21ST CENTURY ASTRONOMY

The Solar System

SIXTH EDITION

21ST CENTURY ASTRONOMY

The Solar System

SIXTH EDITION

Laura Kay
Barnard College

Stacy Palen
Weber State University

George Blumenthal
University of California—Santa Cruz

W. W. NORTON & COMPANY
NEW YORK · LONDON

W. W. Norton & Company has been independent since its founding in 1923, when William Warder Norton and Mary D. Herter Norton first published lectures delivered at the People's Institute, the adult education division of New York City's Cooper Union. The firm soon expanded its program beyond the Institute, publishing books by celebrated academics from America and abroad. By midcentury, the two major pillars of Norton's publishing program—trade books and college texts—were firmly established. In the 1950s, the Norton family transferred control of the company to its employees, and today—with a staff of four hundred and a comparable number of trade, college, and professional titles published each year—W. W. Norton & Company stands as the largest and oldest publishing house owned wholly by its employees.

Copyright © 2019, 2016 by Laura Kay, Stacy Palen, and George Blumenthal.
Copyright © 2013 by Laura Kay, Stacy Palen, Bradford Smith, and George Blumenthal.
Copyright © 2010 by Jeff Hester, Bradford Smith, George Blumenthal, Laura Kay, and Howard G. Voss. © 2007 by Jeff Hester, David Burstein, George Blumenthal, Ronald Greeley, Bradford Smith, and Howard G. Voss. Copyright © 2002 by Jeff Hester, David Burstein, George Blumenthal, Ronald Greeley, Bradford Smith, Howard G. Voss, and Gary Wegner.

All rights reserved
Printed in Canada

Editor: Erik Fahlgren
Project Editor: Diane Cipollone
Editorial Assistant: Sara Bonacum
Managing Editor, College: Marian Johnson
Managing Editor, College Digital Media: Kim Yi
Production Manager: Ashley Horna
Media Editor: Rob Bellinger
Associate Media Editor: Arielle Holstein
Media Project Editor: Danielle Belfiore
Media Editorial Assistant: Kelly Smith
Marketing Manager: Katie Sweeney
Design Director: Rubina Yeh
Photo Editor: Trish Marx
Director of College Permissions: Megan Schindel
Permissions Associate: Elizabeth Trammell
Composition: Graphic World
Manufacturing: Transcontinental

ISBN 978-0-393-64472-2 (pbk)

W. W. Norton & Company, Inc., 500 Fifth Avenue, New York, NY 10110-0017
wwnorton.com
W. W. Norton & Company Ltd., 15 Carlisle Street, London W1D 3BS
1 2 3 4 5 6 7 8 9 0

Aloha and Bon Voyage to our late co-author Brad Smith, who showed the world the outer solar system through his cameras on the Voyager spacecraft.

LAURA KAY thanks her wife, M.P.M. She dedicates this book to her late uncle, Lee Jacobi, for an early introduction to physics, and to her late colleagues at Barnard College, Tally Kampen and Sally Chapman.

STACY PALEN thanks her husband, John Armstrong, for his patient support during this project.

GEORGE BLUMENTHAL gratefully thanks his wife, Kelly Weisberg, and his children, Aaron and Sarah Blumenthal, for their support during this project. He also wants to thank Professor Robert Greenler for stimulating his interest in all things related to physics.

BRIEF CONTENTS

Chapters 15-23 are not included in this edition.

PART I INTRODUCTION TO ASTRONOMY

Chapter 1   Thinking Like an Astronomer   2
        2   Patterns in the Sky—Motions of Earth and the Moon   22
        3   Motion of Astronomical Bodies   58
        4   Gravity and Orbits   82
        5   Light   108
        6   The Tools of the Astronomer   140

PART II THE SOLAR SYSTEM

        7   The Formation of Planetary Systems   170
        8   The Terrestrial Planets and Earth's Moon   196
        9   Atmospheres of the Terrestrial Planets   230
       10   Worlds of Gas and Liquid—The Giant Planets   264
       11   Planetary Moons and Rings   292
       12   Dwarf Planets and Small Solar System Bodies   320

PART III STARS AND STELLAR EVOLUTION

       13   Taking the Measure of Stars   352
       14   Our Star—The Sun   382

PART IV GALAXIES, THE UNIVERSE, AND COSMOLOGY

       24   Life   660

CONTENTS

Chapters 15–23 are not included in this edition.

Preface xxi
About the Authors xxxii

PART I INTRODUCTION TO ASTRONOMY

1 Thinking Like an Astronomer 2

1.1 Earth Occupies a Small Place in the Universe 4
1.2 Science Is a Way of Viewing the Universe 7
Process of Science The Scientific Method 9
1.3 Astronomers Use Mathematics to Find Patterns 12
Working It Out 1.1 Mathematical Tools 13
Working It Out 1.2 Reading a Graph 14
Origins An Introduction 15
Reading Astronomy News A Universe of 2 Trillion Galaxies 16

Summary 17
Unanswered Questions 17
Questions and Problems 17
Exploration: Logical Fallacies 21

2 Patterns in the Sky—Motions of Earth and the Moon 22

2.1 Earth Spins on Its Axis 24
Working It Out 2.1 How to Estimate Earth's Size 31
2.2 Revolution about the Sun Leads to Changes during the Year 33
Process of Science Theories Must Fit All the Known Facts 37

vii

2.3 The Moon's Appearance Changes as It Orbits Earth   40

2.4 Calendars Are Based on the Day, Month, and Year   43

2.5 Eclipses Result from the Alignment of Earth, the Moon, and the Sun   45

Origins The Obliquity of Earth   51

Reading Astronomy News 2,500 Miles of Citizen Scientists   52

Summary   53
Unanswered Question   53
Questions and Problems   54
Exploration: The Phases of the Moon   57

3 Motion of Astronomical Bodies   58

3.1 The Motions of Planets in the Sky   60

Working It Out 3.1 How Copernicus Computed Orbital Periods and Scaled the Solar System   64

3.2 Kepler's Laws Describe Planetary Motion   64

Process of Science Theories Are Falsifiable   67

Working It Out 3.2 Kepler's Third Law   68

3.3 Galileo's Observations Supported the Heliocentric Model   70

3.4 Newton's Three Laws Help Explain How Celestial Bodies Move   71

Working It Out 3.3 Using Newton's Laws   74

Origins Planets and Orbits   75

Reading Astronomy News NASA Spacecraft Take Spring Break at Mars   76

Summary   77
Unanswered Questions   77
Questions and Problems   77
Exploration: Kepler's Laws   81

4 Gravity and Orbits   82

4.1 Gravity Is an Attractive Force between Objects   84

Working It Out 4.1 Playing with Newton's Laws of Motion and Gravitation   87

4.2 An Orbit Is One Body "Falling Around" Another   89

Process of Science Universality   91

Working It Out 4.2 Circular Velocity and Escape Velocity   94

Working It Out 4.3 Calculating Mass from Orbital Periods   95

4.3 Tidal Forces Are Caused by Gravity   95

Working It Out 4.4 Tidal Forces   98

4.4 Tidal Forces Affect Solid Bodies   99

Origins Tidal Forces and Life   101

Reading Astronomy News Exploding Stars Prove Newton's Law of Gravity Unchanged over Cosmic Time   102

Summary 103
Unanswered Question 103
Questions and Problems 103
Exploration: Newton's Laws 107

5 Light 108

5.1 Light Brings Us the News of the Universe 110

Process of Science Agreement between Fields 115

Working It Out 5.1 Working with Electromagnetic Radiation 116

5.2 The Quantum View of Matter Explains Spectral Lines 116

5.3 The Doppler Shift Indicates Motion Toward or Away from Us 124

Working It Out 5.2 Making Use of the Doppler Effect 125

5.4 Temperature Affects the Spectrum of Light That an Object Emits 126

Working It Out 5.3 Working with the Stefan-Boltzmann Law and Wien's Law 130

5.5 The Brightness of Light Depends on the Luminosity and Distance of the Source 130

Working It Out 5.4 Using Radiation Laws to Calculate Equilibrium Temperatures of Planets 132

Origins Temperatures of Planets 133

Reading Astronomy News A Study in Scarlet 134

Summary 135
Unanswered Questions 135
Questions and Problems 136
Exploration: Light as a Wave 139

6 The Tools of the Astronomer 140

6.1 The Optical Telescope Revolutionized Astronomy 142

Working It Out 6.1 Telescope Aperture and Magnification 144

Working It Out 6.2 Diffraction Limit 148

6.2 Optical Detectors and Instruments Used with Telescopes 150

6.3 Astronomers Observe in Wavelengths Beyond the Visible 153

6.4 Planetary Spacecraft Explore the Solar System 157

6.5 Other Tools Contribute to the Study of the Universe 159

Process of Science Technology and Science Are Symbiotic 161

Origins Microwave Telescopes Detect Radiation from the Big Bang 162

Reading Astronomy News Construction Begins on the World's First Super Telescope 164

Summary 165
Unanswered Questions 165
Questions and Problems 166
Exploration: Geometric Optics and Lenses 169

PART II THE SOLAR SYSTEM

7 The Formation of Planetary Systems 170

7.1 Planetary Systems Form around a Star 172

Process of Science Converging Lines of Inquiry 174

7.2 The Solar System Began with a Disk 175

Working It Out 7.1 Angular Momentum 177

7.3 The Inner Disk and Outer Disk Formed at Different Temperatures 178

7.4 The Formation of Our Solar System 181

7.5 Planetary Systems Are Common 183

Working It Out 7.2 Estimating the Size of the Orbit of a Planet 185

Working It Out 7.3 Estimating the Radius of an Exoplanet 187

Origins The Search for Earth-Sized Planets 189

Reading Astronomy News Four Earth-Sized Planets Detected Orbiting the Nearest Sun-like Star 190

Summary 191
Unanswered Questions 191
Questions and Problems 192
Exploration: Exploring Exoplanets 195

8 The Terrestrial Planets and Earth's Moon 196

8.1 Impacts Help Shape the Evolution of the Planets 198

Process of Science Certainty Is Sometimes Out of Reach 202

8.2 Radioactive Dating Tells Us the Age of the Moon and the Solar System 203

Working It Out 8.1 Computing the Ages of Rocks 204

8.3 The Surface of a Terrestrial Planet Is Affected by Processes in the Interior 205

Working It Out 8.2 How Planets Cool 208

8.4 Planetary Surfaces Evolve through Tectonism   210

8.5 Volcanism Signifies a Geologically Active Planet   215

8.6 The Geological Evidence for Water   218

Origins The Death of the Dinosaurs   223

Reading Astronomy News Moon Had a Magnetic Field for at Least a Billion Years Longer than Thought   224

Summary   225
Unanswered Questions   225
Questions and Problems   226
Exploration: Exponential Behavior   229

9 Atmospheres of the Terrestrial Planets   230

9.1 Atmospheres Change over Time   232

9.2 Secondary Atmospheres Evolve   234

Working It Out 9.1 Atmosphere Retention   235

9.3 Earth's Atmosphere Has Detailed Structure   238

9.4 The Atmospheres of Venus and Mars Differ from Earth's   246

9.5 Greenhouse Gases Affect Global Climates   249

Process of Science Thinking about Complexity   254

Origins Our Special Planet   255

Reading Astronomy News NASA's Maven Spacecraft Unlocks the Mystery of Mars's Lost Atmosphere   257

Summary   258
Unanswered Questions   259
Questions and Problems   259
Exploration: Climate Change   263

10 Worlds of Gas and Liquid—The Giant Planets   264

10.1 The Giant Planets Are Large, Cold, and Massive   266

Process of Science Scientific Laws Make Testable Predictions   268

10.2 The Giant Planets Have Clouds and Weather   271

Working It Out 10.1 Measuring Wind Speeds on Different Planets   276

10.3 The Interiors of the Giant Planets Are Hot and Dense   277

Working It Out 10.2 Internal Thermal Energy Heats the Giant Planets   278

10.4 The Giant Planets Are Magnetic Powerhouses   280

10.5 The Planets of Our Solar System Might Not Be Typical   283

Origins Giant Planet Migration and the Inner Solar System   285

Reading Astronomy News *Juno* Spacecraft Spots Jupiter's Great Red Spot   286

Summary   287
Unanswered Questions   287
Questions and Problems   288
Exploration: Estimating Rotation Periods of the Giant Planets   291

11 Planetary Moons and Rings 292

11.1 Many Solar System Planets Have Moons 294

Working It Out 11.1 Using Moons to Compute the Mass of a Planet 296

11.2 Some Moons Have Geological Activity and Water 297

Working It Out 11.2 Tidal Forces on the Moons 299

11.3 Rings Surround the Giant Planets 304

Working It Out 11.3 Feeding the Rings 308

11.4 Ring Systems Have a Complex Structure 308

Process of Science Following Up on the Unexpected 311

Origins Extreme Environments 313

Reading Astronomy NASA Killed *Cassini* to Avoid Contaminating Saturn's Moons 314

Summary 315
Unanswered Questions 315
Questions and Problems 316
Exploration: Measuring Features on Io 319

12 Dwarf Planets and Small Solar System Bodies 320

12.1 Dwarf Planets May Outnumber Planets 322

Process of Science How to Classify Pluto 324

Working It Out 12.1 Eccentric Orbits 326

12.2 Asteroids Are Pieces of the Past 326

12.3 Comets Are Clumps of Ice 331

12.4 Meteorites Are Remnants of the Early Solar System 338

12.5 Comet and Asteroid Collisions Still Happen Today 342

Working It Out 12.2 Impact Energy 344

Origins Comets, Asteroids, Meteoroids, and Life 345

Reading Astronomy News *Rosetta* Spacecraft Finds Water on Earth Didn't Come from Comets 346

Summary 347
Unanswered Questions 347
Questions and Problems 348
Exploration: Asteroid Discovery 351

PART III STARS AND STELLAR EVOLUTION

13 Taking the Measure of Stars   352

13.1 Astronomers Measure the Distance, Brightness, and Luminosity of Stars   354

Working It Out 13.1 Parallax and Distance   357

Working It Out 13.2 The Magnitude System   358

13.2 Astronomers Can Determine the Temperature, Size, and Composition of Stars   359

Working It Out 13.3 Estimating the Sizes of Stars   364

13.3 Measuring the Masses of Stars in Binary Systems   365

Working It Out 13.4 Measuring the Mass of an Eclipsing Binary Pair   368

13.4 The Hertzsprung-Russell Diagram Is the Key to Understanding Stars   369

Process of Science Science Is Collaborative   372

Origins Habitable Zones   375

Reading Astronomy News NASA's Hubble Extends Stellar Tape Measure 10 Times Farther into Space   376

Summary   377
Unanswered Questions   377
Questions and Problems   378
Exploration: The H-R Diagram   381

14 Our Star—The Sun   382

14.1 The Sun Is Powered by Nuclear Fusion   384

Working It Out 14.1 The Source of the Sun's Energy   386

14.2 Energy Is Transferred from the Interior of the Sun   389

Process of Science Learning from Failure   392

14.3 The Atmosphere of the Sun   394

14.4 The Atmosphere of the Sun Is Very Active 396

Working It Out 14.2 Sunspots and Temperature 398

Origins The Solar Wind and Life 403

Reading Astronomy News Carrington-Class CME Narrowly Misses Earth 404

Summary 405
Unanswered Questions 405
Questions and Problems 406
Exploration: The Proton-Proton Chain 409

15 The Interstellar Medium and Star Formation 410

15.1 The Interstellar Medium Fills the Space between the Stars 412

Working It Out 15.1 Dust Glows in the Infrared 415

Process of Science All Branches of Science Are Interconnected 419

15.2 Molecular Clouds Are the Cradles of Star Formation 420

15.3 Formation and Evolution of Protostars 422

15.4 Evolution before the Main Sequence 427

Working It Out 15.2 Luminosity, Surface Temperature, and Radius of Protostars 428

Origins Star Formation, Planets, and Life 431

Reading Astronomy News Interstellar Dust Discovered inside NASA Spacecraft 432

Summary 433
Unanswered Questions 433
Questions and Problems 434
Exploration: The Stellar Thermostat 437

16 Evolution of Low-Mass Stars 438

16.1 The Life of a Main-Sequence Star Depends on Its Mass 440

Working It Out 16.1 Estimating Main-Sequence Lifetimes 442

16.2 The Star Leaves the Main Sequence 443

16.3 Helium Fuses in the Degenerate Core 447

16.4 Dying Stars Shed Their Outer Layers 450

Working It Out 16.2 Escaping the Surface of an Evolved Star 451

16.5 Binary Star Evolution 456

Origins Stellar Lifetimes and Biological Evolution 460

Process of Science Science Is Unfinished 461

Reading Astronomy News Scientists Solve Riddle of Celestial Archaeology 462

Summary 463
Unanswered Questions 463
Questions and Problems 464
Exploration: Low-Mass Stellar Evolution 467

17 Evolution of High-Mass Stars 468

17.1 High-Mass Stars Follow Their Own Path 470

17.2 High-Mass Stars Go Out with a Bang 474

Working It Out 17.1 Binding Energy of Atomic Nuclei 475

17.3 The Spectacle and Legacy of Supernovae 479

Working It Out 17.2 Gravity on a Neutron Star 481

Process of Science Occam's Razor 482

17.4 Star Clusters Are Snapshots of Stellar Evolution 485

Origins Seeding the Universe with New Chemical Elements 489

Reading Astronomy News Astronomers Find Mass Limit for Neutron Stars Before Collapsing into Black Holes 490

Summary 491
Unanswered Questions 491
Questions and Problems 492
Exploration: The CNO Cycle 495

18 Relativity and Black Holes 496

18.1 Relative Motion Affects Measured Velocities 498

18.2 Special Relativity Explains How Time and Space Are Related 500

Working It Out 18.1 Time Dilation 505

18.3 Gravity Is a Distortion of Spacetime 506

Process of Science New Science Can Encompass the Old 510

18.4 Black Holes 512

Working It Out 18.2 Masses in X-ray Binaries 516

Origins Gamma-Ray Bursts 516

Reading Astronomy News LIGO Just Detected the Oldest Gravitational Waves Ever Discovered 518

Summary 519
Unanswered Questions 519
Questions and Problems 520
Exploration: Black Holes 523

PART IV GALAXIES, THE UNIVERSE, AND COSMOLOGY

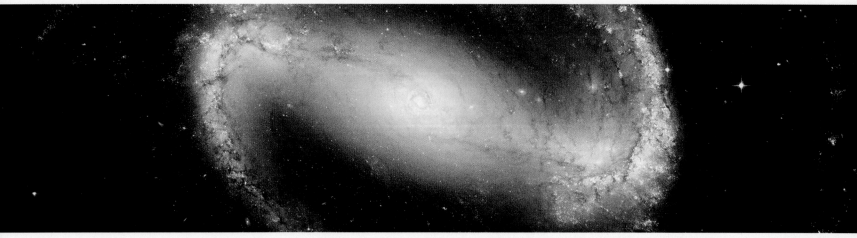

19 Galaxies 524

19.1 Galaxies Come in Different Shapes and Sizes 526

19.2 Astronomers Use Several Methods to Find Distances to Galaxies 532

Working It Out 19.1 Finding the Distance from a Type Ia Supernova 534

Working It Out 19.2 Redshift—Calculating the Recession Velocity and Distance of Galaxies 536

19.3 Galaxies Are Mostly Dark Matter 536

19.4 Most Galaxies Have a Supermassive Black Hole at the Center 539

Working It Out 19.3 The Size, Density, and Power of a Supermassive Black Hole 543

Process of Science Finding the Common Thread 545

Origins Habitability in Galaxies 547

Reading Astronomy News A Lot of Galaxies Need Guarding in This NASA Hubble View 548

Summary 549
Unanswered Questions 549
Questions and Problems 550
Exploration: Galaxy Classification 553

20 The Milky Way—A Normal Spiral Galaxy 554

20.1 Astronomers Have Measured the Size and Structure of the Milky Way 556

20.2 The Components of the Milky Way Reveal Its Evolution 560

Process of Science Unknown Unknowns 561

20.3 Most of the Milky Way Is Unseen 566

Working It Out 20.1 The Mass of the Milky Way inside the Sun's Orbit 568

Working It Out 20.2 The Mass of the Milky Way's Central Black Hole 569

20.4 The History and Future of the Milky Way 570

Origins The Galactic Habitable Zone 573

Reading Astronomy News You Are Here: Scientists Unveil Precise Map of More Than a Billion Stars 574

Summary 575
Unanswered Questions 575
Questions and Problems 576
Exploration: The Center of the Milky Way 579

21 The Expanding Universe 580

21.1 The Cosmological Principle 582

Process of Science Data Are the Ultimate Authority 586

21.2 The Universe Began in the Big Bang 587

Working It Out 21.1 Expansion and the Age of the Universe 588

21.3 Expansion Is Described with a Scale Factor 590

21.4 Astronomers Observe Cosmic Microwave Background Radiation 592

Working It Out 21.2 When Redshift Exceeds 1 593

Origins Big Bang Nucleosynthesis 596

Reading Astronomy News 50th Anniversary of the Big Bang Discovery 598

Summary 599
Unanswered Questions 599
Questions and Problems 599
Exploration: Hubble's Law for Balloons 603

22 Cosmology 604

22.1 Gravity and the Expansion of the Universe 606

Working It Out 22.1 Calculating the Critical Density 608

22.2 The Accelerating Universe 608

Process of Science Never Throw Anything Away 610

22.3 Inflation Solves Several Problems in Cosmology 614

22.4 The Earliest Moments of the Universe Connect the Very Largest Size Scales to the Very Smallest 618

Working It Out 22.2 Pair Production in the Early Universe 620

22.5 String Theory and Multiverses 624

Origins Our Own Universe Must Support Life 626

Reading Astronomy News Astronomers Find Half of the Missing Matter in the Universe 628

Summary 629
Unanswered Questions 629
Questions and Problems 630
Exploration: Studying Particles 633

xvii

23 Large-Scale Structure in the Universe 634

23.1 Galaxies Form Groups, Clusters, and Larger Structures 636

Working It Out 23.1 Mass of a Cluster of Galaxies 638

23.2 Gravity Forms Large-Scale Structure 639

23.3 First Light of Stars and Galaxies 643

Process of Science Multiple Streams of Evidence 644

Working It Out 23.2 Observing High-Redshift Objects 647

23.4 Galaxies Evolve 649

Origins We Are the 4 or 5 Percent 654

Reading Astronomy News Welcome to Laniakea, Your Galactic Supercluster Home 655

Summary 656

Unanswered Questions 656

Questions and Problems 656

Exploration: The Story of a Proton 659

24 Life 660

24.1 Life Evolves on Earth 662

Working It Out 24.1 Exponential Growth 666

24.2 Life Involves Complex Chemical Processes 667

Process of Science All of Science Is Interconnected 668

24.3 Where Do Astronomers Look for Life? 669

24.4 Scientists Are Searching for Signs of Intelligent Life 675

Working It Out 24.2 Putting Numbers into the Drake Equation 676

Origins The Fate of Life on Earth 678

Reading Astronomy News When Reporting News about Aliens, Caution Is Advised 680

Summary 681

Unanswered Questions 681

Questions and Problems 682

Exploration: Fermi Problems and the Drake Equation 685

APPENDIX 1 Mathematical Tools A-1
APPENDIX 2 Physical Constants and Units A-6
APPENDIX 3 Periodic Table of the Elements A-8
APPENDIX 4 Properties of Planets, Dwarf Planets, and Moons A-9
APPENDIX 5 Space Missions A-13
APPENDIX 6 Brightest Stars A-15
APPENDIX 7 Observing the Sky A-16
APPENDIX 8 Uniform Circular Motion and Circular Orbits A-25

Glossary G-1   Selected Answers SA-1   Credits C-1   Index I-1

Working It Out

Chapters 15–23 are not included in this edition.

1.1 Mathematical Tools 13
1.2 Reading a Graph 14
2.1 How to Estimate Earth's Size 31
3.1 How Copernicus Computed Orbital Periods and Scaled the Solar System 64
3.2 Kepler's Third Law 68
3.3 Using Newton's Laws 74
4.1 Playing with Newton's Laws of Motion and Gravitation 87
4.2 Circular Velocity and Escape Velocity 94
4.3 Calculating Mass from Orbital Periods 95
4.4 Tidal Forces 98
5.1 Working with Electromagnetic Radiation 116
5.2 Making Use of the Doppler Effect 125
5.3 Working with the Stefan-Boltzmann Law and Wien's Law 130
5.4 Using Radiation Laws to Calculate Equilibrium Temperatures of Planets 132
6.1 Telescope Aperture and Magnification 144
6.2 Diffraction Limit 148
7.1 Angular Momentum 177
7.2 Estimating the Size of the Orbit of a Planet 185
7.3 Estimating the Radius of an Exoplanet 187
8.1 Computing the Ages of Rocks 204
8.2 How Planets Cool 208
9.1 Atmosphere Retention 235
10.1 Measuring Wind Speeds on Different Planets 276
10.2 Internal Thermal Energy Heats the Giant Planets 278
11.1 Using Moons to Compute the Mass of a Planet 296
11.2 Tidal Forces on the Moons 299
11.3 Feeding the Rings 308
12.1 Eccentric Orbits 326
12.2 Impact Energy 344
13.1 Parallax and Distance 357
13.2 The Magnitude System 358
13.3 Estimating the Sizes of Stars 364
13.4 Measuring the Mass of an Eclipsing Binary Pair 368
14.1 The Source of the Sun's Energy 386
14.2 Sunspots and Temperature 398
15.1 Dust Glows in the Infrared 415
15.2 Luminosity, Surface Temperature, and Radius of Protostars 428
16.1 Estimating Main-Sequence Lifetimes 442
16.2 Escaping the Surface of an Evolved Star 451
17.1 Binding Energy of Atomic Nuclei 475
17.2 Gravity on a Neutron Star 481
18.1 Time Dilation 505
18.2 Masses in X-ray Binaries 516
19.1 Finding the Distance from a Type Ia Supernova 534
19.2 Redshift—Calculating the Recession Velocity and Distance of Galaxies 536
19.3 The Size, Density, and Power of a Supermassive Black Hole 543
20.1 The Mass of the Milky Way inside the Sun's Orbit 568
20.2 The Mass of the Milky Way's Central Black Hole 569
21.1 Expansion and the Age of the Universe 588
21.2 When Redshift Exceeds 1 593
22.1 Calculating the Critical Density 608
22.2 Pair Production in the Early Universe 620
23.1 Mass of a Cluster of Galaxies 638
23.2 Observing High-Redshift Objects 647
24.1 Exponential Growth 666
24.2 Putting Numbers into the Drake Equation 676

AstroTours

Chapters 15-23 are not included in this edition.

AstroTour animations, Interactive Simulations, and Astronomy in Action Videos are all available from the free Student Site at the Digital Resources Site, and they are also integrated into assignable Smartwork5 exercises. **digital.wwnorton.com/astro6**

The Celestial Sphere and the Ecliptic   25
The View from the Poles   26
The Earth Spins and Revolves   35
The Moon's Orbit: Eclipses and Phases   40
Kepler's Laws   66
Velocity, Acceleration, Inertia   70, 75
Newton's Laws and Universal Gravitation   89
Elliptical Orbit   92
Tides and the Moon   95
Light as a Wave, Light as a Photon   116, 139
Atomic Energy Levels and the Bohr Model   117
Atomic Energy Levels and Light Emission and Absorption   121
The Doppler Effect   124
Geometric Optics and Lenses   146, 169

Solar System Formation   172
Traffic Circle Analogy   177
Processes That Shape the Planets   201
Continental Drift   211
Hot Spot Creating a Chain of Islands   216
Atmospheres: Formation and Escape   233
Greenhouse Effect   237
Stellar Spectrum   360
H-R Diagram   370
The Solar Core   387
Star Formation   421
Hubble's Law   534, 587
Dark Matter   536
Active Galactic Nuclei   542
Big Bang Nucleosynthesis   597

Interactive Simulations

Chapters 15-23 are not included in this edition.

Phases of the Moon   40, 57
Planetary Orbits   66, 81, 107
Tides on Earth   95
Radial Velocity   185, 195

H-R Diagram   370, 381, 467
Proton-Proton Chain   387, 409
CNO Cycle   470, 495
H-R Explorer   472

Astronomy in Action Videos

Chapters 15-23 are not included in this edition.

Vocabulary of the Celestial Sphere   26
The Cause of Earth's Seasons   36
The Earth-Moon-Sun System   38
Phases of the Moon   40
Velocity, Force, and Acceleration   75
Center of Mass   94
Tides   95
Emission and Absorption   118
Doppler Shift   124, 185
Changing Equilibrium   126, 236
Wien's Law   129

Inverse Square Law   131
Angular Momentum   175, 421
Charged Particles and Magnetic Fields   244
Parallax   354
Random Walk   389
Type II Supernova   478
Pulsar Rotation   481
Galaxy Shapes   527
Size of Active Galactic Nuclei   542
Expanding Balloon Universe   587
Observable versus Actual Universe   589

Preface

Dear Student

Why is it a good idea to take a science course, and in particular, why is astronomy a course worth taking? Many people choose to learn about astronomy because they are curious about the universe. Your instructor likely has two basic goals in mind for you as you take this course. The first is to understand some basic physical concepts and how they apply to the universe around us. The second is to think like a scientist and learn to use the scientific method not only to answer questions in this course but also to be able to evaluate and understand new information you encounter in your life. We have written the sixth edition of *21st Century Astronomy* with these two goals in mind.

Throughout this book, we emphasize not only the content of astronomy (for example, the differences among the planets, the formation of chemical elements) but also *how* we know what we know. The scientific method is a valuable tool that you can carry with you and use for the rest of your life. One way we highlight the process of science is the **Process of Science Figures**. In each chapter, we have chosen one discovery and provided a visual representation illustrating the discovery or a principle of the process of science. In these figures, we try to illustrate that science is not a tidy process, and that discoveries are sometimes made by different groups, sometimes by accident, but always because people are trying to answer a question and show why or how we think something is the way it is.

The most effective way to learn something is to "do" it. Whether playing an instrument or a sport or becoming a good cook, reading "how" can only take you so far. The same is true of learning astronomy. We have written this book to help you "do" as you learn. Sometimes this means doing an activity to apply a concept and sometimes we provide a tool to make reading a more active process.

- At the beginning of each chapter, we have provided a set of Learning Goals to guide you as you read. There is a lot of information in every chapter, and the Learning Goals should help you focus on the most important points.
- We present a big-picture question in association with the chapter-opening photo at the beginning of each chapter. For each of these, we have tried to pose a question that is not only relevant to its chapter but also something you may have wondered about. We hope that these questions, plus the photographs that accompany them, capture your attention as well as your imagination.

CHECK YOUR UNDERSTANDING 7.1

Which of the following pieces of evidence support the nebular hypothesis? (Choose all that apply.) (a) Planets orbit the Sun in the same direction. (b) The Solar System is relatively flat. (c) Earth has a large Moon. (d) We observe disks of gas and dust around other stars.

- There are **Check Your Understanding** questions at the end of each chapter section. These questions are designed to be answered quickly if you have understood the previous section. The answers are provided in the back of the book so you can check your answer and decide if further review is necessary.

- As a citizen of the world, you make judgments about science, distinguishing between good science and pseudoscience. You use these judgments to make decisions in the grocery store, pharmacy, car dealership, and voting booth. You may base these decisions on the presentation of information you receive through the media, which is very different from the presentation in class. One important skill is the ability to recognize what is credible and to question what is not. To help you hone this skill, we have provided **Reading Astronomy News** sections at the end of every chapter. These features include a news article with questions to help you make sense of how science is presented to you. It is important that you learn to be critical of the information you receive, and these features will help you do that.

- While we know a lot about the universe, science is an ongoing process, and we continue to search for new answers. To give you a glimpse of what we don't know, we provide an **Unanswered Questions** feature near the end of each chapter. Most of these questions represent topics that scientists are currently studying.

Unanswered Questions

- How typical is the Solar System? Only within the past decade have astronomers found other systems containing four or more planets, and so far the observed distributions of large and small planets in those multiplanet systems have looked different from those of the Solar System. Computer simulations of planetary system formation suggest that a system with stable orbits and a planetary distribution like those of the Solar System may develop only rarely. Improved supercomputers can run more complex simulations, which can be compared with the observations to better understand how solar systems are configured.

- How Earth-like must a planet be before scientists declare it to be "another Earth"? An editorial in the science journal *Nature* cautioned that scientists should define "Earth-like" in advance—before multiple discoveries of planets "similar" to Earth are announced and a media frenzy ensues. Must a planet be of similar size and mass, be located in the habitable zone, and have spectroscopic evidence of liquid water before we call it "Earth 2.0"?

- The language of science is mathematics, and it can be as challenging to learn as any other language. The choice to use mathematics as the language of science is not arbitrary; nature "speaks" math. To learn about nature, you will need to speak its language. We don't want the language of math to obscure the concepts, so we have placed this book's mathematics in **Working It Out** boxes to make it clear when we are beginning and ending a mathematical argument, so that you can spend time with the concepts in the chapter text and then revisit the mathematics of the concept to study the formal language of the argument. You will learn to work with data and identify when data aren't quite right. We want you to be comfortable reading, hearing, and speaking the language of science, and we will provide you with tools to make it easier.

7.1 Working It Out Angular Momentum

In its simplest form, the angular momentum (L) of a system is given by

$$L = m \times v \times r$$

where m is the mass, v is the speed at which the mass is moving, and r represents how spread out the mass is.

Let's apply that relationship to the angular momentum of Jupiter in its orbit about the Sun. The angular momentum from one body orbiting another is called *orbital* angular momentum, $L_{orbital}$. The mass (m) of Jupiter is 1.90×10^{27} kilograms (kg), the speed of Jupiter in orbit (v) is 1.31×10^4 meters per second (m/s), and the radius of Jupiter's orbit (r) is 7.79×10^{11} meters. Putting all that together gives

$$L_{orbital} = (1.90 \times 10^{27} \text{ kg}) \times (1.31 \times 10^4 \text{ m/s}) \times (7.79 \times 10^{11} \text{ m})$$

$$L_{orbital} = 1.94 \times 10^{43} \text{ kg m}^2/\text{s}$$

where R is the radius of the sphere and P is the rotation period of its spin.

Let's compare Jupiter's orbital angular momentum with the Sun's spin angular momentum to investigate the distribution of angular momentum in the Solar System. Appendix 2 provides the Sun's radius (6.96×10^8 meters) and mass (1.99×10^{30} kg), and its rotation period is 24.5 days = 2.12×10^6 seconds. If we assume that the Sun is a uniform sphere, the spin angular momentum of the Sun is

$$L_{spin} = \frac{4 \times \pi \times (1.99 \times 10^{30} \text{ kg}) \times (6.96 \times 10^8 \text{ m})^2}{5 \times (2.12 \times 10^6 \text{ s})}$$

$$L_{spin} = 1.14 \times 10^{42} \text{ kg m}^2/\text{s}$$

$L_{orbital}$ of Jupiter is about 17 times greater than L_{spin} of the Sun. Thus, most of the angular momentum of the Solar System now

- Each chapter concludes with an **Origins** section, which relates material or subjects found in the chapter to questions about the origin of the universe and the origin of life in the universe and on Earth. Astrobiologists have made much progress in recent years on understanding how conditions in the universe may have helped or hindered the origin of life, and in each Origins we explore an example from its chapter that relates to how the universe and life formed and evolved.

- At the end of each chapter, we have provided several types of questions, problems, and activities for you to practice your skills. The **Test Your Understanding** questions focus on more detailed facts and concepts from the chapter. **Thinking about the Concepts** questions ask you to synthesize information and explain the "how" or "why" of a situation. **Applying the Concepts** problems give you a chance to practice the quantitative skills you learned in the chapter and to work through a situation mathematically. The **Using the Web** questions represent other opportunities to "learn by doing." Using the Web sends you to websites of space missions, observatories, experiments, or archives to access recent observations, results, or press releases. Other sites are for "citizen science" projects in which you can contribute to the analysis of new data.

- **Explorations** show you how to use the concepts and skills you learned in an interactive way. Some of the book's Explorations ask you to use animations and simulations on the Student Site, while the others are hands-on, paper-and-pencil activities that use everyday objects such as ice cubes or balloons.

The resources outside of the book (at the Student Site) can help you understand and visualize many of the physical concepts described in the book. **AstroTours and Interactive Simulations** are represented by icons in the margins of the book. There is also a series of short **Astronomy in Action** videos that are represented by icons in the margins and available at the Student Site. These videos feature one of the authors (and several students) demonstrating physical concepts at work. Your instructor might assign these videos to you or you might choose to watch them on your own to create a better picture of each concept in your mind.

Astronomy gives you a sense of perspective that no other field of study offers. The universe is vast, fascinating, and beautiful, filled with a wealth of objects that, surprisingly, can be understood using only a handful of principles. By the end of this book, you will have gained a sense of your place in the universe.

USING THE WEB

46. Using the exoplanet catalogs:
    a. Go to the "Catalog" Web page (http://exoplanet.eu/catalog) of the Extrasolar Planets Encyclopedia. Look for a star that has multiple planets. Make a graph showing the distances of the planets from that star, and note the masses and sizes of the planets. Put the Solar System planets on the same axis. How does that extrasolar planet system compare with the Solar System?
    b. Go to the "Exoplanets Data Explorer" website (http://exoplanets.org) and click on "Table." That website lists planets that have detailed orbital data published in scientific journals, and it may have a smaller total count than the website in part (a). Pick a planet discovered recently, as specified in the "First Reference" column. What is the planet's minimum mass? What is its semimajor axis and the period of its orbit? What is the eccentricity of its orbit? Click on the star name in the first column to get more information. Does that planet have a radial velocity curve? Was it observed in transit, and if so, what is the planet's radius and density? Is it more like Jupiter or more like Earth?

Dear Instructor

We wrote this book with a few overarching goals: to inspire students, to make the material interactive, and to create a useful and flexible tool that can support multiple learning styles.

As scientists and as teachers, we are passionate about the work we do. We hope to share that passion with students and inspire them to engage in science on their own. Through our own experience, familiarity with education research, and surveys of instructors, we have come to know a great deal about how students learn and what goals teachers have for their students. We have explicitly addressed many of these goals and learning styles in this book, sometimes in large, immediately visible ways such as the inclusion of features but also through less obvious efforts such as questions and problems that relate astronomical concepts to everyday situations or a fresh approach to organizing material.

For example, many teachers state that they would like their students to become "educated scientific consumers" and "critical thinkers" or that their students should "be able to read a news story about science and understand its significance." We have specifically addressed these goals in our **Reading Astronomy News** feature, which presents a news article and a series of questions that guide a student's critical thinking about the article, the data presented, and the sources.

Education research shows that the most effective way to learn is by doing. **Exploration** activities at the end of each chapter are hands-on, asking students to take the concepts they've learned in the chapter and apply them as they interact with animations and simulations on the Student Site or work through pencil-and-paper activities. Many of these Explorations incorporate everyday objects and can be used either in your classroom or as activities at home. The **Using the Web** problems direct students to "citizen science" projects, where they can contribute to the analysis of new astronomical data. Other problems send students to websites of space missions, observatories, collaborative projects, and catalogs to access the most current observations, results, and news releases. These Web problems can be used for homework, lab exercises, recitations, or "writing across the curriculum" projects.

We also believe students should be exposed to the more formal language of science—mathematics. We have placed the math in **Working It Out** boxes, so it does not interrupt the flow of the text or get in the way of students' understanding of conceptual material. But we've gone further by beginning with fundamental ideas in early Working It Out boxes and slowly building in complexity through the book. We've also worked to remove some of the stumbling blocks that affect student confidence by providing calculator hints, references to earlier Working It Out boxes, and detailed, fully worked examples. Many chapters include problems on reading and interpreting graphs. Appendix 1, "Mathematical Tools," summarizes some math concepts for students.

Discussion of basic physics is contained in Part I to accommodate courses that use the *Solar System* or *Stars and Galaxies* volumes. A "just-in-time" approach to introducing the physics is still possible by bringing in material from Chapters 2–6 as needed. For example, the sections on tidal forces in Chapter 4 can be taught along with the moons of the Solar System in Part II, or with mass transfer in binary stars in Part III, or with galaxy interactions in Part IV. Spectral lines in Chapter 5 can be taught with planetary atmospheres in Part II or with stellar spectral types in Part III, and so on.

In our overall organization, we have made several efforts to encourage students to engage with the material and build confidence in their scientific skills as they proceed through the book. For planets, stars and galaxies, we have organized the material to cover the general case first and then delve into more details with specific examples. Thus, you will find "planetary systems" before our own Solar System, "stars" before the Sun, and "galaxies" before the Milky Way. This allows us to avoid frustrating students by making assumptions about what they know about stars or galaxies or forward-referencing to basic definitions and overarching concepts. This organization also implicitly helps students understand their place in the universe: our galaxy and our star are each one of many. They are specific examples of a physical universe in which the same laws apply everywhere. Planets have been organized comparatively to emphasize that science is a process of studying individual examples that lead to collective conclusions. All of these organizational choices were made with the student perspective in mind and a clear sense of the logical hierarchy of the material.

We begin in Chapter 19 by introducing galaxies as a whole and our measurements of them, including recession velocities. Then we address the Milky Way in Chapter 20—a specific example of a galaxy that we can discuss in detail. This follows the repeating motif of moving from the general to the specific that exists throughout the text and gives students a basic grounding in the concepts of spiral galaxies, supermassive black holes, and dark matter before they need to apply those concepts to the specific example of our own galaxy. Chapter 21, "The Expanding Universe," covers the cosmological principle, the Hubble expansion, and the observational evidence for the Big Bang.

The **Origins** sections illustrate how astrobiologists and other scientists approach the study of a scientific question from the chapter related to the origin of the universe and of life. Material about exoplanets is introduced in Chapter 7, incorporated into other chapters when appropriate, and continued in Chapter 24. We revised each chapter, streamlining some topics, and updating the science to reflect the progress in the field. This includes:

- Updates on new telescopes and space missions
- New images and results from Solar System exploration missions to the Moon, Mars, Jupiter, Saturn, and Pluto
- Updated graphical data on climate change
- Results from Kepler and other missions on populations of exoplanets
- Results from the GAIA mission on stellar distances, the HR diagram, and the Milky Way
- Results from LIGO on the discovery of gravitational waves
- Updated cosmology from the Planck mission
- Many new articles for Reading Astronomy News, such as "2,500 Miles of Citizen Scientists", "NASA killed Cassini to Avoid Contaminating Saturn's Moons", and "LIGO Just Detected the Oldest Gravitational Waves Ever Discovered"
- Revised Process of Science figures for each chapter
- Revised simulations for some Explorations exercises

Many professors find themselves under pressure from accrediting bodies or internal assessment offices to assess their courses in terms of learning goals. Each chapter has Learning Goals and end-of-chapter Summary to correspond to the chapter's Learning Goals. In Smartwork5, questions and problems are tagged and can be sorted by Learning Goal. Smartwork5 contains more than 2,000 questions and problems that are tied directly to this text, including the Check Your Understanding questions and versions of the Reading Astronomy News and Exploration questions. Any of these could be used as a reading quiz to be completed before class or as homework. Every question in Smartwork5 has hints and answer-specific feedback so that students are coached to work toward the correct answer. An instructor can easily modify any of the provided questions, answers, and feedback or can create his or her own questions.

We've also created a series of 23 videos explaining and demonstrating concepts from the text, accompanied by questions in Smartwork5. You might assign these videos prior to lecture—either as part of a flipped modality or as a "reading quiz." In either case, you can use the diagnostic feedback from the questions in Smartwork5 to tailor your in-class discussions. Or you might show them in class, to stimulate discussion. Or you might simply use them as a jumping-off point—to get ideas for activities to do with your own students.

We continue to look for better ways to engage students, so please let us know how these features work for your students.

Digital Media for Students

digital.wwnorton.com/astro6

smartwork5 Online Homework

Steven Desch, Guilford Technical Community College

Diane Friend, University of Montana

Ana M. Larson, University of Washington

Violet Mager, Penn State Wilkes-Barre

Ryan Oelkers, Vanderbilt University

Todd Young, Wayne State College

William Younger, Tidewater Community College

More than 2,000 questions support *21st Century Astronomy, Sixth Edition*—all with answer-specific feedback, hints, and ebook links. Questions include Summary Self-Tests, versions of the Explorations (based on AstroTours and new Interactive Simulations), and questions based on the Reading Astronomy News features. Astronomy in Action video questions focus on getting students to come to class prepared and on overcoming common misconceptions. Process of Science Guided Inquiry Assignments help students apply the scientific method to important questions in astronomy, challenging them to think like scientists. Rounding out the Smartwork5 course are ranking, sorting, and labeling exercises—all classified according to the same Learning Goals used in the textbook and test bank.

Smartwork5 can be set up to work right in your LMS, with student scores flowing directly to your LMS gradebook. Your local Norton representative can help you set up LMS integration.

At Play in the Cosmos: The Videogame

Jeff Bary (Colgate University), Adam Frank (University of Rochester), Learning Games Network (University of Wisconsin and Massachusetts Institute of Technology), Gear Learning at UW-Madison's Wisconsin Center for Education Research

This ground-breaking and award-winning educational videogame is designed to engage introductory astronomy students by asking them to confront challenges using concepts that span the scope of the course. Using the process of science, students complete missions by flying their ship and choosing the right scientific tools for each task. The videogame can be downloaded to PC, Mac, and tablets and can be included with new copies of the Sixth Edition at no additional cost.

A complete videogame instructor's manual provides suggestions for meaningfully integrating the game into your course—for example, as a lab activity or as homework. The manual, written by lead game author Jeff Bary, provides mission overviews, plus multiple-choice and discussion questions for every mission.

Norton Ebook

The 21st Century Astronomy, Sixth Edition ebook provides students an enhanced reading experience at a fraction of the cost of a print textbook. Students are able to have an active reading experience and can take notes, bookmark, search, highlight, and even read offline. Instructors can add notes for students to see as they read the text. Norton Ebooks can be viewed on—and synced among—all computers and mobile devices. Every question in Smartwork5 provides a reference link to the ebook so that students have easy access to their textbook when completing homework assignments.

Student Site

W. W. Norton's student website features the following:

- Thirty AstroTour animations. These animations, some of which are interactive, use art from the text to help students visualize important physical and astronomical concepts. All are now tablet-compatible.
- Seven new Interactive Simulations, authored by Stacy Palen, pair with the Exploration activities in the text, allowing students to explore topics such as Moon phases, Kepler's laws, and the Hertzsprung-Russell diagram.
- Twenty-three Astronomy in Action videos that feature author Stacy Palen demonstrating the most important concepts in a visual, easy to understand, and memorable way.

Learning Astronomy by Doing Astronomy: Collaborative Lecture Activities

Stacy Palen, Weber State University

Ana Larson, University of Washington

Students learn best by doing. Devising, writing, testing, and revising suitable in-class activities that use real astronomical data, illuminate astronomical concepts, and pose probing questions that ask students to confront misconceptions can be challenging and time consuming. In this workbook, the authors draw on their experience teaching thousands of students in many different types of courses (large in-class, small in-class, hybrid, online, flipped, and so forth) to bring 30 field-tested activities that can be used in any classroom today. The activities have been designed to require no special software, materials, or equipment and to take no more than 50 minutes to do.

Starry Night Planetarium Software (College Version) and Workbook

Steven Desch, Guilford Technical Community College
Michael Marks, Bristol Community College

Starry Night is a realistic, user-friendly planetarium simulation program designed to allow students in urban areas to perform observational activities on a computer screen. Norton's unique accompanying workbook offers observation assignments that guide students' virtual explorations and help them apply what they've learned from the text reading assignments.

For Instructors

Instructor's Manual

Ana Larson, University of Washington

This resource includes brief chapter overviews, suggested discussion points, notes on the Process of Science figures, AstroTour animations, Astronomy in Action videos, Interactive Simulations, Reading Astronomy News, Explorations, and worked solutions to Check Your Understanding and end-of-chapter questions and problems. Also included are notes on teaching with *Learning Astronomy by Doing Astronomy: Collaborative Lecture Activities* and answers to the *Starry Night Workbook* exercises.

Norton Interactive Instructor's Guide (IIG)

This new and searchable online resource is designed to help instructors prepare for lecture in real time. It contains the following resources, each tagged by topic and chapter:

- Discussion Points
- Notes on the Process of Science figures
- AstroTour animations
- Astronomy in Action videos
- Interactive Simulations
- Notes on Reading Astronomy News and Explorations
- Notes on teaching with Learning Astronomy by Doing Astronomy: Collaborative Lecture Activites
- Solutions to *Check Your Understanding and end-of-chapter* questions and problems
- Answers to *Starry Night Workbook* exercises
- PowerPoint Lecture Slides
- All art and tables in JPEG and PowerPoint formats
- LMS Coursepacks, available in Blackboard, Canvas, Desire2Learn, and Moodle formats

All of the above content is also available on the traditional Instructor's Resource Site.

Teaching Astronomy by Doing Astronomy blog
tada101.com

This new resource features posts by Stacy Palen on teaching ideas, classroom activities, and current events. Instructors looking for additional ways to incorporate active learning will find this blog to be an invaluable resource.

PowerPoint Lecture Slides

Christa Speights, Northern Kentucky University

These ready-made lecture slides integrate selected textbook art, all Check Your Understanding and Working It Out questions from the text, class questions, and links to the AstroTour animations. These lecture slides are fully editable and are available in Microsoft PowerPoint format.

Test Bank

Matthew Newby, Temple University

The Test Bank has been revised using Bloom's Taxonomy and provides over 2,400 multiple-choice and short-answer problems. Each chapter of the Test Bank consists of five question levels classified according to Bloom's Taxonomy:

- Remembering
- Understanding
- Applying
- Analyzing
- Evaluating

Problems are further classified by section and difficulty level, making it easy to construct tests and quizzes that are meaningful and diagnostic. The Test Bank assesses a common set of Learning Objectives consistent with the textbook and Smartwork5 online homework.

LMS Coursepacks

Norton's Coursepacks, available for use in various Learning Management Systems (LMSs), feature all Test Bank questions, links to the AstroTours, worksheets based on the Explorations and Astronomy in Action videos, and automatically graded versions of selected end-of-chapter Test Your Understanding multiple-choice questions. Coursepacks are available in BlackBoard, Canvas, Desire2Learn, and Moodle formats.

Acknowledgments

The authors would like to acknowledge the extraordinary efforts of the staff at W. W. Norton: Sara Bonacum who kept things flowing smoothly; Jane Miller and Trish Marx for managing the numerous photos; and the copy editor, Gabe Waggoner, who made sure that all the grammar and punctuation survived the multiple rounds of the editing process. We would especially like to thank John Murdzek for the developmental editing process, and Diane Cipollone, who shepherded the manuscript through production.

Erik Fahlgren was the editor. Ashley Horna managed the production and Rubina Yeh was the design director. Rob Bellinger, Arielle Holstein, and Kelly Smith worked on the media and supplements, and Katie Sweeney will help get this book into the hands of people who can use it.

We gratefully acknowledge the contributions of the authors who worked on previous editions of *21st Century Astronomy*: Jeff Hester, Gary Wegner and the late Dave Burstein, Ron Greeley, Brad Smith, and Howard Voss, with special thanks to Dave for starting the project, to Jeff for leading the original authors through the first edition, and to Brad for leading the second and third editions.

Laura Kay

Stacy Palen

George Blumenthal

And we would like to thank the reviewers, whose input at every stage improved the book:

Sixth Edition Reviewers

Manuel Alvarado, El Paso Community College
Gagandeep Anand, Boston University
Simon Balm, Santa Monica College
Raymond Benge, Tarrant College
Ryan Bennett, University of North Texas
Brett Bochner, Hofstra University
Jack Brockway, Radford University
Shea Brown, University of Iowa
Spencer Buckner, Austin Peay State University
C. Austin Campbell, El Paso Community College
Christina Cavalli, Austin Community College
Gerald Cecil, University of North Carolina- Chapel Hill
Scott Cochran, Central Michigan University
Tara Cotton, University of Georgia
Robert Coyne, Texas Tech University
Steve Desche, Arizona State University
Michael Endl, Austin Community College
Duncan Farrah, University of Hawaii
Sunil Fernandes, University of Texas- San Antonio
Ken Gayley, University of Iowa
Christopher Gerardy, University of North Carolina- Charlotte
Guillermo Gonzalez, Ball State University
Ioannis Haranas, Wilfrid Laurier College
Javier Hasbun, University of West Georgia
Jim Higdon, Georgia Southern University
Scott Hildreth, Chabot College
Christopher Hodges, Sacramento State University
Mike Hood, Mt. San Antonio College
James Imamura, University of Oregon
Jennifer Jones, Arapahoe Community College
Bruno Jungweirt, North Carolina State University
Patrick Kelly, Dalhousie University
Bethuel Khamala, El Paso Community College
Keigo Fukumura, James Madison University
Rafael Lang, Purdue University

Kenneth Lanzetta, Stony Brook University
Lee LaRue, Paris Junior College
Denis Leahy, University of Calgary
Hector Leal, University of Texas- Rio Grande Valley
Hyun-chul Lee, University of Texas- Rio Grande Valley
Ludwik Lembryk, University of Toledo
Isaac Lopez, Boston University
Petrus Martens, Georgia State University
Justin Mason, Old Dominion University
Douglas McElroy, Chaffey College
Robert Morehead, Texas Tech University
Hon-kie Ng, Florida State University
Merav Opher, Boston University
Robert Parks, Louisiana State University
Jon Pedicino, College of the Redwoods
Nicolas Pereyra, University of Texas- Rio Grande Valley
Bob Powell, University of West Georgia
Barry Rice, Sierra College
Ruben Sandapen, Acadia University
Eric Schlegel, University of Texas- San Antonio
Michael Schwartz, Santa Monica College
Parampreet Singh, Louisiana State University
James Sy, El Paso Community College
Catherine Tabor, El Paso Community College
Benjamin Team, University of West Georgia
Fiorella Terenzi, Florida International University
Lennart Van Haaften, Texas Tech University
Jasper Wall, University of British Columbia
Colin Wallace, University of North Carolina- Chapel Hill
G. Scott Watson, Syracuse University
James Webb, Florida International University
Julia Wickett, Collin College
Kurt Williams, Texas A&M University
Fred Wilson, Collin College
David Wood, San Antonio College

Reviewers of Previous Editions

Scott Atkins, University of South Dakota
Timothy Barker, Wheaton College
Peter A. Becker, George Mason University
Timothy C. Beers, Michigan State University
David Bennum, University of Nevada-Reno
Edwin Bergin, University of Pittsburgh
William Blass, University of Tennessee
Steve Bloom, Hampden Sydney College
Daniel Boice, University of Texas at San Antonio
Bram Boroson, Clayton State University
David Branning, Trinity College
Julie Bray-Ali, Mt. San Antonio College
Suzanne Bushnell, McNeese State University
Paul Butterworth, George Washington University
Juan E. Cabanela, Minnesota State University-Moorhead
Amy Campbell, Louisiana State University
Michael Carini, West Kentucky University
Supriya Chakrabarti, Boston University
Robert Cicerone, Bridgewater State College
David Cinabro, Wayne State University
Judith Cohen, California Institute of Technology
Eric M. Collins, California State University-Northridge
John Cowan, University of Oklahoma-Norman
Debashis Dasgupta, University of Wisconsin-Milwaukee
Robert Dick, Carleton University
Gregory Dolise, Harrisburg Area Community College
Tom English, Guilford Technical Community College
David Ennis, The Ohio State University
John Finley, Purdue University
Matthew Francis, Lambuth University
Kevin Gannon, College of Saint Rose
Todd Gary, O'More College of Design
Christopher Gay, Santa Fe College
Parviz Ghavamian, Towson University
Martha Gilmore, Wesleyan University
Bill Gutsch, St. Peter's College
Karl Haisch, Utah Valley University
Charles Hawkins, Northern Kentucky University
Sebastian Heinz, University of Wisconsin-Madison
Barry Hillard, Baldwin Wallace College

Paul Hintzen, California State University-Long Beach
Paul Hodge, University of Washington
William A. Hollerman, University of Louisiana at Lafayette
Hal Hollingsworth, Florida International University
Olencka Hubickyj-Cabot, San Jose State University
Kevin M. Huffenberger, University of Miami
Adam Johnston, Weber State University
Steven Kawaler, Iowa State University
Monika Kress, San Jose State University
Jessica Lair, Eastern Kentucky University
Alex Lazarian, University of Wisconsin-Madison
Kevin Lee, University of Nebraska-Lincoln
Matthew Lister, Purdue University
M. A. K. Lodhi, Texas Tech University
Leslie Looney, University of Illinois at Urbana-Champaign
Jack MacConnell, Case Western Reserve University
Kevin Mackay, University of South Florida
Dale Mais, Indiana University-South Bend
Michael Marks, Bristol Community College
Norm Markworth, Stephen F. Austin State University
Kevin Marshall, Bucknell University
Stephan Martin, Bristol Community College
Amanda Maxham, University of Nevada-Las Vegas
Chris McCarthy, San Francisco State University
Ben McGimsey, Georgia State University
Charles McGruder, West Kentucky University
Janet E. McLarty-Schroeder, Cerritos College
Stanimir Metchev, Stony Brook University
Chris Mihos, Case Western University
Milan Mijic, California State University-Los Angeles
J. Scott Miller, University of Louisville
Scott Miller, Sam Houston State University
Kent Montgomery, Texas A&M University-Commerce
Andrew Morrison, Illinois Wesleyan University
Edward M. Murphy, University of Virginia
Kentaro Nagamine, University of Nevada-Las Vegas
Ylva Pihlström, University of New Mexico

Jascha Polet, California State Polytechnic University
Dora Preminger, California State University-Northridge
Daniel Proga, University of Nevada-Las Vegas
Laurie Reed, Saginaw Valley State University
Judit Györgyey Ries, University of Texas
Allen Rogel, Bowling Green State University
Kenneth Rumstay, Valdosta State University
Masao Sako, University of Pennsylvania
Samir Salim, Indiana University-Bloomington
Ata Sarajedini, University of Florida
Paul Schmidtke, Arizona State University
Ann Schmiedekamp, Pennsylvania State University
Jonathan Secaur, Kent State University
Ohad Shemmer, University of North Texas
Caroline Simpson, Florida International University
Paul P. Sipiera, William Rainey Harper College
Ian Skilling, University of Pittsburgh
Tammy Smecker-Hane, University of California-Irvine
Allyn Smith, Austin Peay State University
Jason Smolinski, State University of New York at Oneonta
Roger Stanley, San Antonio College
Ben Sugerman, Goucher College
Neal Sumerlin, Lynchburg College
Angelle Tanner, Mississippi State University
Christopher Taylor, California State University-Sacramento
Donald Terndrup, The Ohio State University
Todd Thompson, The Ohio State University
Glenn Tiede, Bowling Green State University
Frances Timmes, Arizona State University
Trina Van Ausdal, Salt Lake Community College
Walter Van Hamme, Florida International University
Karen Vanlandingham, West Chester University
Nilakshi Veerabathina, University of Texas at Arlington
Paul Voytas, Wittenberg University
Ezekiel Walker, University of North Texas
James Webb, Florida International University
Paul Wiita, Georgia State University
Richard Williamon, Emory University
David Wittman, University of California-Davis

About the Authors

Laura Kay is an award-winning professor and Chair of Physics and Astronomy at Barnard College, where she has taught since 1991. She received a BS degree in physics and an AB degree in feminist studies from Stanford University, and MS and PhD degrees in astronomy and astrophysics from the University of California, Santa Cruz. As a graduate student she spent 13 months at the Amundsen Scott station at the South Pole in Antarctica. She studies active galactic nuclei, using ground-based and space telescopes. She teaches courses in astronomy, astrobiology, women and science, and polar exploration. At Barnard she has served as chair of the Physics & Astronomy Department, chair of the Women's Studies Department, chair of Faculty Governance, and interim associate dean for Curriculum and Governance.

Stacy Palen is an award-winning professor in the physics department and the director of the Ott Planetarium at Weber State University. She received her BS in physics from Rutgers University and her PhD in physics from the University of Iowa. As a lecturer and postdoc at the University of Washington, she taught Introductory Astronomy more than 20 times over 4 years. Since joining Weber State, she has been very active in science outreach activities ranging from star parties to running the state Science Olympiad. Stacy does research in formal and informal astronomy education and the death of Sun-like stars. She spends much of her time thinking, teaching, and writing about the applications of science in everyday life. She then puts that science to use on her small farm in Ogden, Utah.

George Blumenthal has been a professor of astronomy and astrophysics at the University of California, Santa Cruz since 1972, and served as the university's Chancellor from 2006 to 2019. He received his BS degree from the University of Wisconsin–Milwaukee and his PhD in physics from the University of California, San Diego. As a theoretical astrophysicist, George's research encompasses several broad areas, including the nature of the dark matter that constitutes most of the mass in the universe, the origin of galaxies and other large structures in the universe, the earliest moments in the universe, astrophysical radiation processes, and the structure of active galactic nuclei such as quasars. Besides teaching and conducting research, he has served as Chair of the UC Santa Cruz Astronomy and Astrophysics Department, has chaired the Academic Senate for both the UC Santa Cruz campus and the entire University of California system, and has served as the faculty representative to the UC Board of Regents.

21ST CENTURY ASTRONOMY
The Solar System

SIXTH EDITION

1 Thinking Like an Astronomer

Now is a fascinating time to be studying this most ancient of the sciences. Loosely translated, the word **astronomy** means "patterns among the stars." But modern astronomy—the astronomy we talk about in this book—is about far more than looking at the sky and cataloging the visible stars. The contents of the universe, its origin and fate, and the nature of space and time have become the subjects of rigorous scientific investigation. Humans have long speculated about our *origins*. How and when did the Sun, Earth, and Moon form? Are other galaxies, stars, planets, and moons similar to our own? The answers that scientists are finding to those questions are changing our view of not only the cosmos but also ourselves.

LEARNING GOALS

In this chapter, we will begin studying astronomy by exploring our place in the universe and the methods of science. By the end of this chapter, you should be able to:

LG 1 Describe the size and age of the universe and Earth's place in it.

LG 2 Explain how astronomers use the scientific method to study the universe.

LG 3 Show how scientists use mathematics, including graphs, to find patterns in nature.

LG 4 Describe our astronomical origins.

The first view of Earth seen by humans from deep space. In December 1968, *Apollo 8* astronauts photographed Earth above the Moon's limb. ▶▶▶

Why is this considered one of the most famous photos ever taken?

CHAPTER 1 Thinking Like an Astronomer

1.1 Earth Occupies a Small Place in the Universe

Astronomers contemplate our place in the universe by studying Earth's position in space and time. Locating Earth in the larger universe is the first step in learning the science of astronomy. In this section, you will get a feel for the neighborhood in which Earth is located. You will also begin to explore the scale of the universe in space and time.

Our Place in the Universe

Most people receive their postal mail at an address—house or building number, street, city, state, and country. If we expand our view to include the enormously vast universe, however, our "cosmic address" might include our planet, star, galaxy, galaxy group or cluster, and galaxy supercluster.

We reside on a planet called Earth, which is orbiting under the influence of gravity around a star called the **Sun**. The Sun is an ordinary middle-aged star more massive and luminous than some stars but less massive and luminous than others. The Sun is extraordinary only because of its importance to us within our own **Solar System**. Our Solar System consists of eight planets (listed in order of distance from the Sun): Mercury, Venus, Earth, Mars, Jupiter, Saturn, Uranus, and Neptune. It also contains many smaller bodies, such as dwarf planets (for example, Pluto, Ceres, and Eris), asteroids (for example, Ida and Eros), and comets (for example, Halley). All those objects are gravitationally bound to the Sun.

The Sun is located about halfway out from the center of the **Milky Way Galaxy**, a flattened collection of stars, gas, and dust. Our Sun is just one among several hundred billion stars scattered throughout our galaxy, and planets orbit many of those stars.

The Milky Way is a member of a collection of a few dozen galaxies called the **Local Group**. Most galaxies in that group are much smaller than the Milky Way. As we look farther outward, the Local Group is part of a vastly larger collection of thousands of galaxies—a **supercluster**—called the Virgo Supercluster, which is part of an even larger grouping called the Laniakea Supercluster. The observable universe has millions of superclusters.

We can now define our cosmic address—Earth, Solar System, Milky Way Galaxy, Local Group, Virgo Supercluster, Laniakea Supercluster—as illustrated in **Figure 1.1**. Yet even that address is not complete because it encompasses only the *local universe*. The part of the universe that we can see—the *observable universe*—extends to many times the size of Laniakea in every direction. That volume contains about 1 trillion (thousand billion) to 2 trillion galaxies. The entire universe is much larger than the local universe and contains much more than the observed planets, stars, and galaxies. Astronomers estimate that about 95 percent of the universe is made up of matter that does not interact with light, known as *dark matter*, and a form of energy that permeates all space, known as *dark energy*. Dark matter and dark energy aren't well understood, and they are among the many exciting areas of research in astronomy.

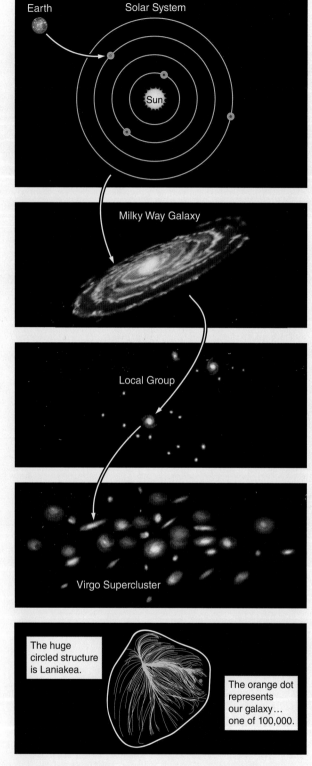

Figure 1.1 Our cosmic address is Earth, Solar System, Milky Way Galaxy, Local Group, Virgo Supercluster, Laniakea Supercluster. We live on Earth, a planet in our Solar System, orbiting the Sun, a star in the Milky Way Galaxy. The Milky Way is a large galaxy within the Local Group of galaxies, located in the Virgo Supercluster, one of four superclusters in the Laniakea Supercluster.

The Scale of the Universe

As you saw in Figure 1.1, the size of the universe completely dwarfs our human experience. We can start by comparing astronomical sizes and distances to something more familiar. For example, the diameter of our Moon (3,474 kilometers [km]) is slightly greater than the distance between New York, New York, and Ogden, Utah (**Figure 1.2a**), where two of the authors of this book work. The distance from Earth to the Moon is about 100 times the Moon's diameter, and the planet Saturn with its majestic rings would fill much of that distance (**Figure 1.2b**). The distance from Earth to the Sun is about 400 times the Earth–Moon distance, and the distance to the planet Neptune is about 30 times the Earth–Sun distance.

As we move out from the Solar System to the stars, however, the enormous distances become difficult to comprehend. The nearest star is about 9,000 times farther from the Sun than the Sun's distance to Neptune. The diameter of our Milky Way Galaxy is 30,000 times the distance to that nearest star. The Andromeda Galaxy, the nearest large galaxy similar to the Milky Way, is about 600,000 times farther than that nearest star. The diameter of the Local Group of galaxies is about 4 times the distance to Andromeda, and the diameter of the Laniakea Supercluster, which includes the Local Group and many other galaxy groups and clusters, is 50 times larger than the Local Group. As noted earlier, that supercluster is just one of millions in the observable universe.

To get a better sense of those distances, imagine a model in which the objects and distances in the universe are 1 billion times smaller than they really are. In that model, Earth is about the size of a marble or a peanut M&M (about 1.3 centimeters [cm], or half an inch), the Moon is 38 cm away, and the Sun is 150 meters away. Neptune is 4.5 km from the Sun, and the nearest star to the Sun is about 40,000 km away (or about the length of the real Earth's circumference). The model Milky Way Galaxy would fill the real Solar System nearly to the orbit of Saturn. The distance between the model Milky Way and Andromeda galaxies would fill the real Solar System 20 times farther, beyond humanity's most distant space probe. The model Laniakea Supercluster would fill the real Solar System and go about one-eighth of the way to the nearest star.

When thinking about distances in the universe, discussing the time it takes to travel to various places can be helpful. If someone asks you how far it is to the nearest city, you might say 100 km or you might say 1 hour. In either case, you will have given that person an idea of how far the city is. In astronomy, the speed of a car on the highway is far too slow to be useful. Instead, we use the fastest speed in the universe—the speed of light. Light travels at 300,000 kilometers per second (km/s). **Figure 1.3** shows how distances that range from the size of Earth to the size of the observable universe can be measured in the time it takes light to travel that far. Light can circle Earth, for example—a distance of 40,000 km—in just under 1/7 of a second. The circumference of Earth, then, is 1/7 of a light-second. Even relatively small distances in astronomy are so vast that they are measured in units of **light-years (ly)**: the distance light travels in 1 year, about 9.5 trillion km, or 6 trillion miles.

Because light takes time to reach us, we see astronomical objects as they were in the past: how far back in time depends on the object's distance from us. Because light from the Moon takes 1 1/4 seconds to reach us, we see the Moon as it was 1 1/4 seconds ago. Because light from the Sun takes 8 1/3 minutes to reach us, we see the Sun as it was 8 1/3 minutes ago. We see the nearest star as it was more than 4 years ago and objects across the Milky Way as they were tens of

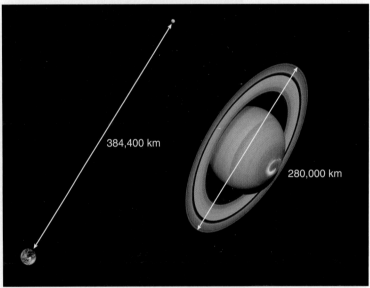

Figure 1.2 (a) The diameter of the Moon is about the same as the distance between New York, New York, and Ogden, Utah. (b) The size of Saturn, including the rings, is about 70 percent of the distance between Earth and the Moon.

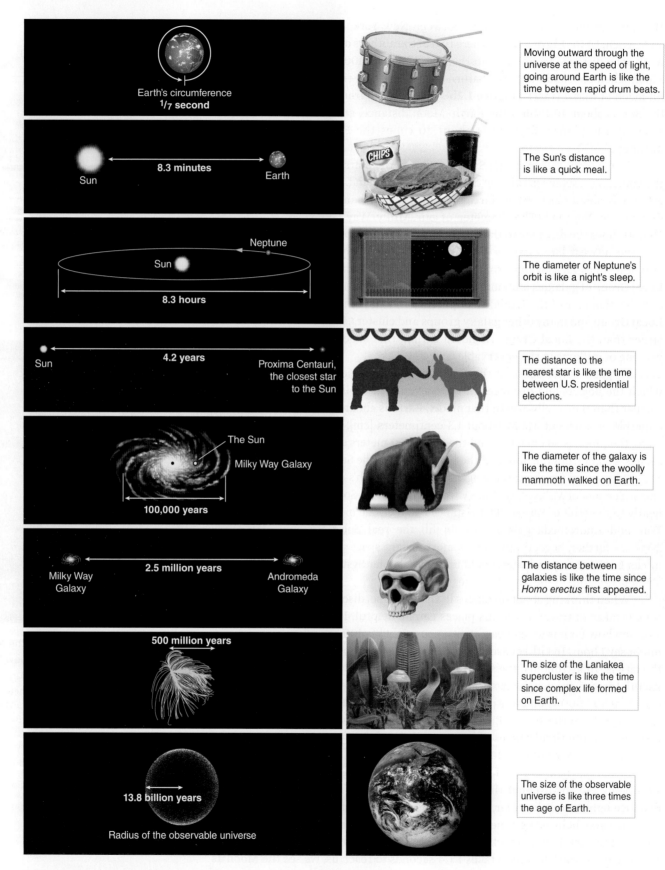

Figure 1.3 Thinking about the time light takes to travel between objects helps us comprehend the vast distances in the observable universe.

thousands of years ago. The light from the Virgo Cluster of galaxies has travelled for 50 million years to reach us. The light from the most distant observable objects has been traveling for almost the age of the universe—nearly 13.8 billion years.

The vast distances from Earth to other objects in the universe tell us that we occupy a very small part of the space in the universe and a very small part of time. Earth and the Solar System are only about one-third as old as the universe. Animals have existed on Earth for even less time. Imagine the age of the universe and the important events in it as though they took place within a single day, as illustrated in **Figure 1.4**. In that timeline, the Big Bang begins the cosmic day at midnight, and the original light chemical elements are created within the first 2 seconds. The first stars and galaxies appear within the first 10 minutes. Our Solar System formed from recycled gas and dust left over from previous generations of stars, at about 4 P.M. on that cosmic clock. The first bacterial life appears on Earth at 5:20 P.M., the first land animals at 11:20 P.M., and modern humans at 11:59:59.8 P.M.—that is, with just 1/5 of a second left in the cosmic day. We humans appeared quite recently in the history of the universe.

CHECK YOUR UNDERSTANDING 1.1

Rank the following from smallest to largest: (a) a light-minute, (b) a light-year, (c) a light-hour, (d) the radius of Earth, (e) the distance from Earth to the Sun, (f) the radius of the Solar System.

1.2 Science Is a Way of Viewing the Universe

Humans have long paid attention to the sky and the stars, making new discoveries and contributing new ideas about the universe. Those discoveries and new ideas still occur often and evolve rapidly, making astronomy a dynamic science. To view the universe through the eyes of an astronomer, you need to understand how science itself works. Throughout this book, we emphasize not only scientific discoveries but also the *process* of science. We begin with this section, which examines the scientific method.

The Scientific Method

The **scientific method** is a systematic way of exploring the world on the basis of developing and then testing new ideas or explanations. Often, the method begins with a fact—an observation or a measurement. For example, you might observe that the weather changes predictably each year and wonder why that happens. You then create a **hypothesis**, a testable explanation of the observation: "I think that it is cold in the winter and warm in the summer because Earth is closer to the Sun in the summer." You come up with a test: if it is cold in the winter and warm in the summer because Earth is closer to the Sun in the summer, then everywhere on the planet will be cold in the winter—Australia should have winter at the same time of year as the United States. You can use that test to check your hypothesis. In January, you travel from the United States to Australia and find that it is summer in Australia. Your hypothesis has just been proved incorrect, so we say that it has been **falsified**. That is different from the word's meaning in common usage, in which "falsified" evidence has been manipulated to misrepresent the truth. Your test has two important elements that all scientific tests share. Your observation is reproducible: anyone who goes to Australia will find the same result. And

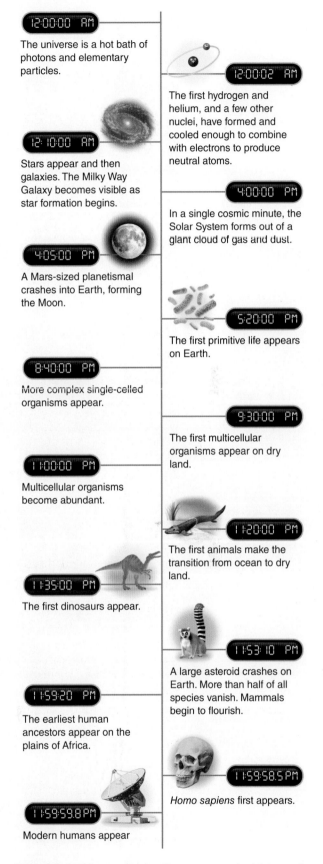

Figure 1.4 This cosmic timeline presents the history of the universe as a 24-hour day.

your result is repeatable: if you conducted a similar test next year or the year after, you would get the same result. Because you have falsified your hypothesis, you must revise or replace it to be consistent with the new data.

Any idea that is not testable—that is, not falsifiable—must be accepted or rejected based on intuition alone, so it is not a scientific idea. A falsifiable hypothesis or idea does not have to be testable using current technology, but we must be able to imagine an experiment or observation that could prove the idea wrong if we could carry it out. As continuing tests support a hypothesis by failing to disprove it, scientists come to accept the hypothesis as a theory. A **theory** is a well-developed idea or group of ideas that must agree with known physical laws and make testable predictions of a given phenomenon or system. As with "falsified," the scientific meaning of "theory" is different from the meaning in common usage. In everyday language, theory may mean a guess: "Do you have a theory about who did it?" In everyday language, a theory can be something we don't take too seriously. "After all," people say, "it's only a theory."

In stark contrast, scientists use the word *theory* to mean a carefully constructed proposition that takes into account every piece of relevant data as well as our entire understanding of how the world works. A theory has been or will be used to make testable predictions, and all those predictions must come true. Every attempt to prove it false fails. A classic example is Einstein's theory of relativity, which Chapter 18 covers in some depth. For more than a century, scientists have tested the predictions of the theory of relativity and have not been able to falsify it. Even after 100 years of verification, if a prediction of the theory of relativity failed tomorrow, the theory would require revision or replacement. As Einstein himself noted, a theory that fails only one test is proved false. In that sense, all scientific knowledge is subject to challenge. In addition, scientific results should be reproducible so that other people get the same results.

In the loosely defined hierarchy of scientific knowledge, an *idea* is a notion about how something might be. Moving up the hierarchy, we come to a *fact*—an observation or measurement. The measured value of Earth's radius is a fact. A *hypothesis* is an idea that leads to testable predictions. A hypothesis may be the forerunner of a scientific theory, may be based on an existing theory, or both. At the top of the hierarchy is a *theory*: an idea that has been examined carefully, is consistent with all existing theoretical and observational knowledge, and makes testable predictions. Ultimately, the success of the predictions is the deciding factor between competing theories. A scientific *law* is a series of observations that enables making predictions but has no underlying explanation of why the phenomenon occurs. A *law* of daytime might say that the Sun rises and sets once each day, whereas a *theory* of daytime might say that the Sun rises and sets once each day because Earth spins on its axis. Scientists themselves can be sloppy about how they use those words, so you may sometimes see them used differently from how we have defined them here.

As the **Process of Science Figure** shows, the scientific method follows a specific sequence. Scientists begin with an observation or idea, followed by analysis, followed by a hypothesis, followed by prediction, followed by further observations or experiments to test the prediction. A hypothesis may lead to a scientific theory, may be based on an existing theory, or both. Ultimately, the basis for deciding among competing theories is the success of their predictions. Scientists can use theories to take their knowledge a step further by building theoretical models. A **theoretical model** describes the properties of a particular object or system in terms of known physical laws or theories, which are used to connect the theoretical model to the behavior of a complex system.

Scientific Principles

Scientific **principles** are general ideas or a sense about the universe that often guides the construction of new theories. Two principles used in studying astronomy are the *cosmological principle* and *Occam's razor*.

The **cosmological principle** is the testable assumption that the physical laws that apply here and now also apply everywhere and at all times. That principle also implies that the universe has no special locations or directions and that matter and energy obey the same physical laws throughout space and time as they do today on Earth. The principle means that the same physical laws that we observe and apply in laboratories on Earth can be used to understand what goes on in the centers of stars or in distant galaxies. Each new theory that comes from applying the cosmological principle to observations of the universe around us adds to our confidence in the validity of that principle, which we discuss in more detail in Chapter 21.

Occam's razor states that when facing two hypotheses that explain all the observations equally well, we should favor the one that requires the fewest assumptions until we have evidence to the contrary. For example, a hypothesis that atoms are constructed differently in the Andromeda Galaxy and in the Milky Way Galaxy would violate the cosmological principle. That hypothesis also would require many assumptions to explain how atoms in the Andromeda Galaxy are constructed differently and yet still appear to behave in the same way as atoms in the Milky Way. For example, suppose you assume that the center of an atom in Andromeda is negatively charged, whereas the center of an atom in the Milky Way is positively charged. That assumption would require you to make additional assumptions about the location of the boundary between Andromeda-like matter and Milky Way–like matter and about why atoms on the boundary between the two regions did not destroy each other. You also would need an assumption about why atoms in the two regions are constructed so differently. If reasonable experimental evidence ever challenges the validity of the cosmological principle, scientists will construct a new description of the universe that takes those new data into account. Until then, the cosmological principle is the hypothesis that has the fewest assumptions, thereby satisfying Occam's razor. To date, the cosmological principle has been repeatedly tested and remains valid.

In many sciences, researchers can conduct controlled experiments to test hypotheses. That experimental method is often unavailable to astronomers: we cannot change the tilt of Earth or the temperature of a star to see what happens. Instead, astronomers work from observations or models. Astronomers make multiple observations by using various methods and then create mathematical and physical models based on established science to explain the observations.

For example, when astronomers first discovered planets orbiting other stars, those new extrasolar planets, or *exoplanets*, were most often giant gaseous planets (similar to Jupiter) in short-period orbits very close to their star. Those planets are the easiest to discover because they have a strong pull on their star, as we discuss in Chapter 7. However, their proximity to their star is completely unlike the situation in our own Solar System, where the giant planets are far from the Sun. Those discoveries challenged existing ideas about how our Solar System formed. As different observers using multiple telescopes found more and more of those planets, astronomers realized they needed new explanations of planet formation to explain how such large planets could wind up so close to their star. Astronomers could not build different solar systems—but they could use the known laws of physics to create computer simulations of planetary systems. When researchers did that, they found that planets can migrate, moving to orbits closer to or farther from their star. Planetary scientists are searching for evidence to test that idea in our own Solar System, where that process may have occurred early in the life of the Sun. In that example, the theory was motivated by the earlier observations.

Alternatively, an astronomer might make predictions from an existing successful mathematical or physical model and then conduct observations and analyze data to test the predictions. One example is the discovery of black holes. In the late 18th century, two scientists hypothesized the existence of "dark stars": massive objects having such strong gravity that light could not escape. At that time, the scientists had no way to test that hypothesis. More than 100 years later, in the early 20th century, Karl Schwarzschild (1873–1916) studied Einstein's relativity equations and calculated that those collapsed dark stars would be very small, with a radius of only a few kilometers. Fifty years later, those objects were named *black holes*. No evidence of their existence was available until the 1970s and 1980s, when the new technology of space-based X-ray telescopes made possible the observations needed to test the hypothesis. Einstein's existing theory made the discovery of black holes possible.

The scientific method provides the rules for testing whether an idea is false, but it offers no insight into where the idea came from in the first place or how an experiment was designed. Scientists discussing their work use words such as *insight*, *intuition*, and *creativity*. Scientists speak of a beautiful theory in the same way that an artist speaks of a beautiful painting or a musician speaks of a beautiful song. Science has an aesthetic that is as human and as profound as any found in the arts.

In one important respect, though, science is different from art or music. Whereas art and music are judged by a human jury alone, the validity of a scientific hypothesis or theory is subject to the natural world. Nature alone is the judge of which theories can be kept and which theories must be discarded. What we want to be true does not matter. In the history of science, many a beautiful and beloved theory has been abandoned.

Scientific Revolutions

Scientific inquiry is necessarily dynamic. Scientists do not have all the answers and must constantly refine their ideas in response to new data and new insights. The vulnerability of knowledge may seem like a weakness. "Gee, you really don't know anything," the cynical person might say. But that vulnerability is actually the scientific process's greatest strength: it means that science self-corrects. New information eventually overturns incorrect ideas. In science, even our most cherished ideas about the nature of the physical world remain fair game, subject to challenge by new evidence. Many of history's best scientists earned their status by falsifying a universally accepted idea. That is a powerful motivation for scientists to challenge old ideas constantly—to formulate and test new explanations for their observations.

For example, the classical physics that Sir Isaac Newton developed in the 17th century to explain motion, forces, and gravity withstood the scrutiny of scientists for more than 200 years. During the late 19th and early 20th centuries, however, a series of scientific revolutions completely changed our understanding of the nature of reality. The work of Albert Einstein (**Figure 1.5**) is representative of those scientific revolutions. Einstein's special and general theories of relativity replaced Newton's mechanics. Einstein did not prove Newton wrong but rather

Figure 1.5 Albert Einstein is perhaps the most famous scientist of the 20th century, and he was *Time* magazine's selection for Person of the Century. Einstein helped usher in two scientific revolutions.

showed that Newton's theories were a special case of a far more general and powerful set of physical theories. Einstein's new ideas unified the concepts of mass and energy and destroyed the conventional notion of space and time as separate concepts.

Throughout this text, you will encounter many other discoveries that forced scientists to abandon accepted theories. Einstein himself never embraced the view of the world offered by *quantum mechanics*—a second revolution he helped start. Yet quantum mechanics, a statistical description of the behavior of particles smaller than atoms, has held up for more than 100 years. In science, all authorities are subject to challenge, even Einstein.

Science is a way of thinking about the world. It is a search for the relationships that make our world what it is. Scientific knowledge is an accumulated collection of ideas about how the universe works, yet scientists are always aware that what is known today may be superseded tomorrow. A scientist assumes that order exists in the universe and that the human mind can grasp the essence of the rules underlying that order. Scientists build on those assumptions to make and then test predictions, finding the underlying rules that allow humanity to solve problems, invent new technologies, or find a new appreciation for the natural world. In the final analysis, science has found such a central place in our civilization because *science works*.

Figure 1.6 Since ancient times, people recognized that patterns in the sky change with the seasons. Those and other patterns shape our lives. These star maps show the sky in the Northern Hemisphere during each season.

CHECK YOUR UNDERSTANDING 1.2

The scientific method is a process by which scientists: (a) prove theories to be known facts; (b) gain confidence in theories by failing to prove them wrong; (c) show all theories to be wrong; (d) survey what most scientists think about a theory.

1.3 Astronomers Use Mathematics to Find Patterns

Scientific thinking allows scientists to make predictions. Once a pattern has been observed, such as the daily rising and setting of the Sun, scientists can predict what will happen next.

Imagine that the patterns in your life became disrupted, so that the world became entirely unpredictable. For example, what would life be like if you could not predict whether an object you dropped would fall up or down? Or what if one morning the Sun rose in the east and set in the west, and the next day it rose in the west and set in the east? In fact, objects do fall toward the ground. The Sun rises, sets, and then rises again at predictable times and in predictable locations. Spring turns into summer, summer turns into autumn, autumn turns into winter, and winter turns into spring again. The rhythms of nature produce patterns in our lives, and those patterns give us clues about the nature of the physical world. Astronomers identify and characterize those patterns and use them to understand the world around us. Some of the most easily identified patterns in nature are those we see in the sky. What in the sky will look different or the same a week from now? A month from now? A year from now? As you can see in **Figure 1.6**, patterns in the sky mark the changing of the seasons. Because the seasons determine planting and harvesting times, astronomy—which studies those patterns—is the oldest of all sciences. We describe many other examples of patterns in the sky in Chapter 2.

Astronomers use mathematics to analyze patterns and to communicate complex material compactly and accurately. Because studying patterns in nature is so important to science, it should come as no surprise that mathematics is the language of science. Many people find mathematics to be a major obstacle that prevents them from appreciating the beauty and elegance of the world as seen through the eyes of a scientist. In this book, we strive to explain any necessary math in everyday language. We will describe what equations mean and help you use them in a way that allows you to connect scientific concepts to the world. Your responsibility is to accept the challenge and make an honest effort to understand the material. **Working It Out 1.1** and **Working It Out 1.2** review some basics of mathematical tools and graphs. At the back of the book, Appendix 1 explains some essential mathematical tools, and Appendix 2 contains physical constants of nature. Other appendixes contain data tables with key information about planets, moons, and stars.

CHECK YOUR UNDERSTANDING 1.3

When you see a pattern in nature, it is usually evidence of: (a) a theory's being displayed; (b) a breakdown of random clustering; (c) an underlying physical law; (d) a coincidence.

1.1 Working It Out Mathematical Tools

Scientists use mathematics to understand the patterns they observe and to communicate that understanding to others. The following mathematical tools will be useful in our study of astronomy:

Scientific notation. Scientific notation is how we handle numbers of vastly different sizes. Writing out 7,540,000,000,000,000,000,000 in standard notation is inefficient. Scientific notation uses the first few digits (the *significant* ones) and counts the number of decimal places to create the condensed form 7.54×10^{21}. Similarly, instead of writing out 0.000000000005, we write 5×10^{-12}. The exponent on the 10 is positive or negative depending on whether the decimal point moves right or left, respectively. For example, the average distance to the Sun is 149,600,000 km, but astronomers usually express that value as 1.496×10^8 km.

Ratios. Ratios are a useful way to compare objects. A star may be "10 times as massive as the Sun" or "10,000 times as luminous as the Sun." Those expressions are ratios.

Proportionality. Often, understanding a concept amounts to understanding the *sense* of the relationships that it predicts or describes. "If you have twice as far to go, getting there will take you twice as long." "If you have half as much money, you will be able to buy only half as much gas." Those are examples of proportionalities.

Appendix 1 further explains the mathematical tools used in this book.

1.2 Working It Out Reading a Graph

Scientists often convey complex information and mathematical patterns in graphical form. Reading graphs is an important skill not only in astronomy but also in life. Economists, social scientists, mortgage brokers, financial analysts, urban planners, historians, medical professionals, and scientists all use graphs to evaluate and communicate important information.

Graphs typically have two axes: a horizontal axis (the x-axis) and a vertical axis (the y-axis). The x-axis usually shows an independent variable, the one a researcher might have control over in an experiment, whereas the y-axis shows the dependent variable, which—in many experiments—is the variable a researcher is studying.

Graphs can take different shapes. Suppose we plot the distance a car travels over time, as shown in **Figure 1.7a**. In a linear graph, each interval on an axis represents the same-sized step. Each step on the horizontal axis of the graph in Figure 1.7a represents 5 minutes. Each step on the vertical axis represents a distance of 5 km traveled by the car. Data are plotted on the graph, with one dot for each observation; for example, after 20 minutes, the car has traveled 20 km.

Drawing a line through those data indicates the trend (or relationship) of the data. To understand what the trend means, scientists often find the slope of the line, which is the relationship of the line's rise along the y-axis to its movement along the x-axis. To find the slope, we look at the change between two points on the vertical axis divided by the change between the same two points on the horizontal axis. For example, finding the slope of the line gives

$$\text{Slope} = \frac{\text{Change in vertical axis}}{\text{Change in horizontal axis}}$$

$$\text{Slope} = \frac{(15-10)\ \text{km}}{(15-10)\ \text{min}}$$

$$= 1\ \text{km/min}$$

There, the trend tells us that the car is traveling at 1 kilometer per minute (km/min), or 60 kilometers per hour (km/h).

Many observations of natural processes do not yield a straight line on a graph—for example, an exponential process. Think about what happens when you catch a cold. You feel fine when you get up in the morning at 7 A.M. At 9 A.M. you feel a little tired. By 11 A.M. you have a bit of a sore throat or a sniffle and think, "I wonder whether I'm getting sick," and by 1 P.M., you have a runny nose, congestion, fever, and chills. That process is exponential because the virus that has infected you reproduces exponentially.

For the sake of this discussion, suppose that the virus produces one copy of itself each time it invades a cell. (Viruses actually produce 1,000–10,000 copies each time they invade a cell, so the exponential curve is much steeper.) One virus infects a cell and multiplies, so now two viruses exist—the original and a copy. Those viruses invade two new cells, and each one produces a copy. Now four viruses are there. After the next cell invasion, we have eight. Then 16, 32, 64, 128, 256, 512, 1,024, 2,048, and so on. That behavior is plotted in **Figure 1.7b**.

Seeing what's happening in the early stages of an exponential curve can be difficult because the later numbers are so much larger than the earlier ones. That's why we sometimes plot that type of data *logarithmically*, by putting the logarithm (the power of 10) of the data on the vertical axis, as shown in **Figure 1.7c**. Now each step on the axis represents 10 times as many viruses as the previous step. Even though we draw all the steps the same size on the page, they represent different-sized steps in the data—the number of viruses, for example. We often use that technique in astronomy because it has a second, related advantage: very large variations in the data can easily fit on the same graph.

Each time you see a graph, you should first understand the axes—what data are plotted on the graph? Then you should check whether the axes are linear or logarithmic. Finally, you can look at the actual data or lines in the graph to understand how the system behaves.

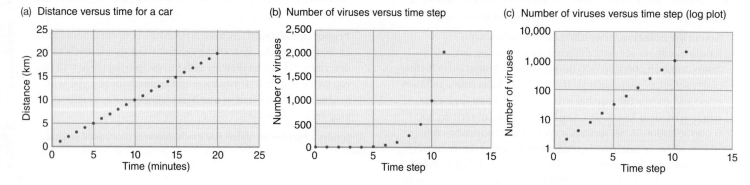

Figure 1.7 Graphs such as these show relationships between quantities. (a) The relationship between time and distance traveled. (b) The relationship between time and number of viruses. (c) The data in part (b) plotted logarithmically.

Origins An Introduction

How and when did the universe begin? What combination of events led to the existence of humans as sentient beings living on a small, rocky planet orbiting a typical middle-aged star? Are others like us scattered throughout the galaxy?

Earlier in the chapter we mentioned the theme of origins. Throughout this book, that theme involves much more than how humans came to be on Earth. In these Origins sections, which conclude each chapter, we look into the origin of the universe and the origin of life on Earth. We also examine the possibilities of life elsewhere in the Solar System and beyond—a subject called **astrobiology**. Our origins theme includes the discovery of planets around other stars and how they compare with the planets of our own Solar System.

Later in the book, we present observational evidence that supports the **Big Bang** theory, which states that the universe started expanding from an infinitesimal size about 13.8 billion years ago. In the early universe, only the lightest chemical elements were found in substantial amounts: hydrogen and helium, and tiny amounts of lithium and beryllium. However, we live on a planet with a central core consisting mostly of heavy elements such as iron and nickel, surrounded by outer layers made up of rocks containing large amounts of silicon and various other elements—all heavier than the original elements. The human body contains carbon, nitrogen, oxygen, calcium, phosphorus, and a host of other chemical elements—except for hydrogen itself, all are heavier than hydrogen and helium. If those heavier elements that make up Earth and our bodies were not present in the early universe, where did they come from?

The answer to that question lies within the stars themselves (**Figure 1.8**). In the core of a star, light elements, such as hydrogen, combine to form more massive atoms, which eventually leads to atoms such as carbon. When a star nears the end of its life, it often loses much of its material—including some of the new atoms formed in its interior—by blasting it back into interstellar space. That material combines with material lost from other stars—some of which produced even more massive atoms as they exploded—to form large clouds of dust and gas. Those clouds go on to make new stars and planets, similar to our Sun and Solar System. Earlier "generations" of stars supplied the building blocks for the chemical processes that we see in the universe, including life. The atoms that make up much of what we see were formed in the cores of stars. The phrase "we are stardust" is not just poetry. We are actually made of recycled stardust.

Studying origins also gives examples of the process of science. Many physical processes in chemistry, planetary science, physics, and astronomy that are seen on Earth or in the Solar System are observed across the Milky Way and throughout the universe. As of this writing, though, the only biology we know about is that which exists on Earth. At this point in human history, then, much of what scientists can say about the origin of life on Earth and the possibility of life elsewhere is reasoned extrapolation and educated speculation. In these Origins sections, we address some of those hypotheses and try to be clear about which are speculative and which have been tested.

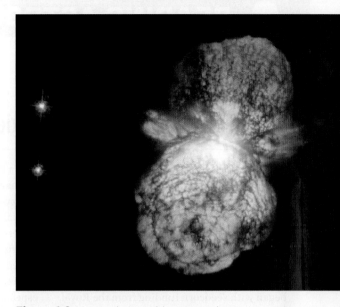

Figure 1.8 You and everything around you are composed of atoms forged in the interiors of stars that lived and died before the Sun and Earth formed. The supermassive star Eta Carinae, shown here, is ejecting a cloud of chemically enriched material just as earlier generations of stars once did to enrich the gas that would become our Solar System.

READING ASTRONOMY NEWS

ARTICLES   QUESTIONS

Astronomers think there are 10 times more galaxies than previously estimated.

A Universe of 2 Trillion Galaxies

Royal Astronomical Society

An international team of astronomers, led by Christopher Conselice, professor of astrophysics at the University of Nottingham, have found that the universe contains at least 2 trillion galaxies, 10 times more than previously thought. The team's work, which began with seedcorn funding from the Royal Astronomical Society, appears in the *Astrophysical Journal*.

Astronomers have long sought to determine how many galaxies there are in the observable universe, the part of the cosmos where light from distant objects has had time to reach us. Over the last 20 years, scientists have used images from the Hubble Space Telescope to estimate that the universe we can see contains around 100–200 billion galaxies. Current astronomical technology allows us to study just 10 percent of these galaxies, and the remaining 90 percent will be only seen once bigger and better telescopes are developed.

Professor Conselice's research is the culmination of 15 years' work, part-funded by a research grant from the Royal Astronomical Society awarded to Aaron Wilkinson, an undergraduate student at the time. Aaron, now a PhD student at the University of Nottingham, began by performing the initial galaxy-counting analysis, work that was crucial for establishing the feasibility of the larger-scale study.

Professor Conselice's team then converted pencil beam images of deep space from telescopes around the world, and especially from the Hubble telescope, into three-dimensional maps. These allowed them to calculate the density of galaxies as well as the volume of one small region of space after another. This painstaking research enabled the team to establish how many galaxies we have missed—much like an intergalactic archaeological dig.

The results of this study are based on the measurements of the number of observed galaxies at different epochs—different instances in time—through the universe's history. When Professor Conselice and his team at Nottingham, in collaboration with scientists from the Leiden Observatory at Leiden University in the Netherlands and the Institute for Astronomy at the University of Edinburgh, examined how many galaxies there were at a given epoch, they found that there were significantly more at earlier times.

It appears that when the universe was only a few billion years old, there were 10 times as many galaxies in a given volume of space as there are within a similar volume today. Most of these galaxies were low-mass systems with masses similar to those of the satellite galaxies surrounding the Milky Way.

Professor Conselice said, "This is very surprising as we know that, over the 13.7 billion years of cosmic evolution since the Big Bang, galaxies have been growing through star formation and mergers with other galaxies. Finding more galaxies in the past implies that significant evolution must have occurred to reduce their number through extensive merging of systems."

He continued, "We are missing the vast majority of galaxies because they are very faint and far away. The number of galaxies in the universe is a fundamental question in astronomy, and it boggles the mind that over 90 percent of the galaxies in the cosmos have yet to be studied. Who knows what interesting properties we will find when we study these galaxies with the next generation of telescopes?"

ARTICLES   QUESTIONS

1. Explain how observing galaxies at different distances tells us about how galaxies existed in different periods.
2. Why does that result suggest that galaxies started small?
3. What is meant by "observable universe"?
4. Watch Hubblecast 96 based on this news story (http://spacetelescope.org/videos/heic1620a/). What else did you learn from the video?
5. Do a search to see if there is a new estimate of the number of observable galaxies.

Source: "A universe of two trillion galaxies," Royal Astronomical Society News Archive, October 13, 2016. Reprinted with permission.

Summary

Astronomy seeks answers to many compelling questions about the universe. Astronomy uses all available tools to follow the scientific method. The process of science is based on objective reality, physical evidence, and testable hypotheses. Scientists continually strive to improve their understanding of the natural world and must be willing to challenge accepted truths as new information becomes available.

LG 1 **Describe the size and age of the universe and Earth's place in it.** We reside on a planet orbiting a star at the center of a solar system in a vast galaxy that is one of many in the universe. We occupy a very tiny part of the universe in space and time.

LG 2 **Explain how astronomers use the scientific method to study the universe.** The scientific method is an approach to learning about the physical universe. The method includes observations, forming hypotheses, making predictions to test and refine those hypotheses, and repeated testing of theories. All scientific knowledge is provisional. Like art, literature, and music, science is a creative human activity; it is also a remarkably powerful, successful, and aesthetically beautiful way of viewing the world.

LG 3 **Show how scientists use mathematics, including graphs, to find patterns in nature.** Mathematics provides many of the tools that astronomers need to understand the patterns we see and to communicate that understanding to others.

LG 4 **Describe our astronomical origins.** We are a product of the universe: The very atoms we're made of formed in stars that died long before the Sun and Earth formed. The atoms of the chemical elements that make up matter come from particles that have been around since the Big Bang, processed through stars, and collected into the Sun and the Solar System.

? Unanswered Questions

- What makes up the universe? We have listed planets, stars, and galaxies as making up the cosmos, but astronomers now have evidence that 95 percent of the universe is composed of dark matter and dark energy, which we do not understand. Scientists are using the largest telescopes and particle colliders on Earth, as well as telescopes and experiments in space, to explore what makes up dark matter and what constitutes dark energy.

- Does life as we know it exist elsewhere in the universe? At the time of this writing, no scientific evidence indicates that life exists on any other planet. Our universe is enormously large and has existed for a great length of time. What if life is too far away or existed too long ago for us ever to "meet"?

Questions and Problems

Test Your Understanding

1. Rank the following in order from smallest to largest.
   a. Local Group
   b. Milky Way
   c. Solar System
   d. universe
   e. Sun
   f. Earth
   g. Laniakea Supercluster
   h. Virgo Supercluster

2. If an event were to take place on the Sun, approximately how long would the light it generates take to reach us?
   a. 8 minutes
   b. 11 hours
   c. 1 second
   d. 1 day
   e. It would reach us instantaneously.

3. *Understanding* in science means that
   a. we have accumulated lots of facts.
   b. we can connect facts through an underlying idea.
   c. we can predict events on the basis of accumulated facts.
   d. we can predict events on the basis of an underlying idea.

4. The cosmological principle states that
   a. on a large scale, the universe is the same everywhere at a given time.
   b. the universe is the same at all times.
   c. our location is special.
   d. all of the above

5. The Sun is part of
   a. the Solar System.
   b. the Milky Way Galaxy.
   c. the universe.
   d. all of the above

6. A light-year is a measure of
   a. distance.
   b. time.
   c. speed.
   d. mass.

7. Occam's razor states that
   a. the universe is expanding in all directions.
   b. the laws of nature are the same everywhere in the universe.
   c. if two hypotheses fit the facts equally well, the simpler one is more likely to apply.
   d. patterns in nature are really manifestations of random occurrences.

8. The circumference of Earth is 1/7 of a light-second. Therefore,
   a. if you were traveling at the speed of light, you would travel around Earth 7 times in 1 second.
   b. light travels a distance equal to Earth's circumference in 1/7 of a second.
   c. neither a nor b
   d. both a and b

9. According to the graphs in Figures 1.7b and c, by how much did the number of viruses increase in four time steps?
   a. It doubled.
   b. It tripled.
   c. It quadrupled.
   d. It went up more than 10 times.

10. Any explanation of a phenomenon that includes a supernatural influence is not scientific because
    a. it is not based on a hypothesis.
    b. it is wrong.
    c. people who believe in the supernatural are not credible.
    d. science is the study of the natural world.

11. "All scientific knowledge is provisional." In that context, *provisional* means
    a. "wrong."
    b. "relative."
    c. "temporary."
    d. "incomplete."

12. When we observe a star that is 10 light-years away, we are seeing that star
    a. as it is today.
    b. as it was 10 days ago.
    c. as it was 10 years ago.
    d. as it was 20 years ago.

13. Which of the following was *not* made in the Big Bang?
    a. hydrogen
    b. lithium
    c. beryllium
    d. carbon

14. "We are stardust" means that
    a. Earth exists because two stars collided.
    b. the atoms in our bodies have passed through (and often formed in) stars.
    c. Earth is formed primarily of material that used to be in the Sun.
    d. Earth and the other planets will eventually form a star.

15. According to our current understanding of the universe, the following astronomical events led to the formation of you. Place them in order of their occurrence over astronomical time.
    a. Stars die and distribute heavy elements into the space between the stars.
    b. Hydrogen and helium are made in the Big Bang.
    c. Enriched dust and gas gather into clouds in interstellar space.
    d. Stars are born and process light elements into heavier ones.
    e. The Sun and planets form from a cloud of interstellar dust and gas.

Thinking about the Concepts

16. Suppose you lived on the planet named "Tau Ceti e" that orbits Tau Ceti, a nearby star in our galaxy. How would you write your cosmic address?

17. Imagine yourself living on a planet orbiting a star in a very distant galaxy. What does the cosmological principle tell you about the physical laws at that distant location?

18. If the Sun suddenly exploded, how soon after the explosion would we know about it?

19. If a star exploded in the Andromeda Galaxy, how long would that information take to reach Earth?

20. Give an example of a scientific theory that a newer theory has superseded. As scientists developed that new theory, where on the Process of Science Figure did a change occur so that the old theory became invalid and the new theory was accepted?

21. Some people have proposed the theory that extraterrestrials (aliens) visited Earth in the remote past. Can you think of any tests that could support or refute that theory? Is it falsifiable?

22. What does the word *falsifiable* mean? Give an example of an idea that is not falsifiable. Give an example of an idea that is falsifiable.

23. Explain how the word *theory* is used differently in the context of science than it is in everyday language.

24. What is the difference between a *hypothesis* and a *theory* in science?

25. Suppose the tabloid newspaper at your local supermarket claimed that children born under a full Moon become better students than children born at other times.
    a. Is that theory falsifiable?
    b. If so, how could it be tested?

26. A textbook published in 1945 stated that light from the Andromeda Galaxy takes 800,000 years to reach Earth. In this book, we assert that it takes 2,500,000 years. What does that difference tell you about a scientific "fact" and how our knowledge evolves?

27. Astrology makes testable predictions. For example, it predicts that the horoscope for your star sign on any day should fit you better than horoscopes for other star signs. Read the daily horoscopes for all the astrological signs in a newspaper or online. How many of them might fit the day you had yesterday? Repeat the experiment every day for a week and record which horoscopes fit your day each day. Did your horoscope sign consistently best describe your experiences?

28. A scientist on television states that it is a known fact that life does not exist beyond Earth. Would you consider that scientist's statement reputable? Explain your answer.

29. Some astrologers use elaborate mathematical formulas and procedures to predict the future. Does doing so show that astrology is a science? Why or why not?

30. Why can it be said that we are made of stardust? Explain why current theory supports that statement.

Applying the Concepts

31. Review Working It Out 1.1., and then convert the following numbers to scientific notation:
    a. 7,000,000,000
    b. 0.00346
    c. 1,238

32. Review Working It Out 1.1, and then convert the following numbers to standard notation:
    a. 5.34×10^8
    b. 4.1×10^3
    c. 6.24×10^{-5}

33. If a car is traveling at 35 km/h, how far does it travel in
    a. 1 hour?
    b. half an hour?
    c. 1 minute?

34. Review Appendix 1. The surface area of a sphere is proportional to the square of its radius. How many times larger is the surface area if the radius is
    a. doubled?
    b. tripled?
    c. halved (divided by 2)?
    d. divided by 3?

35. The average distance from Earth to the Moon is 384,400 km. How many days would it take you, traveling at 800 km/h—the typical speed of jet aircraft—to reach the Moon?

36. The average distance from Earth to the Moon is 384,400 km. In the late 1960s, astronauts reached the Moon in about 3 days. How fast (on average) must they have been traveling (in kilometers per hour) to cover that distance in that time? Compare that speed with the speed of a jet aircraft (800 km/h).

37. (a) If light takes about 8 1/3 minutes to travel from the Sun to Earth, and Neptune is 30 times farther from Earth than the Sun is, how long does light from Neptune take to reach Earth? (b) Radio waves travel at the speed of light. What does that fact imply about the problems you would have if you tried to conduct a two-way conversation between Earth and a spacecraft orbiting Neptune?

38. The distance from Earth to Mars varies from 56 million km to 400 million km. How long does a radio signal traveling at the speed of light take to reach a spacecraft on Mars when Mars is closest and when Mars is farthest away?

39. The surface area of a sphere is proportional to the square of its radius. The radius of the Moon is only about one-quarter that of Earth. How does the surface area of the Moon compare with that of Earth?

40. A remote Web page may sometimes reach your computer by going through a satellite orbiting approximately 3.6×10^4 km above Earth's surface. What is the minimum delay, in seconds, before the Web page shows up on your computer?

41. Imagine that you have become a biologist, studying rats in Indonesia. Usually, Indonesian rats maintain a constant population. Every half century, however, those rats suddenly begin to multiply exponentially. Then the population crashes back to the constant level. Sketch a graph that shows the rat population over two of those episodes.

42. New York is 2,444 miles from Los Angeles. What is that distance in car-hours? In car-days? (Assume a travel speed of 60 mph.)

43. Some people theorize that a tray of hot water will freeze more quickly than a tray of cold water when both are placed in a freezer.
    a. Does that theory make sense to you?
    b. Is the theory falsifiable?
    c. Do the experiment yourself. Note the results. Was your intuition borne out?

44. A pizzeria offers a 9-inch-diameter pizza for $12 and an 18-inch-diameter pizza for $24. Are both offerings equally economical? If not, which is the better deal? Explain your reasoning.

45. The circumference of a circle is given by $C = 2\pi r$, where r is the radius of the circle.
    a. Calculate the approximate circumference of Earth's orbit around the Sun, assuming that the orbit is a circle with a radius of 1.5×10^8 km.
    b. A year has 8,766 hours, so how fast, in kilometers per hour, does Earth move in its orbit?
    c. How far along in its orbit does Earth move in 1 day?

USING THE WEB

46. Go to the interactive "The Scale of the Universe 2" (Web page at https://apod.nasa.gov/apod/ap140112.html, or download the app or find on YouTube). Start at 10° (human size) and scale upward; clicking on an object gives you its exact size. What astronomical bodies are about the size of the United States? What objects are about the size of Earth? What stars are larger than the distance from Earth to the Sun? How many light-days is the distance of *Voyager 1* to Earth? What objects are about the size of the distance from the Sun to the nearest star? How many times larger is the Milky Way than the Solar System? How many times larger is the Local Group than the Milky Way? How many times larger is the observable universe than the Local Group?

47. a. For a video representation of the scale of the universe, view the short video *The Known Universe* at the Hayden Planetarium website (https://amnh.org/our-research/hayden-planetarium/digital-universe), which takes the viewer on a journey from the Himalayan mountains to the most distant galaxies. How far have broadcast radio programs from Earth traveled? Is the Sun a particularly luminous star in comparison with others? Do you think the video is effective for showing the size and scale of the universe?

    b. A similar film produced in 1996 in IMAX format, *Cosmic Voyage*, can be found online on YouTube. Watch the "powers of 10" zoom out to the cosmos, starting at the 7-minute mark, for about 5 minutes. Do the "powers of 10" circles add to your understanding of the size and scale of the universe?

48. Go to the Astronomy Picture of the Day (APOD) app or website (https://apod.nasa.gov/apod) and click on "Archive" to look at the recent pictures and videos. Submissions to the website come from all around the world. Pick one and read the explanation. Was the image or video taken from Earth or from space? Is it a combination of several images? Does it show Earth, our Solar System, objects in our Milky Way Galaxy, more distant galaxies, or something else? Would someone who has not studied astronomy understand the explanation? Do you think the website promotes a general interest in astronomy?

49. Throughout this book, we examine how the media cover discoveries in astronomy and space. On your favorite news website (or one assigned by your instructor), find a recent article about astronomy or space. Does the website have a separate section for science? Is the article you selected based on a press release, on interviews with scientists, or on an article in a scientific journal? Use Google News or the equivalent to see how widespread the coverage of the story is. Have many newspapers carried it? Has it been picked up internationally? Has it been discussed in blogs? Do you think the story was interesting enough to be covered?

50. Go to a blog about astronomy or space. Is the blogger a scientist, a science writer, a student, or an enthusiastic amateur astronomer? What is the current topic of interest? Is it controversial? Are readers making many comments? Is the blog something you would want to read again?

EXPLORATION

Logical Fallacies

digital.wwnorton/astro6

Logic is fundamental to studying science and to scientific thinking. A logical fallacy is an error in reasoning, which good scientific thinking avoids. For example, "because Einstein said so" is not an adequate argument. No matter how famous the scientist is (even if he is Einstein), he or she must still supply a logical argument and evidence to support a claim. Anyone who claims that something must be true because Einstein said it has committed the logical fallacy known as an *appeal to authority*. Many types of logical fallacies exist, but a few crop up often enough in discussions about science that you should be aware of them.

Ad hominem. In that fallacy, you attack the person making the argument instead of the argument itself. For example, "A well-known actor says dogs are better pets than cats, but I think that actor is an idiot, so cats must be better."

Appeal to belief. That fallacy has the general pattern "Most people believe X is true; therefore, X is true." For example, "Most people believe Earth orbits the Sun. Therefore, Earth orbits the Sun." Even if your conclusion is correct, you may still have committed a logical fallacy in your argument.

Begging the question. In that fallacy, also known as circular reasoning, you assume that the claim is true and then use that assumption as evidence to prove the claim is true. For example, "I am trustworthy; therefore, I must be telling the truth." No real evidence is presented for the conclusion.

Biased sample. If a sample drawn from a smaller pool has a bias, conclusions about the sample cannot be applied to a larger pool. For example, imagine you poll students at your university and find that 30 percent of them visit the library at least once per week. Then you conclude that 30 percent of Americans visit the library at least once per week. You have committed the biased sample fallacy because university students are not a representative sample of the American public.

Post hoc ergo propter hoc. That term is Latin for "after this, therefore because of this." Just because one thing follows another doesn't mean that one caused the other. For example, "An eclipse occurred, and then the king died. Therefore, the eclipse killed the king." That fallacy is often connected to related inverse reasoning: "If we can prevent an eclipse, the king won't die."

Slippery slope. In that fallacy, you claim that a chain reaction of events will take place, inevitably leading to a conclusion that no one could want. For example, "If I don't get A's in all my classes, I will not ever be able to get into graduate school, and then I won't ever be able to get a good job, and then I will be living in a van down by the river until I'm old and starve to death." None of those steps actually follows inevitably from the one before.

The following are examples of logical fallacies. Identify the fallacy represented. Each fallacy just discussed is represented once.

1 You get a chain email threatening terrible consequences if you break the chain. You move the message to your spam box. Later that day you get in a car accident. The next morning, you retrieve the chain email and send it along.

_____
_____
_____

2 If I get question 1 on the assignment wrong, then I'll get question 2 wrong as well, and before you know it, I will never catch up in the class.

_____
_____
_____

3 All my friends love the band Degenerate Electrons. Therefore, all people my age love that band.

_____
_____
_____

4 Eighty percent of Americans believe in the tooth fairy. Therefore, the tooth fairy exists.

_____
_____
_____

5 My professor says that the universe is expanding. But my professor is a geek, and I don't like geeks, so the universe can't be expanding.

_____
_____
_____

6 When applying for a job, you use a friend as a reference. Your prospective employer asks you how she can be sure your friend is trustworthy, and you say, "I can vouch for him."

_____
_____
_____

2

Patterns in the Sky—Motions of Earth and the Moon

Ancient peoples learned to use the patterns they observed in the sky to predict how the length of day, the seasons, and the appearance of the Moon would change. Some people understood those patterns well enough to create complicated calendars and predict rare eclipses. But now we can see those patterns with the perspective of centuries of modern science, and we can explain those changes as being a consequence of the motions of Earth and the Moon. Discovering what causes those patterns has shown us the way outward into the universe.

LEARNING GOALS

Here in Chapter 2, we examine the patterns in the sky and on Earth and the underlying motions that cause them. By the end of this chapter, you should be able to:

LG 1 Describe how Earth's rotation about its axis and its revolution around the Sun affect how we perceive celestial motions from different places on Earth.

LG 2 Explain why seasons change throughout the year.

LG 3 Describe the factors that create the phases of the Moon.

LG 4 Sketch the alignment of Earth, the Moon, and the Sun during eclipses of the Sun and the Moon.

Total eclipse of the Sun on August 21, 2017. ▶▶▶

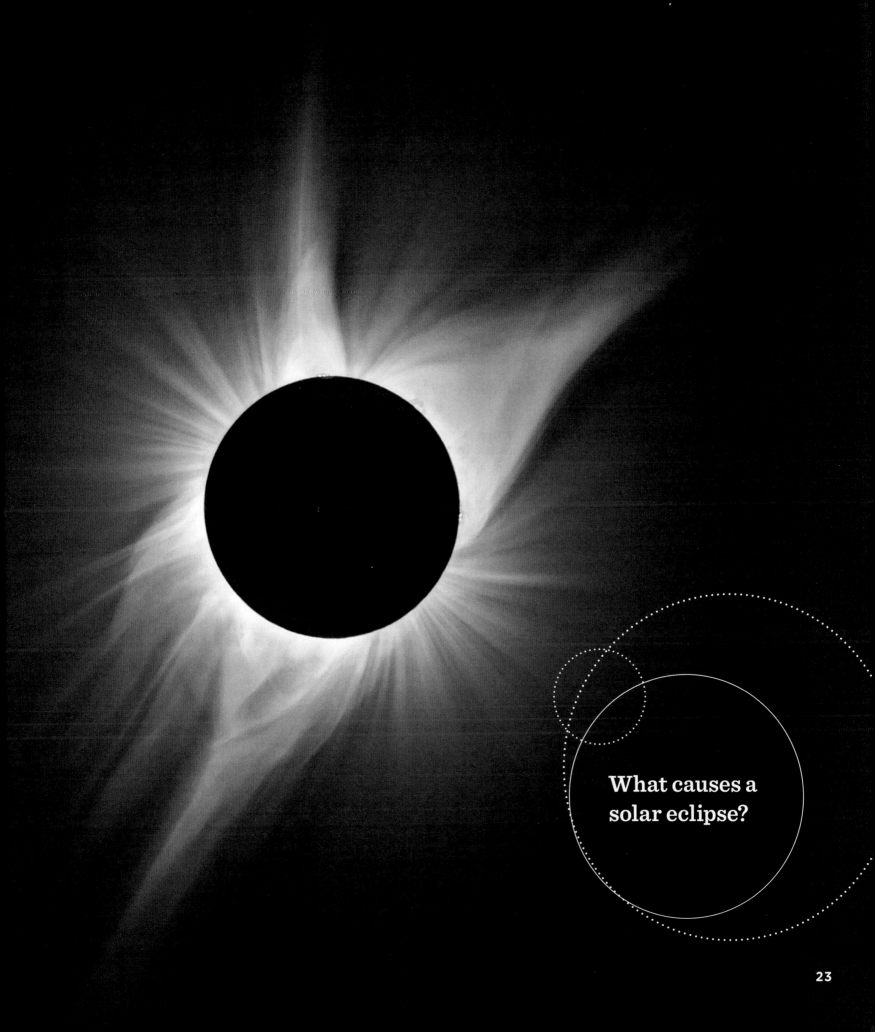

What causes a solar eclipse?

Figure 2.1 (a) One suspected use of Stonehenge 4,000 years ago was to keep track of celestial events. (b) The Mayan El Caracol at Chichén Itzá in Mexico (906 CE) is believed to have been designed to align with Venus. (c) The Beijing Ancient Observatory in China (1442 CE) includes ancient astronomical instruments as well as some brought by European Jesuits in the 17th and 18th centuries.

2.1 Earth Spins on Its Axis

Ancient humans may not have known that they were "stardust," but they did sense a connection between their lives on Earth and the sky above. Before our modern technological civilization, people's lives were more attuned to the ebb and flow of nature, which includes the patterns in the sky. By watching the repeating patterns of the Sun, Moon, and stars in the sky, people could predict when the seasons would change and when the rains would come and the crops would grow. Ancient astronomers with knowledge of the sky—priests and priestesses, natural philosophers, and explorers—all had knowledge of the world that others did not, and knowledge was power. Some of those early observations and ideas live on today in the names of stars and the apparent grouping of stars we call **constellations**, in calendars based on the Moon and Sun that many cultures still use, and in the astronomical names of the days of the week.

From Mesopotamia to Africa, from Europe to Asia, from the Americas to the British Isles, the archaeological record holds evidence that ancient cultures built structures to study astronomical positions and events. **Figure 2.1** shows some examples. From the 8th through 17th centuries, people used pre-telescopic astronomical observatories to study the sky for timekeeping and navigation. Many of those structures and observatories are now national historical or UNESCO World Heritage sites.

The Celestial Sphere

Long before Christopher Columbus sailed west from Spain, Aristotle and other Greek philosophers knew that Earth was a sphere. However, because Earth seems stationary, they did not realize that its motion caused the changes they observed in the sky from day to day and year to year. As we explain here, Earth's rotation on its axis determines the rising and setting of the Sun, Moon, and stars—one of the rhythms that governs life on Earth (**Figure 2.2**).

One reason the ancients did not suspect that Earth rotates was that they could not perceive its spinning motion. As Earth rotates about its axis, the planet's surface is moving quite fast—about 1,674 kilometers per hour (km/h) at the equator. We do not feel that motion any more than we would feel the motion of a car

Figure 2.2 Motions in the Solar System, as viewed from above Earth's North Pole. (Not drawn to scale.)

with a perfectly smooth ride cruising down a straight highway. Nor do we feel the direction of Earth's spin, although the hourly motion of the Sun, Moon, and stars across the sky reveals that direction. Earth's **North Pole** is at the north end of Earth's rotation axis. Imagine that you are in space far above Earth's North Pole. From there you would see Earth complete a counterclockwise rotation, once each 24-hour period, as Figure 2.2 shows. As the rotating Earth carries an observer on the surface from west to east, objects in the sky appear to move in the opposite direction, from east to west. As seen from Earth's surface, each object seems to move across the sky in a path called its **apparent daily motion**.

To help visualize the apparent daily motions of the Sun and stars, think of the sky as a huge sphere with the stars painted on its surface and Earth at its center. From ancient Greek times to the Renaissance, many people believed that to be a true representation of the heavens. Astronomers call that imaginary sphere the **celestial sphere** (**Figure 2.3**). The celestial sphere is a useful concept because it is easy to visualize, but never forget that it is, in fact, imaginary.

Each point on the celestial sphere indicates a direction in space. Directly above Earth's North Pole is the **north celestial pole** (**NCP**). Directly above Earth's **South Pole**, which is at the south end of Earth's rotation axis, is the **south celestial pole** (**SCP**). Directly above Earth's **equator** is the **celestial equator**, an imaginary circle that divides the sky into a northern half and a southern half. Just as the north celestial pole is the projection of the direction of Earth's North Pole into the sky, the celestial equator is the projection of the plane of Earth's equator into the sky. Just as Earth's North Pole is 90° away from Earth's equator, the north celestial pole is 90° away from the celestial equator. If you are in the Northern Hemisphere and you point one arm toward the celestial equator and one arm toward the north celestial pole, your arms will always form a right angle, so the north celestial pole is 90° away from the celestial equator. The same holds true for the Southern Hemisphere: the angle between the celestial equator and the south celestial pole is always 90° as well.

Between the celestial poles and the equator, objects have positions on the celestial sphere with coordinates analogous to latitude and longitude on Earth. **Latitude** indicates distance north or south from Earth's equator. On the celestial sphere, **declination** similarly indicates the distance of an object north or south of the celestial equator (from 0° to ±90°). On Earth, **longitude** measures how far east or west you are from the Royal Observatory in Greenwich, England. On the celestial sphere, **right ascension** is similar to longitude on Earth. Right ascension measures a celestial body's angular distance eastward along the celestial equator from the point where the Sun's path crosses the celestial equator from south to north. Those coordinates are used to quickly locate objects in the sky. The **ecliptic** is the Sun's path in the sky throughout the year (Figure 2.3). In Appendix 7, we discuss detailed descriptions and illustrations of latitude and longitude as well as coordinates used with the celestial sphere.

The **zenith** is the point in the sky directly above you, wherever you are, as **Figure 2.4a** shows. You can find the **horizon** by standing up and pointing your right hand at the zenith and your left hand straight out from your side. Turn in a complete circle. Your left hand has traced out the entire horizon. You can divide the sky into an eastern half and a western half with a line that runs from the horizon at due north through the zenith to the horizon at due south. That imaginary north–south line is the **meridian**, shown as a dashed line in Figure 2.4a. **Figure 2.4b** shows those locations on the celestial sphere. The meridian line continues around the far side of the celestial sphere, through the **nadir** (the point directly below you), and back to the starting point due north.

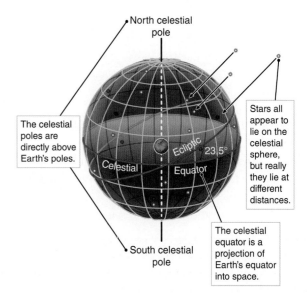

Figure 2.3 The celestial sphere is a useful fiction in thinking about the stars' appearance and apparent motion in the sky.

AstroTour: The Celestial Sphere and the Ecliptic

Figure 2.4 (a) The meridian is a line on the celestial sphere that runs from north to south, dividing the sky into an east half and a west half. (b) At any location on Earth, the sky is divided into an east half and a west half by an imaginary meridian projected onto the celestial sphere.

Astronomy in Action: Vocabulary of the Celestial Sphere

Take a moment to visualize all those locations in space. To see how to use the celestial sphere, consider the Sun at noon and at midnight. Local noon occurs when the Sun crosses the meridian at your location. That is the highest point above the horizon that the Sun will reach on any given day. The highest point is almost never the zenith. You have to be in a specific place on a specific day for the Sun to be directly overhead at noon. One such place and time is at latitude 23.5° north of the equator on June 20. From our perspective on Earth, the celestial sphere appears to rotate, carrying the Sun across the sky to its highest point at noon, over toward the west to set in the evening. In *reality*, the Sun remains in the same place in space through the entire 24-hour period, and Earth rotates so that any given location on Earth faces a different direction at every moment. When it is noon where you live, Earth has rotated so that you face most directly toward the Sun. Half a day later, at midnight, your location on Earth has rotated to face most directly away from the Sun. Local midnight occurs when the Sun is precisely opposite from its position at local noon.

The View from the Poles

AstroTour: The View from the Poles

The apparent daily motions of the stars and the Sun depend on where you live. For example, the apparent daily motions of celestial objects in a northern place such as Alaska are different from the apparent daily motions seen from a tropical island such as Hawaii. To understand why your location matters, let's examine the view of the stars from the poles—and then use them to guide our thinking about the view of the stars from other latitudes.

Imagine that you are the person in **Figure 2.5a** who is standing on the North Pole watching the sky. There, the north celestial pole is directly overhead at the zenith. Ignore the Sun for the moment and pretend that you can always see stars in the sky. You are standing where Earth's axis of rotation intersects its surface, which is like standing at the center of a rotating carousel. As Earth rotates, the spot directly above you remains fixed over your head while everything else in the sky appears to revolve counterclockwise around that spot. **Figure 2.5b** depicts that overhead view.

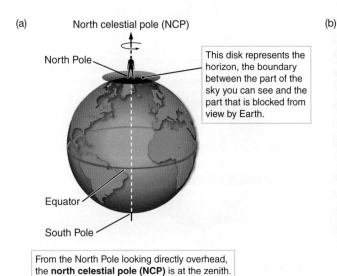

From the North Pole looking directly overhead, the **north celestial pole (NCP)** is at the zenith.

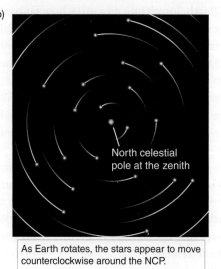

As Earth rotates, the stars appear to move counterclockwise around the NCP.

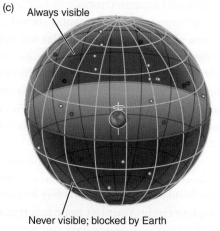

From the North Pole, you always see the same half of the sky.

Figure 2.5 As viewed from (a) Earth's North Pole, (b) stars move throughout the night in counterclockwise, circular paths about the zenith. (c) The same half of the sky is always visible from the North Pole.

Wherever you are on Earth, you can see only half the sky. The horizon is the boundary between the part of the sky you can see and the other half of the sky that is blocked by Earth. Except at the poles, the visible half of the sky changes constantly as Earth rotates because the zenith points to different locations in the sky as Earth carries you around. If you are standing at the North Pole, however, the zenith is always in the same location in space, so objects visible from the North Pole follow circular paths that always have the same **altitude**, or angle above the horizon. Objects close to the zenith appear to follow small circles, whereas objects near the horizon follow the largest circles (Figure 2.5b). The view from the North Pole is special: you will always see the *same* half of the celestial sphere from there because nothing rises or sets each day as Earth turns (**Figure 2.5c**).

The view from Earth's South Pole is much the same—with two major differences. First, the South Pole is on the opposite side of Earth from the North Pole, so the visible half of the sky at the South Pole is precisely the half hidden from view at the North Pole. Second, stars appear to move clockwise around the south celestial pole rather than counterclockwise as they do at the north celestial pole. To visualize why those motions are different, stand up and spin around from right to left. As you look at the ceiling, things appear to move counterclockwise, but as you look at the floor, they appear to be moving clockwise.

CHECK YOUR UNDERSTANDING 2.1A

No matter where you are on Earth, stars appear to rotate about a point called the:
(a) zenith; (b) celestial pole; (c) nadir; (d) meridian.

The View Away from the Poles

Suppose that you leave the North Pole to travel south to lower latitudes. Imagine a line from the center of Earth to your location on the surface, as in **Figure 2.6**. Now imagine a second line from Earth's center to the point on the equator closest to you. The angle between those two lines is your latitude. At the North Pole, for example, those two imaginary lines form a 90° angle. At the equator, they form a 0° angle. So the latitude of the North Pole is 90° north, and the latitude of the equator is 0°. The South Pole is at latitude 90° south.

Your latitude determines the part of the sky that you can see throughout the year. As you move south from the North Pole, your zenith moves away from the north celestial pole, and so the horizon moves as well. At the North Pole, the horizon makes a 90° angle with the north celestial pole, which is at the zenith. At a latitude of 60° north, as in Figure 2.6, your horizon is tilted 60° from the north celestial pole. No matter where you are on Earth, the angle between your horizon and the north celestial pole is equal to your latitude. The situation is the same in the Southern Hemisphere—your latitude is the altitude of the south celestial pole. At the equator, at a latitude of 0°, the north and south celestial poles would be at the northern and southern horizons, respectively.

One way to solidify your understanding of the view of the sky at different latitudes is to draw pictures like the one in Figure 2.6. If you can draw a picture like that for any latitude—filling in the values for each angle in the drawing and imagining what the sky looks like from there—you are developing a working knowledge of the appearance of the sky. That knowledge will prove useful later when we discuss a variety of phenomena, such as the changing of the seasons.

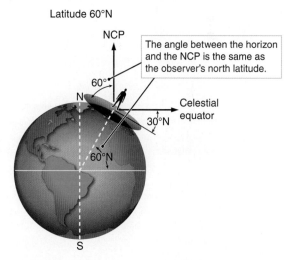

Figure 2.6 Your perspective on the sky depends on your location on Earth. The locations of the celestial poles and the celestial equator in an observer's sky depend on the observer's latitude. Here, an observer at latitude 60° north sees the north celestial pole at an altitude of 60° above the northern horizon and the celestial equator 30° above the southern horizon.

Figure 2.7 Time exposures of the sky showing the apparent motions of stars through the night. Note the difference in the circumpolar portion of the sky as seen from the two northern latitudes.

From a location in the Canadian woods, the north celestial pole appears high in the sky…

…but at a lower latitude in Utah, the north celestial pole appears closer to the horizon.

MOTIONS OF THE STARS AND THE CELESTIAL POLES **Figure 2.7** shows two time-lapse views of the sky from different latitudes. The stars' apparent motions about the celestial poles also differ from latitude to latitude. The visible part of the sky constantly changes as stars rise and set with Earth's rotation. From that perspective the horizon appears fixed, whereas the stars appear to move. From those latitudes, if we focus our attention on the north celestial pole, we see much the same thing we saw from Earth's North Pole. The north celestial pole remains fixed in the sky, and all the stars appear to move throughout the night in counterclockwise, circular paths around that point. The north celestial pole is no longer directly overhead as it was at the North Pole, however, so the apparent circular paths of the stars are now tipped with respect to the horizon. (More correctly, your horizon is now tipped with respect to the apparent circular paths of the stars.)

From the vantage point of an observer in the Northern Hemisphere, stars located near the north celestial pole in the sky never dip below the horizon and thus stay visible all night. Recall from Figure 2.6 that the latitude is equal to the altitude of the north celestial pole. Stars closer to the north celestial pole than that angle never dip below the horizon as they complete their apparent paths around the pole. Those stars are **circumpolar** ("around the pole"). Another group of stars, near the south celestial pole, never rises above the horizon in the Northern Hemisphere. Stars between those that never rise and those in the circumpolar region (that is, the ones that never set) can be seen for *only part* of each 24-hour day. Those stars appear to rise and set as Earth turns. The equator is the only place on Earth where you can see the entire sky over the course of 24 hours. From there, the north and south celestial poles sit on the northern and southern horizons, respectively, and all the stars move through the sky each 24-hour day. (Even though the Sun lights the sky for roughly half that time, the stars are still there.)

Figure 2.8 shows the orientation of the sky as seen by observers at four latitudes. For an observer at the North Pole (Figure 2.8a), the celestial equator lies exactly along the horizon. The north celestial pole is at the zenith, and the southern half of the sky is never visible. Stars neither rise nor set; their paths form circles parallel to the horizon. For an observer at the equator (Figure 2.8b), the celestial poles are both at the horizon, and all the stars are visible in a 24-hour period, rising straight up and setting straight down each day.

At other latitudes, the celestial equator intersects the horizon due east and due west. Therefore, a star on the celestial equator rises due east and sets due west. Stars located north of the celestial equator rise north of east and set north of west. Stars located south of the celestial equator rise south of east and set south of west.

From everywhere else on Earth (except at the poles), half of the celestial equator is always visible above the horizon. Therefore, any object located on the celestial equator is visible half the time—above the horizon for 12 hours each day. Objects not on the celestial equator are above the horizon for different amounts of time. Figures 2.8c and d show that stars in the observer's hemisphere are visible

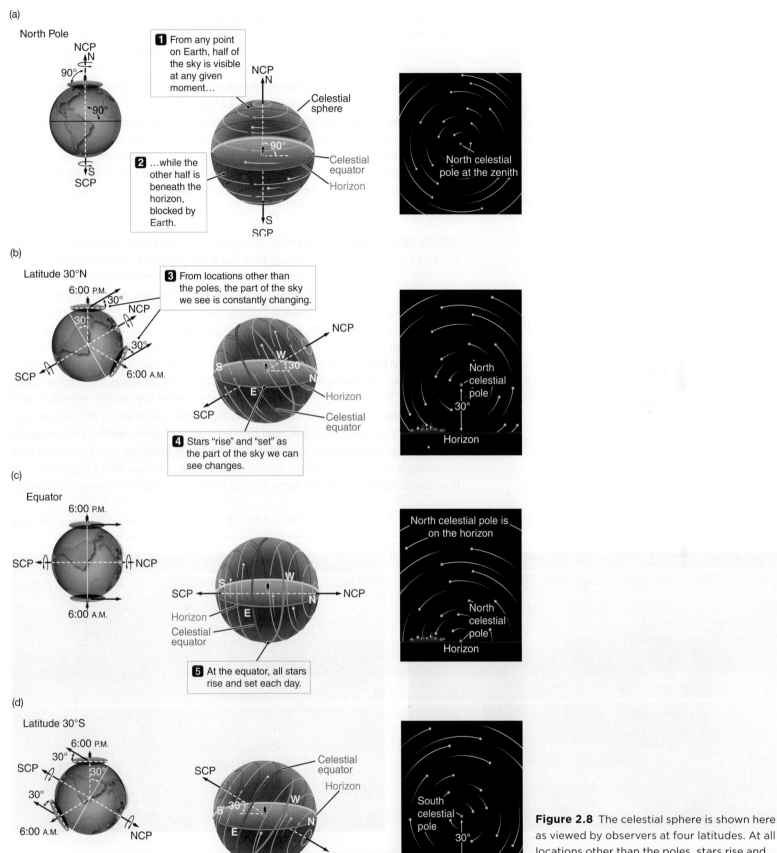

Figure 2.8 The celestial sphere is shown here as viewed by observers at four latitudes. At all locations other than the poles, stars rise and set as the part of the celestial sphere that we see changes during the day.

for more than half the day because more than half of each star's path in the sky is above the horizon. In contrast, stars in the opposite hemisphere are visible for less than half the day because less than half of each star's path in the sky is above the horizon.

As seen from the Northern Hemisphere, stars north of the celestial equator remain above the horizon for more than 12 hours each day. The farther north the star is, the longer it stays up. Circumpolar stars are the extreme example of that phenomenon; they are always above the horizon. In contrast, stars south of the celestial equator are above the horizon for less than 12 hours each day. The farther south a star is, the less time it is visible. For an observer in the Northern Hemisphere, stars located close to the south celestial pole never rise above the horizon.

USING THE STARS TO NAVIGATE Since ancient times, travelers have used the stars to navigate. They would find the north or south celestial poles by recognizing the stars that surround them. In the Northern Hemisphere, a moderately bright star happens to be located close to the north celestial pole (**Figure 2.9a**). That star is called Polaris, the "North Star." Polaris's altitude in the sky is nearly equal to your latitude. If you are in Phoenix, Arizona, for example (latitude 33.5° north), Polaris has an altitude very close to 33.5°. In Fairbanks, Alaska (latitude 64.6° north), Polaris sits much higher, with an altitude of 64.6°. In the Southern Hemisphere, the constellation Crux (commonly called the Southern Cross) points to a star near the south celestial pole (**Figure 2.9b**). A navigator who has located a pole star can identify north and south and therefore east, west, and her latitude. Doing so enables the navigator to determine which direction to travel. The location of the north celestial pole in the sky was used to measure Earth's size, as described in **Working It Out 2.1**. Because of Earth's rotation, determining your longitude by astronomical meth-

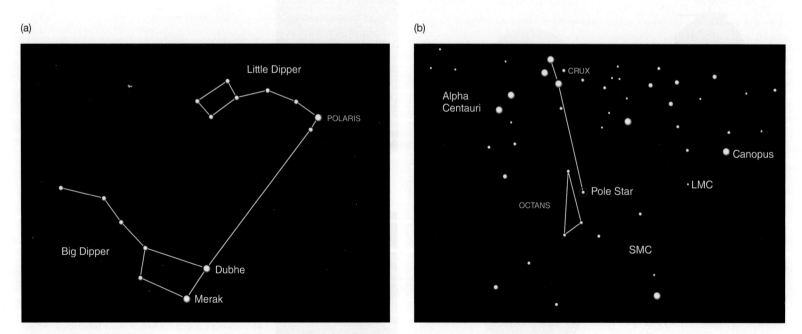

Figure 2.9 Groups of stars near the pole stars in the sky can be used to locate a pole star. (a) Two bright "pointer stars" in the cup of the Big Dipper point toward Polaris, the "North Star." (b) In the Southern Hemisphere, the constellation Crux and two of its bright pointer stars can be used to locate the relatively faint southern pole star.

2.1 Working It Out How to Estimate Earth's Size

We can use the location of the north celestial pole in the sky to estimate Earth's size. Suppose we start out in Phoenix, Arizona, and we observe the north celestial pole to be 33.5° above the horizon. If we head north, by the time we reach the Grand Canyon, about 290 km from Phoenix, we notice that the north celestial pole has risen to about 36° above the horizon. That difference between 33.5° and 36° (2.5°) is 1/144 of the way around a circle. (A circle is 360°, and 2.5°/360° = 1/144.)

That means we must have traveled 1/144 of the way around Earth's circumference, C, by traveling the 290 km between Phoenix and the Grand Canyon. In other words,

$$\frac{1}{144} \times C = 290 \text{ km}$$

When we rearrange the expression, Earth's circumference is given by

$$C = 144 \times 290 \text{ km} \approx 42,000 \text{ km}$$

Earth's actual circumference is just over 40,000 km, so our simple calculation was close. The circumference of a circle is equal to $2\pi r$, where r is its radius. Earth's radius, then, is given by

$$\text{Radius} = \frac{C}{2\pi} = \frac{40,000 \text{ km}}{2\pi} = 6,400 \text{ km}$$

In about 230 BCE, using much the same method, the Greek astronomer Eratosthenes (276–194 BCE) was the first to accurately measure Earth's size. As **Figure 2.10** illustrates, Eratosthenes used the distance between his home city of Alexandria, Egypt, and the city of Syene (now Aswân, also in Egypt), which was 5,000 "stadia." He noticed that on the first day of summer in Syene, the sunlight reflected directly off the water in a deep well, so the Sun must have been nearly at the zenith. By measuring the shadow of the Sun from an upright stick in Alexandria, he saw that the Sun was about 7.2° south of the zenith on the same date. Assuming that Earth was spherical and that Syene was directly south of Alexandria, he determined the distance between the two cities to be 7.2° divided by 360, or 1/50 of Earth's circumference.

Accurately estimating distances in those days was hard. Although historians know Eratosthenes concluded that the cities were 5,000 stadia apart, they are still not at all sure of the value of his stadion unit. If the stadion was 185 meters, the size of Greek stadia, Eratosthenes would have worked the math similarly:

$$\frac{1}{50} \times C = 5,000 \text{ stadia} \times 185 \text{ meters/stadion}$$
$$= 925,000 \text{ meters} = 925 \text{ km}$$

Eratosthenes would have found Earth's circumference to be

$$C = 50 \times 925 \text{ km} = 46,250 \text{ km}$$

That result is only about 16 percent higher than the modern value.

Figure 2.10 Eratosthenes used observations and basic calculations to estimate Earth's size.

---

ods is much more complicated. Longitude cannot easily be determined from astronomical observation alone.

Relative Motions and Frame of Reference

Why don't we feel motion as Earth spins on its axis and moves through space in its orbit around the Sun? The answer lies in the concept of a **frame of reference**, a coordinate system within which an observer measures positions and motions. The difference in motion between two individual frames of reference is called the

(a) Frame of reference: Viewer on the street

(b) Frame of reference: Viewer in faster car

(c) Frame of reference: Viewer in slower car

Figure 2.11 The motion of an object is always measured relative to the frame of reference of the observer.

relative motion. For example, imagine that you are riding in a car traveling down a straight section of highway at a constant speed. Under those conditions, without looking out the window or feeling road vibrations, you cannot easily tell the difference between being in the moving car and sitting in a parked car. Because everything in the car is moving together, you can measure only the relative motions between objects in the car, and they are all zero—no motion is observed. Similarly, the resulting relative motions between objects that are near each other on Earth are zero because they have the same rotational motion. That is why we do not notice Earth's motion.

Now imagine two cars driving down the road at different speeds, as **Figure 2.11a** shows. For the moment, ignore any real-world complications, such as wind resistance. If you were to throw a ball from the faster-moving car directly out the side window at the slower-moving car as the two cars passed, you would miss. The ball shares the forward motion of the faster car, so the ball outruns the forward motion of the slower car. From your perspective in the faster car (**Figure 2.11b**), the slower car lagged behind the ball. From the slower car's perspective (**Figure 2.11c**), your car and the ball sped on ahead.

Although we cannot feel Earth's rotation, it influences things as diverse as the motion of weather patterns and how an artillery gunner must aim at a distant target. An object on Earth's surface moves in a circle each day around the planet's rotation axis. That circle is larger for objects near the equator and smaller for objects closer to one of the poles, but all objects complete their circular motion in exactly 1 day. As you can see in **Figure 2.12**, Earth's surface moves faster at the equator than at higher latitudes. An object closer to the equator has farther to go each day than one nearer a pole does. Therefore, the object nearer the equator must be moving faster than the one at higher latitude. If an object starts out at one latitude and then moves to another, that difference in speed influences its apparent motion over the surface.

Now consider two locations at different latitudes. Imagine a cannonball launched directly north from a point in the Northern Hemisphere, as in Figure 2.12a. Because the cannon is closer to the equator than its target is, the cannonball is moving toward the east faster than its target. Even though the cannonball is fired toward the north, it shares the eastward velocity of the cannon itself, just like the ball thrown from the faster car in Figure 2.11. Therefore, the cannonball is *also* moving toward the east faster than its target. As the cannonball flies north, it too moves toward the east faster than the ground underneath it does. To an observer on the ground, the cannonball appears to curve toward the east as it outruns the eastward motion of the ground it is crossing. The farther north the cannonball flies, the greater the difference between its eastward velocity and the ground's. Thus, the cannonball follows a path that appears to curve more and more to the east the farther north it goes. If you are located in the Northern Hemisphere and fire a cannonball *south* toward the equator, as in Figure 2.12b, the opposite effect will occur. Now the cannon is moving toward the east more slowly than its target. As the cannonball flies toward the south, its eastward motion lags behind that of the ground underneath it, and the cannonball appears to curve toward the west.

That curving motion of objects as a result of the difference in Earth's rotation speeds at different latitudes is called the **Coriolis effect**. In the Northern Hemisphere, the Coriolis effect causes a cannonball fired north to drift to the east as seen from the ground. In other words, the cannonball appears to curve to the

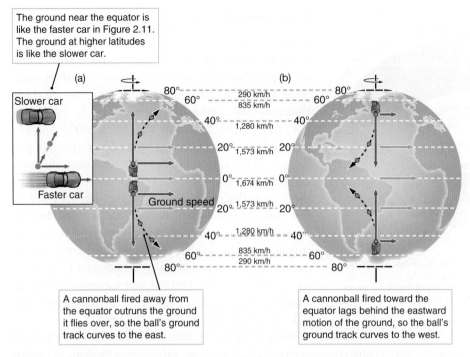

Figure 2.12 The Coriolis effect causes objects to be deflected as they move across Earth's surface. The green dashed curve shows the cannonball's path as measured by a local observer.

right. A cannonball fired south appears to curve to the west, which also gives it the appearance of curving to the right. In the Northern Hemisphere, the Coriolis effect seems to deflect things to the *right*. In the Southern Hemisphere, it seems to deflect things to the *left*. In between, at the equator itself, the Coriolis effect vanishes.

On Earth, the effect is enough to deflect a fly ball hit north or south into deep left field in a stadium in the northern United States by about a half a centimeter. At some time or other, the Coriolis effect has probably determined the outcome of a baseball game.

CHECK YOUR UNDERSTANDING 2.1B

If the star Polaris has an altitude of 35°, we know that: (a) our longitude is 55° east; (b) our latitude is 55° north; (c) our longitude is 35° west; (d) our latitude is 35° north.

2.2 Revolution about the Sun Leads to Changes during the Year

Earth orbits (or **revolves**) around the Sun in the same direction that Earth spins about its axis—counterclockwise as viewed from above Earth's North Pole (see Figure 2.2). A **year** is the time Earth takes to complete one revolution around the Sun. Earth's motion around the Sun is responsible for many of the patterns of change we see in the sky and on Earth, including changes in the stars we see overhead. Because of that motion, the stars in the night sky change throughout the year, and Earth experiences seasons.

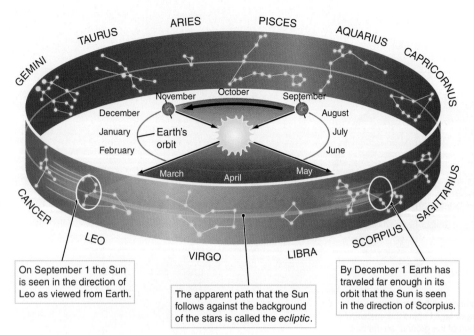

Figure 2.13 As Earth orbits the Sun, different stars appear in the night sky, and the Sun's apparent position against the background of stars changes. The imaginary circle that the Sun's annual path traces is called the ecliptic. Constellations along the ecliptic form the zodiac.

Constellations and the Zodiac

As Earth orbits the Sun, our view of the night sky changes (**Figure 2.13**). Six months from now, Earth will be on the other side of the Sun. The stars overhead at midnight 6 months from now will be those that are near overhead at noon today. To follow the patterns of the Sun and the stars, early humans grouped together stars that formed recognizable patterns, called constellations. People from different cultures, though, saw different patterns and projected ideas from their own cultures onto what they saw in the sky. Constellations named for winged horses, dragons, and other imaginary things—and the stories that go with them—are creations of the human imagination. If you look at the sky, no obvious pictures of those things emerge. Instead, you see only the random pattern of stars—about 9,000 of them visible to the naked eye—spread out across the sky.

Modern constellations visible from the Northern Hemisphere draw heavily from the list of constellations that the Alexandrian astronomer Ptolemy compiled 2,000 years ago. Modern constellation names in the southern sky come from European explorers who visited the Southern Hemisphere during the 17th and 18th centuries. Today, astronomers use an officially sanctioned set of 88 constellations to serve as a road map of the sky. Every star in the sky lies within the borders of a single constellation, and the names of constellations are used in naming their component stars. For example, Sirius, the brightest star in the sky, lies within the constellation Canis Major (meaning "big dog"). Following the Greek alphabet, astronomers call the brightest star in a constellation *alpha*; the second brightest, *beta*; and so forth. The official name of Sirius is Alpha Canis Majoris, indicating that it is the brightest star in Canis Major. Appendix 7 includes sky maps showing the constellations.

If you noted the position of the Sun in relation to the stars each day for a year, you would find that the Sun traces out a great circle against the background of the stars. On September 1, the Sun appears to be in the direction of the constellation Leo. Six months later, on March 1, Earth is on the other side of the Sun, and the Sun appears from our perspective on Earth to be in the direction of the constellation Aquarius. Recall that the Sun's apparent path against the background of the stars is called the ecliptic (the blue band in Figure 2.13). The constellations that lie along the ecliptic through which the Sun appears to move are called the constellations of the **zodiac**.

Earth's Axial Tilt and the Seasons

To understand why the seasons change, we need to consider the combined effects of Earth's rotation on its axis and revolution around the Sun. Many people believe that seasons change because Earth is closer to the Sun in summer and farther away in winter. How can we falsify that hypothesis? We can predict that if the distance between Earth and the Sun caused the seasons, then all of Earth should experience summer at the same time of year. But the United States has summer in June, whereas Australia has summer in December. In modern times, we can directly measure the distance: Earth is actually closest to the Sun at the beginning of January. We have just falsified the hypothesis tying seasonal change to distance, and we need to look for another explanation that explains all the available facts.

As Earth orbits the Sun, the Sun appears to move along the ecliptic, which is tilted 23.5° with respect to the celestial equator. That tilt occurs because Earth's axis of rotation is tilted 23.5° from the perpendicular to Earth's orbital plane. The days are longer in summer than in winter, and the Sun is higher in the sky as it crosses the meridian in summer than in winter.

Figure 2.14 shows that as Earth moves around the Sun, its axis always points in the same direction in space (toward Polaris). As it orbits, Earth is sometimes on one side of the Sun and sometimes on the other. As a result, Earth's North Pole is sometimes tilted more toward the Sun, and other times the South Pole is tilted more toward the Sun. When Earth's North Pole is tilted toward the Sun, an observer on Earth views the Sun north of the celestial equator. For observers in the Northern Hemisphere, the Sun is above the horizon more than 12 hours each day, thus making the days longer than 12 hours. Six months later, when Earth's North Pole is tilted away from the Sun, an observer in the same place views the Sun south of the celestial equator and the days are shorter.

AstroTour: The Earth Spins and Revolves

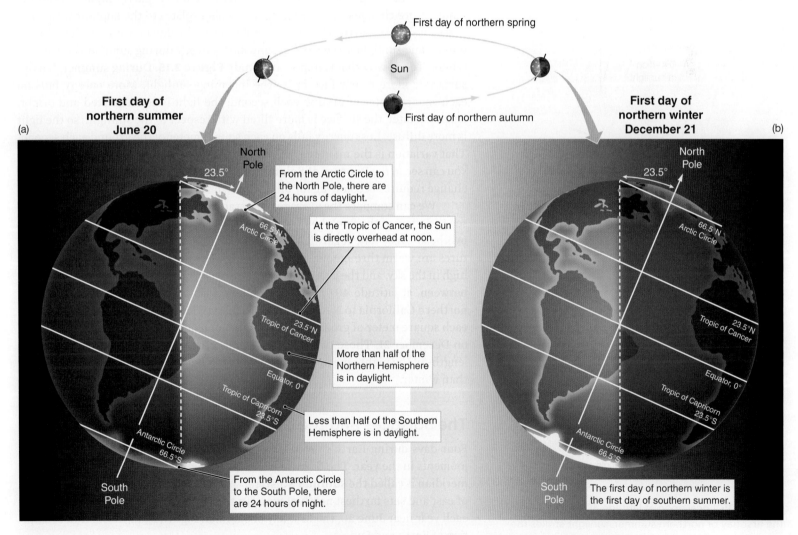

Figure 2.14 (a) On the first day of the northern summer (around June 20, the summer solstice), the northern end of Earth's axis is tilted most nearly toward the Sun, whereas the Southern Hemisphere is tipped away. (b) Six months later, on the first day of the northern winter (around December 21, the winter solstice), the situation is reversed. Seasons are opposite in the two hemispheres.

Astronomy in Action: The Cause of Earth's Seasons

Figure 2.15 Local noon at latitude 40° north. (a) On the first day of northern summer, sunlight strikes the ground almost face-on. (b) On the first day of northern winter, sunlight strikes the ground more obliquely, and less than half as much sunlight falls on each square meter of ground each second.

In the preceding paragraph, we specified the *Northern* Hemisphere because seasons are opposite in the Southern Hemisphere. Look again at Figure 2.14. Around June 20, while the Northern Hemisphere is enjoying the long days and short nights of summer, Earth's South Pole is tilted away from the Sun. It is winter in the Southern Hemisphere; less than half of the Southern Hemisphere is illuminated by the Sun, and the days are shorter than 12 hours. On December 21, Earth's South Pole is tilted toward the Sun. It is summer in the Southern Hemisphere, so its days are long and the nights are short.

To understand how the combination of Earth's axial tilt and its path around the Sun creates seasons, consider a limiting case. If Earth's spin axis were exactly perpendicular to the plane of Earth's orbit (the **ecliptic plane**), the Sun would always be on the celestial equator. At every latitude (except the poles), the Sun would follow the same path through the sky every day, rising due east each morning and setting due west each evening. The Sun would be above the horizon exactly half the time, and days and nights would always be exactly 12 hours long everywhere on Earth. In short, each day would be just like the last, and Earth would have no seasons.

The changing length of days through the year only partly explains seasonal temperature changes. Another important effect relates to the angle at which the Sun's rays strike Earth. The Sun is higher in the sky during summer than during winter, and sunlight strikes the ground *more directly* during summer than during winter. To see why that is important, study **Figure 2.15**. During summer, Earth's surface is more nearly face-on to the incoming sunlight. More energy falls on each square meter of ground each second; the light is concentrated and bright. During winter, the surface is more tilted with respect to the sunlight, so the light is more diffuse. Less energy falls on each square meter of the ground each second. That variation is the main reason why summer is hotter and winter is colder. As you can see in the **Process of Science Figure**, determining the causes of seasonal change requires accounting for all the known facts.

We can compare the average temperatures at different latitudes on Earth to see the effect of the Sun's height in the sky. Near the equator, the Sun passes high overhead every day, regardless of the season. As a result, the average temperatures are warm throughout the year. At high latitudes, however, the Sun is *never* high in the sky, and the average temperatures can be cold even during summer. In between, at latitude 40° north, which stretches across the United States from northern California to New Jersey, more than twice as much solar energy falls on each square meter of ground per second at noon on June 20 as falls there at noon on December 21. Those two effects—the directness of sunlight and the different lengths of the night—mean that the Sun heats a hemisphere more during summer than winter.

The Solstices and the Equinoxes

Four days during Earth's orbit (two solstices and two equinoxes) mark unique moments in the year. The day when the Sun is highest in the sky as it crosses the meridian is called the **summer solstice**. On that day, the Sun rises farthest north of east and sets farthest north of west. In the Northern Hemisphere, that occurs each year near June 20, the first day of summer. Figure 2.14a shows that orientation of Earth and Sun.

Six months after the summer solstice, around December 21, the North Pole is tilted away from the Sun (Figure 2.14b). That day is the **winter solstice** in the Northern Hemisphere—the shortest day of the year and the first day of winter in

PROCESS OF SCIENCE

Theories Must Fit All the Known Facts

Many people misunderstand the phenomenon of changing seasons because they do not account for all the relevant facts.

HYPOTHESIS 1

We have seasons because Earth is closer to the Sun in the summer and farther away in winter.

THE TEST If this were true, both the Northern and Southern Hemispheres would have summer in July. However, the Northern and Southern Hemispheres experience opposite seasons.

THE CONCLUSION The hypothesis is falsified.

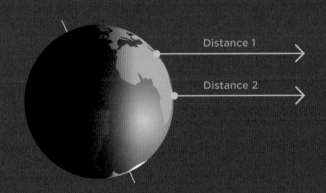

HYPOTHESIS 2

We have seasons because the tilt of Earth's axis causes one hemisphere to be significantly closer to the Sun than the other.

THE TEST If this were true, the distances would have to be very different to cause such a large effect. Earth is tiny compared to its distance from the Sun: The distances between each hemisphere and the Sun differ by less than 0.004 percent.

THE CONCLUSION The hypothesis is falsified.

HYPOTHESIS 3

We have seasons because Earth's tilt changes the distribution of energy—one hemisphere receives more intense light than the other.

THE TEST If this were true, the amount of sunlight striking the ground in the summer would have to be more than in the winter, and the days would have to be longer in summer.

THE CONCLUSION Seasons are caused primarily by a change in illumination due to Earth's tilt. During winter, less energy falls on each square meter of ground per second.

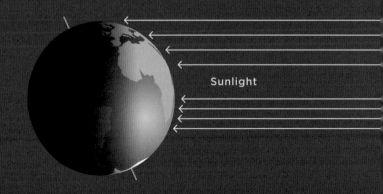

New information often challenges misconceptions. Incorporating this new information may alter current theories to improve the explanation of observed phenomena.

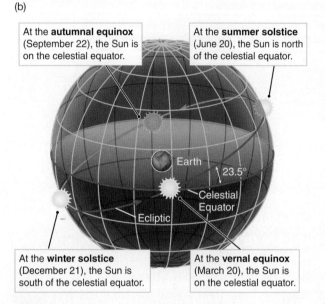

Figure 2.16 Earth's motion around the Sun from the frame of reference of (a) the Sun and (b) Earth.

the Northern Hemisphere. Almost all cultural traditions in the Northern Hemisphere include some sort of major celebration in late December. Those winter festivals celebrate the return of the source of Earth's light and warmth. The days have stopped growing shorter and are beginning to get longer, and spring will come again.

Between the two solstices, the ecliptic crosses the celestial equator on two days. On those days, the Sun lies directly above Earth's equator. We call those days *equinoxes*, which means "equal night," because the entire Earth experiences 12 hours of daylight and 12 hours of darkness (except at the poles). Halfway between summer solstice and winter solstice, the **autumnal equinox** marks the beginning of fall in the Northern Hemisphere; it occurs around September 22. Halfway between winter solstice and summer solstice, the **vernal equinox** marks the beginning of spring in the Northern Hemisphere; it occurs around March 20. In the Southern Hemisphere, the dates for the summer and winter solstices, and the autumnal and vernal equinoxes, are reversed from those in the Northern Hemisphere.

Figure 2.16 shows the solstices and equinoxes from two perspectives. Figure 2.16a shows Earth in orbit around a stationary Sun, whereas Figure 2.16b shows the Sun's apparent motion along the celestial sphere, which is how it appears to observers on Earth. In both cases, we are looking at the plane of Earth's orbit from the side, so that it is shown in perspective and looks quite flattened. We also have tilted the images so that Earth's North Pole points straight up. The equinoxes correspond to points in the sky where the celestial equator meets the ecliptic. Practice shifting between those perspectives. You will know that you understand them when you can look at a position in either panel and predict the corresponding positions of the Sun and Earth in the other panel.

Just as a pot of water on a stove takes time to heat up when the burner is turned on and to cool when the burner is turned off, Earth takes time to respond to changes in heating from the Sun. The hottest months of northern summer are usually July and August, which come after the summer solstice, when days are growing shorter. Similarly, the coldest months of northern winter are usually January and February, which occur after the winter solstice, when days are growing longer. Temperature changes on Earth lag behind changes in the amount of heating we receive from the Sun.

That picture of the seasons must be modified somewhat near Earth's poles. At latitudes north of 66.5° north and south of 66.5° south, the Sun is circumpolar

Astronomy in Action: The Earth-Moon-Sun System

Figure 2.17 This composite photo shows the midnight Sun, which is visible in latitudes above 66.5° north (or south). In the 360-degree panoramic view, the Sun moves 15° each hour.

for a part of the year surrounding the first day of summer. Those lines of latitude are the **Arctic Circle** and the **Antarctic Circle**, respectively (see Figure 2.14). When the Sun is circumpolar, it is above the horizon 24 hours per day, earning the polar regions the nickname "land of the midnight Sun" (**Figure 2.17**). An equally long period surrounds the first day of winter, when the Sun never rises and the nights are 24 hours long. The Sun never rises high in the Arctic or Antarctic sky, so the sunlight is never very direct. Even with the long days at the height of summer, the Arctic and Antarctic regions remain relatively cool.

In contrast, on the equator, *all* stars, including the Sun, are above the horizon approximately 12 hours per day. Days and nights there are 12 hours long throughout the year. The Sun passes directly overhead on the first day of spring and the first day of autumn because on those days the Sun is on the celestial equator. Sunlight is most direct, perpendicular to the ground, at the equator on those days. On the summer solstice, the Sun is at its northernmost point along the ecliptic. On that day, and on the winter solstice, the Sun is farthest from the zenith at noon, and therefore sunlight is least direct at the equator.

Latitude 23.5° north is called the Tropic of Cancer, and latitude 23.5° south is called the Tropic of Capricorn (Figure 2.14). The band between those latitudes is called the **Tropics**. If you live in the tropics—in Rio de Janeiro or Honolulu, for example—the Sun will be directly overhead at noon twice during the year.

Precession of the Equinoxes

Two thousand years ago, when Ptolemy and his associates were formalizing their knowledge of the positions and motions of objects in the sky, the Sun appeared in the constellation Cancer on the first day of northern summer and in the constellation Capricornus on the first day of northern winter. Today, the Sun is in Taurus on the first day of northern summer and in Sagittarius on the first day of northern winter. Why have the constellations in which solstices appear changed? Two motions are actually associated with Earth and its axis: Earth spins on its axis, but its axis also wobbles like the axis of a spinning top slowing down (**Figure 2.18a**). The wobble is very slow: the north celestial pole takes about 26,000 years to complete one trip around a large circle centered on the north ecliptic pole. Polaris is the star we now see near the north celestial pole. However, if you could travel several thousand years into the past or the future, the point about which the northern sky appears to rotate would no longer be near Polaris. Instead, the stars would rotate about another point on the path shown in Figure 2.18b. That figure shows the path of the north celestial pole through the sky during one cycle of that wobble.

The celestial equator is perpendicular to Earth's axis. Therefore, as Earth's axis wobbles, the celestial equator also must wobble. As it does so, the locations where the celestial equator crosses the ecliptic—the equinoxes—change as well. During each 26,000-year wobble of Earth's axis, the locations of the equinoxes make one complete circuit around the celestial equator. That change of the position of the equinox, due to the wobble of Earth's axis, is called the **precession of the equinoxes**.

CHECK YOUR UNDERSTANDING 2.2

If Earth's axis were tilted by 45°, instead of its actual tilt of 23.5°, how (if at all) would the seasons be different from what they are now? (a) The seasons would remain the same. (b) Summers would be colder. (c) Winters would be shorter. (d) Winters would be colder.

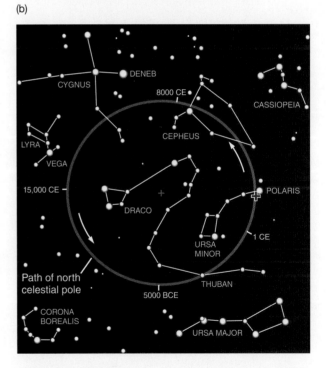

Figure 2.18 (a) Earth's axis of rotation changes orientation in the same way that the axis of a spinning top changes orientation. Polaris is the current North Pole star, Thuban was the North Pole star in 3000 BCE, and Vega will be the Pole star in 14,000 CE. (b) That precession causes the projection of Earth's rotation axis to move in a circle, centered on the north ecliptic pole (orange cross in the center). The red cross on the circle marks the projection of Earth's axis on the sky in the early 21st century.

2.3 The Moon's Appearance Changes as It Orbits Earth

After the Sun, the most prominent object in our sky is the Moon. Just as Earth orbits around the Sun, the Moon orbits around Earth once every 27.32 days. In this section, we discuss the phases of the Moon as seen from Earth.

The Changing Phases of the Moon

The Moon and its changing aspects have been the frequent subject of mythology, art, literature, and music. In mythology, the Moon was the Greek goddess Artemis, the Roman goddess Diana, and the Inuit god Igaluk. We speak of the "man in the Moon," the "harvest Moon," and sometimes a "blue Moon."

Unlike the Sun, the Moon has no light source of its own; instead, the Moon shines by reflected sunlight. As the Moon orbits Earth, our view of the illuminated portion of the Moon constantly changes. Those different appearances of the Moon are called **phases**. During a new Moon, when the Moon is between Earth and the Sun, the side facing away from us is illuminated, and during a full Moon, when Earth is between the Sun and the Moon, the side facing toward us is illuminated. The rest of the time, we can see only part of the illuminated portion from Earth. Sometimes the Moon appears as a circular disk in the sky. Other times it is nothing more than a thin sliver, or its face appears dark.

To help you visualize the changing phases of the Moon, use an orange, a lamp, and your head (**Figure 2.19**). Your head is Earth, the orange is the Moon, and the lamp is the Sun. Turn off all the other lights in the room, and step back as far from the lamp as you can. Hold up the orange slightly above your head so that the lamp illuminates it from one side. Move the orange clockwise around your head and watch how the appearance of the orange changes. When you are between the orange and the lamp, the face of the orange that is toward you is fully illuminated. The orange appears to be a bright, circular disk. As the orange moves around its circle, you will see a progression of lighted shapes, depending on how much of the bright side and how much of the dark side of the orange you can see. That progression of shapes mimics the changing phases of the Moon.

Figure 2.20 shows the changing phases of the Moon. The **new Moon** occurs when the Moon is between Earth and the Sun. The far side is illuminated, but the near side is in darkness and we cannot see it. The new Moon appears close to the Sun in the sky, so it is up in the daytime with the Sun: it rises in the east at sunrise, crosses the meridian near noon, and sets in the west near sunset. A new Moon is never above the horizon in the nighttime sky.

A few days after a new Moon, as the Moon orbits Earth, a sliver of its illuminated half, called a **waxing crescent Moon**, becomes visible. *Waxing* here means "growing in size and brilliance"; the name refers to the fact that the Moon appears to be "filling out" from night to night at that time. From our perspective, the Moon has also moved away from the Sun in the sky. Because the Moon travels around Earth in the same direction in which Earth rotates, we now see the Moon trailing the Sun, so the Moon is east of the Sun in the sky. A waxing crescent Moon is visible in the western sky in the evening, near the setting Sun but remaining above the horizon after the Sun sets. The "horns" of the crescent always point directly away from the Sun.

Figure 2.19 An orange and a lamp can help you visualize the changing phases of the Moon.

Astronomy in Action: Phases of the Moon

AstroTour: The Moon's Orbit: Eclipses and Phases

Simulation: Phases of the Moon

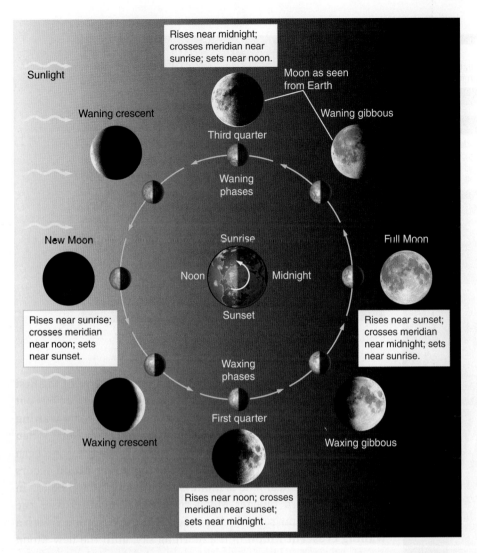

Figure 2.20 The inner circle of images (connected by blue arrows) shows the Moon as it orbits Earth, as seen by an observer far above Earth's North Pole. The Sun is on the left. The outer ring of images shows the corresponding phases of the Moon as seen from the northern hemisphere of Earth.

As the Moon moves farther along in its orbit, and the angle between the Sun and Moon grows larger, more and more of its near side becomes illuminated. About a week after the new Moon, half of the near side of the Moon is illuminated and half is in darkness. That phase is called a **first quarter Moon** because the Moon has moved a quarter of the way around Earth and has completed the first quarter of its cycle from new Moon to new Moon (Figure 2.20). The first quarter Moon rises at noon, crosses the meridian at sunset, and sets at midnight.

As the Moon moves beyond first quarter, more than half of its near side is illuminated. That phase is called a **waxing gibbous Moon**, from the Latin *gibbus*, meaning "hump." The waxing gibbous Moon continues nightly to "grow" until finally we see the entire near side of the Moon illuminated—a **full Moon**. Earth is now between the Sun and the Moon, which appear opposite each other in the sky when viewed from Earth. The full Moon rises as the Sun sets, crosses the meridian at midnight, and sets in the morning as the Sun rises.

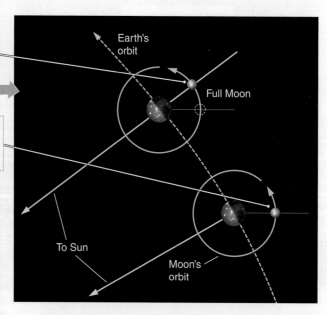

Figure 2.21 (a) The Moon completes one sidereal orbit in 27.32 days, but the synodic period (the period between phases seen from Earth) from one full Moon to the next is 29.53 days. The horizontal orange line to the right of the Moon indicates a fixed direction in space. (b) Here the orbits of Earth and the Moon are shown to scale, but the sizes of Earth and the Moon are not.

The second half of the Moon's orbit is the reverse of the first half. The Moon continues in its orbit, again appearing gibbous but now becoming smaller each night. That phase is called a **waning gibbous Moon**, where *waning* means "becoming smaller." When the Moon is waning, the left side—as viewed from the Northern Hemisphere—appears illuminated. A **third quarter Moon** occurs when half of the near side is illuminated by sunlight and half is in darkness. A third quarter Moon rises at midnight, crosses the meridian near sunrise, and sets at noon. The cycle continues with a **waning crescent Moon** in the morning sky, until the new Moon once again rises and sets with the Sun, and the cycle begins again. Notice that when the Moon is farther from the Sun than Earth is, it is in gibbous (or full) phases, whereas when the Moon is closer to the Sun than Earth is, it is in crescent (or new) phases.

You can always tell a waxing Moon from a waning Moon because the illuminated side is always the side facing the Sun. When the Moon is waxing, it appears in the evening sky, so its western side is illuminated. That is the right side as viewed from the Northern Hemisphere. Conversely, when the Moon is waning, it appears in the morning sky, so the eastern side appears bright. That is the left side as viewed from the Northern Hemisphere.

Figure 2.21 illustrates two types of lunar periods: the first one based on the Moon's orbit in space and the second based on the alignment of the moving Moon, Earth, and Sun. The Moon completes one orbit around Earth in 27.32 days: that **sidereal period** is how long the Moon takes to return to the same location in its orbit. However, because the relationships among Earth, the Moon, and the Sun change as a result of Earth's orbital motion, going from one full Moon to the next takes 29.53 days. That cycle is known as the Moon's **synodic period** and is the basis for our "month," because it is what we can easily observe from Earth.

Do not try to memorize all possible combinations of where the Moon is in the sky at what phase and at what time of day. Instead, work on understanding the motion and phases of the Moon, and then use your understanding to figure out the specifics of any given case. To study the phases of the Moon, draw a picture like Figure 2.20, and use it to follow the Moon around its orbit. From your drawing, figure out what phase you would see and where it would appear in the sky at a given time of day. You might also try the simulations described in "Exploration: The Phases of the Moon" at the end of the chapter.

The Moon's Visible Face

Although the Moon's illumination varies as it orbits, one aspect of the Moon's appearance does *not* change: the face that we see. If we were to go outside next week, next month, 20 years from now, or 200 centuries from now, we would still see the same side of the Moon that we see tonight. That is true because the Moon rotates on its axis exactly once for each revolution that it makes around Earth.

Imagine walking around the Washington Monument while always keeping your face toward the monument. By the time you complete one circle around the monument, your head has turned completely around once. When you were south of the monument, you were facing north; when you were east of the monument, you were facing west; and so on. But someone looking at you from the monument would always see your face. The Moon does the same thing, rotating on its axis once per revolution around Earth, always keeping the same face toward Earth (**Figure 2.22**). Objects whose revolution and rotation are synchronized (or "in sync") with each other are said to be in **synchronous rotation**. That effect occurs as a result of the gravitational and tidal forces between Earth and the Moon (Section 4.4). In later chapters we discuss other examples of synchronous rotation.

The Moon's *far side*, facing away from Earth, is often called the "dark side of the Moon." In fact, no side of the Moon is always dark. At any given time, half of the Moon is in sunlight and half is in darkness—just as for Earth. The side of the Moon that faces away from Earth, the "far side," spends just as much time in sunlight as the side of the Moon that faces Earth.

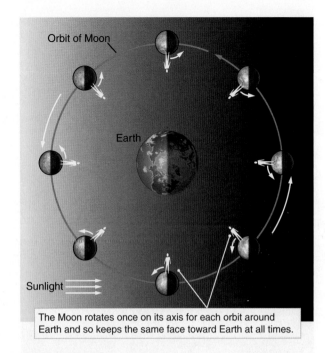

Figure 2.22 The Moon rotates once on its axis for each orbit around Earth—an effect called synchronous rotation. Here, the Sun is far to the left of the Earth-Moon system.

CHECK YOUR UNDERSTANDING 2.3

If you see the Moon rising just as the Sun is setting, what phase is the Moon in? What is the phase if the Moon is setting at noon?

2.4 Calendars Are Based on the Day, Month, and Year

Archaeologists tell us that developing agriculture was crucial for the rise of human civilization, and keeping track of the seasons and best times of the year to plant and harvest was critical to successful farming. Records going back to the dawn of humanity suggest that people kept track of time by following the patterns in the sky, especially those of the Sun, Moon, and stars. Some anthropologists have speculated that notches on fragments of bone found in southern France represent a 33,000-year-old lunar calendar. In this section, we examine some different calendars.

Lunar and Lunisolar Calendars

As civilizations developed around the globe, different cultures tried to solve the "problem" of the calendar. The rates of rotation of Earth and the revolutions of the Moon around Earth and of Earth around the Sun are not multiples of one another. A lunar cycle (from full Moon to full Moon) is 29.5 days, and a solar cycle—the time the Sun takes to appear to move from and return to its highest possible point in the sky at noon on the summer solstice—is 365.24 days. One solar cycle, then, has 12.38 lunar cycles. Those fractions of days and months are what make calendars complicated.

Some of the oldest known calendars come from the Egyptians, the Babylonians, and the Chinese. The ancient Egyptians used a system of 12 months of 30 days each—which added up to 360 days—and then added five "festival days" to the end of the year (**Figure 2.23**). Without leap years, the seasons in that 365-day year drifted, so an extra month was added when necessary. When we consider how some Western countries celebrate the days between the modern December holidays and the new year, an end-of-year calendar break for festivals seems like a good solution!

The Babylonians started the 24-hour day and 7-day week—7 for the Sun, the Moon, and the 5 planets visible with the naked eye. They created the first **lunisolar calendar**, in which a month began with the first sighting of the lunar crescent, and a 13th month was added when needed to catch up to the solar year. As the Babylonians developed mathematics, they discovered that 235 lunar months equals 19 solar years (and 6,940 days). Then they created a calendar cycle that consisted of 19 years, 12 of which have 12 months and 7 of which have 13 months, and then the cycle repeats. The ancient Hebrew calendar adopted that lunisolar cycle, and the Jewish calendar still uses it today. That type of calendar keeps holidays in the same season from year to year, even though the dates change.

The ancient Chinese calendar, dating back several thousand years, also is lunisolar. By about 500 BCE, the Chinese were using a year of 365.25 days and a system similar to that of the Babylonians of adding a 13th month to some years. To mark out a year, a few other cultures used stellar calendars, which followed the position of a bright star such as Sirius or certain prominent groups of stars in their sky, such as the Pleiades or the Big Dipper.

The Islamic calendar is purely lunar, with no added 13th lunar month. Their 12 months of 29 or 30 days each add up to 354 days—11.24 days short of the solar year. For that reason, the Islamic New Year and all other holidays drift earlier in each solar year. In the Islamic calendar, a holiday may fall in the winter in some years, and then a few years later it will have moved back to autumn.

The Modern Civil Calendar

The international civil calendar used today is a solar calendar known as the **Gregorian calendar**. It is based on the **tropical year**, which measures the 365.242 solar days between vernal (spring) equinoxes. A **solar day** is the 24-hour period of Earth's rotation that brings the Sun back to the same local meridian. That measure is in contrast to the **sidereal day**, the time Earth takes to make one rotation and face the same star on the meridian. The sidereal day is about 23 hours 56 minutes and is slightly shorter than the solar day because of Earth's motion around the Sun.

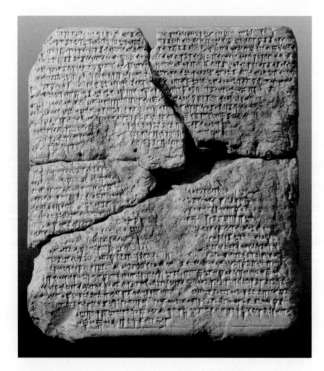

Figure 2.23 The ancient Egyptian calendar used a system of 12 months, plus festival days.

The Gregorian calendar includes a system of **leap years**—years in which a 29th day is added to February—decreed by Julius Caesar in 45 BCE to make up for the extra fraction of a day. Leap years prevent the seasons from slowly sliding through the year to become increasingly out of sync with the months, so that we don't end up experiencing winter in December one year and in August in later years.

The Gregorian calendar is named for Pope Gregory XIII. He was concerned that the Easter holiday—the first Sunday after the first full Moon after March 21—was drifting away from the vernal equinox. Julius Caesar's rule of one leap year every 4 years resulted in an average year of 365.25 days, but the actual year is 365.242 days. That difference of 0.008 day is about 11.5 minutes per year, or about 3 days every 400 years, and by Gregory's time it had caused the date of the vernal equinox to drift in the Julian calendar by about 10 days. In 1582, therefore, Pope Gregory decreed that 10 days would be deleted from the calendar to move the vernal equinox back to March 21. To make that system work out better in the future, he declared that only century years divisible by 400 are leap years, thereby deleting 3 leap years (and the 3 days) every 400 years. Catholic countries followed that system immediately, but Protestant countries did not adopt it until the 1700s. Eastern Orthodox countries, including Russia, did not switch from the Julian to the Gregorian calendar until the 1900s. One slight further revision—making years divisible by 4,000 into common 365-day years—has been proposed so the modern Gregorian calendar will slip by about only 1 day in 20,000 years.

Despite international adoption of the Gregorian calendar, billions of people still celebrate holidays and festivals according to a lunisolar or lunar calendar. Chinese New Year, Passover, Easter, Ramadan, Rosh Hashanah, and Diwali, among others, have dates that change from one year to the next because they are based on lunar months from lunisolar or lunar calendars. The astronomy of people from long ago is still in use today.

CHECK YOUR UNDERSTANDING 2.4

Suppose that the astronomical cycles were even multiples of each other, so that a cycle of lunar phases was precisely 30 days and Earth's orbital period was precisely 12 months. In such a system, how would the dates of "wandering" holidays, such as Chinese New Year or Ramadan, change?

2.5 Eclipses Result from the Alignment of Earth, the Moon, and the Sun

Ancient peoples attuned to the patterns of the sky must have been terrified to watch the Sun being eaten away as if by a giant dragon or to see the full Moon turning the color of blood. An **eclipse** is the total or partial obscuration of one celestial body, or the light from that body, by another celestial body. Archaeological evidence suggests that ancient peoples tried hard to figure out the pattern of eclipses and thereby bring them into the orderly scheme of the heavens. Stonehenge, pictured in Figure 2.1a, may have enabled its builders to predict when eclipses might occur. Ancient Chinese, Babylonian, and Greek astronomers had figured out that eclipses occur in cycles, and they used that knowledge to predict when and where eclipses would occur. In this section, we describe the types of eclipses and their frequency.

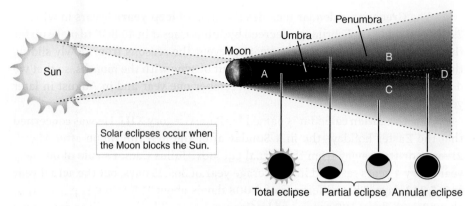

Figure 2.24 Different parts of the Sun are blocked at different places within the Moon's shadow. An observer on Earth in the umbra (point A) sees a total solar eclipse, observers in the penumbra (points B and C) see a partially eclipsed Sun, and observers at point D see an annular solar eclipse. Not to scale.

Solar Eclipses

A **solar eclipse** occurs when the Moon passes between Earth and the Sun; observers on Earth in the shadow of the Moon will see the eclipse. Three types of solar eclipses exist: *total*, *partial*, and *annular*. Consider the structure of the shadow of the Sun cast by a round object such as the Moon, as shown in **Figure 2.24**. An observer at point A would see the Sun blocked. That darkest, inner part of the shadow is called the **umbra**. If a location on Earth passes through the Moon's umbra, the Moon blocks all the Sun's light, and a **total solar eclipse** will be observed (Chapter Opening Figure and **Figures 2.25** and **2.26a**). At points B and C in Figure 2.24, an observer can see one side of the disk of the Sun but not the other. Only partially in shadow, that outer region is the **penumbra**. If a location on Earth passes through the Moon's penumbra, viewers there will observe a **partial solar eclipse**, in which the Moon's disk blocks the light from a portion of the Sun's disk.

In the third type of solar eclipse, called an **annular solar eclipse**, the Sun appears as a bright ring surrounding the dark disk of the Moon (**Figure 2.26b**). An observer at point D in Figure 2.24 is far enough from the Moon that it appears smaller than the Sun. An object's apparent size in the sky depends on the object's actual size and its distance from us. The Sun is about 400 times the diameter of the Moon, and the distance between the Sun and Earth is about 400 times more than the distance between the Moon and Earth. As a result, the Moon and Sun have almost the same apparent size in the sky. In addition, the Moon's orbit is not a perfect circle. When the Moon and Earth are a bit closer together than average, the Moon appears slightly larger in the sky than the Sun. An eclipse occurring then will be total for some observers. When the Moon and Earth are farther apart than average, the Moon appears smaller than the Sun, so eclipses occurring then will be annular for some observers. Among all solar eclipses, one-third are total at some location on Earth, one-third are annular, and one-third are only partial.

Figure 2.27 shows the geometry of a solar eclipse when the Moon's shadow falls on Earth's surface. Figures like that usually show Earth and the Moon much closer together than they really are. The page is too small to draw them to scale and still show the critical details. The relative sizes and distances between Earth

Figure 2.25 The full spectacle of a total solar eclipse.

Figure 2.26 Time-lapse images of the Sun taken (a) during a total solar eclipse and (b) during the annular solar eclipse of May 20, 2012. The Sun set during the ending phases.

and the Moon are roughly equivalent to the difference between a basketball and a tennis ball, respectively, placed 7 meters apart. Figure 2.27b shows the geometry of a solar eclipse with Earth, the Moon, and the separation between them drawn to scale. Compare that drawing with Figure 2.27a to see why drawings of Earth and the Moon are rarely drawn to scale. If the Sun were drawn to scale in Figure 2.27a, it would be bigger than your head and located almost 64 meters off the left side of the page.

From any particular location, you are more likely to observe a partial solar eclipse than a total solar eclipse. Where the Moon's penumbra touches Earth, it has a diameter of almost 7,000 km—large enough to cover a substantial fraction of Earth. Thus, a partial solar eclipse is often visible from many locations on Earth.

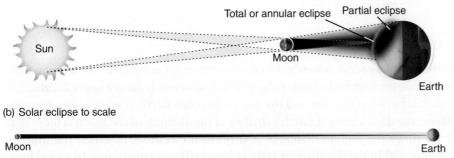

Figure 2.27 (a, b) A solar eclipse occurs when the shadow of the Moon falls on the surface of Earth. Note that (b) is drawn to scale.

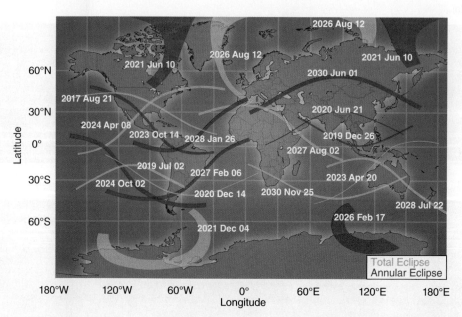

Figure 2.28 The paths of total solar eclipses through 2030. Solar eclipses occurring in Earth's polar regions cover more territory because the Moon's shadow hits the ground obliquely.

In contrast, the path along which a total solar eclipse is visible, shown in **Figure 2.28**, covers only a tiny fraction of Earth's surface. Even when the distance between Earth and the Moon is at a minimum, the umbra is only 270 km wide on Earth. As the Moon moves along in its orbit, that tiny shadow sweeps across Earth at a few thousand kilometers per hour. In addition, the Moon's shadow falls on Earth's curved surface, causing the region shaded by the Moon during a solar eclipse to be elongated by various amounts. The curvature can even cause an eclipse that started out as annular to become total.

As a result of those factors, a total solar eclipse can never last longer than 7 1/2 minutes and is usually significantly shorter. Even so, it is one of the most amazing sights in nature. People all over the world flock to the most remote corners of Earth to witness the fleeting spectacle of the bright disk of the Sun blotted out of the daytime sky. Perhaps you saw some of the total eclipse visible from much of the United States in August 2017. The next total solar eclipse visible in the continental United States will take place in 2024. Annular eclipses will be visible from parts of the United States in 2021 and 2023. Viewing a solar eclipse should be on your lifetime to-do list!

Lunar Eclipses

Lunar eclipses occur when the Moon moves through Earth's shadow. **Figure 2.29a** shows the geometry of a lunar eclipse and is drawn to scale in **Figure 2.29b**. Here Earth is between the Sun and the Moon. Because Earth is much larger than the Moon, the dark umbra of Earth's shadow at the distance of the Moon is more than 2 1/2 times the Moon's diameter. A **total lunar eclipse**, in which the Moon is entirely within Earth's shadow, lasts as long as 1 hour 40 minutes. In a total lunar eclipse, the Moon often appears red (**Figure 2.30a**). That "blood-red Moon," as literature and poetry have called it, occurs because red light from the Sun is bent

2.5 Eclipses Result from the Alignment of Earth, the Moon, and the Sun

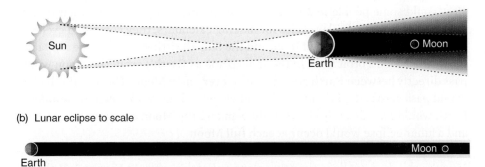

(a) Lunar eclipse geometry (not to scale)

(b) Lunar eclipse to scale

Figure 2.29 (a, b) A lunar eclipse occurs when the Moon passes through Earth's shadow. Note that (b) is drawn to scale.

as it travels through Earth's atmosphere and then illuminates the Moon. Earth's atmosphere absorbs or scatters other colors of sunlight away from the Moon and therefore does not illuminate it.

A **penumbral lunar eclipse** occurs when the Moon passes through the penumbra of Earth's shadow; those are noticeable only from a very dark location or when the Moon passes within about 1,000 km of the umbra. If Earth's shadow incompletely covers the Moon, some of the Moon's disk remains bright and some of it is in shadow. The result is a **partial lunar eclipse**. **Figure 2.30b** shows a composite of images taken at different times during a partial lunar eclipse. In the middle frame, Earth's shadow nearly completely eclipses the Moon.

Many more people have observed a total lunar eclipse than have observed a total solar eclipse. To see a total solar eclipse, you must be located within that very narrow band of the Moon's shadow as it moves across Earth's surface. In contrast, when the Moon is immersed in Earth's shadow, anyone located in the hemisphere facing the Moon can see it. As a result, from any location, total eclipses of the Moon are relatively common, and you may have seen at least one.

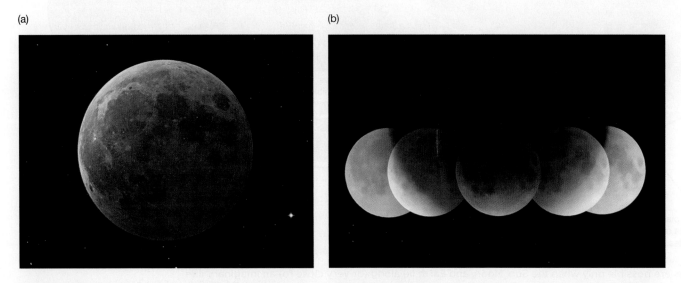

Figure 2.30 (a) During a total lunar eclipse, the Moon often appears blood red. (b) A time-lapse series of photographs of a partial lunar eclipse clearly shows Earth's shadow. Note the size of Earth's shadow in comparison with the size of the Moon.

Frequency of Eclipse Seasons

How could some people in ancient cultures predict eclipses? From their understanding of lunar and solar cycles for making calendars, they computed cycles of eclipses. Imagine Earth, the Moon, and the Sun all sitting on the same flat tabletop. If the Moon's orbit were in the same plane as Earth's orbit, the Moon would pass directly between Earth and the Sun at every new Moon. The Moon's shadow would pass across the face of Earth, and we would see a solar eclipse. Similarly, Earth would pass directly between the Sun and the Moon every synodic month, and a lunar eclipse would occur at each full Moon.

Solar and lunar eclipses do *not* happen every month, however, because the Moon's orbit does *not* lie in the same plane as Earth's orbit. As **Figure 2.31** shows, the plane of the Moon's orbit around Earth is inclined by about 5.2° with respect to the plane of Earth's orbit around the Sun. The line along which the orbital planes of the Sun and the Moon intersect is called the **line of nodes**. For part of the year, the line of nodes points generally toward the Sun. During those times, called **eclipse seasons**, a new Moon passes directly between the Sun and Earth, casting its shadow on Earth's surface and causing a solar eclipse. Similarly, a full Moon occurring during an eclipse season passes through Earth's shadow, causing a lunar eclipse. An eclipse season lasts only 38 days. That's how long the Sun is close enough to the line of nodes for eclipses to occur. Usually, the line of nodes

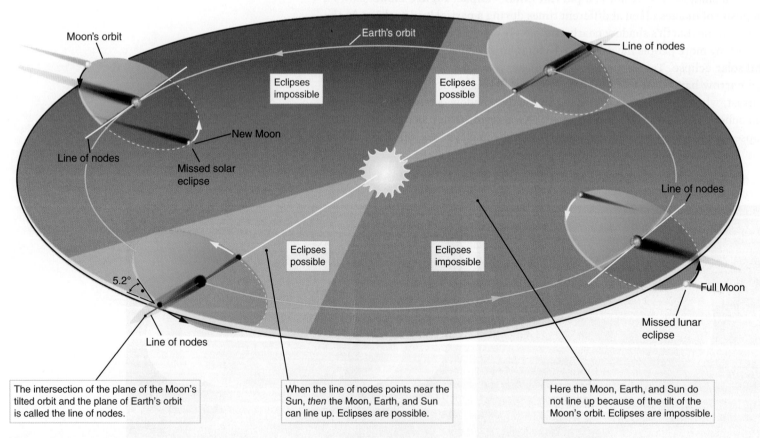

Figure 2.31 Eclipses are possible only when the Sun, Moon, and Earth lie along (or very close to) an imaginary line known as the line of nodes. When the Sun does not lie along the line of nodes, Earth passes under or over the shadow of a new Moon, and a full Moon passes under or over the shadow of Earth.

points farther away from the Sun, and Earth, Moon, and Sun cannot line up closely enough for an eclipse to occur. At those times, a solar eclipse cannot take place because the shadow of a new Moon passes "above" or "below" Earth. Similarly, no lunar eclipse can occur because a full Moon passes "above" or "below" Earth's shadow.

If the plane of the Moon's orbit always had the same orientation, eclipse seasons would occur twice a year, as suggested in Figure 2.31. In actuality, eclipse seasons occur about every 5 months 20 days. The roughly 10-day difference occurs because the plane of the Moon's orbit slowly wobbles, much like the wobble of a spinning plate balanced on the end of a circus performer's stick. As the Moon's orbital plane wobbles, the line of nodes changes direction. That wobble rotates in the direction opposite the direction of the Moon's motion in its orbit. That is, the line of nodes moves clockwise as viewed from above Earth's orbital plane. One wobble of the Moon's orbit takes 18.6 years, so we say that the line of nodes regresses by 360° every 18.6 years, or 19.4° per year. That rate amounts to about a 20-day regression each year. If January 1 marks the middle of an eclipse season, the next eclipse season will be centered around June 20, and the one after that around December 10.

CHECK YOUR UNDERSTANDING 2.5

If the Moon were in its same orbital plane but twice as far from Earth, which of the following would happen? (a) The Moon would not go through phases. (b) Total eclipses of the Sun would not be possible. (c) Annular eclipses of the Sun would not be possible. (d) Total eclipses of the Moon would not be possible.

Origins The Obliquity of Earth

Earth's various motions give rise to the most basic of patterns faced by life on Earth. Earth's rotation is responsible for the cycle of night and day. Earth's axial tilt and its passage around the Sun bring the change of the seasons. As life evolved on Earth, it had to adapt to those patterns.

Earth's range of climate, based on distance from the equator, probably has contributed to the broad diversity of life on our planet. Earth's biodiversity includes life that adapted to the long, cold polar nights and to the much higher temperatures of the tropics. Earth's life adapted to seasonal patterns in rain and drought, leading to acquired seasonal patterns of migration and reproduction. If Earth had no axial tilt, the poles would continuously be in winter and probably too cold for humans. Latitudes near the equator would be consistently even warmer than they are now.

Periodic changes in Earth's axial tilt might have affected life. If Earth's tilt were larger than 23.5°, the seasonal variation would be even stronger. If the tilt were smaller, the seasonal variation would be weaker. Chinese, Indian, Greek, and Arabic records going back 3,000 years indicate that people estimated the tilt by measuring the length of the shadow from a vertical pole on the day of the solstice. Earth's axial tilt actually varies from 22.1° to 24.5° over a 41,000-year cycle. Currently, the tilt is about midway between the two extremes and getting smaller. It will reach its minimum value of 22.1° in about 10,000 years. Scientists are studying whether that variation in tilt correlates with periods of temperature change on Earth, especially the times of ice ages.

READING ASTRONOMY NEWS

ARTICLES | QUESTIONS

Citizen scientists across the nation captured images of the Sun's corona to create the largest dataset of its kind.

2,500 Miles of Citizen Scientists

LAURA REED, sciencenode.org

On August 21 many of us stood outside and marveled at the solar eclipse.

But for some, the eclipse was more than just a show. It was an opportunity to participate in the National Solar Observatory (NSO) Citizen Continental-America Telescopic Eclipse (CATE) Experiment.

The Citizen CATE Experiment is funded by federal, corporate, and private sources, including several companies that donated equipment to the project. Volunteer observers and schools get to keep the equipment used in the study.

A grand experiment

Scientists, students, and volunteers assembled along the 2,500-mile path of eclipse totality tracked the Sun using 68 identical telescopes.

Using specialized software and instrument packages, they produced more than 1,000 images from the start of the partial solar eclipse until totality.

The goal was to capture 90 minutes of continuous, high-resolution, rapid-cadence images detailing a difficult-to-capture region of the solar atmosphere: the Sun's inner corona.

This is the first time scientists have collected research-quality observations of the corona during the eclipse's entire transit across the U.S.

"This dataset is extraordinary," says Matt Penn, principal investigator for the project. "Normally during a solar eclipse, we get about 2 minutes of data in the region closest to the photosphere. But Citizen CATE allows us to get an hour and a half of data."

My corona

The corona is the Sun's outer atmosphere. It is difficult for scientists to study because the photosphere, or solar surface, is so bright that it overpowers the faint corona. We can only see the corona when something obscures the photosphere.

(Think of a light so bright that it makes it difficult to see an object close to you. Blocking the light with your hand allows a better view of the object.)

Scientists can create artificial eclipses using an instrument called a coronagraph that covers the Sun's bright disk. The proximity of the instrument to the observer distorts the Sun's edge, making precise observation and measurement difficult.

During a real eclipse, the Moon blocks the Sun. The Moon's great distance lets scientists measure and study the corona in greater detail.

The process

On the day of the event, skies were clear for 58 of the 68 observation sites.

Observers, spaced about 50 miles from each other, started observations when the Moon's shadow appeared on their horizon. They captured images every 10 seconds during totality.

The rapid cadence of imagery along with a 2-arcsecond pixel resolution should help scientists understand the intensity of the corona over an extended period as well as the motions of prominences, coronal inflows, coronal mass ejections, and other active regions.

The next total solar eclipse in the United States will be April 8, 2024. The path of totality will begin in Texas, moving north from Mexico, and will exit the U.S. via Maine.

Start making plans to join the next team of citizen scientists.

ARTICLES | **QUESTIONS**

1. Why does a solar eclipse move across the surface of the Earth?
2. Why did scientists want data from along the whole track of the eclipse?
3. Why can the corona be seen only during an eclipse?
4. Would you have expected a lunar eclipse near the date of the solar eclipse? Explain.
5. See whether any results are available on the project website (http://citizencate.org/). What did participants learn from this experiment?

Source: Laura Reed, "2,500 miles of citizen scientists," sciencenode.org, August 22, 2017. © Trustees of Indiana University. Under a CC BY-NC-ND 4.0 license. Reprinted by permission of Science Node.

Summary

The motions of Earth and the Moon are responsible for many of the repeating patterns that can be observed in the sky. Calendars keep track of time by using those patterns. Earth's rotation about its axis causes daily patterns of rising and setting, and Earth's revolution around the Sun causes yearly patterns of the stars in the sky and marks the passage of the seasons. Earth's tilt on its axis changes both the length of daytime and the intensity of sunlight, causing the seasons. Life on Earth adapted to those seasonal variations. Earth's axial tilt varies slightly over tens of thousands of years. The Moon's revolution around Earth causes the monthlong pattern of the phases of the Moon. Occasionally, alignments of Earth, the Moon, and the Sun cause eclipses.

LG 1 **Describe how Earth's rotation about its axis and its revolution around the Sun affect how we perceive celestial motions from different places on Earth.** Earth's daily rotation on its axis causes the apparent daily motion of the Sun, Moon, and stars. Our location on Earth and Earth's location in its orbit around the Sun determine which stars we see at night. You can determine your latitude from the altitude of the pole star. When observing objects in our sky, we need to consider how Earth moves in relation to other objects. The ecliptic is the path that the Sun appears to take through the stars.

LG 2 **Explain why seasons change throughout the year.** A year is the time Earth takes to complete one revolution around the Sun. Constellations are patterns of stars that reappear in the same place in the sky at the same time of night each year. The tilt of Earth's axis determines the seasons by changing the angle at which sunlight strikes Earth's surface in different locations. The changing angle of sunlight and the differing length of the day cause the seasonal variations on Earth. Equinoxes and solstices mark the changing seasons.

LG 3 **Describe the factors that create the phases of the Moon.** The relative locations of the Sun, Earth, and Moon determine the phases of the Moon. The Moon takes one sidereal month to complete one revolution around Earth and one synodic month to go through a cycle of phases. The Moon's motion around Earth causes it to be illuminated differently at different times. When farther from the Sun than Earth is, the Moon is in gibbous phases. When closer to the Sun than Earth is, the Moon is in crescent phases.

LG 4 **Sketch the alignment of Earth, the Moon, and the Sun during eclipses of the Sun and the Moon.** A solar eclipse occurs when the new Moon is in the plane of Earth and the Sun, and the shadow of the Moon falls on Earth. A lunar eclipse occurs when the full Moon is in the plane of Earth and the Sun, and the shadow of Earth falls on the Moon. Twice a year, at new or at full Moon, the Moon is exactly in line between Earth and the Sun. At those times, eclipses occur.

? Unanswered Questions

- How long will Earth continue to have total solar eclipses? Those occur because the Moon and the Sun are coincidentally the same size in our sky, but will that always be the case? An object's observed size in the sky depends on its actual diameter and its distance from us. One or both of those can change. The Moon is slowly moving away from Earth by about 4 meters per century. Over time, the Moon will appear smaller in the sky, and it won't be able to cover the full disk of the Sun. Although we can measure how fast the Moon is moving away from Earth now, we are less certain of how that rate may change with time. A lesser and more uncertain effect comes from the Sun—which will continue to brighten slowly, as it has throughout its history. With that brightening, the Sun's actual diameter will slightly increase, and it will appear larger in our sky. A more distant Moon and a larger Sun will eventually (in hundreds of millions of years) end total eclipses on Earth.

Questions and Problems

Test Your Understanding

1. Constellations are groups of stars that
   a. are close to one another in space.
   b. are bound to each other by gravity.
   c. are relatively close to one another in Earth's sky.
   d. all have the same composition.

2. Where on Earth can you stand and, over the course of a year, see the entire sky?
   a. only at the North Pole
   b. at either pole
   c. at the equator
   d. anywhere

3. Day and night are caused by
   a. the tilt of Earth on its axis.
   b. the rotation of Earth on its axis.
   c. the revolution of Earth around the Sun.
   d. the revolution of the Sun around Earth.

4. Polaris, the North Star, is unique because
   a. it is the brightest star in the night sky.
   b. it is the only star in the sky that doesn't move throughout the night.
   c. it is always located at the zenith, for any observer.
   d. it has a longer path above the horizon than any other star.

5. An angle exists between the ecliptic and the celestial equator because
   a. Earth's axis is tilted with respect to its orbit.
   b. Earth's orbit is tilted with respect to the orbits of other planets.
   c. the Sun follows a rising and falling path through space.
   d. the Sun's orbit is tilted with respect to Earth's.

6. Earth's axial tilt causes the seasons because
   a. one hemisphere of Earth is closer to the Sun in summer.
   b. the days are longer in summer.
   c. the rays of light strike the ground more directly in summer.
   d. both a and b
   e. both b and c

7. Which is *not* true on the vernal and autumnal equinoxes?
   a. Every nonpolar place on Earth has 12 hours of daylight and 12 hours of darkness.
   b. The Sun rises due east and sets due west.
   c. The Sun is located on the celestial equator.
   d. The motion of the stars in the sky is different from that on other days.

8. We always see the same side of the Moon because
   a. the Moon does not rotate on its axis.
   b. the Moon rotates on its axis once for each revolution around Earth.
   c. the other side of the Moon is unlit when it is facing Earth.
   d. the other side of the Moon is on the opposite side of Earth when it is facing Earth.

9. If you see the Moon on the meridian at sunrise, the phase of the Moon is
   a. waxing gibbous.
   b. full.
   c. first quarter.
   d. third quarter.

10. A lunar eclipse occurs when _____ shadow falls on _____.
    a. Earth's; the Moon
    b. the Moon's; Earth
    c. the Sun's; the Moon
    d. the Sun's; Earth

11. Different cultures created different calendars because
    a. they had measured different lengths of the day, month, and year.
    b. they used different definitions of the day, month, and year.
    c. the number of days in a month and the number of days and months in a year are not integers.
    d. calendars are completely arbitrary.

12. Which stars we see at night depends on
    a. our location on Earth.
    b. Earth's location in its orbit.
    c. the time of the observation.
    d. all of the above.

13. On the summer solstice in June, the Sun will be directly above _____ and all locations north of _____ will experience daylight all day.
    a. the Tropic of Cancer; the Antarctic Circle
    b. the Tropic of Capricorn; the Arctic Circle
    c. the Tropic of Cancer; the Arctic Circle
    d. the Tropic of Capricorn; the Antarctic Circle

14. The Sun, Moon, and stars
    a. appear to move each day because the celestial sphere rotates about Earth.
    b. change their relative positions over time.
    c. rise north or south of west and set north or south of east, depending on their location on the celestial sphere.
    d. always remain in the same position with respect to one another.

15. If you see the first quarter Moon on the meridian, the Sun must be
    a. on the western horizon.
    b. on the eastern horizon.
    c. below the horizon.
    d. on the meridian.

Thinking about the Concepts

16. Seafarers such as Columbus used Polaris to navigate as they sailed from Europe to North America. When Magellan sailed the South Seas, he could not use Polaris. Explain why.

17. If you were standing at Earth's North Pole, where would you see the north celestial pole in relation to your zenith?

18. Observers in the Northern Hemisphere see the zodiacal constellation Gemini in the winter. Why don't they see it in the summer?

19. Imagine that you are flying along in a jetliner.
    a. Describe ways to tell that you are moving.
    b. If you look down at a building, which way is it moving relative to you?

20. Astronomers are sometimes asked to serve as expert witnesses in court cases. Suppose you are called as an expert witness, and the defendant states that he could not see the pedestrian because the full Moon was casting long shadows across the street at midnight. Is that claim credible? Why or why not?

21. Imagine that one person was developing a theory of seasons as described in the three hypotheses in this chapter's Process of Science Figure. Compare that process with the flowchart of Chapter 1's Process of Science Figure. How would the development of that theory look on the diagram?

22. Why is the winter solstice *not* the coldest time of year?

23. Earth spins on its axis and wobbles like a top.
    a. How long does it take to complete one spin?
    b. How long does it take to complete one wobble?

24. At what approximate time of day do you see the full Moon near the meridian? At what time is the first quarter (waxing) Moon on the eastern horizon? Use a sketch to help explain your answers.

25. Assume that the Moon's orbit is circular, and imagine that you are standing on the side of the Moon that faces Earth.
    a. How would Earth appear to move in the sky as the Moon made one revolution around Earth?
    b. How would the "phases of Earth" appear to you compared with the phases of the Moon as seen from Earth?

26. If people on Earth were observing a lunar eclipse, what would you see from the Moon?

27. From any given location, why are you more likely to witness a partial eclipse of the Sun than a total eclipse?

28. Why don't we see a lunar eclipse each time the Moon is full or witness a solar eclipse each time the Moon is new?

29. How would Earth's temperature variation be different if it were tilted 90° on its axis (like the planet Uranus)?

30. Explain how a cyclic change in Earth's tilt could affect the planet's seasonal temperatures.

Applying the Concepts

31. Earth spins along at 1,674 km/h at the equator. Use that fact, along with the length of the day, to calculate Earth's equatorial diameter.

32. Determine the latitude where you live. Draw and label a diagram showing that your latitude is the same as (a) the altitude of the north celestial pole and (b) the angle (along the meridian) between the celestial equator and your local zenith. What is the Sun's altitude at noon as seen from your home at the winter solstice and at the summer solstice?

33. If the angle between your zenith and Polaris is 40°, what is the altitude of Polaris? What is your latitude?

34. The southernmost star in a group of stars known as the Southern Cross lies approximately 65° south of the celestial equator. What is the farthest-north latitude from which the entire Southern Cross is visible? Can it be seen in any U.S. states? If so, which ones?

35. Imagine that you are standing on the South Pole at the time of the southern summer solstice.
    a. How far above the horizon will the Sun be at noon?
    b. How far above (or below) the horizon will the Sun be at midnight?

36. Suppose the tilt of Earth's equator with respect to its orbit were 10° instead of 23.5°. At what latitudes would the Arctic and Antarctic circles and the Tropics of Cancer and Capricorn be located?

37. The Moon's orbit is tilted by about 5° with respect to Earth's orbit around the Sun. What is the highest altitude in the sky that the Moon can reach, as seen in Philadelphia (latitude 40° north)?

38. Suppose you would like to witness the midnight Sun (when the Sun appears just above the northern horizon at midnight), but you don't want to travel any farther north than necessary.
    a. How far north (that is, to which latitude) would you have to go?
    b. At what time of year would you make that trip?

39. If, as some historians believe, Eratosthenes used the Egyptian stadion, about 157.5 meters, what would he have computed for the size of Earth? How close is that result to the modern value?

40. a. The vernal equinox is now in the zodiacal constellation Pisces. The precession of Earth's axis will eventually cause the vernal equinox to move into Aquarius. How long, on average, does the vernal equinox spend in each zodiacal constellation?
    b. Stonehenge was erected roughly 4,000 years ago. Referring to the zodiacal constellations shown in Figure 2.13, identify the constellation in which those ancient builders saw the vernal equinox.

41. Referring to Figure 2.18, estimate when Vega, the fifth-brightest star in our sky (excluding the Sun), will once again be the northern pole star.

42. The Moon's apparent diameter in the sky is approximately ½°. How long does the Moon take to move 360°? About how long does the Moon take to move a distance equal to its own diameter across the sky?

43. An object's apparent size in the sky is proportional to its actual diameter divided by its distance. The Moon has a radius of 1,737 km, with an average distance of 3.780×10^5 km from Earth. The Sun has a radius of 696,000 km, with an average distance of 1.496×10^8 km from Earth. Show that the apparent sizes of the Moon and Sun in our sky are approximately the same.

44. Earth has an average radius of 6,371 km. If you were standing on the Moon, how much larger would Earth appear in the lunar sky than the Moon appears in our sky?

45. How would the length of the eclipse season change if the plane of the Moon's orbit were inclined less than its current 5.2° to the plane of Earth's orbit? Explain your answer.

USING THE WEB

46. Go to the U.S. Naval Observatory website (USNO "Data Services," at http://aa.usno.navy.mil/data). Look up the times for sunrise and sunset for your location for the current week. (You can change the dates one at a time or bring up a table for the entire month.) How are the times changing from one day to the next? Bring up the "Duration of Daylight/Darkness Table for One Year" page for your location. Are the days getting longer or shorter? When do the shortest and the longest days occur? Look up a location in the opposite hemisphere (Northern or Southern). When are the days shortest and longest?

47. Go to the "Earth and Moon Viewer" website (http://fourmilab.ch/earthview). Under "Viewing the Earth," click on "latitude, longitude and altitude" and enter your approximate latitude and longitude, and 40,000 for altitude; then select "View Earth." Are you in daytime or nighttime? Now play with the locations; keep the same latitude but change to the opposite hemisphere (Northern or Southern). Is it still night or day? Go back to your latitude, and this time enter 180° minus your longitude, and change from west to east, or from east to west, so that you are looking at the opposite side of Earth. Is it night or day there? What do you see at the North Pole (latitude 90° north) and the South Pole (latitude 90° south)? At the bottom of your screen you can play with the time. Move back 12 hours. What do you observe at your location and at the poles?

48. Go to the U.S. Naval Observatory website (USNO "Data Services," at http://aa.usno.navy.mil/data). Look up the Moon data for today. When will it rise and set? What is the phase? How will it change over the next 4 weeks? Enter one day at a time or look at the yearlong tables for moonrise and moonset and for the dates of primary phases. What time of day does a third quarter Moon rise? When (and in what phases) can you see the Moon in the daytime?

49. Using the times of moonrise and moonset that you found in question 48, make a plan to observe the Moon directly for a few weeks. Take a picture of the Moon (or make a sketch) every other day. How is the brightness of the Moon changing? When it's daytime, estimate how far the Moon is from the Sun in the sky. When it's nighttime, use your pictures to see whether the stars near the Moon in the sky are the same every night.

50. Go to the "NASA Eclipse" website (https://eclipse.gsfc.nasa.gov/eclipse.html). When is the next lunar eclipse? Will it be visible at your location if the skies are clear? Is it a total or partial eclipse? When is the next solar eclipse? Will it be visible at your location? Compare the fraction of Earth that the solar eclipse will affect with the fraction for the lunar eclipse. Why are lunar eclipses visible in so many more locations?

EXPLORATION

The Phases of the Moon

digital.wwnorton.com/astro6

Visit the Digital Resources Page and on the Student Site open the "Phases of the Moon" Interactive Simulation in Chapter 2.

Study the diagrams shown in the simulator. The largest window shows a view of the Earth-Moon system as seen from above Earth's North Pole. The Sun is far off the screen to the left. An observer stands on Earth. The small window at upper right shows how the Moon looks from the Northern Hemisphere. The small window at lower right shows the observer's location, with the Sun and Moon pictured as flat disks in the sky.

1 Given the relative positions of the observer and the Sun, approximately what time is it for this observer: 6:00 A.M., noon, 6:00 P.M., or midnight?

_____

2 Where is the Moon in the observer's sky: on the eastern horizon, on the western horizon, below the horizon, or crossing the meridian?

_____

3 What is the phase of the Moon?

_____

4 Imagine yourself on the Moon in the image shown in the larger window. If you looked toward Earth, what phase of Earth would you see?

_____

Now select "start animation." Allow the animation to run until the Moon is 20 percent illuminated (as shown in the upper small window on the right).

5 In which direction does the Moon orbit Earth: clockwise or counterclockwise?

_____

6 For observers in the Northern Hemisphere, which side of the Moon is illuminated first after a new Moon: right or left?

_____

7 If you observe a crescent Moon with the horns of the crescent pointing right, is the Moon waxing or waning?

_____

Grab the Moon with your mouse and drag it to first quarter. Drag the observer so that her local time is approximately midnight.

8 Where is the first quarter Moon in the observer's sky: on the eastern horizon, on the western horizon, below the horizon, or crossing the meridian?

_____

The time, the phase of the Moon, and the Moon's location in the sky are related.

9 Arrange the observer and the Moon so that the Moon is full and crossing the meridian. What time is it for the observer: 6:00 A.M., noon, 6:00 P.M., or midnight?

_____

10 Arrange the observer and the Moon so that the Moon is in third quarter and the time for the observer is approximately noon (the Sun is on the meridian). Where is the Moon in the observer's sky: on the eastern horizon, on the western horizon, below the horizon, or crossing the meridian?

_____

11 Arrange the observer and the Moon so that it is approximately 6:00 P.M. (sunset) for the observer, and the Moon is just rising on the eastern horizon. What is the phase of the Moon?

_____

Many other combinations of the time, phase of the Moon, and Moon's location are available to play with. Challenge yourself to be able to set up any two of those and find the third. When you can do that without the simulator, just by making the picture in your head, you will really understand the phases of the Moon.

3

Motion of Astronomical Bodies

The birth of modern astronomy dates back to when astronomers and mathematicians discovered regular patterns in the motions of the planets. A successful theory of how Earth moved and how it fit in with its neighbors in the Solar System was the first step toward understanding our planet's place in the universe.

LEARNING GOALS

Here in Chapter 3, we examine how astronomers came to understand that Earth and other planets orbit the Sun. By the end of this chapter, you should be able to:

LG 1 Compare the geocentric and heliocentric models of the Solar System.

LG 2 Use Kepler's laws to describe how objects in the Solar System move.

LG 3 Explain how Galileo's astronomical discoveries convinced him that the heliocentric model was correct.

LG 4 Describe the physical laws of motion discovered by Galileo and Newton.

The shadows of a few of Jupiter's Galilean moons fall on the planet. ▶▶▶

Who discovered Jupiter's moons?

3.1 The Motions of Planets in the Sky

When people in ancient times looked up at the sky, they saw the Sun, Moon, and stars rise in the east and set in the west and *appear* to move around Earth. The ancient peoples were aware of five planets (*planet* means "wandering star") because from one night to the next they moved in a generally eastward direction among the stars, whose positions were fixed on the celestial sphere. The ancients did *not* know, however, that Earth was similar to those planets. A successful theory of how Earth and the planets move and how Earth fits in with its neighbors in the Solar System was the first step to understanding Earth's place in the universe. The history of how those ideas evolved—from Earth at the center of all things to Earth as just another planet—is a prime example of science's self-correcting nature.

The Geocentric Model

As discussed in Chapter 2, hard evidence of the motions of Earth was remarkably difficult to come by. Largely because we do not feel Earth moving through space, early astronomers developed a **geocentric** (Earth-centered) **model** of the Solar System to explain what they observed in the sky. Looking up at the sky, people saw that the Sun, Moon, planets, and stars appeared to move around Earth. Before the 17th century, Western educated people believed that the Sun, Moon, and known planets all moved in circles around a stationary Earth. **Figure 3.1** illustrates the geocentric model of the Alexandrian astronomer Ptolemy (Claudius Ptolemaeus, 90–168 CE).

The geocentric model, though, did not account for all observations. Ancient astronomers knew that the planets usually have an eastward **prograde motion**, in which each night they move a little eastward with respect to the background stars. Sometimes, however, the ancients observed apparent **retrograde motion**, in which the planets appear to move westward for a time before resuming their normal eastward travel. **Figure 3.2** shows retrograde motion for Mars.

That odd behavior of the five "naked eye" planets—Mercury, Venus, Mars, Jupiter, and Saturn—created a puzzling problem for the geocentric model as

Figure 3.1 In the Ptolemaic view of the heavens, Earth is at the center, orbited by the Moon, Mercury, Venus, the Sun, Mars, Jupiter, and Saturn.

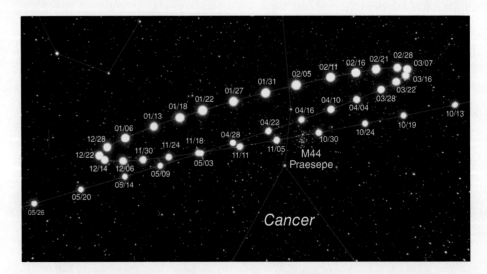

Figure 3.2 This time-lapse photographic series shows Mars moving in apparent retrograde motion.

Ptolemy summarized it in 150 CE. Some astronomers and philosophers in ancient times, such as Aristarchus of Samos (310–230 BCE), hypothesized that the Sun might be the center of the Solar System, but they lacked the tools to test that hypothesis or the mathematical insight to formulate a more complete and testable model. But most other astronomers of the time were skeptical because they thought that if Earth moved around the Sun, they should feel Earth's motion. Therefore, they preferred the geocentric model, in which the Sun, Moon, and planets all moved in perfect circles around a stationary Earth, with the "fixed stars" located somewhere way out beyond the planets.

Because the geocentric model in its simplest form failed to explain the retrograde motion of the planets, Ptolemy added an embellishment called an *epicycle*—a small circle superimposed on each planet's larger circle (**Figure 3.3**). Thus, as the planet travels along its larger circle around Earth, it also moves periodically along its epicycle. When moving along the smaller circle in the opposite direction of the forward motion of the larger circle, the planet's forward motion would be reversed and it would appear to move backward in the sky. Ptolemy's model had many epicycles, but they made the model *work*, in the sense that this model was reasonably successful at predicting the positions of planets in the sky. For nearly 1,500 years, Ptolemy's model of the heavens was the accepted paradigm in the Western world.

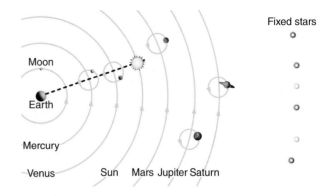

Figure 3.3 To reconcile retrograde motion with the geocentric model of the Solar System, Ptolemy added loops called epicycles to each planet's circular orbit around Earth.

CHECK YOUR UNDERSTANDING 3.1A

How did the ancients know the planets were different from the stars?

Copernicus Proposes a Heliocentric Model

Nicolaus Copernicus (1473–1543—**Figure 3.4**) is famous for placing the Sun, rather than Earth, at the center of the Solar System. He was not the first person to suggest that Earth orbited the Sun; a few ancient Greek and medieval Arab astronomers had briefly considered the idea when they saw problems with the Earth-centered model. However, Copernicus was the first to develop a comprehensive mathematical model with the Sun at the center that later astronomers could test. That model was the beginning of the Copernican Revolution. Through the work of 16th- and 17th-century scientists such as Tycho Brahe, Galileo Galilei, Johannes Kepler, and Isaac Newton, the **heliocentric** (Sun-centered) **model** of the Solar System became one of the best-corroborated theories in all of science.

Copernicus was multilingual and highly educated: he studied philosophy, canon (Catholic) law, medicine, economics, mathematics, and astronomy in his native Poland and in Italy. He conducted astronomical observations from a small tower, and sometime around 1514 he started writing about heliocentricity. Eighteen years later, he completed his manuscript. He did not publish the book because he knew his ideas would be controversial: philosophical and religious views of the time held that humanity, and thus Earth, must be the center of the universe. Late in his life, Copernicus was finally persuaded to publish his ideas, and his great work *De revolutionibus orbium coelestium* ("On the Revolutions of the Heavenly Spheres") appeared in 1543, the year he died.

Figure 3.5 shows Copernicus's model with the planets orbiting around the Sun. That model explained the observed motions of Earth, the Moon, and the planets, including retrograde motion, much more simply than the geocentric model did. It's like when you are in a car or train and you pass a slower-moving

Figure 3.4 Nicolaus Copernicus rejected the ancient Greek model of an Earth-centered universe and replaced it with a model that centered on the Sun.

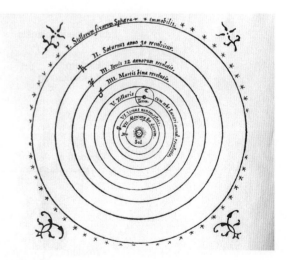

Figure 3.5 The Copernican heliocentric view of the Solar System (II–VII) and the fixed stars (I). The Sun is at the center, orbited by Mercury, Venus, Earth, Mars, Jupiter, and Saturn. The Moon orbits Earth.

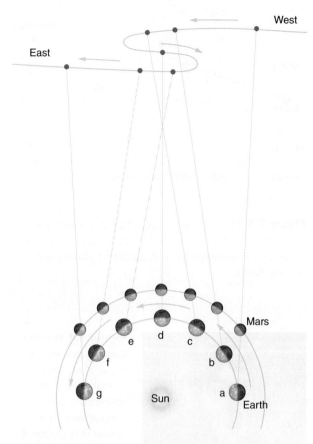

Figure 3.6 The Copernican model explains the apparent retrograde motion of Mars (see Figure 3.2) as seen in Earth's sky when Earth passes Mars in its orbit. (Not drawn to scale.)

car or train: the other vehicle seems as though it is moving backward. Without an external frame of reference, telling which vehicle is moving and in which direction can be hard. Through his model, Copernicus provided that frame of reference for the Sun and its planets.

In the Copernican model, planets farther from the Sun undergo apparent retrograde motion when Earth overtakes them in their orbits. **Figure 3.6** illustrates that effect for Mars. Conversely, planets closer to the Sun than Earth is—Mercury and Venus—move in apparent retrograde motion when overtaking Earth. Except for the Sun, all Solar System objects exhibit apparent retrograde motion. The effect diminishes with increasing distance from Earth. Retrograde motion is an illusion caused by the relative motion between Earth and the other planets.

Copernicus still conceived of the planets as moving in perfectly circular orbits at constant speeds, and so he needed to use some epicycles to match the observations. His model made testable predictions of the location of each planet on a given night. Those predictions were at least as accurate as those of the geocentric model. But the heliocentric model was, overall, simpler than the geocentric model and became the basis for further refinements in understanding how Earth moved. As copies of *De Revolutionibus* and Copernicus's ideas slowly spread across Europe, other scientists began to consider and then accept the heliocentric model, and a scientific revolution began.

Scaling the Solar System

From his observations, Copernicus deduced the correct order of the planets and concluded that planets closer to the Sun traveled faster than planets farther out. He also realized he needed to consider two categories of planets: **inferior** planets, which are closer to the Sun than Earth is, and **superior** planets, which are farther from the Sun than Earth is.

In Copernicus's model, Earth, another planet, and the Sun periodically align in space to form either a line or a right triangle. **Figure 3.7a** shows a superior planet in line with the Sun and Earth but on the other side of the Sun from Earth—a configuration called a *conjunction*. A superior planet in conjunction rises

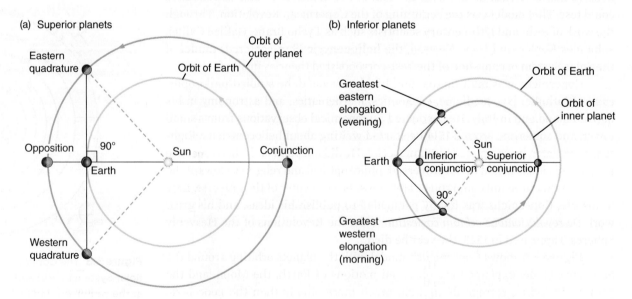

Figure 3.7 Planetary configurations for (a) superior (outer) planets and (b) inferior (inner) planets.

and sets in the sky with the Sun. When a superior planet is in conjunction, it is at its farthest from Earth, so it is at its faintest, too. You won't see the planet at all exactly at conjunction because it's on the far side of the Sun.

In contrast, when a superior planet is in line with the Sun and Earth on the same side of the Sun as Earth, the configuration is an *opposition* (Figure 3.7a). At opposition, the superior planet is "opposite" the Sun in the sky: like a full Moon, it rises when the Sun sets and sets when the Sun rises. When the superior planet is in opposition, it is at its closest to Earth and thus at its brightest; therefore, opposition is the best time to observe the planet in the sky. Opposition is also the time when the planet exhibits retrograde motion because that is exactly when Earth is overtaking the planet in its orbit. *Quadrature* is said to occur when Earth, the Sun, and a superior planet form a right triangle in space (Figure 3.7a).

For an inferior planet, the configurations are slightly different (**Figure 3.7b**). If the inferior planet is between Earth and the Sun, we call the configuration an *inferior conjunction*—when the planet is closest to Earth. If the inferior planet is on the other side of the Sun from Earth, we call the configuration a *superior conjunction*—when the planet is farthest from Earth. When the inferior planet forms a right triangle with Earth and the Sun, and thus is the farthest it gets from the Sun in the sky, we call the configuration the *greatest elongation*. When you are standing on Earth and looking toward the inner planets, you always see them close to the Sun in the sky, so you see Mercury and Venus only within a few hours of sunrise or sunset. The best time to observe those planets is at greatest elongation because they will have the greatest separation from the Sun in the sky.

Copernicus used the geometry of those alignments along with his observations of the positions of the planets in the sky, including their altitudes and the times they rose and set, to estimate the planet–Sun distances in multiples of the Earth–Sun distance. He realized that two types of orbital periods existed. Similar to the terms used for lunar orbits, a planet's *sidereal period* is how long the planet takes to make one orbit around the Sun with respect to the stars and return to the same point in space. A planet's *synodic period* is how long the planet takes to return to the same configuration with the Sun and Earth, such as inferior conjunction to inferior conjunction or opposition to opposition. The synodic period is what can be observed from Earth.

In **Figure 3.8a**, Earth and the superior planet are in opposition at point A. Superior planets move around the Sun more slowly than Earth does, so Earth orbits the Sun once and then catches up to the superior planet to form the next opposition at point B. In **Figure 3.8b**, Earth and the inferior planet are in inferior conjunction at point A. An inferior planet moves around the Sun faster than Earth does, so it completes one sidereal period and then must continue in its orbit to catch up to Earth for the next inferior conjunction at point B. **Table 3.1** shows that those relative distances Copernicus calculated are remarkably close to the distances modern methods yield. **Working It Out 3.1** shows the numerical details. Copernicus's model not only predicted planetary positions in the sky but also could be used to compare the distances between the planets and the Sun accurately and thus to set the scale of the Solar System.

CHECK YOUR UNDERSTANDING 3.1B

The planet Uranus will be observed in retrograde motion when: (a) Uranus is closest to the Sun; (b) Uranus is farthest from the Sun; (c) Earth overtakes Uranus in its orbit; (d) Uranus overtakes Earth in its orbit.

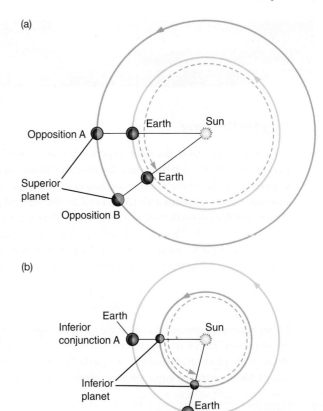

Figure 3.8 The synodic periods of planets indicate how long they take to return to the same configuration with Earth and the Sun. (a) Earth completes one orbit around the Sun first and then catches up to the superior planet. (b) Inferior planets complete a full orbit around the Sun first and then catch up to Earth.

TABLE 3.1 Copernicus's Scale of the Solar System

Planet	Copernicus's Value (AU)	Modern Value (AU)
Mercury	0.38	0.39
Venus	0.72	0.72
Earth	1.00	1.00
Mars	1.52	1.52
Jupiter	5.22	5.20
Saturn	9.17	9.58

AU = astronomical unit.

3.1 Working It Out: How Copernicus Computed Orbital Periods and Scaled the Solar System

Orbital Periods

Copernicus calculated the sidereal period by observing the synodic period. He didn't know the actual Earth–Sun distance in miles or kilometers, so he let the Earth–Sun distance equal 1. We call that distance 1 **astronomical unit** (**AU**). Let P be the sidereal period of a planet and S, its synodic period. E is Earth's sidereal period, which equals 1 year, or 365.25 days. By thinking about the distance that Earth and the planet move in one synodic period, and noting that an inferior planet orbits the Sun faster than Earth does, we can show that

$$\frac{1}{P} = \frac{1}{E} + \frac{1}{S}$$

for an inferior planet, with P, E, and S all in the same units of days or years. Similarly, Earth orbits the Sun faster than a superior planet does, so the planet has traveled only part of its orbit around the Sun after 1 Earth year. The equation for a superior planet is

$$\frac{1}{P} = \frac{1}{E} - \frac{1}{S}$$

For Saturn, the time that passes between oppositions—the date of maximum brightness—shows that Saturn's synodic period (S) is 378 days, or 378 ÷ 365.25 = 1.035 years. Then, to compute Saturn's sidereal period (P) in years, we use $S = 1.035$ yr and $E = 1$ yr in the equation for a superior planet:

$$\frac{1}{P} = \frac{1}{1\,\text{yr}} - \frac{1}{1.035\,\text{yr}} = 1 - 0.966\,\text{yr}^{-1} = 0.034\,\text{yr}^{-1}$$

Thus,

$$P = \frac{1}{0.034\,\text{yr}^{-1}} = 29.4\,\text{yr}$$

Saturn's sidereal period is 29.4 years, which means that Saturn takes 29.4 years to travel around the Sun and return to where it started in space.

Scaling the Solar System

Copernicus used the configurations of the planets shown in Figure 3.7 along with their sidereal periods to compute the relative distances of the planets. For the superior planets, he measured the fraction of the circular orbit that the planet completed in the time between opposition and quadrature, and then he used trigonometry to solve for the planet–Sun distance in astronomical units (see Figure 3.7a). For the inferior planets, he had a right triangle at the point of greatest elongation, and then he used right-triangle trigonometry to solve for the planet–Sun distance in astronomical units (see Figure 3.7b). Copernicus's values are impressively similar to modern values (see Table 3.1). Copernicus still did not know the actual value of the astronomical unit in miles, but he was the first to accurately compute the relative distances of the planets from the Sun.

3.2 Kepler's Laws Describe Planetary Motion

Copernicus did not understand *why* the planets move about the Sun, but he realized that his heliocentric picture offered a way to compute the planets' relative distances. His theory is an example of **empirical science**, which seeks to describe patterns in nature as accurately as possible. Copernicus's work was revolutionary because he challenged the accepted geocentric model and proposed that Earth is one planet among many. His conclusions paved the way for other great empiricists, including Tycho Brahe and Johannes Kepler.

Tycho Brahe's Observations

Tycho Brahe (1546–1601—**Figure 3.9**) was a Danish astronomer of noble birth who entered university at age 13 to study philosophy and law. After seeing a partial solar eclipse in 1560, Tycho (conventionally referred to by his first name) became interested in astronomy. A few years later, he observed Jupiter and Saturn near each other in the sky, though not exactly where one would expect

Figure 3.9 Tycho Brahe, known commonly as Tycho, was one of the greatest astronomical observers before the invention of the telescope.

from the astronomical tables based on Ptolemy's model. Tycho gave up studying law and devoted himself to making better tables of the planets' positions in the sky.

The king of Denmark granted Tycho the island of Hven, between Sweden and Denmark, to build an observatory. Tycho designed and built new instruments, operated a printing press, and taught students and others how to conduct observations. With the help of his sister Sophie, Tycho carefully measured the precise positions of planets in the sky over several decades, developing the most comprehensive set of planetary data then available. He created his own geo-heliocentric model, shown in **Figure 3.10**. In Tycho's model, the planets orbit the Sun, and the Sun and planets orbit Earth. His model gained some acceptance among people who preferred to keep Earth at the center for philosophical or religious reasons. Tycho lost his financial support when the king died, and in 1600 he relocated to Prague.

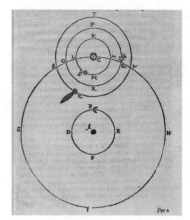

Figure 3.10 Tycho's geo-heliocentric model, showing the Moon and Sun orbiting Earth, with the other planets orbiting the Sun.

Kepler's Laws

In 1600, Tycho hired a more mathematically inclined astronomer, Johannes Kepler (1571–1630—**Figure 3.11**), as his assistant. Kepler, who had studied Copernicus's ideas, was responsible for the next major step toward understanding the planets' motions. Kepler inherited the records of Tycho's observations when he died. Working first with Tycho's observations of Mars, Kepler deduced three rules, now generally referred to as **Kepler's laws**, which accurately describe how the planets move. Kepler's laws are empirical: they use existing data to make predictions about future behavior but do not include an underlying theory of why the objects behave as they do.

KEPLER'S FIRST LAW Comparing Tycho's extensive planetary observations with predictions from Copernicus's heliocentric model, Kepler expected the data to confirm circular orbits for planets orbiting the Sun. Instead, he found disagreements between his predictions and Tycho's observations. He was not the first to notice such discrepancies. Rather than discard Copernicus's model, Kepler revised it.

By replacing Copernicus's circular orbits with *elliptical* orbits, Kepler could predict the positions of planets for any day, and his predictions fit Tycho's observations almost perfectly. Looking like an elongated circle, an **ellipse** is symmetric from right to left and from top to bottom. As shown in **Figure 3.12a**, you can draw an ellipse by attaching the two ends of a piece of string to a piece of paper, stretching the string tight with the tip of a pencil, and then drawing around those two points while keeping the string tight. Each point at which the string is attached is a **focus** (plural: **foci**) of the ellipse. Notice in **Figure 3.12b** that the ellipse becomes more circular as the two foci move closer together. As the two foci move farther apart, the ellipse becomes more and more elongated. The **eccentricity** (*e*) of an ellipse measures that elongation; it is determined by the separation between the

Figure 3.11 Johannes Kepler explained the motions of the planets with three empirically determined laws.

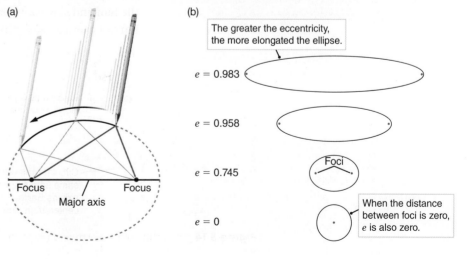

Figure 3.12 (a) We can draw an ellipse by attaching a length of string to a piece of paper at two points (called foci) and then pulling the string around as shown. The long axis is called the major axis. (b) Ellipses range from circles (*e* = 0) to elongated eccentric shapes. *e* = eccentricity.

Figure 3.13 According to Kepler's first law, planets move in elliptical orbits with the Sun at one focus. (Nothing is at the other focus.) The orbit's eccentricity is the center-to-focus distance divided by the semimajor axis.

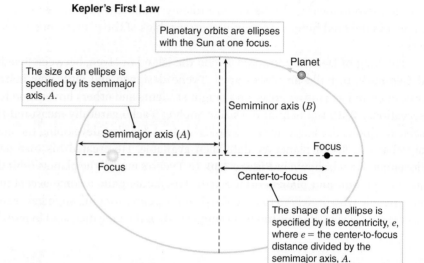

AstroTour: Kepler's Laws

Simulation: Planetary Orbits

two foci divided by the length of the long axis, called the **major axis**. A circle has an eccentricity of 0 because the two foci coincide at the center. The more elongated the ellipse becomes, the closer its eccentricity gets to 1.

Kepler's first law of planetary motion states that the orbit of each planet is an ellipse with the Sun located at one focus (see the **Process of Science Figure**). Nothing but empty space is at the other focus. **Figure 3.13** illustrates Kepler's first law and shows the features of an ellipse that can be measured. The dashed lines in Figure 3.13 represent the major (long) and minor (short) axes of the ellipse. Half the length of the major axis of the ellipse is called the **semimajor axis**, A. The average distance between the Sun and a planet equals the semimajor axis of the planet's orbit.

The eccentricities of planetary orbits vary widely, but most planetary objects in our Solar System have nearly circular orbits. As **Figure 3.14a** shows, Earth's orbit, with an eccentricity of 0.017, is very nearly a circle centered on the Sun; therefore the distance variation is small. In contrast, dwarf planet Pluto's orbit, depicted in **Figure 3.14b**, has an eccentricity of 0.249. The orbit is noticeably elongated, with the Sun offset from center.

KEPLER'S SECOND LAW By analyzing Tycho's observational data of changes in the planets' positions, Kepler found that a planet moves fastest when closest to the Sun and slowest when farthest from the Sun. For example, we now measure Earth's average speed in its orbit around the Sun at 29.8 kilometers per second

Figure 3.14 The orbits of (a) Earth and (b) Pluto in comparison with circles around the Sun. e = eccentricity.

PROCESS OF SCIENCE

Theories Are Falsifiable

Early astronomers studied the motions of the planets, but did not understand why the planets behaved as they do.

Copernicus
1473–1543

The Hypothesis
Copernicus proposed that planets moved in circular orbits (with epicycles) around the Sun.

Tycho
1546–1601

The Observation
Tycho, born three years after Copernicus died, observed and collected enormous amounts of data about planet positions.

Kepler
1571–1630

The Prediction
Kepler, 25 years younger than Tycho, used Copernicus's model and Tycho's data to predict where the planets should be.

The Test
When Kepler compared the predictions with actual data, they disagreed. Copernicus's idea was falsified!

The New Hypothesis
Planet orbits are not circular. They are elliptical.

Newton
1643–1726

This new hypothesis gained acceptance a century later when Newton showed that planetary orbits are an inevitable consequence of gravitational attraction.

In order for a theory to be scientific, it must be falsifiable, even if the test can't be carried out until decades or centuries later. Disproving an old theory is part of the self-correcting nature of science: It always leads to deeper understanding.

Kepler's Second Law

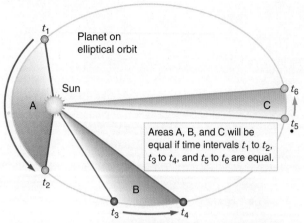

Figure 3.15 An imaginary line between a planet and the Sun sweeps out an area as the planet orbits. According to Kepler's second law, if the three time intervals shown are equal, the three areas A, B, and C will be the same.

(km/s). When closest to the Sun, Earth travels at 30.3 km/s. At its farthest from the Sun, Earth travels at 29.3 km/s.

Kepler found a way to elegantly describe the changing speed of a planet in its orbit around the Sun. **Figure 3.15** shows a planet at six points along the ellipse of its orbit (t_1 to t_6). Imagine a straight line connecting the Sun with that planet. We can think of the line as "sweeping out" an area as it moves with the planet from one point to another. Area A (in orange) is swept out between times t_1 and t_2, area B (in blue) is swept out between times t_3 and t_4, and area C (in green) is swept out between times t_5 and t_6. When the planet is closest to the Sun (area A), it is moving the fastest, but the distance between the planet and the Sun is small. Kepler realized that changes in the distance from the Sun to a planet and changes in a planet's speed work together so that the area a planet sweeps out in the same amount of time is always the same, regardless of where the planet is in its orbit. Therefore, if the three time intervals in the figure are equal (that is, $t_1 \rightarrow t_2 = t_3 \rightarrow t_4 = t_5 \rightarrow t_6$), the three areas A, B, and C must be equal as well.

Kepler's second law, also called Kepler's **law of equal areas**, states that the imaginary line connecting a planet to the Sun sweeps out equal areas in equal times, regardless of where the planet is along the ellipse of its orbit. That law applies to only one planet at a time. The area swept out by Earth in a given time interval is always the same. Likewise, the area swept out by Mars in a given time is always the same. But the area swept out by Earth and the area swept out by Mars in a given time are *not* the same. That law can be used to find the speed of a planet anywhere in its orbit.

KEPLER'S THIRD LAW Kepler looked for patterns in the planets' orbital periods. Compared with planets closer to the Sun, he found that planets farther from the Sun have longer orbits and orbit the Sun more slowly. Kepler discovered a mathematical relationship between a planet's sidereal period—how many years the planet takes to go around the Sun and return to the same position in the Solar System—and its average distance from the Sun in astronomical units. **Kepler's third law** states that in those units, the square of the sidereal

---

3.2 Working It Out Kepler's Third Law

Kepler's third law states that the square of the period of a planet's orbit measured in years, P_{years}, is equal to the cube of the semimajor axis of the planet's orbit measured in astronomical units, A_{AU}. As an equation, the law says

$$(P_{\text{years}})^2 = (A_{\text{AU}})^3$$

For mathematical convenience, Kepler used units based on Earth's orbit—astronomical units and years. If other units were used, such as kilometers and days, P^2 would still be proportional to A^3, but the constant of proportionality would not be 1.

To calculate the average size of Neptune's orbit in astronomical units, you first need to find out how long Neptune's period is in Earth years, which you can determine by observing the synodic period and from that computing its sidereal period (using Working It Out 3.1). Neptune's sidereal period is 165 years. Plugging that number into Kepler's third law gives the following result:

$$(P_{\text{years}})^2 = (165)^2 = 27{,}225 = (A_{\text{AU}})^3$$

To solve that equation, you must first square 165 to get 27,225 and then take its cube root (see Appendix 1 for calculator hints). Then

$$A_{\text{AU}} = \sqrt[3]{27{,}225} = 30.1$$

The semimajor axis of Neptune's orbit—that is, the average distance between Neptune and the Sun—is 30.1 AU.

TABLE 3.2 Kepler's Third Law: $P^2 = A^3$

Planet	Period P (years)	Semimajor Axis A (AU)	$\dfrac{P^2}{A^3}$
Mercury	0.241	0.387	$\dfrac{0.241^2}{0.387^3} = 1.00$
Venus	0.615	0.723	$\dfrac{0.615^2}{0.723^3} = 1.00$
Earth	1.000	1.000	$\dfrac{1.000^2}{1.000^3} = 1.00$
Mars	1.881	1.524	$\dfrac{1.881^2}{1.524^3} = 1.00$
Ceres	4.599	2.765	$\dfrac{4.599^2}{2.765^3} = 1.00$
Jupiter	11.86	5.204	$\dfrac{11.86^2}{5.204^3} = 1.00$
Saturn	29.46	9.582	$\dfrac{29.46^2}{9.582^3} = 0.99^*$
Uranus	84.01	19.201	$\dfrac{84.01^2}{19.201^3} = 1.00$
Neptune	164.79	30.047	$\dfrac{164.79^2}{30.047^3} = 1.00$
Pluto	247.92	39.482	$\dfrac{247.92^2}{39.482^3} = 1.00$
Eris	557.00	67.696	$\dfrac{557.00^2}{67.696^3} = 1.00$

*Ratios are not exactly 1.00 because of slight perturbations from the gravity of other planets.

period (P) of a planet's orbit is equal to the cube of the semimajor axis (A) of the planet's orbit.

Table 3.2 lists the periods and semimajor axes of the orbits of the eight classical planets and three of the dwarf planets, along with the values of the ratio P^2 divided by A^3. Those data are also plotted in **Figure 3.16**. Kepler referred to that relationship as his **harmonic law** or, more poetically, as the "Harmony of the Worlds." Kepler's third law is explored further in **Working It Out 3.2**. Kepler's laws enhanced the heliocentric mathematical model of Copernicus and led to its greater acceptance.

CHECK YOUR UNDERSTANDING 3.2

Order the following from largest to smallest semimajor axis: (a) a planet with a period of 84 Earth days; (b) a planet with a period of 1 Earth year; (c) a planet with a period of 2 Earth years; (d) a planet with a period of 0.5 Earth years.

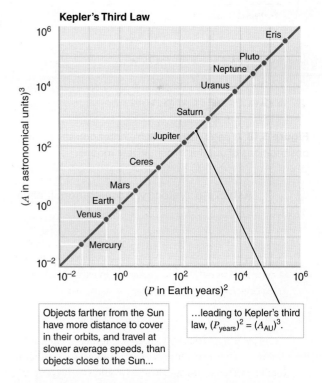

Figure 3.16 A plot of A^3 versus P^2 for objects in our Solar System shows that they obey Kepler's third law. (By plotting powers of 10 on each axis, we can fit both large and small values on the same plot. We will use that approach often.)

Figure 3.17 Galileo Galilei laid the physical framework for Newton's laws.

3.3 Galileo's Observations Supported the Heliocentric Model

Galileo Galilei (1564–1642—**Figure 3.17**) was the first to use a telescope to make and report significant discoveries about astronomical objects. Galileo's telescopes were relatively small, yet sufficient for him to observe spots on the Sun, the uneven surface and craters of the Moon, and the many stars in the band of light in the sky called the Milky Way.

Galileo's Observations

Galileo provided the first observational evidence that some objects in the sky do not orbit Earth. When Galileo turned his telescope on Jupiter, he observed four "stars" in a line near the planet. Over time he observed that the objects changed position from night to night (**Figure 3.18**). Galileo hypothesized that those objects were moons orbiting Jupiter. Those four moons are the largest of Jupiter's many moons and are still called the Galilean moons. Galileo also estimated the relative distance of each moon from Jupiter and the periods of their orbits, and he showed that they followed Kepler's third law.

Galileo also observed that Venus went through phases like the Moon. He noticed that the phases of Venus were correlated with how large Venus appeared in his telescope. In a geocentric model in which Venus orbits Earth like the Moon does, the apparent size of Venus would change only slightly, when it looped an epicycle. In the heliocentric model, however, the Earth–Venus distance varies by a lot, and Venus's size changes accordingly. When Venus is in its gibbous to full phases, it is farther away, on the other side of the Sun from Earth, and smaller in the sky. When Venus is in its crescent to new phases, it is closer, on the same side of the Sun as Earth, and larger in the sky (**Figure 3.19**). His observations of Jupiter's moons and the phases of Venus in particular convinced Galileo that Copernicus was correct to place the Sun at the center of the Solar System.

In addition to his astronomical observations, Galileo did important work on the motion of objects. Unlike natural philosophers, who thought about objects in motion but did not actually experiment with them, Galileo conducted experiments with falling and rolling objects. As with his telescopes, Galileo improved or developed new technology to enable him to conduct those experiments. For example, by carefully rolling balls down an inclined plane and by dropping various objects from a height, he found that a falling object travels a distance proportional to the square of the time it has been falling. If he simultaneously dropped two objects of different masses, they reached the ground at the same time, showing that all objects falling to Earth accelerate at the same rate, independent of their mass.

Galileo's observations and experiments with many types of moving objects, such as carts and balls, led him to disagree with the Greek philosophers about when and why objects continue to move or come to rest. Before Galileo, it was thought that an object's natural state was to be at rest. But he found that an object naturally does what it was doing until a force acts on it. That is, an object in motion continues moving along a straight line with a constant speed until a force acts on it to change that state of motion. That idea of *inertia*, which Newton later adopted as his first law of motion, has implications for not only the motion of carts and balls but also the orbits of planets.

Figure 3.18 A page from Galileo's notebook shows his observations of Jupiter's four largest moons.

AstroTour: Velocity, Acceleration, Inertia

Dialogue Concerning the Two Chief World Systems

Galileo faced considerable danger because of his work. His later life was consumed by conflict with the Catholic Church over his support of the Copernican system. In 1632, Galileo published his best-selling book, *Dialogo sopra i due massimi sistemi del mondo* ("Dialogue Concerning the Two Chief World Systems"). The *Dialogo* presents a brilliant philosopher named Salviati as the champion of the Copernican heliocentric view of the universe. The defender of an Earth-centered universe, Simplicio—who uses arguments made by the classical Greek philosophers and the pope—sounds silly and ignorant.

Galileo, a religious man with two daughters in a convent, thought he had the Catholic Church's tacit approval for his book. But when he placed several of the pope's geocentric arguments in the unflattering mouth of Simplicio, the perceived attack on the pope got the church's attention. Galileo was put on trial for heresy, sentenced to prison, and eventually placed under house arrest. To escape a harsher sentence, Galileo was forced to publicly recant his belief in the Copernican theory that Earth moves around the Sun. According to one story, as he left the courtroom, Galileo stamped his foot on the ground and muttered, "And yet it moves!"

The *Dialogo* was placed on the pope's Index of Prohibited Books, along with Copernicus's *De Revolutionibus*. Yet Galileo's work traveled across Europe, was translated into other languages, and was read by other scientists. Galileo spent his final years compiling his research on inertia and other ideas into the book *Discourses and Mathematical Demonstrations Relating to Two New Sciences*, which was published in 1638 in Holland, outside the Catholic Church's jurisdiction. (In 1992, Pope John Paul II apologized for the "Galileo Case.")

Figure 3.19 Modern photographs of the phases of Venus show that when we see Venus more illuminated, it also appears smaller, implying that Venus is farther away then.

CHECK YOUR UNDERSTANDING 3.3

Which of Galileo's astronomical observations did Copernicus's model explain better than Ptolemy's? (a) sunspots; (b) craters on the Moon; (c) moons of Jupiter; (d) apparent size and phases of Venus

3.4 Newton's Three Laws Help Explain How Celestial Bodies Move

Empirical laws, such as Kepler's laws, describe what happens, but they do not explain why. Kepler described the orbits of planets as ellipses, but he did not explain why they should be so. To take that next step in the scientific process, scientists use basic physical principles and the tools of mathematics to derive the empirically determined laws. Alternatively, a scientist might start with physical laws and predict relationships, which are then verified or falsified through experiment and observation. If those predictions are verified, the scientist may have determined something fundamental about how the universe works.

Sir Isaac Newton (1642–1727—**Figure 3.20**) took that next step in explaining the nature of motion. Newton was a student of mathematics at Cambridge University when it closed because of the Great Plague and students were sent home to the safer countryside. Over the next 2 years, he studied on his own, and at the age of 23 he invented calculus, which would become crucial to his development of the physics of motion. (The German mathematician Gottfried Leibniz independently developed calculus around the same time.)

Figure 3.20 Sir Isaac Newton formulated three laws of motion.

Figure 3.21 (a) An object moving in a straight line at a constant speed is at rest in its own inertial frame of reference. (b) Any change in an object's velocity is an acceleration. When you are driving, for example, any time your speed changes or you follow a curve in the road, you are accelerating. (Throughout the text, velocity arrows are red and acceleration arrows are green.)

Building on the work of Kepler, Galileo, and others, Newton proposed three physical laws that govern the motions of all objects in the sky and on Earth. To understand how the planets and all other celestial bodies move, you must understand these three laws.

Newton's First Law: Objects at Rest Stay at Rest; Objects in Motion Stay in Motion

A **force** (*F*) is a push or a pull on an object. Two or more forces can oppose one another such that they are perfectly balanced and cancel out. For example, gravity pulls down on you as you sit in your chair, but the chair pushes up on you with an exactly equal and opposite force. As a result, you stay motionless. Forces that cancel out do not affect an object's motion. When forces combine to produce an effect, we often use the term *net force*, or sometimes just *force*.

Imagine that you are driving a car, and your book is on the seat next to you. A rabbit runs across the road in front of you, and you hit the brakes hard. You feel the seat belt tighten to restrain you. At the same time, your book flies off the seat and hits the dashboard. You have just experienced what Newton describes in his first law of motion. **Inertia** is an object's tendency to maintain its state—either of uniform motion or of rest—until a net force pushes or pulls it. In the stopping car, you did not hit the dashboard because the force of the seat belt slowed you down. The book, however, hit the dashboard because no such force acted upon it.

Newton's first law of motion describes inertia and states that an object in motion tends to stay in motion, in the same direction, until a net force acts upon it; and an object at rest tends to stay at rest until a net force acts upon it. Galileo's law of inertia became the cornerstone of physics as Newton's first law.

Recall from Section 2.1 the concept of a frame of reference. Within a frame of reference, only the relative motions between objects have any meaning. Without external clues, you cannot tell the difference between sitting still and traveling at constant speed in a straight line. For example, if you close your eyes while riding in the passenger seat of a quiet car on a smooth road, you would feel as though you were sitting still. In the earlier example in this chapter, your book was "at rest" beside you on the front seat of your car, but a person standing by the side of the road would see the book moving past at the same speed as the car. People in a car approaching you would see the book moving quite fast—at the speed they are traveling plus the speed you are traveling! All those perspectives are equally valid, and all those speeds of the book are correct when measured in the appropriate reference frame.

A reference frame moving in a straight line at a constant speed is called an **inertial frame of reference**. Any inertial frame of reference is as good as another. In the inertial frame of reference of a cup of coffee, for example, **Figure 3.21a** shows that the cup is at rest in its own frame even if the car is moving quickly down the road.

Newton's Second Law: Motion Is Changed by Forces

What if a net force does act? In the earlier example, you were traveling in the car, and your motion slowed when the force of the seat belt acted upon you. Forces change an object's motion—by changing either the speed or the direction. That effect reflects **Newton's second law of motion**: if a net force acts on an object, the object's motion changes.

In the driver's seat of a car, you have several controls, including an accelerator and a brake pedal, which you use to make the car speed up or slow down. A *change in speed* is one way the car's motion can change. But you also have the steering wheel in your hands. When you are moving down the road and you turn the wheel, your speed does not necessarily change, but the direction of your motion does. A *change in direction* also is a kind of change in motion.

Together, an object's speed and direction are called **velocity** (*v*). "Traveling at 50 kilometers per hour (km/h)" indicates speed; "traveling north at 50 km/h" indicates velocity. **Acceleration** (*a*) describes the changes in an object's velocity. For example, if you go from 0 to 100 km/h in 4 seconds, you feel a strong push from the seat back as it shoves your body forward, causing you to accelerate along with the car. However, if you take 2 minutes to go from 0 to 100 km/h, the acceleration is so slight that you hardly notice it.

Partly because a car's gas pedal is often called the accelerator, some people think *acceleration* always means that an object is speeding up. But in physics, *any change in speed or direction is an acceleration*. **Figure 3.21b** illustrates that point by showing what happens to the coffee in a cup as the car speeds up, slows down, or turns. Slamming on your brakes and going from 100 to 0 km/h in 4 seconds is just as much an acceleration as going from 0 to 100 km/h in 4 seconds. Similarly, the acceleration you experience as you go through a fast, tight turn at a constant speed is every bit as real as the acceleration you feel when you slam your foot on the accelerator or the brake. Whether speeding up, slowing down, turning left, or turning right—if you are not moving in a straight line at a constant speed, you are accelerating.

Newton's second law of motion says that a net force causes acceleration. An object's acceleration depends on two things. First, as shown in **Figure 3.22**, the acceleration depends on the strength of the net force acting on the object to change its motion. If the forces acting on the object do *not* add up to zero, a net force is present and the object accelerates (Figure 3.22a). The stronger the net force, the greater the acceleration (Figure 3.22b). Push on something twice as hard and it experiences twice as much acceleration. Push on something 3 times as hard and its acceleration is 3 times as great. The resulting change in motion occurs in the direction the net force points. Push an object away from you, and it will accelerate away from you.

An object's acceleration also depends on its inertia. You can push some objects easily, such as the empty box from a new refrigerator. But you can't easily shove the actual refrigerator around, however, even though it is about the same size as the box. Figure 3.22c shows that the greater the mass, the greater the inertia, and the *less* acceleration that will occur in response to the same net force. That relationship among acceleration, force, and mass is expressed mathematically in **Working It Out 3.3**.

Newton's Third Law: Whatever Gets Pushed Pushes Back

Imagine that you are standing on a skateboard and pushing yourself along with your foot. Each shove of your foot against the ground sends you faster along your way. But why? You accelerate because as you push on the ground, the ground pushes back on you.

Part of Newton's genius was his ability to see patterns in such everyday events. Newton realized that *every* time one object exerts a force on another, the second object exerts a matching force on the first. That second force is as strong as the first force but is in the opposite direction. When you accelerate yourself on the skateboard, you push backward on Earth, and Earth pushes you forward. As

Figure 3.22 According to Newton's second law of motion, an object's acceleration is the force acting on the object divided by the object's mass. (Throughout the text, force arrows are blue.)

3.3 Working It Out Using Newton's Laws

Your acceleration is determined by how much your velocity changes, divided by how long that change takes to happen:

$$\text{Acceleration} = \frac{\text{How much velocity changes}}{\text{How long the change takes to happen}}$$

For example, if an object's speed goes from 5 to 15 meters per second (m/s), the change in velocity is 10 m/s. If that change happens in 2 seconds, the acceleration is given by

$$a = \frac{15\,\text{m/s} - 5\,\text{m/s}}{2\,\text{s}} = 5\,\text{m/s}^2$$

To determine how an object's motion is changing, we need to know two things: what net force is acting on the object and the object's resistance to that force. We can put that idea into equation form as follows:

$$\begin{pmatrix}\text{An object's}\\ \text{Acceleration}\end{pmatrix} = \frac{\text{The force acting to change the object's motion}}{\text{The object's resistance to that change}} = \frac{\text{Force}}{\text{Mass}}$$

Newton's second law above is often written as Force = mass × acceleration, or $F = ma$. The units of force are called **newtons (N)**, so $1\,\text{N} = 1\,\text{kg m/s}^2$.

Suppose you are holding two blocks of the same size, the block in your right hand has twice the mass of the block in your left hand. When you drop the blocks, they both fall with the same acceleration, as Galileo showed, and they hit your two feet at the same time. Which will hit with more force: the block falling onto your right foot or the one falling onto your left foot? The block in your right hand, with twice the mass, will hit your right foot with twice the force that the other block hits your left foot.

To see how Newton's three laws of motion work together, study **Figure 3.23**. An astronaut is adrift in space, motionless with respect to the nearby space shuttle. With no tether to pull on, how can the astronaut get back to the ship? Suppose the 100-kg astronaut throws a 1-kg wrench directly away from the shuttle at a speed of 10 m/s. Newton's second law says that to change the wrench's motion, the astronaut must apply a force to it in the direction away from the shuttle. Newton's third law says that the wrench must therefore push back on the astronaut with as much force but in the opposite direction. The force of the wrench on the astronaut causes the astronaut to begin drifting toward the shuttle. How fast will the astronaut move? Turn to Newton's second law again. Because the astronaut has more mass, he will accelerate less than the wrench will. A force that causes the 1-kg wrench to accelerate to 10 m/s will have much less effect

Figure 3.23 According to Newton's laws, if an astronaut adrift in space throws a wrench, the two will move in opposite directions. Their speeds will depend on their masses: the same force will produce a smaller acceleration of a more massive object than of a less massive object. (Acceleration and velocity arrows not drawn to scale.)

on the 100-kg astronaut. Because acceleration equals force divided by mass, the 100-kg astronaut will experience only 1/100 as much acceleration as the 1-kg wrench. The astronaut will drift toward the shuttle, but only at the leisurely rate of 1/100 × 10 m/s, or 0.1 m/s.

shown in **Figure 3.24**, when a woman moves a load on a cart by pulling a rope, the rope pulls back, and when a car tire pushes back on the road, the road pushes forward on the tire. Similarly, when Earth pulls on the Moon, the Moon pulls on Earth, and when a rocket engine pushes hot gases out of its nozzle, those hot gases push back on the rocket, propelling it into space.

All those force pairs are examples of **Newton's third law of motion**, which says that forces always come in pairs, with those two forces always equal in strength but opposite in direction. The forces in those pairs always act on two objects. Your weight pushes down on the floor, and the floor pushes back up on your feet with the same amount of force. For every force, an equal force in the opposite direction is *always* there.

CHECK YOUR UNDERSTANDING 3.4

Imagine a planet moving in a perfectly circular orbit around the Sun. Is that planet accelerating? (a) Yes, because it is constantly changing its speed. (b) Yes, because it is constantly changing its direction. (c) No, because its speed is not constantly changing. (d) No, because planets do not accelerate.

Origins Planets and Orbits

In addition to the planets in our own Solar System, researchers have detected thousands of planets orbiting stars other than our own Sun. Those planets' orbits can be calculated and understood by applying the same three Kepler's laws that we have discussed here. Astrobiologists think that a planet's orbit around its star affects the chances of life developing.

Consider the average distance of the planet from its star. A planet close to its star receives more energy than a planet farther out. If Earth were closer to the Sun, it would be hotter throughout the year—probably so hot that water would evaporate and no longer exist as a liquid. If Earth were farther from the Sun, it would be colder and all surface water would probably freeze. Liquid water was crucial for life to form on Earth, so some astronomers look for planets at a distance from their star such that liquid water can exist (that distance varies depending on the star's temperature and size).

Next consider the eccentricity of a planet's orbit. Recall from Figure 3.14a that Earth's orbit differs from a circle by less than 2 percent. Thus, Earth's distance from the Sun varies by only about 5 million km throughout the year; and as we saw in Chapter 2, seasons change on Earth because of the tilt of Earth's axis, not because of slight changes in distance from the Sun. Mars, however, has about the same axial tilt as Earth. But because the orbit of Mars is more eccentric, Mars has greater seasonal variation. The distance between Mars and the Sun varies from 1.38 AU (207 million km) to 1.67 AU (249 million km)—an eccentricity of 9 percent (or ~40 million km). As a result, the seasons on Mars are unequal. They are shorter when Mars is closer to the Sun and moving faster, and longer when Mars is farther from the Sun and moving slower. The inequality of the martian seasons affects the overall stability of the planet's temperature and climate. When we look at planets orbiting other stars, we see that many have orbital eccentricities even higher than that of Mars and therefore large variations in temperature.

Earth is at the right distance from the Sun to have temperatures that permit water to be liquid, and its orbital eccentricity is low enough that the average planetary temperature does not change much during its orbit. Those orbital characteristics have contributed to making the conditions on Earth suitable for life to develop.

Figure 3.24 Newton's third law states that for every force, an equal and opposite force is always present. Those opposing forces always act on the two objects in the same pair.

Astronomy in Action: Velocity, Force, and Acceleration

AstroTour: Velocity, Acceleration, Inertia

READING ASTRONOMY NEWS

ARTICLES | QUESTIONS

> In this article we see that planetary alignments have practical implications. A spacecraft on the other side of the Sun will lose radio contact with Earth.

NASA Spacecraft Take Spring Break at Mars

MIKE WALL, Space.com

NASA's robotic Mars explorers are taking a cosmic break for the next few weeks, thanks to an unfavorable planetary alignment of Mars, the Earth, and the Sun.

Mission controllers won't send any commands to the agency's *Opportunity* rover, *Mars Reconnaissance Orbiter* (*MRO*), or *Mars Odyssey* orbiter from today (April 9) through April 26. The blackout is even longer for NASA's car-size *Curiosity* rover, which is slated to go solo from April 4 through May 1.

The cause of the communications moratorium is a phenomenon called a Mars solar conjunction, during which the Sun comes between Earth and the Red Planet (**Figure 3.25**). Our star can disrupt and degrade interplanetary signals in this formation, so mission teams won't be taking any chances.

"Receiving a partial command could confuse the spacecraft, putting them in grave danger," NASA officials explain in a video posted last month by the agency's Jet Propulsion Laboratory (JPL) in Pasadena, California.

Opportunity and *Curiosity* will continue performing stationary science work, using commands already beamed to the rovers. *Curiosity* will focus on gathering weather data, assessing the martian radiation environment, and searching for signs of subsurface water and hydrated minerals, officials said Monday (April 8).

MRO and *Odyssey* will also keep studying the Red Planet from above, and they'll continue to serve as communications links between the rovers and Earth. The conjunction will also affect the European Space Agency's *Mars Express* orbiter, officials have said.

Odyssey will send rover data home as usual during conjunction, though the orbiter may have to relay information multiple times due to dropouts. *MRO*, on the other hand, entered record-only mode on April 4. The spacecraft will probably have about 52 gigabits of data to relay when it's ready to start transmitting again on May 1, *MRO* officials have said.

Mars solar conjunctions occur every 26 months, so NASA's Red Planet veterans have dealt with them before. This is the fifth conjunction for *Opportunity*, in fact, and the sixth for *Odyssey*, which began orbiting Mars in 2001.

But it'll be the first for *Curiosity*, which touched down on August 5, kicking off a 2-year surface mission to determine if the Red Planet has ever been capable of supporting microbial life.

"The biggest difference for this 2013 conjunction is having *Curiosity* on Mars," *Odyssey* mission manager Chris Potts, of NASA's Jet Propulsion Laboratory in Pasadena, California, said in a statement last month.

Figure 3.25 Every 26 months, Mars and Earth are on opposite sides of the Sun, preventing communication between the planets.

ARTICLES | **QUESTIONS**

1. How often do Mars solar conjunctions occur? Does that interval correspond to the sidereal period or to the synodic period of Mars?
2. Give two reasons that conjunction is a bad time to view Mars in the sky.
3. Make a sketch of the Mars solar conjunction, showing Mars, Earth, and the Sun. Using the orbital period data in Table 3.2, add the positions of Mars and Earth at the next Mars conjunction. Will it take place at the same location in space?
4. View this short video from NASA: https://www.jpl.nasa.gov/video/details.php?id=1204. Do you think it explains the concepts well to someone who is not studying astronomy?
5. If people were on a mission on Mars, losing contact would be troubling. How might NASA plan to avoid losing contact with astronauts on Mars during conjunctions?

Source: "NASA spacecraft take spring break at Mars," by Mike Wall. Space.com, April 9, 2013. Copyrighted 2013. Purch. 261743:0918AT. Permission conveyed through Wright's Media.

Summary

Early astronomers hypothesized that Earth was stationary at the center of the Solar System. Later astronomers realized that a Sun-centered Solar System was much simpler and could explain what people observed in the sky. Planets, such as Jupiter, orbit the Sun, not Earth. Kepler's laws describe the planets' elliptical orbits around the Sun, including details about how fast a planet travels at various points in its orbit. Those laws helped Newton develop his laws of motion, which govern how all objects move (not just orbiting ones). Orbital semimajor axis, eccentricity, and stability may affect a planet's suitability to foster life.

LG 1 **Compare the geocentric and heliocentric models of the Solar System.** Earth's motion is hard to detect, so before the Copernican Revolution, most people accepted a geocentric model of the Solar System, in which all objects orbit Earth. But in that model, the planets' apparent retrograde motion was hard to understand. Copernicus created the first comprehensive mathematical model of the Solar System with the Sun at the center, called a heliocentric model. His model explained apparent retrograde motion as a visual illusion seen when an inner planet passes an outer planet in their orbits.

LG 2 **Use Kepler's laws to describe how objects in the Solar System move.** Using Tycho's observational data, Kepler developed empirical rules to describe the motions of the planets. Kepler's three laws state that (1) planets move in elliptical orbits around the Sun; (2) planets move fastest when closest to the Sun and slowest when farthest from the Sun, so that the planets sweep out equal areas in equal times; and (3) the orbital period (P) of a planet squared equals the semimajor axis (A) of its orbit cubed: $P^2 = A^3$.

LG 3 **Explain how Galileo's astronomical discoveries convinced him that the heliocentric model was correct.** Galileo invented astronomical telescopes and used them to observe sunspots, craters on the Moon, and moons orbiting Jupiter. He also saw Venus going through phases, like the Moon, but changing its apparent size in each phase. The observations of Venus were hard to explain with Ptolemy's model.

LG 4 **Describe the physical laws of motion discovered by Galileo and Newton.** Galileo studied the physics of falling objects and discovered the principle of inertia. Newton's laws state that (1) objects do not change their motion unless they experience a net force, (2) force = mass × acceleration, and (3) every force has an equal and opposite force. Net forces cause accelerations (changes in motion). Inertia resists changes in motion.

? Unanswered Questions

- Would the history of scientific discoveries in physics and astronomy have been different if the Inquisition had not prosecuted Galileo? Galileo wrote the *Dialogo* after the Catholic Church in 1616 ordered him not to "hold or defend" the idea that Earth moves and the Sun is still. And he wrote his equally famous *Discorsi e Dimostrazioni Matematiche* (often shortened in English to "Two New Sciences") while under house arrest after his trial. However undeterred Galileo appeared to be, the effects of the decrees, prohibitions, and prosecutions might have dissuaded other scientists in Catholic countries from pursuing that type of work. Indeed, after Galileo's experiences, the center of the scientific revolution moved north to Protestant Europe.

- What percentage of planets are in unstable orbits? In younger planetary systems, planets might migrate in their orbits because of the presence of other massive planets nearby. We explain in Chapter 7 that Uranus and Neptune might have migrated in that way. Some planets have been discovered moving through the galaxy without any obvious orbit around a star—and therefore are not in the stable orbits we see in our own Solar System.

Questions and Problems

Test Your Understanding

1. An *empirical science* is one based on
   a. hypothesis.
   b. calculus.
   c. computer models.
   d. observed data.

2. When Earth catches up to a slower-moving outer planet and passes it in its orbit, the outer planet
   a. exhibits retrograde motion.
   b. slows down because it feels Earth's gravitational pull.
   c. becomes dimmer as it passes through Earth's shadow.
   d. moves into a more elliptical orbit.

3. Copernicus's model of the Solar System was superior to Ptolemy's because
   a. it had a mathematical basis and computed the spacing of the planets.
   b. it was much more accurate.
   c. it did not require epicycles.
   d. it fit the telescopic data better.

4. An inferior planet is one that is
   a. smaller than Earth.
   b. larger than Earth.
   c. closer to the Sun than Earth.
   d. farther from the Sun than Earth.

5. The time a planet takes to come back to the same position in space in relation to the Sun is called its _____ period.
   a. synodic
   b. sidereal
   c. heliocentric
   d. geocentric

6. Suppose a planet is discovered orbiting a star in a highly elliptical orbit. While the planet is close to the star, it moves _____, but while it is far away, it moves _____.
   a. faster; slower
   b. slower; faster
   c. retrograde; prograde
   d. prograde; retrograde

7. If a superior planet is observed from Earth to have a synodic period of 1.2 years, what is its sidereal period?
   a. 0.54 years
   b. 1.8 years
   c. 4.0 years
   d. 6.0 years

8. A net force must be acting when an object
   a. accelerates.
   b. changes direction but not speed.
   c. changes speed but not direction.
   d. all of the above

9. For Earth, $P^2/A^3 = 1.0$ (in appropriate units). Suppose a new dwarf planet is discovered that is 14 times farther from the Sun than Earth is. For that planet,
   a. $P^2/A^3 = 1.0$.
   b. $P^2/A^3 > 1.0$.
   c. $P^2/A^3 < 1.0$.
   d. we can't know the value of P^2/A^3 without more information.

10. Galileo observed that Venus had phases that correlated with the size of its image in his telescope. From that information, you may conclude that Venus
    a. is the center of the Solar System.
    b. orbits the Sun.
    c. orbits Earth.
    d. orbits the Moon.

11. Kepler's second law says that
    a. planetary orbits are ellipses with the Sun at one focus.
    b. the square of a planet's orbital period equals the cube of its semimajor axis.
    c. net forces cause changes in motion.
    d. planets move fastest when closest to the Sun.

12. The average speed of a new planet in orbit has been reported to be 33 km/s. When it is closest to its star it moves at 31 km/s, and when it is farthest from its star it moves at 35 km/s. Those data must be in error because
    a. the average speed is far too fast.
    b. Kepler's third law says the planet has to sweep out equal areas in equal times, so the speed of the planet cannot change.
    c. Kepler's second law says the planet must move fastest when closest to its star, not when farthest away.
    d. with those numbers, the square of the orbital period will not be equal to the cube of the semimajor axis.

13. Galileo observed that Jupiter has moons. From that information, you may conclude that
    a. Jupiter is the center of the Solar System.
    b. Jupiter orbits the Sun.
    c. Jupiter orbits Earth.
    d. some things do not orbit Earth.

14. If you start from rest and accelerate at 10 mph/s and end up traveling at 60 mph, how long did it take?
    a. 1 second
    b. 6 seconds
    c. 60 seconds
    d. 0.6 seconds

15. Planets with high eccentricity may be unlikely candidates for life because
    a. the speed varies too much.
    b. the period varies too much.
    c. the temperature varies too much.
    d. the orbit varies too much.

Thinking about the Concepts

16. Study Figure 3.2. During normal motion, does Mars move toward the east or west? Which direction does it travel when moving retrogradely? For how many days did Mars move retrogradely? If one of the martian missions were photographing *Earth* in the sky during those days, what would it have observed?

17. Copernicus and Kepler engaged in what is called empirical science. What do we mean by *empirical*?

18. Explain why Saturn's synodic period is very close to 1 Earth year (a sketch may help).

19. Make a sketch of Earth, Venus, and the Sun in a geocentric model and in a heliocentric model. Label the Sun and Earth, and then show the changing phases of Venus during one of its orbits. What would we observe in each model? Why was the invention of the telescope necessary to distinguish between those models?

20. Experiment with falling objects as Galileo did. When you drop pairs of objects with different masses, do they reach the ground at the same time? Do they hit the ground with the same force? Does that approach work with a sheet of paper or a tissue—why or why not?

21. A planet's orbital speed around the Sun varies. When is the planet moving fastest in its orbit? Slowest?

22. The Moon's orbit around Earth is elliptical, too, with an eccentricity of 0.05. How does that value compare with the eccentricity of Earth's orbit? How do those elliptical orbits explain the types of solar eclipses discussed in Chapter 2?

23. Galileo came up with the concept of inertia. What do we mean by *inertia*?

24. If Kepler had lived on Mars, would he have deduced the same empirical laws for the motion of the planets? Explain.

25. What is the difference between speed and velocity? Between velocity and acceleration?

26. When involved in an automobile collision, a person not wearing a seat belt will move through the car and often strike the windshield directly. Which of Newton's laws explains why the person continues forward, even though the car stopped?

27. When riding in a car, we can sense changes in speed or direction through the forces that the car applies on us. Do we wear seat belts in cars and airplanes to protect us from speed or from acceleration? Explain your answer.

28. An astronaut standing on Earth can easily lift a wrench having a mass of 1 kg, but not a scientific instrument with a mass of 100 kg. On the International Space Station, he can manipulate both, although the scientific instrument responds much more slowly than the wrench. Explain why.

29. The Process of Science Figure illustrates that scientific ideas are always open to challenge. Construct an argument that the constant process of challenging and falsifying ideas is a strength of science, rather than a weakness.

30. How might you expect conditions on Earth to be different if the eccentricity of its orbit were 0.17 instead of 0.017?

Applying the Concepts

31. Study the graph in Figure 3.16. Is that graph linear or logarithmic? From the data on the graph, find the approximate semimajor axis and period of Saturn. Show your work.

32. Study Figure 3.19, which shows that Venus's apparent size changes as it goes through phases. Approximately how many times larger is Venus in the sky at the thinnest crescent than at the gibbous phase shown? Approximately how many times closer is Venus to us at the phase of that thinnest crescent than at the gibbous phase?

33. Suppose a dwarf planet is discovered orbiting the Sun with a semimajor axis of 50 AU. What would be the planet's orbital period?

34. The orbital period of Uranus is 84 years. Compute the semimajor axis of its orbit. How much time passes between oppositions of Uranus?

35. Dwarf planet Ceres is 2.77 AU from the Sun. Its synodic period is 1.278 years.
    a. Use Working It Out 3.1 to find the sidereal period in years.
    b. Use Kepler's law to find the sidereal period in years.
    c. Compare your results for (a) and (b).

36. Suppose you read online that "experts have discovered a new planet with a distance from the Sun of 2 AU and a period of 3 years." Use Kepler's third law to argue that this is impossible.

37. Show, as Galileo did, that Kepler's third law applies to the four Galilean moons of Jupiter by calculating P^2 divided by A^3 for each moon. (Data on the moons can be found in Appendix 4.)

38. In a period of 3 months, a planet travels 30,000 km with an average speed of 3.8 m/s. Some time later, the same planet travels 65,000 km in 3 months. How fast is the planet traveling at that later time? During which period is the planet closer to the Sun?

39. If you were on Mars, how often would you see retrograde motion of *Earth* in the martian night sky? (You can view a picture at https://mars.jpl.nasa.gov/allaboutmars/nightsky/retrograde.)

40. The elliptical orbit of a comet that a spacecraft recently visited is 1.24 AU from the Sun at its closest approach and 5.68 AU from the Sun at its farthest.
    a. Sketch the comet's orbit. When is it moving fastest? Slowest?
    b. What is the semimajor axis of its orbit? How long does the comet take to go around the Sun?
    c. What is the distance from the Sun to the "center" of the ellipse? What is the eccentricity of the comet's orbit?

41. You are driving down a straight road at a speed of 90 km/h, and you see another car approaching you at a speed of 110 km/h along the road.
    a. With respect to your own frame of reference, how fast is the other car approaching you?
    b. With respect to the other driver's frame of reference, how fast are you approaching the other driver's car?

42. During the latter half of the 19th century, a few astronomers thought a planet might be circling the Sun inside Mercury's orbit. They even gave it a name: Vulcan. We now know that Vulcan does not exist. If a planet with an orbit one-fourth the size of Mercury's actually existed, what would its orbital period be in comparison with that of Mercury?

43. Suppose you are pushing a small refrigerator of mass 50 kg on wheels. You push with a force of 100 N.
    a. What is the refrigerator's acceleration?
    b. Assume the refrigerator starts at rest. How long will the refrigerator accelerate at that rate before it gets away from you (that is, before it is moving faster than you can run—on the order of 10 m/s)?

44. If a 100-kg astronaut pushes on a 5,000-kg satellite and the satellite accelerates at 0.1 m/s^2, what acceleration does the astronaut experience in the opposite direction?

45. Sketch the orbit of Mars by using the information provided in the "Origins: Planets and Orbits" section of the chapter for the closest and farthest distances of Mars from the Sun.
    a. What is the orbit's major axis? Semimajor axis?
    b. What is the distance from the "center" of the orbit to the Sun? Compute the orbit's eccentricity. Compare that value with the eccentricity of Earth's orbit.

USING THE WEB

46. Go to the Web page "This Week's Sky at a Glance" (https://www.skyandtelescope.com/observing/sky-at-a-glance/) at the *Sky & Telescope* magazine website. Which planets are visible in your sky this week? Why are Mercury and Venus visible only in the morning before sunrise or in the evening just after sunset? Before telescopes, how did people know the planets were different from the stars?

47. Look up the dates for the next opposition of Mars, Jupiter, or Saturn. One source is the NASA "Sky Events Calendar" (https://eclipse.gsfc.nasa.gov/SKYCAL/SKYCAL.html). Check only the "Planet Events" box in "Section 2: Sky Events"; and in Section 3, generate a Sky Events Calendar or Table for the year. If you are coming up on an opposition (Opp), take pictures of the planet over the next few weeks. Can you see its position move in retrograde motion with respect to the background stars?

48. Refer to the Web page from question 47 to find the current observational positions of all the planets.
    a. Which ones are in or near to conjunction, opposition, or greatest elongation?
    b. Which are visible in the morning sky? In the evening sky?
    c. To connect the time we can see the planets on Earth with their physical alignments with Earth and the Sun in space, sketch the Solar System with Earth, the Sun, and the planets as it looks today from "above." Check your result with NASA's "Solar System Simulator" (https://space.jpl.nasa.gov): set it for "Show Me Solar System" as seen from above, and set the field of view at 2°, then 20°, and then 45° to see the inner and then the outer planets. Does your sketch agree with the simulator?

49. Go to the Museo Galileo website and view the Room VII video and exhibit at https://catalogue.museogalileo.it/room/RoomVII.html. What did his telescope look like? What other instruments did he use? From the museum page you can link to short videos (in English) on his science and his trial (http://catalogue.museogalileo.it/index/VideoIndexByThematicArea.html#s7). For example, click on "Galileo's micrometer." How did he measure the separation of the moons from Jupiter? How did that measurement allow him to show that the moons obeyed Kepler's third law? Why is Galileo often considered the first modern scientist? Why is his middle finger on display in the museum?

50. Go to the online "Extrasolar Planets Encyclopedia" (http://exoplanet.eu/catalog).
    a. Find a planet with an orbital period similar to that of Earth (click on "Period" twice and it will show decreasing values in days). What is the semimajor axis of the exoplanet's orbit? If it is very different from 1 AU, the star's mass is different from the Sun's. Click on the star name in the first column to see the star's mass. What is the planet's orbital eccentricity?
    b. Click on "Planet" to sort by name, and select a star with multiple planets. Verify that Kepler's third law applies by showing that the value of P^2/A^3 is about the same for each planet of that star. How eccentric are the orbits of the multiple planets?

EXPLORATION

Kepler's Laws

digital.wwnorton.com/astro6

In this Exploration, we examine how Kepler's laws apply to the orbit of Mercury. Visit the Digital Resources Page and on the Student Site open the "Planetary Orbits" Interactive Simulation in Chapter 3. That simulator animates the orbits of the planets, enabling you to control the simulation speed, as well as several other parameters. Here we focus on exploring Mercury's orbit, but you may wish to spend some time examining the orbits of other planets as well.

Kepler's First Law

In the "Orbit Settings" panel, use the drop-down menu next to "set parameters for" to select "Mercury" and then click "OK." Click the "Kepler's 1st Law" tab at the bottom of the control panel. Use the radio buttons to select "show empty focus" and "show center."

1 How would you describe the shape of Mercury's orbit?

_____

_____

Deselect "show empty focus" and "show center," and select "show semiminor axis" and "show semimajor axis." Under "Visualization Options," select "show grid."

2 Use the grid markings to estimate the ratio of the semiminor axis to the semimajor axis.

_____

_____

3 Calculate the eccentricity of Mercury's orbit from that ratio by using $e = [1 - (\text{Ratio})^2]^{1/2}$.

_____

_____

Kepler's Second Law

Click on "reset" near the top of the control panel, set parameters for Mercury, and click "OK." Then click on the "Kepler's 2nd Law" tab at the bottom of the control panel. Slide the "adjust size" slider to the right until the fractional sweep size is 1/8.

Click on "start sweeping." The planet moves around its orbit, and the simulation fills in area until one-eighth of the ellipse is filled. Click on "start sweeping" again as the planet arrives at the rightmost point in its orbit (that is, at the point in its orbit farthest from the Sun). You may need to slow the animation rate by using the slider under "Animation Controls." Click on "show grid" under the visualization options. (If the moving planet annoys you, pause the animation.) One easy way to estimate an area is to count the squares.

4 Count the squares in the yellow area and in the red area. You will need to decide what to do with fractional squares. Are the areas the same? Should they be?

_____

_____

Kepler's Third Law

Click on "reset" near the top of the control panel, set parameters for Mercury, and then click on the "Kepler's 3rd Law" tab at the bottom of the control panel. Select "show solar system orbits" in the "Visualization Options" panel. Study the graph. Use the eccentricity slider to change the simulated planet's eccentricity. Make the eccentricity first smaller and then larger.

5 Did anything in the graph change?

_____

_____

6 What do your observations of the graph tell you about the dependence of the period on the eccentricity?

_____

_____

Set the parameters back to those for Mercury. Now use the semimajor axis slider to change the semimajor axis of the simulated planet.

7 What happens to the period when you make the semimajor axis smaller?

_____

_____

8 What happens when you make it larger?

_____

_____

9 What do those results tell you about the dependence of the period on the semimajor axis?

_____

_____

4 Gravity and Orbits

Here in Chapter 4, we explore the physical laws that explain the regular patterns in the motions of the planets. Because the Sun is far more massive than all the other parts of the Solar System combined, its gravity shapes the motions of every object in its vicinity, from the almost circular orbits of some planets to the extremely elongated orbits of comets.

LEARNING GOALS

By the end of this chapter, you should be able to:

LG 1 Explain the elements of Newton's universal law of gravitation.

LG 2 Use the laws of motion and gravitation to explain planetary orbits.

LG 3 Explain how tidal forces from the Sun and Moon create Earth's tides.

LG 4 Describe how tidal forces affect solid bodies.

The International Space Station in orbit around Earth. ▶▶▶

What keeps a space station in orbit?

4.1 Gravity Is an Attractive Force between Objects

In Chapter 3, we explored Kepler's work on the movement of the planets around the Sun and Newton's laws of motion. Here in Chapter 4, we build on those concepts to understand Newton's universal law of gravitation. Although some properties of gravity were observed before Newton, his work connected the everyday phenomenon of falling objects to the motion of the planets around the Sun. Newton's law of gravity combined Kepler's empirical laws and Newton's own laws of motion.

Gravity, Mass, and Weight

Many forces that we see in everyday life involve direct contact between objects. The cue ball on a pool table slams into the eight ball, knocking it into the pocket. The shoe of the child pushing a scooter presses directly against the pavement. When physical contact occurs between two objects, the source of the forces between them is easy to see.

If you drop a ball, it falls toward the ground. The ball picks up speed as it falls, accelerating downward toward Earth. According to Newton's second law, where there is acceleration, there is force. But where is the force that causes the ball to accelerate? The ball falling toward Earth is an example of a different kind of force, one that acts at a distance across the intervening void of space. The ball falling toward Earth is accelerating in response to the force of gravity. **Gravity**, one of the fundamental forces of nature, is the mutually attractive force between objects with mass.

Recall from Chapter 3 that Galileo discovered that all freely falling objects accelerate toward Earth at the same rate, regardless of their mass. Drop a marble and a book at the same time and from the same height, and they will hit the ground together. Note that air resistance becomes a factor at higher speeds, but the effect is negligible for dense, slow objects. The acceleration of falling objects due to gravity near the surface of Earth, which Galileo also measured experimentally, is usually written as g (lowercase) and has an average value across the surface of Earth of 9.8 meters per second squared (m/s^2). The value of g varies slightly across Earth's surface, ranging from 9.78 m/s^2 at the equator to 9.83 m/s^2 at the poles. That variation exists because Earth is not a perfect sphere: its rotation makes it flatter at the poles, so the radius of Earth is smaller there.

After working out the laws governing the motion of objects, Newton realized that if all objects, regardless of mass, fall with the same acceleration, the gravitational *force* on an object must be determined by the object's *mass*.

Newton's second law (Section 3.4) states that acceleration equals force divided by mass, or $a = F/m$. The acceleration due to gravity can be the same for all objects only if the value of the force divided by the mass is the same for all objects. In other words, an object twice as massive has double the gravitational force acting on it; an object three times as massive has triple the gravitational force acting on it, and so on.

The gravitational force acting on an object attracted by a planet is called the object's **weight**. On the surface of Earth, weight equals mass multiplied by the acceleration of gravity at Earth's surface, g. In common language, we often use weight and mass interchangeably. To be more scientifically precise, astronomers use *mass* to refer to the amount of matter in an object and *weight* to refer to the force that the planet's gravitational pull exerts on that object. Your mass is the same no matter what planet or moon you are on, but your weight will be different:

$$F_{\text{weight}} = m \times g$$

where F_{weight} is an object's weight in newtons (N), the metric unit of force; m is the object's mass in kilograms (kg); and g is Earth's constant for acceleration due to gravity, 9.8 m/s². On Earth, then, an object with a *mass* of 1 kg has a *weight* of 9.8 N (**Figure 4.1**). On the Moon, the acceleration due to gravity is 1.6 m/s², which is about one-sixth of its value on Earth, so a 1-kg mass has a weight of 1.6 N on the Moon. Similarly, your weight on the Moon would be about one-sixth of your weight on Earth.

Newton's Law of Gravity

As Newton told the story, he saw an apple fall from a tree to the ground. He reasoned that if gravity is a force that depends on mass, a gravitational force should exist between *any* two masses, including between a falling apple and Earth. That great insight came from applying his third law of motion to gravity. Recall that Newton's third law states that for every force, an equal and opposite force is present. Therefore, if Earth exerts a force of 9.8 N on a 1-kg mass sitting on its surface, that 1-kg mass must also exert a force of 9.8 N on Earth. Drop a 7-kg bowling ball and it falls toward Earth, but at the same time Earth falls toward the 7-kg bowling ball. We don't notice the planet's motion because Earth is very massive, so it has a lot of resistance to changes in its motion. In the time a 7-kg bowling ball takes to fall to the ground from a height of 1 kilometer (km), Earth has "fallen" toward the bowling ball by only a tiny fraction of the size of an atom.

Newton reasoned that if doubling the mass of any object doubles the gravitational force between the object and Earth, doubling the mass of Earth ought to do the same thing. In short, the gravitational force between Earth and an object equals the product of the two masses multiplied by *something*:

$$\text{Gravitational force} = \text{Something} \times \text{Mass of Earth} \times \text{Mass of object}$$

If the mass of the object is 2 times greater, the force of gravity will be 2 times greater. Likewise, if the mass of Earth happened to be 3 times what it is, the force of gravity also would have to be 3 times greater. If the mass of both Earth and the object were greater by those amounts, the gravitational force would increase by a factor of 2×3, or 6 times. Because objects fall toward the center of Earth, we know that gravity is an attractive force acting along a line between the two masses.

If gravity is a force that depends on mass, a gravitational force should exist between *any* two masses. Suppose we have two masses—call them mass 1 and mass 2, or m_1 and m_2 for short. The gravitational force between them is *something* multiplied by the product of the masses:

$$\text{Gravitational force between two objects} = \text{Something} \times m_1 \times m_2$$

We have gotten this far just by combining Galileo's observations of falling objects with (1) Newton's laws of motion and (2) Newton's belief that Earth is a mass just like any other mass. But what about that "something" in the previous expression?

Kepler had already thought about that question. He reasoned that because the Sun is the focal point for planetary orbits, the Sun must exert an influence over the motions of the planets. Kepler speculated that the influence must grow

Figure 4.1 On the Moon, a mass of 1 kg has 1/6 the weight (displayed in newtons) that it has on Earth.

weaker with distance from the Sun because the planets closer to the Sun moved much faster than the farther ones. Kepler did not know about forces or inertia or gravity as the cause of celestial motion, but he thought that geometry alone suggested how that solar "influence" might change for planets progressively farther from the Sun.

To see why the influence must become weaker, imagine you have a certain amount of paint to spread over the surface of a sphere. If the sphere is small, you can apply a thick coat of paint. If the sphere is larger, however, the paint has to be spread farther, so the coat will be thinner. The surface area of a sphere depends on the square of the sphere's radius. Double the radius of a sphere, and the sphere's surface area becomes 4 times what it was. If you paint that new, larger sphere, the paint must cover 4 times as much area, and the thickness of the paint will be only a fourth of what it was on the smaller sphere. If you triple the radius of the sphere, the sphere's surface will be 9 times larger and the coat of paint will be only one-ninth as thick.

Kepler reasoned that as the influence of the Sun extended farther and farther into space, it would have to spread out to cover the surface of a larger and larger imaginary sphere centered on the Sun. The Sun's influence should diminish with the square of the distance from the Sun—a relationship known as an **inverse square law**.

Kepler had an interesting idea, but not a scientific hypothesis with testable predictions. He couldn't explain how the Sun influences the planets, and he lacked the mathematical tools to calculate how an object would move under such an influence. Newton, however, had both. If gravity is a force between *any* two objects, a gravitational force should then exist between the Sun and each of the planets. If that gravitational force were the same as Kepler's "influence," gravity might behave according to an inverse square law.

Newton's expression for gravity came to look like this:

$$\text{Gravitational force between two objects} = \text{Something} \times \frac{m_1 \times m_2}{(\text{Distance between objects})^2}$$

The expression still has a "something," and that something is a constant of proportionality. That constant determines the strength of gravity between objects, and it is the same for all pairs of objects. Newton named it the **universal gravitational constant**. Newton estimated that gravitational constant, written as G (uppercase), by using Galileo's measurement of g, estimates of Earth's radius, and a guess at the mass of Earth by assuming it had about the same density as typical rocks. Not until many years later was the actual value of G first measured. Today the value of G is accepted as 6.67×10^{-11} N m^2/kg^2 or its equivalents: 6.67×10^{-11} m^3/(kg s^2) or 6.67×10^{-20} km^3/(kg s^2).

A Universal Law of Gravitation

Newton's **universal law of gravitation**, illustrated in **Figure 4.2**, states that gravity is a force between any two objects having mass. The force due to gravity has the following properties:

1. It is an attractive force acting along a straight line between the two objects.
2. It is proportional to the mass of one object (m_1) multiplied by the mass of the other object (m_2). If we double m_1, the force (F) increases by a factor of 2. Likewise, if we double m_2, F increases by a factor of 2.

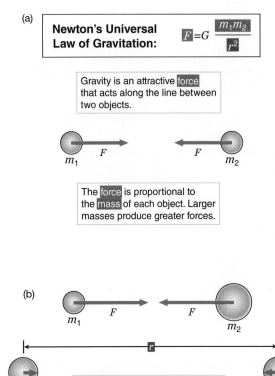

Figure 4.2 Gravity is an attractive force between two objects. The force of gravity depends on the masses of the objects, m_1 and m_2, and the distance, r, between them.

3. It is inversely proportional to the square of the distance r between the centers of the two objects. The graph in **Figure 4.3** shows that if we double r, F decreases by a factor of 4. If we triple r, F falls by a factor of 9.

Written as a mathematical formula, the universal law of gravitation is

$$F_{grav} = G \times \frac{m_1 \times m_2}{r^2}$$

where F_{grav} is the force of gravity between two objects; m_1 and m_2 are the masses of objects 1 and 2, respectively; r is the distance between the centers of mass of the two objects; and G is the universal gravitational constant. The relationship between the force of gravity and the masses of and distance between two objects is further explored in **Working It Out 4.1**.

Gravity pulls you toward the center of Earth. Gravity holds the planets and stars together and keeps the thin blanket of air we breathe close to Earth's surface. The planets, including Earth, orbit the Sun, and gravity holds them in orbit. Gravity caused a vast

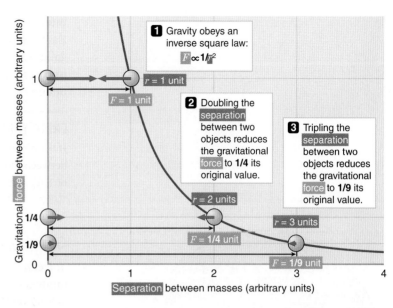

Figure 4.3 As two objects move apart, the gravitational force between them decreases by the inverse square of the distance between them.

4.1 Working It Out Playing with Newton's Laws of Motion and Gravitation

For any two objects, the force of gravity is directly proportional to the masses and inversely proportional to the *square* of the distance between them. Let's look at a few examples of how to use this equation:

Changing the Distance

How would the gravitational force between Earth and the Moon change if the distance between them were doubled? In that example, the masses of Earth and the Moon stay the same, and r becomes $2r$. We can calculate how the force changes by writing the equation for distance r and again for distance $2r$, and then taking a ratio to compare them:

$$\frac{F_{grav\ at\ distance\ 2r}}{F_{grav\ at\ distance\ r}} = \frac{G \times \frac{M_{Earth} M_{Moon}}{(2r)^2}}{G \times \frac{M_{Earth} M_{Moon}}{r^2}}$$

We can cancel out the constant G and the masses of Earth and the Moon, which do not change. Then we need to multiply both the numerator and denominator of the fraction by r^2, remembering that both the 2 and the r get squared in $(2r)^2 = 4r^2$. The equation becomes

$$\frac{F_{grav\ at\ distance\ 2r}}{F_{grav\ at\ distance\ r}} = \frac{\cancel{G} \times \cancel{M_{Earth} M_{Moon}}}{\cancel{G} \times \cancel{M_{Earth} M_{Moon}}} \times \frac{r^2}{(2r)^2}$$

$$\frac{F_{grav\ at\ distance\ 2r}}{F_{grav\ at\ distance\ r}} = \frac{r^2}{4r^2} = \frac{1}{4}$$

Doubling the distance reduced the force by a factor of 4; that is, the force is 1/4 as strong.

Gravitational Acceleration

We have two ways to think about the gravitational force that Earth exerts on an object with mass m located on the surface of Earth. Recall Newton's second law of motion: $F = m \times a$. Here we are considering the gravitational force and the acceleration due to gravity, or $F_{grav} = m \times g$. The other way to think about the force is from the perspective of the universal law of gravitation:

$$F_{grav} = G \times \frac{M_{Earth} \times m}{R_{Earth}^2}$$

Here, M_{Earth} is the mass of Earth, and R_{Earth} is the radius of Earth. The two expressions describing that force must be equal. Therefore,

$$m \times g = G \times \frac{M_{Earth} \times m}{R_{Earth}^2}$$

The mass m is on both sides of the equation, so we can cancel it out. The equation then becomes

$$g = G \times \frac{M_{Earth}}{R_{Earth}^2}$$

The expression shows that the gravitational acceleration (g) experienced by an object of mass m on the surface of Earth is determined by Earth's mass and radius. The mass of the object itself (m) appears nowhere in the expression, so changing m does not affect the gravitational *acceleration* of an object on Earth.

interstellar cloud of gas and dust to collapse 4.5 billion years ago to form the Sun, Earth, and the rest of the Solar System. Gravity binds colossal groups of stars into galaxies. Gravity shapes space and time, and it can affect the ultimate fate of the universe. We return often to the concept of gravity because it is central to an understanding of the universe.

CHECK YOUR UNDERSTANDING 4.1A

If the distance between Earth and the Sun were cut in half, the gravitational force between those two objects would: (a) decrease by a factor of 4; (b) decrease by a factor of 2; (c) increase by a factor of 2; (d) increase by a factor of 4.

Gravity Differs from Place to Place within an Object

You can think of Earth as a collection of small masses, each of which feels a gravitational attraction toward every other small part of Earth. The mutual gravitational attraction that occurs among all parts of the same object is called *self-gravity*.

As you sit reading this book, you are exerting a gravitational attraction on every other fragment of Earth, and every other fragment of Earth is exerting a gravitational attraction on you. Your gravitational interaction is strongest with the parts of Earth closest to you. The parts of Earth on the other side of our planet are much farther from you, so their pull on you is correspondingly less.

The net effect of all those forces is to pull you (or any other object) toward Earth's center. If you drop a hammer, it falls directly toward the ground. Because Earth is nearly spherical, for every piece of Earth pulling you toward your right, a corresponding piece of Earth is pulling you toward your left with just as much force. For every piece of Earth pulling you forward, a corresponding piece of Earth is pulling you backward. Because Earth is almost spherically symmetric, all those "sideways" forces (see the blue arrows in **Figure 4.4a**) cancel out, leaving behind an overall force that points toward Earth's center (**Figure 4.4b**).

Some parts of Earth are closer to you and others are farther away, but an average distance exists between you and all the small fragments of Earth pulling on you. That average distance is the distance between you and the center of Earth. As illustrated in Figure 4.4b, the overall pull that you feel is the same as it would be if all the mass of Earth were concentrated at a single point located at the very center of the planet.

That relationship is true for any spherically symmetric object. Outside the object, its gravity behaves as though all the mass of that object were concentrated at a point at its center. That relationship is important in many applications. For example, when you estimate your weight on another planet, you are calculating the force of gravity between you and the planet. The "distance" in the gravitational equation will be the distance between you and the center of the planet, which is just the radius of the planet plus your altitude.

CHECK YOUR UNDERSTANDING 4.1B

If Earth shrank to a smaller radius but kept the same mass, would the gravitational force between Earth and the Moon: (a) decrease; (b) increase; or (c) stay the same? Would everyone's weight at Earth's surface: (a) increase; (b) decrease; or (c) stay the same?

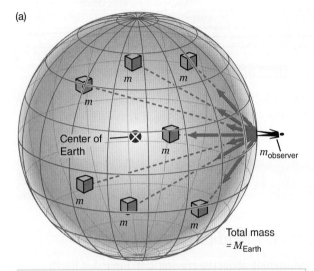

An object on the surface of a spherical mass (such as Earth) feels a gravitational attraction toward each small part of the sphere.

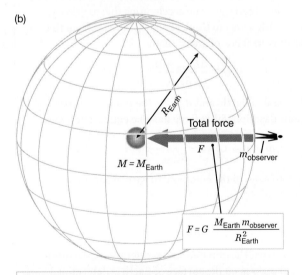

The net force is the same as if we scooped up the mass of the entire sphere and concentrated it at a point at the center.

Figure 4.4 Outside a sphere, the net gravitational force due to a spherical mass is the same as the gravitational force from the same mass concentrated at a point at the center of the sphere.

4.2 An Orbit Is One Body "Falling Around" Another

Kepler's laws on the motions of planets enabled astronomers to predict the positions of the planets accurately, but those laws did not explain *why* the planets behave as they do. Newton's work explained why planets orbit the Sun.

Newton used his laws of motion and his proposed law of gravity to predict the paths of planetary orbits. His calculations showed that those orbits should be ellipses with the Sun at one focus, that planets should move faster when closer to the Sun, and that the square of the period of a planet's orbit should vary as the cube of the semimajor axis of that elliptical orbit. Newton's universal law of gravitation *predicted* that planets should orbit the Sun in just the way that Kepler's empirical laws described. By explaining Kepler's laws, Newton found important corroboration for his law of gravitation.

Gravity and Orbits

Newton's laws tell us how forces change an object's motion and how objects interact with one another through gravity. To know where an object will be at any given time, we have to "add up" the object's motion over time. Newton invented calculus to do that, but here we aim just for a conceptual understanding.

In Newton's time, the closest thing to making a heavy object fly was shooting cannonballs out of a cannon, so he used cannonballs in "thought experiments" about planetary motions. If you drop a cannonball, it falls directly to the ground, like any other mass does. A cannonball fired out of a cannon that is level with the ground behaves differently, however, as shown in **Figure 4.5a**. The cannonball still falls to the ground in the same amount of time as it does when it is dropped, but while falling it also travels over the ground, following a curved path that carries it a horizontal distance before it finally lands. The faster the cannonball moves when it is fired from the cannon, the farther it will go before it hits the ground (**Figure 4.5b**).

In the real world, that experiment reaches a natural limit. To travel through air, the cannonball must push air out of its way—an effect normally referred to as *air resistance*—which slows it down. But we can ignore such real-world complications in this thought experiment. Instead imagine that, having inertia, the cannonball continues along its course until it runs into something. The faster the cannonball moves when it is fired, the farther it goes before hitting the ground. If the cannonball flies far enough, Earth's surface curves out from under it, as shown in **Figure 4.5c**. Eventually a point is reached (**Figure 4.5d**) at which the cannonball is flying so fast that the surface of Earth curves away from the cannonball at the same rate at which the cannonball is falling toward Earth. When that occurs, the cannonball, which always falls toward the center of Earth, is, in a sense, falling around the world. An **orbit** is the path of one object that freely falls around another.

Why do astronauts appear to float freely about the cabin of a spacecraft? It is not because they have escaped Earth's gravity; Earth's gravity is what holds them in their orbit. Instead, the answer lies in Galileo's early observation that all objects fall with the same acceleration, regardless of their mass. The astronauts and the spacecraft are both in orbit around Earth, moving in the same direction, at the same speed, and experiencing the same gravitational acceleration, so they fall around Earth together. **Figure 4.6** demonstrates that point. The astronaut is orbiting Earth just as the spacecraft is orbiting Earth. On the surface of Earth,

Figure 4.5 Newton realized that a cannonball fired at the right speed would fall around Earth in a circle. Velocity (*v*) is indicated by a red arrow and acceleration (*a*), by a green arrow.

AstroTour: Newton's Laws and Universal Gravitation

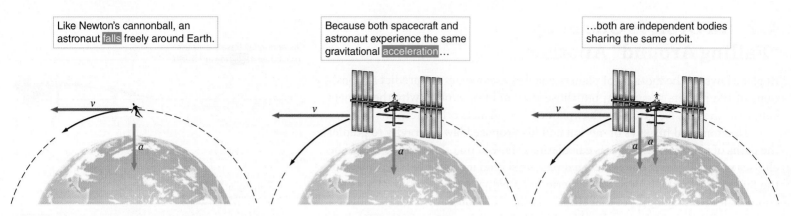

Figure 4.6 A "weightless" astronaut has not escaped Earth's gravity. Rather, an astronaut and a spacecraft share the same orbit as they fall around Earth together.

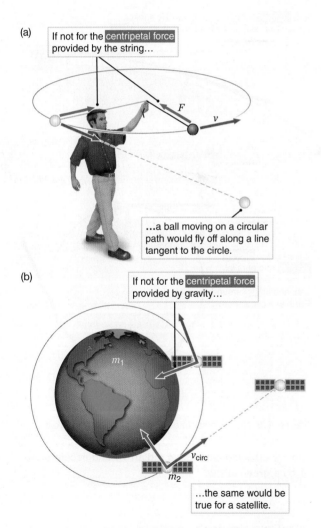

Figure 4.7 (a) A string provides the centripetal force that keeps a ball moving in a circle. (We are ignoring the smaller force of gravity that also acts on the ball.) (b) Gravity provides the centripetal force that holds a satellite in a circular orbit about Earth.

your body tries to fall toward the center of Earth, but the ground gets in the way. You feel your weight when you are standing on Earth because the ground pushes on you to oppose the force of gravity, which pulls you downward. In the spacecraft, however, nothing interrupts the astronaut's fall because the spacecraft is falling around Earth in the same orbit as the astronaut. The astronaut is in **free fall**, falling freely in Earth's gravity. The **Process of Science Figure** illustrates the universality of Newton's law of gravitation.

What Velocity Is Needed to Reach Orbit?

How fast must Newton's cannonball be fired for it to fall around the world? The cannonball would be in **uniform circular motion**, meaning it moves along a circular path at constant speed. That type of motion is discussed in more depth in Appendix 8. Another example of uniform circular motion is a ball whirling around your head on a string, as shown in **Figure 4.7a**. If you let go of the string, the ball will fly off in a straight line in whatever direction it is traveling at the time, just as Newton's first law predicts for an object in motion. The string prevents the ball from flying off by constantly changing the direction the ball is traveling. The central force of the string on the ball is called a **centripetal force**: a force toward the center of the circle. Using a more massive ball, speeding up its motion, and making the string shorter so that the turn is tighter all are ways to increase the force needed to keep a ball moving in a circle.

For Newton's cannonball (or a satellite), no string holds the ball in its circular motion. Instead, gravity supplies the force, as illustrated in **Figure 4.7b**. The force of gravity must be just enough to keep the satellite moving on its circular path. Because that force has a specific strength, the satellite must therefore be moving at a particular speed around the circle, which we call its **circular velocity** (v_{circ}). If the satellite were moving at any other velocity, it would not be moving in a circular orbit. Remember the cannonball: If it is moving too slowly, it will drop below the circular path and hit the ground. Similarly, if the cannonball is moving too fast, its motion will carry it above the circular orbit. Only a cannonball moving at just the right velocity—the circular velocity—will fall around Earth on a circular path (see Figure 4.5d).

Newton's thought experiment became a reality in 1957, when the Soviet Union launched the first artificial object to orbit Earth. They used a rocket to lift Sputnik 1, an object about the size of a basketball, high enough above Earth's upper atmosphere that air resistance wasn't an issue. Sputnik 1 achieved a high enough speed that it fell around Earth, just like Newton's imaginary cannonball.

PROCESS OF SCIENCE

Universality

The laws of physics are the same everywhere and at all times.
The principle of universality underlies our understanding of the natural world.

1604 — Galileo

In 1604, Galileo determined that all objects on Earth have the same gravitational acceleration, *a*. Universality means that all objects on any world share the same gravitational acceleration with each other.

$a_{large} = a_{small}$

1971 — David Scott and *Apollo 15*

In 1971, *Apollo 15* commander David Scott tested the gravitational law with a hammer and a falcon feather on the Moon. With no air resistance, the objects had the same gravitational acceleration.

$a_{hammer} = a_{feather}$

The physical laws that apply to falling objects also apply to planets orbiting the Sun, to stars orbiting within the galaxy, and to galaxies orbiting each other. This universality allows us to draw conclusions about very distant objects.

92 CHAPTER 4 Gravity and Orbits

When one object is falling around a much more massive object, we say that the less massive object is a **satellite** of the more massive object. Planets are satellites of the Sun, and moons are natural satellites of planets. Newton's imaginary cannonball and Sputnik 1 were satellites (*sputnik* means "satellite" in Russian). An Earth-orbiting spacecraft and the astronauts inside it are independent satellites of Earth that conveniently happen to share the same orbit.

The Shape of Orbits

Some Earth satellites travel a circular path at constant speed. Just like the ball on the string, satellites traveling at the circular velocity always stay the same distance from Earth, neither speeding up nor slowing down in orbit. But what if the satellite then fired its rockets and started traveling *faster* than the circular velocity? Earth's pull is as strong as ever, but because the satellite has greater speed, Earth's gravity does not bend the satellite's path sharply enough to hold it in a circle. When that happens, the satellite begins to climb above a circular orbit.

AstroTour: Elliptical Orbit

As the distance between the satellite and Earth increases, the satellite slows down. Think about a ball thrown upward into the air, as shown in **Figure 4.8a**. As the ball climbs higher, the pull of Earth's gravity opposes its motion, slowing the ball down. The ball climbs more and more slowly until its vertical motion stops for an instant and then is reversed; the ball then begins to fall back toward Earth, speeding up along the way. A satellite does the same thing. As the satellite climbs above a circular orbit and begins to move away from Earth, Earth's gravity opposes the satellite's outward motion, slowing the satellite down. The farther the satellite is from Earth, the more slowly the satellite moves—just like the ball thrown into the air. Also just like the ball, the satellite reaches a maximum height on its curving path and then begins falling back toward Earth. When it does, Earth's gravity speeds the satellite up as it gets closer and closer to Earth. The satellite's orbit has changed from circular to elliptical.

Any object in an elliptical orbit, including a planet orbiting the Sun (**Figure 4.8b**), will therefore lose speed as it pulls away from what it is orbiting

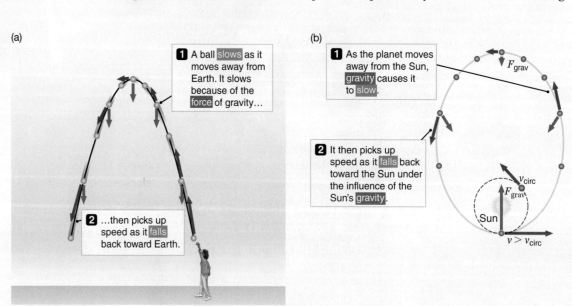

Figure 4.8 (a) A ball thrown into the air slows as it climbs away from Earth and then speeds up as it heads back toward Earth. (b) A planet in an elliptical orbit around the Sun does the same thing. (Although no planet has an orbit as eccentric as the one shown here, the orbits of comets can be far more eccentric.)

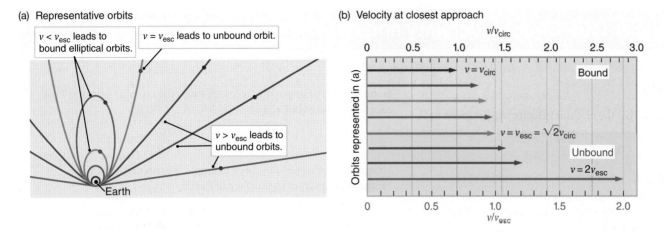

Figure 4.9 (a) Various orbits that share the same point of closest approach but differ in velocity at that point. (b) Closest-approach velocities for the orbits in (a). An object's velocity at closest approach determines the orbit's shape and whether the orbit is bound or unbound. v_{circ} = circular velocity; v_{esc} = escape velocity.

and then gain that speed back again as gravity causes it to fall inward toward what it is orbiting. Gravity, therefore, helps explain Kepler's second law from Section 3.2—namely, that a planet moves fastest when closest to the Sun and slowest when farthest from the Sun.

Newton's laws do more than explain Kepler's laws: they predict different types of orbits beyond Kepler's empirical experience. **Figure 4.9a** shows a series of satellite orbits, each with the same point of closest approach to Earth but with different velocities at that point, as indicated in **Figure 4.9b**. The more speed a satellite has at its closest approach to Earth, the farther the satellite can pull away from Earth, and the more eccentric its orbit becomes. As long as it remains elliptical, no matter how eccentric, the orbit is called a **bound orbit** because the satellite is gravitationally bound to the object it is orbiting.

In that sequence of faster and faster satellites, a point comes at which the satellite is moving so fast that gravity cannot reverse its outward motion, so the object travels away from Earth, never to return. The lowest speed at which an object can permanently leave the gravitational grasp of another mass is called the **escape velocity**, v_{esc}. Once a satellite's velocity at closest approach equals or exceeds v_{esc}, and it is no longer gravitationally bound to the object it was orbiting, we say it is in an **unbound orbit**.

As Figure 4.9a shows, an object with a velocity *less* than the escape velocity (v_{esc}) has an elliptically shaped orbit and follows the same path over and over again, whereas unbound orbits do not close like an ellipse. An object such as a comet on an unbound orbit makes only a single pass around the Sun and then continues away from the Sun into deep space, never to return. Circular velocity and escape velocity are further explored in **Working It Out 4.2**.

Estimating Mass by Using Newton's Version of Kepler's Laws

Newton's calculations opened up a new way to investigate the universe. He showed that the same physical laws that describe the flight of a cannonball on Earth—or the legendary apple falling from a tree—also describe the motions of the planets through the heavens. His laws of motion and gravitation predict all three of Kepler's empirical laws of planetary motion. Newton's version of Kepler's

4.2 Working It Out Circular Velocity and Escape Velocity

Circular Velocity

In Appendix 8, we show that the circular velocity (v_{circ}) is given by

$$v_{circ} = \sqrt{\frac{GM}{r}}$$

where M is the mass of the orbited object, r is the radius of the circular orbit, and G is the universal gravitational constant. A cannonball moving at just the right velocity—the circular velocity—will fall around Earth on a circular path.

We can use that equation to show how fast Newton's cannonball would have to travel to stay in its circular orbit. The average radius of Earth is 6,370 km, the mass of Earth is 5.97×10^{24} kg, and the gravitational constant is 6.67×10^{-20} km³/(kg s²). Inserting those values into the expression for v_{circ}, we get

$$v_{circ} = \sqrt{\frac{[6.67 \times 10^{-20} \text{ km}^3/(\text{kg s}^2)] \times (5.97 \times 10^{24} \text{ kg})}{6,370 \text{ km}}} = 7.91 \text{ km/s}$$

To stay in its circular orbit, Newton's cannonball would have to be traveling about 8 kilometers per second (km/s)—more than 28,000 kilometers per hour (km/h). That's well beyond the range of a typical cannon, but rockets routinely reach those speeds.

Now let's compare that speed with that needed to launch a satellite from the Moon into orbit just above the lunar surface. The radius of the Moon is 1,740 km, and its mass is 7.35×10^{22} kg. Those values give a considerably lower circular velocity than the one for Earth:

$$v_{circ} = \sqrt{\frac{(6.67 \times 10^{-20} \text{ km}^3/\text{kg s}^2) \times (7.35 \times 10^{22} \text{ kg})}{1,740 \text{ km}}} = 1.68 \text{ km/s}$$

Escape Velocity

Sending a spacecraft to another planet requires launching it with a velocity greater than Earth's escape velocity. The escape velocity is a factor of $\sqrt{2}$, or approximately 1.41, times the circular velocity. That relation can be expressed as

$$v_{esc} = \sqrt{2} \times v_{circ} = \sqrt{\frac{2GM}{R}}$$

Using the numbers in the above example, we can calculate the escape velocity from the surface of Earth:

$$v_{esc} = \sqrt{2} \times v_{circ} = 1.41 \times 7.91 \text{ km/s} = 11.15 \text{ km/s}$$

To leave Earth, a rocket must have a speed of at least 11.15 km/s, or 40,140 km/h.

As with weight, the escape velocity from other astronomical objects is different from the escape velocity from Earth. Ida is a small asteroid orbiting the Sun between the orbits of Mars and Jupiter. Ida has an average radius of 15.7 km and a mass of 4.2×10^{16} kg. Therefore,

$$v_{esc} = \sqrt{\frac{2 \times [6.67 \times 10^{-20} \text{ km}^3/(\text{kg s}^2)] \times (4.2 \times 10^{16} \text{ kg})}{15.7 \text{ km}}}$$

$$v_{esc} = 0.019 \text{ km/s} = 68 \text{ km/h}$$

A baseball thrown at about 130 km/h (81 mph) would easily escape from Ida's surface and fly off into interplanetary space.

When two objects are closer to having the same mass, however—such as dwarf planet Pluto and its moon Charon, or a large planet and a star, or two stars—both objects experience significant accelerations in response to their mutual gravitational attraction. The two objects are both orbiting about a common point located between them, called the **center of mass**, so we now must think of them as falling around each other. Each mass is moving on its own elliptical orbit around the two objects' mutual center of mass. From measuring the size and period of any orbit, we can calculate the *sum* of the masses of the orbiting objects. Almost all knowledge about the masses of astronomical objects comes directly from applying Newton's version of Kepler's third law.

Astronomy in Action: Center of Mass

laws can be used to estimate the mass of the Sun from the orbit of Earth, as seen in **Working It Out 4.3**.

When Earth, a much less massive object, is orbiting the Sun, a much more massive object, the Sun's gravity has a strong influence on Earth, but Earth's gravity has little effect on the Sun. Therefore, the Sun remains nearly motionless from Earth's orbit around it.

CHECK YOUR UNDERSTANDING 4.2

Rank the circular velocities of the following objects from smallest to largest: (a) a 5-kg object orbiting Earth halfway to the Moon; (b) a 10-kg object orbiting Earth just above Earth's surface; (c) a 15-kg object orbiting Earth at the same distance as the Moon.

4.3 Working It Out Calculating Mass from Orbital Periods

Newton's Version of Kepler's Third Law

The time a planet takes to complete one orbit around the Sun (the orbital period, P) equals the distance traveled divided by the planet's speed. For simplicity, let's assume the orbit is circular. Thus, the time an object takes to make one trip around the Sun is the circumference of the circle ($2\pi r$) divided by the object's speed (v). The speed of the planet must be equal to the circular velocity discussed in Working It Out 4.2. Putting those relationships together, we have

$$\text{Orbital period } (P) = \frac{\text{Circumference of orbit}}{\text{Circular velocity}} = \frac{2\pi r}{\sqrt{\frac{GM_{\text{Sun}}}{r}}}$$

Squaring both sides of the equation gives

$$P^2 = \frac{4\pi^2 r^2}{\frac{GM_{\text{Sun}}}{r}} = \frac{4\pi^2}{GM_{\text{Sun}}} \times r^3$$

The square of the period of an orbit is equal to a constant ($4\pi^2/GM_{\text{Sun}}$) multiplied by the cube of the radius of the orbit. That is Kepler's third law ($P^2 = \text{constants} \times A^3$) applied to circular orbits. When Kepler used Earth units of years and astronomical units for the planets, he was taking a ratio of their periods and orbital radii with those for Earth, so the constants canceled out. Using calculus, Newton similarly could derive Kepler's third law for elliptical orbits with semimajor axis A instead of radius r:

$$P^2 = \frac{4\pi^2}{GM_{\text{Sun}}} \times A^3$$

Mass of the Sun

If we can measure the size and period of any orbit, we can use Newton's universal law of gravitation to calculate the mass of the object being orbited. To do so, we rearrange Newton's form of Kepler's third law above to read

$$M = \frac{4\pi^2}{G} \times \frac{A^3}{P^2}$$

Everything on the right side of that equation is either a constant (4, π, and G) or a measurable quantity (the semimajor axis A and period P of an orbit). The left side of the equation is the mass of the object at the focus of the ellipse. For example, we can find the mass of the Sun by noting the period and semimajor axis of the orbit of a planet around the Sun. Let's use the numbers for Earth. Whenever we have an equation with G, putting everything else into the same units as G (km, kg, s) is best. So first we must compute the number of seconds in 1 year: P = 1 yr = 365.24 days/yr × 24 h/day × 60 min/h × 60 s/min = 3.16×10^7 s. The semimajor axis A = 1 AU = 1.5×10^8 km. Then the mass of the Sun can be computed:

$$M_{\text{Sun}} = \frac{4\pi^2}{G} \times \frac{A^3}{P^2} = \frac{4\pi^2}{6.67 \times 10^{-20} \text{ km}^3/(\text{kg s}^2)} \times \frac{(1.5 \times 10^8 \text{ km})^3}{(3.16 \times 10^7 \text{ s})^2}$$

$$M_{\text{Sun}} = 2.0 \times 10^{30} \text{ kg}$$

We could have used the period and semimajor axis of another Solar System planet to get the same result for the mass of the Sun.

4.3 Tidal Forces Are Caused by Gravity

The rise and the fall of the oceans are called Earth's **tides**. Coastal dwellers long ago noted that the strength of the tides varies with the phase of the Moon. Tides are strongest during a new or a full Moon and are weakest during first quarter or third quarter Moon. Here in Section 4.3, we see how tides result from differences between the strength of the gravitational pull of the Moon and Sun on one part of Earth in comparison with their pull on other parts of Earth.

Astronomy in Action: Tides

Tides and the Moon

Recall from Figure 4.4 that each small part of an object feels a gravitational attraction toward every other small part of the object, and that self-gravity differs from place to place. In addition, each small part of an object feels a gravitational attraction toward every other mass in the universe, and those external forces also differ from place to place within the object.

Simulation: Tides on Earth

The Moon's gravity pulls on Earth as though the Moon's mass were concentrated at its center. The side of Earth that faces the Moon is closer to the Moon

AstroTour: Tides and the Moon

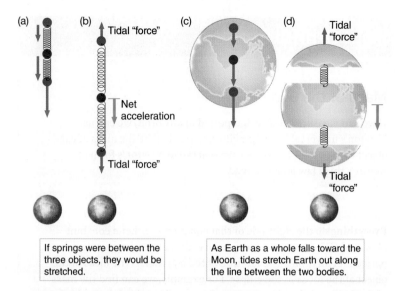

If springs were between the three objects, they would be stretched.

As Earth as a whole falls toward the Moon, tides stretch Earth out along the line between the two bodies.

Figure 4.10 (a) Imagine three objects connected by springs. (b) The springs are stretched as though forces were pulling outward on each end of the chain. (c) Similarly, three locations on Earth experience different gravitational attractions toward the Moon. (d) The difference in the Moon's gravitational attraction across Earth causes Earth's tides.

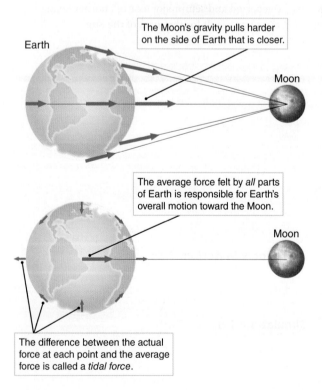

The Moon's gravity pulls harder on the side of Earth that is closer.

The average force felt by *all* parts of Earth is responsible for Earth's overall motion toward the Moon.

The difference between the actual force at each point and the average force is called a *tidal force*.

Figure 4.11 Tidal forces stretch Earth along the line between Earth and the Moon but compress Earth perpendicular to that line.

than the rest of Earth is, so it feels a stronger-than-average gravitational attraction toward the Moon. In contrast, the side of Earth facing away from the Moon is farther than average from the Moon, so it feels a weaker-than-average attraction toward the Moon. The pull of the Moon on the near side of Earth is about 7 percent greater than its pull on the far side of Earth.

To understand the consequence of that variation in the pull of the Moon, imagine three rocks being pulled by gravity toward the Moon. A rock closer to the Moon feels a stronger force than a rock farther from the Moon. Now suppose the three rocks are connected by springs (**Figure 4.10a**). As the rocks are pulled toward the Moon, the purple rock pulls away from the red rock, and the red rock pulls away from the blue rock. Therefore, the differences in the gravitational forces they feel will stretch *both* springs (**Figure 4.10b**). Now, instead of springs, imagine that the rocks are at different places on Earth (**Figure 4.10c**). On the side of Earth away from the Moon, the force is smaller (as indicated by the shorter arrow), so that part gets left behind (**Figure 4.10d**). Those differences in the Moon's gravitational attraction on different parts of Earth are called **tidal forces**.

Figure 4.11 shows how Earth is stretched, causing a tidal bulge. The Moon is not pushing the far side of Earth away; rather, the Moon simply is not pulling on the far side of Earth as hard as it is pulling on the planet as a whole. The far side of Earth is "left behind" as the rest of the planet is pulled more strongly toward the Moon. Figure 4.11 shows that a net force also is squeezing inward on Earth in the direction perpendicular to the line between Earth and the Moon. Together, the stretching by tidal forces along the line between Earth and the Moon and the squeezing by tidal forces perpendicular to that line distort Earth's shape, like a rubber ball caught in the middle of a tug-of-war.

If the surface of Earth were perfectly smooth and covered with an ocean of uniform depth and Earth did not rotate, the Moon would pull our oceans into an elongated **tidal bulge** like that shown in **Figure 4.12a**. The water would be at its deepest on the side toward the Moon and on the side away from the Moon, and the water would be at its shallowest midway between. But Earth is *not* covered with perfectly uniform oceans, and Earth *does* rotate. As any point on Earth rotates through the ocean's tidal bulges, that point experiences the ebb and flow of the tides. In addition, friction between the spinning Earth and its tidal bulge drags the oceanic tidal bulge around in the direction of Earth's rotation, as illustrated in **Figure 4.12b**.

Figure 4.12c shows the tides you would experience throughout a day. Begin as the rotating Earth carries you through the tidal bulge on the Moonward side of the planet. Because Earth's rotation drags the tidal bulge, the Moon is not exactly overhead but is instead high in the western sky. When you are at the high point in the tidal bulge, the ocean around you has risen higher than average—called a *high tide*. The Moon rises 50 minutes later than it did on the previous day, so the time between the Moon's return to the same position in the sky from the previous day is 24h 50m. Each high and low tide is 1/4 of that, or about 6 1/4 hours later. So, 6 1/4 hours after the high tide, somewhat after the Moon has settled beneath the western horizon, the rotation of Earth carries you through a point where the ocean is lower than average—called a *low tide*. If you wait another 6 1/4 hours, it is again high tide. You are now passing through the region where ocean water is "left behind" (with respect to Earth as a whole) in the tidal bulge on the side of Earth that is away from the Moon. The Moon at that time is hidden from view on

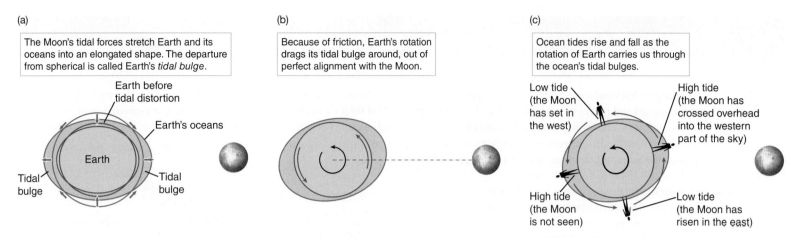

Figure 4.12 (a) Tidal forces pull Earth and its oceans into a tidal bulge. (b) Earth's rotation pulls its tidal bulge slightly out of alignment with the Moon. (c) As Earth's rotation carries us through those bulges, we experience the ocean tides. The magnitude of the tides has been exaggerated here for clarity. In these figures, the observer is looking down from above Earth's North Pole. Sizes and distances are not to scale.

the far side of Earth. About 6 1/4 hours later, sometime after the Moon has risen above the eastern horizon, it is low tide. About 25 hours after you started this journey—the amount of time the Moon takes to return to the same point in the sky from which it started—you again pass through the tidal bulge on the near side of the planet. That is the age-old pattern by which mariners have lived their lives for millennia: the twice-daily coming and going of high tide, shifting through the day in lockstep with the passing of the Moon.

Ocean currents and local geography, such as the shapes of Earth's shorelines and ocean basins, complicates the simple picture of tides. In addition, oceanwide oscillations occur, similar to water sloshing around in a basin. As the oceans respond to the tidal forces from the Moon, they flow around the various landmasses that break up the water covering our planet. Some places, such as the Mediterranean Sea and the Baltic Sea, are protected from tides by their relatively small sizes and the narrow passages connecting those bodies of water to the larger ocean. In other places, the shape of the land funnels the tidal surge from a large region of ocean into a relatively small area, concentrating its effect, as at eastern Canada's Bay of Fundy (**Figure 4.13**).

Solar Tides

Tides resulting from the pull of the Moon are called **lunar tides**, but the Sun also influences Earth's tides. The gravitational pull of the Sun causes Earth to stretch along a line pointing approximately in the direction of the Sun. The side of Earth closer to the Sun is pulled toward the Sun more strongly than the side of Earth away from the Sun, just as the side of Earth closest to the Moon is pulled more strongly toward the Moon. Tides on Earth resulting from differences in the gravitational pull of the Sun are called **solar tides**. The absolute strength of the Sun's pull on Earth is nearly 200 times greater than the strength of the Moon's pull on Earth. However, the Sun's gravitational attraction does not change by much from one side of Earth to the other because the Sun is much farther away than the Moon. As a result, solar tides are only about half as strong as lunar tides (**Working It Out 4.4**).

Figure 4.13 The world's most extreme tides are found in the Bay of Fundy in eastern Canada. Water rocks back and forth in the bay with a period of about 13 hours, close to the 12.5-hour period of the tides. The shape of the basin amplifies the tides so that the difference in water depth between low tide (a) and high tide (b) is extreme—typically about 14.5 meters and as much as 16.6 meters.

4.4 Working It Out   Tidal Forces

The strength of the gravitational force between two bodies is proportional to their masses and inversely proportional to the square of the distance between them. The strength of tidal forces caused by one body acting on another is also proportional to the mass of the body that is raising the tides, but the strength is inversely proportional to the *cube* of the distance between them.

The equation for the tidal force comes from the difference of the gravitational force on one side of a body compared with the force on the other side. We can approximate the tidal force of the Moon acting on a mass m on the surface of Earth with

$$F_{\text{tidal}}(\text{Moon}) = \frac{2GM_{\text{Moon}}R_{\text{Earth}}m}{d^3_{\text{Earth-Moon}}}$$

where R_{Earth} is Earth's radius and $d_{\text{Earth-Moon}}$ is the distance between Earth and the Moon.

Let's compare the Moon's tidal force acting on Earth with the Sun's tidal force acting on Earth. F_{tidal} from the Moon is given in the preceding equation; F_{tidal} from the Sun is given by

$$F_{\text{tidal}}(\text{Sun}) = \frac{2GM_{\text{Sun}}R_{\text{Earth}}m}{d^3_{\text{Earth-Sun}}}$$

To compare the lunar and solar tides, we can take a ratio of the tidal forces and proceed in a similar way to our comparison of gravitational forces in Working It Out 4.1:

$$\frac{F_{\text{tidal}}(\text{Moon})}{F_{\text{tidal}}(\text{Sun})} = \frac{\dfrac{2GM_{\text{Moon}}R_{\text{Earth}}m}{d^3_{\text{Earth-Moon}}}}{\dfrac{2GM_{\text{Sun}}R_{\text{Earth}}m}{d^3_{\text{Earth-Sun}}}}$$

Canceling out the constant G and the terms common in both equations (m and R_{Earth}) gives

$$\frac{F_{\text{tidal}}(\text{Moon})}{F_{\text{tidal}}(\text{Sun})} = \frac{M_{\text{Moon}}}{M_{\text{Sun}}} \times \frac{d^3_{\text{Earth-Sun}}}{d^3_{\text{Earth-Moon}}} = \frac{M_{\text{Moon}}}{M_{\text{Sun}}} \times \left(\frac{d_{\text{Earth-Sun}}}{d_{\text{Earth-Moon}}}\right)^3$$

Using the values from Appendix 4, $M_{\text{Moon}} = 7.35 \times 10^{22}$ kg, $M_{\text{Sun}} = 2 \times 10^{30}$ kg, $d_{\text{Earth-Moon}} = 384{,}400$ km, and $d_{\text{Earth-Sun}} = 1.5 \times 10^8$ km, gives

$$\frac{F_{\text{tidal}}(\text{Moon})}{F_{\text{tidal}}(\text{Sun})} = \frac{7.35 \times 10^{22} \text{ kg}}{2 \times 10^{30} \text{ kg}} \times \left(\frac{1.5 \times 10^8 \text{ km}}{384{,}400 \text{ km}}\right)^3 = 2.2$$

Thus, the tidal force between Earth and the Moon is 2.2 times stronger than the tidal force between Earth and the Sun. The Sun is an important factor, too, though, and that's why the tides change depending on the alignment of the Moon and the Sun with Earth.

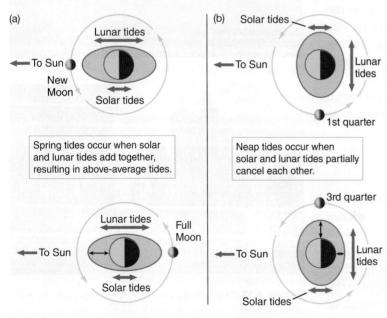

Figure 4.14 Solar tides are about half as strong as lunar tides. The interactions of solar and lunar tides result in either (a) spring tides when they are added together or (b) neap tides when they partially cancel each other.

Solar and lunar tides interact. When the Moon and the Sun are lined up with Earth (**Figure 4.14a**), at either new or full Moon, the lunar and solar tides on Earth overlap. That alignment creates more extreme tides that range from higher high tides to lower low tides. The extreme tides near the new or full Moon are called **spring tides**—not because of the season but because the water appears to spring out of the sea. Conversely, when the Moon, Earth, and Sun make a right angle (**Figure 4.14b**), at the Moon's first and third quarters, the lunar and solar tidal forces stretch Earth in different directions, creating less extreme tides known as **neap tides**. The word *neap* is derived from the Saxon word *neafte*, which means "scarcity": at those times of the month, shellfish and other food gathered in the tidal region are less accessible because the low tide is higher than at other times. Neap tides are only about half as strong as average tides and only a third as strong as spring tides.

CHECK YOUR UNDERSTANDING 4.3

Rank the following in order of weakest to strongest total tides: (a) new Moon in July; (b) first quarter Moon in July; (c) full Moon in January; (d) third quarter Moon in January.

4.4 Tidal Forces Affect Solid Bodies

In Section 4.3, we focused on how the liquid of Earth's oceans moves in response to the tidal forces from the Moon and Sun. But those tidal forces also affect the solid body of Earth. As Earth rotates through its tidal bulge, the solid body of the planet is constantly being deformed by tidal forces. Earth is somewhat elastic (like a rubber ball), and tidal stresses cause a vertical displacement of about 30 centimeters (cm) between high tide and low tide, or roughly a third of the displacement of the oceans. Deforming the shape of a solid object requires energy. (For a practical demonstration of that fact, hold a rubber ball in your hand and squeeze and release it a few dozen times.) That energy from the deformation is converted into thermal energy by friction in Earth's interior. That friction opposes and takes energy from the rotation of Earth, causing Earth to gradually slow. Earth's internal friction adds to the slowing caused by friction between Earth and its oceans as the planet rotates through the tidal bulge of the oceans. As a result, Earth's days are currently getting about 1.7 milliseconds (ms) longer every century. That sounds insignificant, but it adds up: when dinosaurs ruled 200 million years ago, the day was closer to 23 hours long, and 200 million years into the future, the day will be close to 25 hours.

Tidal Locking

Other solid bodies besides Earth experience tidal forces. For example, the Moon has no bodies of liquid to make tidal forces obvious, but its shape is distorted in the same manner as Earth's. Because of Earth's much greater mass and the Moon's smaller radius, Earth's tidal effects on the Moon are stronger than the Moon's tidal effects on Earth. Given that the average tidal deformation of Earth is about 30 cm, the average tidal deformation of the Moon should be about 6 meters. However, what we actually observe on the Moon is a tidal bulge of about 20 meters. That unexpectedly large displacement exists because the Moon's tidal bulge was "frozen" into its relatively rigid crust at an earlier time, when the Moon was closer to Earth and tidal forces were much stronger than they are today. Planetary scientists sometimes call that deformation the Moon's *fossil tidal bulge*.

Recall that the Moon's rotation period equals its orbital period (see Figure 2.22). That synchronous rotation of the Moon is a result of **tidal locking**. Early in the Moon's history, the period of its rotation was almost certainly different from its orbital period. As the Moon rotated through its extreme tidal bulge, however, friction within the Moon's crust was tremendous, rapidly slowing the Moon's rotation. After a fairly short time, the period of the Moon's rotation equaled the period of its orbit. When its orbital and rotation periods became equalized, the Moon no longer rotated with respect to its tidal bulge. Instead, the Moon and its tidal bulge rotated *together*, in lockstep with the Moon's orbit around Earth. As illustrated in **Figure 4.15**, that scenario continues today as the tidally distorted Moon orbits Earth, always keeping the same face and the long axis of its tidal bulge toward Earth.

Tidal forces affect not only the rotations of the Moon and Earth but also their orbits. Because of its tidal bulge, Earth is not perfectly spherical. Therefore, the material in Earth's tidal bulge on the side nearer the Moon pulls on the Moon more strongly than does material in the tidal bulge on the back side of Earth. Because the tidal bulge on the Moonward side of Earth "leads" the Moon

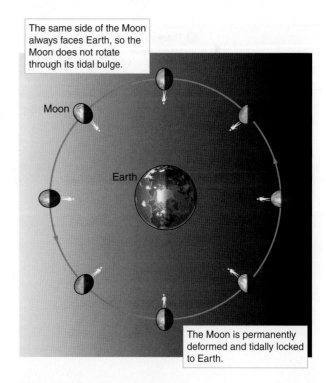

Figure 4.15 Tidal forces due to Earth's gravity lock the Moon's rotation to its orbital period.

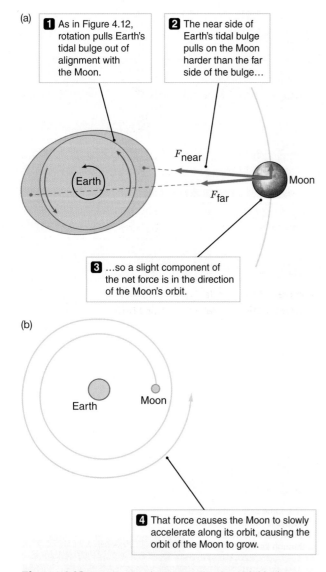

Figure 4.16 Interaction between Earth's tidal bulge and the Moon causes the Moon to accelerate in its orbit and the Moon's orbit to grow.

somewhat, as shown in **Figure 4.16**, the gravitational attraction of the bulge causes the Moon to accelerate slightly along the direction of its orbit around Earth. The rotation of Earth is dragging the Moon along with it. The acceleration of the Moon in the direction of its orbital motion causes the orbit of the Moon to grow larger. At present, the Moon is drifting away from Earth at a rate of 3.83 cm per year.

As the Moon grows more distant, the lunar month gets about 0.014 second longer each century. If that increase in the radius of the Moon's orbit were to continue long enough (about 50 billion years), Earth would become tidally locked to the Moon, just as the Moon is now tidally locked to Earth. At that point, the Earth's period of rotation, the Moon's period of rotation, and the Moon's orbital period would all be the same—about 47 of our current days—and the Moon would be about 43 percent farther from Earth than it is today. That situation will never actually occur, however, because the Sun will have burned itself out long before then.

The effects of tidal forces are apparent throughout the Solar System. Most of the moons in the Solar System are tidally locked to their parent planets, and for dwarf planet Pluto and its largest moon, Charon, each is tidally locked to the other.

Tidal locking is only one way that orbits and rotations can be coupled. Tidal forces have coupled the planet Mercury's rotation to its very elliptical orbit around the Sun. Unlike the Moon's synchronous rotation, however, Mercury spins on its axis three times for every two trips around the Sun. The period of Mercury's orbit—87.97 Earth days—is exactly 1 1/2 times the 58.64 days that Mercury takes to spin once on its axis. When Mercury comes to the point in its orbit that is closest to the Sun, one hemisphere faces the Sun, and then in the next orbit, the other hemisphere faces the Sun.

Tidal Forces on Many Scales

We normally think of the effects of tidal forces as small in comparison with the force of gravity holding an object together, yet tidal effects can be extremely destructive. Consider for a moment the fate of a small moon, asteroid, or comet that wanders too close to a massive planet such as Jupiter or Saturn. All objects in the Solar System larger than about a kilometer in diameter are held together by their self-gravity. However, the self-gravity of a small object such as an asteroid, a comet, or a small moon is feeble. In contrast, the tidal forces close to a massive object such as Jupiter can be very strong. If the tidal forces trying to tear an object apart become greater than the self-gravity trying to hold the object together, the object will break into pieces.

The **Roche limit** is the distance at which a planet's tidal forces become greater than the self-gravity of a smaller object—such as a moon, asteroid, or comet—causing the object to break apart. For a smaller object having the same density as the planet, the Roche limit is about 2.45 times the planet's radius. Such an object bound together solely by its own gravity can remain intact when it is outside a planet's Roche limit, but not when it is inside the limit. Objects such as the International Space Station and other Earth satellites are not torn apart, even though they orbit well within Earth's Roche limit, because chemical and physical bonds hold them together, not just self-gravity.

We have concentrated on the role that tidal forces play on Earth and the Moon, but tidal forces exist throughout the Solar System and the universe.

Anytime two objects of significant size or two collections of objects interact gravitationally, the gravitational forces will differ from one place to another within the objects, giving rise to tidal effects. Tidal disruption of small bodies is the source of the particles that make up the rings of the giant planets. Tidal interactions can cause material from one star in a binary pair to be pulled onto the other star. Tidal effects can strip stars from clusters consisting of thousands of stars. Galaxies can pass close enough together to strongly interact gravitationally. When that happens, as in **Figure 4.17**, tidal effects can grossly distort both galaxies. Tidal forces even play a role in shaping huge collections of galaxies—the largest known structures in the universe.

CHECK YOUR UNDERSTANDING 4.4
The Moon always keeps the same face toward Earth because of: (a) tidal locking; (b) tidal forces from the Sun; (c) tidal forces from the Earth and Sun.

Figure 4.17 The tidal "tails" seen here are characteristic of tidal interactions between galaxies.

Origins Tidal Forces and Life

Earth's rotation is slowing down as the Moon slowly moves away into a larger orbit, so in the distant past the Moon was closer and Earth rotated faster. In that time, tides would have been stronger and the interval between high tides would have been shorter. Tides are also affected by the configuration of the continents and oceans on Earth, so precisely how much faster Earth rotated billions of years ago is unknown. What is known, though, is that the stronger and more frequent tides would have provided additional energy to the oceans of the young Earth.

Scientists debate whether life on Earth originated deep in the ocean, on the surface of the ocean, or in hot springs on land (see Chapter 24). The tides shaped the regions in the margins between land and ocean, such as tide pools and coastal flats. Some researchers think that those border regions, which alternate between wet and dry with the tides, could have been places where concentrations of biochemicals periodically became high enough for more complex reactions to take place. Those complex reactions were important to the earliest life. Later, those border regions may have been important as advanced life moved from the sea to the land.

Elsewhere in the Solar System, the giant planets Jupiter and Saturn are far from the Sun and thus very cold. Both Jupiter and Saturn have many moons, and the closest moons would experience strong tidal forces from their respective planets. Several of those moons are thought to have a liquid ocean underneath an icy surface. Tidal forces from Jupiter or Saturn supply the heat to keep the water in a liquid state. Astrobiologists think that those subsurface liquid oceans are perhaps the most probable location for life elsewhere in the Solar System.

On Earth, the solar tides are about half as strong as lunar tides. A planet with a closer orbit would experience much stronger tidal forces from its star. Many planets detected outside the Solar System have orbits very close to their stars (see Chapter 7), so those exoplanets experience strong tidal forces. The planets might be tidally locked, too, so they have a synchronous rotation like the Moon does around Earth, with one side of the planet always facing the star and one side facing away. How might life on Earth have evolved differently if half the planet was in perpetual night and half in perpetual day?

READING ASTRONOMY NEWS

> Astronomers and physicists have wondered whether the value of Newton's gravitational constant, G, has changed over time. This press release reports on one experiment to measure the value of G.

Exploding Stars Prove Newton's Law of Gravity Unchanged over Cosmic Time

By Swinburne Media Centre

Australian astronomers have combined all observations of supernovae ever made to determine that the strength of gravity has remained unchanged over the last 9 billion years.

Newton's gravitational constant, known as G, describes the attractive force between two objects, together with the separation between them and their masses. It had previously been suggested that G could have been slowly changing over the 13.8 billion years since the Big Bang.

If G had been decreasing over time, for example, this would mean the Earth's distance to the Sun was slightly smaller in the past, meaning that we would experience longer seasons now compared to much earlier points in the Earth's history.

But researchers at Swinburne University of Technology in Melbourne have now analysed the light given off by 580 supernova explosions in the nearby and far universe and have shown that the strength of gravity has not changed.

"Looking back in cosmic time to find out how the laws of physics may have changed is not new," said Professor Jeremy Mould. "But supernova cosmology now allows us to do this with gravity."

A Type Ia supernova marks the violent death of a star called a white dwarf, which is as massive as our Sun but packed into a ball the size of our Earth.

Telescopes can detect the light from this explosion and use its brightness as a "standard candle" to measure distances in the universe, a tool that helped Australian astronomer Professor Brian Schmidt in his 2011 Nobel Prize–winning work discovering the mysterious force Dark Energy.

Professor Mould and his PhD student Syed Uddin at the Swinburne Centre for Astrophysics and Supercomputing and the ARC Centre of Excellence for All-sky Astrophysics (CAASTRO) assumed that these supernova explosions happen when a white dwarf reaches a critical mass or after colliding with other stars to "tip it over the edge."

"This critical mass depends on Newton's gravitational constant G and allows us to monitor it over billions of years of cosmic time—instead of only decades as was the case in previous studies," Professor Mould said.

Despite these vastly different time spans, their results agree with findings from the Lunar Laser Ranging Experiment that has been measuring the distance between the Earth and the Moon since NASA's *Apollo* missions in the 1960s and has been able to monitor possible variations in G at very high precision.

"Our cosmological analysis complements experimental efforts to describe and constrain the laws of physics in a new way and over cosmic time," Mr. Uddin said.

In their current publication, the Swinburne researchers were able to set an upper limit on the change in Newton's gravitational constant of 1 part in 10 billion per year over the past 9 billion years.

---

1. Explain how observing distant supernovae means observing back in time.
2. Why are physicists still measuring G, centuries after Newton described that constant?
3. What quantities that you studied here in Chapter 4 would change if the value of G changed over time?
4. Why is it important to the process of science that this result agrees with a result from a totally different experiment?
5. If the average distance to the Sun were slightly less in the past, would solar tides have been weaker or stronger? Would you expect that change to have been as important to tides on Earth as changes in the distance to the Moon?

Source: "Exploding stars prove Newton's law of gravity unchanged over cosmic time," Swinburne University of Technology Media Center News Release, March 24, 2014. Reprinted by permission of Swinburne University of Technology.

Summary

Objects stay in orbit because of gravity. Newton's laws of motion and his proposed law of gravity predict the paths of planetary orbits and explain Kepler's laws. Newton's calculations showed that those orbits should be ellipses with the Sun at one focus, that planets should move faster when closer to the Sun, and that the square of the period of a planet's orbit should vary as the cube of the semimajor axis of that elliptical orbit. Newton also mathematically confirmed Galileo's conclusion that falling objects have an accelerated motion independent of their mass. Tidal forces provide energy to Earth's oceans. Tide pools on Earth may have been a site of early biochemical reactions. Some moons in the outer Solar System might have liquid water because of tidal heating from their respective planets.

LG 1 **Explain the elements of Newton's universal law of gravitation.** Gravity is a force between any two objects with mass. Gravity is one of the fundamental forces of nature—it binds the universe together. The force of gravity is proportional to the mass of each object and inversely proportional to the square of the distance between them.

LG 2 **Use the laws of motion and gravitation to explain planetary orbits.** An orbit is one body "falling around" another. Planets orbit the Sun in elliptical orbits. All objects affected by gravity move either in bound elliptical orbits or unbound paths. Orbits are ultimately given their shape by the gravitational attraction of the objects involved, which in turn reflects the masses of those objects.

LG 3 **Explain how tidal forces from the Sun and Moon create Earth's tides.** Tides on Earth are the result of differences between how hard the Moon and Sun pull on one part of Earth in comparison with their pull on other parts of Earth. The primary cause of tides is the Moon, which stretches out Earth. The tides are the strongest when the Sun, Moon, and Earth are aligned. As Earth rotates, tides rise and fall twice each day.

LG 4 **Describe how tidal forces affect solid bodies.** Tidal forces lock the Moon's rotation to its orbit around Earth. Tidal forces can break up an object if its gets too close to a more massive object. Tidal forces are observed throughout the universe—in planets and moons, pairs of stars, and interacting galaxies.

? Unanswered Question

- What range of gravities will support human life? Humans have evolved to live on Earth's surface, but what happens when humans go elsewhere? What are the limits for our hearts, lungs, eyes, and bones? At the higher end of human tolerance, fighter pilots have been trained to experience about 10 times the normal surface gravity on Earth for very short periods (too long and they black out). Astronauts who spend several months in near-weightless conditions experience medical problems such as bone loss. On the Moon or Mars, humans will weigh much less than on Earth. Many science fiction tales have been written about what happens to children born on a space station or on another planet or moon with low surface gravity: would their hearts and bodies ever be able to adjust to the higher surface gravity of Earth, or must they stay in space forever?

Questions and Problems

Test Your Understanding

1. In Newton's universal law of gravitation, the force is
   a. proportional to both masses.
   b. proportional to the radius.
   c. proportional to the radius squared.
   d. inversely proportional to the orbiting mass.

2. If we wanted to increase a satellite's altitude above Earth and keep it in a stable orbit, we also would need to
   a. increase its orbital speed.
   b. increase its weight.
   c. decrease its weight.
   d. decrease its orbital speed.

3. An object in a(n) _____ orbit in the Solar System will remain in its orbit forever. An object in a(n) _____ orbit will escape from the Solar System.
   a. unbound; bound
   b. circular; elliptical
   c. bound; unbound
   d. elliptical; circular

4. Compared with your mass on Earth, your mass on the Moon would be
   a. lower because the Moon is smaller than Earth.
   b. lower because the Moon has less mass than Earth.
   c. higher because of the combination of the Moon's mass and size.
   d. the same—mass doesn't change.

5. If you went to Mars, your weight would be
   a. higher because you are closer to the center of the planet.
   b. lower because Mars has two small moons instead of one big moon, so less tidal force is present.
   c. lower because the lower mass and smaller radius of Mars combine to a lower surface gravity.
   d. the same as on Earth.

6. Venus has about 80 percent of Earth's mass and about 95 percent of Earth's radius. Your weight on Venus will be
   a. 20 percent more than on Earth.
   b. 20 percent less than on Earth.
   c. 10 percent more than on Earth.
   d. 10 percent less than on Earth.

7. The connection between gravity and orbits enables astronomers to measure the _____ of stars and planets.
   a. distances
   b. sizes
   c. masses
   d. compositions

8. If the Moon were twice as massive, how would the strength of lunar tides change?
   a. The highs would be higher, and the lows would be lower.
   b. Both the highs and the lows would be higher.
   c. The highs would be lower, and the lows would be higher.
   d. Nothing would change.

9. If Earth had half its current radius, how would the strength of lunar tides change?
   a. The highs would be higher, and the lows would be lower.
   b. Both the highs and the lows would be higher.
   c. The highs would be lower, and the lows would be higher.
   d. Both the highs and the lows would be lower.

10. If the Moon were 2 times closer to Earth than it is now, the gravitational force between Earth and the Moon would be
    a. 2 times stronger.
    b. 4 times stronger.
    c. 8 times stronger.
    d. 16 times stronger.

11. If the Moon were 2 times closer to Earth than it is now, the tides would be
    a. 2 times stronger.
    b. 4 times stronger.
    c. 8 times stronger.
    d. 16 times stronger.

12. If two objects are tidally locked to each other,
    a. the tides always stay on the same place on each object.
    b. the objects always remain in the same place in each other's sky.
    c. the objects are falling together.
    d. both a and b

13. Spring tides occur only when
    a. the Sun is near the vernal equinox.
    b. the Moon's phase is new or full.
    c. the Moon's phase is first quarter or third quarter.
    d. it is either spring or fall.

14. If an object crosses from farther to closer than the Roche limit, it
    a. can no longer be seen.
    b. begins to accelerate very quickly.
    c. slows down.
    d. may be torn apart.

15. Self-gravity is
    a. the gravitational pull of a person.
    b. the force that holds objects such as people and lamps together.
    c. the gravitational interaction of all the parts of a body.
    d. the force that holds objects on Earth.

Thinking about the Concepts

16. Both Kepler's laws and Newton's laws tell us something about the motion of the planets, but they have a fundamental difference. What is that difference?

17. Explain the difference between circular velocity and escape velocity. Which of those must be larger? Why?

18. Explain the difference between weight and mass.

19. Weight on Earth is proportional to mass. On the Moon, too, weight is proportional to mass, but the constant of proportionality is different on the Moon from what it is on Earth. Why? Explain why that difference does not violate the universality of physical law, as described in the Process of Science Figure.

20. Two comets are leaving the vicinity of the Sun, one traveling in an elliptical orbit and the other in an unbound orbit. What can you say about the future of those two comets? Would you expect either of them to return eventually?

21. What is the advantage of launching satellites from spaceports located near the equator? Would you expect satellites to be launched to the east or to the west? Why?

22. Explain how to use celestial orbits to estimate an object's mass. What observational quantities do you need to make that estimation?

23. Suppose astronomers discovered an object approaching the Sun in an unbound orbit. What would that say about the origin of the object?

24. What determines the strength of gravity at various radii between Earth's center and its surface?

25. The best time to dig for clams along the seashore is when the ocean tide is at its lowest. What phases of the Moon would be best for clam digging? What would be the best times of day during those phases?

26. The Moon is on the meridian at your seaside home, but your tide calendar does not show that it is high tide. What might explain that apparent discrepancy?

27. We may have an intuitive feeling for why lunar tides raise sea level on the side of Earth facing the Moon, but why is sea level also raised on the side facing away from the Moon?

28. Tides raise and lower the level of Earth's oceans. Can they do the same for Earth's landmasses? Explain your answer.

29. Lunar tides raise the ocean surface by less than 1 meter. How can tides as large as 5–10 meters occur?

30. Most commercial satellites are well inside the Roche limit as they orbit Earth. Why are they not torn apart?

Applying the Concepts

31. Mars has about one-tenth the mass of Earth and about half of Earth's radius. What is the value of gravitational acceleration on the surface of Mars in comparison with that on Earth? Compare your estimated mass and weight on Mars with your mass and weight on Earth. Do Hollywood movies showing people on Mars accurately portray that difference in weight?

32. Earth speeds along at 29.8 km/s in its orbit. Neptune's nearly circular orbit has a radius of 4.5×10^9 km, and the planet takes 164.8 years to make one trip around the Sun. Calculate how fast Neptune moves along in its orbit.

33. Venus's circular velocity is 35.03 km/s, and its orbital radius is 1.082×10^8 km. Use that information to calculate the mass of the Sun.

34. At the surface of Earth, the escape velocity is 11.2 km/s. What would be the escape velocity at the surface of a very small asteroid having a radius 10^{-4} that of Earth and a mass 10^{-12} that of Earth?

35. How long does Newton's cannonball, moving at 7.9 km/s just above Earth's surface, take to complete one orbit around Earth?

36. When a spacecraft is sent to Mars, it is first launched into an Earth orbit with circular velocity.
    a. Describe the shape of that orbit.
    b. What minimum velocity must we give the spacecraft to send it on its way to Mars?

37. Earth's average radius is 6,370 km and its mass is 5.97×10^{24} kg. Show that the acceleration of gravity at the surface of Earth is 9.81 m/s^2.

38. Using 6,370 km for Earth's radius, compare the gravitational force acting on a NASA rocket when it is sitting on its launchpad with the gravitational force acting on it when it is orbiting 350 km above Earth's surface.

39. The International Space Station travels on a nearly circular orbit 350 km above Earth's surface. What is its orbital speed?

40. Using the last equation in Working It Out 4.1, calculate the value of g on Earth. Compare that result with the measured value.

41. As described in Working It Out 4.4, tidal force is proportional to the masses of the two objects and is inversely proportional to the cube of the distance between them. Some astrologers claim that your destiny is determined by the "influence" of the planets that are rising above the horizon at the moment of your birth. Compare the tidal force of Jupiter (mass, 1.9×10^{27} kg; distance, 7.8×10^8 km) with that of the doctor in attendance at your birth (mass, 80 kg; distance, 1 meter).

42. The asteroid Ida (mass, 4.2×10^{16} kg) is attended by a tiny asteroidal moon, Dactyl, which orbits Ida at an average distance of 90 km. If you ignore the mass of the tiny moon, what is Dactyl's orbital period in hours?

43. Suppose you go skydiving.
    a. Just as you fall out of the airplane, what is your gravitational acceleration?
    b. Would that acceleration be bigger, smaller, or the same if you were strapped to a flight instructor and so had twice the mass?
    c. Just as you fall out of the airplane, what is the gravitational force on you? (Assume that your mass is 70 kg.)
    d. Would that force be bigger, smaller, or the same if you were strapped to a flight instructor and so had twice the mass?

44. Assume that a planet just like Earth is orbiting the bright star Vega at a distance of 1 astronomical unit (AU). Vega has twice the mass of the Sun.
    a. How long in Earth years will that planet take to complete one orbit around Vega?
    b. How fast is the Earth-like planet traveling in its orbit around Vega?

45. Suppose that in the past the Moon was 80 percent of the distance from Earth that it is now. Calculate how much stronger the lunar tides would have been. How would the neap and spring tides be different from now?

USING THE WEB

46. Go to NASA's "Apollo 15 Hammer-Feather Drop" Web page (https://nssdc.gsfc.nasa.gov/planetary/lunar/apollo_15_feather_drop.html or https://youtube.com/watch?v=oYEgdZ3iEKA) and watch the video from *Apollo 15* of astronaut David Scott dropping the hammer and falcon feather on the Moon. What did that experiment show? What would happen if you tried that on Earth with a feather and a hammer? Would it work? Suppose instead you dropped the hammer and a big nail. How would they fall? How does the acceleration of falling objects on the Moon compare with the acceleration of falling objects on Earth?

47. Go to the Exploratorium's "Your Weight on Other Worlds" Web page (http://exploratorium.edu/ronh/weight), which will calculate your weight on other planets and moons in our Solar System. On which objects would your weight be higher than it is on Earth? What difficulties would human bodies have in a higher-gravity environment? For example, could you easily get up out of bed and walk? What are the possible short-term and long-term effects of lower gravity on the human body? What types of life on Earth might adapt well to a different gravity?

48. a. Watch the first 12 minutes of the episode "Gravity" in season 2 of the series *The Universe* (http://history.com/shows/the-universe/season-2/episode-17 or http://dailymotion.com/video/x1p90qs or on YouTube) to see several illustrated examples of gravity. Why does the tennis ball appear to float when dropped at the top of the roller coaster? How could Newton imagine a satellite orbiting Earth centuries before it was possible? Why does the speed of the cannonball determine whether it goes into orbit? What was the technical difficulty in launching a satellite?
    b. In the same video as in part (a), watch the trip on the zero-G plane (the "vomit comet"), at 23–29 minutes. How does the plane simulate zero-G? Why does it last for only 20–30 seconds? How is that similar to the roller coaster in part (a)?

49. Go to a website that will show you the times for high and low tides—for example, http://saltwatertides.com. Pick a location and bring up the tide table for today and the next 14 days. Why do two high tides and two low tides occur every day? What is the difference in the height of the water between high and low tides? In the last few columns of the table, the times of moonrise and moonset are indicated, as well as the percentage of lunar illumination. Does the time of the high tide lead or follow the highest position of the Moon in the sky? Compare with Figure 4.12c: what phases of the Moon have the greatest differences in the height of high and low tides?

50. Figure 4.17 shows two galaxies pulling at each other, most likely after they have already collided. Watch "The Mice—Interacting Galaxies" on YouTube (https://youtube.com/watch?v=pR4pzthwqkA) to see a simulation of that interaction. Galaxies are not solid objects: they contain stars and gas and dust and a lot of empty space. Explain how those tidal tails can result from that type of interaction.

EXPLORATION

Newton's Laws

digital.wwnorton.com/astro6

In the Exploration of Chapter 3, we used the "Planetary Orbits" Interactive Simulation to explore Kepler's laws for Mercury. Now that we know how Newton's laws explain why Kepler's laws describe orbits, we will revisit the simulator to explore the Newtonian features of Mercury's orbit. Visit the Digital Resources Page and on the Student Site open the "Orbits" Interactive Simulation in Chapter 3.

Acceleration

To begin exploring the simulation, set parameters for "Mercury" in the "Orbit Settings" panel and then click "OK." Click the "Newtonian Features" tab at the bottom of the control panel. Select "show solar system orbits" and "show grid" under "Visualization Options." Change the animation rate to 0.01, and select the "start animation" button.

Examine the graph at the bottom of the panel.

1 Where is Mercury in its orbit when the acceleration is smallest?

2 Where is Mercury in its orbit when the acceleration is largest?

3 What are the values of the largest and smallest accelerations?

In the "Newtonian Features" graph, mark the boxes for "vector" and "line" that correspond to the acceleration. Specifying those parameters will insert an arrow that shows the direction of the acceleration and a line that extends the arrow.

4 To what Solar System object does the arrow point?

5 In what direction is the force on the planet?

Velocity

Examine the graph at the bottom of the panel again.

6 Where is Mercury in its orbit when the velocity is smallest?

7 Where is Mercury in its orbit when the velocity is largest?

8 What are the values of the largest and smallest velocities?

Add the velocity vector and line to the simulation by clicking on the boxes in the graph window. Study the resulting arrows carefully.

9 Are the velocity and the acceleration always perpendicular (is the angle between them always 90°)?

10 If the orbit were a perfect circle, what would be the angle between the velocity and the acceleration?

Hypothetical Planet

In the "Orbit Settings" panel, change the semimajor axis to 0.8 AU.

11 How does that imaginary planet's orbital period now compare with Mercury's?

Now change the semimajor axis to 0.1 AU.

12 How does the planet's orbital period now compare with Mercury's?

13 Summarize your observations of the relationship between the speed of an orbiting object and the semimajor axis.

5 Light

Our knowledge of the universe beyond Earth comes from light emitted, absorbed, or reflected by astronomical objects. Light carries information about the temperature, composition, and speed of the objects. Light also tells us about the nature of the material that the light passed through on its way to Earth. Yet light plays a far larger role in astronomy than that of being a messenger. Light is one of the primary means by which energy is transported throughout the universe. Stars, planets, and vast clouds of gas and dust filling the space between the stars heat up as they absorb light and cool off as they emit light. Light carries energy generated in the heart of a star outward through the star and off into space. Light transports energy from the Sun outward through the Solar System, heating the planets; and light carries energy away from each planet, allowing each one to cool. The balance between those two processes establishes each planet's temperature and therefore a planet's possible suitability for life.

LEARNING GOALS

An astronomer must try to understand the universe by studying the light and other particles that reach Earth from distant objects. By the end of this chapter, you should be able to:

LG 1 Describe the wave and particle properties of light, and describe the electromagnetic spectrum.

LG 2 Describe how to measure the chemical composition of distant objects by using the unique spectral lines of different types of atoms.

LG 3 Describe the Doppler effect and how it can be used to measure the motion of distant objects.

LG 4 Explain how the spectrum of light that an object emits depends on its temperature.

LG 5 Differentiate luminosity from brightness, and illustrate how distance affects each.

The visible part of the electromagnetic spectrum is laid out in the colors of this rainbow. ▶▶▶

What is light?

5.1 Light Brings Us the News of the Universe

Is light a **wave**—a disturbance that travels from one point to another—or is it made up of particles? Scientists have come to understand that light sometimes acts like a wave and sometimes acts like a particle. We begin with a discussion of how fast light travels. We then discuss its wavelike and particle-like properties.

The Speed of Light

In the 1670s, Danish astronomer Ole Rømer (1644–1710) studied the movement of the moons of Jupiter, measuring the times when each moon disappeared behind the planet. To his amazement, the observed times did not follow the regular schedule that he predicted from Kepler's laws. Sometimes the moons disappeared behind Jupiter sooner than expected, whereas at other times they disappeared behind Jupiter later than expected. Rømer realized that the difference depended on where Earth was in its orbit. If he began tracking the moons when Earth was closest to Jupiter, the moons were almost 17 minutes "late" by the time Earth was farthest from Jupiter. When Earth was once again closest to Jupiter, the moons again passed behind Jupiter at the predicted times.

Rømer correctly concluded that his observations did not represent a failure of Kepler's laws. Instead, he was seeing the first clear evidence that light travels at a finite speed. As shown in **Figure 5.1**, the moons appeared "late" when Earth was farther from Jupiter because of the time needed for light to travel the extra distance between the two planets. Over the course of Earth's yearly trip around the Sun, the distance between Earth and Jupiter changes by 2 astronomical units (AU). The speed of light equals that distance divided by Rømer's 16.7-minute delay, or about 3×10^5 kilometers per second (km/s). The value that Rømer actually announced in 1676 was a bit on the low side—2.25×10^5 km/s—because the size of Earth's orbit was not well known. Modern measurements of the speed of light give a value of 2.99792458×10^5 km/s in a **vacuum** (a region of space devoid of matter). The speed of light in a vacuum is one of nature's fundamental constants, usually written as c (lowercase). The speed of light through any medium, such as air or glass, is always less than c.

The International Space Station moves around Earth at a speed of about 28,000 kilometers per hour (km/h), taking 91 minutes to complete one orbit. Light travels almost 40,000 times faster than that and can circle Earth in only

Figure 5.1 Danish astronomer Ole Rømer realized that apparent differences between the predicted and observed orbital motions of Jupiter's moons depend on the distance between Earth and Jupiter. He used those observations to measure the speed of light. (The superscript letters in "$16^m\ 40^s$" stand for minutes and seconds of time, respectively.)

1/7 of a second. Because light is so fast, its travel time is a convenient way to express cosmic distances. Light takes 1 1/4 seconds to travel between Earth and the Moon, so we say that the Moon is 1 1/4 light-*seconds* from Earth. The Sun is 8 1/3 light-*minutes* away, and the next-nearest star is 4 1/3 light-*years* distant. Thus, a *light-year* is defined as the distance light travels in 1 year, or about 9.5 trillion km. The light-year is a measure of distance.

While traveling at that high speed, light carries energy from place to place. **Energy** is the ability to do work, and it comes in many forms. **Kinetic energy** is the energy of moving objects. **Thermal energy** is closely related to kinetic energy and is the sum of all the random motion of atoms, molecules, and particles, by which we measure their temperature. For example, when light from the Sun strikes the pavement, the pavement heats up. Light carried that energy from the Sun to the pavement. Rømer knew how long light took to travel a given distance, but physicists would take more than 200 years to figure out what light actually is.

Figure 5.2 An electromagnetic wave consists of oscillating electric and magnetic fields that are perpendicular both to each other and to the direction in which the wave travels.

Light as an Electromagnetic Wave

In the late 19th century, the Scottish physicist James Clerk Maxwell (1831–1879) introduced the concept that electricity and magnetism are two components of the same physical phenomenon. An **electric force** is the push or pull between electrically charged particles that make up atoms, such as protons and electrons, arising from their electric charges. Particles with opposite charges attract, and those with like charges repel. A **magnetic force** is a force between electrically charged particles arising from their motion.

To describe those electric and magnetic forces, Maxwell considered what happens when charged particles move in *electric fields* and *magnetic fields*. An **electric field** is a measure of the electric force on a charge at any point in space. Similarly, a **magnetic field** is a measure of the magnetic force acting on a small magnet at any point in space.

Maxwell summarized the behavior of electric fields and magnetic fields in four elegant equations. Among other things, those equations say that a changing electric field causes a magnetic field and that a changing magnetic field causes an electric field. A change in the motion of a charged particle causes a changing electric field, which causes a changing magnetic field, which causes a changing electric field, and so on. You can see that interaction in **Figure 5.2**. Once the process starts, a self-sustaining procession of oscillating electric and magnetic fields moves out in all directions through space. In other words, an accelerating charged particle gives rise to an **electromagnetic wave**. Those electromagnetic waves, and the accelerating charges that generate them, are the sources of electromagnetic radiation. Maxwell's equations also predict the speed at which an electromagnetic wave should travel, which agrees with the measured speed of light (*c*).

Maxwell's wave description of light also suggests how light originates and how it interacts with matter. When a drop of water falls from the faucet into a sink full of water, it causes a disturbance, or wave, like the one shown in **Figure 5.3a**. The wave moves outward as a ripple on the surface of the water. **Figure 5.3b** shows that electromagnetic waves resulting from periodic changes in the strength of the electric and magnetic fields move out through space away from their source in much the same way. The ripples in the sink, however, are distortions of the water's surface, and they require a **medium**: a substance to travel through. Light waves move through empty space—what we call a vacuum—in the absence of a medium.

Figure 5.3 (a) A drop falling into water generates waves that move outward across the water's surface. (b) Similarly, an oscillating (accelerated) electric charge generates electromagnetic waves that move away at the speed of light.

Figure 5.4 (a) When waves moving across the surface of water reach a cork, they cause the cork to bob up and down. (b) Similarly, a passing electromagnetic wave causes an electric charge to oscillate in response to the wave.

Now imagine that a cork is floating in the sink (**Figure 5.4a**). The cork remains stationary until the ripple from the dripping faucet reaches it. As the ripple passes by, the rising and falling water causes the cork to rise and fall. That can happen only if the wave is carrying energy—a conserved quantity that gives objects and particles the ability to do work. Light waves similarly carry energy through space and cause electrically charged particles to vibrate, as in **Figure 5.4b**.

Characterizing Waves

We discuss several kinds of waves in this book, including electromagnetic waves crossing the vast expanse of the universe and seismic waves traveling through Earth. Waves are generally characterized by four quantities: *amplitude*, *speed*, *frequency*, and *wavelength*. Each of those quantities is illustrated in **Figure 5.5**. The **amplitude** of a wave is the height of the wave above the undisturbed position (Figure 5.5a). For water waves, the amplitude is how far the water is lifted up by the wave. For light waves, the amplitude is related to the brightness of the light. A water wave travels at a particular speed, v (Figure 5.5b), through the water. The water itself doesn't travel; it just moves up and down at the same location. For waves such as those in water, that speed varies and depends on the density of the substance the wave moves through, among other things. Light, in contrast, always moves through a vacuum at the same speed, $c \approx 300,000$ km/s.

The distance from one crest of a wave to the next is the **wavelength**, usually denoted by the Greek letter lambda, λ (Figure 5.5c). The number of wave crests passing a point in space each second is called the wave's **frequency**, f. The unit of frequency is cycles per second, which is called **hertz** (**Hz**) after the 19th-century physicist Heinrich Hertz (1857–1894), who was the first to experimentally confirm Maxwell's predictions about electromagnetic radiation.

Figures 5.5c and 5.5d show that waves with longer (larger) wavelengths have lower (smaller) frequencies, whereas waves with shorter (smaller) wavelengths have higher (larger) frequencies. Higher-frequency waves carry more energy. Imagine standing on an ocean beach with the waves lapping up against you: the amount of energy you feel from the waves increases as the frequency of the ocean waves increases.

Waves travel a distance of one wavelength each cycle, so the speed of a wave can be found by multiplying the frequency and the wavelength. Translating that idea into math, we have $v = \lambda f$. The speed of light in a vacuum is always c, so once the wavelength of a wave of light is known, its frequency is known, and vice versa. Because light travels at constant speed, its wavelength and frequency are inversely proportional to each other: if the wavelength increases, the frequency decreases. Waves can carry a tremendous amount of information, such as complex and beautiful music, which travels by sound waves. As you continue your study of the universe, time and time again you will find that the information we receive, whether about the interior of Earth or about a distant star or galaxy, rides in on a wave.

The Electromagnetic Spectrum

Most light signals are made up of many wavelengths. A rainbow, such as the one shown in the chapter-opening figure, is created when white light interacts

Figure 5.5 A wave is characterized by the distance from one peak to the next (wavelength, λ), the frequency of the peaks (f), the maximum deviations from the medium's undisturbed state (amplitude), and the velocity (v) at which the wave pattern travels from one place to another. In an electromagnetic wave, the amplitude is the maximum strength of the electric field, and the speed of light is written as c.

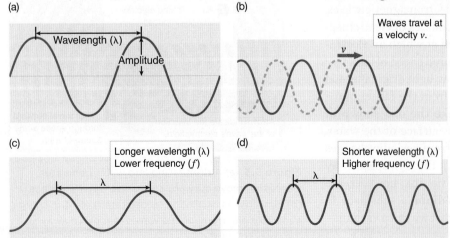

with water droplets and is spread out into its component colors. Light spread out by wavelength is called a **spectrum**. At the long-wavelength (and therefore low-frequency) end of the visible spectrum is red light. At the short-wavelength (high-frequency) end is violet light. A commonly used unit for the wavelength of visible light is the **nanometer** (**nm**). A nanometer is one-billionth (10^{-9}) of a meter. Human eyes can see light between violet (about 380 nm) and red (750 nm). Stretched out between the two, in a rainbow, is the rest of the visible spectrum.

The light-sensitive cells in our eyes respond to visible light, but that is only a small sample of the range of possible wavelengths for electromagnetic radiation. Light can have wavelengths much shorter or much longer than your eyes can perceive. The whole range of wavelengths of light—collectively called the **electromagnetic spectrum**—is shown in **Figure 5.6**. Most of the electromagnetic spectrum—and therefore most of the information in the universe—is invisible to the human eye. To detect light outside the visible range, we must use specialized detectors of various kinds, as we explain in Chapter 6.

Refer to Figure 5.6 as we tour the electromagnetic spectrum, beginning with the shortest wavelengths and working our way to the longest ones. The very shortest wavelengths of light are called **gamma rays**, or sometimes gamma radiation. Because that light has the shortest wavelengths, it has the highest frequency and the highest energy, so it penetrates matter easily. Wavelengths between 0.1 and 40 nm are called **X-rays**. You have probably encountered X-rays at the dentist's office or in a hospital's emergency room—X-ray light has enough energy to penetrate skin and muscle but is stopped by denser bone. **Ultraviolet** (**UV**) **radiation** has wavelengths between 40 and about 380 nm—longer than X-rays but shorter than visible light. You are familiar with that type of light from sunburns: UV light has enough energy to penetrate your skin, but not much deeper.

Infrared (**IR**) **radiation** has longer wavelengths than the reddest wavelengths in the visible range. You are familiar with infrared radiation because you often feel it as heat. When you hold your hand next to a hot stove, some of the heat you feel is carried to your hand by infrared radiation emitted from the stove. In that sense, you could think of your skin as being a giant infrared eyeball—it is

Figure 5.6 By convention, the electromagnetic spectrum is divided into loosely defined regions ranging from gamma rays to radio waves. Throughout the book, we use the following labels to indicate the form of radiation used to produce astronomical images, with an icon to remind you: G = gamma rays; X = X-rays; U = ultraviolet; V = visible; I = infrared; R = radio. If more than one region is represented, multiple labels are highlighted.

sensitive to infrared wavelengths. Infrared radiation also is used in television remote controls, and night vision goggles detect infrared radiation from warm objects such as animals. A useful unit for infrared light is the **micron** (μm, where μ is the Greek letter mu). One micron is 1,000 nm, or one-millionth (10^{-6}) of a meter. Infrared wavelengths are longer than red light and shorter than 500 microns.

Microwave radiation has even longer wavelengths than those of infrared radiation. The microwave in your kitchen heats the water in food by using light of those wavelengths. The longest-wavelength light, which has wavelengths longer than a few centimeters, is called **radio waves**. Light of those wavelengths in the form of FM, AM, television, and cell phone signals is used to transmit information.

CHECK YOUR UNDERSTANDING 5.1A

Rank the following in order from longest wavelength to shortest wavelength: (a) gamma rays; (b) visible light; (c) infrared light; (d) ultraviolet light; (e) radio waves

Light as a Particle

Although the wave theory of light describes many observations, it does not provide a complete picture of the properties of light. Many of the difficulties with the wave model of light have to do with how light interacts with small particles such as atoms and molecules.

The work of Albert Einstein and other scientists modified our understanding of light to show that light both acts like a wave and acts like a particle. In 1905, Einstein explained the *photoelectric effect*, in which subatomic particles called electrons are emitted when surfaces are illuminated by electromagnetic radiation greater than a certain frequency. He showed that the rate at which electrons are emitted depends only on the amount of incoming light and that the speed of the electrons depends only on the frequency of the incoming radiation. That work earned Einstein the Nobel Prize in Physics in 1921. In the particle model, light is made up of massless particles called **photons** (*phot-* means "light," as in *photograph*; and *-on* signifies a particle). Photons always travel at the speed of light (**Process of Science Figure**), and they carry energy. A photon is a *quantum* of light; *quantized* means that something is subdivided into individual units. **Quantum mechanics** is a branch of physics that deals with the quantization of energy and of other properties of matter.

The energy of a photon and the frequency of the electromagnetic wave are directly proportional to each other: the higher the frequency of the light, the greater the energy each photon carries. That relationship connects the particle and the wave concepts of light. For example, photons with higher frequencies carry more energy than that carried by photons with lower frequencies. The constant of proportionality between the energy (E) and the frequency (f) is called Planck's constant (h), which is equal to 6.63×10^{-34} joule-seconds (a joule is a unit of energy):

$$E = hf$$

Because the wavelength (λ) and frequency (f) of electromagnetic waves are inversely proportional, that also means that the photon energy is inversely proportional to the wavelength: $E = hc/\lambda$. In visible light, high-energy light is blue and low-energy light is red. **Working It Out 5.1** explores the relationships among the wavelength, frequency, and energy of light.

PROCESS OF SCIENCE

Agreement between Fields

Scientists working on very different problems in different fields all found the same results: light has a speed that can be measured.

Ole Rømer (1644–1710) studied eclipses of Jupiter's moons.
Rømer calculated the speed of light from the eclipse delays of Jupiter's moons.

In 1727, James Bradley (1693–1762) studied the apparent motions of stars, which appeared to make small circles because of the relative motion of Earth.
A half century after Rømer's measurement, Bradley's motion studies led to a more accurate measurement of the speed of light.

James Clerk Maxwell (1831–1879) studied electricity and magnetism.
Maxwell determined that the speed of electromagnetic waves must be a constant, and is equal to the speed of light.

$$c = 1/(\sqrt{\mu_0 \varepsilon_0})$$

Einstein (1879–1955) explored the consequences of the fact that the speed of light is the same for all observers.
Predictions from this theory have been tested repeatedly since 1905.

$$t' = \frac{t}{\sqrt{1 - \frac{v^2}{c^2}}}$$

Astronomers and physicists converged on an understanding that photons travel at the "speed of light"—a fundamental constant of the universe that is the same for all observers.

5.1 Working It Out Working with Electromagnetic Radiation

Wavelength and Frequency

When you tune to a radio station to, say, 770 AM, you are receiving an electromagnetic signal that travels at the speed of light and is broadcast at a frequency of 770 kilohertz (kHz), or 7.7×10^5 Hz. We can use the relationship between wavelength and frequency, $c = \lambda f$, to calculate the wavelength of the AM signal:

$$\lambda = \frac{c}{f} = \left(\frac{3 \times 10^8 \text{ m/s}}{7.7 \times 10^5/\text{s}}\right) = 390 \text{ m}$$

FM wavelengths are much shorter than AM wavelengths. For example, a station at 89.5 FM broadcasts signals with a wavelength of 3.4 m.

The human eye is most sensitive to light in green and yellow wavelengths, about 500–590 nm. If we examine green light with a wavelength of 530 nm, we can compute its frequency:

$$f = \frac{c}{\lambda} = \left(\frac{3 \times 10^8 \text{ m/s}}{530 \times 10^{-9} \text{ m}}\right) = 5.66 \times 10^{14}/\text{s} = 5.66 \times 10^{14} \text{ Hz}$$

That frequency corresponds to 566 *trillion* wave crests passing by each second.

Photon Energy

How does the energy of an X-ray photon with a wavelength of 1 nm compare with the energy of a visible-light photon with a wavelength of 530 nm (the same green light used in the previous calculation)? The equation for the energy of a photon is $E = hf$. Because $f = c/\lambda$, substituting c/λ for f yields the inverse relationship, $E = hc/\lambda$. Because we are making a *comparison*, we can take a ratio, and then the constants h and c cancel out:

$$\frac{E_{\text{X-ray photon}}}{E_{\text{visible photon}}} = \frac{hc/\lambda_{1\text{ nm}}}{hc/\lambda_{530\text{ nm}}} = \frac{h\cancel{c}}{h\cancel{c}} \times \frac{\lambda_{500\text{ nm}}}{\lambda_{1\text{ nm}}} = \frac{530 \text{ nm}}{1 \text{ nm}} = 530$$

The X-ray photon has 530 times the energy of the visible-light photon.

In the particle description of light, the electromagnetic spectrum is a spectrum of photon energies. The higher the frequency of the electromagnetic wave, the greater the energy carried by each photon. Photons of shorter wavelength (higher frequency) carry more energy than that carried by photons of longer wavelength (lower frequency). For example, photons of blue light carry more energy than that carried by photons of longer-wavelength red light. Ultraviolet photons carry more energy than that carried by photons of visible light, and X-ray photons carry more energy than that carried by ultraviolet photons. The lowest-energy photons are radio wave photons. A beam of red light can carry just as much energy as a beam of blue light, but the red beam will have more photons than the blue beam.

AstroTour: Light as a Wave, Light as a Photon

CHECK YOUR UNDERSTANDING 5.1B

As wavelength increases, the energy of a photon _____ and its frequency _____. (a) increases; decreases (b) increases; increases (c) decreases; decreases (d) decreases; increases

5.2 The Quantum View of Matter Explains Spectral Lines

Light and matter interact, and that interaction allows us to detect matter even at great distances in space. To understand that interaction, we must first understand the building blocks of matter. Here in Section 5.2, we review atomic structure and the process by which astronomers identify the chemical elements in astronomical objects.

Atomic Structure

Matter is anything that occupies space and has mass. Atoms are composed of a central massive **nucleus**, which contains **protons** with a positive charge and **neutrons**, which have no charge. A cloud of negatively charged **electrons** surrounds the nucleus. Atoms with the same number of protons are all of the same type of chemical **element**. For example, an atom with two protons, shown in **Figure 5.7a**, is the element helium. An atom with six protons is the element carbon, one with eight protons is the element oxygen, and so forth. An element may have many **isotopes**: atoms with the same number of protons but different numbers of neutrons. **Molecules** are groups of atoms bound together by shared electrons. A single teaspoon of water contains about 10^{23} atoms—about as many atoms as the number of stars in the observable universe.

For an atom to be electrically neutral, it must have the same number of electrons as protons. Electrons have much less mass than protons or neutrons, so almost all the mass of an atom is found in its nucleus. The Danish physicist Niels Bohr (1885–1962) proposed a model of the atom in 1913 in which a massive nucleus sits in the center and the smaller electrons orbit around it, much as planets orbit around the Sun (**Figure 5.7b**).

The **Bohr model**, however, is an incomplete description of the atom. Just as waves of light have particle-like properties, particles of matter also have wavelike properties. Once that was understood, the Bohr model was modified so that the positively charged nucleus is surrounded by electron "clouds" or "waves," as shown in **Figure 5.7c**. In that quantum mechanical model, it is not possible to know precisely where the electron is in its orbit. The wave characteristics of particles make it impossible to pin down simultaneously their exact location and their exact velocity; some uncertainty will always exist. That is why a featureless cloud is used to represent electrons in orbit around an atomic nucleus.

Atomic Energy Levels

Because of their wavelike properties, electrons in an atom can take on only certain specific energies that depend on the energy states of the atom. The form that the electron waves take depends on the possible energy states of atoms. The energy states of atoms are like the shelves of the bookcase depicted in **Figure 5.8a**. The energy of an atom might correspond to the energy of one state or to the energy of the next state, but the energy of the atom is never found between the two states, just as a book can be on only one shelf at a time and cannot be partly on one shelf and partly on another. A given atom may have many energy states available to it, but those stable states are *discrete*.

Astronomers keep track of the allowed states of an atom by using energy level diagrams, as shown in **Figure 5.8b**. Both the bookcase and the energy level diagram are simplifications of the

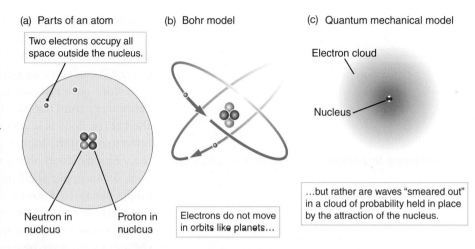

Figure 5.7 (a) An atom (here, helium) is made up of a nucleus consisting of positively charged protons and electrically neutral neutrons and is surrounded by much less massive negatively charged electrons. (b) Atoms are often drawn as miniature "solar systems," but this Bohr model is incorrect. (c) Electrons are actually smeared out around the nucleus in quantum mechanical clouds of probability.

AstroTour: Atomic Energy Levels and the Bohr Model

Figure 5.8 (a) Energy states of an atom are analogous to shelves in a bookcase. You can move a book from one shelf to another, but books can never be placed between shelves. (b) Atoms exist in one allowed energy state or another but never in between. The ground state is the lowest possible level.

Astronomy in Action: Emission and Absorption

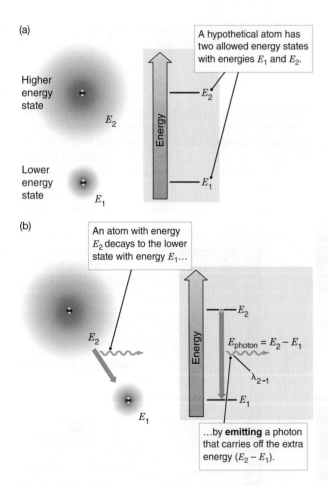

Figure 5.9 (a) The energy levels of a hypothetical two-level atom. (b) A photon with energy $E_{photon} = E_2 - E_1$ is emitted when an atom in the higher energy state (E_2) decays to the lower energy state (E_1).

possible energies of a three-dimensional system. The lowest possible energy state for a system (or part of a system) such as an atom is called the **ground state**. When the atom is in the ground state, the electron has its minimum energy. It can't give up any more energy to move to a lower state because a lower state does not exist. An atom will remain in its ground state forever unless it gets energy from outside. In the bookcase analogy, a book sitting on the bottom shelf at the floor is in its ground state. It has nowhere left to fall, and it cannot jump to one of the higher shelves on its own.

Energy levels above the ground state are called **excited states**. Just as a book on an upper shelf might fall to a lower shelf, an atom in an excited state might **decay** to a lower state by getting rid of some of its extra energy. An important difference between the atom and the book on the shelf, however, is that whereas a snapshot might catch the book falling between the two shelves, the atom will never be caught between two energy states. When the transition occurs from a higher state to a lower one, the difference in energy between the two states is carried off all at once. A common way for an atom to do that is to emit a photon. The photon emitted by the atom carries away exactly the amount of energy that atom loses as it goes from the higher energy state to the lower energy state. Similarly, atoms moving from a lower energy state to a higher energy state can absorb only certain specific energies.

To make an analogy with money, suppose you have a penny (1 cent), a nickel (5 cents), and a dime (10 cents), totaling 16 cents. Now imagine that you give away the nickel and are left with 11 cents. You never had exactly 13 cents or 13.6 cents. You had 16 cents and then 11 cents. Atoms don't accept and give away money to change energy states, but they do accept and give away photons with well-defined energies.

EMISSION SPECTRA Imagine a hypothetical atom that has only two available energy states. The energy of the lower energy state (the ground state) is E_1, and the energy of the higher energy state (the excited state) is E_2. The energy levels of that atom can be represented in an energy level diagram such as the one in **Figure 5.9a**. An atom in the excited state moves to the ground state by getting rid of the "extra" energy all at once. In that **emission** process, a photon is released. The atom goes from one energy state to another, but it never has an amount of energy in between.

In **Figure 5.9b**, the green arrow pointing straight down indicates that the atom went from the higher state (E_2) to the lower state (E_1). The atom lost an amount of energy equal to $E_2 - E_1$, the difference between the two states. Because energy is never truly lost or created, the energy the atom lost has to show up somewhere. Here, the energy shows up in the form of a photon emitted by the atom. The energy of the emitted photon matches the energy lost by the atom; that is, $E_{photon} = E_2 - E_1$.

An atom can emit photons with energies corresponding *only* to the difference between two of its allowed energy states. Because the energy of a photon is related to the frequency or wavelength of electromagnetic radiation, in the transition from level 2 to level 1, a photon of energy $E_{photon} = E_2 - E_1$ has a specific wavelength $\lambda_{2\to1}$ and a specific frequency $f_{2\to1}$. Therefore, those emitted photons have a specific color, and every photon emitted in any transition from E_2 to E_1 will have that same color defined by a specific wavelength. The energy level structure of an atom determines the wavelengths of the photons it emits—the color of the light that the atom gives off.

Figure 5.10 A cloud of gas containing atoms with two energy states, E_1 and E_2, emits photons with an energy $E_{photon} = E_2 - E_1$, which appear in the spectrum (right) as a single bright *emission line*.

Figure 5.10 illustrates the light coming from a cloud of gas consisting of the hypothetical two-state atoms in Figure 5.9. Any atom in the higher energy state (E_2) quickly decays and emits a photon in a random direction, and an enormous number of photons come pouring out of the cloud of gas. Instead of containing photons of all different energies—light of all different colors—that light contains only photons with the specific energy $E_2 - E_1$ and wavelength $\lambda_{2 \to 1}$. In other words, all the light coming from those atoms in the cloud is the same color. If you spread the light out into its component colors, only one color would be present—a single bright line called an **emission line**.

Why was the atom in the excited state E_2 in the first place? An atom sitting in its ground state will remain there unless it absorbs just the right amount of energy to kick it up to an excited state. In general, the atom either absorbs the energy of a photon or it collides with another atom or an unattached electron and absorbs some of the other particle's energy. In a neon sign, an alternating electric field inside the glass tube pushes electrons in the gas back and forth through the neon gas inside the tube. Some of those electrons crash into atoms of the gas, knocking them into excited states. The atoms then drop back down to their ground states by emitting photons, causing the gas inside the tube to glow a color characteristic of the element neon.

ABSORPTION SPECTRA An atom in a low energy state can absorb the energy of a passing photon and jump up to a higher energy state, as shown in **Figure 5.11**. Once again, the energy required to go from E_1 to E_2 is the difference in energy between the two states, $E_2 - E_1$. For a photon to cause an atom to jump from E_1 to E_2, it must provide exactly that much energy. The only photons that can excite atoms from E_1 to E_2 are photons with $E_{photon} = E_2 - E_1$. As with emission, those photons have a corresponding frequency ($f_{1 \to 2} = E_{photon}/h$) and wavelength

Figure 5.11 An atom in a lower energy state (E_1) may absorb a photon of energy $E_{photon} = E_2 - E_1$, leaving the atom in a higher energy state (E_2).

($\lambda_{1\rightarrow 2} = hc/E_{photon}$). Those photons have the same energy—the same color of light—emitted by the atoms when they decay from E_2 to E_1. That is not a coincidence. The energy difference between the two levels is the same whether the atom is emitting a photon or absorbing one, so the energy of the photon involved is the same in either case.

In **Figure 5.12a**, white light (with all wavelengths of photons in it) passes directly through a glass prism, which spreads out the light into a rainbow of colors. That is called a **continuous spectrum**. When the white light passes through a cool cloud composed of our hypothetical gas of two-state atoms (**Figure 5.12b**), however, some photons will be absorbed. Almost all the photons will pass through the cloud of gas unaffected because they do not have the right energy ($E_2 - E_1$) to be absorbed by atoms of the gas. Photons with just the right amount of energy can be absorbed, however, and as a result, those photons will be missing from the light passing through the prism. When that occurs, a sharp, dark line appears at the wavelength corresponding to the energy of the missing photons. That process by which atoms capture the energy of passing photons is called

Figure 5.12 (a) When passed through a prism, white light produces a spectrum containing all colors. (b) When light of all colors passes through a cloud of hypothetical two-state atoms, photons with energy $E_{photon} = E_2 - E_1$ may be absorbed, leading to the dark absorption line in the spectrum.

absorption, and the dark line seen in the spectrum is called an **absorption line**. **Figure 5.13a** shows the absorption lines in the spectrum of a star. The same spectrum is shown again in **Figure 5.13b** as a graph of the brightness at every wavelength. The brightness in Figure 5.13b drops abruptly at the wavelengths corresponding to the dark lines in Figure 5.13a. Places between the dark lines are brighter and therefore higher on the graph than the absorption lines.

When an atom absorbs a photon, it may quickly decay to its previous lower energy state, emitting a photon with the same energy as the photon it just absorbed. If the atom reemits a photon just like the one it absorbed, why does the absorption matter? The photon that was taken out of the passing light was replaced, but all the absorbed photons were originally traveling in the same direction, whereas the emitted photons are traveling in random directions. In other words, some photons with energies equal to $E_2 - E_1$ are diverted from their original paths by their interaction with atoms. If you look at a white light through the cloud in Figure 5.12b, you will observe an absorption line at a wavelength of $\lambda_{2 \to 1}$, but if you look at the cloud from another direction, you will observe an emission line at that same wavelength.

Figure 5.13 Absorption lines in the spectrum of a star as an image (a) and a graph (b).

AstroTour: Atomic Energy Levels and Light Emission and Absorption

Spectral Fingerprints of Atoms

In the mid-19th century, Gustav Kirchoff (1824–1887) first observed emission, continuous, and absorption spectra from the three types of sources shown in Figures 5.10 and 5.12. He summarized his findings as three empirical laws—but he did *not* know about energy levels in atoms. Kirchoff, together with Robert Bunsen (1811–1899), concluded that the dark-line (absorption) spectrum of the Sun was the "reverse" of the bright-line (emission) spectrum that would be produced by the Sun's atmosphere alone and identified some of the elements on the Sun. (A few years later the element helium was discovered in the solar spectrum.) Others observed the stars. Noteworthy among them were William (1824–1910) and Margaret Lindsay (1848–1915) Huggins, who published an atlas of stellar spectra in 1899 and who showed that the types of atoms seen in the stars are the same as those found on Earth.

The study of spectra of astronomical objects is fundamental to our understanding of the universe. Astronomers who study spectra will say that "a spectrum is worth a thousand pictures" because of the wealth of information that can come from it. We refer to spectra in every chapter in this book.

Real atoms can occupy many more than just two possible energy states; therefore, any given type of atom can emit and absorb photons at many wavelengths. An atom with three energy states, for example, might jump from state 3 to state 2, or from state 3 to state 1, or from state 2 to state 1. The three distinct emission lines in the spectrum from a gas made up of those atoms would have wavelengths of $hc/(E_3 - E_2)$, $hc/(E_3 - E_1)$, and $hc/(E_2 - E_1)$, respectively.

An atom's allowed energy states are determined by the interactions among the electrons and the nucleus. Every neutral hydrogen atom consists of a nucleus containing one proton, plus a single electron in a cloud surrounding the nucleus. Therefore, every hydrogen atom has the same energy states available to it, and all hydrogen atoms have the same emission and absorption lines. **Figure 5.14a** shows the energy level diagram of hydrogen. **Figure 5.14b** shows the visible emission spectrum from hydrogen, and **Figure 5.14c** displays that same information as a graph.

Each type of atom—that is, each chemical element—has a unique set of available energy states and therefore a unique set of wavelengths at which it can emit or absorb radiation. **Figure 5.14d** shows the emission spectra of the elements sodium, helium, neon, and mercury. Those unique sets of wavelengths serve as unmistakable spectral "fingerprints" for each chemical element.

For astronomers, spectral fingerprints are crucial. They let astronomers figure out what types of atoms (or molecules) are present in distant objects by simply looking at the spectrum of light from those objects. If the spectral lines of hydrogen, helium, carbon, oxygen, or any other element are visible in the light from a distant object, we know that element is present in that object. The strength of a line is determined in part by how many atoms of that type are present in the source. By measuring the strength of the lines from different types of atoms in the spectrum of a distant object, astronomers can often infer the relative amounts of elements that make up the object. In addition, by looking at the relative strength of lines from the same element, astronomers often can determine the temperature, density, and pressure of the material.

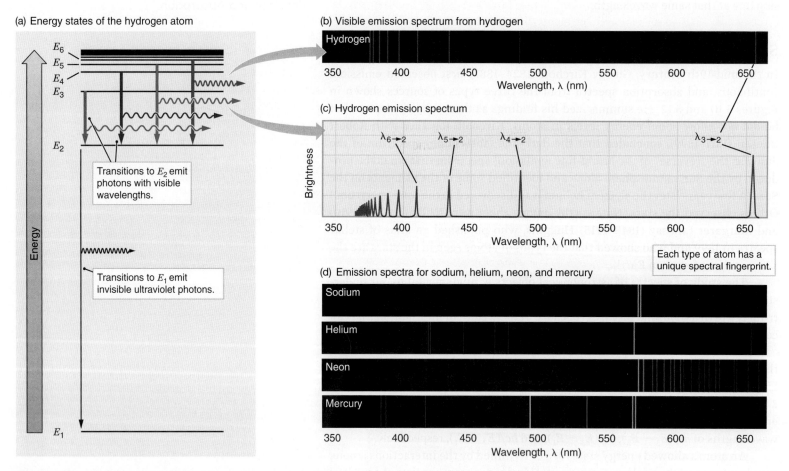

Figure 5.14 (a) The energy states of the hydrogen atom. Decays to level E_2 emit photons in the visible part of the spectrum. (b) This visible emission spectrum is what you might see if you looked at the light from a hydrogen lamp projected through a prism onto a screen. (c) The brightness at every wavelength can be measured to produce a graph of the brightness of spectral lines versus their wavelength. (d) Emission spectra from gaseous sodium, helium, neon, and mercury.

Spectra are how we know what makes up the stars and planets, and that we on Earth are composed of the same elements. Astronomers use the relative abundance of the elements in the Sun (called the **solar abundance**) as a standard reference. Hydrogen (H) is the most abundant element in the Sun (**Figure 5.15**), followed by helium (He) and 13 others. Those 15 elements make up 99.99 percent of the mass of the Sun. Most of the elements on the regular periodic table (in the lower right) make up less than 0.01 percent of the mass of the Sun.

Excitement and Decay

If a book on a level shelf is not disturbed, it will sit there forever; something must *cause* the book to fall off the shelf. Similarly, physicists have wondered what causes an atom in an excited state to jump down to a lower energy state and emit a photon. Sometimes an atom in a higher energy state can be "stimulated" into emitting a photon—but usually nothing causes the atom to jump to the lower energy state. Instead, the atom decays *spontaneously*. Although scientists can determine on average how long the atom is likely to remain in the excited state, they can't know exactly when a particular atom will decay until after the decay has happened. An atom decays at a random moment that is not influenced by anything in the universe and cannot be predicted.

Toys that glow in the dark are an example of that phenomenon. Photons in sunlight or from a lightbulb are absorbed by certain phosphorescent atoms in the toy, knocking those atoms into excited energy states. The excited states of the atoms in the toy last for many seconds, unlike the excited energy states of many atoms that tend to decay in a small fraction of a second. If on average those atoms tend to

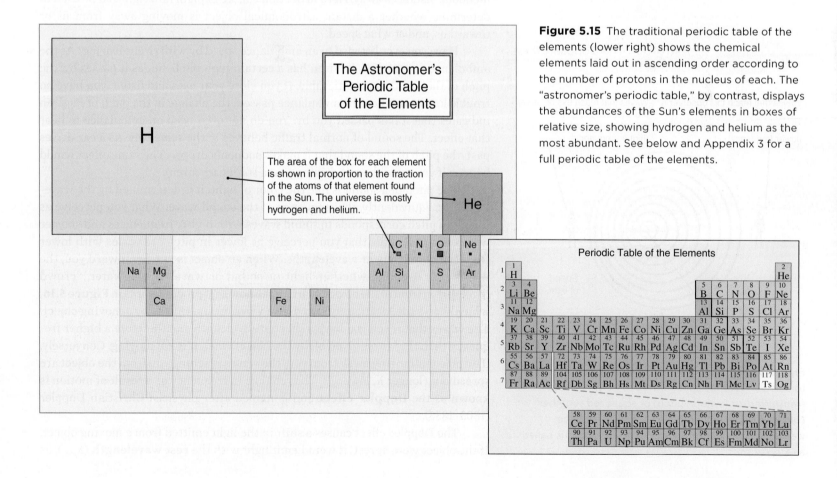

Figure 5.15 The traditional periodic table of the elements (lower right) shows the chemical elements laid out in ascending order according to the number of protons in the nucleus of each. The "astronomer's periodic table," by contrast, displays the abundances of the Sun's elements in boxes of relative size, showing hydrogen and helium as the most abundant. See below and Appendix 3 for a full periodic table of the elements.

remain in their excited state for 1 minute before decaying and emitting a photon, then after 1 minute a 50-50 chance exists that any particular atom in the toy will have decayed and a 50-50 chance that the atom will remain in its excited state. Although it is impossible to say exactly which atoms will decay, about half of the trillions and trillions of atoms in the toy will decay within 1 minute, and the brightness of the glow from the toy will have dropped to half of what it was. After each minute, half of the remaining excited atoms decay, and the glow from the toy drops to half of what it was 1 minute earlier. Thus, the glow from the toy slowly fades away.

In deep space, where atoms can remain undisturbed for long periods, certain excited states of atoms last, on average, for tens of millions of years or even longer. An atom may have been in such an excited energy state for a few seconds, a few hours, or 50 million years when, in an instant, it decays to the lower energy state without anything having caused it to do so. Physicists can only calculate the *probabilities* that certain events will take place.

CHECK YOUR UNDERSTANDING 5.2

How can spectra tell us the chemical composition of a distant star?

5.3 The Doppler Shift Indicates Motion Toward or Away from Us

Astronomy in Action: Doppler Shift

AstroTour: The Doppler Effect

You have already seen that light is a tightly packed bundle of information that can reveal a wealth of information about the physical state of material located tremendous distances away. Here in Section 5.3, we explain how light can be used to determine whether a distant astronomical object is moving away from us or toward us, and at what speed.

Have you ever listened to an ambulance speed by with sirens blaring? As the ambulance approaches, its siren has a certain high pitch, but as it passes by, the pitch of the siren drops noticeably. If you close your eyes and listen, you have no trouble knowing when the ambulance passed; the change in the pitch of its siren indicates that it has passed you by. You do not even need an ambulance to hear that effect. The sound of normal traffic behaves in the same way. As a car drives past, the pitch of the sound that it makes suddenly drops. (The same effect would happen if you were the one driving past a stationary siren).

The pitch of a sound is like the color of light: it is determined by the wavelength or, equivalently, the frequency of the sound wave. What you perceive as higher in pitch corresponds to sound waves with higher frequencies and shorter wavelengths. Sounds that you perceive as lower in pitch are waves with lower frequencies and longer wavelengths. When an object is moving toward you, the waves that it emits, whether light or sound or waves in the water, "crowd together" in front of the object. You can see how that process works in **Figure 5.16**, which shows the locations of successive wave crests emitted by a moving object. The waves that reach you have a shorter wavelength and therefore a higher frequency than the waves given off by the object when it is not moving. Conversely, if an object is moving away from you, the waves reaching you from the object are spread out (longer λ, lower f). That change in frequency as a result of motion is known as the **Doppler effect** and is named after physicist Christian Doppler (1803–1853).

The Doppler effect causes a shift in the light emitted from a moving object. If the object were at rest, it would emit light with the **rest wavelength** (λ_{rest}), as

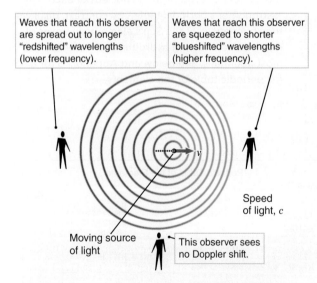

Figure 5.16 Motion of a light or sound source relative to an observer may cause waves to be spread out (*redshifted*, or lower in pitch) or squeezed together (*blueshifted*, or higher in pitch). A change in the wavelength of light or the frequency of sound is called a *Doppler shift*.

shown in **Figure 5.17a**. If an object such as a star is moving *toward* you, the light reaching you from the object has a shorter wavelength than its rest wavelength—the light is "bluer" than the rest wavelength, and the light is described as **blueshifted**, as shown by the blue waves in **Figure 5.17b**. In contrast, light from a source moving *away* from you is shifted to longer wavelengths. The light that you see is "redder" than if the source were not moving away from you and is described as **redshifted**, as shown by the red waves in **Figure 5.17c**. The faster the object is moving with respect to you, the larger the shift. The amount by which the wavelength of light is shifted by the Doppler effect is called the light's **Doppler shift**, which depends on the speed of the object emitting the light.

The Doppler shift provides information only about the **radial velocity** (v_r) of the object, which is the part of the motion that is toward you or away from you. The radial velocity is the rate at which the distance between you and the object is changing: if v_r is positive, the object is getting farther away from you; if v_r is negative, the object is getting closer. At the moment the ambulance is passing you, it is getting neither closer nor farther away, so the pitch you hear is the same as the pitch heard by the crew riding on the truck. Similarly, a distant object moving slowly across the sky does not move toward or away from you, and so its light will not be Doppler shifted from your point of view.

Doppler shifts become especially useful when you are looking at an object that has emission or absorption lines in its spectrum. Those spectral lines enable astronomers to determine how rapidly the object is moving toward or away from Earth. To determine that velocity, astronomers first identify the spectral line as being from a certain chemical element, which has a unique rest wavelength (λ_{rest}) measured in a lab on Earth. They then measure the observed wavelength (λ_{obs}) in the spectrum of the distant object. The difference between the rest wavelength and the observed wavelength indicates the object's radial velocity. See **Working It Out 5.2**.

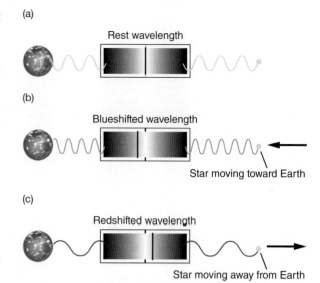

Figure 5.17 From their rest wavelength (a), spectral lines of astronomical objects are blueshifted if they are moving toward the observer (b) and redshifted if they are moving away from the observer (c).

5.2 Working It Out Making Use of the Doppler Effect

The Doppler formula for objects moving at a radial velocity (v_r) much less than the speed of light is given by

$$v_r = \frac{\lambda_{obs} - \lambda_{rest}}{\lambda_{rest}} \times c$$

A prominent spectral line of hydrogen atoms has a rest wavelength, λ_{rest}, of 656.3 nm (see Figure 5.14b). Suppose that you measure the wavelength of that line in the spectrum of a distant object and find that instead of seeing the line at 656.3 nm, you see the line at a wavelength, λ_{obs}, of 659.0 nm. What is its radial velocity? Using the above equation,

$$v_r = \frac{659.0 \text{ nm} - 656.3 \text{ nm}}{656.3 \text{ nm}} \times (3 \times 10^5 \text{ km/s})$$

$$v_r = 1{,}200 \text{ km/s}$$

Thus, the object is moving away from you with a radial velocity of 1,200 km/s.

Suppose instead that you know the velocity and want to compute the wavelength at which you would observe the spectral line. Earth's nearest stellar neighbor, Proxima Centauri, is moving toward us at a radial velocity of −21.6 km/s. What is the observed wavelength, λ_{obs}, of a magnesium line in Proxima Centauri's spectrum that has a rest wavelength, λ_{rest}, of 517.27 nm? We can rearrange the Doppler formula to solve for λ_{obs}:

$$\lambda_{obs} = \left(1 + \frac{v_r}{c}\right) \times \lambda_{rest}$$

$$\lambda_{obs} = \left(1 + \frac{-21.6 \text{ km/s}}{3 \times 10^5 \text{ km/s}}\right) \times 517.27 \text{ nm} = 517.23 \text{ nm}$$

Although the observed Doppler blueshift ($\lambda_{obs} - \lambda_{rest} = 517.23 - 517.27$) is only −0.04 nm, it is easily measured with modern instrumentation.

CHECK YOUR UNDERSTANDING 5.3

Which of the following Doppler shifts indicates the fastest-approaching object?
(a) 0.04 nm; (b) 0.06 nm; (c) −0.04 nm; (d) −0.06 nm

5.4 Temperature Affects the Spectrum of Light That an Object Emits

The balance between heating and cooling in an object determines its temperature. If an object's temperature is constant, those two must be in balance. Here in Section 5.4, we examine that balance and see how we can use it to predict the temperatures of planets and stars.

Equilibrium and Balance

Your body is heated by the release of chemical energy inside it. Sometimes your body is also heated by energy from your surroundings. If you are standing in sunshine on a hot day, the hot air around you and the sunlight falling on you both heat you. In response to that heating, your body cools itself off by perspiring: water seeps from the pores in your skin and evaporates. The energy to evaporate the water comes from your body. As the perspiration evaporates, it cools you down because it removes heat energy from your body. For your body temperature to remain stable, the heating must be balanced by the cooling. If more heating than cooling occurs, your body temperature increases; if more cooling than heating, your body temperature decreases.

In a tug-of-war contest between two perfectly matched teams, each team pulls on the rope, but the force of one team's pull is only enough to match, not overcome, the force exerted by the other team. A picture taken now and another taken 5 minutes from now would not differ in any significant way. In that static equilibrium, opposing forces balance each other exactly. Static equilibrium can be stable, unstable, or neutral. A marble in a bowl is in a stable equilibrium: if it moves, it will return to its original position at the bottom of the bowl. A book standing on its edge, unsupported on either side, is in an unstable equilibrium: if you nudge it, it will fall over, not settle back into its original position. When an unstable equilibrium is disturbed, it moves further away from equilibrium rather than back toward it.

Equilibrium also can be dynamic, which means the system is constantly changing so that one source of change is exactly balanced by another source of change, and the configuration of the system remains the same. In **Figure 5.18**, placing a can with a hole cut in the bottom under an open water faucet provides a simple example of dynamic equilibrium. The depth of the water in the can determines how fast water pours out through the hole in the bottom of the can. When the water reaches just the right depth (Figure 5.18a), water pours out of the hole in the bottom of the can at the same rate it pours into the top of the can from the faucet. The water leaving the can balances the water entering, and equilibrium is established. If you took a picture now and another picture in a few minutes, little of the water in the can would be the same, but the pictures would be indistinguishable.

If a system is not in equilibrium, its configuration will change. If the level of the water in the can is too low (Figure 5.18b), water will not flow out of the bottom of the can fast enough to balance the water flowing in. The water level will begin to rise until the amount coming in equals the amount going out. A picture taken

Astronomy in Action: Changing Equilibrium

Figure 5.18 The relative rates at which water flows into and out of a can determine the water level in the can. This is an example of dynamic equilibrium.

now and another taken a short time later would not look the same. Conversely, if the water level in the can is too high (Figure 5.18c), water will flow out of the can faster than it flows in. The water level will begin to fall until the amount coming in equals the amount going out. Once again, if the system is not in equilibrium, its configuration will change.

When heating is balanced by cooling, we call it **thermal equilibrium**. Planets have a dynamic but stable thermal equilibrium, and electromagnetic radiation plays a crucial role in maintaining that balance. Energy from sunlight heats the surface of a planet, driving its temperature up, and the planet emits thermal radiation into space, cooling it down. For a planet to remain at the same average temperature over time, the energy it radiates into space must exactly balance the energy it absorbs from the Sun. **Figure 5.19** shows how the equilibrium temperature of a planet is analogous to the water level in Figure 5.18. We return to planetary equilibrium later in the chapter. Many kinds of equilibrium exist besides thermal equilibrium, some of which we encounter later in the book.

Temperature

In everyday life, we define hot and cold subjectively: something is hot or cold when it feels hot or cold. When we measure *temperature*, we specify degrees on a thermometer, but the way we define a degree is arbitrary. If you grew up in the United States, you probably think of temperatures in degrees Fahrenheit (°F), whereas if you grew up almost anywhere else in the world, you think of temperatures in degrees Celsius (°C).

Temperature is a measurement of how energetically the atoms that make up an object are moving about. The air around us is composed of vast numbers of atoms and molecules. Those molecules are moving about every which way. Some move slowly; some move more rapidly. All atoms and molecules are constantly in motion. The average kinetic energy (E_K) is given by $E_K = 1/2\, mv^2$, where m is the mass of an atom or molecule and v is its velocity. The more energetically the atoms or molecules are bouncing about, the higher the object's temperature. In fact, the random motions of atoms and molecules are often called their **thermal motions** to emphasize the connection between those motions and temperature. **Figure 5.20** shows that when the temperature of a gas is increased, the kinetic energy is increased, in which case the atoms move faster.

The atoms and molecules in a solid body (like you) cannot move about freely like a gas, but they still move back and forth around their average location, and temperature measures the amount of that movement. If an object is hotter than you are, thermal energy flows from that object into you. At the atomic level, that means the object's atoms are bouncing more energetically than the atoms in your body are, so if you touch the object, its atoms collide with your atoms, causing the atoms in your body to move faster. Your body gets hotter as thermal energy flows from the object to you. At the same time, those collisions rob the particles in the object of some of their energy. Their motions slow down, and the hotter object cools. Heating processes increase the average thermal energy of an object's particles, whereas cooling processes decrease the average thermal energy of those particles.

Figure 5.19 Planets are heated by absorbing sunlight (and sometimes by internal heat sources) and cooled by emitting thermal radiation into space. In the absence of other sources of heating or means of cooling, the equilibrium between those processes determines the temperature of the planet.

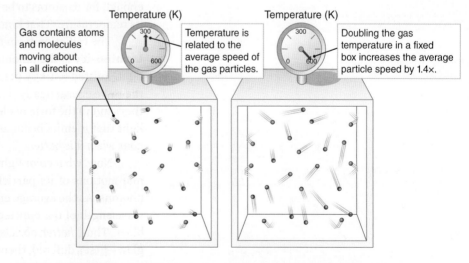

Figure 5.20 Higher gas temperatures correspond to faster-moving atoms.

Figure 5.21 Temperatures on the Fahrenheit, Celsius, and Kelvin scales.

The three most common temperature scales are shown in **Figure 5.21**. On the Fahrenheit scale, water at sea level has a freezing point of 32°F and a boiling point of 212°F. On the Celsius scale, water freezes at 0°C and boils at 100°C. Because the number of degrees between freezing and boiling on those two scales is different (180°F vs. 100°C), a 1-degree change measured in °F is not the same as a 1-degree change measured in °C.

As the motions of the particles in an object slow down, the temperature decreases more and more. The lowest possible temperature, at which all thermal motions would stop, is called **absolute zero**. Absolute zero corresponds to −273.15°C and −459.57°F. It also is the starting point of the **Kelvin temperature scale**. The size of one unit on the Kelvin scale, called a **kelvin (K)**, is the same as the Celsius degree. To convert between °C and K, just add 273.15 to the temperature in °C. Thus, water freezes at 273.15 K and water boils at 373.15 K. The Kelvin scale has no negative temperatures.

When temperatures are measured in kelvins, the average thermal energy of particles is proportional to the measured temperature. Thus, the average thermal energy of the atoms in an object with a temperature of 200 K is twice the average thermal energy of the atoms in an object with a temperature of 100 K.

Temperature, Luminosity, and Color

We have seen the way discrete atoms emit and absorb radiation, which leads to a useful understanding of emission lines and absorption lines that tell us about the physical state and motion of distant objects. But not all objects have spectra dominated by discrete spectral lines. As you saw in Figure 5.12a, if you pass the light from a lightbulb through a prism, instead of discrete bright and dark bands you will see light spread out smoothly from the blue end of the spectrum to the red. Similarly, if you look closely at the spectrum of the Sun, you will see absorption lines, and you will see light smoothly spread out across all colors of the spectrum—the continuous spectrum noted earlier.

We can think of a dense material as being composed of a collection of charged particles being jostled as their thermal motions cause them to run into their neighbors. The hotter the material is, the more violently its particles are being jostled. A charged particle radiates anytime it accelerates, so the jostling of particles that results from their thermal motions causes them to give off a continuous spectrum of electromagnetic radiation. That is why any material dense enough for its atoms to be jostled by their neighbors emits light simply because of its temperature. That kind of radiation is called **thermal radiation**.

The radiation from an object changes as the object heats up or cools down. **Luminosity** is the amount of light *leaving* a source—that is, the total amount of light emitted (energy per second, measured in watts, W). The hotter the object, the more energetically the charged particles within it move, and the more energy they emit in the form of electromagnetic radiation. So, as an object gets hotter, the light that it emits becomes more intense. In other words, *an object is more luminous when it is hotter*.

Now, what color light does an object emit? As the object gets hotter, the thermal motions of its particles become more energetic, producing more energetic photons. As the average energy of the photons that it emits increases, the average wavelength of the emitted photons decreases, and the light from the object gets bluer. Thus, *hotter objects are bluer*. If you heat a piece of metal, the metal will glow—first a dull red, then orange, and then yellow. The hotter the metal becomes, the more the highly energetic blue photons become mixed with the less energetic red photons, and the color of the light shifts from red toward blue. The light becomes more intense and bluer as the metal becomes hotter.

Blackbodies are objects that emit electromagnetic radiation only because of their temperature, not their composition. Blackbodies emit just as much thermal radiation as they absorb from their surroundings. Physicist Max Planck (1858–1947) graphed the intensity of the emitted radiation across all wavelengths and obtained the characteristic curves that we now call **Planck spectra** or **blackbody spectra**. **Figure 5.22** shows blackbody spectra for objects at several temperatures.

Blackbody Laws

In the real world, the light from stars such as the Sun and the thermal radiation from a planet often come close to having blackbody spectra. Those objects follow the Stefan-Boltzmann law, which relates luminosity with temperature, and Wien's law, which relates temperature with color.

STEFAN-BOLTZMANN LAW As the temperature of an object increases, the object gives off more radiation at every wavelength, so the luminosity of the object should increase. Adding up all the energy in a blackbody spectrum shows that the increase in luminosity is proportional to the fourth power of the temperature. That relationship, luminosity $\propto T^4$, is known as the **Stefan-Boltzmann law**. It was discovered in the laboratory by physicist Josef Stefan (1835–1893) and derived mathematically by his student Ludwig Boltzmann (1844–1906).

Measuring all the photons emitted by Earth or the Sun in all possible directions is hard, but measuring the *flux* is easier. The **flux** ($\mathcal{F}$) is the amount of energy radiated by each square meter of the surface of an object each second. Flux is proportional to luminosity. You can find the luminosity by multiplying the flux by the total surface area. According to the Stefan-Boltzmann law, the flux is given by $\mathcal{F} = \sigma T^4$, where σ (the Greek letter sigma) is the **Stefan-Boltzmann constant**. It equals 5.67×10^{-8} W/(m² K⁴), where 1 watt = 1 joule per second (J/s).

The Stefan-Boltzmann law says that an object rapidly becomes more luminous as its temperature increases. If the temperature of an object doubles, the amount of energy being radiated each second increases by a factor of $2^4 = 16$. If the temperature of an object increases by a factor of 3, the energy being radiated by the object each second increases by a factor of $3^4 = 81$. A lightbulb with a filament temperature of 3000 K radiates 16 times as much light as it would if the filament temperature were 1500 K. Even modest changes in temperature can result in large changes in the luminosity of an object.

WIEN'S LAW Look again at Figure 5.22. The wavelength where the blackbody spectrum is at its peak, λ_{peak}, corresponds to where the electromagnetic radiation from an object is greatest. As the temperature, T, increases, the peak of the spectrum shifts toward shorter wavelengths. For example, $\lambda_{peak} = 970$ nm for a 3000 K object, but only 480 nm for a 6000 K object. Photon energy and wavelength are inversely related; thus, as the peak wavelength decreases, the average photon energy increases. The object becomes bluer. The physicist Wilhelm Wien (1864–1928) found that the peak wavelength in the spectrum is inversely proportional to the temperature of the object ($\lambda_{peak} \propto 1/T$). **Wien's law** states that if you double the temperature, the peak wavelength becomes half of what it was. If you increase the temperature by a factor of 3, the peak wavelength becomes a third

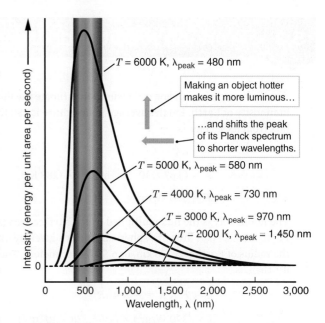

Figure 5.22 This graph shows blackbody spectra emitted by sources with temperatures of 2000, 3000, 4000, 5000, and 6000 K. At higher temperatures, the peak of the spectrum shifts toward shorter wavelengths, and the amount of energy radiated per second from each square meter of the source increases.

Astronomy in Action: Wien's Law

5.3 Working It Out: Working with the Stefan-Boltzmann Law and Wien's Law

The Stefan-Boltzmann law can be used to estimate the flux and luminosity of Earth. Earth's average temperature is 288 K, so the flux from its surface is

$$\mathcal{F} = \sigma T^4$$
$$\mathcal{F} = (5.67 \times 10^{-8} \text{ W/m}^2 \text{ K}^4) \times (288 \text{ K})^4$$
$$\mathcal{F} = 390 \text{ W/m}^2$$

The luminosity is the flux multiplied by the surface area (A) of Earth. The surface area of a spherical object is given by $4\pi R^2$, and the radius of Earth is 6,378 km, or 6.378×10^6 meters. The luminosity, then, is

$$L = \mathcal{F} \times A = \mathcal{F} \times 4\pi R^2$$
$$L = (390 \text{ W/m}^2) \times [4\pi (6.378 \times 10^6 \text{ m})^2]$$
$$L \approx 2 \times 10^{17} \text{ W}$$

Earth emits the equivalent of the energy used by 2,000,000,000,000,000 (2 million billion) hundred-watt lightbulbs. That value is still not anywhere close to the amount emitted by the Sun.

If astronomers measure the spectrum of an object emitting thermal radiation and find where the peak in the spectrum is, Wien's law can be used to calculate the temperature of the object. Wien's law can be written as

$$T = \frac{2{,}900{,}000 \text{ nm K}}{\lambda_{\text{peak}}}$$

What, for example, is the surface temperature of the Sun? The spectrum of the light coming from the Sun peaks at a wavelength of $\lambda_{\text{peak}} = 500$ nm, so

$$T = \frac{2{,}900{,}000 \text{ nm K}}{500 \text{ nm}} = 5800 \text{ K}$$

What is the peak wavelength at which Earth radiates? Using Earth's average temperature of 288 K in Wien's law gives

$$\lambda_{\text{peak}} = \frac{2{,}900{,}000 \text{ nm K}}{288 \text{ K}} = 10{,}100 \text{ nm} = 10.1 \text{ μm}$$

Thus, Earth's radiation peaks in the infrared region of the spectrum.

---

of what it was. Stefan-Boltzmann's law and Wien's law are further explored in **Working It Out 5.3**. We use those laws later in the chapter to estimate the temperatures of the planets.

CHECK YOUR UNDERSTANDING 5.4

When you look at the sky on a dark night you see stars of different colors. Rank them from coolest to hottest. (a) orange; (b) red-orange; (c) yellow; (d) red; (e) blue

5.5 The Brightness of Light Depends on the Luminosity and Distance of the Source

Whereas luminosity is the amount of light *leaving* a source, the **brightness** of electromagnetic radiation is the amount of light *arriving* at a particular location. Therefore, the observed brightness depends on the luminosity and the distance of the light source. For example, replacing a 50-W lightbulb with a 100-W bulb makes a room twice as bright because it doubles the light reaching any point in the room. But brightness also depends on the distance from a source of electromagnetic radiation. For example, if you needed more light to read this book, you could replace the bulb in your lamp with a more luminous bulb or you could move the book closer to the light. Conversely, if a light were too bright, you could move away from it. Our everyday experience teaches us that as we move away from a light, its brightness decreases.

The particle description of light offers another way to think about the brightness of radiation and how brightness depends on distance. Suppose you had a piece of cardboard that measured 1 meter by 1 meter. To make the light falling on the cardboard twice as bright, you would need to double the number of photons that hit the cardboard each second. Tripling the brightness of the light would mean increasing the number of photons hitting the cardboard each second by a factor of 3, and so on. Brightness depends on the number of photons falling on each square meter of a surface each second.

Now imagine a lightbulb sitting at the center of a spherical shell (**Figure 5.23**). Photons from the bulb travel in all directions and land on the inside of the shell. To find the number of photons landing on each square meter of the shell during each second—that is, to determine the brightness of the light—take the *total* number of photons given off by the lightbulb each second and divide by the number of square meters over which those photons have to be spread. The surface area of a spherical shell is given by the formula $A = 4\pi d^2$, where d is the distance between the bulb and the surface of the sphere. The number of photons striking one square meter each second is equal to the total number of photons emitted each second divided by the surface area $4\pi d^2$.

Now increase the size of the spherical shell while keeping the total number of photons given off by the lightbulb each second the same. As the shell becomes larger, the photons from the lightbulb must spread out to cover a larger surface area. Each square meter of the shell receives fewer photons each second, so the brightness of the light decreases. If the shell's surface is moved twice as far from the light, the area over which the light must spread increases by a factor of $2^2 = 2 \times 2 = 4$. The photons from the bulb spread out over 4 times as much area, so the number of photons falling on each square meter each second becomes 1/4 of what it was. If the surface of the sphere is 3 times as far from the light, the area over which the light must spread increases by a factor of $3^2 = 3 \times 3 = 9$, and the number of photons per second falling on each square meter becomes 1/9 of what it was originally. That is the same kind of inverse square relationship you saw for gravity in Chapter 4. The brightness of the light from an object is inversely proportional to the square of the distance from the object. Twice as far means one-fourth as bright.

The idea of photons streaming and spreading onto a surface from a light explains why brightness follows an inverse square law. In practice, however, talking about the average *energy* falling on a surface each second, rather than the number of photons, is usually more more useful to an astronomer.

The luminosity of an object is the total number of photons given off by the object multiplied by the energy of each photon. So, instead of thinking about how the number of photons must spread out to cover the surface of a sphere, we can think about how the energy carried by the photons must spread out to cover the surface of a sphere. The brightness of the light is the amount of energy falling on a square meter in a second, and it equals the luminosity L divided by the area of the sphere, which depends on the radius squared. That tells us, for example, that the brightness of the Sun depends on the inverse square of the planet's distance from the Sun. That relationship factors in when we estimate the equilibrium temperatures of the planets in **Working It Out 5.4**.

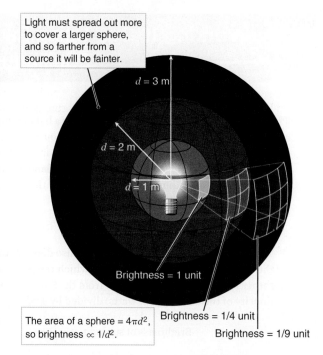

Figure 5.23 Light obeys an inverse square law as it spreads away from a source. Twice as far means one-fourth as bright, three times as far means one-ninth as bright, and so on.

Astronomy in Action: Inverse Square Law

CHECK YOUR UNDERSTANDING 5.5

The average distance of Mars from the Sun is 1.4 AU. How bright is the Sun on Mars compared with its brightness on Earth? (a) 1.4 times brighter; (b) about 2 times brighter; (c) about 2 times fainter; (d) 1.4 times fainter

5.4 Working It Out: Using Radiation Laws to Calculate Equilibrium Temperatures of Planets

The temperature of a planet is determined by a balance between the amount of sunlight being absorbed and the amount of energy being radiated back into space. We begin with the amount of sunlight being absorbed. When viewed from the Sun, a planet looks like a circular disk with a radius equal to the radius of the planet, R_{planet}. The area of the planet that is lit by the Sun is

$$\text{Absorbing area of planet} = \pi R_{planet}^2$$

The amount of energy striking a planet also depends on the brightness of sunlight at the distance at which the planet orbits. The brightness of sunlight at a distance d from the Sun is equal to the luminosity of the Sun (L_{Sun}, in watts) divided by $4\pi d^2$:

$$\text{Brightness of sunlight} = \frac{L_{Sun}}{4\pi d^2}$$

A planet does not absorb *all* the sunlight that falls on it. **Albedo**, a, is the fraction of the sunlight that reflects from a planet. The corresponding fraction of the sunlight absorbed by the planet is 1 minus the albedo $(1 - a)$. A planet covered in snow would have a high albedo (close to 1), whereas a planet covered by black rocks would have a low albedo, close to 0:

$$\text{Fraction of sunlight absorbed} = 1 - a$$

We can now calculate the energy absorbed by the planet each second. Writing that relationship as an equation, we say that

$$\begin{pmatrix} \text{Energy absorbed} \\ \text{by the planet} \\ \text{each second} \end{pmatrix} = \begin{pmatrix} \text{Absorbing} \\ \text{area of} \\ \text{the planet} \end{pmatrix} \times \begin{pmatrix} \text{Brightness} \\ \text{of sunlight} \end{pmatrix} \times \begin{pmatrix} \text{Fraction} \\ \text{of sunlight} \\ \text{absorbed} \end{pmatrix}$$

$$= \pi R_{planet}^2 \times \frac{L_{Sun}}{4\pi d^2} \times (1 - a)$$

Now let's turn to the other piece of the equilibrium: the amount of energy that the planet radiates away into space each second. We can calculate that amount by multiplying the number of square meters of the planet's total surface area by the energy radiated by each square meter each second. The surface area for the planet is given by $4\pi R_{planet}^2$. According to the Stefan-Boltzmann law, the energy radiated by each square meter each second is given by σT^4. Thus,

$$\begin{pmatrix} \text{Energy radiated} \\ \text{by the planet} \\ \text{each second} \end{pmatrix} = \begin{pmatrix} \text{Surface} \\ \text{area of} \\ \text{the planet} \end{pmatrix} \times \begin{pmatrix} \text{Energy radiated} \\ \text{per square meter} \\ \text{per second} \end{pmatrix}$$

$$= 4\pi R_{planet}^2 \times \sigma T^4$$

If the planet's temperature is to remain stable—not heating up or cooling down—then each second the "energy radiated" must be equal to "energy absorbed":

$$\begin{pmatrix} \text{Energy radiated} \\ \text{by the planet} \\ \text{each second} \end{pmatrix} = \begin{pmatrix} \text{Energy absorbed} \\ \text{by the planet} \\ \text{each second} \end{pmatrix}$$

When we set those two quantities equal to each other, we arrive at the following expression:

$$4\pi R_{planet}^2 \sigma T^4 = \pi R_{planet}^2 \frac{L_{Sun}}{4\pi d^2}(1 - a)$$

Canceling out πR_{planet}^2 on both sides, and rearranging the equation to put T on one side and everything else on the other, gives

$$T^4 = \frac{L_{Sun}(1 - a)}{16\sigma\pi d^2}$$

If we take the fourth root of each side, we get

$$T = \left(\frac{L_{Sun}(1 - a)}{16\sigma\pi d^2}\right)^{1/4}$$

Putting in the appropriate numbers for the known luminosity of the Sun, L_{Sun}, and the constants π and σ yields a simpler equation:

$$T = 279\,\text{K} \times \left(\frac{1 - a}{d_{AU}^2}\right)^{1/4}$$

where d_{AU} is the distance of the planet from the Sun in astronomical units.

To use that equation, we need to know a planet's distance from the Sun and its average albedo. For a blackbody ($a = 0$) at 1 AU from the Sun, the temperature is 279 K. For Earth, with an albedo of 0.3 and a distance from the Sun of 1 AU, the temperature is

$$T = 279\,\text{K} \times \left(\frac{1 - 0.3}{1^2}\right)^{1/4} = 255\,\text{K}$$

(Calculator hint: To take a fourth root, you can take the square root twice, or use the x^y button with $y = 0.25$.)

Earth is cooler than a blackbody at 1 AU from the Sun because its average albedo is greater than zero. If Earth's albedo changed or the Sun's luminosity changed, that would affect the result. When we examine planets around other stars, we must use the luminosity of the particular star in the equation, instead of the Sun's luminosity, so the temperature at 1 AU will be different from what it is for Earth.

Origins Temperatures of Planets

In the previous chapters, we discussed how a planet's axial tilt and its orbital shape affect its temperature and thus its prospects for life. Now let's get more specific about the temperatures of planets, using what you learned in this chapter about thermal radiation. For a planet at an equilibrium temperature, the energy radiated by a planet exactly balances the energy absorbed by the planet. If the planet is hotter than that equilibrium temperature, it will radiate energy faster than it absorbs sunlight, and its temperature will decrease. If the planet is cooler than that temperature, it will radiate energy slower than it absorbs sunlight, and its temperature will increase.

Planets at different distances from the Sun will have different temperatures, and the temperature should be inversely proportional to the square root of the distance, as you saw in Working It Out 5.4. **Figure 5.24** plots the actual and predicted temperatures of nine solar system objects. Each vertical orange bar shows the range of temperatures found on the surface of the planet or, for the giant planets, at the top of the planet's clouds. The black dots show the predictions made using the equation in Working It Out 5.4. For most planets, the predictions are not too far off, indicating that our basic understanding of *why* planets have the temperatures they have is probably pretty good. The data for Mercury, Mars, and Pluto agree particularly well.

Sometimes, however, the predictions are wrong. For Earth, the actual measured temperature is a bit higher than the predicted temperature, and for Venus the actual surface temperature is much higher than the prediction.

The predicted values assume that the temperature of the planet is the same everywhere, but planets are likely to be hotter on the day side than on the night side. The predictions also assume that a planet's only source of energy is sunlight and that the fraction of sunlight reflected is constant over the surface of each planet. The values also incorporate the assumption that the planets absorb and radiate energy into space as blackbodies.

The discrepancies between the calculated and the measured temperatures of some of the planets indicate that for those planets, some or all of those assumptions are incorrect. For example, the planet may have its own source of energy besides sunlight, or it may have an atmosphere that traps more solar energy and increases the temperature of the planet. Understanding the temperatures of planets makes it possible to hypothesize which planets are more suitable for life.

Figure 5.24 Predicted temperatures for the planets and dwarf planet Pluto are based on the equilibrium between absorbed sunlight and thermal radiation into space. Those temperatures are compared with ranges of observed surface temperatures.

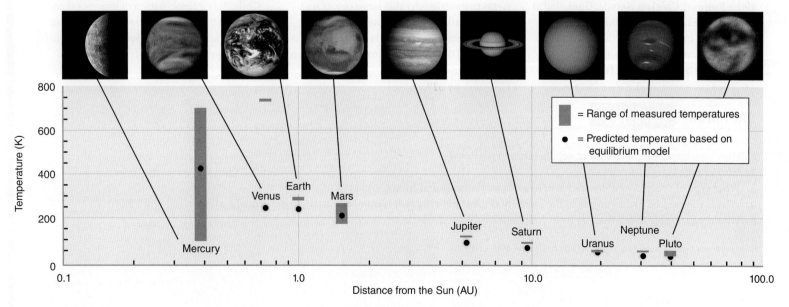

READING ASTRONOMY NEWS

ARTICLES | QUESTIONS

This press release from the European Southern Observatory (ESO) in Chile was widely reprinted in news sites around the world.

A Study in Scarlet

ESO

This new image from ESO's La Silla Observatory in Chile reveals a cloud of hydrogen called Gum 41 (**Figure 5.25**). In the middle of this little-known nebula, brilliant hot young stars are giving off energetic radiation that causes the surrounding hydrogen to glow with a characteristic red hue.

This area of the southern sky, in the constellation of Centaurus (The Centaur), is home to many bright nebulae, each associated with hot newborn stars that formed out of the clouds of hydrogen gas. The intense radiation from the stellar newborns excites the remaining hydrogen around them, making the gas glow in the distinctive shade of red typical of star-forming regions. Another famous example of this phenomenon is the Lagoon Nebula, a vast cloud that glows in similar bright shades of scarlet.

The nebula in this picture is located some 7,300 light-years from Earth. Australian astronomer Colin Gum discovered it on photographs taken at the Mount Stromlo Observatory near Canberra, and included it in his catalog of 84 emission nebulae, published in 1955. Gum 41 is actually one small part of a bigger structure called the Lambda Centauri Nebula, also known by the more exotic name of the Running Chicken Nebula. Gum died at a tragically early age in a skiing accident in Switzerland in 1960.

In this picture of Gum 41, the clouds appear to be quite thick and bright, but this is actually misleading. If a hypothetical human space traveler could pass through this nebula, it is likely that they would not notice it as—even at close quarters—it would be too faint for the human eye to see. This helps to explain why this large object had to wait until the mid-twentieth century to be discovered—its light is spread very thinly and the red glow cannot be well seen visually.

This new portrait of Gum 41—likely one of the best so far of this elusive object—has been created using data from the Wide Field Imager (WFI) on the MPG/ESO 2.2-meter telescope at the La Silla Observatory in Chile. It is a combination of images taken through blue, green, and red filters, along with an image using a special filter designed to pick out the red glow from hydrogen.

Figure 5.25 The Gum 41 Nebula.

ARTICLES | **QUESTIONS**

1. How long has the light from that nebula taken to reach us?
2. Why are the young stars blue?
3. What type of spectra would you expect to get from the stars and from the gas?
4. Why is the excited hydrogen gas in the image of Gum 41 glowing red? (Hint: Look at the spectrum of hydrogen in Figure 5.14.)
5. Would you be able to see Gum 41 from your location? Why or why not?

Source: "Photo Release: A Study in Scarlet," ESO.org, April 16, 2014. CC by 4.0.

Summary

Light carries both information and energy throughout the universe. The speed of light in a vacuum is 300,000 km/s; nothing can travel faster. Visible light is only a tiny portion of the entire electromagnetic spectrum. Atoms absorb and emit radiation at unique wavelengths, giving them spectral fingerprints. A planet's temperature depends on its distance from its star, its albedo, and the luminosity of its star.

LG 1 **Describe the wave and particle properties of light, and describe the electromagnetic spectrum.** Light is simultaneously a stream of particles called photons and an electromagnetic wave. Different types of electromagnetic radiation, from gamma rays to visible light to radio waves, are electromagnetic waves that differ in frequency and wavelength.

LG 2 **Describe how to measure the chemical composition of distant objects by using the unique spectral lines of different types of atoms.** Nearly all matter is composed of atoms, and light can reveal the identity of the chemical elements present in matter. The electron energy levels of each element have different (unique) spacings, and the wavelengths of the photons each element emits correspond to the differences in those levels. As a result, we can identify different chemical elements and molecules in distant objects.

LG 3 **Describe the Doppler effect and how it can be used to measure the motion of distant objects.** Because of the Doppler effect, light from receding objects is redshifted to longer wavelengths, and light from approaching objects is blueshifted to shorter wavelengths. The wavelength shifts of the spectral lines indicate how fast an astronomical object is moving toward or away from Earth.

LG 4 **Explain how the spectrum of light that an object emits depends on its temperature.** Temperature is a measure of how energetically particles are moving in an object. A light source that emits electromagnetic radiation because of its temperature is called a blackbody. A blackbody emits a continuous spectrum. The total amount of energy emitted is proportional to the temperature to the fourth power, and the peak wavelength, which determines its color, is inversely proportional to the temperature.

LG 5 **Differentiate luminosity from brightness, and illustrate how distance affects each.** The luminosity of an object is the amount of light it emits. The brightness of an object is proportional to its luminosity divided by its distance squared. Thus, the brightness of the Sun is different when measured from each Solar System planet, but the luminosity is the same.

? Unanswered Questions

- Has the speed of light always been 300,000 km/s? Some theoretical physicists have questioned whether light traveled much faster earlier in the history of our universe. The observational evidence that may test that idea comes from studying the spectra of the most distant objects—whose light has been traveling for billions of years—and determining whether billions of years ago chemical elements absorbed light differently from how they do today. So far, no evidence exists that the speed of light changes.

- Will traveling faster than the speed of light ever be possible? Our current understanding of the science says no. A staple of science fiction films and stories is spaceships that go into "warp speed" or "hyperdrive"—moving faster than light—to traverse the huge distances of space (and visit a different planetary system every week). If that premise is simply fictional and the speed of light is a true universal limit, travel between the stars will take many years. Because all electromagnetic radiation travels at the speed of light, even an electromagnetic signal sent to another planetary system would take many years to get there. Interstellar visits (and interstellar conversations) will be quite prolonged.

Questions and Problems

Test Your Understanding

1. If the Sun instantaneously stopped giving off light, what would happen on Earth?
   a. Earth would immediately get dark.
   b. Earth would get dark 8 minutes later.
   c. Earth would get dark 27 minutes later.
   d. Earth would get dark 1 hour later.

2. Why is an iron atom a different element from a sodium atom?
   a. A sodium atom has fewer neutrons in its nucleus than an iron atom has.
   b. An iron atom has more protons in its nucleus than a sodium atom has.
   c. A sodium atom is bigger than an iron atom.
   d. A sodium atom has more electrons.

3. Suppose an atom has three energy levels, specified in arbitrary units as 10, 7, and 5. In those units, which of the following energies might an emitted photon have? (Select all that apply.)
   a. 3
   b. 2
   c. 5
   d. 4

4. When a boat moves through the water, the waves in front of the boat bunch up, whereas the waves behind the boat spread out. That is an example of
   a. the Bohr model.
   b. the wave nature of light.
   c. emission and absorption.
   d. the Doppler effect.

5. As a blackbody becomes hotter, it also becomes _____ and _____.
   a. more luminous; redder
   b. more luminous; bluer
   c. less luminous; redder
   d. less luminous; bluer

6. Which of the following factors does *not* directly influence the temperature of a planet?
   a. the luminosity of the Sun
   b. the distance from the planet to the Sun
   c. the albedo of the planet
   d. the size of the planet

7. Two stars are of equal luminosity. Star A is 3 times as far from you as star B. Star A appears _____ star B.
   a. 9 times brighter than
   b. 3 times brighter than
   c. the same brightness as
   d. 1/3 as bright as
   e. 1/9 as bright as

8. When less energy radiates from a planet, its _____ increases until a new _____ is achieved.
   a. temperature; equilibrium
   b. size; temperature
   c. equilibrium; size
   d. temperature; size

9. How does the speed of light in a medium compare with the speed in a vacuum?
   a. The speed is the same in both a medium and a vacuum because the speed of light is a constant.
   b. The speed in the medium is always faster than the speed in a vacuum.
   c. The speed in the medium is always slower than the speed in a vacuum.
   d. The speed in the medium may be faster or slower, depending on the medium.

10. When an electron moves from a higher energy level in an atom to a lower energy level,
    a. a continuous spectrum is emitted.
    b. a photon is emitted.
    c. a photon is absorbed.
    d. a redshifted spectrum is emitted.

11. In Figure 5.14, the red photons come from the transition from E_3 to E_2. Those photons will have the _____ wavelengths because they have the _____ energy compared with that of the other photons.
    a. shortest; least
    b. shortest; most
    c. longest; least
    d. longest; most

12. Star A and star B appear equally bright in the sky. Star A is twice as far away from Earth as star B. How do the luminosities of stars A and B compare?
    a. Star A is 4 times as luminous as star B.
    b. Star A is 2 times as luminous as star B.
    c. Star B is 2 times as luminous as star A.
    d. Star B is 4 times as luminous as star A.

13. What is the surface temperature of a star that has a peak wavelength of 290 nm?
    a. 1000 K
    b. 2000 K
    c. 5000 K
    d. 10,000 K
    e. 100,000 K

14. If a planet is in thermal equilibrium,
    a. no energy is leaving the planet.
    b. no energy is arriving on the planet.
    c. the amount of energy leaving equals the amount of energy arriving.
    d. the temperature is very low.

15. The temperature of an object has a specific meaning as it relates to the object's atoms. A high temperature means that the atoms
    a. are very large.
    b. are moving very fast.
    c. are all moving together.
    d. have a lot of energy.

Thinking about the Concepts

16. The speed of light in a vacuum is 3×10^5 km/s. Can light travel at a lower speed? Explain your answer.

17. Is light a wave or a particle or both? Explain your answer.

18. If any of the experiments mentioned in the Process of Science Figure had *not* agreed with the others, what would that mean for the conclusion that light has a finite, constant speed?

19. If photons of blue light have more energy than photons of red light, how can a beam of red light carry as much energy as a beam of blue light?

20. Patterns of emission or absorption lines in spectra can uniquely identify individual atomic elements. How can the positive identification of atomic elements be used to test the validity of the cosmological principle discussed in Chapter 1?

21. An atom in an excited state can drop to a lower energy state by emitting a photon. Can we predict exactly how long the atom will remain in the higher energy state? Explain your answer.

22. Spectra of astronomical objects show both bright and dark lines. Describe what those lines indicate about the atoms responsible for the spectral lines.

23. Astronomers describe certain celestial objects as being *redshifted* or *blueshifted*. What do those terms indicate about the objects?

24. An object somewhere near you is emitting a pure tone at middle C on the octave scale (262 Hz). You, having perfect pitch, hear the tone as A above middle C (440 Hz). Describe the motion of that object relative to where you are standing.

25. During a popular art exhibition, the museum staff finds it necessary to protect the artwork by limiting the total number of viewers in the museum at any particular time. New viewers are admitted at the same rate that others leave. Is that system an example of static equilibrium or of dynamic equilibrium? Explain.

26. A favorite object for amateur astronomers is the double star Albireo, with one of its components a golden yellow and the other a bright blue. What do those colors tell you about the relative temperatures of the two stars?

27. The stars you see in the night sky cover a large range of brightness. What does that range tell you about the distances of the various stars? Explain your answer.

28. Why is it not surprising that sunlight peaks in the "visible"?

29. In Figure 5.24, why is the range of temperatures so much greater for Mercury than for the other planets?

30. Suppose you want to find a planet with the same temperature as Earth. What could you say about the size of the orbit of such a planet if it is orbiting a cooler red star? A yellow star like the Sun? A hotter blue star?

Applying the Concepts

31. You are tuned to 790 on AM radio. That station is broadcasting at a frequency of 790 kHz (7.90×10^5 Hz). You switch to 98.3 on FM radio. That station is broadcasting at a frequency of 98.3 MHz (9.83×10^7 Hz).
    a. What are the wavelengths of the AM and FM radio signals?
    b. Which broadcasts at higher frequencies, AM or FM?
    c. What are the photon energies of the two broadcasts?

32. Your microwave oven cooks by vibrating water molecules at a frequency of 2.45 gigahertz (GHz), or 2.45×10^9 Hz. What is the wavelength, in centimeters, of the microwave's electromagnetic radiation?

33. Assume that an object emitting a pure tone of 440 Hz is on a vehicle approaching you at a speed of 25 m/s. If the speed of sound at this particular atmospheric temperature and pressure is 340 m/s, what will be the frequency of the sound that you hear? (Hint: Keep in mind that frequency is inversely proportional to wavelength.)

34. You observe a spectral line of hydrogen at a wavelength of 502.3 nm in a distant galaxy. The rest wavelength of that line is 486.1 nm. What is the radial velocity of that galaxy? Is it moving toward you or away from you?

35. If half of the phosphorescent atoms in a glow-in-the-dark toy give up a photon every 30 minutes, how bright (relative to its original brightness) will the toy be after 2 hours?

36. How bright would the Sun appear from Neptune, 30 AU from the Sun, compared with its brightness as seen from Earth? The spacecraft *Voyager 1* is now about 140 AU from the Sun and heading out of the Solar System. Compare the brightness of the Sun as seen from *Voyager 1* with that seen from Earth.

37. On a dark night you notice that a distant lightbulb happens to have the same brightness as a firefly 5 meters away from you. If the lightbulb is a million times more luminous than the firefly, how far away is the lightbulb?

38. Two stars appear to have the same brightness, but one star is 3 times more distant than the other. How much more luminous is the more distant star?

39. A panel with an area of 1 square meter (m^2) is heated to a temperature of 500 K. How many watts is the panel radiating into its surroundings?

40. The Sun has a radius of 6.96×10^5 km and a blackbody temperature of 5780 K. Calculate the Sun's luminosity.

41. Some of the hottest stars known have a blackbody temperature of 100,000 K. What is the peak wavelength of their radiation? What type of radiation is it?

42. Your body emits radiation at a temperature of about 98.6°F.
    a. What is that temperature in kelvins? What is the peak wavelength, in microns, of your emitted radiation? What region of the spectrum is this?
    b. If you assume an exposed body surface area of 0.25 m^2, how many watts of power do you radiate?

43. A planet with no atmosphere at 1 AU from the Sun would have an average blackbody surface temperature of 279 K if it absorbed all the Sun's electromagnetic energy falling on it (albedo = 0).
    a. What would the average temperature on that planet be if its albedo were 0.1, typical of a rock-covered surface?
    b. What would the average temperature be if its albedo were 0.9, typical of a snow-covered surface?

44. The orbit of Eris, a dwarf planet, carries it out to a maximum distance of 97.7 AU from the Sun. If you assume an albedo of 0.8, what is the average temperature of Eris when it is farthest from the Sun?

45. Suppose our Sun had 10 times its current luminosity. What would the average blackbody surface temperature of Earth be if Earth had the same albedo?

USING THE WEB

46. a. Go to the website for NASA's Astronomy Picture of the Day (https://apod.nasa.gov/apod/ap101027.html) and study the picture of the Andromeda Galaxy in visible light and in ultraviolet light (slide the mouse on and off to flip between the two pictures). Which light represents a hotter temperature? What differences do you see in the two images?
    b. Go to the APOD archive (https://apod.nasa.gov/cgi-bin/apod/apod_search) and enter "false color" in the search box. Examine a few images that come up in the search. What does *false color* mean in that context? What wavelength(s) were the pictures exposed in? What is the color coding; that is, what wavelength does each color in the image represent? You can read more about false color at http://chandra.harvard.edu/photo/false_color.html.

47. Crime scene investigators may use different types of light to examine a crime scene. Search for "forensic lighting" in your browser. What wavelengths of light are used to search for blood and saliva? For fingerprints? Why is having access to different kinds of light useful for an investigator? Search on "forensic spectroscopy" and select a recent report. How is spectroscopy being used in crime scene investigations?

48. Using Google Images or an equivalent website, search for "night vision imaging" and "thermal imaging." How do night vision goggles and thermal-imaging devices work differently from regular binoculars or cameras? When are night vision or thermal-imaging tools useful?

49. The Transportation Security Administration (TSA) uses several types of imaging devices to screen passengers in airports. Search for "TSA imaging" in your browser. What wavelengths of light are being used in those devices? What concerns do passengers have about some of those imaging devices?

50. Go to the NASA Earth Observations website (https://neo.sci.gsfc.nasa.gov) and look at a map of Earth's albedo (click on "albedo" in the menu for "energy" or "land" if it didn't come up). Compare that map with those of 2, 4, 6, 8, and 10 months earlier. Which parts of Earth have the lowest and highest albedos? In which parts do the albedos seem to change the most with the time of the year? Would you expect ice, snow, oceans, clouds, forests, and deserts to add or subtract in each case from the total Earth albedo? Which parts of Earth are not showing up on this map?

EXPLORATION

Light as a Wave

digital.wwnorton.com/astro6

Visit the Digital Resources Page and on the Student Site open the "Light as a Wave, Light as a Photon" AstroTour in Chapter 5. Watch the first section and then click through, using the "Play" button, until you reach "Section 2 of 3."

Here we explore the following questions: How many properties does a wave have? Are any of those properties related to each other?

Work your way to the experimental section, where you can adjust the properties of the wave. Watch the simulation for a moment to see how fast the frequency counter increases.

1 Increase the wavelength by pressing the arrow key. What happens to the rate of the frequency counter?

_____

_____

2 Reset the simulation and then decrease the wavelength. What happens to the rate of the frequency counter?

_____

_____

3 How are the wavelength and frequency related to each other?

_____

_____

4 Imagine that you increase the frequency instead of the wavelength. How should the wavelength change when you increase the frequency?

_____

_____

5 Reset the simulation, and increase the frequency. Did the wavelength change in the way you expected?

_____

_____

_____

6 Reset the simulation, and increase the amplitude. What happens to the wavelength and the frequency counter?

_____

_____

7 Decrease the amplitude. What happens to the wavelength and the frequency counter?

_____

_____

8 Is the amplitude related to the wavelength or frequency?

_____

_____

_____

9 Why can't you change the speed of this wave?

_____

_____

6 The Tools of the Astronomer

In Chapter 5, we explained that astronomers learn about the physical and chemical properties of distant planets, stars, and galaxies by studying the light they emit. That electromagnetic radiation, though, must first be collected and processed before it can be analyzed and converted to useful knowledge. Here in Chapter 6, we describe the tools that astronomers use to capture and scrutinize that information.

LEARNING GOALS

By the end of this chapter, you should be able to:

LG 1 Compare how the two main types of optical telescopes gather and focus light.

LG 2 Summarize the main types of detectors that are used on telescopes.

LG 3 Explain why some wavelengths of radiation must be observed from high, dry, and remote observatories on Earth, or from space.

LG 4 Explain the benefits of sending spacecraft to study the planets and moons of our Solar System.

LG 5 Describe other astronomical tools that contribute to the study of the universe.

The twin 10-meter Keck reflectors on Mauna Kea, Hawaii, have a multiple-mirror, compact design. ▶▶▶

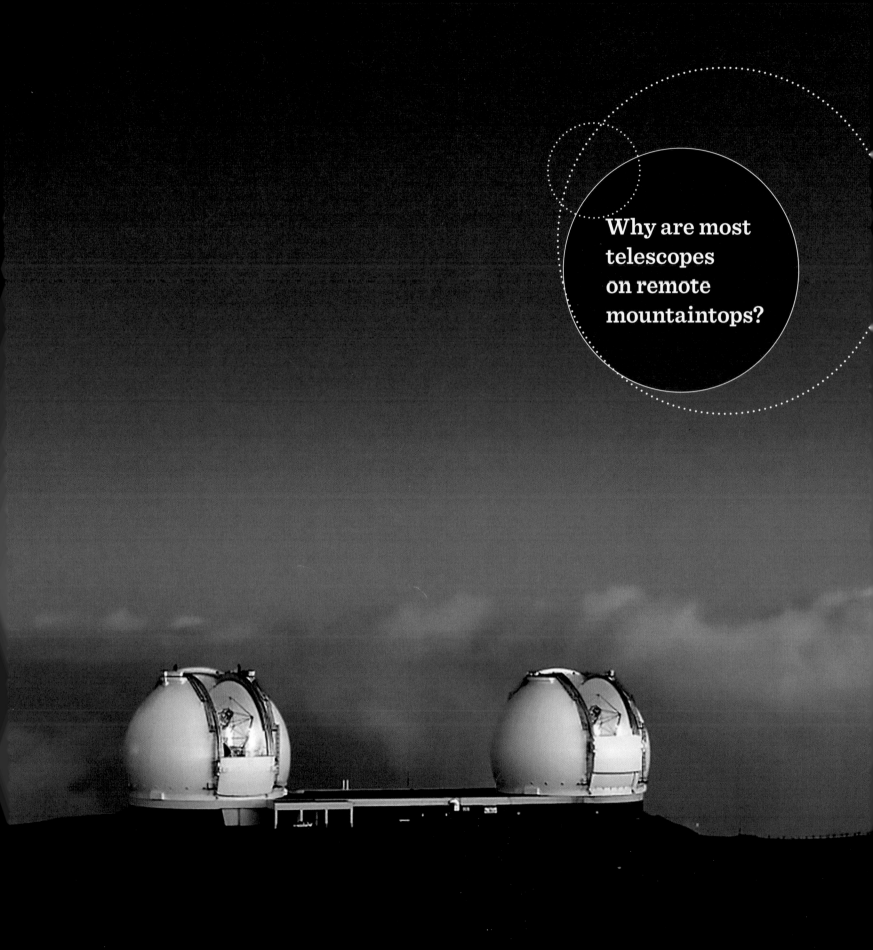

Why are most telescopes on remote mountaintops?

142 CHAPTER 6 The Tools of the Astronomer

Figure 6.1 A schematic view of the human eye creating an image of an object (the blue arrow).

6.1 The Optical Telescope Revolutionized Astronomy

Astronomical observations began with the human eye—information about the overall colors of stars and their brightness in the night sky is apparent even to the naked eye—that is, unassisted by binoculars or telescopes or filters. The development of **telescopes**—devices for collecting and focusing light—in the 17th century greatly increased the amount of light that can be collected from astronomical objects. With modern telescopes, astronomers can detect light that has been traveling across space for billions of years—even electromagnetic radiation from soon after the Big Bang, the beginning of the universe itself.

The Eye

Human eyes are sensitive to light with wavelengths ranging from about 380 nanometers (nm) (deep violet) to 750 nm (far red). A simplified schematic of the human eye is shown in **Figure 6.1**. The part of the human eye that detects light is called the retina, and the individual receptor cells that respond to light falling on the retina are called rods and cones. The center of the human retina consists solely of cones, which detect color and provide the greatest visual acuity. Away from the center, rods and cones intermingle, with rods dominating far from the center, where they are responsible for peripheral vision.

Our vision is limited by the eye's angular **resolution**, which refers to how close two points of light can be to each other before we can no longer distinguish them. Unaided, the best human eyes can resolve objects separated by 1 arcminute (1/60 of a degree), an angular distance of about 1/30 the diameter of the full Moon. (A more in-depth description of angular units—radians, degrees, arcminutes, and arcseconds—can be found in Appendix 1.4.) That may seem small, but thousands of stars and galaxies may reside within a patch of sky with that diameter.

Refracting Telescopes

Optical telescopes come in two primary types: **refracting telescopes**, which use lenses, and **reflecting telescopes**, which use mirrors. For all telescopes, the "size" of the telescope refers to the diameter of the largest mirror or lens, which determines the light-collecting area. That diameter is called the **aperture**. The light-gathering power of a telescope is proportional to its opening—that is, to the square of its diameter. The larger the aperture, the more light the telescope can collect. A "1-meter telescope" has a primary mirror (or lens) that is 1 meter in diameter. The aperture of the human eye is 6–7 millimeters (mm).

In the late 13th century, craftsmen in Venice were making small lentil-shaped disks of glass that could be mounted in frames and worn over the eyes to improve vision. More than 300 years later, Hans Lippershey (1570–1619), a spectacle maker living in the Netherlands, put two of his lenses together in a tube. With that new instrument, he saw distant objects magnified and could see farther. Galileo Galilei heard news of that invention and constructed one of his own. Recall from Chapter 3 that by 1610, Galileo had become the first to see the phases of Venus and the moons of Jupiter and was among the first to see craters on the Moon. He also was the first to realize that the Milky Way is made up of many individual stars. The refracting telescope—one that uses lenses—quickly revolutionized the science of astronomy.

Figure 6.2 (a) When wavefronts enter a new medium, they bend in a new direction relative to a line perpendicular to the surface (black lines). (b) An actual light ray entering and leaving a medium.

Refraction is the basis for the refracting telescope. Recall from Chapter 5 that the speed of light is constant in a vacuum, but through a medium such as air or glass, the speed of light is always lower. As light enters a new medium, its speed changes. If the light strikes the surface at an angle, some of the crest of the wave arrives at the surface earlier and some arrives later. You can see that happen in **Figure 6.2a**, a schematic diagram of wave crests (red lines) striking a medium at an angle. **Figure 6.2b** shows an actual light ray passing into and out of a medium (here, glass). The ray bends each time the medium changes. That bending of light when it changes the medium through which it travels is called **refraction**.

The amount of refraction depends on the properties of the medium. A medium's index of refraction (n) is equal to the ratio of the speed of light in a vacuum (c) to its speed in the medium (v). That relationship can be expressed by the equation $n = c/v$. For example, most glass has an index of refraction of approximately 1.5, so the speed of light in glass is 300,000 kilometers per second (km/s) divided by 1.5, or 200,000 km/s. The light bends by an amount that depends on the index of refraction of the materials involved and the angle at which the light strikes.

The primary lens in a refracting telescope is a simple convex lens, called the **objective lens** (**Figure 6.3**), whose curved surfaces refract the light from a distant object. That refracted light forms an image on the telescope's **focal plane**, which is perpendicular to the *optical axis*—the path that light takes through the center of the lens. Because the telescope's glass lens is curved, light at the outer edges of the lens strikes the surface more obliquely than light near the center. As a result, light at the outer edges of the lens is refracted more than light near its center. The lens concentrates the light rays entering the telescope, bringing them to a sharp focus at a distance called the **focal length**. Sometimes focal length is specified on a telescope as *focal ratio*, which equals focal length divided by aperture; that term may be familiar to you from the lenses used in photography.

Figure 6.4 illustrates how a telescope uses the light that passes through its lenses. Figure 6.4a shows the light from two stars passing through a lens and converging at the focal plane of the lens. Figure 6.4b shows the same situation for a lens with a longer focal length. Longer focal lengths increase the size and separation of objects in the focal plane. Aperture and focal length are the two most important parameters of a telescope. The image can be viewed with an **eyepiece**—a changeable lens whose focal length determines the magnification (**Working It Out 6.1**). In modern research, however, the images are sent directly to a camera or other detector.

Refracting telescopes have two major shortcomings. First, physical limits constrain the size of refracting telescopes. The larger the area of the objective lens, the more light-gathering power it has and the fainter the stars we can observe. As objective lenses get larger, however, they get heavier, and a massive piece of glass at the end of a very long tube sags too much under the force of gravity. Refracting telescopes grew in size until the 1897 completion of the Yerkes 1-meter refractor (**Figure 6.5**), the world's largest operational refracting telescope. Located in Williams Bay, Wisconsin, the Yerkes telescope consists of a 450-kilogram (kg) objective lens mounted at the end of a 19.2-meter tube.

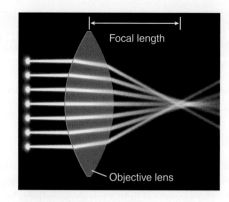

Figure 6.3 For a curved lens such as the one shown, refraction causes the light to focus to a point. That point is in a slightly different location for different wavelengths (colors) of light.

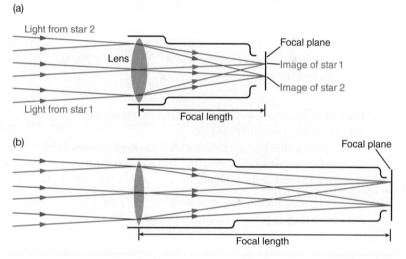

Figure 6.4 (a) A refracting telescope uses a lens to collect and focus light from two stars, forming images of the stars on its focal plane. (b) A telescope with a longer focal length produces larger images.

Figure 6.5 (a) The Yerkes 1-meter refractor is the world's largest refracting telescope. (b) The main parts of a refractor.

6.1 Working It Out Telescope Aperture and Magnification

If you are shopping for a telescope, you need to consider its aperture and magnification.

Aperture

The light-gathering power of a telescope is proportional to the area of its lens or mirror, $\pi \times (D/2)^2$, and thus to the square of the aperture (D). The amount of light a telescope collects increases as the aperture increases. How does the light-gathering power of a telescope with a diameter of 200 mm, or 8 inches, compare with that of the pupil of your eye, which is about 6 mm in the dark?

$$\text{Light-gathering power of telescope} = \frac{\pi}{4} \times (200\,\text{mm})^2$$

and

$$\text{Light-gathering power of eye} = \frac{\pi}{4} \times (6\,\text{mm})^2$$

So, to compare:

$$\frac{\text{Light-gathering power of telescope}}{\text{Light-gathering power of eye}} = \frac{\frac{\cancel{\pi}}{\cancel{4}}(200\,\text{mm})^2}{\frac{\cancel{\pi}}{\cancel{4}}(6\,\text{mm})^2} = \left(\frac{200}{6}\right)^2 = 1{,}111$$

Thus, an 8-inch telescope has more than 1,000 times the light-gathering power of your eye.

Comparing that 8-inch telescope with the Keck 10-meter telescope shows why bigger is better: 200 mm = 0.2 meter, and we cancel out the $\pi/4$ again to obtain

$$\frac{\text{Light-gathering power of Keck}}{\text{Light-gathering power of 8-inch telescope}} = \left(\frac{10\,\text{m}}{0.2\,\text{m}}\right)^2 = 2{,}500$$

Even larger telescopes, 25–40 meters in diameter, are under construction.

Magnification

Most telescopes have a set focal length and come with a collection of eyepieces. The magnification of the image in the telescope is given by

$$\text{Magnification} = \frac{\text{Telescope focal length}}{\text{Eyepiece focal length}}$$

Suppose the focal length of the 200-mm telescope in the preceding example is 2,000 mm. Combined with the focal length of a standard eyepiece, 25 mm, that telescope will give the following magnification:

$$\text{Magnification} = \frac{2{,}000\,\text{mm}}{25\,\text{mm}} = 80$$

That telescope and eyepiece combination has a magnifying power of 80, meaning that a crater on the Moon will appear 80 times larger in the telescope's eyepiece than it does when viewed by the naked eye. An eyepiece that has a focal length of 8 mm will have about 3 times more magnifying power, or 250 times.

A higher magnification will not necessarily let you see an object better. A faint and fuzzy image will not look clearer when magnified.

The second major shortcoming of refracting telescopes is **chromatic aberration** (**Figure 6.6**). Starlight is made up of all the colors of the rainbow, and each color refracts at a slightly different angle because the index of refraction depends on the wavelength of the light. Shorter (bluer) wavelengths are refracted more than longer (redder) wavelengths (Figure 6.6a). That wavelength-dependent difference in refraction, which spreads the white light out into its spectral colors, is

Figure 6.6 (a) Different wavelengths of light come to different foci along the optical axis of a simple lens, causing chromatic aberration. (b) A compound lens using two types of glass with different indices of refraction can compensate for much of the chromatic aberration, so different colors of light all come to a focus at the same point.

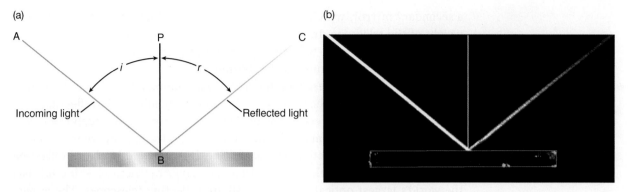

Figure 6.7 (a) When a ray of incoming light (AB) shines on a flat surface, it reflects from the surface, becoming the reflected ray BC. The angle between AB and PB, the perpendicular to the surface, is the angle of incidence (i). The angle between BC and PB is the angle of reflection (r). The angles of incidence and reflection are always equal. (b) Light from a laser beam is reflected from a flat glass surface.

called **dispersion**. Dispersion causes bluer light to come to a shorter focus than that of the longer visible wavelengths, creating chromatic aberration. In a refracting telescope with a simple convex lens, chromatic aberration produces haloed images around the star. That issue was addressed in early telescopes by increasing the focal length, but then the telescope requires a longer tube. Now manufacturers of quality cameras and telescopes use a **compound lens** composed of two types of glass to correct for chromatic aberration (Figure 6.6b).

Reflecting Telescopes

Another property of light, called **reflection**, is the basis for reflecting telescopes. When light encounters a different medium—here, going from air to glass—some of it will be reflected from the surface of the new medium. In other words, some of the light will reverse its direction of travel. The most common example occurs when light encounters an ordinary flat mirror (**Figure 6.7**). The angle of the incoming light and the angle of outgoing light are always equal. A reflected image in a mirror looks like the original image, although left and right are interchanged.

In 1668, Isaac Newton designed a reflecting telescope, which uses mirrors instead of lenses (**Figure 6.8a**). The angle of reflection does not depend on the wavelength of light, so chromatic aberration is *not* a problem in reflecting telescopes. The important parts of Newton's reflecting telescope are shown schematically in **Figure 6.8b**. To make that reflecting telescope, Newton cast a 2-inch primary mirror made of copper and tin and polished it to a special curvature. He then placed the primary mirror at the bottom of a tube with a secondary flat mirror mounted above it at a 45° angle. The second mirror directed the focused light to an eyepiece on the outside of the tube.

Astronomers use mirrors with a surface that curves inward toward the incoming light, called *concave* mirrors. The same rules of incidence and reflection for flat mirrors hold here for each ray of light, but for concave mirrors the reflected rays do not maintain the same angle with respect to each other as they do with a flat mirror. Concave mirrors will reflect the rays so that they converge to form an image (**Figure 6.9**). If the incoming light rays are parallel, as from a distant source such as a star, the reflected light rays cross at the focal length of the mirror. The light path from the primary mirror to the focal plane can be "folded" by using

Figure 6.8 (a) Newton's reflecting telescope. (b) A Newtonian focus telescope has a curved primary and a flat secondary mirror, as shown in the sketch. The eyepiece is on the side of the tube.

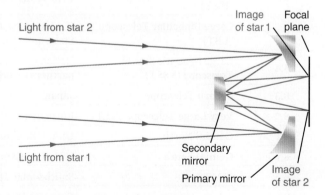

Figure 6.9 Large reflecting telescopes often use Cassegrain focus, in which a secondary mirror directs the light back through a hole in the concave parabolic primary mirror to an accessible focal plane behind the primary mirror.

a secondary mirror, which enables a significant reduction in the length and weight of the telescope. In many modern telescopes, the primary mirror has a hole so that light can pass back through it; the eyepiece is on the back of the tube of the telescope, and the tube can be shortened.

AstroTour: Geometric Optics and Lenses

Large reflecting telescopes did not become common until the latter half of the 18th century. Since then, the size of the primary mirrors in reflecting telescopes has grown larger every decade. Primary mirrors can be supported from the back, and they can be made thinner and therefore less massive than the objective lenses found in refracting telescopes. The limitation on the size of reflecting telescopes is the cost of their fabrication and support structure. **Table 6.1** lists the world's largest optical telescopes. All are reflecting telescopes. The largest

TABLE 6.1 The World's Largest Optical Telescopes

Mirror Diameter (meters)	Telescope	Sponsor(s)	Location	Operational Date
39.3	Extremely Large Telescope (ELT)	European Southern Observatory (Europe, Chile, Brazil)	Cerro Armazones, Chile	Under construction
30.0	Thirty Meter Telescope (TMT)	International collaboration led by Caltech, U. of California, U. of Hawaii, China, Japan, India, and Canada	Mauna Kea, Hawaii, or Canary Islands, Spain	TBD
24.5	Giant Magellan Telescope (GMT)	Carnegie Institution, Arizona State U., Harvard U., Smithsonian Institution, U. of Arizona, U. of Texas, Texas A&M U., U. of Chicago, São Paulo Research Foundation (FAPESP), Australian National U., Astronomy Australia Ltd., and Korea Astronomy and Space Science Institute	Cerro Las Campanas, Chile	Under construction
10.4	Gran Telescopio Canarias (GTC)	Spain, Mexico, U. of Florida	Canary Islands	2007
10	Keck I	Caltech, U. of California, NASA	Mauna Kea, Hawaii	1993
10	Keck II	Caltech, U. of California, NASA	Mauna Kea, Hawaii	1996
10	South African Large Telescope (SALT)	South Africa, USA, UK, Germany, Poland, New Zealand, India	Sutherland, South Africa	2005
10	Hobby-Eberly Telescope (HET)	U. of Texas, Penn State U., Stanford U., Germany	Mount Fowlkes, Texas	1999
8.4 × 2	Large Binocular Telescope (LBT)	U. of Arizona, Ohio State U., Italy, Germany, Arizona State, and others	Mount Graham, Arizona	2008
8.4	Large Synoptic Survey Telescope (LSST)	National Science Foundation, Dept. of Energy, and other partners through the private LSST corporation	Cerro Pachón, Chile	Under construction
8.3	Subaru Telescope	Japan	Mauna Kea, Hawaii	1999
8.2 × 4	Very Large Telescope (VLT)	European Southern Observatory	Cerro Paranal, Chile	2000
8.1	Gemini North	USA, UK, Canada, Chile, Brazil, Argentina, Australia	Mauna Kea, Hawaii	1999
8.1	Gemini South	USA, UK, Canada, Chile, Brazil, Argentina, Australia	Cerro Pachón, Chile	2000
6.5	MMT	Smithsonian Institution, U. of Arizona	Tucson, Arizona	2000
6.5	Magellan I	Carnegie Institution, U. of Arizona, Harvard U., U. of Michigan, MIT	Cerro Las Campanas, Chile	2000
6.5	Magellan II	Carnegie Institution, U. of Arizona, Harvard U., U. of Michigan, MIT	Cerro Las Campanas, Chile	2002

single mirrors constructed today are 8 meters in diameter, but reflecting telescopes even bigger than that are designed to make use of an array of smaller segments. The primary mirror of each of the 10-meter twin Keck telescopes is made up of 36 hexagon-shaped segments that are 1.8 meters in diameter (**Figure 6.10**). Located on 4,100-meter-high Mauna Kea in Hawaii, the Keck telescopes are among the world's largest reflecting telescopes. Each one has 4 million times the light-gathering power of the human eye.

CHECK YOUR UNDERSTANDING 6.1A

Why are all large astronomical telescopes reflectors (choose all reasons that apply)? (a) chromatic aberration is minimized; (b) they are not as heavy; (c) they can be shorter; (d) the glass doesn't need to be curved.

Optical and Atmospheric Limitations

The *resolution* of a telescope determines how close two points of light can be to each other before they are indistinguishable (**Figure 6.11**). Review Figure 6.4a to see the path followed by rays of light from two distant stars as they pass through the lens of a refracting telescope. Figure 6.4b illustrated that increasing the focal length increases the size of and separation between the images that a telescope produces. That is one reason why telescopes provide a much clearer view of the stars than that obtained with the naked eye. The focal length of a human eye is typically about 20 mm, whereas telescopes used by professional astronomers often have focal lengths of tens or even hundreds of meters. Such telescopes make images that are far larger than those formed by the human eye, and consequently they contain far more detail.

Focal length explains only one difference between the resolution of telescopes and that of the unaided eye. The other difference results from the wave nature of light. **Figure 6.12** shows that light waves spread out from the edges of the lens or mirror when they pass through the aperture of a telescope. The distortion that occurs as light passes the edge of an opaque object is called **diffraction**. Diffraction "diverts" some of the light from its path, slightly blurring the image

Figure 6.10 Each of the Keck 10-meter reflectors uses an aligned group of 36 hexagonal mirrors to collect light.

Figure 6.11 Resolution is the ability to separate two images that appear close together. When resolution is lower (a), the two images blend together. When resolution is higher (b), individual images can be distinguished.

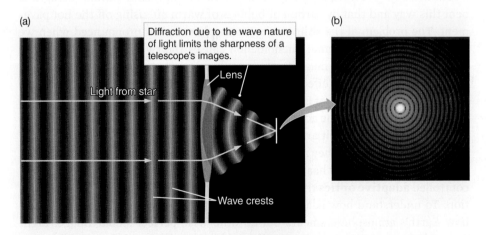

Figure 6.12 (a) Light waves from a star are diffracted by the edges of a telescope's lens or mirror. (b) That diffraction causes the stellar image to be blurred, limiting a telescope's ability to resolve objects.

6.2 Working It Out Diffraction Limit

The practical limit on the angular resolution of a telescope (θ) is called the diffraction limit. That limit is determined by the ratio of the wavelength of light passing through the telescope (λ) to the diameter of the aperture (D):

$$\theta = 2.06 \times 10^5 \left(\frac{\lambda}{D}\right) \text{arcsec}$$

The constant, 2.06×10^5, has units of arcseconds (arcsec). An **arcsecond** is a tiny angular measure found by first dividing the sky into 360 degrees, then dividing each degree by 60 to get arcminutes, and then dividing each arcminute by 60 to get arcseconds. An arcsecond is 1/1,800 of the size of the Moon in the sky, or about the size of a tennis ball if you could see it from 13 km (8 miles) away.

Both λ and D must be expressed in the same units—usually meters. The smaller the ratio of λ/D, the better the resolution. For example, the size of the human pupil (see Figure 6.1) ranges from about 2 mm in bright light to 8 mm (0.008 meter) in the dark. Visible (green) light has a wavelength (λ) of 550 nm—that is, 550×10^{-9} meter, or 5.5×10^{-7} meter. Using those values for the aperture and the wavelength gives

$$\theta = 2.06 \times 10^5 \left(\frac{5.5 \times 10^{-7}\,\text{m}}{0.008\,\text{m}}\right) \text{arcsec} = 14\,\text{arcsec}$$

or about 1/4 arcmin for the theoretical resolution. The typical actual resolution of the human eye is 1 arcmin. We do not achieve the theoretical resolution with our eyes because the physical properties of our eyes are not perfect.

How does the resolution of the human eye compare with that of the Hubble Space Telescope when it is operating in the visible part of the spectrum? Its primary mirror has a diameter of 2.4 meters. Substituting that value for D and again using visible (green) light gives

$$\theta = 2.06 \times 10^5 \left(\frac{5.5 \times 10^{-7}\,\text{m}}{2.4\,\text{m}}\right) \text{arcsec} = 0.047\,\text{arcsec}$$

or about 300 times better than the theoretical resolving power of the human eye.

---

made by the telescope. The degree of blurring depends on the wavelength of the light and the telescope's aperture. The larger the aperture, the smaller the problem diffraction poses. The best resolution that a given telescope can achieve is known as the **diffraction limit** (**Working It Out 6.2**).

Larger telescopes have better resolution and can distinguish objects that appear closer together. Theoretically, the 10-meter Keck telescopes have a diffraction-limited resolution of 0.0113 arcsec in visible light, which would be good enough for you to read newspaper headlines 60 km away. But for telescopes with apertures larger than about a meter, Earth's atmosphere stands in the way of better resolution. If you have ever looked out across a large asphalt parking lot on a summer day, you have probably seen the distant horizon shimmer as light is bent this way and that by turbulent bubbles of warm air rising off the hot pavement. The problem of the shimmering atmosphere is less pronounced when we look overhead, but the same phenomenon causes the twinkling of stars in the night sky. As telescopes magnify the angular diameter of an object, they also magnify the shimmering effects of the atmosphere. The limit on the resolution of a telescope on the surface of Earth caused by that atmospheric distortion is called **astronomical seeing**. One advantage of launching telescopes such as the Hubble Space Telescope into orbit around Earth is that they are not hindered by astronomical seeing because they are above the atmosphere.

Modern technology has improved ground-based telescopes with computer-controlled **adaptive optics** that compensate for much of the atmosphere's distortion. To understand how adaptive optics work, we need to look more closely at how Earth's atmosphere smears out an otherwise perfect stellar image. Light from a distant star arrives at the top of Earth's atmosphere with flat, parallel wave crests. If Earth's atmosphere were perfectly uniform, the crests would remain flat as they reached the objective lens or primary mirror of a ground-based telescope. After making their way through the telescope's optical system, the crests would produce a tiny diffraction disk in the focal plane, as shown in Figure 6.12b. But Earth's atmosphere is not uniform. It is filled with bubbles of air that have slightly different temperatures from those of their surroundings. Different temperatures mean different densities, and different densities mean different refractive properties, so each bubble bends light differently.

Those air bubbles act as weak lenses, and by the time the waves reach the telescope they are far from flat, as shown in **Figure 6.13**. Instead of a tiny diffraction disk, the image in the telescope's focal plane is distorted and swollen, degrading the resolution. Adaptive optics flatten out that distortion. First, an optical device within the telescope constantly measures the wave crests. Then, before reaching the telescope's focal plane, the light is reflected off yet another mirror, which has a flexible surface. A computer analyzes the light and bends the flexible mirror so that it accurately corrects for the distortion caused by the air bubbles. **Figure 6.14** shows an example of an image corrected by adaptive optics. To calibrate those systems, a laser beam is shone into the atmosphere to create an artificial "star" in the same direction as the telescope is pointing. The widespread use of adaptive optics has made the image quality of ground-based telescopes competitive with the quality of Hubble images from space at some wavelengths. Telescopes also can have active optics, in which actuators on the rear of the mirror correct for factors such as wind, temperature, and sagging from gravity.

Observatory Locations

What makes a good location for a telescope on Earth? According to Table 6.1, astronomers look for sites that are high, dry, and dark. The best sites are far away from the lights of cities; in locations with little moisture, humidity, or rain; and where the atmosphere is relatively still. Telescopes are located as high as possible so that they get above a significant part of Earth's atmosphere, which distorts images and blocks infrared and microwave light. Many telescopes are situated on remote, high mountaintops surrounded by desert or ocean. Recall from Chapter 2 that the stars that can be seen throughout the year depend on latitude, and only at the equator would a telescope have access to all the stars in the sky. But because equatorial latitudes have tropical weather—wet, humid, and stormy—they are poor locations for telescopes. To cover the entire sky, astronomers have built telescopes in both northern and southern locations. In the United States, large telescopes are located in California, Arizona, New Mexico, Texas, and Hawaii. The largest southern-sky observatories are found in Chile, South Africa, and Australia. The twin Gemini telescopes, designed to be a matched pair, are located in Hawaii in the Northern Hemisphere and in Chile in the Southern Hemisphere.

Newer and larger telescopes are planned for many of the same locations listed in Table 6.1. The 8-meter Large Synoptic Survey Telescope (LSST) is headed for Cerro Pachón in Chile, current site of the Gemini South telescope. The Giant Magellan Telescope (GMT), consisting of seven 8-meter mirrors in a pattern equivalent to a 24.5-meter mirror, will be constructed at Cerro Las Campanas in Chile. The Thirty Meter Telescope (TMT) is planned for Mauna Kea in Hawaii or the Canary Islands, and the European Southern Observatory (ESO) is building the 39-meter Extremely Large Telescope (ELT) at

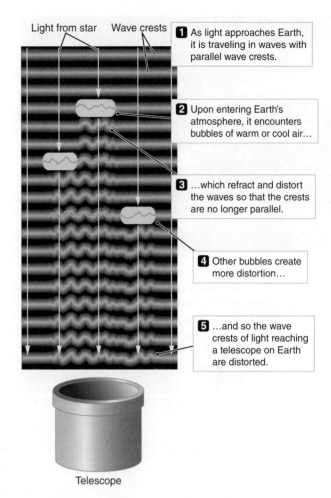

Figure 6.13 Bubbles of warmer or cooler air in Earth's atmosphere distort the wavefront of light from a distant object.

Figure 6.14 These images of the Cat's Eye Nebula from the Palomar Observatory telescope without (left) and with (right) adaptive optics show the benefit of the technique.

Figure 6.15 An artist's rendering of the European Extremely Large Telescope, a reflecting telescope under construction in Chile.

Cerro Armazones in Chile (**Figure 6.15**). As telescopes get larger—and more expensive—international collaboration becomes even more important.

Today's professional astronomers rarely look through the eyepiece of a telescope because they learn much more and make better use of observing time by permanently recording an object's image at a variety of wavelengths or seeing its light spread out into a revealing spectrum. Some astronomers no longer travel to telescopes at all, instead observing remotely from the base of the mountain or far away at their own institutions.

Professional and amateur astronomers alike are concerned about loss of the dark sky. As cities and suburbs around the world grow and expand, the use of outdoor artificial light becomes more widespread. Pictures from space show how bright many areas of Earth are at night (**Figure 6.16**). In the United States, two-thirds of the population resides in an area too bright to see the Milky Way in the sky at night, and it has been estimated that by 2025 the continental United States will have almost no dark skies. Increased air pollution also dims the view of the night sky in many locations. The U.S. National Park Service now advertises evening astronomy programs in natural, unpolluted dark skies as one of the reasons to visit some parks. Several international astronomy associations are working with UNESCO (the United Nations Educational, Scientific and Cultural Organization) to promote the "right to starlight," arguing that for historical, cultural, and scientific reasons, it would be a huge loss if humanity could no longer view the stars. Those organizations are encouraging countries to create starlight reserves and starlight parks where people can experience increasingly rare dark skies and a natural nocturnal environment.

CHECK YOUR UNDERSTANDING 6.1B

In practice, the smallest angular size that you can resolve with a 10-inch telescope is governed by the: (a) blurring caused by Earth's atmosphere; (b) diffraction limit of the telescope; (c) size of the primary mirror; (d) magnification of the telescope.

6.2 Optical Detectors and Instruments Used with Telescopes

Beginning in the 1800s, the development of film photography, and later digital photography, revolutionized astronomy, allowing astronomers to detect fainter and more distant objects than possible to detect with the eye alone. Here in Section 6.2, we examine some of the more common types of detectors.

Integration Time and Quantum Efficiency

Originally, the retina of the human eye was the only astronomical detector. The limit of the faintest stars we can see with our unaided eyes is determined in part by two factors that are characteristic of all detectors: *integration time* and *quantum efficiency*.

Integration time is the limited time interval during which the eye can add up photons—that is analogous to leaving the shutter open on a camera. The brain "reads out" the information gathered by the eye about every 100 milliseconds (ms). Anything that happens faster than that appears to happen all at once. If two images on a computer screen appear 30 ms apart, you will see them as a single image

Figure 6.16 This satellite image of Earth at night shows that few populated areas are free from light pollution.

because your eyes will add up (or integrate) whatever they see over an interval of 100 ms or less. If the images occur 200 ms apart, however, you will see them as separate images. That relatively brief integration time is the most important factor limiting our nighttime vision. Stars too faint to be seen with the unaided eye are those from which you receive too few photons for your eyes to process in 100 ms.

Quantum efficiency determines how many responses occur for each photon received. For the human eye, 10 photons must strike a cone within 100 ms to activate a single response. So the quantum efficiency of our eyes is about 10 percent: for every 10 events, the eye sends one signal to the brain. Together, integration time and quantum efficiency determine the rate at which photons must arrive at the retina before the brain says, "I see something." Astronomers seek to use detectors with longer integration times and higher quantum efficiency than our eyes can achieve.

Figure 6.17 A photograph of the Moon taken by John W. Draper in 1840.

From Photographic Plates to Charge-Coupled Devices

For more than two centuries after the invention of the telescope, astronomers struggled with the problem of surface brightness. Only *point sources* such as stars appear brighter in a telescope. Extended astronomical objects, such as the Moon, appear bigger in the eyepiece, but the light is spread over that larger image, so their surfaces are no brighter than they appear to the unaided eye. Even when astronomers built larger telescopes, nebulae and galaxies appeared larger, but the details of these faint objects remained elusive. The problem was not with the telescopes but with the limitations of optics and the human eye. Only with the longer exposure times made possible by the invention of photography and the later development of electronic cameras could astronomers finally discern intricate details in faint objects.

In 1840, John W. Draper (1811–1882), a New York chemistry professor, created the earliest known astronomical photograph (**Figure 6.17**). By the late 1800s, astronomers had created thousands of photographic plates with permanent images of planets, nebulae, and galaxies. The quantum efficiency of most photographic systems used in astronomy was poorer than that of the human eye—typically 1–3 percent. But unlike the eye, photography can overcome poor quantum efficiency by leaving the shutter open on the camera, increasing the integration time to many hours of exposure. Photography enabled astronomers to record and study objects invisible to the human eye. However, one problem is that the response of photography to light is not linear, especially at long exposures, so if you doubled the exposure time, you did not get twice as much light on your image. By the middle of the 20th century, the search was on for electronic detectors that would overcome photography's problems in sensitivity, spectral range, and nonlinearity.

In 1969, scientists at Bell Laboratories invented a detector called a **charge-coupled device**, or **CCD**. By the late 1970s, the CCD had become the detector of choice in almost all astronomical-imaging applications. CCDs are linear, so doubling the exposure means you record twice as much light. As a result, they are good for measuring objects that vary in brightness, as well as for faint objects that require long exposures. CCDs have a quantum efficiency far superior to that of photography or the eye—up to 80 percent at some wavelengths. That improvement dramatically increases the ability to view faint objects with short exposure times.

A CCD is an ultrathin wafer of silicon—less than the thickness of a human hair—that is divided into a two-dimensional array of picture elements, or **pixels** (**Figure 6.18a**). When a photon strikes a pixel, it creates a small electric charge

(a)

(b)

Figure 6.18 (a) In this simplified diagram of a charge-coupled device (CCD), photons from a star land on pixels (represented by gray squares) and produce free electrons within the silicon. The electron charges are electronically moved sequentially to the collecting register at the bottom. Each row is then moved out to the right to an electronic amplifier, which converts the electric charge of each pixel into a digital signal. (b) This large CCD (about 6 inches across) contains 12,288 × 8,192 pixels.

within the silicon. As each CCD pixel is read out, the digital signal that flows to the computer is nearly proportional to the accumulated charge. That is what we mean when we say that the CCD is a linear device. If a CCD is exposed to too much light, however, it can lose its linearity.

Liquid nitrogen or helium is used to cool the CCDs down to very low temperatures to reduce noise caused by the movement of the charge-carrying electrons within the silicon wafer. The first astronomical CCDs were small arrays containing a few hundred thousand pixels. The larger CCDs used in astronomy today may contain more than 100 million pixels (**Figure 6.18b**). Still larger arrays are under development as ever-faster computing power keeps up with image-processing demands.

The output from a CCD is a digital signal that can be sent directly from the telescope to image-processing software or stored electronically for later analysis. Nearly every spectacular astronomical image in ultraviolet, visible, or infrared wavelength that you find online was recorded by a CCD in a telescope either on the ground or in space. CCDs are found in many everyday devices, too, such as digital cameras, digital video cameras, and camera phones.

Your cell phone takes color pictures by using a grid of CCD pixels arranged in groups of three. Each pixel in a group is constructed to respond to only a particular range of colors—to only red light, for example. That also is true for digital image displays. You can see for yourself if you place a small drop of water on the screen of your smartphone or tablet and turn it on. The water magnifies the grid of pixels so that you can see them individually. The grid degrades the angular resolution of the camera because each spot in the final image requires three pixels of information. Astronomers instead use all the pixels on the camera to determine how many photons fall on each pixel, without regard to color. They put filters in front of the camera to allow light of only particular wavelengths to pass through, such as the light of a specific spectral line. Color pictures such as those from the Hubble Space Telescope are constructed by taking multiple pictures, coloring each one, and then carefully aligning and overlapping them to produce beautiful and informative images. Sometimes the colors are "true"; that is, they are close to the colors you would see if you were actually looking at the object with your eyes. Other times "false" colors represent different portions of the electromagnetic spectrum and tell you the temperature or composition of different parts of the object. Using changeable filters instead of designated color pixels gives astronomers greater flexibility and greater resolution.

Spectrographs

Spectroscopy is the study of an object's *spectrum* (plural: *spectra*)—its electromagnetic radiation split into component wavelengths. **Spectrographs** (sometimes called **spectrometers**) are instruments that take the spectrum of an object and then record it. The first spectrographs used prisms to disperse the light. Modern spectrographs use a **diffraction grating**, which is made by engraving closely spaced lines on glass to disperse incoming light into its constituent wavelengths. **Figure 6.19a** shows the light reflected from a CD or DVD; the closely spaced tracks act as a grating and create a spectrum. **Figure 6.19b** shows a grating spectrograph: Light from an astronomical object enters a telescope and passes through a slit. The light is reflected onto the diffraction grating, which creates a spectrum like the one shown in Figure 5.13. The spectra are recorded and then

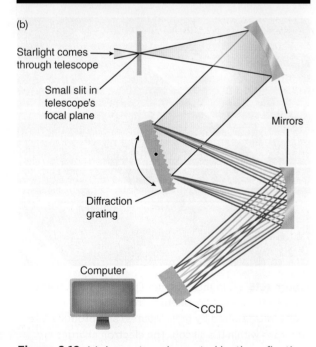

Figure 6.19 (a) A spectrum is created by the reflection of light from the closely spaced tracks of a CD. (b) In a grating spectrograph, light goes through the telescope and then a slit, where it is reflected to the diffraction grating and split into components. The spectrum is recorded on the CCD.

analyzed. Some modern spectrographs use bundles of optical fibers, or masks with multiple slits, to obtain spectra simultaneously from multiple objects in the field of view of the telescope.

CHECK YOUR UNDERSTANDING 6.2

CCD cameras have much higher quantum efficiency than that of other detectors. Therefore, CCD cameras: (a) can collect photons for longer times; (b) can collect photons of different energies; (c) can generate a signal from fewer photons; (d) can split light into different colors.

6.3 Astronomers Observe in Wavelengths Beyond the Visible

Recall from Section 5.4 (and Wien's law) that an object's temperature can be found from the peak wavelength of its continuous spectrum. Extending beyond visible light, radio or infrared telescopes are used to study cool objects, such as clouds of dust, whereas X-ray or gamma-ray telescopes are used to study violently hot gas. Thus, astronomers use telescopes that observe at all the wavelengths of the electromagnetic spectrum. **Figure 6.20** shows Earth's **atmospheric windows** and which parts of the spectrum they let in. The largest window is in radio wavelengths, including microwaves at the short-wavelength end of the radio window. Those telescopes can be built on the ground. However, gamma-ray,

Figure 6.20 Earth's atmosphere blocks most electromagnetic radiation. Fermi = Fermi Gamma-ray Space Telescope (orbiting); Chandra = Chandra X-ray Observatory (orbiting); HST = Hubble Space Telescope (orbiting); Keck = Keck Observatory (Hawaii); Spitzer = Spitzer Space Telescope (orbiting); JCMT = James Clerk Maxwell Telescope (Hawaii); VLA = Very Large Array (New Mexico); Green Bank = Robert C. Byrd Green Bank Telescope (West Virginia); Arecibo = Arecibo Observatory (Puerto Rico).

(a)

(b)

Figure 6.21 (a) The Green Bank telescope in West Virginia is the world's largest steerable radio telescope. (b) The FAST radio telescope in China is the world's largest single-dish telescope.

Figure 6.22 The VLA in New Mexico combines signals from 27 telescopes so that they act as one "very large" telescope.

X-ray, ultraviolet, and most of the infrared light arriving at Earth fails to reach the ground because it is partially or completely absorbed by ozone, water vapor, carbon dioxide, and other molecules in Earth's atmosphere. Light at those wavelengths has to be observed from space.

Radio Telescopes

Karl Jansky (1905–1950), a young physicist working for Bell Laboratories in the early 1930s, identified a radio source in the Milky Way in the direction of the galactic center, in the constellation Sagittarius. Jansky's discovery marked the birth of radio astronomy, and in his honor, the basic unit for the strength of a radio source is called the **jansky** (Jy). A few years later, Grote Reber (1911–2002), a radio engineer and ham radio operator, built his own radio telescope and conducted the first survey of the sky at radio frequencies. Reber was largely responsible for the rapid advancement of radio astronomy in the post–World War II era.

Most radio telescopes are large, steerable dishes, typically tens of meters in diameter, such as the one shown in **Figure 6.21a**. The world's largest single-dish radio telescopes are the new FAST (Five-hundred-meter Aperture Spherical Telescope) dish in China (**Figure 6.21b**) and the 305-meter Arecibo dish built into a natural bowl-shaped depression in Puerto Rico. Those large single-dish telescopes are not steerable, so they can observe only sources that pass within 20° of the zenith as Earth's rotation carries them overhead.

As large as radio telescopes are, they have relatively poor angular resolution. Recall that a telescope's angular resolution is determined by the ratio λ/D, so the resolution decreases as that ratio increases. Radio telescopes have diameters (D) much larger than the apertures of most optical telescopes. However, the wavelengths (λ) of radio waves range from about 1 centimeter (cm) to 10 meters, or up to several hundred thousand times greater than the wavelengths of visible light, which makes the ratio larger. Radio telescopes are thus limited by the very long wavelengths they are designed to receive. The resolution of the Arecibo dish is typically about 1 arcmin—similar to that of the unaided human eye.

Radio astronomers have had to develop ways to improve resolution. Mathematically combining the signals from two radio telescopes turns them into a telescope with a diameter equal to the separation between them. For example, if two 10-meter telescopes are located 1,000 meters apart, the D in λ/D is 1,000, not 10. That combination of two (or more) telescopes is called an **interferometer**, and it makes use of the wavelike properties of light. Usually, several telescopes are used in an arrangement called an **interferometric array**. Through the use of very large arrays, radio astronomers can better observe bright sources and exceed the angular resolution possible with optical telescopes.

The Very Large Array (VLA) in New Mexico (**Figure 6.22**) is an interferometric array made up of 27 movable dishes spread out in a Y-shaped configuration up to 36 km across. At a wavelength of 10 cm, that array reaches resolutions of less than 1 arcsec. The Very Long Baseline Array (VLBA) uses 10 radio telescopes spread out over more than 8,000 km from the Virgin Islands in the Caribbean to Hawaii in the Pacific. At a wavelength of 10 cm, that array can attain resolutions of better than 0.003 arcsec. A radio telescope put into near-Earth orbit as part of a Space Very Long Baseline Interferometer (SVLBI) overcomes even that limit. The new Event Horizon Telescope combines many of the most advanced existing radio telescopes, from Greenland to the South Pole, to make an *Earth-sized* interferometer. For a few nights each year, all the telescopes observe the same object, with a combined resolution that may be good enough to image objects near the center of the Milky Way.

Some radio telescopes use many small dishes. The Atacama Large Millimeter/submillimeter Array (ALMA; **Figure 6.23**), located at an elevation of 5,000 meters in the Atacama Desert in Chile, was completed in 2013. That project, an international collaboration of astronomers from Europe, North America, East Asia, and Chile, consists of 66 dishes with diameters of 12 and 7 meters for observations in the 0.3- to 9.6-mm-wavelength range. The Square Kilometre Array (SKA) is designed to have *thousands* of small radio dishes, which together will act as one dish with a collecting area of 1 square kilometer (km^2). Twenty countries are supporting that telescope, which will be located in Australia and South Africa.

Optical telescopes also can be combined in an array to yield resolutions greater than those of single telescopes, although for technical reasons the individual units cannot be spread as far apart as radio telescopes. The Very Large Telescope Interferometer (VLTI) in Chile, operated by ESO, combines four 8-meter telescopes with four movable 1.8-meter auxiliary telescopes. It has a baseline of up to 200 meters, yielding angular resolution of about 0.001 arcsec. The six-telescope Center for High Angular Resolution Astronomy (CHARA) array in California works in the visible and near-infrared regions. It has a baseline of 330 meters with resolution of 0.0003 arcsec.

Figure 6.23 The new Atacama Large Millimeter/submillimeter Array (ALMA) telescope in the Atacama Desert in northern Chile has many international partners.

Infrared Telescopes

Molecules such as water vapor in Earth's atmosphere block infrared (IR) photons from reaching astronomical telescopes on the ground, so telescopes that observe in the infrared [0.75–30 microns (μm)] are at the highest locations. Mauna Kea, a dormant volcano and home of the Mauna Kea Observatories (MKO), rises 4,200 meters above the Pacific Ocean. At that altitude, the MKO telescopes sit above 40 percent of Earth's atmosphere; but more important, 90 percent of Earth's atmospheric water vapor lies below. Still, for the infrared astronomer, the remaining 10 percent is troublesome.

Airborne observatories overcome atmospheric absorption of infrared light by placing telescopes above most of the water vapor in the atmosphere. NASA's Stratospheric Observatory for Infrared Astronomy (SOFIA) (**Figure 6.24**), a joint project with the German Aerospace Center (DLR), is a modified 747 airplane that carries a 2.5-meter telescope and works in the far-infrared region of the spectrum, from 1 to 650 μm. It flies in the stratosphere at an altitude of about 12 km, above 99 percent of the water vapor in Earth's lower atmosphere. Because airplanes are highly mobile, SOFIA can observe in both the Northern and Southern Hemispheres. Other infrared wavelengths must be observed from space.

Orbiting Observatories

Gaining full access to the complete electromagnetic spectrum requires getting completely above Earth's atmosphere. The first astronomical satellite was the British Ariel 1, launched in 1962 to study solar UV and X-ray radiation. Today, a

Figure 6.24 SOFIA is a 2.5-meter infrared telescope mounted in a Boeing 747 aircraft.

TABLE 6.2 Selected Space Observatories

Telescope	Sponsor(s)	Description	Year
Hubble Space Telescope (HST)	NASA, ESA	Optical, infrared, ultraviolet observations	1990–
Chandra X-ray Observatory	NASA	X-ray imaging and spectroscopy	1999–
X-ray Multi-Mirror Mission (XMM-Newton)	ESA	X-ray spectroscopy	1999–
Galaxy Evolution Explorer (GALEX)	NASA	Ultraviolet observations	2003–2012
Spitzer Space Telescope	NASA	Infrared observations	2004–
Neil Gehrels Swift Observatory	NASA	Gamma-ray bursts	2004–
Convection Rotation and Planetary Transits (COROT) space telescope	CNES (France)	Planet finder	2006–2013
Fermi Gamma-ray Space Telescope	NASA, European partners	Gamma-ray imaging and gamma-ray bursts	2008–
Planck telescope	ESA	Cosmic microwave background radiation	2009–2013
Herschel Space Observatory	ESA	Far-infrared and submillimeter observations	2009–2013
Kepler telescope	NASA	Planet finder	2009–2018
Solar Dynamics Observatory (SDO)	NASA	Sun, solar weather	2010–
RadioAstron	Russia	Very-long-baseline interferometry in space	2011–
Nuclear Spectroscopic Telescopic Array (NuSTAR)	NASA	High-energy X-ray	2012–
Gaia	ESA	Optical, digital 3D space camera	2013–
Transiting Exoplanet Survey Satellite (TESS)	NASA, MIT, SAO	Survey entire sky for transiting planets around nearby and bright stars	2018–
CHEOPS (CHaracterising ExOPlanet Satellite)	ESA	High-precision brightness measurements of stars known to have planets	2019 (scheduled)
James Webb Space Telescope (JWST)	NASA, ESA, Canadian Space Agency	Optical and infrared; replacement for HST	2021 (scheduled)

multitude of orbiting astronomical telescopes cover the electromagnetic spectrum from gamma rays to microwaves, with more in the planning stage (**Table 6.2**). Optical telescopes, such as the 2.4-meter Hubble Space Telescope (HST; **Figure 6.25a**), operate successfully at low Earth orbit, 600 km above Earth's surface. Launched in 1990, HST has been the workhorse for UV, visible, and IR space astronomy for more than 25 years. Low Earth orbit is also the region where the International Space Station and many scientific satellites orbit.

For certain other satellites and space telescopes, however, 600 km is not high enough. The Chandra X-ray Observatory (**Figure 6.25b**), NASA's X-ray telescope, cannot see through even the tiniest traces of atmosphere and therefore orbits more than 16,000 km above Earth's surface. NASA's Spitzer Space Telescope, an infrared telescope, is so sensitive that it needs to be completely free from Earth's own infrared radiation. The solution was to put it into a *solar* orbit, trailing tens of millions of kilometers behind Earth. The James Webb Space Telescope, scheduled to replace the HST, will observe primarily in infrared wavelengths. It will be located 1.5 million miles away from Earth, orbiting the Sun at a fixed distance from the Sun and Earth.

Orbiting telescopes located above the atmosphere are unaffected by atmospheric image distortions, weather, or brightening night skies. But space observatories are much more expensive than ground-based observatories and can be difficult or impossible to repair. The HST required several servicing missions, but such missions are impossible for the observatories in more distant Earth orbits. Ground-based telescopes at even the most remote mountaintop locations can receive shipments of replacement parts in a few days; space telescopes cannot. Some wavelengths can be observed from space only, but issues of cost and repair are why ground-based telescopes are much more prevalent.

CHECK YOUR UNDERSTANDING 6.3

Which of the following is the biggest disadvantage of putting a telescope in space?
(a) Astronomers don't have as much control in choosing what to observe.
(b) Astronomers have to wait until the telescopes come back to Earth to get their images.
(c) Space telescopes can observe only in certain parts of the electromagnetic spectrum.
(d) Space telescopes are much more expensive and riskier than similar ground-based telescopes.

6.4 Planetary Spacecraft Explore the Solar System

Recall from Chapter 2 that everyone always sees the same face of the Moon from Earth because the Moon's orbital and rotational periods are equal. The first view of the "far" side was in 1959, when the Soviet flyby mission *Luna 3* sent back pictures showing that the far side of the Moon was very different from its Earth-facing half. No matter how powerful our ground-based or Earth-orbiting telescopes, sometimes we need to send a spacecraft for a different view.

Only in the past half century has the technology existed to explore the Solar System. Spacecraft have now visited all the planets and some of their moons, as well as a few comets and asteroids, providing the first close-up views of those distant worlds. The study of the Solar System from space is an international collaboration involving NASA, the European Space Agency (ESA), the Russian Federal Space Agency (Roscosmos), the Japan Aerospace Exploration Agency (JAXA), the China National Space Administration (CNSA), and the Indian Space Research Organisation (ISRO). Other countries may soon join the endeavor. In this section, we look at the types of spacecraft used to explore our Solar System.

Flybys and Orbiters

Exploration of the Solar System began with a reconnaissance phase, using spacecraft to fly by or orbit a planet or other body. A **flyby** is a spacecraft that first approaches and then continues flying past the target. As those spacecraft speed by, instruments aboard them briefly probe the physical and chemical properties of the target and its environment.

Flyby missions are the most common first phase of exploration. They cost less than orbiters or landers and are easier to design and execute. Flyby spacecraft,

(a)

(b)

Figure 6.25 (a) The Hubble Space Telescope (HST) observes in the ultraviolet, visible, and infrared regions of the electromagnetic spectrum. (b) NASA's Chandra X-ray Observatory, shown in this artist's conception, collects X-rays. Different telescopes in space allow astronomers to observe at all wavelengths of the electromagnetic spectrum.

(a)

(b)

Figure 6.26 The two *Voyager* spacecraft (a) flew past the outer planets and (b) have recently passed the boundary of our Solar System.

Figure 6.27 The robotic rover *Curiosity* took this selfie on the surface of Mars.

such as *Voyager 1* and *Voyager 2*, can sometimes visit several worlds during their travels (**Figure 6.26**). The downside of flyby missions is that because of the physics of orbits, the spacecraft must move by very swiftly. They are limited to just a few hours or at most a few days in which to conduct close-up studies of their targets. Flyby spacecraft give astronomers their first close-up views of Solar System objects, and sometimes the data obtained are then used to plan follow-up studies.

More detailed reconnaissance work is done by spacecraft known as **orbiters** because they orbit around their target. Such missions are intrinsically more difficult than flyby missions because they have to make risky maneuvers and use their limited fuel to change their speed to enter an orbit. But orbiters can linger, looking in detail at more of the surfaces of the objects they are orbiting and studying things that change with time, such as planetary weather.

Orbiters use remote-sensing instrumentation like that used by Earth-orbiting satellites to study our own planet. Those instruments include tools such as cameras that take images at different wavelength ranges, radar that can map surfaces hidden beneath obscuring layers of clouds, and spectrographs that analyze the electromagnetic spectrum. Those instruments enable planetary scientists to map other worlds, measure the heights of mountains, identify geological features and rock types, watch weather patterns develop, measure the composition of atmospheres, and get a general sense of the place. Additional instruments make measurements of the extended atmospheres and space environment through which they travel.

Landers, Rovers, and Atmospheric Probes

Reconnaissance spacecraft provide a wealth of information about a planet, but no better way exists to explore a planet than doing so within a planet's atmosphere or on solid ground. Spacecraft have landed on the Moon, Mars, Venus, Saturn's large moon Titan, several asteroids, and a comet. Those spacecraft have taken pictures of planetary surfaces, measured surface chemistry, and conducted experiments to determine the physical properties of the surface rocks and soils.

Using **landers**—spacecraft that touch down and remain on the surface—has several disadvantages. Because of the expense, only a few landings in limited areas are practical. With that limitation, the results may apply only to the small area around the landing site. Imagine, for example, what a different picture of Earth you might get from a spacecraft that landed in Antarctica, as opposed to a spacecraft that landed in a volcano or on the floor of a dry riverbed. Sites to be explored with landed spacecraft must be carefully chosen on the basis of reconnaissance data. Some landers have wheels and can explore the vicinity of the landing site. Such remote-controlled vehicles, called **rovers**, were used first by the Soviet Union on the Moon four decades ago and more recently by the United States on Mars. **Figure 6.27** shows a self-portrait of the *Curiosity* rover on Mars.

Atmospheric probes descend into the atmospheres of planets and continually measure and send back data on temperature, pressure, and wind speed, along with other properties, such as chemical composition. Atmospheric probes have survived all the way to the solid surfaces of Venus and of Saturn's moon Titan (**Figure 6.28**), sending back streams of data during their descent. An atmospheric probe sent into Jupiter's atmosphere never reached that planet's surface because, as we discuss later in the book, Jupiter does not have a solid surface in the same sense that terrestrial planets and moons do. After sending back its data, the Jupiter probe eventually melted and vaporized as it descended into the hotter layers of the planet's atmosphere.

Sample Returns

If you pick up a rock from the side of a road, you might learn a lot from the rock by using tools that you could carry in your pocket or in your car. Much better, though, would be to pick up a few samples and carry them back to a laboratory equipped with a full range of state-of-the-art instruments that can measure chemical composition, mineral type, age, and other information needed to reconstruct the story of your rock sample's origin and evolution. The same is true of Solar System exploration. One of the most powerful methods for investigating remote objects is to collect samples of the objects and bring them back to Earth for study. So far, only samples of the Moon, a comet, and streams of charged particles from the Sun have been collected and returned to Earth. Scientists have found meteorites on Earth that are pieces of Mars that were blasted loose when objects crashed into that world. In the next 10 years, robotic "sample and return" missions may go to Mars.

The missions discussed so far in this section have all been conducted with robotic spacecraft. The only spacecraft that took people to another world were the *Apollo* missions to the Moon. That program ran from 1961 to 1972 and included several missions before the actual Moon landings. The *Apollo 8* astronauts brought back the famous picture of Earth viewed over the surface of the Moon (see the opening figure of Chapter 1). Each mission from *Apollo 11* through *Apollo 17* had three astronauts—two to land on the Moon and one to remain in orbit. *Apollo 13* did not reach the Moon but returned to Earth safely. Twelve American astronauts walked on the Moon between 1969 and 1972 and brought back a total of 382 kg of rocks and other material (**Figure 6.29**).

The return of extraterrestrial samples to Earth is governed by international treaties and standards to ensure that the samples do not contaminate our world. For example, before the lunar samples that the *Apollo* missions brought back could be studied, they (and the astronauts) had to be quarantined and tested for the presence of alien life-forms. The same international standards apply to spacecraft landing on other planets. The goal of those standards is to avoid transporting life-forms from Earth to another planet; we do not want to "discover" life that we, in fact, introduced. In addition, we do not want to potentially harm life that may exist on other planets.

With many missions under way and others on the horizon, robotic exploration of the Solar System is an ongoing, dynamic activity. Appendix 5 summarizes some recent and current missions. We will discuss some of those missions in the relevant chapter. Information on the latest discoveries can be found on mission websites and in science news sources.

Figure 6.28 Artist's illustration showing the Huygens space probe descending and landing on Saturn's moon Titan in 2005.

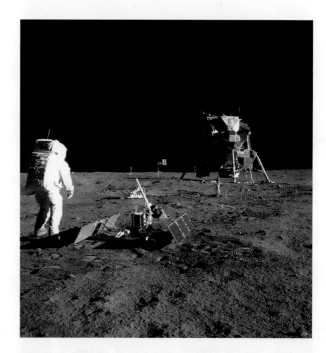

Figure 6.29 *Apollo 11* astronaut Buzz Aldrin on the moon, as photographed by mission commander Neil Armstrong, in 1969.

CHECK YOUR UNDERSTANDING 6.4

Spacecraft are the most effective way to study planets in our Solar System because: (a) planets move too fast across the sky for us to image them well from Earth; (b) planets cannot be imaged from Earth; (c) spacecraft can collect more information than is available just from images from Earth; (d) space missions are easier than long observing campaigns.

6.5 Other Tools Contribute to the Study of the Universe

High-profile space missions have sent back stunning images and data from across the electromagnetic spectrum. But astronomers use other tools as well, including particle accelerators and colliders, neutrino and gravitational-wave detectors, and super-computers.

Figure 6.30 The ATLAS particle detector at CERN's Large Hadron Collider near Geneva, Switzerland. The instrument's enormous size is evident from the person standing near the bottom center of the picture.

Particle Accelerators

Ever since the early years of the 20th century, physicists have been peering into the structure of the atom by observing what happens when small particles collide. By the 1930s, physicists had developed the technology to accelerate charged subatomic particles such as protons to very high speeds and then observe what happens when they slam into a target. From such experiments, physicists have discovered many kinds of subatomic particles and learned about their physical properties. High-energy particle colliders have proved to be an essential tool for physicists studying the basic building blocks of matter.

Astronomers have realized that to understand the very largest structures seen in the universe, it is important to understand the physics that took place during the earliest moments in the universe, when everything was extremely hot and dense. High-energy particle colliders that physicists use today are designed to approach the energies of the early universe. The effectiveness of particle accelerators is determined by the energy they can achieve and the number of particles they can accelerate. Modern particle colliders such as the Large Hadron Collider near Geneva, Switzerland (**Figure 6.30**), reach very high energies. Particles also can be studied from space. The Alpha Magnetic Spectrometer, installed on the International Space Station in 2011, searches for some of the most exotic forms of matter, such as dark matter, antimatter, and high-energy particles called cosmic rays.

Neutrinos and Gravitational Waves

The **neutrino** is an elusive elementary particle that plays a major role in the physics of the interiors of stars. Neutrinos are extremely difficult to detect. In less time than you take to read this sentence, a thousand trillion (10^{15}) solar neutrinos from the Sun are passing through your body, even during the night. Neutrinos are so nonreactive with matter that they can pass right through Earth (and you) as though it (or you) weren't there at all. To be observed, a neutrino has to interact with a detector. Neutrino detectors typically record only one of every 10^{22} (10 billion trillion) neutrinos passing through them, but that's enough to reveal processes deep within the Sun or the violent death of a star 160,000 light-years away.

Experiments designed to look for neutrinos originating outside Earth are buried deep underground in mines or caverns or under the ocean or ice to ensure that only neutrinos are detected. For example, the ANTARES experiment uses the Mediterranean Sea as a neutrino telescope. Detectors located 2.5 km under the sea, off the coast of France, observe neutrinos that originated in objects visible in southern skies and passed through Earth. In the IceCube neutrino observatory located at the South Pole in Antarctica, the neutrino detectors are 1.5–2.5 km under the ice, and they observe neutrinos that originated in objects visible in northern skies (**Figure 6.31**).

Another elusive phenomenon is the **gravitational wave**. Gravitational waves are disturbances in a gravitational field, similar to the waves that spread out from the disturbance you create when you toss a pebble onto the quiet surface of a pond (**Process of Science Figure**). Several facilities, including the Laser Interferometer Gravitational-Wave Observatory (LIGO) and the European VIRGO interferometer, have been constructed to detect gravitational waves. Recently, LIGO has detected several pairs of coalescing binary black holes and neutron stars, as we explore in Chapter 18.

Figure 6.31 The IceCube neutrino telescope at the South Pole, Antarctica.

PROCESS OF SCIENCE

Technology and Science Are Symbiotic

Scientists have been searching for waves that carry gravitational information for nearly 100 years, but the accuracy of their measurements has been limited by the available technology.

1916
Einstein predicts that changing gravitational fields should produce gravitational waves. He calculates the size of the effect, and concludes that it will be impossible to detect them.

c. 1970
Joseph Weber constructs precision-machined bars of metal that should "ring" as a gravitational wave passes by. His results were never confirmed.

1990s
Construction begins on LIGO, an observatory which uses lasers to precisely measure the change in the length of long tubes as gravitational waves pass by.

2015
LIGO detects gravitational waves for the first time, announcing the discovery in early 2016, 100 years after Einstein's prediction. This first detection was followed by several more in the next two years.

Future (2034?)
LISA, a gravitational wave observatory in space, will be more sensitive; it will be able to detect much weaker events than LIGO.

Technology and science develop together. New technologies enable humans to ask new scientific questions. Asking new scientific questions pushes the development of better instrumentation and new technologies.

Computers

Astronomers use powerful computers to gather, analyze, and interpret data. A single CCD image may contain millions of pixels, with each pixel displaying roughly 30,000 levels of brightness. That adds up to several trillion pieces of information in each image. To analyze their data, astronomers typically do calculations for *every single pixel* of an image to remove unwanted contributions from Earth's atmosphere or to correct for instrumental effects. Astronomers conduct many types of sky surveys—in which one or more telescopes survey a specific part of the sky—yielding thousands of images that need to be analyzed.

High-performance computers also play an essential role in generating and testing theoretical models of astronomical objects. Even when we completely understand the underlying physical laws that govern the behavior of a particular object, often the object is so complex that calculating its properties and behavior would be impossible without the assistance of high-performance computers. As discussed in Chapter 4, for example, you can use Newton's laws to compute the orbits of two stars that are gravitationally bound to each other because their orbits take the form of simple ellipses. However, understanding the orbits of the several hundred billion stars that make up the Milky Way Galaxy is not so easy, even though the underlying physical laws are the same.

Computer modeling is used to determine the interior properties of stars and planets, including Earth. Although astronomers cannot see beneath the surfaces of those bodies, they have a surprisingly good understanding of their interiors, which we describe in later chapters. Astronomers start a model by assigning well-understood physical properties to tiny volumes within a planet or star. The computer assembles an enormous number of those individual elements into an overall representation. The result is a rather good picture of what the interior of the star or planet is like.

Astronomers also use high-performance computers to study how astronomical objects, systems of objects, and the universe as a whole evolve. For example, astronomers create models of galaxies and then run computer simulations to study how those galaxies might change over billions of years. **Figure 6.32** shows a simulation of the collision of two galaxies. The results of the computer simulations are then compared with telescopic observations. If the simulations do not match the observations, the model is adjusted and the simulations are run again until general agreement exists between them.

CHECK YOUR UNDERSTANDING 6.5

High-performance computers have become one of an astronomer's most important tools. Which of the following require the use of that type of computer? (Choose all that apply.) (a) analyzing images taken with very large CCDs; (b) generating and testing theoretical models; (c) pointing a telescope from object to object; (d) studying how astronomical objects or systems evolve.

Origins Microwave Telescopes Detect Radiation from the Big Bang

In this chapter, we explored the tools of the astronomer, from basic optical telescopes to instruments that observe in different wavelengths. Now let's examine in more detail one type of telescope that has aided the study of the origin of the

Figure 6.32 These images show supercomputer simulations of the collision of two galaxies. Astronomers compare simulations such as those with telescopic observations.

universe. Recall from Chapter 1 that astronomers think the universe originated with a hot Big Bang. The multiple strands of evidence for that conclusion are discussed in Chapter 21. Here, we look at one piece: the observation of faint microwave radiation left over from the early hot universe. Two Bell Laboratories physicists, Arno Penzias (1933–) and Robert Wilson (1936–), were working on satellite communications when they first detected that radiation in 1964 with a microwave dish antenna in New Jersey. Today, we routinely use cell phones and handheld GPS devices that communicate directly with satellites, but at the time, that capability was at the limit of technology.

Penzias and Wilson needed a very sensitive microwave telescope for the work they were doing for Bell Labs because any spurious signals coming from the telescope itself might wash out the faint signals bounced off a satellite. To that end, they were working hard to eliminate all possible sources of interference originating from within their instrument, including keeping the telescope free of bird droppings. No matter how carefully they tried to eliminate sources of extraneous noise, they always still detected a faint signal at microwave wavelengths. That faint signal was the same in every direction and turned out to be from the Big Bang. Penzias and Wilson shared the 1978 Nobel Prize in Physics for discovering the **cosmic microwave background radiation (CMB)** left over from the Big Bang itself.

Since 1964, astronomers from around the world have designed increasingly precise instruments to measure that radiation from the ground, from high-altitude balloons, from rockets, and from satellites. The Russian experiment RELIKT-1, launched in 1983, found some limits on the variation of the CMB. The COBE (Cosmic Background Explorer) satellite, launched in 1989, showed that the spectrum of that radiation precisely matched that of a blackbody with a temperature of 2.73 K—exactly what was predicted for the radiation left over from the Big Bang. (Compare **Figure 6.33** with the curves in Figure 5.22.) The data also showed some slight differences in temperature—small fractions of a degree—over the map of the sky. Those slight variations tell us about how the universe evolved from one dominated by radiation to one that contains structures such as galaxies, stars, planets, and us. John Mather and George Smoot shared the 2006 Nobel Prize in Physics for that work.

In 1998 and 2003, a high-altitude balloon experiment called BOOMERANG (short for "balloon observations of millimetric extragalactic radiation and geophysics") flew over Antarctica at an altitude of 42 km to study CMB variations and estimate the overall geometry of the universe. The *WMAP* (Wilkinson Microwave Anisotropy Probe) satellite, launched in 2001, created an even more detailed map of the temperature variations in that radiation, yielding more precise values for the age and shape of the universe and the presence of dark matter and dark energy. The Planck space telescope, operated from 2009 to 2013 by the European Space Agency, was much more sensitive than *WMAP* and studied those CMB variations in even more precise detail. The Atacama Cosmology Telescope and Simons Array (Chile) and the South Pole Telescope (Antarctica) study that radiation to look for evidence of when galaxy clusters formed. Those experiments and observations have opened up the current era of precision cosmology, in which astronomers can make detailed models of how the universe was born, eventually leading to stars, planets, and us.

Figure 6.33 This graph shows the spectrum of the cosmic microwave background radiation (CMB) as measured by the COBE satellite (red dots). A steradian is a unit of solid angle. The uncertainty in the measurement at each wavelength is much less than the size of a dot. The line running through the data is a Planck blackbody spectrum with a temperature of 2.73 K.

Construction Begins on the World's First Super Telescope

University of Oxford

Scientists are a step closer to understanding the inner-workings of the universe following the laying of the first stone, and construction starting on the world's largest optical and infrared telescope.

With a main mirror 39 metres in diameter, the Extremely Large Telescope (ELT) is going to be, as its name suggests, enormous. Unlike any other before it, ELT is also designed to be an adaptive telescope and has the ability to correct atmospheric turbulence, taking telescope engineering to another level.

To mark the construction's milestone, a ceremony was held at ESO's Paranal residencia in northern Chile, close to the site of the future giant telescope which will be on top of Cerro Armazones, a 3,046-metre peak mountain.

Among many other representatives from industry, the significance of the project was highlighted by the attendance of the Director General of ESO, Tim de Zeeuw, and President of the Republic of Chile, Michelle Bachelet Jeria.

The ELT is being built by the European Southern Observatory (ESO), an international collaboration supported by the UK's Science and Technology Facilities Council (STFC). Oxford University scientists are playing a key role in the project, and are responsible for the design and construction of its spectrograph, 'HARMONI', an instrument designed to simultaneously take 4,000 images, each in a slightly different colour. The visible and near-infrared instrument will harness the telescope's adaptive optics to provide extremely sharp images.

'HARMONI' will enable scientists to form a more detailed picture of the formation and evolution of objects in the Universe. Supporting researchers to view everything from the planets in our own solar system and stars in our own and nearby galaxies with unprecedented depth and precision, to the formation and evolution of distant galaxies that have never been observed before.

Niranjan Thatte, Principal Investigator for 'HARMONI' and Professor of Astrophysics at Oxford's Department of Physics, said: 'For me, the ELT represents a big leap forward in capability, and that means that we will use it to find many interesting things about the Universe that we have no knowledge of today.

'It is the element of "exploring the unknown" that most excites me about the ELT. It will be an engineering feat, and its sheer size and light grasp will dwarf all other telescopes that we have built to date.'

A time capsule, created by members of the ESO team and sealed at the event, will serve as a lasting memory of the research and the scale of ambition and commitment behind it. Contents include a copy of a book describing the original scientific aims of the telescope, images of the staff that have and will play a role in its construction and a poster of an ELT visualisation. The cover of the time capsule is engraved with a hexagon made of Zerodur, a one-fifth scale model of one of the ELT's primary mirror segments.

Tim de Zeeuw, Director General of ESO, said: 'The ELT will produce discoveries that we simply cannot imagine today, and it will surely inspire numerous people around the world to think about science, technology and our place in the Universe. This will bring great benefit to the ESO member states, to Chile, and to the rest of the world.'

1. Why do astronomers want to build at this particular location?
2. What are the advantages of larger telescopes?
3. How can adaptive optics make images sharper?
4. Why will this telescope observe only in optical and infrared wavelengths?
5. Do a Web search to see the status of this project. What countries are partners? Does any local opposition to the project exist in the host country?

Source: "Construction begins on the world's first super telescope," by the University of Oxford, May 26, 2017. Reprinted with permission.

Summary

Earth's atmosphere blocks many spectral regions and distorts telescopic images. Telescopes are sited to be above as much of the atmosphere as possible. Telescopes are matched to the wavelengths of observation, with different technologies required for each region of the spectrum. The aperture of a telescope both determines its light-gathering power and limits its resolution; larger telescopes are better in both measures. Modern CCD cameras have better quantum efficiency and longer integration times, allowing astronomers to study fainter and more distant objects than were observable with earlier detectors. Telescopes observing at microwave wavelengths have detected radiation left over from the Big Bang.

LG 1 **Compare how the two main types of optical telescopes gather and focus light.** The telescope is the astronomer's most important tool. Ground-based telescopes that observe in visible wavelengths are either refractors (lenses) or reflectors (mirrors). All large astronomical telescopes are reflectors. Large telescopes collect more light and have greater resolution. The diffraction limit is the limiting resolution of a telescope.

LG 2 **Summarize the main types of detectors that are used on telescopes.** Photography improved the ability of astronomers to record details of faint objects seen in telescopes. CCDs are today's astronomical detector of choice because they are much more linear, have a broader spectral response, and can send electronic images directly to a computer. Spectrographs are specialized instruments that take the spectrum of an object to reveal what the object is made of and many other physical properties.

LG 3 **Explain why some wavelengths of radiation must be observed from high, dry, and remote observatories on Earth, or from space.** Radio, near-infrared, and optical telescopes can see through our atmosphere. Those types of telescopes can be arrayed to greatly increase angular resolution. Putting telescopes in space solves problems created by Earth's atmosphere.

LG 4 **Explain the benefits of sending spacecraft to study the planets and moons of our Solar System.** Most of what is known about the planets and their moons comes from observations by spacecraft. Flyby and orbiting missions obtain data from space, and landers and rovers collect data from the ground.

LG 5 **Describe other astronomical tools that contribute to the study of the universe.** Astronomers also use particle accelerators, neutrino detectors, and gravitational-wave detectors to study the universe. High-performance computers are essential to acquiring, analyzing, and interpreting astronomical data.

? Unanswered Questions

- Will telescopes be placed on the Moon? The Moon has no atmosphere to make stars twinkle, cause weather, or block certain wavelengths of light from reaching its surface. All parts of the Moon have days and nights that last for 2 Earth weeks each. China had a small (15 cm) ultraviolet telescope operated from Earth on their *Chang'e 3* lander. One proposal is for a small array of radio telescopes on the side of the Moon facing Earth that would study the Sun and the solar wind. Eventually, telescopes would go on the far side of the Moon, which faces away from the light and radio radiation of Earth. On the far side one might put an array of hundreds of radio telescopes that would be deployed to study the earliest formation of stars and galaxies. Another proposal is for the Lunar Liquid Mirror Telescope (LLMT), with a diameter of 20–100 meters, to be located at one of the Moon's poles. Gravity would settle the rotating liquid into the necessary parabolic shape, and those liquid mirror telescopes are much simpler than are arrays of telescopes with large glass mirrors. Astronomers debate whether telescopes on the Moon would be easier to service and repair than those in space and whether problems caused by lunar dust would outweigh any advantages.

- Will human exploration of the Solar System occur within your lifetime? Since the *Apollo* program, humans have not returned to the Moon or traveled to other planets or moons in the Solar System. Sending humans to the worlds of the Solar System is much more complicated, risky, and expensive than sending robotic spacecraft. Humans need life support such as air, water, and food. Radiation in space can be dangerous. Furthermore, human explorers would expect to return to Earth, whereas most spacecraft do not come back. Astronomers and space scientists have heated debates about human spaceflight versus robotic exploration. Some argue that true exploration requires that human eyes and brains actually go there; others argue that the costs and risks are too high for the potential additional scientific knowledge. Beyond basic exploration, we also do not know whether humans will ever permanently colonize space.

Questions and Problems

Test Your Understanding

1. If one telescope has an aperture of 20 cm, and another has an aperture of 30 cm, and if aperture size is the only difference, then which should you choose, and why?
   a. The 20 cm, because the light-gathering power will be better.
   b. The 20 cm, because the image size will be larger.
   c. The 30 cm, because the light-gathering power will be better.
   d. The 30 cm, because the image size will be larger.

2. Which of the following can be observed from Earth's surface? (Choose all that apply.)
   a. radio waves
   b. gamma radiation
   c. far UV light
   d. X-ray light
   e. visible light

3. Match the following properties of telescopes (lettered) with their corresponding definitions (numbered).
   a. aperture
   b. resolution
   c. focal length
   d. chromatic aberration
   e. diffraction
   f. interferometer
   g. adaptive optics

   (1) two or more telescopes connected to act as one
   (2) distance from lens to focal plane
   (3) diameter
   (4) ability to distinguish close objects
   (5) computer-controlled correction for atmospheric distortion
   (6) color-separating effect
   (7) smearing effect due to sharp edge

4. Two 10-meter telescopes, separated by 85 meters, can operate as an interferometer. What is its resolution when it observes in the infrared at a wavelength of 2 microns?
   a. 0.01 arcsec
   b. 0.005 arcsec
   c. 0.2 arcsec
   d. 0.05 arcsec

5. Arrays of radio telescopes can produce much better resolution than single-dish telescopes can because they work based on the principle of
   a. reflection.
   b. refraction.
   c. diffraction.
   d. interference.

6. Refraction is caused by
   a. light bouncing off a surface.
   b. light changing colors as it enters a new medium.
   c. light changing speed as it enters a new medium.
   d. two light beams interfering.

7. The light-gathering power of a 4-meter telescope is _____ than that of a 2-meter telescope.
   a. 4 times larger
   b. 8 times larger
   c. 16 times smaller
   d. 2 times smaller

8. Improved resolution is helpful to astronomers because
   a. they often want to look in detail at small features of an object.
   b. they often want to look at very distant objects.
   c. they often want to look at many objects close together.
   d. all of the above

9. The part of the human eye that acts as the detector is the
   a. retina.
   b. pupil.
   c. lens.
   d. iris.

10. Cameras that use adaptive optics provide higher-spatial-resolution images primarily because
    a. they operate above Earth's atmosphere.
    b. deformable mirrors are used to correct the blurring due to Earth's atmosphere.
    c. composite lenses correct for chromatic aberration.
    d. they simulate a much larger telescope.

11. The advantage of an interferometer is that
    a. the resolution is dramatically improved.
    b. the focal length is dramatically increased.
    c. the light-gathering power is dramatically increased.
    d. diffraction effects are dramatically decreased.
    e. chromatic aberration is dramatically decreased.

12. The angular resolution of a ground-based telescope is usually determined by
    a. diffraction.
    b. the focal length.
    c. refraction.
    d. atmospheric seeing.

13. A grating can spread white light out into a spectrum of colors because of the property of
    a. reflection.
    b. interference.
    c. dispersion.
    d. diffraction.

14. Why would astronomers put telescopes in airplanes?
    a. to get the telescopes closer to the stars
    b. to get the telescopes above most of the water vapor in Earth's atmosphere
    c. to be able to observe one object for more than 24 hours without stopping
    d. to allow the telescopes to observe the full spectrum of light

15. If we could increase the quantum efficiency of the human eye, doing so would
    a. allow humans to see a larger range of wavelengths.
    b. allow humans to see better at night or in other low-light conditions.
    c. increase the resolution of the human eye.
    d. decrease the resolution of the human eye.

Thinking about the Concepts

16. Galileo's telescope used simple lenses. What is the primary disadvantage of using a simple lens in a refracting telescope?

17. The largest astronomical refractor has an aperture of 1 meter. List several reasons why building a larger refractor with twice that aperture would be impractical.

18. Your camera may have a zoom lens, ranging between wide angle (short focal length) and telephoto (long focal length). How does the size of an object in the camera's focal plane differ between wide angle and telephoto?

19. Optical telescopes reveal much about the nature of astronomical objects. Why do astronomers also need information provided by gamma-ray, X-ray, infrared, and radio telescopes?

20. For light reflecting from a flat surface, the angles of incidence and reflection are the same. That is also true for light reflecting from the curved surface of a reflecting telescope's primary mirror. Sketch a curved mirror and several of those reflecting rays.

21. Consider two optically perfect telescopes having different diameters but the same focal length. Is the image of a star larger or smaller in the focal plane of the larger telescope? Explain your answer.

22. Study the Process of Science Figure. Make a flowchart for the symbiosis between technology and science that led to the detection of gravitational waves.

23. Explain adaptive optics and how they improve a telescope's image quality.

24. Explain integration time and quantum efficiency and how each contributes to the detection of faint astronomical objects.

25. Some people believe that we put astronomical telescopes on high mountaintops or in orbit because doing so gets them closer to the objects they are observing. Explain what is wrong with that popular misconception, and give the actual reason telescopes are located in those places.

26. Humans have sent various kinds of spacecraft—including flybys, orbiters, and landers—to all the planets in our Solar System. Explain the advantages and disadvantages of each of those types of spacecraft.

27. If Earth has meteorites that are pieces of Mars, why is going to Mars and bringing back samples of the martian surface so important?

28. Humans had a first look at the far side of the Moon as recently as 1959. Why had we not seen it earlier—when Galileo first observed the Moon with his telescope in 1610?

29. Where are neutrino detectors located? Why are neutrinos so difficult to detect?

30. Why do telescopes in space give a better picture of the leftover radiation from the Big Bang?

Applying the Concepts

31. Compare the light-gathering power of the Thirty Meter Telescope with that of the dark-adapted human eye (aperture, 8 mm) and with that of one of the 10-meter Keck telescopes.

32. Study the photograph of light entering and leaving a block of refractive material in Figure 6.2b. Use a protractor to measure the angles of the green light as it enters the block and as it leaves the block. How are those angles related?

33. Many amateur astronomers start out with a 4-inch (aperture) telescope and then graduate to a 16-inch telescope. By what factor does the light-gathering power of the telescope increase with that upgrade? How much fainter are the faintest stars that can be seen in the larger telescope?

34. The resolution of the human eye is about 1.5 arcmin. What would the aperture of a radio telescope (observing at 21 cm) have to be to have that resolution? Even though the atmosphere is transparent at radio wavelengths, humans do not see light in the radio range. Using your calculations and logic, explain why.

35. Assume that you have a telescope with an aperture of 1 meter. Compare the telescope's theoretical resolution when you are observing in the near-infrared region of the spectrum ($\lambda = 1{,}000$ nm) with that when you are observing in the violet region of the spectrum ($\lambda = 400$ nm).

36. Assume that the maximum aperture of the human eye, D, is approximately 8 mm and the average wavelength of visible light, λ, is 5.5×10^{-4} mm.
    a. Calculate the diffraction limit of the human eye in visible light.
    b. How does the diffraction limit compare with the actual resolution of 1–2 arcmin (60–120 arcsec)?
    c. To what do you attribute the difference?

37. The diameter of the full Moon in the focal plane of an average amateur's telescope (focal length, 1.5 meters) is 13.8 mm. How big would the Moon be in the focal plane of a very large astronomical telescope (focal length, 250 meters)?

38. One of the earliest astronomical CCDs had 160,000 pixels, each recording 8 bits (256 levels of brightness). A new generation of astronomical CCDs may contain a billion pixels, each recording 15 bits (32,768 levels of brightness). Compare the number of bits of data that each of those two CCD types produces in a single image.

39. Consider a CCD with a quantum efficiency of 80 percent and a photographic plate with a quantum efficiency of 1 percent. If an exposure time of 1 hour is required to photograph a celestial object with a given telescope, how much observing time would be saved by substituting a CCD for the photographic plate?

40. The VLBA uses an array of radio telescopes ranging across 8,000 km of Earth's surface from the Virgin Islands to Hawaii.
    a. Calculate the angular resolution of the array when radio astronomers are observing interstellar water molecules at a microwave wavelength of 1.35 cm.
    b. How does that resolution compare with the angular resolution of two large optical telescopes separated by 100 meters and operating as an interferometer at a visible wavelength of 550 nm?

41. When operational, the Space VLBI may have a baseline of 100,000 km. What will be the angular resolution when studying interstellar molecules emitting at a wavelength of 17 mm from a distant galaxy?

42. The *Mars Reconnaissance Orbiter* (*MRO*) flies at an average altitude of 280 km above the martian surface. If its cameras have an angular resolution of 0.2 arcsec, what is the size of the smallest objects that the *MRO* can detect on the martian surface?

43. At this writing, *Voyager 1* is about 140 astronomical units (AU) from Earth, continuing to record its environment as it departed the boundary of our Solar System.
    a. How far away is *Voyager 1*, in kilometers?
    b. How long do observational data take to come back to us from *Voyager 1*?
    c. How does *Voyager 1*'s distance from Earth compare with that of the nearest star (other than the Sun)?

44. Gravitational waves travel at the speed of light. Their speed, wavelength, and frequency are related as $c = \lambda \times f$. If we were to observe a gravitational wave from a distant cosmic event with a frequency of 10 hertz (Hz), what would be the wavelength of the gravitational wave?

45. Compute the peak of the blackbody spectrum with a temperature of 2.73 K. What region of the spectrum is this?

USING THE WEB

46. A webcast for the International Year of Astronomy 2009 called "Around the World in 80 Telescopes" can be accessed at http://eso.org/public/events/special-evt/100ha.html. The 80 telescopes are situated all over, including Antarctica and space. Pick two of the telescopes and watch the videos. Do you think those videos are effective for public outreach for the observatory in question? For astronomy in general? For each telescope you choose, answer the following questions: Does the telescope observe in the Northern Hemisphere or the Southern Hemisphere? What wavelengths does the telescope observe? What are some of the key science projects at the telescope?

47. Most major observatories have their own websites. Use the link in Question 46 to find a master list of telescopes, and click on a telescope name to link to an observatory website (or run a search on names from Tables 6.1 and 6.2). For the telescope you choose, answer the following questions: What is that telescope's "claim to fame"—is it the largest? At the highest altitude? At the driest location? With the darkest skies? The newest? Does the observatory website have news releases? What is a recent discovery from that telescope?

48. Go to the website for the International Dark Sky Association (http://darksky.org/). Watch the short "Losing the Dark" film (available under the "Resources" tab). Is a "dark sky place" located near you? What are the ecological arguments against too much light at night?

49. What is the current status of the James Webb Space Telescope (https://jwst.nasa.gov)? How is it different from the Hubble Space Telescope? What are some of the instruments for the JWST and its planned projects? What is the current estimated cost of the JWST?

50. Pick a mission from Appendix 5, go to its website, and see what's new. For the mission you choose, answer the following questions: Is the spacecraft still active? Is it sending images? What new science is coming from this mission?

EXPLORATION

Geometric Optics and Lenses

digital.wwnorton.com/astro6

Visit the Digital Resources Page and on the Student Site open the "Geometric Optics and Lenses" AstroTour in Chapter 6. Read through the animation until you reach the optics simulation, pictured in **Figure 6.34**. The simulator shows a converging lens and a pencil. Rays come from the pencil on the left of the converging lens, pass through the lens, and make an image to the right of the lens. The view that would be seen by an observer at the position of the eye is shown in the circle at upper right. Initially, when the pencil is at position 2.3 and the eye is at position 2.0, the pencil is out of focus and blurry.

1 Is the eraser at the top or the bottom of the actual pencil? (That answer becomes important later.)

_____

Using the red slider in the upper left of the window, try moving the pencil to the right. Pause when the observer's eye sees a recognizable pencil (even if it's still blurry).

2 Does the eye see the pencil right side up or upside down?

_____

3 This is somewhat analogous to the view through a telescope. The objects are very far from the lenses, and the observer sees things upside down in the telescope. If an object in your field of view is at the top of the field and you want it in the center, should you move the telescope up or down?

_____

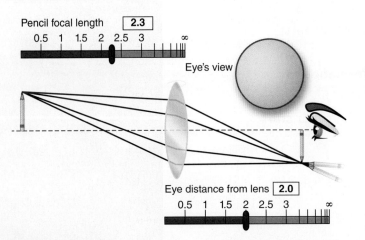

Figure 6.34 Use this simulation to change the position of the object of the eye to explore what will be seen for various configurations.

Now return the pencil to position 2.3. Use the red slider in the lower right of the window to move the eye closer to the lens (to the left).

4 At what distance does the image of the pencil first become crisp and clear?

_____

5 Is the pencil right side up or upside down?

_____

6 In practice at the telescope, we do not move the observer back (away from the eyepiece) to bring the image into focus. Why not?

_____

7 Instead of moving the observer, we use a focusing knob to move the lens in the eyepiece, which brings the image into focus. Imagine that you are looking through the eyepiece of a telescope and the image is blurry. You turn the focusing knob and things get blurrier! What should you try next?

_____

8 Now imagine that you get the image focused just right, so it is crisp and sharp. The next person to use the telescope wears glasses and insists that the image is blurry. When you look through the telescope again, though, the image is still crisp. Explain why your experiences differ.

_____

Step through the animation to the next picture. Carefully study the two telescopes shown and the path the light takes through them.

9 Which telescope has a longer focal length: the top one or the bottom one?

_____

10 Which telescope produces an image with the red and the blue stars more separated: the top one or the bottom one?

_____

11 A longer focal length is an advantage in one sense, but it's not the entire story. What are some disadvantages of a telescope with a very long focal length?

_____

7

The Formation of Planetary Systems

The planetary system containing Earth—our Solar System—is a by-product of the birth of the Sun, but the physical processes that shaped the formation of the Solar System are not unique to it. The same processes have formed many other multiplanet systems. Here in Chapter 7, we examine how planetary systems are born and evolve.

LEARNING GOALS

By the end of this chapter, you should be able to:

LG 1 Describe how our understanding of planetary system formation developed from the work of both planetary and stellar scientists.

LG 2 Discuss the role of gravity and angular momentum in explaining why planets orbit the Sun in a plane and why they revolve in the same direction that the Sun rotates.

LG 3 Explain how temperature at different locations in the protoplanetary disk affects the composition of planets, moons, and other bodies.

LG 4 Discuss the processes that resulted in the formation of planets and other objects in our Solar System.

LG 5 Describe how astronomers find planets around other stars and derive exoplanet properties.

From clouds of gas and dust, planetary systems are born. ▶▶▶

How did our Solar System form?

7.1 Planetary Systems Form around a Star

Earth is part of a collection of **planets**—large, round isolated bodies that orbit a star. Astronomers call a system of planets surrounding a star a **planetary system**. The Solar System, shown in **Figure 7.1**, is the planetary system that includes Earth, seven other planets, and the Sun. The system also includes moons that orbit planets and small bodies that occupy particular regions of the Solar System, such as the asteroid belt or the Kuiper Belt. Our Solar System is a tiny part of our galaxy, which is a tiny part of the universe. Review Figure 1.3 to remind yourself of the size scales involved. Light takes about 4 hours to travel to Earth from Neptune, the outermost known planet in the Solar System, but light from the most distant galaxies has taken more than 13 *billion* years to reach Earth.

Until the latter part of the 20th century, the origin of the Solar System remained speculative. Over the past century, with the aid of spectroscopy, astronomers have determined that the Sun is an ordinary star, one of hundreds of billions in its galaxy, the Milky Way, and that the Milky Way is an ordinary galaxy, one of hundreds of billions in the universe. In the past few decades, stellar astronomers studying the formation of stars and planetary scientists analyzing clues about the history of the Solar System have arrived at the same picture of the early Solar System—but from two very different directions. That unified understanding provides the foundation for the way astronomers now think about the Sun and the objects that orbit it. In this section, we look at how the work of stellar and planetary scientists converged to inform our understanding of planetary system formation.

AstroTour: Solar System Formation

The Nebular Hypothesis

The first plausible theory for the formation of the Solar System, the **nebular hypothesis**, was proposed in 1755 by the German philosopher Immanuel Kant (1724–1804) and conceived independently in 1796 by the French astronomer Pierre-Simon Laplace (1749–1827). Kant and Laplace argued that a rotating cloud of interstellar gas, or **nebula** (Latin for "cloud"), gradually collapsed and flattened to form a disk with the Sun at its center. Surrounding the Sun were rings of material from which the planets formed. That configuration would explain why the planets orbit the Sun in the same direction in the same plane. The nebular hypothesis remained popular throughout the 19th century, and those basic principles of the hypothesis are still retained today.

Our modern theory of planetary system formation calculates the conditions required for a cloud of interstellar gas to collapse under the force of their own self-gravity to form stars. Recall from Chapter 4 that self-gravity is the gravitational attraction between the parts of an object, such as a planet or star, that pulls all the parts toward the object's center. That inward force is opposed by either structural strength (for rocks that make up terrestrial planets) or the outward force resulting from gas pressure and radiation pressure within a star. If the outward force is less than self-gravity, the object contracts; if it is greater,

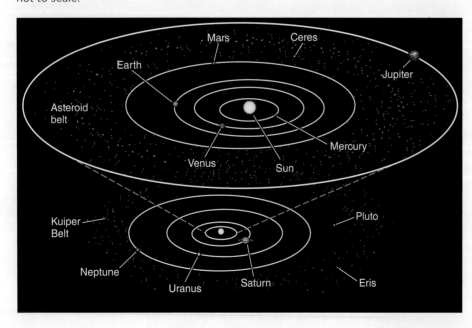

Figure 7.1 Our Solar System includes planets, moons, and other small bodies. Sizes and distances shown are not to scale.

the object expands. In a stable object, the inward and outward forces are balanced.

In support of the nebular hypothesis, disks of gas and dust have been observed surrounding young stellar objects (**Figure 7.2**). From that observational evidence, stellar astronomers have shown that, much like a spinning ball of pizza dough spreads out to form a flat crust, the cloud that produces a star—the Sun, for example—collapses first into a rotating disk. Material in the disk eventually suffers one of three fates: it travels inward onto the forming star at its center, it remains in the disk itself to form planets and other objects, or it is ejected back into interstellar space.

Planetary Scientists and the Convergence of Evidence

While astronomers were working to understand star formation, other groups of scientists with very different backgrounds were piecing together the history of the Solar System. Planetary scientists, geochemists, and geologists looking at the current structure of the Solar System inferred what some of its early characteristics must have been. The orbits of all the planets lie very close to a single plane, so the early Solar System must have been flat. In addition, all the planets orbit the Sun in the same direction, so the material from which the planets formed must have been orbiting the Sun in the same direction as well.

To find out more, scientists study samples of the very early Solar System. Rocks that fall to Earth from space, known as **meteorites**, include pieces of material left over from the Solar System's youth. Many meteorites, such as the one in **Figure 7.3**, resemble a piece of concrete in which pebbles and sand are mixed with a much finer filler, suggesting that the larger bodies in the Solar System must have grown from the aggregation of smaller bodies. That finding suggests an early Solar System in which the young Sun was surrounded by a flattened disk of both gaseous and solid material. Our Solar System formed from that swirling disk of gas and dust.

As astronomers and planetary scientists compared notes, they realized they had arrived at the same picture of the early Solar System from two completely different directions. The rotating disk from which the planets formed was the remains of the disk that had accompanied the formation of the Sun. Earth, along with all the other orbiting bodies that make up the Solar System, formed from the remnants of an *interstellar cloud* that collapsed to form the local star, the Sun. The connection between the formation of stars and the origin and later evolution of the Solar System is one of the cornerstones of both astronomy and planetary science—a central theme of our understanding of our Solar System (see the **Process of Science Figure**).

CHECK YOUR UNDERSTANDING 7.1

Which of the following pieces of evidence support the nebular hypothesis? (Choose all that apply.) (a) Planets orbit the Sun in the same direction. (b) The Solar System is relatively flat. (c) Earth has a large Moon. (d) We observe disks of gas and dust around other stars.

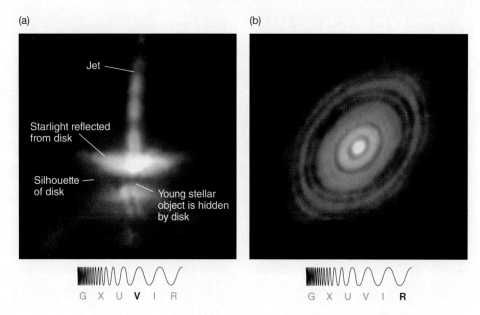

Figure 7.2 (a) Hubble Space Telescope image of a disk around newly formed stars. The dark band is the silhouette of the disk seen edge on. Bright regions are dust illuminated by the star's light. The jets are shown in green. (b) The Atacama Large Millimeter/submillimeter Array (ALMA) obtained this image of a protoplanetary disk around the star HL Tau. This image shows substructures and possibly planets in the system's dark patches.

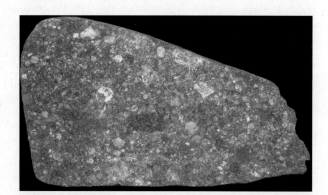

Figure 7.3 Meteorites are the surviving pieces of Solar System fragments that land on planets. This meteorite formed from many smaller components that stuck together.

PROCESS OF SCIENCE

Converging Lines of Inquiry

Why is the Solar System a disk, with all planets orbiting in the same direction? Scientists from different disciplines often contribute to the solution to a problem. These scientists approach the problem from different directions and perspectives, so when they arrive at the same conclusion, that is compelling evidence that they are all on the right track.

Mathematicians:
- Suggested the nebular hypothesis—that a collapsing rotating cloud formed the Solar System.

Stellar Astronomers:
- Tested the nebular hypothesis, seeking evidence for or against.
- Found dust and gas around young stars.
- Observed this gas and dust to be in the shape of disks.

Planetary Scientists:
- Tested the nebular hypothesis, seeking evidence for or against.
- Studied meteorites, which showed that the planets formed from many smaller bodies.

Beginning from the same fundamental observations about the shape of the Solar System, theorists (the mathematicians), stellar astronomers, and planetary scientists **converged on the nebular theory** that stars and planets form together from a collapsing cloud of gas and dust.

7.2 The Solar System Began with a Disk

Planets form in a disk around young stars, but what are some of the specifics of the process? **Figure 7.4** illustrates the young Solar System as it appeared roughly 5 billion years ago. At that time, the Sun was still a **protostar**—a large ball of gas but not yet hot enough in its center to be a star. As the cloud of interstellar gas collapsed to form the protostar, its gravitational energy was converted into heat energy and radiation. Surrounding the protostellar Sun was a flat, orbiting disk of gas and dust. Each bit of the material in that thin disk orbited the Sun in accordance with the same laws of motion and gravitation that govern the orbits of the planets. The disk around the Sun—like the disks that astronomers see today surrounding protostars elsewhere in our galaxy—is called a **protoplanetary disk**. The disk probably contained less than 1 percent of the mass of the star forming at its center, but that amount was more than enough to account for the bodies that make up the Solar System today.

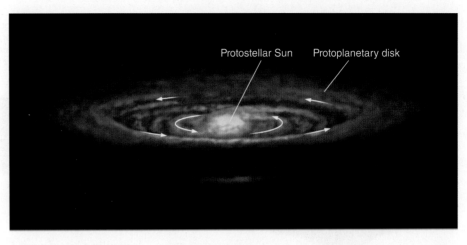

Figure 7.4 Think of the young Sun as being surrounded by a flat, rotating disk of gas and dust that was flared at its outer edge.

The Collapsing Cloud and Angular Momentum

Conservation of *angular momentum* causes protoplanetary disks to form. **Angular momentum** is a conserved quantity associated with a revolving or rotating system and depends on both the velocity and distribution of the system's mass. The angular momentum of an isolated system is always conserved; that is, it remains unchanged unless acted on by an external force. A figure-skater spinning on the ice (**Figure 7.5**), like any other rotating object, has some amount of angular momentum. Unless frictional forces act to reduce her angular momentum, she will always have the same amount of angular momentum.

The amount of angular momentum depends on three factors:

1. How fast the object is rotating. The faster an object is rotating, the more angular momentum it has.
2. The mass of the object. If a similarly sized bowling ball and a basketball are spinning at the same speed, the bowling ball has more angular momentum because it has more mass.
3. How the mass of the object is distributed relative to the spin axis—that is, how spread out the object is. For an object of a given mass and rate of rotation, the more spread out it is, the more angular momentum it has. A spread-out object rotating slowly might have the same angular momentum as a compact object rotating rapidly.

Both an ice-skater and a collapsing interstellar cloud are subject to the **conservation of angular momentum**: the angular momentum must remain the same in the absence of an external force. For angular momentum to be conserved, a change in one of the above quantities (the rate of spin, mass, or distribution of mass) must be accompanied by a change in another quantity. Because an ice-skater's mass doesn't change, for example, she can control how rapidly she spins

Astronomy in Action: Angular Momentum

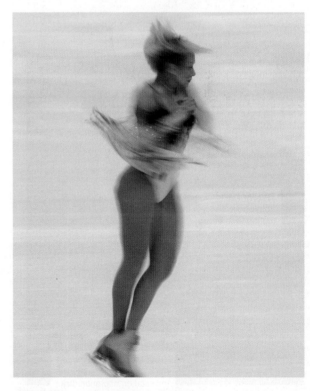

Figure 7.5 A figure-skater relies on the principle of conservation of angular momentum to change the speed at which she spins.

Figure 7.6 A rotating interstellar cloud collapses in a direction parallel to its axis of rotation, thus forming an accretion disk.

by pulling in or extending her arms or legs. As she pulls in her arms to become more compact, she changes her distribution of mass and must spin faster to maintain the same angular momentum. When her arms are held tightly in front of her and one leg is wrapped around the other, the skater's spin becomes a blur. She finishes with a flourish by throwing her arms and leg out—an action that abruptly slows her spin by spreading out her mass. The skater's angular momentum remains constant throughout the maneuver. Similarly—as shown in **Figure 7.6**— the cloud that formed our Sun rotated faster and faster as it collapsed, just as the ice-skater speeds up when she pulls in her arms.

That description, however, presents a puzzle. Suppose the Sun formed from a typical cloud—one about a light-year across and rotating so slowly that completing one rotation took a million years. By the time the cloud collapsed to the size of the Sun today, it would have been spinning so fast that one rotation would occur every 0.6 second. That rate is more than 3 million times faster than our Sun actually spins. At that rate of rotation, the Sun would tear itself apart. It *appears* that angular momentum was *not* conserved in the actual formation of the Sun—but that can't be right because angular momentum *must* be conserved. We must be missing something. Where did the angular momentum go?

The Formation of an Accretion Disk

To understand how angular momentum is conserved in disk formation, we must think in three dimensions. Imagine that the ice-skater bends her knees, compressing herself downward instead of bringing her arms toward her body. As she does so, she again makes herself less spread out, but her rate of spin does not change because no part of her body has become any closer to the axis of spin. Similarly, as shown in Figure 7.6, a clump of a molecular cloud can flatten out without speeding up by collapsing parallel to its axis of rotation. Instead of collapsing into a ball, the interstellar cloud flattens into a disk. As the cloud collapses, its self-gravity increases and the inner parts begin to fall freely inward, raining down on the growing object at the center. The outer portions of the cloud lose the support of the collapsed inner portion, and they start falling inward, too. As that material makes its final inward plunge, it lands on a thin, rotating disk—called an **accretion disk**—that forms from the accretion of material around a massive object.

The formation of accretion disks is common in the universe. The process is shown in **Figure 7.7**. As material falls onto the disk, its motion perpendicular to the disk stops abruptly, but the motion of its mass *parallel* to the surface of the

Figure 7.7 Gas from a rotating cloud falls inward from opposite sides, piling up onto a rotating disk.

7.1 Working It Out Angular Momentum

In its simplest form, the angular momentum (L) of a system is given by

$$L = m \times v \times r$$

where m is the mass, v is the speed at which the mass is moving, and r represents how spread out the mass is.

Let's apply that relationship to the angular momentum of Jupiter in its orbit about the Sun. The angular momentum from one body orbiting another is called *orbital* angular momentum, $L_{orbital}$. The mass (m) of Jupiter is 1.90×10^{27} kilograms (kg), the speed of Jupiter in orbit (v) is 1.31×10^4 meters per second (m/s), and the radius of Jupiter's orbit (r) is 7.79×10^{11} meters. Putting all that together gives

$$L_{orbital} = (1.90 \times 10^{27} \text{ kg}) \times (1.31 \times 10^4 \text{ m/s}) \times (7.79 \times 10^{11} \text{ m})$$

$$L_{orbital} = 1.94 \times 10^{43} \text{ kg m}^2/\text{s}$$

Calculating the *spin* angular momentum of a spinning object, such as a skater, a planet, a star, or an interstellar cloud, is more complicated. Here, we must add up the individual angular momenta of *every tiny mass element* within the object. For a uniform sphere, the spin angular momentum is

$$L_{spin} = \frac{4\pi m R^2}{5P}$$

where R is the radius of the sphere and P is the rotation period of its spin.

Let's compare Jupiter's orbital angular momentum with the Sun's spin angular momentum to investigate the distribution of angular momentum in the Solar System. Appendix 2 provides the Sun's radius (6.96×10^8 meters) and mass (1.99×10^{30} kg), and its rotation period is 24.5 days = 2.12×10^6 seconds. If we assume that the Sun is a uniform sphere, the spin angular momentum of the Sun is

$$L_{spin} = \frac{4 \times \pi \times (1.99 \times 10^{30} \text{ kg}) \times (6.96 \times 10^8 \text{ m})^2}{5 \times (2.12 \times 10^6 \text{ s})}$$

$$L_{spin} = 1.14 \times 10^{42} \text{ kg m}^2/\text{s}$$

$L_{orbital}$ of Jupiter is about 17 times greater than L_{spin} of the Sun. Thus, most of the angular momentum of the Solar System now resides in the orbits of its major planets.

For a collapsing sphere to conserve L_{spin}, its rotation period P must be proportional to R^2. As with the skater, when a sphere decreases in radius, its rotation period decreases; that is, it spins faster.

---

disk adds to the disk's total angular momentum. Thus, the angular momentum of the infalling material is transferred to the accretion disk. The rotating accretion disk has a radius of hundreds of astronomical units, and that is *thousands* of times greater than the radius of the star that will eventually form at its center. Therefore, most of the angular momentum in the original interstellar cloud ends up in the accretion disk rather than in the central protostar (see **Working It Out 7.1** for an example of the relevant calculation).

Most of the matter that lands on the accretion disk either becomes part of the star or is ejected back into interstellar space in the form of jets or other outflows (Figure 7.2a). Those jets are bipolar—they come in pairs aligned along an axis. Material swirling in the bipolar jets carries angular momentum away from the accretion disk in the general direction of the poles of the rotation axis. However, a small amount of material is left behind in the disk. The objects in that leftover disk—the dregs of the process of star formation—form planets and other objects that orbit the star.

AstroTour Traffic Circle Analogy

Formation of Large Objects

Random motions of the gas within the protoplanetary disk eventually push the smaller grains of solid material toward larger grains. As that happens, the smaller grains stick to the larger grains. The "sticking" process among smaller grains is due to the same static electricity that causes dust bunnies to grow under your bed.

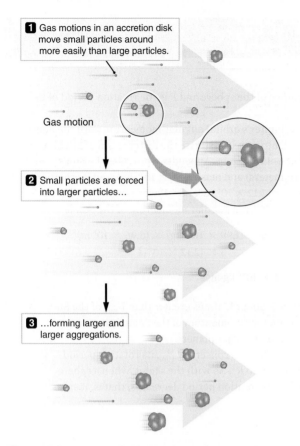

Figure 7.8 Motions of gas in a protoplanetary disk blow smaller particles of dust into larger particles, making the larger particles larger still. That process continues, eventually creating objects many meters in size.

Starting out at only a few microns (μm) across—about the size of particles in smoke—the slightly larger bits of dust grow to the size of pebbles and then to clumps the size of boulders, which are not as easily pushed around by gas (**Figure 7.8**). When clumps grow to about 100 meters across, the objects are so far apart that they collide less often, and their growth rate slows down but does not stop. Within a protoplanetary disk, the larger dust grains become larger at the expense of the smaller grains.

For two large clumps to stick together rather than explode into many small pieces, they must bump into each other very gently: collision speeds must be about 0.1 m/s or less. If your stride is about a meter long, a collision speed of 0.1 m/s would correspond to only one step every 10 seconds. The process does not yield larger and larger bodies with every collision. When violent collisions occur in an accretion disk, larger clumps break back into smaller pieces. But over a long time, large bodies do form.

Objects continue to grow by "sweeping up" smaller objects that get in their way. Those objects can eventually measure up to several hundred meters across. As the clumps reach the size of about a kilometer, a different process becomes important. Those kilometer-sized objects are massive enough that their gravity pulls on nearby bodies, as shown in **Figure 7.9**. Those bodies of rock and ice are known as **planetesimals** ("tiny planets") and eventually combine with one another to form planets. Besides chance collisions with other objects, a planetesimal's gravity can now pull in and capture small objects outside its direct path. That process speeds up the growth of planetesimals, so larger planetesimals quickly consume most of the remaining bodies near their orbits. The final survivors of that process are large enough to be called planets. As with the major bodies in orbit around the Sun, some of the planets may be relatively small and others quite large.

CHECK YOUR UNDERSTANDING 7.2

Where does most of the angular momentum of the original cloud go? (a) into the orbital angular momentum of planets; (b) into the star; (c) into the spin of the planets; (d) lost along the jets from the star

7.3 The Inner Disk and Outer Disk Formed at Different Temperatures

In the Solar System, the inner planets are small and mostly rocky, whereas the outer planets are very large and mostly gaseous. That distinct difference between the inner and outer Solar System can be explained by how the local disk environment affects the formation process. In this section, we examine these differences.

Energy in the Disk

The accretion disks surrounding young stars form from interstellar material that may have a temperature of only a few kelvins, but the disks themselves reach temperatures of hundreds of kelvins or more. Astronomers want to understand what heats up the disk around a forming star so that we can calculate how hot those disks get.

The law of **conservation of energy** states that unless energy is added to or taken away from a system from the outside, the total amount of energy in the

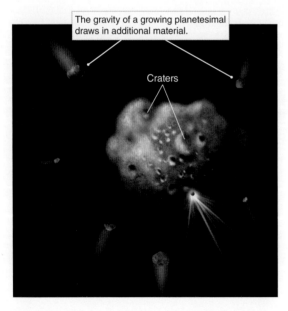

Figure 7.9 The gravity of a planetesimal is strong enough to attract surrounding material, which causes the planetesimal to grow more rapidly.

system must remain constant. The form the energy takes, however, can change. With that concept in mind, why does gas falling on the disk make the disk hot? Imagine you are working against gravity by lifting a heavy object, such as a bowling ball. Lifting the bowling ball takes energy, and the law of conservation of energy states that energy is never lost. Where does that energy go? The energy is stored and changed into a form called **gravitational potential energy**. If you drop the bowling ball, it falls, and as it falls, it speeds up. The gravitational potential energy that was stored is converted to energy of motion, which is called **kinetic energy**. When the bowling ball hits the floor, it stops suddenly. The bowling ball loses its energy of motion, so what form does that energy take now? Some of the energy is converted into the sound the bowling ball makes when it hits the floor, too, and some goes into heating and distorting the floor. But much of the energy is converted into thermal energy. The atoms and molecules that make up the bowling ball are moving around within the bowling ball a bit faster than they were before the bowling ball hit, so the bowling ball and its surroundings, including the floor, grow a tiny bit warmer. Similarly, as gas falls toward the disk surrounding a protostar, gravitational potential energy is converted first to kinetic energy, causing the gas to pick up speed. When the gas hits the disk and stops suddenly, that kinetic energy is turned into thermal energy.

Figure 7.10 Atoms in a gas fall together until they hit the accretion disk, at which point their motions become randomized, raising the temperature of the gas.

Material falling onto the accretion disk around a forming star causes the disk to heat up too (**Figure 7.10**). The amount of heating depends on *where* the material hits the disk. Material hitting the inner part of the disk (the *inner disk*) has fallen farther and picked up greater speed within the gravitational field of the forming star than material hitting the disk farther out. Like a brick dropped from a tall building, material striking the inner disk is moving quite rapidly when it hits, so it heats the inner disk to high temperatures. In contrast, material falling onto the outer part of the disk (the *outer disk*) is moving much more slowly, like a brick dropped from just a foot or so above the ground. As a result, the temperature at the outermost parts of the disk is not much higher than that of the original interstellar cloud. Stated another way: material falling onto the inner disk converts more gravitational potential energy into thermal energy than does material falling onto the outer disk.

The energy released as material falls onto the disk is not the only source of thermal energy in the disk. Even before the nuclear reactions that will one day power the new star have ignited, the conversion of gravitational energy into thermal energy drives the temperature at the surface of the protostar to several thousand kelvins, and it drives the luminosity of the huge ball of glowing gas to many times the luminosity of the present-day Sun. For the same reasons that Mercury is hot whereas Pluto is not (see Chapter 5), the radiation streaming outward from the protostar at the center of the disk drives the temperature in the inner parts of the disk even higher, increasing the difference in temperature between the inner and outer parts of the disk.

The Compositions of Planets

Temperature affects whether a material exists in a solid form. On a hot summer day, ice melts and water quickly evaporates; on a cold winter night, water in your breath freezes into tiny ice crystals. Metals and rocky materials, such as iron, **silicates** (minerals containing silicon and oxygen), and carbon, remain solid even at high temperatures. Substances that can withstand high temperatures without melting or being vaporized are called **refractory materials**. Other materials,

Figure 7.11 Differences in temperature within a protoplanetary disk determine the composition of dust grains that then evolve into planetesimals and planets. The colored bars show that refractory materials are found throughout the disk, whereas water ice is found only outside Jupiter's orbit, and highly volatile materials are found only outside Saturn's orbit. Shown here are the proto-Sun and the orbits of Venus (V), Earth (E), Mars (M), Jupiter, Saturn, and Uranus.

such as water, ammonia, and methane, remain in a solid form only if their temperature is very low. Those materials, which become gases at moderate temperatures, are called **volatile materials** (or *volatiles* for short). Astronomers generally call the solid form of any volatile material an **ice**.

Differences in temperature from place to place within the protoplanetary disk significantly affect the makeup of the dust grains in the disk. **Figure 7.11** illustrates that only refractory substances exist in the hottest parts of the disk—the area closest to the protostar. In the inner disk, dust grains are composed almost entirely of refractory materials. Some substances can survive in solid form somewhat farther out, including some hardier volatiles, such as water ice and certain chemical compounds that are **organic** (meaning that they contain molecules with a carbon-hydrogen bond). Those solids add to the materials that make up dust grains. In the coldest, outermost parts of the accretion disk, far from the central protostar, highly volatile components such as methane, ammonia, and carbon monoxide ices and other organic molecules survive only in solid form. The differences in composition of dust grains within the disk are reflected in the composition of the planets formed from that dust. Planets that form closer to the central star tend to be made up mostly of refractory materials, such as rock and metals, but are deficient in volatiles. Planets that form farther from the central star contain not only refractory materials but also large quantities of ices and organic materials.

In the Solar System, the inner planets are composed of rocky material surrounding metallic cores of iron and nickel. Objects in the outer Solar System, including moons, giant planets, and comets, are composed largely of ices of various types. But not all planetary systems are so neatly organized as our Solar System. When planets around other stars were first discovered, they appeared to be very different, with large planets close to their respective stars. Astronomers now think that **chaotic** encounters—in which a small change in the initial state of a system can lead to a large change in the final state of the system—may change the organization of planetary compositions. In a process called **planet migration**, the force of gravity from all the nearby objects can move some planets so that they end up far from the place of their birth. In our Solar System, for example, Uranus and Neptune originally may have formed nearer to the orbits of Jupiter and Saturn but were then driven outward to their current locations by gravitational encounters with Jupiter and Saturn. A planet also can migrate when it gives up some of its orbital angular momentum to the disk material that surrounds it. Such a loss of angular momentum causes the planet to slowly spiral inward toward the central star. Thus, the order of planets in a system can change.

Formation of an Atmosphere

Once a solid planet has formed, it may continue to grow by capturing gas from the protoplanetary disk. To do so, it must act quickly. Young stars and protostars emit fast-moving particles and intense radiation that can quickly disperse the gaseous remains of the accretion disk. Gaseous planets such as Jupiter probably have only about 10 million years to form and to grab whatever gas they can. Because of their strong gravitational fields, more massive young planets can capture more of the hydrogen and helium gas that makes up the bulk of the disk. What follows is much like the formation of a star and protoplanetary disk, but on a smaller scale—namely, gas from a mini accretion disk moves inward and falls onto the planet.

The gas that a planet captures when it forms—primarily hydrogen and helium—is called the planet's **primary atmosphere**. The primary atmosphere of a large planet can be more massive than the solid body, as with Jupiter. Some of the solid material in the mini accretion disk might stay behind to coalesce into larger bodies in much the same way that particles of dust in the protoplanetary disk came together to form planets. The result is a mini "solar system"—a group of moons that orbit about the planet.

A less massive planet also may capture some gas from the protoplanetary disk, only to lose it later. The gravity of small planets may be too weak to hold low-mass gases such as hydrogen or helium. Even if a small planet can gather some hydrogen and helium from its surroundings, that primary atmosphere will not last long. In the inner solar system, the temperatures are higher, so the hydrogen and helium atoms are moving faster than in the outer solar system and will escape from a small planet. The atmosphere that remains around a small planet such as Earth is a **secondary atmosphere**, which forms later in the life of a planet. Volcanism is one important source of a secondary atmosphere because it releases heavier and thus slower-moving gases, such as carbon dioxide, water vapor, and other gases trapped in the planet's interior. In addition, volatile-rich comets that formed in the outer parts of the disk fall inward toward the new star long after its planets have formed, and they sometimes collide with planets. **Comets** are icy planetesimals that survive planetary accretion. They may serve as a significant source of water, organic compounds, and other volatile materials on planets close to the central star.

CHECK YOUR UNDERSTANDING 7.3

In our Solar System, the inner planets are rocky because: (a) the original cloud had more rocky material near the center; (b) warm temperatures in the inner disk caused the inner planetesimals to be formed of only rocky material; (c) the inner disk filled a smaller volume, and so it was denser; (d) the hydrogen and helium atoms were too low mass to remain in the inner disk.

7.4 The Formation of Our Solar System

Nearly 5 billion years ago, the Sun was still a protostar surrounded by a protoplanetary disk of gas and dust. During the next few hundred thousand years, much of the dust in the disk had collected into planetesimals—clumps of rock and metal near the emerging Sun, and aggregates of rock, metal, ice, and organic materials farther from the Sun. In this section, we look at the formation of the types of planets in our own Solar System.

The Terrestrial Planets

Within the inner 5 astronomical units (AU) of the disk, several rock and metal planetesimals quickly grew larger to become the dominant masses in their orbits. With their ever-strengthening gravitational fields, they either captured most of the remaining planetesimals or ejected them from the inner part of the disk. **Figure 7.12** shows some results from a computer simulation of how that might have happened. The dominant planetesimals became planet-sized bodies with masses ranging between 5 percent and 100 percent of Earth's mass. Those dominant planetesimals evolved into the **terrestrial planets**, which are rocky, Earth-like planets. Today, the surviving terrestrial planets are Mercury, Venus,

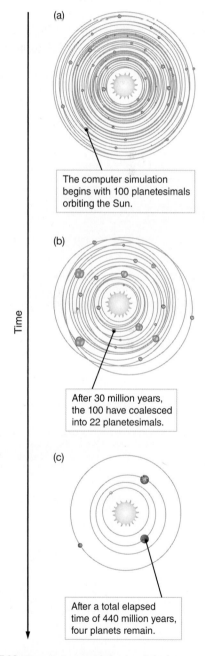

Figure 7.12 Computer models simulate how material in the protoplanetary disk became clumped into the planets over time. Only a few planets remain at the end.

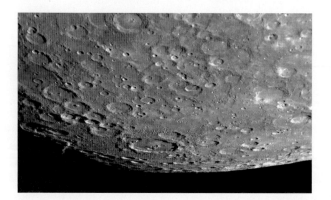

Figure 7.13 Large impact craters on Mercury (and on other solid bodies throughout the Solar System) record the final days of the Solar System's youth, when planets and planetesimals grew as smaller planetesimals rained down on their surfaces.

Earth, and Mars. Earth's Moon is often grouped with those terrestrial planets because of its similar physical and geological properties, even though it is not a planet itself and formed differently. It is possible that one or two other planets or large moons formed in the young Solar System but were later destroyed.

For several hundred million years after the four surviving terrestrial planets formed, leftover pieces of debris still in orbit around the Sun continued to rain down on the surfaces of those planets. Today, we can still see the scars of those early impacts on the cratered surfaces of all the terrestrial planets (**Figure 7.13**). That rain of debris continues even today, but at a much lower rate.

Before the proto-Sun became a true star, gas in the inner part of the protoplanetary disk was still plentiful. During that early period the two larger terrestrial planets, Earth and Venus, may have held on to weak primary atmospheres of hydrogen and helium, but those thin atmospheres were soon lost to space. The terrestrial planets did not develop thick atmospheres until the formation of the secondary atmospheres that now surround Venus, Earth, and Mars. Mercury's size and proximity to the Sun and the Moon's small mass prevented those bodies from retaining significant secondary atmospheres.

The Giant Planets

Beyond 5 AU from the Sun, in a much colder part of the accretion disk, planetesimals combined to form several bodies with masses about 5–20 times that of Earth (5–20 M_{Earth}). Those planet-sized objects formed from planetesimals containing volatile ices and organic compounds in addition to rock and metal. In a process astronomers call *core accretion–gas capture*, mini accretion disks formed around those planetary cores, capturing massive amounts of hydrogen and helium and funneling that material onto the planets. Four such massive bodies became the cores of the **giant planets**—Jupiter, Saturn, Uranus, and Neptune. Those giant planets are many times the mass of any terrestrial planet.

Jupiter's massive solid core captured and retained the most gas—roughly 300 M_{Earth}. The other outer planetary cores captured less hydrogen and helium, perhaps because their cores were less massive or because less gas was available to them. Saturn ended up with less than 100 M_{Earth} of gas, and Uranus and Neptune grabbed less than 20 M_{Earth} of gas.

The core accretion model indicates that a Jupiter-like planet could take up to 10 million years to accumulate. Some planetary scientists think that our protoplanetary disk could not have survived long enough to form gas giants such as Jupiter through the general process of core accretion. All the gas may have dispersed in roughly half that time, cutting off Jupiter's supply of hydrogen and helium. An alternative explanation is a process called *disk instability*, in which the protoplanetary disk suddenly and quickly fragments into massive clumps equivalent to those of a large planet. Both core accretion and disk instability may have played a role in the formation of our own and other planetary systems.

During the formation of the planets, gravitational energy was converted into thermal energy as the individual atoms and molecules moved faster. That conversion warmed the gas surrounding the cores of the giant planets. Proto-Jupiter and proto-Saturn probably became so hot that they glowed a deep red color, similar to the heating element on an electric stove. Their internal temperatures may have been even higher.

In the mini accretion disks surrounding the giant planets, some of the remaining material combined into small bodies, which became moons. A **moon** is any natural satellite in orbit about a planet or asteroid. The composition of the moons that formed around the giant planets followed the same trend as that of the planets that formed around the Sun: the innermost moons formed under the hottest conditions and therefore contained the smallest amounts of volatile material. For example, the closest of Jupiter's many moons may have experienced high temperatures from nearby Jupiter's glowing so intensely that it would have evaporated most of the volatile substances in the inner part of its mini accretion disk.

Remaining Planetesimals

Not all planetesimals in the disk became planets. For example, dwarf planets orbit the Sun but have not cleared other, smaller bodies from their orbits. Ceres and Pluto (Figure 7.1) are dwarf planets. More dwarf planets, along with many smaller bodies, are found in the Kuiper Belt, beyond Pluto's orbit. Asteroids are small bodies found inside Jupiter's orbit around the Sun; most are located in the main asteroid belt between the orbits of Mars and Jupiter. Jupiter's gravity kept the region between Jupiter and Mars so stirred up that most planetesimals there never formed a large planet.

Planetesimals persist to this day in the outermost part of the Solar System as well. Formed in a deep freeze, those objects have retained most of the highly volatile materials found in the grains present when the accretion disk formed. Unlike the crowded inner part of the disk, the outermost parts had planetesimals too sparsely distributed for large planets to grow. Icy planetesimals in the outer Solar System that survived planetary accretion remain today as **comet nuclei**. The frozen, distant dwarf planets Pluto and Eris are especially large examples of those residents of the outer Solar System.

Many Solar System objects show evidence of cataclysmic impacts that reshaped worlds, suggesting that the early Solar System must have been a remarkably violent and chaotic place. The dramatic difference in the terrain of the northern and southern hemispheres on Mars, for example, has been interpreted as the result of one or more colossal collisions. The leading theory for the origin of our Moon is that it resulted from the collision of an object with Earth. Mercury has a crater on its surface from an impact so devastating that it caused the crust to buckle on the opposite side of the planet. In the outer Solar System, one of Saturn's moons, Mimas, has a crater roughly one-third the diameter of the moon itself. Uranus suffered one or more collisions violent enough to knock it on its side. As a result, its equatorial plane is tilted at almost a right angle to its orbital plane. Other examples are discussed in later chapters.

CHECK YOUR UNDERSTANDING 7.4

Suppose that astronomers found a rocky, terrestrial planet beyond the orbit of Neptune. What is the most likely explanation for its origin? (a) It formed close to the Sun and migrated outward. (b) It formed in that location and was not disturbed by migration. (c) It formed later in the Sun's history than other planets. (d) It is a captured planet that formed around another star.

7.5 Planetary Systems Are Common

When astronomers turn their telescopes to young nearby stars, they see disks of the same type from which the Solar System formed. When the light from the central star is blocked (**Figure 7.14**), evidence of the planetary disk is observed. The

Figure 7.14 An edge-on circumstellar dust disk is seen extending outward to 60 AU from the young (12-million-year-old) star AU Microscopii. The star itself, whose brilliance would otherwise overpower the circumstellar disk, is hidden behind an occulting disk (opaque mask) placed in the telescope's focal plane. The star's position is represented by the dot.

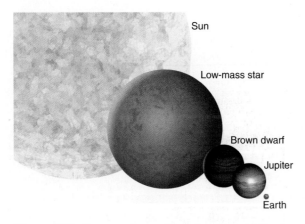

Figure 7.15 A comparison of the diameters of the Sun, a low-mass star, a brown dwarf, Jupiter, and Earth.

Figure 7.16 Doppler shifts observed in the spectrum of a star as it moves around its common center of gravity with its planet. The spectral lines are blueshifted as the star moves toward us and redshifted as it moves away from us.

Figure 7.17 Radial velocity data for a star with a planet. A positive velocity is motion away from the observer, whereas a negative velocity is motion toward the observer. The plot repeats as the planet completes another orbit around the star.

physical processes that led to the formation of the Solar System should be commonplace wherever new stars are being born. Compared with stars, however, planets are small and dim objects. They shine primarily by reflection and therefore are millions to billions of times fainter than their host stars. Thus, they were difficult to find until advances in telescope detector technology in the 1990s enabled astronomers to discover them through indirect methods. In 1995, astronomers announced the first confirmed **extrasolar planet**, usually shortened to **exoplanet**—a planet orbiting around a star other than the Sun. Today, the number of known exoplanets has grown to the thousands, and new discoveries occur almost daily.

The IAU defines an exoplanet as an object that orbits a star other than the Sun and has a mass less than 10–13 Jupiter masses (10–13 M_{Jup}). Objects more massive than 10–13 M_{Jup} but less massive than 0.08 solar masses (0.8 M_{Sun}; about 80 M_{Jup}) are **brown dwarfs**. Objects more massive than 0.08 M_{Sun} are defined as stars. **Figure 7.15** compares the diameters of typical objects of each class.

The Search for Exoplanets

The first planets were discovered indirectly, by observing their gravitational tug on the central star. As technology has improved, other methods have become more productive. Astronomers now have direct images of planets orbiting stars and have taken spectra of some exoplanets to observe the composition of their atmospheres. Almost certainly, between the time we write this and the time you read it, new discoveries will have been made. The field is advancing extremely quickly: more than 100 projects are searching for exoplanets from the ground and from space. We will now look at each discovery method.

THE RADIAL VELOCITY METHOD As a planet orbits a star, the planet's gravity tugs the star around ever so slightly. If the star has a radial velocity (Chapter 5), an observable Doppler shift may appear in the spectrum of the star. **Figure 7.16** illustrates that motion. When the star is moving toward us (negative radial velocity), the light is blueshifted; when the star is moving away from us (positive radial velocity), the light is redshifted. That pattern of radial velocity repeats. After detecting those changes in the radial velocity (**Figure 7.17**), astronomers can infer the existence of the planet and find the planet's mass and distance from the star.

The smallest planet that can be found depends on the precision of the observations. For example, in our Solar System, Jupiter's mass is greater than the mass of all the other planets, asteroids, and comets combined, so Jupiter is the planet with the largest effect on the Sun. Both the Sun and Jupiter orbit a common center of gravity (sometimes called center of mass—the location where the effect of one mass balances the other) that lies just outside the surface of the Sun, as shown in **Figure 7.18**. Jupiter tugs the Sun around in a circle with a speed of 12 m/s. Alien astronomers would find that the Sun's radial velocity varies by ±12 m/s, with a period equal to Jupiter's orbital period of 11.86 years. From that information, the astronomers would rightly conclude that the Sun has at least one planet with a mass comparable to Jupiter's. Without greater precision, the observers would be unaware of the other, less massive planets. To detect Saturn, they would need to improve the precision of their measurements to 2.7 m/s. Earth would not be detectable unless the aliens could detect motions as small as 0.09 m/s.

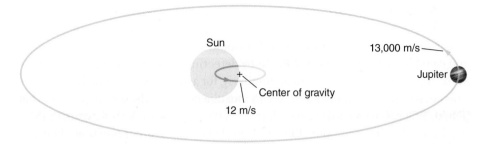

Figure 7.18 The Sun and Jupiter orbit around a common center of gravity (+), which lies just outside the Sun's surface. Spectroscopic measurements made by an extrasolar astronomer would reveal the Sun's radial velocity varying by ±12 m/s over 11.86 years, which is Jupiter's orbital period. Jupiter travels around its orbit at a speed of 13,000 m/s.

The precision of radial velocity instruments has been about 0.3 m/s. The technique enabled astronomers to detect giant planets around solar-type stars, but not yet to find planets with masses similar to Earth's. Finding the signal of the Doppler shift in the noise of the observation requires the star to be quite bright in our sky. A new instrument at the Very Large Telescope is expected to improve precision and be able to detect smaller planets. **Working It Out 7.2** explains more about the spectroscopic radial velocity method.

Astronomy in Action: Doppler Shift

Simulation: Radial Velocity

7.2 Working It Out Estimating the Size of the Orbit of a Planet

In the spectroscopic radial velocity method, the star is moving about its center of mass, and its spectral lines are Doppler-shifted accordingly. Recall from Figure 7.18 that an alien astronomer looking toward the Solar System would observe a shift in the wavelengths of the Sun's spectral lines—caused by the presence of Jupiter—of ~12 m/s.

Figure 7.17 showed the radial velocity data for a star with a planet discovered with that method. How do astronomers use that method to estimate the distance (A) of the planet from the star? Recall from Chapter 4 that Newton generalized Kepler's law relating the period of an object's orbit to the orbital semimajor axis:

$$P^2 = \frac{4\pi^2}{G} \times \frac{A^3}{M}$$

where A is the semimajor axis of the orbit, P is its period, and M is the combined mass of the two objects. To find A, we rearrange the equation as follows:

$$A^3 = \frac{G}{4\pi^2} \times M \times P^2$$

According to the graph of radial velocity observations in Figure 7.17, the period of the orbit is 5.7 years. A year has 3.16×10^7 seconds, so

$$P = 5.7 \text{ yr} \times (3.16 \times 10^7 \text{ s/yr}) = 1.8 \times 10^8 \text{ s}$$

The mass of the star is much greater than the mass of the planet, so the combined masses of the star and the planet can be approximated as the mass of the star, which here is about equal to the mass of the Sun, 2×10^{30} kg. (Stellar masses can be estimated from their spectra.) The gravitational constant G is 6.67×10^{-20} km^3/(kg s^2). Plugging in the numbers gives

$$A^3 = \frac{6.67 \times 10^{-20} \frac{\text{km}^3}{\text{kg s}^2}}{4\pi^2} \times (2 \times 10^{30} \text{ kg}) \times (1.8 \times 10^8 \text{ s})^2$$

$$A^3 = 1.1 \times 10^{26} \text{ km}^3$$

taking the cube root,

$$A = 4.8 \times 10^8 \text{ km}$$

If we convert that number into astronomical units (where 1 AU = 1.5×10^8 km), the semimajor axis of the orbit of this planet is

$$A = \frac{4.8 \times 10^8 \text{ km}}{1.5 \times 10^8 \text{ km/AU}} = 3.2 \text{ AU}$$

The planet is 3.2 times farther from its star than Earth is from the Sun.

THE TRANSIT METHOD From Earth it is sometimes possible to see the inner planets Mercury and Venus transit, or pass in front of, the Sun. An alien located somewhere in the plane of Earth's orbit would see Earth pass in front of the Sun and could infer the existence of Earth by detecting the 0.009 percent drop in the Sun's brightness during the transit. Similarly, for astronomers on Earth to observe a planet passing in front of a star, Earth must lie nearly in the orbital plane of that planet. When an exoplanet passes in front of its parent star, the light from the star diminishes by a tiny amount (**Figure 7.19**). This is the **transit method** of detecting exoplanets, in which we observe the dimming of starlight as a planet passes in front of its parent star. Whereas the radial velocity method gives us the mass of the planet and its orbital distance from a star, the transit method provides the radius of a planet. **Working It Out 7.3** shows how the radii are estimated.

Current ground-based technology limits the sensitivity of the transit method to about 0.1 percent of a star's brightness. Telescopes in space improve the sensitivity because smaller dips in brightness can be measured. The small Fench COROT telescope (27 cm) discovered 32 planets during its 6 years of operation (2007–2013). NASA's 0.95-meter Kepler telescope has discovered many planets and has found thousands more candidates that are being investigated further. **Figure 7.20** illustrates how systems of multiple planets are identified with that method: if one planet is found, then multiple sets of transits can indicate that other planets are orbiting the same star. Several thousand exoplanets have been detected from ground-based and space telescopes by using the transit method.

OTHER METHODS The gravitational field of an unseen planet can act like a lens, bending the light from a distant star in such a way that it causes the star to brighten temporarily while the planet is passing in front of it. Because the effect is small, it is usually called *microlensing*. Like the radial velocity method, microlensing provides an estimate of the mass of the planet. To date, about 60 exoplanets have been found with that technique.

Planets also may be detected by *astrometry*—precisely measuring the position of a star in the sky. If the system is viewed from "above," the star moves in a mini-orbit as the planet pulls it around. That motion is generally tiny and therefore very difficult to measure. For systems viewed from above the plane of the planet's orbit, however, none of the prior methods will work because the planet neither passes in front of the star nor causes a shift in its speed along the line of sight. Space missions such as the Gaia observatory, launched in 2013 by the European Space Agency, conduct such observations.

Direct imaging involves taking a picture of the planet directly. The technique is conceptually straightforward but is technically difficult because it involves searching for a relatively faint planet in the overpowering glare of a bright star—a challenge far more difficult than looking for a star in a clear, bright daytime sky. Even when an object is detected by direct imaging, an astronomer must still determine whether the observed object is actually a planet. Suppose we detect a faint object near a bright star. Could it be a more distant star that just happens to be in the line of sight? Future observations could tell whether the object shares the bright star's motion through space, but it also could be a brown dwarf rather than a true planet. An astronomer would need to make further observations to determine the object's mass.

Figure 7.19 A planet that passes in front of a star blocks some of the light coming from the star's surface, causing the brightness of the star to decrease slightly. (The decrease in brightness is exaggerated here.)

Figure 7.20 Multiple planets can be detected through multiple transits with different changes in brightness. The arrows point to the changes in the total light as the three planets transit the star.

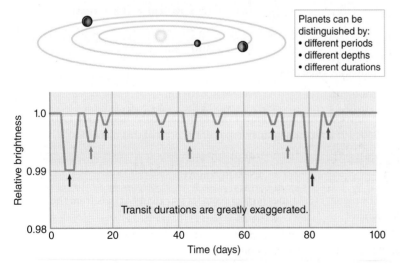

7.3 Working It Out Estimating the Radius of an Exoplanet

The masses of exoplanets can often be estimated using Kepler's laws and the conservation of angular momentum. When planets are detected with the transit method, astronomers can estimate the radius of an exoplanet. In that method, astronomers look for planets that eclipse their stars and then observe how much the star's light decreases during that eclipse (see Figure 7.19). When Venus or Mercury transits the Sun, a black circular disk is visible on the face of the circular Sun. During the transit, the amount of light from the transited star is reduced by the area of the circular disk of the planet divided by the area of the circular disk of the star:

$$\% \text{ reduction in light} = \frac{\text{Area of disk of planet}}{\text{Area of disk of star}} = \frac{\pi R_{planet}^2}{\pi R_{star}^2} = \frac{R_{planet}^2}{R_{star}^2}$$

Then, to solve for the radius of the planet, astronomers need an estimate of the radius of the star and a measurement of the percentage of reduction in light during the transit. The radius of a star is estimated from the surface temperature and the luminosity of the star.

Kepler-11, for example, is a system of at least six planets that transit a star. The radius of the star, R_{star}, is estimated to be 1.1 times the radius of the Sun, or $1.1 \times (7.0 \times 10^5 \text{ km}) = 7.7 \times 10^5$ km. The light from the Kepler-11 star is observed to decrease by 0.077 percent, or 0.00077 (see Figure 7.19), from planet Kepler-11c. What is the radius of Kepler-11c?

$$0.00077 = \frac{R_{Kepler-11c}^2}{R_{star}^2} = \frac{R_{Kepler-11c}^2}{(7.7 \times 10^5 \text{ km})^2}$$

$$R_{Kepler-11c}^2 = 4.5 \times 10^8 \text{ km}^2 \quad \text{thus} \quad R_{Kepler-11c} = 2.1 \times 10^4 \text{ km}$$

Dividing $R_{Kepler-11c}$ by R_{Earth} (6,400 km) shows that $R_{Kepler-11c} = 3.3\, R_{Earth}$.

Some planets have been discovered through that method with large ground-based telescopes operating in the infrared region of the spectrum, using adaptive optics. **Figure 7.21** is an infrared image of Beta Pictoris b. A related form of direct observation involves separating the spectrum of a planet from the spectrum of its star to obtain information about the planet directly. Large ground-based telescopes have obtained spectra of the atmospheres of some exoplanets and have found, for example, carbon monoxide and water in those atmospheres.

Types of Exoplanets

Searches for exoplanets have been remarkably successful. Between the discovery of the first (in 1995) and this writing, nearly 4,000 more have been confirmed, and thousands more candidates are under investigation. As the number of observed systems with single and multiple planets increases, astronomers can compare those worlds with Solar System planets, finding more variation than they expected. The field is changing so fast that the most up-to-date information can be found only online.

The first discoveries included many **hot Jupiters**—Jupiter-sized planets orbiting solar-type stars in circular or highly eccentric orbits that bring them closer to their parent stars than Mercury is to our own Sun. Those planets were among the first to be detected because they are relatively easy targets for the spectroscopic radial velocity method. The large mass of a nearby hot Jupiter tugs the star very hard, creating large radial velocity variations in the star. In addition, those large planets orbiting close to their parent stars are more likely to pass in front of the star periodically and reveal themselves via the transit method.

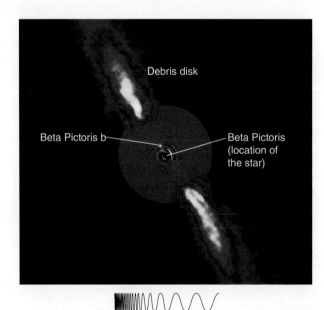

Figure 7.21 The planet Beta Pictoris b is seen orbiting within a dusty debris disk that surrounds the bright naked-eye star Beta Pictoris. The planet's estimated mass is 8 times that of Jupiter. The star is hidden behind an opaque mask, and the planet appears through a semitransparent mask used to subdue the brightness of the dusty disk.

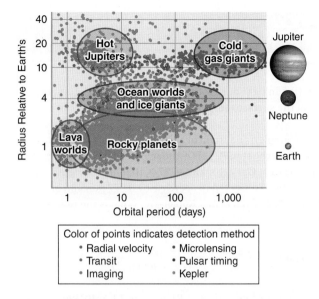

Figure 7.22 Different populations of exoplanets. Color of points indicates detection method. Jupiter-sized planets can be close to their star and hot, or they can be far from their star and cold. Neptune-sized planets are probably composed of water and ice. Earth-sized planets have higher density, indicating that they are made of rock.

Therefore, those hot Jupiter systems are easier to find than smaller, more distant planets. Astronomers realized that those hot Jupiter systems are not representative of most planetary systems; they were just easier to find. Scientists call that type of bias a *selection effect*. **Figure 7.22** illustrates the types of exoplanet populations.

Astronomers were surprised by the hot Jupiters because, according to the theory of planet formation described earlier in the chapter, those giant, volatile-rich planets should not have been able to form so close to their parent stars. From theories based on the Solar System, astronomers expected that Jupiter-type planets should form in the more distant, cooler regions of the protoplanetary disk, where the volatiles that make up much of their composition can survive. Hot Jupiters may form much farther from their parent stars and later migrate inward to a closer orbit. That migration may be caused by an interaction with gas or planetesimals in which orbital angular momentum is transferred from the planet to its surroundings, allowing it to spiral inward.

The radial velocity method yields an estimate of the mass of a planet (see Working It Out 7.2), and the transit method yields an estimate of the size of a planet (see Working It Out 7.3). With both pieces of information, the density (mass divided by the volume) of the planet can be computed to determine whether the planet is mostly gaseous (low density) or mostly rock (high density). Many of the new planets being discovered by Kepler are mini-Neptunes (gaseous planets with masses of 2–10 M_{Earth}) or super-Earths (rocky planets more massive than Earth). **Figure 7.23** illustrates the formation of those two types.

Planets with longer orbital periods, and therefore larger orbits, can be discovered only when the observations have gone on long enough to observe more than one complete orbit. Some of the exoplanets have highly elliptical orbits compared with those in the Solar System. Planets have been found with orbits that are highly tilted with respect to the plane of the rotation of their star, and some planets move in orbits whose direction is opposite that of their star's rotation. Multiple-planet systems have been observed in which the larger mini-Neptunes alternate with smaller super-Earths. The multiple-planet systems that have been found with the transit method reside in flat systems like our own, offering further evidence that the planets formed in a flat protoplanetary disk around a young star. But the current hypothesis to explain the Solar System's inner, small rocky planets and outer, large gaseous planets may not apply to all other planetary systems.

In addition, some planets detected through microlensing seem to be wandering freely through the Milky Way. Those planets may have been ejected from their solar systems after they formed and are no longer in gravitationally bound orbits around their stars.

We return to exoplanets at several points in this book—namely, when we review the planets in our own Solar System, when we consider the types of stars, and when we discuss the search for Earth-like planets in Chapter 24. The frequent new discoveries requiring revisions of existing theories make exoplanets one of the most exciting topics in astronomy today.

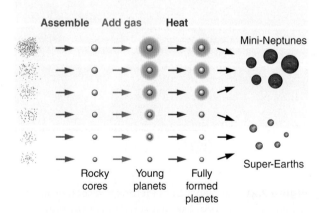

Figure 7.23 Many exoplanets are between the sizes of Earth and Neptune (~4 R_E). "Super-Earths" began with smaller rocky cores and less material, whereas "mini-Neptunes" had larger cores and collected more gas while forming.

CHECK YOUR UNDERSTANDING 7.5

What is the most common method for discovering an Earth-mass planet around a star?
(a) Doppler spectroscopy; (b) direct imaging; (c) transit; (d) astrometric

Origins The Search for Earth-Sized Planets

The discovery of planetary systems, many different from the Solar System, shows us that the formation of planets often, and perhaps always, accompanies the formation of stars. The implications of that conclusion are profound. Planets are a common by-product of star formation. In a galaxy of 200 billion stars and a universe of hundreds of billions of galaxies, how many planets (and moons) might exist? And with all those planets in the universe, how many might have conditions suitable for the particular category of chemical reactions that we refer to as "life"? (We return to that point in Chapter 24.)

The Kepler Mission was developed by NASA to find Earth-sized and larger planets in orbit about a variety of stars. Kepler is a 0.95-meter telescope with 42 CCD detectors that observes approximately 150,000 stars in 100 square degrees of sky to look for planetary transits. To confirm a planetary detection, transits need to be observed three times with repeatable changes in brightness, duration of transit times, and computed orbital period. Kepler could detect a dip of 0.01 percent in the brightness of a star—sensitive enough to detect an Earth-sized planet. Kepler identified the first Earth-sized planets in 2011. Stars with transiting planets detected by Kepler also are observed spectroscopically to obtain radial velocity measurements that can lead to an estimate of a planet's mass. If a planet's radius and mass are known, the planet's density (mass per volume) can be estimated, too. From the density, astronomers can get a sense of whether the planet is composed primarily of gas, rock, ice, water, or some mixture of those.

On Earth, liquid water was essential for life to form and evolve. Because life on Earth is the only example of life for which we have evidence, we do not know whether liquid water is a cosmic requirement, but it is a place to start. The Kepler Mission searched for rocky planets at the right distance from their stars to permit the existence of liquid water, a distance known as the **habitable zone**. The location and width of the zone will vary depending on the temperature of the star. If a planet is too close to its star, water will exist only as a vapor; if too far, water will be frozen as ice. In the Solar System, Earth is the only planet currently in the habitable zone. Although announcements of new planets often state whether the planet is in the habitable zone, just being in the zone doesn't guarantee that the planet actually has liquid water—or that the planet is inhabited! An example of an Earth-sized planet in a habitable zone is shown in **Figure 7.24**.

Kepler identified thousands of planet candidates, some in the habitable zones of their respective stars. The candidates must be confirmed by follow-up observations of more transits or of radial velocities before they are officially announced as planet detections. Citizen scientists contribute to that search through the Exoplanet Explorers project; recently, the project announced the discovery of a system with five planets.

Astronomers are looking forward to results from new missions such as the Transiting Exoplanet Survey Satellite (TESS), launched in 2018, and the James Webb Space Telescope (JWST), scheduled for launch in 2020. TESS will monitor 200,000 nearby stars for dimming caused by planetary transits, whereas JWST will observe in the infrared and should be able to detect gases in the atmospheres of exoplanets.

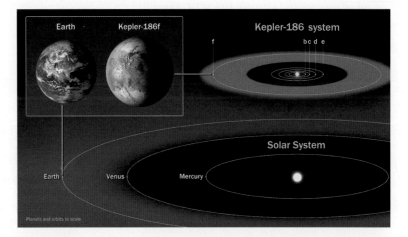

Figure 7.24 Artist's conception of the Kepler-186 system. Located about 500 light-years from Earth, that system has a planet in its habitable zone. The Solar System is shown for comparison. The Kepler-186 star is cooler than the Sun; thus, its habitable zone is closer to the star.

READING ASTRONOMY NEWS

A multiplanet system discovered around a relatively nearby star

Four Earth-Sized Planets Detected Orbiting the Nearest Sun-like Star

August 9, 2017, by TIM STEPHENS

A new study by an international team of astronomers reveals that four Earth-sized planets orbit the nearest Sun-like star, Tau Ceti, which is about 12 light years away and visible to the naked eye. These planets have masses as low as 1.7 Earth mass, making them among the smallest planets ever detected around nearby Sun-like stars. Two of them are super-Earths located in the habitable zone of the star, meaning they could support liquid surface water.

The planets were detected by observing the wobbles in the movement of Tau Ceti. This required techniques sensitive enough to detect variations in the movement of the star as small as 30 centimeters per second.

"We are now finally crossing a threshold where, through very sophisticated modeling of large combined data sets from multiple independent observers, we can disentangle the noise due to stellar surface activity from the very tiny signals generated by the gravitational tugs from Earth-sized orbiting planets," said coauthor Steven Vogt, professor of astronomy and astrophysics at UC Santa Cruz.

According to lead author Fabo Feng of the University of Hertfordshire, UK, the researchers are getting tantalizingly close to the 10-centimeter-per-second limit required for detecting Earth analogs. "Our detection of such weak wobbles is a milestone in the search for Earth analogs and the understanding of the Earth's habitability through comparison with these analogs," Feng said. "We have introduced new methods to remove the noise in the data in order to reveal the weak planetary signals."

The outer two planets around Tau Ceti are likely to be candidate habitable worlds, although a massive debris disk around the star probably reduces their habitability due to intensive bombardment by asteroids and comets.

The same team also investigated Tau Ceti four years ago in 2013, when coauthor Mikko Tuomi of the University of Hertfordshire led an effort in developing data analysis techniques and using the star as a benchmark case. "We came up with an ingenious way of telling the difference between signals caused by planets and those caused by star's activity. We realized that we could see how star's activity differed at different wavelengths and use that information to separate this activity from signals of planets," Tuomi said.

The researchers painstakingly improved the sensitivity of their techniques and were able to rule out two of the signals the team had identified in 2013 as planets. "But no matter how we look at the star, there seem to be at least four rocky planets orbiting it," Tuomi said. "We are slowly learning to tell the difference between wobbles caused by planets and those caused by stellar active surface. This enabled us to essentially verify the existence of the two outer, potentially habitable planets in the system."

Sun-like stars are thought to be the best targets in the search for habitable Earth-like planets due to their similarity to the Sun. Unlike more common smaller stars, such as the red dwarf stars Proxima Centauri and Trappist-1, they are not so faint that planets would be tidally locked, showing the same side to the star at all times. Tau Ceti is very similar to the sun in its size and brightness, and both stars host multi-planet systems.

1. How do planets cause their stars to "wobble"?
2. With the spectroscopic method, why is finding Earth-sized planets harder than finding giant planets?
3. How did the astronomers improve their ability to find small planets?
4. What is a debris disk?
5. What would be the disadvantage to a planet's being tidally locked to its star? (Hint: Refer to Chapter 2.)

Source: The Regents of the University of California, Santa Cruz, Writer: Tim Stephens, "Four Earth-sized planets detected orbiting the nearest sun-like star," University of California, Santa Cruz, News Center, August 8, 2017 ©. Reprinted with permission.

Summary

Stars and their planetary systems form from collapsing interstellar clouds of gas and dust, following the laws of gravity and conservation of angular momentum. Conservation of angular momentum produces an accretion disk around a protostar that often fragments to form multiple planets, as well as smaller objects such as asteroids and dwarf planets, through the gradual accumulation of material into larger and larger objects. Multiple methods exist for finding planets around other stars, and those planets are now thought to be very common. This field of study is evolving very quickly as technology advances.

LG 1 **Describe how our understanding of planetary system formation developed from the work of both planetary and stellar scientists.** Planets are a common by-product of star formation, and many stars are surrounded by planetary systems. Gravity pulls clumps of gas and dust together, causing them to shrink and heat up. Angular momentum must be conserved, leading to both a spinning central star and an accretion disk that rotates and revolves in the same direction as the central star. Solar System meteorites show that larger objects build up from smaller objects.

LG 2 **Discuss the role of gravity and angular momentum in explaining why planets orbit the Sun in a plane and why they revolve in the same direction that the Sun rotates.** As particles orbit the forming star, the cloud of dust and gas flattens into a plane. Conservation of angular momentum determines both the speed and the direction of the revolution of the objects in the forming system. Dust grains in the protoplanetary disk first stick together because of collisions and static electricity. As those objects grow, they eventually have enough mass to attract other objects gravitationally. Once that occurs, they begin emptying the space around them. Collisions of planetesimals lead to the formation of planets.

LG 3 **Explain how temperature at different locations in the protoplanetary disk affects the composition of planets, moons, and other bodies.** The temperature is higher near the central protostar, forcing volatile elements, such as water, to evaporate and leave the inner part of the disk. Planets in the inner part of the disk will have fewer volatiles than those in the outer part of the disk. The gas that a planet captures when it forms is the planet's primary atmosphere. Less massive planets lose their primary atmospheres and then form secondary atmospheres.

LG 4 **Discuss the processes that resulted in the formation of planets and other objects in our Solar System.** In the current model of the formation of the Solar System, solid terrestrial planets formed in the inner disk, where temperatures were high, whereas giant gaseous planets formed in the outer disk, where temperatures were low. Dwarf planets such as Pluto formed in the asteroid belt and in the region beyond the orbit of Neptune. Asteroids and comet nuclei remain today as leftover debris.

LG 5 **Describe how astronomers find planets around other stars and derive exoplanet properties.** Astronomers find planets around other stars by using the radial velocity method, the transit method, microlensing, astrometry, and direct imaging. As technology has improved, the number and variety of known exoplanets has increased dramatically, with thousands of planets and planet candidates discovered orbiting other stars near the Sun within the Milky Way Galaxy in just the past few years.

? Unanswered Questions

- How typical is the Solar System? Only within the past decade have astronomers found other systems containing four or more planets, and so far the observed distributions of large and small planets in those multiplanet systems have looked different from those of the Solar System. Computer simulations of planetary system formation suggest that a system with stable orbits and a planetary distribution like those of the Solar System may develop only rarely. Improved supercomputers can run more complex simulations, which can be compared with the observations to better understand how solar systems are configured.

- How Earth-like must a planet be before scientists declare it to be "another Earth"? An editorial in the science journal *Nature* cautioned that scientists should define "Earth-like" in advance—before multiple discoveries of planets "similar" to Earth are announced and a media frenzy ensues. Must a planet be of similar size and mass, be located in the habitable zone, and have spectroscopic evidence of liquid water before we call it "Earth 2.0"?

Questions and Problems

Test Your Understanding

1. Place the following events in the order that corresponds to the formation of a planetary system.
   a. Gravity collapses a cloud of interstellar gas.
   b. A rotating disk forms.
   c. Small bodies collide to form larger bodies.
   d. A stellar wind "turns on" and sweeps away gas and dust.
   e. Primary atmospheres form.
   f. Primary atmospheres are lost.
   g. Secondary atmospheres form.
   h. Dust grains stick together by static electricity.

2. If the radius of an object's orbit is halved, and angular momentum is conserved, what must happen to the object's speed?
   a. It must be halved.
   b. It must stay the same.
   c. It must be doubled.
   d. It must be squared.

3. Unlike the giant planets, the terrestrial planets formed when
   a. the inner Solar System was richer in heavy elements than the outer Solar System.
   b. the inner Solar System was hotter than the outer Solar System.
   c. the outer Solar System took up more volume than the inner Solar System, so more material was available to form planets.
   d. the inner Solar System was moving faster than the outer Solar System.

4. The terrestrial planets and the giant planets have different compositions because
   a. the giant planets are much larger.
   b. the terrestrial planets formed closer to the Sun.
   c. the giant planets are made mostly of solids.
   d. the terrestrial planets have few moons.

5. The spectroscopic radial velocity method preferentially detects
   a. large planets close to the central star.
   b. small planets close to the central star.
   c. large planets far from the central star.
   d. small planets far from the central star.
   e. The method detects all those planets equally well.

6. The concept of disk instability was developed to solve the problem that
   a. Jupiter-like planets migrate after formation.
   b. not enough gas was in the Solar System to form Jupiter.
   c. the early solar nebula probably dispersed too soon to form Jupiter.
   d. Jupiter consists mostly of volatiles.

7. Because angular momentum is conserved, an ice-skater who throws her arms out will
   a. rotate more slowly.
   b. rotate more quickly.
   c. rotate at the same rate.
   d. stop rotating.

8. Clumps grow into planetesimals by
   a. gravitationally pulling in other clumps.
   b. colliding with other clumps.
   c. attracting other clumps with opposite charge.
   d. conserving angular momentum.

9. The transit method preferentially detects
   a. large planets close to the central star.
   b. small planets close to the central star.
   c. large planets far from the central star.
   d. small planets far from the central star.
   e. The method detects all those planets equally well.

10. If the radius of a spherical object is halved, what must happen to the period so that the spin angular momentum is conserved?
    a. It must be divided by 4.
    b. It must be halved.
    c. It must stay the same.
    d. It must double.
    e. It must be multiplied by 4.

11. The amount of angular momentum in a spherical object does *not* depend on
    a. its radius.
    b. its mass.
    c. its rotation speed.
    d. its temperature.

12. The planets in the inner part of the Solar System are made primarily of refractory materials; the planets in the outer Solar System are made primarily of volatiles. That difference occurs because
    a. refractory materials are heavier than volatiles, so they sank farther into the nebula.
    b. no volatiles were in the inner part of the accretion disk.
    c. the volatiles on the inner planets were lost soon after the planets formed.
    d. the outer Solar System has gained more volatiles from space since formation.

13. If scientists want to find out about the composition of the early Solar System, the best objects to study are
    a. the terrestrial planets.
    b. the giant planets.
    c. the Sun.
    d. asteroids and comets.

14. The direction of revolution in the plane of the Solar System was determined by
    a. the plane of the galaxy in which the Solar System sits.
    b. the direction of the gravitational force within the original cloud.
    c. the direction of rotation of the original cloud.
    d. the amount of material in the original cloud.

15. A planet in the "habitable zone"
    a. is close to the central star.
    b. is far from the central star.
    c. is the same distance from its star as Earth is from the Sun.
    d. is at a distance where liquid water can exist on the surface.

Thinking about the Concepts

16. What is the source of the material that now makes up the Sun and the rest of the Solar System?

17. Describe the different ways by which stellar astronomers and planetary scientists each came to the same conclusion about how planetary systems form.

18. What is a protoplanetary disk? What are two reasons that the inner part of the disk is hotter than the outer part?

19. Physicists describe certain properties, such as angular momentum and energy, as being *conserved*. What does that mean? Do conservation laws imply that an individual object can never lose or gain angular momentum or energy? Explain your reasoning.

20. The Process of Science Figure in this chapter makes the point that different areas of science must agree with one another. Suppose that a few new exoplanets are discovered that appear not to have formed from the collapse of a stellar nebula (for example, the planetary orbits might be in random orientations). What will scientists do with that new information?

21. How does the law of conservation of angular momentum control a figure-skater's rate of spin?

22. What is an accretion disk?

23. Describe how tiny grains of dust grow to become massive planets.

24. Look under your bed, the refrigerator, or any similar place for dust bunnies. Once you find them, blow one toward another. Watch carefully and describe what happens as they meet. What happens if you repeat that action with additional dust bunnies? Will those dust bunnies ever have enough gravity to begin pulling themselves together? If they were in space instead of on the floor, might that happen? What force prevents their mutual gravity from drawing them together into a "bunny-tesimal" under your bed?

25. Why do we find rocky material everywhere in the Solar System but find large amounts of volatile material only in the outer regions?

26. Why could the four giant planets collect massive gaseous atmospheres, whereas the terrestrial planets could not? Explain the source of the secondary atmospheres surrounding the terrestrial planets.

27. Describe four methods that astronomers use to search for exoplanets. What are the limitations of each method; that is, what circumstances are necessary to detect a planet by each method?

28. Why is it difficult to obtain an image of an exoplanet?

29. Many of the first exoplanets that astronomers found orbiting other stars were giant planets with Jupiter-like masses and with orbits located very close to their parent stars. Explain why those characteristics are a selection effect of the discovery method.

30. How has the Kepler telescope found Earth-like planets, and what do astronomers mean by "Earth-like"?

Applying the Concepts

31. In Figure 7.17, what is the maximum radial velocity of HD 70642 b in meters per second? Convert that number to miles per hour (mph). How does that value compare with the speed at which Earth orbits the Sun (67,000 mph)?

32. Use Appendix 4 to answer the following:
    a. What is the total mass of all the planets in the Solar System, expressed in Earth masses (M_{Earth})?
    b. What fraction of that total planetary mass is Jupiter?
    c. What fraction does Earth represent?

33. Compare Earth's orbital angular momentum with its spin angular momentum by using the following values: $m = 5.97 \times 10^{24}$ kg, $v = 29.8$ kilometers per second (km/s), $r = 1$ AU, $R = 6,378$ km, and $P = 1$ day. Assume that Earth is a uniform body. What fraction does each component (orbital and spin) contribute to Earth's total angular momentum? Refer to Working It Out 7.1.

34. Venus has a radius 0.949 times that of Earth and a mass 0.815 times that of Earth. Venus's rotation period is 243 days. What is the ratio of Venus's spin angular momentum to that of Earth? Assume that Venus and Earth are uniform spheres.

35. Jupiter has a mass equal to 318 times Earth's mass, an orbital radius of 5.2 AU, and an orbital velocity of 13.1 km/s. Earth's orbital velocity is 29.8 km/s. What is the ratio of Jupiter's orbital angular momentum to that of Earth?

36. The asteroid Vesta has a diameter of 530 km and a mass of 2.7×10^{20} kg.
    a. Calculate the density (mass/volume) of Vesta.
    b. The density of water is 1,000 kg/m^3, whereas that of rock is about 2,500 kg/m^3. What does that difference tell you about the composition of Vesta?

37. Using data in Appendix 4, compute the densities of Venus, Jupiter, and Neptune. Compare with the densities of rock, water, and gas.

38. Recalling Kepler's laws, put the three planets in Figure 7.20 in order from fastest to slowest. Compare the duration of the transits in Figure 7.20. Why does the outermost planet have the longest duration?

39. Suppose you can measure radial velocities of about 0.3 m/s. Suppose you are observing a spectral line with a wavelength of 575 nanometers (nm). How large a shift in wavelength would a radial velocity of 0.3 m/s produce?

40. Earth tugs the Sun around as it orbits, but that effect (only 0.09 m/s) is much smaller than that of any known exoplanet. How large a shift in wavelength does that effect cause in the Sun's spectrum at 500 nm?

41. If an alien astronomer observed a plot of the light curve as Jupiter passed in front of the Sun, by how much would the Sun's brightness drop during the transit?

42. A planet has been found to orbit a 1-M_{Sun} star in 200 days.
    a. What is the orbital radius of that exoplanet?
    b. Compare its orbit with that of the planets around our own Sun. What temperatures must that planet experience?

43. One of the planets orbiting the star Kepler-11 has an orbital radius of 1.1 R_{Sun}. If the planet has a radius of 4.5 R_{Earth}, by how much does the brightness of Kepler-11 decrease when that planet transits the star?

44. Kepler detected a planet with a diameter of 1.7 Earth (D_{Earth}).
    a. How much larger is the volume of that planet than Earth's?
    b. Assume that the density of the planet is the same as Earth's. How much more massive is it than Earth?

45. The planet COROT-11b was discovered using the transit method, and astronomers have followed up with radial velocity measurements, so both its radius (1.43 R_{Jup}) and its mass (2.33 M_{Jup}) are known. The density provides a clue about whether the object is gaseous or rocky.
    a. What is the mass of the planet in kilograms?
    b. What is the planet's radius in meters?
    c. What is the planet's volume?
    d. What is the planet's density? How does that density compare with the density of water (1,000 kg/m³)? Is the planet likely to be rocky or gaseous?

USING THE WEB

46. Using the exoplanet catalogs:
    a. Go to the "Catalog" Web page (http://exoplanet.eu/catalog) of the Extrasolar Planets Encyclopedia. Look for a star that has multiple planets. Make a graph showing the distances of the planets from that star, and note the masses and sizes of the planets. Put the Solar System planets on the same axis. How does that extrasolar planet system compare with the Solar System?
    b. Go to the "Exoplanets Data Explorer" website (http://exoplanets.org) and click on "Table." That website lists planets that have detailed orbital data published in scientific journals, and it may have a smaller total count than the website in part (a). Pick a planet discovered recently, as specified in the "First Reference" column. What is the planet's minimum mass? What is its semimajor axis and the period of its orbit? What is the eccentricity of its orbit? Click on the star name in the first column to get more information. Does that planet have a radial velocity curve? Was it observed in transit, and if so, what is the planet's radius and density? Is it more like Jupiter or more like Earth?

47. Space missions:
    a. Go to the website for the Kepler Mission (https://nasa.gov/mission_pages/kepler/main/). How many confirmed planets has Kepler discovered? How many planet candidates are there? When did the mission end? Search for the latest version of the "Kepler Orrery," an animation that shows multiplanet systems discovered by Kepler. Do most of those systems look like our own?
    b. Go to the website for the TESS mission (https://tess.gsfc.nasa.gov/). Are there any results?

48. Citizen science projects I:
    a. Go to the "Exoplanet Explorers" website (https://zooniverse.org/projects/ianc2/exoplanet-explorers) Exoplanet Explorers is part of the Zooniverse, a citizen science project that invites individuals to participate in a major science project by using their own computers. To participate in that or any of the other Zooniverse projects mentioned in later chapters, you will need to sign up for an account. Read through the sections under "About," including the FAQ. What are some of the advantages to crowdsourcing Kepler data analysis? Back on the Exoplanet Explorers home page, click on "Classify" and then "Show the project tutorial" under the example on the right. Also click on the "Field Guide" to see more examples. Then look at the data presented and classify.
    b. Go to the "Planet Hunters" website (http://planethunters.org), an older Zooniverse project. Read through the sections under "Science," including the FAQ. Back on the Planet Hunters home page, click on "Start Classifying" and go through the tutorial. Classify a few images.

49. Citizen science projects II: Go to the "Disk Detective" website at http://diskdetective.org/, another Zooniverse project for which you will need to open an account as in problem 48. In that project, you will look at observations of young stars to see whether evidence exists for a planetary disk. Under "Menu," read "Science," "About," "FAQ," and then "Classify." Work through an example and then classify a few images.

50. Go to the "Super Planet Crash" Web page (http://stefanom.org/spc/) or the App. Read "Help" to see the rules. First build a system like ours with four Earth-sized planets in the inner 2 AU. Is that system stable? What happens if you add in super-Earths or "ice giants"? Build up a few completely different planetary systems and see what happens. What types of situations cause instability in the inner 2 AU of those systems?

EXPLORATION

Exploring Exoplanets

digital.wwnorton.com/astro6

Visit the Digital Resources Page and on the Student Site open the "Radial Velocity" Interactive Simulation in Chapter 7. That applet has several panels that allow you to experiment with the variables important for measuring radial velocities. First, in the window labeled "Visualization Controls," check the box to show multiple views. Compare the views shown in panels 1–3 with the colored arrows in the last panel to see where an observer would stand to see the view shown. Start the animation (in the "Animation Controls" panel), and allow it to run while you watch the planet orbit its star from each of the views shown. Stop the animation, and in the "Presets" panel, select "Option A" and then click "set."

1 Is Earth's view of that system most nearly like the "side view" or most nearly like the "orbit view"?

2 Is the orbit of that planet circular or elongated?

3 Study the radial velocity graph in the upper right panel. The blue curve shows the radial velocity of the star over a full period. What is the maximum radial velocity of the star?

4 The horizontal axis of the graph shows the "phase," or fraction of the period. A phase of 0.5 is halfway through a period. The vertical red line indicates the phase shown in views in the upper left panel. Start the animation to see how the red line sweeps across the graph as the planet orbits the star. The period of the planet is 365 days. How many days pass between the minimum radial velocity and the maximum radial velocity?

5 When the planet moves away from Earth, the star moves toward Earth. The sign of the radial velocity tells the direction of the motion (toward or away). Is the radial velocity of the star positive or negative at that time in the orbit? If you could graph the radial velocity of the planet at that point in the orbit, would it be positive or negative?

In the "Presets" window, select "Option B" and then click "set."

6 What has changed about the orbit of the planet as shown in the views in the upper left panel?

7 When is the planet moving fastest: when it is close to the star or when it is far from the star?

8 When is the star moving fastest: when the planet is close to it or when the planet is far away?

9 Explain how an astronomer would determine, from a radial velocity graph of the star's motion, whether the orbit of the planet was circular or elongated.

10 Study the "Earth view" panel at the top of the window. Would that planet be a good candidate for a transit observation? Why or why not?

In the "System Orientation" panel, change the inclination to 0.0.

11 Now is Earth's view of the system most nearly like the "side view" or most nearly like the "orbit view"?

12 How does the radial velocity of the star change as the planet orbits?

13 Click the box that says "show simulated measurements," and change the "noise" to 1.0 m/s. The gray dots are simulated data, and the blue line is the theoretical curve. Use the slider bar to change the inclination. What happens to the radial velocity as the inclination increases? (Hint: Pay attention to the vertical axis as you move the slider, not just the blue line.)

14 What is the smallest inclination for which you would find the data convincing? That is, what is the smallest inclination for which the theoretical curve is in good agreement with the data?

8

The Terrestrial Planets and Earth's Moon

The objects that formed in the inner part of the protoplanetary disk around the Sun are relatively small, rocky worlds, one of which is Earth. A comparison of those worlds reveals the forces that shape a planet. The past six decades have been an exciting time for the exploration of and discovery about Earth and the other planets in the Solar System. Robotic probes have visited every planet, and astronauts walked on the Moon. In addition to discoveries from new space missions and telescopes, improved analytical techniques applied to the rocks and soil brought back from the Moon 50 years ago have led to surprising new results. The information from those missions has revolutionized the understanding of the Solar System, offering insights into the current state of each of the neighboring planets and clues about their histories.

LEARNING GOALS

By the end of this chapter, you should be able to look at an image of a planet and identify which geological features occurred early in the history of that world and which occurred late. You also should be able to:

- **LG 1** Describe how impacts have affected the evolution of the terrestrial planets.
- **LG 2** Explain how radiometric dating is used to measure the ages of rocks and terrestrial planetary surfaces.
- **LG 3** Explain how scientists use both theory and observation to determine the structure of terrestrial planetary interiors.
- **LG 4** Describe tectonism and volcanism and the forms they take on different terrestrial planets.
- **LG 5** Summarize what is known about the presence of water on the terrestrial planets.

Mars Reconnaissance Orbiter image of Newton Crater on Mars. The dark streaks may indicate flowing water. ▶▶▶

Does water exist on other planets?

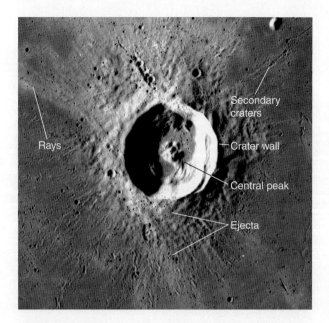

Figure 8.1 A lunar crater showing the crater wall and central peak surrounded by ejected material (*ejecta*), rays, and secondary craters—all typical features associated with impact craters.

8.1 Impacts Help Shape the Evolution of the Planets

Four principal geological processes constantly reshape the terrestrial planets: *impact cratering, tectonism, volcanism,* and *erosion.* Some geological processes originate in a planet's interior, whereas other processes are external. The relative importance of each of those processes varies among the planets. Planetary scientists can learn about the evolutionary history of the Solar System by comparing those geological processes on the terrestrial planets and the Moon. In this section, we examine **impact cratering**, which occurs when large collisions by other Solar System objects leave distinctive scars on the outer layer of a planet.

Comparative Planetology

The four innermost planets in the Solar System—Mercury, Venus, Earth, and Mars—are collectively known as the *terrestrial planets.* Although the Moon is Earth's lone natural satellite, we include it in this chapter because of its similarity to the terrestrial planets.

When comparing planets, we first compare the basic physical characteristics, such as distance from the Sun, size and density, and gravitational pull at the surface. Those characteristics reveal what a planet is made of, what its surface temperature is likely to be, and how well it can hold an atmosphere (planetary atmospheres are discussed in Chapter 9). By comparing the planets, scientists can sort out the vast quantity of information returned by space probes. The correct explanation for a particular aspect of one planet must be consistent with what is known about the other planets. For example, an analysis of why the Moon is covered with craters must allow for the fact that preserved craters are rare on Earth.

TABLE 8.1 Comparison of Physical Properties of the Terrestrial Planets and the Moon

Property	Mercury	Venus	Earth	Mars	Moon
Orbital radius	0.387 AU	0.723 AU	1.000 AU	1.524 AU	384,400 km
Orbital period	0.241 yr	0.615 yr	1.000 yr	1.881 yr	27.32 days
Orbital velocity (km/s)	47.4	35.0	29.8	24.1	1.02
Mass/M_{Earth}	0.055	0.815	1.000	0.107	0.012
Equatorial diameter (km)	4,880	12,104	12,756	6,794	3,476
Equatorial diameter (D/D_{Earth})	0.383	0.950	1.000	0.532	0.273
Density/density of water	5.43	5.24	5.51	3.93	3.34
Sidereal rotation period*	58.65^d	243.02^d	$23^h 56^m$	$24^h 37^m$	27.32^d
Obliquity (degrees)†	0.03	177.36	23.44	25.19	6.68
Surface gravity (m/s^2)	3.70	8.87	9.81	3.71	1.62
Escape velocity (km/s)	4.25	10.36	11.19	5.03	2.38

*The superscript letters *d, h,* and *m* stand for days, hours, and minutes, respectively.
†An obliquity greater than 90° indicates that the planet rotates in a retrograde, or backward, direction.

Figure 8.2 Stages in the formation of an impact crater.

An explanation for why Venus has such a massive atmosphere should point to reasons that Earth and Mars do not. Such comparisons are key to an approach called **comparative planetology**. Some of the basic physical properties of the terrestrial planets are compared in **Table 8.1**.

Impacts and Craters

Of those four geological processes, impact cratering causes the most concentrated and sudden release of energy. Planets and other objects orbit the Sun at very high speeds. As seen in Table 8.1, for example, Earth orbits the Sun at an average speed of around 30 kilometers per second (km/s), or about 67,000 miles per hour (mph). Collisions between orbiting bodies can release huge amounts of energy and produce craters such as the one in **Figure 8.1**. **Figure 8.2** shows the process of impact cratering. When an object hits a planet, the object's kinetic energy heats and compresses the surface that it strikes and throws material far from the resulting impact crater. Sometimes, material thrown from the crater, called **ejecta**, falls back to the surface of the planet with enough energy to cause **secondary craters**. The rebound of heated and compressed material can also lead to the formation of a central peak or a ring of mountains on the crater floor, as shown in the lunar crater in Figure 8.1. Those processes are similar to what happens when a drop lands in liquid (**Figure 8.3**).

The energy of an impact can be great enough to melt or even vaporize rock. The floors of some craters are the cooled surfaces of melted rock that flowed as lava. The energy released in an impact also can lead to the formation of new minerals. Because some minerals form only during an impact, they are evidence of ancient impacts on Earth's surface. The space rocks that cause such impacts are defined by three related terms: **Meteoroids** are small (less than 100 meters in diameter) cometary or asteroid fragments in space. A meteoroid that enters and burns up in a planetary atmosphere is called a **meteor**. Any meteoroids that survive to hit the ground are known as **meteorites** (see Figure 7.3).

Figure 8.3 A drop (a) hitting a pool of liquid illustrates the formation of features in an impact crater, including crater walls (b, c), secondary craters (d, e), and a central peak (f).

Figure 8.4 Meteor Crater (also known as Barringer Crater), located in northern Arizona, is an impact crater 1.2 km in diameter that formed some 50,000 years ago when a nickel-iron meteoroid collided with Earth.

One of the best-preserved impact structures on Earth is Meteor Crater in Arizona (**Figure 8.4**). That impact occurred about 50,000 years ago. From the crater's size and shape and from the remaining pieces of the impacting body, we know that the nickel-iron asteroid fragment was about 50 meters across, had a mass of about 300 million kilograms (kg), and was traveling at 13 km/s relative to Earth when it hit Earth's upper atmosphere. Approximately half of the original mass was vaporized in the atmosphere before the rest hit the ground. That collision released about 300 times as much total energy as the first atom bomb. At only 1.2 km in diameter, Meteor Crater is tiny in comparison with impact craters seen on the Moon or ancient impact scars on Earth.

Impact craters cover the surfaces of Mercury, Mars, and the Moon. The Moon has hundreds of millions of craters of at least 10 meters in diameter and even more smaller ones. As seen outlined by yellow circles in **Figure 8.5**, craters of different sizes exist, one on top of another. Nearly all those craters are the result of impacts. On Earth and Venus, by contrast, most impact craters have been destroyed. Fewer than 200 impact craters have been identified on Earth, and about 1,000 have been found on Venus. Earth's crater shortage is primarily due to two processes we discuss later in the chapter—plate tectonics in Earth's ocean basins and erosion on land. On Venus, lava flows have destroyed most of the craters.

The atmospheres of Earth and Venus provide another explanation for their low number of small craters. The surfaces of the Moon and Mercury are directly exposed to bombardment from space, whereas the surfaces of Earth and Venus are partly protected by their atmospheres. Rock samples from the Moon show craters smaller than a pinhead, formed by micrometeoroids. In contrast, most meteoroids smaller than 100 meters in diameter that enter Earth's atmosphere are either burned up or broken up by friction before they reach the surface. Small meteorites found on the ground on Earth are probably pieces of much larger bodies that broke apart on entering the atmosphere. With an atmosphere far thicker than that of Earth, Venus is even better protected.

Planetary scientists can tell a lot about the surface of a planet by studying its craters because the characteristics of a crater depend on the properties of the planetary surface. An impact in a deep ocean on Earth might create an impressive wave but leave no lasting crater. In contrast, an impact scar formed in an ancient rocky area can be preserved for billions of years. For example, craters on the Moon's pristine surface are often surrounded by strings of smaller secondary craters formed from material thrown out by the impact, such as those shown in Figure 8.1.

Some craters on Mars look very different. They are surrounded by structures that look much like the pattern you might see if you threw a rock into mud (**Figure 8.6**). The flows appear to indicate that the martian surface rocks contained water or ice at the time of the impact. Features resembling canyons and dry riverbeds are further evidence that liquid water may have existed on the surface of Mars. Not all martian craters have that feature, so the water or ice must have been concentrated in only some areas. Those icy locations might have changed—for example, perhaps the meteoroid impact heated the surface enough to temporarily liquefy the frozen water in the ground.

Today, the surface of Mars is dry in some regions and frozen in others, which suggests that water once on the surface has evaporated, or soaked into the ground, much like water frozen in the ground in Earth's polar regions. The energy released by an impact would have melted that ice, turning the surface

Figure 8.5 *Lunar Reconnaissance Orbiter* false-color map of craters on the Moon.

material into a slurry with a consistency much like that of wet concrete. When thrown from the crater by the force of the impact, that slurry would hit the surrounding terrain and slide across the surface, forming the mudlike craters we see today.

Giant Impacts Reshape Planets

Because many planetesimals were roaming around the early Solar System, every young planet experienced heavy bombardment early in its history. The last major bombardment is called *the late heavy bombardment*, which took place from 4.2 billion to 3.8 billion years ago. Observations of uneven thicknesses of the crusts of inner planets have led to theories that some or all of the terrestrial planets were disrupted by at least one giant impact—a collision with an object the size of a large asteroid. A giant impact with Mercury could have removed some of the lighter material of its outside layer, leaving behind an overall denser planet. A giant impact with Venus may have led to its retrograde (backward) rotation. A giant impact probably explains the large differences between the northern and southern hemispheres of Mars. The southern highlands have a thicker crust and formed early in martian history. The northern lowlands have a thinner and younger crust and may have formed when melting occurred after the impact. Early giant impacts on Mars also may be responsible for its loss of a magnetic field. Impacts with smaller icy comets from the outer Solar System brought water, atmospheric gases, and possibly organic molecules to the inner planets. As we discuss at the end of this chapter, an impact with Earth about 65 million years ago might have been a crucial event for the evolution of our species, *Homo sapiens*.

Although several theories have tried to account for how the Moon formed, the leading theory involves a giant impact. According to that impact theory, a Mars-sized protoplanet collided with proto-Earth about 4.5 billion years ago, blasting off and vaporizing parts of proto-Earth's outer layers. The debris from the impact condensed into many "moonlets" in orbit around Earth, and those collected and formed our Moon. That theory accounts for the similarities in composition between the Moon and Earth's outer layers. It also explains the smaller amounts of volatiles on the Moon: during the vaporization stage of the collision, most gases were lost to space, leaving primarily the nonvolatiles to condense as the Moon. Earth, in contrast, was large enough to keep more of its volatiles, which continued to be released from its interior after the collision. Because of the stronger gravity of Earth, those gases were retained as part of Earth's atmosphere. If that account of the Moon's formation is correct, scientists expect to observe chemical evidence of material from the colliding protoplanet that was incorporated into the Moon's composition along with material from Earth. Scientists still debate whether the collision was glancing or head-on, how the Earth-Moon system lost angular momentum, and whether chemical evidence of the colliding object has been found (**Process of Science Figure**). In some models, the *inner* part of the Moon would still contain volatiles. As explained in Section 8.6, debate persists about evidence of water in the Moon's interior.

Figure 8.6 Some craters on Mars look like those formed by rocks thrown into mud, suggesting that material ejected from the crater contained large amounts of water. This crater is about 20 km across.

AstroTour: Processes That Shape the Planets

CHECK YOUR UNDERSTANDING 8.1

Geologists can find the relative age of impact craters on a world because: (a) the ones on top must be older; (b) the ones on top must be younger; (c) the larger ones must be older; (d) the larger ones must be younger.

PROCESS OF SCIENCE

Certainty Is Sometimes Out of Reach

There are several hypotheses for how the Moon formed. One of these fits the data better than the others, but none has been absolutely ruled out.

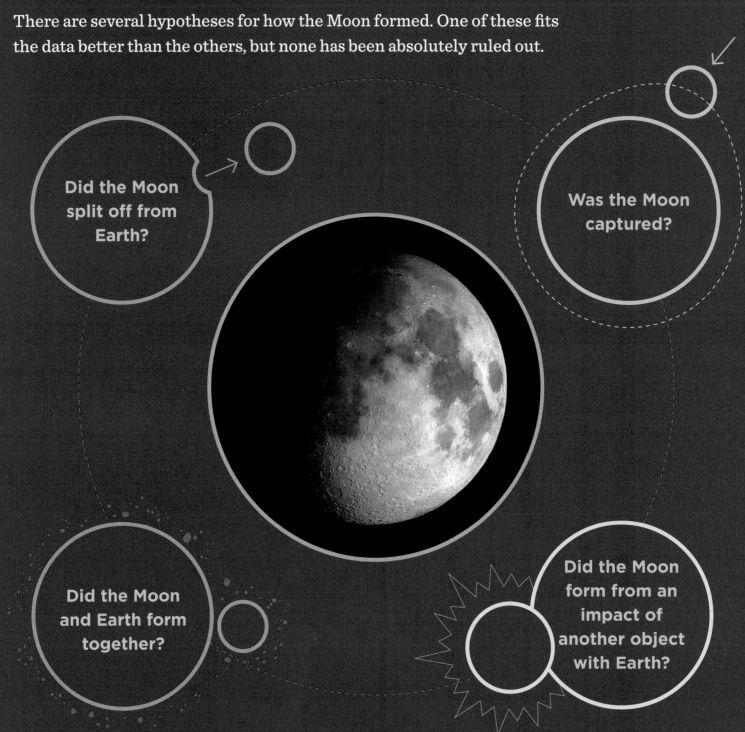

In some cases, hypotheses cannot be definitively falsified (at least not yet). The working hypothesis, then, is the one that best fits the data, but other ideas are kept in mind.

8.2 Radioactive Dating Tells Us the Age of the Moon and the Solar System

The number of visible craters on a planet is determined by the rate at which those craters are destroyed. Geological activity on Earth, Mars, and Venus erased most evidence of early impacts. By contrast, the Moon's surface still preserves the scars of craters dating from about 4 billion years ago. The lunar surface has remained essentially unchanged for more than a billion years because the Moon has no atmosphere or surface water and a cold, geologically dead interior. Mercury also has well-preserved craters, although recent evidence from the *Messenger* mission shows tilted crater floors that are higher on one side than the other—evidence that internal forces lifted the floors unevenly after the craters formed.

Planetary scientists use that cratering record to estimate the ages and geological histories of planetary surfaces: extensive cratering indicates an older planetary surface that remains relatively unchanged because of minimal geological activity. The amount of cratering can serve as a clock to measure the relative ages of surfaces, but to determine the exact age of a surface from the number of craters, we need to know how fast the clock runs. In other words, we need to "calibrate the cratering clock."

To assign real dates to the layers in rock, scientists use a technique called radioactive (or radiometric) dating. A geologist can find the age of a rock by measuring the relative amounts of a radioactive element, known as a **radioisotope**, and the decay products it turns into. Recall that the nuclei of atoms contain neutrons and protons. An isotope is an atom that has the same number of protons as other atoms of the same chemical element but has a different number of neutrons. The radioactive element is known as the **parent element**, and the decay products are called **daughter products**. Chemical analysis of a rock containing radioactive elements immediately after its formation would reveal the presence of the radioactive parents, but the daughter products of the radioactive decay would be absent because they would not have formed yet. As radioactive atoms decay, however, the amount of parent elements decreases and the amount of daughter products builds up. Chemical analysis reveals both the remaining radioactive parent atoms and the daughter products trapped within the structure of the mineral.

The time interval over which a radioactive isotope decays to half its original amount is called that isotope's **half-life**. With every half-life that passes, the remaining amount of the radioisotope decreases by a factor of 2. For example, after 3 half-lives, the remaining amount of a parent radioisotope will be $1/2 \times 1/2 \times 1/2 = 1/8$ of its original amount (**Figure 8.7**). At formation, 100 percent of the material is radioactive parent isotope (in red), with no daughter isotope (in blue) yet. After 1 half-life has passed, half of the parent isotope has decayed, and parent and daughter isotopes are present in equal numbers. After another half-life has passed, the sample is now only 1/4 parent and 3/4 daughter isotopes, and so on. By comparing the percentages of parent and daughter isotopes in a mineral, scientists can determine the number of half-lives that have passed and thus the age of the mineral. Some numerical examples are discussed in **Working It Out 8.1**.

The age of the Solar System is estimated from radioactive dating of meteorites found on Earth that are 4.5 billion to 4.6 billion years old. Earth may be as young as 4.4 billion years. The age of the Moon is determined from radioactive dating of lunar rocks. Between 1969 and 1976, *Apollo* astronauts and Soviet unmanned probes brought back samples from nine locations on the lunar surface.

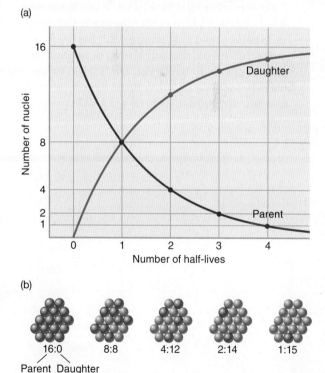

Figure 8.7 The concept of half-life. A parent population of 16 radioactive nuclei decays over several half-lives. That information can be presented (a) graphically or (b) as collections of particles.

8.1 Working It Out Computing the Ages of Rocks

With every half-life that passes, the remaining amount of radioactive isotopes will decrease by a factor of 2. If we express the number of half-lives more generally as n, we can translate that relationship into math:

$$\frac{P_F}{P_O} = \left(\frac{1}{2}\right)^n$$

where P_O and P_F are the original and final amounts, respectively, of a parent radioisotope, and n is the number of half-lives that have gone by, which equals the time interval of decay (its age) divided by the half-life of the isotope.

For example, the most abundant isotope of the element uranium (uranium-238, or ^{238}U—the parent) decays through a series of intermediate daughters to an isotope of the element lead (lead-206, or ^{206}Pb—its final, stable daughter). The half-life of ^{238}U is 4.5 billion years. Therefore, after 4.5 billion years, a sample that originally contained the uranium isotope (the parent) but no lead (its final daughter) would be found instead to contain equal numbers of uranium and lead atoms. If we were to find a mineral with such composition, we would know that half the uranium atoms had turned to lead and that the mineral formed 4.5 billion years ago.

Another isotope of uranium (^{235}U) decays to a different lead isotope (^{207}Pb) with a half-life of 700 million years. Suppose that a lunar mineral brought back by astronauts has 15 times as much ^{207}Pb (the daughter product) as ^{235}U (the parent radioisotope). Therefore, 15/16 of the parent radioisotope (^{235}U) has decayed to the daughter product (^{207}Pb), leaving only 1/16 of the parent remaining in the mineral sample. By noting that 1/16 is $(1/2)^4$, we see that 4 half-lives have elapsed since the mineral was formed, in which case the lunar sample is 4 × 700 million years = 2.8 billion years old.

Because the measured quantity of the isotope is not always a neat power of 2, looking at how we would solve the equation mathematically is worthwhile. We do that by taking the logarithm on both sides:

$$\log_{10}\frac{P_F}{P_O} = \log_{10}\left(\frac{1}{2}\right)^n$$

$$\log_{10}\frac{P_F}{P_O} = n\log_{10}\left(\frac{1}{2}\right)$$

$$\log_{10}\frac{P_F}{P_O} = -0.3n$$

Putting that back into words, we can write the relationship as

$$\log_{10}\left(\frac{\text{Actual measured quantity of isotope}}{\text{Original quantity of isotope}}\right) = -0.3 \times \left(\frac{\text{Time it has been decaying (age)}}{\text{Half-life}}\right)$$

Solving for age,

$$\text{Age} = -3.3 \times \text{Half-life} \times \log_{10}\left(\frac{\text{Actual measured quantity of isotope}}{\text{Original quantity of isotope}}\right)$$

(Most calculators have a button called "log" or "$\log_{10}$" for calculating such numbers.)

For the ^{235}U that decays to ^{207}Pb with a half-life of 700 million years, the lunar mineral is measured to have 15 times as much lead as uranium, so the mineral now contains only 1/16 of the original quantity of uranium:

$$\text{Age of mineral} = -3.3 \times (700 \times 10^6 \text{ yr}) \times \log_{10}\left(\frac{1}{16}\right) = 2.8 \times 10^9 \text{ yr}$$

The mineral is 2.8 billion years old.

By measuring relative amounts of various radioactive elements and the elements into which they decay, scientists assigned ages to those lunar regions. The oldest, most heavily cratered regions on the Moon date back to about 4.4 billion years ago, whereas most of the smoother parts of the lunar surface are typically 3.1 billion to 3.9 billion years old. That finding suggests that the Moon formed after Earth, and the heavy cratering suggests heavy bombardment at that time. As the graph in **Figure 8.8** shows, almost all the major cratering in the Solar System took place within its first billion years.

CHECK YOUR UNDERSTANDING 8.2

If radioactive element A decays into radioactive element B with a half-life of 20 seconds, then after 40 seconds: (a) none of element A will remain; (b) none of element B will remain; (c) half of element A will remain; (d) one-quarter of element A will remain.

8.3 The Surface of a Terrestrial Planet Is Affected by Processes in the Interior

Whereas impact cratering is driven by forces outside a planet, two other important processes, tectonism and volcanism, are determined by conditions in the interior of the planet. To understand those processes, we must understand the structure and composition of the interiors of planets. But how do we know what the interiors of planets are like? On Earth, the deepest holes drilled are about 12 km deep; that is tiny in comparison with Earth's radius of 6,378 km. It is impossible to drill down into Earth's core to observe Earth's interior structure directly. Scientists have determined a lot about the interior of Earth but less about the interiors of the other terrestrial planets.

Probing the Interior of Earth

The composition of Earth's interior can be determined in two ways. In one approach, Kepler's or Newton's laws are used to find the mass of Earth—for example, by applying Kepler's third law to a satellite orbiting Earth. Dividing the mass by the volume of Earth gives an average density of 5,500 kilograms per cubic meter (kg/m^3), or 5.5 times the density of water. But rocky surface material averages only 2,900 kg/m^3. Because the density of the whole planet is greater than the density of the surface, the interior must contain material denser than surface rocks. Another approach to determine the composition of Earth's interior comes from studies of meteorites. Because meteorites are left over from a time when the Solar System was young and Earth was forming from similar materials, the overall composition of Earth should resemble the composition of meteorite material. That material includes minerals with large amounts of iron, which has a density of nearly 8,000 kg/m^3. From those considerations, planetary scientists can determine the composition of Earth's interior.

The most important source of information about the structure of Earth's interior comes from monitoring the vibrations from earthquakes. When an earthquake occurs, vibrations spread out through and across the planet as **seismic waves**.

Surface waves travel across the surface of a planet, much like waves on the ocean. If conditions are right, surface waves from earthquakes can be seen rolling across the countryside like ripples on water. Those waves are responsible for much of the heaving of Earth's surface during an earthquake, causing damage such as the buckling of roadways.

The other types of seismic waves travel *through* Earth, probing the interior of the planet, faster than surface waves travel. **Primary waves** (P waves) are a type of **longitudinal wave** resulting from alternating compression and decompression of a material. Imagine a stretched-out spring (**Figure 8.9a**). A quick push along its length will make a longitudinal wave. P waves distort the material they travel through, much as compression waves do when they move along the length of a spring. **Secondary waves** (S waves) are a type of **transverse wave** that result from the sideways motion of material (**Figure 8.9b**).

The progress of seismic waves through Earth's interior depends on the characteristics of the material

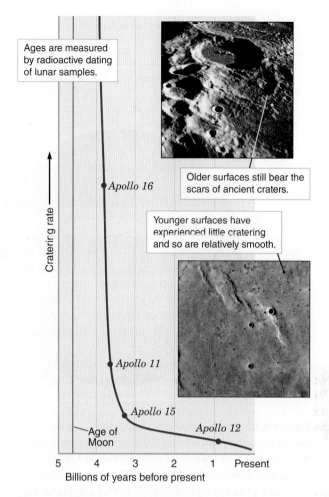

Figure 8.8 Radiometric dating of lunar samples returned from *Apollo* missions that landed at different sites was used to determine how the cratering rate has changed. Cratering records can then be used to establish the age of other parts of the lunar surface.

Figure 8.9 (a) A longitudinal wave involves oscillations along the direction of travel of the wave. (b) A transverse wave involves oscillations perpendicular to the direction in which the wave travels. Primary seismic waves are longitudinal; secondary seismic waves are transverse.

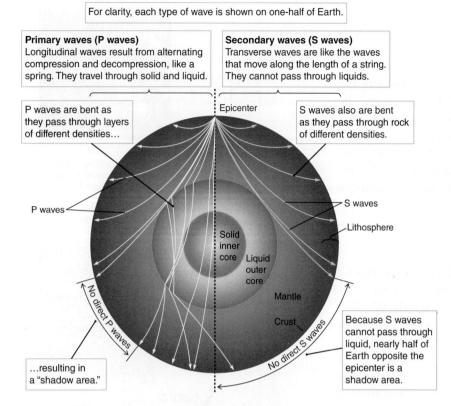

Figure 8.10 Primary and secondary seismic waves move through the interior of Earth in distinctive ways. Measurements of when and where different types of seismic waves arrive after an earthquake enable scientists to test predictions from detailed models of Earth's interior. Note the "shadow areas" caused by the refraction of primary waves (yellow) at the outer boundary of the liquid outer core and the inability of secondary waves (blue) to pass through the liquid outer core.

through which they are moving. Primary waves (white; **Figure 8.10**) can travel through either solids or liquids, but secondary waves (blue) cannot travel through liquids. Seismic waves travel at different speeds, depending on the density and composition of the rocks they encounter. As a result, seismic waves moving through rocks of various densities or composition are bent in much the same way that waves of light are bent when they enter or leave glass. The speed of seismic waves provides additional information about Earth's interior. The refraction of primary waves at the outer edge of Earth's liquid outer core and the inability of secondary waves to penetrate the liquid outer core create "shadows" of the liquid core on the side of Earth opposite an earthquake's epicenter (Figure 8.10). Much of scientists' knowledge of Earth's liquid outer core comes from studies of those waves.

Scientists use instruments called seismometers to measure the distinctive patterns of seismic waves. For more than 100 years, thousands of seismometers scattered around the globe have measured the vibrations from countless earthquakes and other seismic events, such as volcanic eruptions and nuclear explosions. A single seismometer can record ground motion at only one place on Earth, but when combined with the recordings of many other seismometers placed all over Earth, scientists can use the data to get a comprehensive picture of the planet's interior.

Building a Model of Earth's Interior

To model the structure of Earth's interior, geologists use the laws of physics and the properties of materials and how they behave at different temperatures and pressures. The pressure at any point in Earth's interior must be just high enough that the outward forces balance the inward force of the weight of all the material above that point. If the outward pressure at some point within a planet were *less* than the weight per unit area of the overlying material, that material would fall inward, crushing what was underneath it. If the pressure at some point within a planet were *greater* than the weight per unit area of the overlying material, the material would expand and push outward, lifting the overlying material. The situation is stable only when the weight of matter above is just balanced by the pressure within the whole interior of the planet. The balance between pressure and weight is known as **hydrostatic equilibrium**, which is important to the structure of planetary interiors, planetary atmospheres, and the structure and evolution of stars.

From considerations of hydrostatic equilibrium and from seismic wave measurements, scientists construct a layered model of Earth's interior. They then test their model by comparing its predictions of how seismic waves would propagate through Earth with actual observations of seismic waves from real earthquakes. The extent to which the predictions agree with observations identifies both strengths and weaknesses of the model. Geologists adjust the model—always remaining consistent with the known physical properties of materials—until a good match is found between prediction and observation.

That is the method geologists used to arrive at the current picture of the interior of Earth (Figure 8.10). The innermost region of Earth's interior consists of a **core**. Earth's solid inner core is at a temperature of about 6000 K and is composed primarily of iron, nickel, and other dense metals. The liquid outer core is cooler, about 4000 K, and is composed of liquid metals. Outside the outer core is Earth's **mantle**, a rocky shell made of solid, medium-density materials such as silicates. That hot, rocky shell is under pressure, and over long timescales—hundreds of years—it flows like a viscous fluid. Covering the mantle is the **crust**, a thin, hard layer of lower-density materials that is chemically distinct from the interior.

The cross sections in **Figure 8.11** show the interior structures of each terrestrial planet and Earth's Moon. Earth's interior is not uniform. The materials have been separated according to density, a process known as **differentiation**. When rocks of different types are mixed, they tend to stay mixed. Once that rock melts, however, the denser materials sink to the center and the less dense materials float toward the surface. Today, little of Earth's interior is molten, but the differentiated structure shows that Earth was once much hotter, and its interior was liquid throughout. The cores of all the terrestrial planets and the core of the Moon were once molten. Using new, improved methods to reanalyze 8 years of data from seismometers left on the Moon by the *Apollo* astronauts, planetary scientists found that the Moon has a solid inner core, possibly a liquid outer core, and a partially melted layer between the core and the mantle.

The Evolution of Planetary Interiors

The balance between energy received and energy produced and emitted governs the temperature within a planet. The interiors of planets evolve as their temperatures change. Factors that influence how the temperature changes include the size of the planet, the composition of the material, and heating from various sources. Here, we are concerned with thermal energy—the kinetic energy of particles within a substance that determines the temperature. In general, the interior of a planet cools as heat is emitted from the surface. Because heat takes time to travel through rock, the temperature increases the deeper we go within a planet. That trend is similar to the effect of taking a hot pie out of the oven. The pie radiates heat from the surface and eventually cools, but the filling takes much longer to cool than the crust.

Planets lose thermal energy from their surfaces primarily through radiation. Recall from Chapter 5 that the hotter an object is, the more energy it radiates. The relative proportions of the type of energy radiated (infrared, optical, ultraviolet, and so forth) depend on the temperature of the object. The rate at which a planet cools depends on its size. A larger planet has a larger volume of matter and more thermal energy trapped inside. Thermal energy has to escape through the planet's surface, so the planet's surface area determines the rate at which energy is lost. Smaller planets have more surface area in comparison with their small volumes, so they cool faster, whereas larger planets have a smaller surface-area-to-volume ratio and cool more slowly (**Working It Out 8.2**). Because geological activity is powered by heat, smaller objects become geologically inactive sooner. Major geological activity ended on Mercury and the Moon first, but it continued on the larger terrestrial planets—Venus, Earth, and Mars.

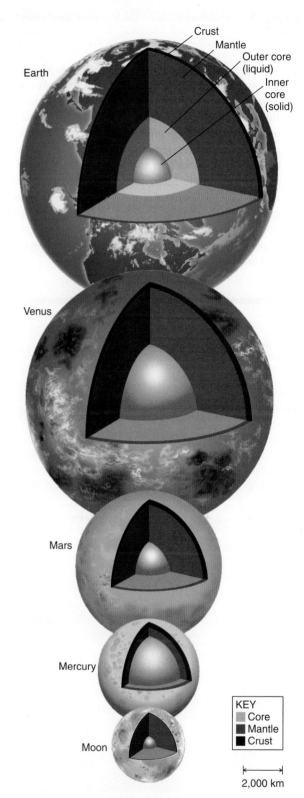

Figure 8.11 A comparison of the interiors of the terrestrial planets and Earth's Moon. Some fractions of the cores of Mercury, Venus, the Moon, and Mars are probably liquid.

8.2 Working It Out How Planets Cool

If we assumed that all the terrestrial planets formed with the same percentage of radioactive materials in their bulk composition and that those radioactive materials are their sole source of internal thermal energy, a planet's volume would determine the total amount of the thermal energy–producing material it contains. The volume of a spherical planet is proportional to the *cube* of the planet's radius (volume = $4/3 \times \pi R^3$).

A planet loses its internal energy by radiating it away at its surface, so a planet's surface area determines the rate at which it can get rid of its thermal energy. The surface area of the planet is proportional to the *square* of the radius (surface area = $4\pi R^2$). The ratio of the two—the volume divided by the surface area through which thermal energy can escape—is given by

$$\frac{\text{Volume}}{\text{Surface area}} = \frac{\frac{4}{3} \times \pi R^3}{4\pi R^2} = \frac{R}{3}$$

so the time it takes to lose internal energy is proportional to the planet's radius.

A planet's ability to transfer internal energy from its hot core to its cooling surface also depends on some of its own internal properties. Nevertheless, all things being equal, planets with larger radii retain their internal energy longer than smaller planets do. For example, Mars has a radius about half that of Earth, so it has been losing its internal thermal energy to space about twice as fast as Earth has. That is one reason that Mars is less geologically active than Earth.

Some of the thermal energy in the interior of Earth is left over from when Earth formed. The tremendous energy of collisions and the energy from short-lived radioactive elements melted the planet, leading to the differentiated structure. As the surface of Earth radiated energy into space, it cooled rapidly. A solid crust formed above a molten interior. Because a solid crust does not conduct thermal energy well, it helped to retain the remaining heat. Over a long time, energy from the interior of the planet continued to leak through the crust and radiate into space. As a result, the interior of the planet cooled slowly, and the mantle and the inner core solidified.

If the thermal energy from Earth's formation were the only source of heating in Earth's interior, Earth would have long ago solidified completely. Most of the rest of the thermal energy in Earth's interior comes from long-lived radioactive elements trapped in the mantle. As those radioactive elements decay, they release energy, which heats the planet's interior. Today, the temperature of Earth's interior is determined by a dynamic equilibrium between the radioactive heating of the interior and the loss of energy to space. As the radioactive elements decay, the amount of thermal energy generated declines, and Earth's interior cools as it ages. A small amount of additional heating of Earth's interior is due to friction generated by tidal effects of the Moon and Sun.

Although temperature plays an important role in a planet's interior structure, whether a material is solid or liquid also depends on pressure. Higher pressure forces atoms and molecules closer together and makes the material more likely to become a solid. Toward the center of Earth, the effects of temperature and pressure oppose each other: the higher temperatures make it more likely that material will melt, but the higher pressure favors a solid form. In the outer core of Earth, the high temperature wins, allowing the material to exist in a molten state. At the center of Earth, even though the temperature is higher, the pressure is so great that Earth's inner core is solid.

CHECK YOUR UNDERSTANDING 8.3A

Differentiation refers to materials that are separated on the basis of their: (a) weight; (b) mass; (c) volume; (d) density.

Magnetic Fields

A magnetic field is created by moving charges and exerts a force on magnetically reactive objects, such as iron and other charged particles. In a navigation compass, for example, the compass needle lines up with Earth's magnetic field and points "north" and "south," as shown in **Figure 8.12a**. In the north, a compass needle points to a location in the Arctic Ocean off the coast of northern Canada, near but not at the geographic North Pole (about which Earth spins). In the south, a compass needle points to a location off the coast of Antarctica, 2,800 km from the geographic South Pole. Earth behaves as though it contained a giant bar magnet that were slightly tilted with respect to the planet's rotation axis and had its two endpoints near the two magnetic poles (**Figure 8.12b**). Earth also has a **magnetosphere**, the region surrounding a planet that is filled with relatively intense magnetic fields and charged particles.

Earth's magnetic field is not actually the result of a bar magnet buried within the planet. A magnetic field is the result of moving electric charges. Earth's magnetic field is created by the combination of Earth's rotation about its axis and a liquid, electrically conducting, circulating outer core. From that combination, Earth converts mechanical energy into magnetic energy. The magnetic field of a planet is an important probe into its internal structure.

Earth's magnetic field is constantly changing. At the moment, the north magnetic pole is traveling several tens of kilometers per year toward the northwest. If that rate and direction continue, the north magnetic pole could be in Siberia before the end of the century. The magnetic pole tends to wander, constantly changing direction as a result of changes in the core.

The geological record shows that much more dramatic changes in the magnetic field have occurred over the history of our planet. When a magnet made of material such as iron gets hot enough, it loses its magnetization. As the material cools, it again becomes magnetized by any magnetic field surrounding it. Thus, iron-bearing minerals record the direction of Earth's magnetic field at the time that they cooled. A memory of that magnetic field thus becomes "frozen" into the material. For example, lava extruded from a volcano carries a record of Earth's magnetic field at the moment the lava cooled. By using radiometric techniques to date those materials, geologists obtain a record of how Earth's magnetic field has changed. Although Earth's magnetic field has probably existed for at least 3.5 billion years, the north and south magnetic poles switch from time to time—on average, about every half-million years.

The general idea of how Earth's magnetic field (and those of other planets) originates is called the *dynamo theory*. In general, magnetic fields result from electric currents, which are moving electric charges. Earth's magnetic field is thought to be a side effect of three factors: Earth's rotation about its axis; an electrically conducting, liquid outer core; and fluid motions within the outer core. That model has been tested with computer simulations, which also produce the pole reversals. The theory suggests that any rotating planet with an internal heat source will have a magnetic field.

During the *Apollo* program, astronauts measured the Moon's local magnetic fields, and small satellites have searched for global magnetism. The Moon has a

Figure 8.12 (a) Earth's magnetic field can be visualized as though it were a giant bar magnet tilted relative to Earth's axis of rotation. Compass needles line up along magnetic-field lines and point toward Earth's north magnetic pole. (b) Iron filings sprinkled around a bar magnet help us visualize such a magnetic field.

very weak field, possibly none at all, because the Moon is very small and therefore has a solid (not liquid and rotating) inner core. The Moon also has a very small core. However, remnant magnetism is preserved in lunar rocks from an earlier time before the lunar surface and rocks solidified. Recent analysis of the oldest lunar rocks brought back on *Apollo 17* suggests that 4.2 billion years ago, the Moon had a liquid core with a generated magnetic field that lasted a billion years or more. The Moon no longer generates a global magnetic field, so what is detected is the "fossil" remains of its early field. Data from India's *Chandrayaan-1* spacecraft suggest that the Moon has a very weak and localized mini-magnetosphere on its far side.

Other than Earth, Mercury is the only terrestrial planet with a significant global magnetic field today—although its field is only about 1 percent as strong as Earth's. Slow rotation and a large iron core, parts of which are molten and circulating, cause Mercury's magnetic field. Because the field is so weak, Mercury's magnetosphere is small, meeting the solar wind about 1,700 km above its surface. At that boundary, twisted bundles of magnetic fields transfer magnetic energy from the planet to space.

Planetary scientists expected that Venus would have a magnetic field because the planet's mass and distance from the Sun imply an iron-rich core and partly molten interior like Earth's. Venus's lack of a magnetic field might be attributed to its extremely slow rotation (see Table 8.1) or to its not having a giant impact early in its history as Earth did. Or perhaps Venus's magnetic field is temporarily dormant—a condition that Earth is believed to have experienced at times of magnetic field reversals.

Mars has a weak magnetic field, presumably frozen in place early in its history. The magnetic signature occurs only in the ancient crustal rocks, showing that early in the history of Mars, some sort of an internally generated magnetic field must have existed. Geologically younger rocks lack that residual magnetism, so the planet's original magnetic field has long since disappeared. The lack of a strong magnetic field today on Mars might be the result of its small core. Or Mars might have lost its ability to generate a magnetic field after a series of giant impacts early in the planet's history, which could have heated the mantle of Mars enough to reduce the flow of heat out of the core to the mantle. The oldest large impact basins on Mars appear to be magnetized; newer ones are not.

CHECK YOUR UNDERSTANDING 8.3B
According to the dynamo theory, a planet will have a strong magnetic field if it has: (a) fast rotation and a solid core; (b) slow rotation and a liquid core; (c) fast rotation and a liquid core; (d) slow rotation and a solid core; (e) fast rotation and a gaseous core.

8.4 Planetary Surfaces Evolve through Tectonism

Now that we have looked at planetary interiors, we can connect the interior conditions to the processes that shape the surface. The crust and part of the upper mantle form the **lithosphere** of a planet. **Tectonism**, the deformation of a planet's lithosphere, warps, twists, and shifts the lithosphere to form visible surface features. If you have driven through mountainous or hilly terrain, you may have seen places like the one shown in **Figure 8.13**, where the roadway has

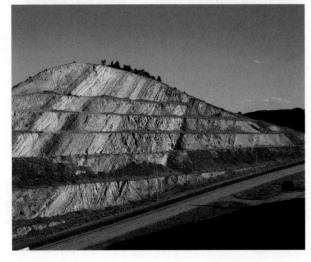

Figure 8.13 Tectonic processes fold and warp Earth's crust, as seen in these rocks along a roadside north of Denver, Colorado.

been cut through rock. The exposed layers tell the story of Earth through the vast expanse of geological time. In this section, we look at tectonic processes that create those layers and play an important part in shaping the surface of a planet.

AstroTour: Continental Drift

The Theory of Plate Tectonics

Early in the 20th century, some scientists recognized that Earth's continents could be fit together like pieces of a giant jigsaw puzzle. In addition, the layers in the rock and the fossil records they hold on the east coast of South America match those on the west coast of Africa. On the basis of that evidence, Alfred Wegener (1880–1930) proposed a hypothesis that the continents were originally joined in one large landmass that broke apart as the continents began to "drift" away from one another over millions of years. That hypothesis was further developed into the theory known today as **plate tectonics**. Geologists now recognize that Earth's outer shell is composed of several relatively brittle segments, or **lithospheric plates**. About seven major plates and about a half dozen smaller plates are floating on the mantle. The motion of those plates is constantly changing the surface of Earth.

Originally, the idea of plate tectonics was met with great skepticism among geologists because they could not imagine a mechanism that could move such huge landmasses. In the late 1950s and early 1960s, however, studies of the ocean floor provided compelling evidence for plate tectonics. Those surveys showed surprising characteristics in bands of basalt—a type of rock formed from cooled lava—that were found on both sides of the ocean rifts. Ocean floor rifts such as the Mid-Atlantic Ridge are **spreading centers**. As **Figure 8.14** shows, hot material in those rifts rises toward Earth's surface, becoming new ocean floor. When that hot material cools, it becomes magnetized along the direction of Earth's magnetic field, thus recording the direction of Earth's magnetic field at that time. Greater distance from the rift indicates the ocean floor is older and formed earlier. Combined with radiometric dates for the rocks, that magnetic record proved that the spreading of the seafloor and the motions of the plates have continued over long geological time spans.

Precise surveying techniques and global positioning systems (GPS) can now determine locations on Earth to within a few centimeters. Those measurements confirm that Earth's lithosphere is moving. Some areas are being pulled apart by more than 15 centimeters (cm) each year. Over millions of years, such motions add up. Over 10 million years—a short time by geological standards—15 cm/yr becomes 1,500 km, and maps definitely need to be redrawn.

The theory of plate tectonics is perhaps the greatest advance in 20th century geology. Plate tectonics is responsible for a variety of geological features on our planet, including the continental drift that Wegener hypothesized.

The Role of Convection

Moving lithospheric plates requires immense forces. Those forces are the result of thermal energy escaping from the interior of Earth. The transport of thermal energy by the movement of packets of gas or liquid is known as **convection**.

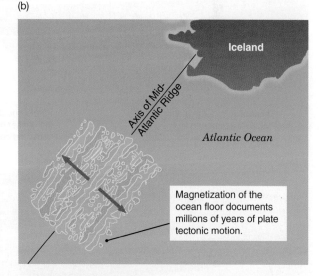

Figure 8.14 (a) New seafloor is formed at a spreading center. The cooling rock becomes magnetized and is then carried away by tectonic motions. (b) Maps such as this one of banded magnetic structure in the seafloor near Iceland support the theory of plate tectonics.

Figure 8.15 (a) Convection occurs when a fluid is heated from below. (b) Convection in Earth's mantle drives plate tectonics.

Figure 8.15a illustrates the process. If you have ever watched water in a heated pot on a stovetop, you have observed convection. Thermal energy from the stove warms water at the bottom of the pot. The warm water expands slightly, becoming less dense than the cooler water above it, and the cooler water with higher density sinks, displacing the warmer water upward. When the lower-density water reaches the surface, it gives up part of its energy to the air and cools; as the water cools, it then becomes denser and sinks back toward the bottom of the pot. Water rises in some locations and sinks in others, forming convection cells. As discussed in later chapters, convection also plays an important role in planetary atmospheres and in the structure of the Sun and stars.

Figure 8.15b shows how convection works in Earth's mantle, where radioactive decay supplies the heat to drive convection. We know the mantle is not molten—if it were, secondary seismic waves could not travel through it—but the mantle is viscous, having the consistency of hot molten glass. That consistency allows convection to take place very slowly. Convection cells in Earth's mantle drive the plates, carrying both continents and ocean crust along with them. Convection also creates crust along rift zones in the ocean basins, where mantle material rises up, cools, and slowly spreads out.

Figure 8.16 illustrates plate tectonics and some of its consequences. If material rises and spreads in one location, it must converge and sink in another. Locations where plates converge and convection currents turn downward are called **subduction zones**. In a subduction zone, one plate slides beneath the other, and convection drags the submerged lithospheric material back down into the mantle. The

Figure 8.16 Divergence and collision of tectonic plates create a variety of geological features.

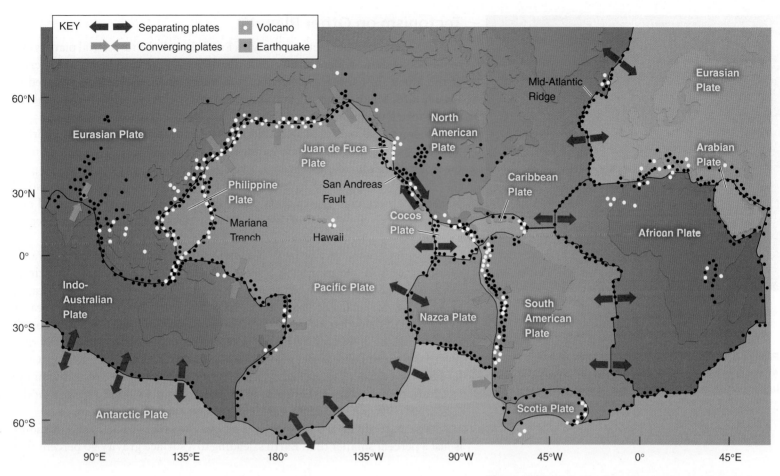

Figure 8.17 Major earthquakes and volcanic activity are often concentrated along the boundaries of Earth's principal tectonic plates.

Mariana Trench—the deepest part (11 km) of Earth's ocean floor—is a subduction zone. Much of the ocean floor lies between spreading centers and subduction zones, so the ocean floor is the youngest portion of Earth's crust. In fact, the *oldest* seafloor rocks are less than 200 million years old.

In some places, the plates are not sinking but instead are colliding and, consequently, shoved upward. The highest mountains on Earth, the Himalayas, grow a half-meter per century as the Indo-Australian subcontinental plate collides with the Eurasian Plate. In other places, plates meet at oblique angles and slide along past each other. One such place is the San Andreas Fault in California, where the Pacific Plate slides past the North American Plate. A **fault** is a fracture in a planet's crust along which material can slide.

Locations where plates meet tend to be very active geologically. One of the best ways to see the outline of Earth's plates is to look at a map of where earthquakes and volcanism occur (**Figure 8.17**). Where plates meet, enormous stresses build up. Earthquakes occur when a portion of the boundary between two plates suddenly slips, relieving the stress. Volcanoes are created when friction between plates melts rock, which is then pushed up through cracks to the surface. Earth also has many **hot spots**, where hot deep-mantle material rises, releasing thermal energy. As plates shift, some parts move more rapidly than others, causing the plates to stretch, buckle, or fracture. Those effects are seen on the surface as folded and faulted rocks. Mountain chains are common near converging plate boundaries, where plates buckle and break.

Tectonism on Other Planets

We have observed plate tectonics only on Earth, but all the terrestrial planets and some moons show evidence of tectonic disruptions. Fractures have cut the crust of the Moon in many areas, leaving fault valleys such as the one pictured in **Figure 8.18**. Many of those features are the result of large impacts that cracked and distorted the lunar crust.

Mercury has fractures and faults similar to those on the Moon. In addition, Mercury has many cliffs that are hundreds of kilometers long. They appear to be the result of the shrinking of Mercury; recent observations by *Messenger* suggest that the planet has shrunk by about 10–14 km across. Like the other terrestrial planets, Mercury was once molten. As it shrank, Mercury's crust cracked and buckled in much the same way that a grape skin wrinkles as it shrinks to become a raisin.

Possibly the most impressive tectonic feature in the Solar System is Valles Marineris on Mars (**Figure 8.19**). Stretching nearly 4,000 km, and nearly 4 times as deep as the Grand Canyon, that canyon system is as long as the distance between San Francisco and New York. Valles Marineris includes massive cracks in the crust of Mars that formed as local forces, perhaps related to mantle convection, pushed the crust upward from below. The surface could not be equally supported by the interior everywhere, and unsupported segments fell in. Once formed, the cracks were eroded by wind, water, and landslides, resulting in the structure we see today. Other parts of Mars have faults similar to those on the Moon, but cliffs as high and long as those seen on Mercury are absent.

The mass of Venus is only 20 percent less than that of Earth, and its radius is just 5 percent smaller than Earth's, leading to a surface gravity 90 percent that

Figure 8.18 This *Apollo 10* photograph shows Rima Ariadaeus, a 2-km-wide valley between two tectonic faults on the Moon.

Figure 8.19 (a) A mosaic of *Viking Orbiter* images shows Valles Marineris, the major tectonic feature on Mars, stretching across the center of the image from left to right. This canyon system is more than 4,000 km long. The dark spots on the left are huge shield volcanoes. (b) This close-up perspective view of the canyon wall was photographed by the European Space Agency's *Mars Express* orbiting spacecraft.

of Earth. Because of the similarities between the planets, many scientists predicted that Venus might also show evidence of plate tectonics. The NASA *Magellan* mission orbited Venus from 1990 to 1994, and *Magellan* mapped nearly the entire surface of Venus, providing the first high-resolution radar views of the planet's surface (**Figure 8.20**).

Venus Express, launched by the European Space Agency (ESA), orbited Venus from 2006 to 2014. The probe mapped Venus in the infrared, which can penetrate the clouds to enable a view of the surface. The impact craters on Venus seem to be evenly distributed, suggesting that the surface is all about the same age—about a billion years old. Venus is mostly covered with smooth lava, but the planet has two highland regions: Ishtar Terra in the north and Aphrodite Terra in the south. The highland rocks are less smooth and older than those on the rest of Venus and may be similar to granite rocks on Earth. Because granite results from plate tectonics and water, the data hint at the possibility that those highlands on Venus are ancient continents, created by volcanic activity, on a planet that once had liquid water oceans.

Because of the similarities between Venus and Earth, the interior of Venus should be very much like the interior of Earth, and convection should be occurring in its mantle. On Earth, mantle convection and plate tectonism release the most thermal energy from the interior. On Venus, however, hot spots may be the principal way that thermal energy escapes from the planet's interior. Circular fractures called coronae on the surface of Venus, ranging from a few hundred kilometers to more than 2,500 km across, may be the result of upwelling plumes of hot mantle that have fractured Venus's lithosphere. Alternatively, energy may build up in the interior until large chunks of the lithosphere melt and overturn, releasing an enormous amount of energy. Then, the surface cools and solidifies. It is uncertain why Venus and Earth are so different with regard to plate tectonics.

Figure 8.20 The atmosphere of Venus blocks our view of the surface in visible light. This false-color view of Venus is a radar image made by the *Magellan* spacecraft. Bright yellow and white areas are mostly fractures and ridges in the crust. Some circular features seen in the image may be regions of mantle upwelling, or *hot spots*. Most of the surface is formed by lava flows, shown in orange.

CHECK YOUR UNDERSTANDING 8.4

On which of the following does plate tectonics occur now? (Select all that apply.)
(a) Mercury; (b) Venus; (c) Earth; (d) the Moon; (e) Mars

8.5 Volcanism Signifies a Geologically Active Planet

The molten rock that a volcano spews onto the surface of Earth is known as **magma**. It originates deep in the crust and in the upper mantle, where sources of thermal energy combine. Those sources include rising convection cells in the mantle, heating by friction from movement in the crust, and concentrations of radioactive elements. In this section, we look at the occurrence of volcanic activity on a planet or moon, called **volcanism**. Volcanism not only shapes planetary surfaces but also is a key indicator of a geologically active planet.

Terrestrial Volcanism Is Related to Tectonism

Volcanoes are usually located along plate boundaries and over hot spots. Maps of geological activity such as the one in Figure 8.17 leave little doubt that most terrestrial volcanism is linked to the same forces responsible for plate motions. A tremendous amount of friction is generated as plates slide under each other. That friction raises the temperature of rock toward its melting point.

Material at the base of a lithospheric plate is under a great deal of pressure because of the weight of the plate pushing down on it. That pressure increases the melting temperature of the material, forcing it to remain solid even at high temperature. As the material is forced up through the crust, its pressure drops, in which case the material's melting temperature drops, too. Material that was solid at the base of a plate becomes molten as it nears the surface. Places where convection carries hot mantle material toward the surface are frequent sites of eruptions. Iceland, one of the most volcanically active regions in the world, sits astride one such spreading center—the Mid-Atlantic Ridge (see Figure 8.17). In recent years, volcano eruptions in Iceland have disrupted travel plans for thousands of people as the airborne volcanic ash made it unsafe for airplanes to fly near the eruption.

Once lava reaches the surface of Earth, it can form many types of structures. Flows often form vast sheets, especially if the eruptions come from long fractures called fissures. If very fluid lava flows from a single *point source*, it spreads out over the surrounding terrain or ocean floor, forming a **shield volcano** (**Figure 8.21a**). A **composite volcano** (**Figure 8.21b**) forms when thick lava flows alternating with explosively generated rock deposits build a steep-sided structure.

Terrestrial volcanism also occurs where convective plumes rise toward the surface in the interiors of lithospheric plates, creating local hot spots. Volcanism over hot spots works much like volcanism elsewhere at a spreading center, except that the convective upwelling occurs at a single spot rather than along the edge of a plate. Those hot spots force mantle and lithospheric material toward the surface, where the material emerges as liquid lava.

Earth has many hot spots, including the regions around Yellowstone Park and the Hawaiian Islands (**Figure 8.21c**). The Hawaiian Islands are a chain of shield volcanoes that formed as the lithospheric plate moved across a hot spot. Volcanoes erupt over a hot spot, building an island. The island ceases to grow as the plate motion carries the island away from the hot spot. Meanwhile, a new island grows over the hot spot. Today, the Hawaiian hot spot is located off the southeast coast of the Big Island of Hawaii, where it continues to power the active volcanoes. On top of the hot spot, the newest Hawaiian island, Loihi, is forming. Loihi is already a massive shield volcano, rising more than 3 km above the ocean floor. Loihi will eventually break the surface of the ocean and merge with the Big Island of Hawaii—but not for another 100,000 years.

AstroTour: Hot Spot Creating a Chain of Islands

Figure 8.21 Magma reaching Earth's surface commonly forms (a) shield volcanoes, such as Mauna Loa, which have gently sloped sides built up by fluid lava flows; and (b) composite volcanoes, such as Mt. Fuji, which have steeply symmetric sides built up by viscous lava flows. (c) Hot spots are convective plumes of lava that can form a successive series of volcanoes as the plate above them slides by.

Volcanism in the Solar System

THE MOON Although Earth is the only planet on which plate tectonics is an important process, evidence of volcanism appears throughout the Solar System, including several moons of the outer planets. Some of the first observers to use telescopes to view the Moon noted dark areas that looked like bodies of water—thus they were named **maria** (singular: *mare*), Latin for "seas." Early photographs showed flowlike features in the dark regions of the Moon. We now know that the maria are actually vast, hardened lava flows, similar to volcanic rocks known as basalts on Earth. Because the maria contain relatively few craters, those volcanic flows must have occurred after the period of heavy bombardment ceased.

Many of the rock samples that the *Apollo* astronauts brought back from the lunar maria contained gas bubbles typical of volcanic materials (**Figure 8.22**). The lava that flowed across the lunar surface must have been relatively fluid. That fluidity, due partly to the lava's chemical composition, explains why lunar basalts form vast sheets that fill low-lying areas such as impact basins (**Figure 8.23**). The fluidity also partly explains the Moon's lack of classic volcanoes: the lava was too fluid to pile up—the way motor oil spreads out when poured from a container.

The lunar rock samples also showed that most of the lunar lava flows are older than 3 billion years. Samples from the heavily cratered terrain of the Moon also originated from magma, indicating that the young Moon went through a molten stage. Those rocks cooled from a "magma ocean" and are more than 4 billion years old, thus preserving the early history of the Solar System. Most of the sources of heating and volcanic activity on the Moon must have shut down some 3 billion years ago—unlike on Earth, where volcanism continues. That conclusion is consistent with the idea that smaller objects and planets cool more efficiently (see Working It Out 8.2) and are thus less active than larger planets.

Only in a few limited areas of the Moon are younger lavas thought to exist, but most of those have not been sampled directly. The *Lunar Reconnaissance Orbiter* observed volcanic cones that were probably built up from volcanic rocks erupting from the surface. Those volcanic rocks are far different from the mare basalt rocks and contain silica and thorium. Those domes could have been formed as recently as 800 million years ago, which would make them the result of the most recent volcanic activity found on the Moon.

MERCURY Mercury also shows evidence of past volcanism. The *Mariner 10* and *Messenger* missions revealed smooth plains on Mercury that looked similar to the lunar maria. Those sparsely cratered plains are the youngest areas on Mercury and, like those on the Moon, are almost certainly volcanic in origin, created when fluid lavas flowed into and filled huge impact basins. Many of the volcanic plains on Mercury also are associated with impact scars. The volcanic activity that created the plains probably ceased 3.8 billion years ago, possibly from the shrinking of the planet as it cooled. The ending of the late heavy bombardment might also have been a factor. High-resolution imaging by *Messenger* has also identified several volcanoes. Vents that could be from explosive volcanism have been found around the large, old impact basin Caloris (**Figure 8.24**) and may be as young as 1 billion to 2 billion years.

MARS Mars also has been volcanically active. More than half the surface of Mars is covered with volcanic rocks. Lavas covered huge regions of Mars, flooding the older, cratered terrain. Most of the vents or long cracks that created those flows are buried under the lava that poured forth from them. Among the most impressive features on Mars are its enormous shield volcanoes—the largest mountains in the Solar System. Olympus Mons, standing 27 km high at its peak

Figure 8.22 This rock sample from the Moon, collected by the *Apollo 15* astronauts from a lunar lava flow, shows gas bubbles typical of gas-rich volcanic materials. The rock is about 6 × 12 cm.

Figure 8.23 The lava flowing across the surface of Mare Imbrium on the Moon must have been extremely fluid to spread out for hundreds of kilometers in sheets only tens of meters thick.

Figure 8.24 The Caloris Basin on Mercury (yellow) is one of the largest impact basins in the Solar System, with a span of about 1,500 km. The orange regions may be volcanic vents. The false color is enhanced to show more detail.

and 550 km wide at its base (**Figure 8.25**), would tower over Earth's largest mountains. Olympus Mons and its neighbors grew as the result of hundreds of thousands of individual eruptions. In the absence of plate tectonics on Mars, its volcanoes have remained over their hot spots for billions of years, growing ever taller and broader in the lower surface gravity with each successive eruption.

Lava flows and other volcanic landforms span nearly the entire history of Mars, estimated to extend from the formation of the crust some 4.4 billion years ago to geologically recent times and to cover more than half of the red planet's surface. Although some "fresh appearing" lava flows have been identified on Mars, we will not know the age of those latest eruptions until rock samples are radiometrically dated. Mars could, in principle, experience eruptions today.

VENUS Of the terrestrial planets, Venus has the most volcanoes. Radar images reveal a variety of volcanic landforms. They include highly fluid flood lavas covering thousands of square kilometers, shield volcanoes approaching the size and complexity of those on Mars, and lava channels thousands of kilometers long. Those lavas must have been extremely hot and fluid to flow for such long distances. Some of the volcanic eruptions on Venus are thought to have been associated with deformation of Venus's lithosphere above hot spots such as the circular features mentioned earlier.

Figure 8.25 The largest known volcano in the Solar System, Olympus Mons is a 27-km-high shield-type volcano on Mars, similar to but much larger than Hawaii's Mauna Loa. This partial view of Olympus Mons was taken by the *Mars Global Surveyor*.

Lavas on Venus are basalts, much like the lavas on Earth, the Moon, Mars, and possibly Mercury. The *Venus Express* spacecraft imaged three of the nine Hawaii-like hot spots in infrared wavelengths (**Figure 8.26**). Each hot spot has several volcanoes, with altitudes of 500–1,200 meters above the nearby plains. Some of the volcanic regions were radiating heat more efficiently than the regions nearby. That finding suggests that those volcanic regions have younger material and that volcanic activity had taken place within the past 2.5 million years—perhaps as recently as a few thousand years ago. As we describe in Chapter 9, Venus has some of the volcanic gas sulfur dioxide in its atmosphere, so Venus may still be cooling its interior through volcanic activity.

(a)

A geological timescale for Venus has not yet been devised. But from its relative lack of impact craters, most of the surface is considered less than 1 billion years old, and some of it may be much more recent. When volcanism began on Venus and how much active volcanism exists today remain unanswered questions.

CHECK YOUR UNDERSTANDING 8.5

Which is *not* a reason for the large size of volcanoes on Mars in comparison with the smaller ones on Earth? (a) absence of plate tectonics; (b) distance from the Sun; (c) lower surface gravity than Earth's; (d) many repeated eruptions

(b)

8.6 The Geological Evidence for Water

Today, Earth is the only planet in the Solar System where the temperature and atmospheric conditions allow extensive liquid surface water to exist. The other inner planets do not have extensive liquid surface water, but evidence exists for water ice in deep craters or in the polar regions and for water below the surface in permafrost, subsurface glaciers, or possibly as liquid in the core.

Figure 8.26 The volcanic peak Idunn Mons on Venus is 200 km in diameter and 2.5 km high. (a) Radar data on top of the topographic data from NASA's *Magellan* mission. (b) Visible and infrared data from ESA's *Venus Express* orbiter. The red-orange area at the summit is the warmest, suggesting the presence of some volcanic outflow activity.

Life, as we know it on Earth, requires water as a solvent and as a delivery mechanism for essential chemistry. Therefore, the search for water is central to the search for life in the Solar System. In addition, if humans are ever going to live on another terrestrial planet, they will need a source of water. In this section, we look at how water modifies the surface of a planet and then discuss the search for water in the Solar System.

Water and Erosion

Tectonism, volcanism, and impact cratering affect Earth's surface by creating variations in the height of the surface. **Erosion** is the wearing away of a planet's surface by mechanical action. The term *erosion* covers a variety of processes. Erosion, not only by running water but also by wind and by the actions of living organisms, wears down hills, mountains, and craters; the resulting debris fills in valleys, lakes, and canyons. If erosion were the only geological process operating, it would eventually smooth out the surface of the planet completely. Because Earth is a geologically and biologically active world, however, its surface is an ever-changing battleground between processes that build up topography and those that tear it down.

Weathering is the first step in erosion. During weathering, rocks are broken into smaller pieces and may be chemically altered. For example, rocks on Earth are physically weathered along shorelines, where the pounding waves break them into beach sand. Other weathering processes include chemical reactions, such as when oxygen in the air combines with iron in rocks to form a type of rust. One of the most efficient forms of weathering is caused by water: liquid water runs into crevices and then freezes. As the water freezes, it expands and shatters the rock.

After weathering, the resulting debris can be carried away by flowing water, glacial ice, or blowing wind and deposited in other areas as sediment. Where material is eroded, we can see features such as river valleys, wind-sculpted hills, or mountains carved by glaciers. Where eroded material is deposited, we see features such as river deltas, sand dunes, or piles of rock at the bases of mountains and cliffs. Erosion is most efficient on planets with water and wind. On Earth, where water and wind are prevalent, most impact craters on land are worn down and filled in.

Even though the Moon and Mercury have almost no atmosphere and no running water, a type of slow erosion is still at work. Radiation from the Sun and from deep space slowly decomposes some types of minerals, effectively weathering the rock. Such effects are only a few millimeters deep at most. Impacts of micrometeoroids also chip away at rocks. In addition, landslides can occur wherever gravity and differences in elevation are present. Although water enhances landslide activity, landslides also are seen on dry bodies such as Mercury and the Moon.

As we discuss in Chapter 9, Earth, Mars, and Venus have atmospheres, and all three planets show the effects of windstorms. Images of Mars and Venus returned by spacecraft landers show surfaces that have been subjected to the forces of wind. Sand dunes are common on Earth and Mars (**Figure 8.27**), and some have been identified on Venus. Orbiting spacecraft also have found wind-eroded hills and surface patterns called wind streaks. Those surface patterns appear, disappear, and change in response to winds blowing sediments around hills, craters, and cliffs. The surface patterns serve as local weather vanes, telling planetary scientists about the direction of local prevailing surface winds. Planet-encompassing windy dust storms have been seen on Mars.

The Search for Water in the Solar System

A priority of recent planetary exploration missions is the search for water on the terrestrial planets and the Moon. Some of the evidence for past water comes from the geological processes discussed earlier in the chapter. Water was brought in by impacts in the early Solar System and is affected by geological and atmospheric activity on a planet. The search for water includes examination of images of the

Figure 8.27 A *Mars Reconnaissance Orbiter* (*MRO*) image of the Nili Patera dune field on Mars. The dunes change over months because of winds on Mars.

MARS The search for water on Mars goes back almost a century and a half. Even small telescopes show polar ice caps, which change with the martian seasons. In 1877, Italian astronomer Giovanni Schiaparelli (1835–1910) observed what appeared to be linear features on Mars and dubbed them *canali* (Italian for "channels"). Unfortunately, some other observers, including the U.S. astronomer Percival Lowell (1855–1916), incorrectly translated Schiaparelli's *canali* as "canals," implying that they were artificially constructed by intelligent life rather than naturally formed as a result of geological processes. Lowell strongly advocated the theory that those "canals" were built to move water around a drying planet Mars. Other observers of the time disputed that idea, arguing that the features were optical illusions seen in telescopes. Larger telescopes and astrophotographs did not show canals, astronomical spectroscopy did not find water vapor, and the idea of artificial canals went out of favor after Lowell died.

Scientists debate about how recently significant liquid water flow has occurred on the surface of Mars. The geological evidence suggests that water once flowed across the surface of Mars in vast quantities. Canyons and huge, dry riverbeds attest to tremendous floods that poured across the martian surface. In addition, many regions on Mars show small networks of valleys that probably were carved by flowing water (**Figure 8.28**). Large deposits of ice have been detected under the surface. Mars may have contained oceans at one time, including one ocean that might have covered a third of the planet's surface.

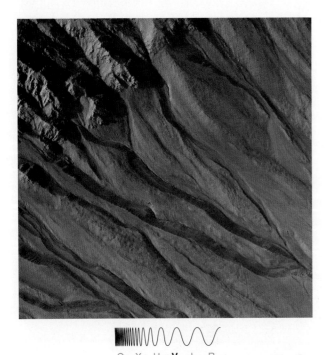

Figure 8.28 A photograph of gully channels in a crater on Mars taken by the *MRO*. The gullies coming from the rocky cliffs near the crater's rim (out of the image, to the upper left) show meandering and braided patterns similar to those of water-carved channels on Earth.

In 2004, NASA sent two instrument-equipped roving vehicles, *Opportunity* and *Spirit*, to search for evidence of water on Mars. *Opportunity* landed inside a crater. For the first time, martian rocks were available for study in the original order in which they had been laid down. Previously, the only rocks that landers and rovers had come across were those that had been dislodged from their original settings by either impacts or river floods. The layered rocks at the *Opportunity* site revealed that they had once been soaked in or transported by water. The form of the layers was typical of layered sandy deposits laid down by gentle currents of water. Magnified images of the rocks showed "blueberries"—small spheres a few millimeters across that probably formed in place among the layered rocks. Analysis of the spheres revealed abundant hematite, an iron-rich mineral that forms in the presence of water.

Observations by ESA's *Mars Express* and NASA's *Mars Odyssey* and *Mars Reconnaissance Orbiter* (*MRO*) have shown the hematite signature and the presence of sulfur-rich compounds in a vast area surrounding *Opportunity*'s landing site. Those observations suggest the existence of an ancient martian sea larger than the combined area of the Great Lakes and as deep as 500 meters.

In 2012, NASA's Mars rover *Curiosity* landed in Gale Crater, a large (150 km), old (3.8 billion years) crater just south of the equator of Mars. *Curiosity* found evidence of a stream that flowed at a rate of about 1 meter per second and was as much as 2/3 meter deep. The streambed is identified by water-worn gravel (**Figure 8.29**). The rover, about the size of a car, includes cameras, a drill, and an instrument to measure chemical composition. When the rover drilled into a rock, it found sulfur, nitrogen, hydrogen, oxygen, phosphorus, and carbon, together with clay minerals that formed in a water-rich environment that was not very

Figure 8.29 The water-worn gravel embedded in sand shown in this Mars image from the *Curiosity* rover is evidence of an ancient streambed.

salty. That crater may well have once been a large lake that held water in the ground long after the lake evaporated.

Where did the water go? Some escaped into the thin atmosphere of Mars, and some is locked up as ice in the polar regions, just as the ice caps on Earth hold much of its water. Unlike Earth's polar caps of frozen water, those on Mars are a mixture of frozen carbon dioxide and frozen water. Water must be hiding elsewhere on Mars. Small amounts of water can be found on the surface, and in 2008 NASA's *Phoenix* lander found water ice just a centimeter or so beneath surface soils at high northern latitudes (**Figure 8.30**). However, most of the water on Mars appears to be trapped well below the surface. Radar imaging by *Mars Express* and the *MRO* indicates huge quantities of subsurface water ice, not only in the polar areas as expected but also at lower latitudes under craters, even as far south as the equator. In addition, *MRO* images suggest that seasonal salt water might flow on the surface far from the poles. Salt water freezes at a lower temperature, so some sites could be warm enough to have temporary liquid salt water. Other *MRO* images show underground ice exposed on a steep slope (**Figure 8.31**). Thus, Mars may have had conditions suitable to support Earth-like microbial life in the distant past.

VENUS Evidence for liquid water on Venus comes primarily from water vapor in its atmosphere, but some geological indications of past water are apparent, such as color differences between the highland and lowland regions. As noted earlier, on Earth such a difference indicates the presence of granite, which requires water for its formation. We return to the subject of what happened to the water on Venus and Mars in Chapter 9, when we discuss their atmospheres.

THE MOON Infrared measurements of the Moon by the U.S. *Clementine* mission in 1994 supported the possibility of ice at the lunar poles. In 1998, NASA's

Figure 8.30 Water ice appears a few centimeters below the surface in this trench dug by a robotic arm on the *Phoenix* lander. The trench measures about 20 × 30 cm.

Figure 8.31 Color-enhanced *MRO* image of a steep slope on Mars in which underground water ice is exposed (and appears bright blue). *MRO* observed eight of these slopes, suggesting that Mars has places where water is accessible.

Lunar Prospector observations suggested subsurface water ice in the polar regions. When the lunar orbiter's primary mission was completed, NASA crashed *Lunar Prospector* into a crater near the Moon's south pole while ground-based telescopes searched for evidence of water vapor above the impact site, but none was seen. Since 2007, several new missions have been sent to the Moon, including Japan's *Kaguya*, India's *Chandrayaan*, and NASA's *Lunar Reconnaissance Orbiter* (*LRO*) and its companion *Lunar Crater Observation and Sensing Satellite* (*LCROSS*).

In 2009, NASA crashed the *LCROSS* launch vehicle into the Cabeus Crater at the lunar south pole, which sent up dust and vapor that the *LCROSS* and *LRO* spacecraft analyzed. More than 5 percent of the resulting plume was water, which makes that part of the Moon wetter than many Earth deserts. Other volatiles also were detected. Measurements of hydrogen by the *LRO* suggest that a fair amount of buried water ice exists in the cold southern polar region. In mid-2018, data from an instrument on *Chandrayaan* definitively confirmed that there was water ice in the shadows of craters in the polar regions (**Figure 8.32**).

Those space observations of lunar ice sent planetary scientists back to the collections of lunar rocks and soil returned to Earth decades ago by the *Apollo* mission astronauts. New analysis of volcanic glass beads in lunar soil suggests that the interior of the Moon may have a much larger amount of volatiles than previously believed. Reanalysis of the *Apollo* lunar rocks also found evidence of water. One way to distinguish whether water on the Moon originated from its interior or from impacts is to look at the ratio of water molecules composed of regular hydrogen (with one proton) and oxygen to water molecules composed of oxygen, hydrogen, and an isotope of hydrogen called deuterium (hydrogen with one proton and one neutron). The ratio is higher in the water in lunar rocks than in the water on Earth. Water that originated in comets or in meteorites rich in water has a different ratio. Alternatively, protons from the solar wind or from high-energy cosmic rays in space could have combined with oxygen on the lunar surface, yielding a different ratio. Findings from a recent study suggest that the Moon's water came from the interactions between solar wind protons and oxygen.

Those results on lunar water are preliminary. Considerable debate continues among scientists about exactly how much water ice exists on the lunar surface and how much liquid water is in the interior, how the Moon acquired that water, and how the presence and origin of water affect the current theories of lunar formation. Several countries—and some private companies—are considering proposals to send robotic spacecraft to the Moon to collect additional lunar material and bring it back to Earth for analysis.

MERCURY Water ice also has been detected in the polar regions of Mercury. Some deep craters in the polar regions of Mercury have floors that are in perpetual shadow and thus receive no sunlight. Temperatures in those permanently shadowed areas remain very cold—below 180 K. For many years, planetary scientists had speculated that ice could be found in those polar craters, with a possible detection by radar occurring in the early 1990s. The *Messenger* spacecraft, which orbited Mercury from 2011 to 2015, found deposits of ice at craters at the planet's north pole. Other frozen volatiles also were seen in the polar craters. The icy areas have sharp boundaries, which indicates they are relatively recent, either from comet impacts or from some ongoing process on the planet.

(a)

(b)

Figure 8.32 NASA's Moon Minerology Mapping instrument found surface ice at the poles of the moon. In this image, the blue areas show the location of the ice, in the coldest and darkest locations at the South Pole (top) and North Pole.

CHECK YOUR UNDERSTANDING 8.6

Which of the following worlds show evidence of the current presence of liquid or frozen water? (Choose all that apply.) (a) Mercury; (b) Venus; (c) Earth; (d) the Moon; (e) Mars

Origins The Death of the Dinosaurs

When large impacts happen on Earth, they can have far-reaching consequences for Earth's climate and for terrestrial life. One of the biggest and most significant impacts happened at the end of the Cretaceous Period, which lasted from 146 million years ago to 65 million years ago. At the end of the Cretaceous Period, more than half of all living species, including the dinosaurs, became extinct. That mass extinction is marked in Earth's fossil record by the Cretaceous-Paleogene boundary, or *K-Pg boundary* (the *K* comes from *Kreide*, German for "Cretaceous"). Fossils of dinosaurs and other now-extinct life-forms are found in older layers below the K-Pg boundary. Fossils in the newer rocks above the K-Pg boundary lack more than half of all previous species but contain a record of many other newly evolving species. Big winners in the new order were the mammals—distant ancestors of humans—that moved into the ecological niches that extinct species vacated.

How do scientists know that an impact was involved? The K-Pg boundary is marked in the fossil record in many areas by a layer of clay. Studies at more than 100 locations around the world have found large amounts of the element iridium in that layer, as well as traces of soot. Iridium is very rare in Earth's crust but is common in meteorites. The soot at the K-Pg boundary possibly indicates that widespread fires burned the world over. The thickness of the layer of clay at the K-Pg boundary and the concentration of iridium increases toward what is today the Yucatán Peninsula in Mexico. Although erosion has largely erased the original crater, geophysical surveys and rocks from drill holes in the area show a deeply deformed subsurface rock structure, similar to that seen at known impact sites. Those results provide compelling evidence that 65 million years ago an asteroid about 10 km in diameter struck the area, throwing great clouds of red-hot dust and other debris into the atmosphere (**Figure 8.33**) and possibly igniting a worldwide conflagration. The energy of the impact is estimated to have been more than that released by 5 billion nuclear bombs.

Figure 8.33 This artist's rendition depicts an asteroid or comet, perhaps 10 km across, striking Earth 65 million years ago in what is now the Yucatán Peninsula in Mexico. The lasting effects of the impact might have killed off most forms of terrestrial life, including the dinosaurs.

An impact of that energy would have devastated terrestrial life. In addition to the possible firestorm ignited by the impact, computer models suggest that earthquakes and tsunamis would have occurred. Dust from the collision and soot from the firestorms thrown into Earth's upper atmosphere would have remained there for years, blocking sunlight and plunging Earth into decades of a cold and dark "impact winter." Recent measurements of ancient microbes in ocean sediments suggest that Earth may have cooled by 7°C. The firestorms, temperature changes, and decreased food supplies could have led to a mass starvation that would have been especially hard on large animals such as the dinosaurs.

However, not all paleontologists believe that mass extinction event was the result of an impact; some think volcanic activity was important as well. However, the evidence is compelling that a great impact did occur at the end of the Cretaceous Period. Life on our planet has had its course altered by sudden and cataclysmic events when asteroids and comets have slammed into Earth. We quite possibly owe our existence to the luck of our remote ancestors—small rodent-like mammals—that could live amid the destruction after such an impact 65 million years ago.

READING ASTRONOMY NEWS

ARTICLES | QUESTIONS

The Moon had a magnetic field for at least 2 billion years, or maybe longer.

Moon Had a Magnetic Field for at Least a Billion Years Longer than Thought

By ASHLEY YEAGER, *ScienceNews*

Analysis of a relatively young rock collected by *Apollo* astronauts reveals the moon had a weak magnetic field until 1 billion to 2.5 billion years ago, at least a billion years later than previous data showed. Extending this lifetime offers insights into how small bodies generate magnetic fields, researchers report August 9 in *Science Advances*. The result may also suggest how life could survive on tiny planets or moons.

"A magnetic field protects the atmosphere of a planet or moon, and the atmosphere protects the surface," says study coauthor Sonia Tikoo, a planetary scientist at Rutgers University in New Brunswick, N.J. Together, the two protect the potential habitability of the planet or moon, possibly those far beyond our solar system.

The moon does not currently have a global magnetic field. Whether one ever existed was a question debated for decades. On Earth, molten rock sloshes around the outer core of the planet over time, causing electrically conductive fluid moving inside to form a magnetic field. This setup is called a dynamo. At 1 percent of Earth's mass, the moon would have cooled too quickly to generate a long-lived roiling interior.

Magnetized rocks brought back by *Apollo* astronauts, however, revealed that the moon must have had some magnetizing force. The rocks suggested that the magnetic field was strong at least 4.25 billion years ago, early on in the moon's history, but then dwindled and maybe even got cut off about 3.1 billion years ago.

Tikoo and colleagues analyzed fragments of a lunar rock collected along the southern rim of the moon's Dune Crater during the *Apollo 15* mission in 1971. The team determined the rock was 1 billion to 2.5 billion years old and found it was magnetized. The finding suggests the moon had a magnetic field, albeit a weak one, when the rock formed, the researchers conclude.

A drop in the magnetic field strength suggests the dynamo driving it was generated in two distinct ways, Tikoo says. Early on, Earth and the moon would have sat much closer together, allowing Earth's gravity to tug on and spin the rocky exterior of the moon. That outer layer would have dragged against the liquid interior, generating friction and a very strong magnetic field.

Then slowly, starting about 3.5 billion years ago, the moon moved away from Earth, weakening the dynamo. But by that point, the moon would have started to cool, causing less dense, hotter material in the core to rise and denser, cooler material to sink, as in Earth's core. This roiling of material would have sustained a weak field that lasted for at least a billion years, until the moon's interior cooled, causing the dynamo to die completely, the team suggests.

The two-pronged explanation for the moon's dynamo is "an entirely plausible idea," says planetary scientist Ian Garrick-Bethell of the University of California, Santa Cruz. But researchers are just starting to create computer simulations of the strength of magnetic fields to understand how such weaker fields might arise. So it is hard to say exactly what generated the lunar dynamo, he says.

If the idea is correct, it may mean other small planets and moons could have similarly weak, long-lived magnetic fields. Having such an enduring shield could protect those bodies from harmful radiation, boosting the chances for life to survive.

ARTICLES | QUESTIONS

1. Refer to Working It Out 8.2. Is the smaller *mass* of the Moon the primary reason it cooled quicker? Explain.
2. What is the other cause of the dynamo's weakening?
3. How might the presence of a magnetic field affect the possibility of habitability?
4. Why did the Moon move away from the Earth billions of years ago? (Hint: see Chapter 4.)

Source: Ashley Yeager, "The Moon had a magnetic field for at least billion years longer than thought," Science News, August 9, 2017. © Society for Science & the Public. Reprinted with permission.

Summary

The terrestrial planets in the Solar System are Mercury, Venus, Earth, and Mars, all of which have evidence of past or present water. The Moon is usually included in discussions of the terrestrial worlds because it is similar to them in many ways. Comparative planetology is the key to understanding the planets. Four geological processes—impacts, volcanism, tectonism, and erosion—are responsible for the topography on the terrestrial planets. Active volcanism and tectonics are the results of a "living" planetary interior: one that is still hot inside. Over time, the interiors cool, and tectonics and volcanism weaken. On Earth, radioactive decay and tidal effects from the Moon help heat the interior. Erosion is a surface phenomenon that results from weathering by wind or water. Surface features on the terrestrial planets, such as tectonic plates, volcanoes, mountain ranges, or canyons, are the result of the interplay among those four processes. Evolution on Earth may have been affected by impacts, such as the impact of an asteroid 65 million years ago that might have led to the death of the dinosaurs.

LG 1 **Describe how impacts have affected the evolution of the terrestrial planets.** Impact cratering is the result of a direct interaction of an astronomical object with the surface of a planet. The layering of craters gives their relative ages, with more recent craters found superimposed on older ones. Crater densities can be used to find the relative ages of regions on a surface; more heavily cratered regions are older than less cratered ones. Planets protected by atmospheres, such as Earth and Venus, have fewer small impact craters. The Moon was probably created when a Mars-sized protoplanet collided with Earth.

LG 2 **Explain how radiometric dating is used to measure the ages of rocks and terrestrial planetary surfaces.** Radioactive isotopes found in rocks can be used to measure their age. The oldest rocks measured, from the Moon and from meteorites, give the Solar System an age of 4.5 billion to 4.6 billion years.

LG 3 **Explain how scientists use both theory and observation to determine the structure of terrestrial planetary interiors.** Models of Earth's interior are used to predict how seismic waves should propagate through the interior, and those predictions are compared with observations of seismic waves. The interiors of other planets are modeled using physical principles, along with observational data of their magnetic fields. Earth has a strong magnetic field, but Venus and Mars do not. The cause for that difference between the terrestrial planets is still uncertain.

LG 4 **Describe tectonism and volcanism and the forms they take on different terrestrial planets.** Tectonism folds, twists, and cracks the outer surface of a planet. Plate tectonics is unique to Earth currently, although other types of tectonic disruptions are observed on the other terrestrial planets, such as cracking and buckling on the surface. Smooth areas on the Moon and Mercury are ancient lava flows. While Venus has the most volcanoes, the largest mountains in the Solar System are volcanoes on Mars. Earth's surface is still changing as volcanic hot-spot activity forms new members of island chains.

LG 5 **Summarize what is known about the presence of water on the terrestrial planets.** Geological features observed in orbiting and surface missions to Mars suggest that liquid water was once on the surface. Mars today has large amounts of subsurface water ice. Venus might have had liquid oceans early in its history. Space mission data indicate that water ice exists near the poles of the Moon and Mercury. The search for water is important to both the search for extraterrestrial life and the possibilities of human colonization of space.

? Unanswered Questions

- Why is Venus so different from Earth? The two planets have similar size, mass, and composition, but they are very different geologically with respect to magnetic fields, plate tectonics, and recent activity, and it is not yet known why. In addition, how did Venus end up rotating in the direction opposite that of its revolution around the Sun? Did it form with a different orbit or rotation? Was it the result of an impact early in its history? Did it change very slowly because of tidal effects from other planets?

- Will humans someday "live off the land" on the Moon? Recent space missions have provided evidence of some water ice on the Moon. If the Moon has water, that would certainly make living there more practical than if water had to be brought from Earth or synthesized from hydrogen and oxygen extracted from the lunar soil. Scientists and engineers have been studying several methods to see whether oxygen can be extracted from lunar rocks to make breathable air for people. Other researchers have looked at using lunar rock as a building material (to make concrete, for example). But the most valuable material on the Moon might turn out to be an isotope of helium, helium-3 (^{3}He, which is helium with two protons and one neutron). On Earth, that isotope exists only as a by-product of nuclear weapons, but up to a million tons of it may be on the Moon. Some scientists and engineers think that helium-3 could be used in a "clean" type of nuclear energy. That helium-3 could be brought back from the Moon for use on Earth or possibly even used in a power plant on the Moon to create energy for a lunar colony.

Questions and Problems

Test Your Understanding

1. _____, _____, and _____ build up structures on the terrestrial planets, whereas _____ tears them down.
   a. impacts, erosion, volcanism; tectonism
   b. impacts, tectonism, volcanism; erosion
   c. tectonism, volcanism, erosion; impacts
   d. tectonism, impacts, erosion; volcanism

2. Geologists can determine the relative age of features on a planet because
   a. the ones on top must be older.
   b. the ones on top must be younger.
   c. the larger ones must be older.
   d. the larger ones must be younger.

3. Scientists can learn about the interiors of the terrestrial planets from
   a. seismic waves.
   b. satellite observations of gravitational fields.
   c. physical arguments about cooling.
   d. satellite observations of magnetic fields.
   e. all of the above

4. Earth's interior is heated by
   a. angular momentum and gravity.
   b. radioactive decay and gravity.
   c. radioactive decay and tidal effects.
   d. angular momentum and tidal effects.
   e. gravity and tidal effects.

5. If a radioactive element has a half-life of 10,000 years, what fraction of it is left in a rock after 40,000 years?
   a. 1/2   c. 1/8   e. 1/32
   b. 1/4   d. 1/16

6. Lava flows on the Moon and Mercury created large, smooth plains. We don't see similar features on Earth because
   a. Earth has less lava.
   b. Earth had fewer large impacts in the past.
   c. Earth has plate tectonics that recycle the surface.
   d. Earth is large with respect to the size of those plains, so they are not as noticeable.
   e. Earth rotates much faster than either of those other worlds.

7. Scientists know the history of Earth's magnetic field because
   a. the magnetic field hasn't changed since Earth formed.
   b. they see today's changes and project backward in time.
   c. the magnetic field becomes frozen into rocks, and plate tectonics spreads those rocks apart.
   d. they compare the magnetic fields on other planets to Earth's.

8. Suppose an earthquake occurs on an imaginary planet. Scientists on the other side of the planet detect primary waves but not secondary waves after the quake. That suggests that
   a. part of the planet's interior is liquid.
   b. the planet's entire interior is solid.
   c. the planet has an iron core.
   d. the planet's interior consists entirely of rocky materials.
   e. the planet's mantle is liquid.

9. Geologists can determine the actual age of features on a planet by
   a. radiometric dating of rocks retrieved from the planet.
   b. comparing cratering rates on one planet with those on another.
   c. assuming that all features on a planetary surface are the same age.
   d. both a and b
   e. both b and c

10. Impacts on the terrestrial planets and the Moon
    a. are more common than they used to be.
    b. have occurred at approximately the same rate since the Solar System formed.
    c. are less common than they used to be.
    d. periodically become more common and then less common.
    e. never occur anymore.

11. Earth has fewer craters than Venus because
    a. Earth's atmosphere protects better than Venus's.
    b. Earth is a smaller target than Venus.
    c. Earth is closer to the asteroid belt.
    d. Earth's surface experiences more erosion.

12. Scientists propose an early period of heavy bombardment in the Solar System because
    a. the Moon is heavily cratered.
    b. all the craters on the Moon are old.
    c. the smooth part of the Moon is nearly as old as the heavily cratered part.
    d. all the craters on the Moon are young.

13. Scientists know that Earth was once completely molten because
    a. the surface is smooth.
    b. the interior layers are denser.
    c. the chemical composition indicates that.
    d. volcanoes exist today.

14. What is the main reason that Earth's inner core is liquid today?
    a. tidal force of the Moon on Earth
    b. seismic waves that travel through Earth's interior
    c. decay of radioactive elements
    d. convective motions in the mantle
    e. pressure on the core from Earth's outer layers

15. Mars has about half of Earth's diameter. If the interiors of those planets are heated by radioactive decays, how does the heating rate of the interior of Mars compare with that of Earth?
    a. The heating rate of Mars is 0.125 times that of Earth.
    b. The heating rate of Mars is 8 times that of Earth.
    c. The heating rate of Mars is 0.5 times that of Earth.
    d. The heating rate of Mars is 4 times that of Earth.
    e. The heating rates are about the same.

Thinking about the Concepts

16. When discussing the terrestrial planets, why do we include Earth's Moon?

17. Can all rocks be dated with radiometric methods? Explain.

18. Explain how scientists know that rock layers at the bottom of the Grand Canyon are older than those found on the rim.

19. Describe the sources of heating responsible for generating Earth's magma.

20. Explain why the Moon's core is cooler than Earth's.

21. Explain the difference between longitudinal waves and transverse waves.

22. How do we know that Earth's core includes a liquid zone?

23. Study the Process of Science Figure. What evidence makes the impactor theory the currently preferred explanation for the origin of the Moon? What evidence remains to be found to rule out the competing theories?

24. Compare tectonism on Venus, Earth, and Mercury.

25. Explain plate tectonics and identify the only planet on which that process has been observed.

26. Volcanoes have been found on all the terrestrial planets. Where are the largest volcanoes in the inner Solar System?

27. Explain the criteria you would apply to images (assume adequate resolution) to distinguish between a crater formed by an impact and one formed by a volcanic eruption.

28. What are the primary reasons scientists have decided that the surfaces of Venus, Earth, and Mars are younger than those of Mercury and the Moon?

29. Explain some of the geological evidence suggesting that Mars once had liquid water on its surface.

30. What evidence supports the theory suggesting that a mass extinction occurred as a consequence of an enormous impact on Earth 65 million years ago?

Applying the Concepts

31. Use Figure 8.8 to answer the following questions.
    a. How has the cratering rate changed? Has it fallen off gradually or abruptly?
    b. At present, what is the cratering rate compared with that about 4 billion years ago?
    c. Why does that falloff in cratering rate fit nicely into the theory of planet formation?

32. Are the vertical and horizontal axes in Figure 8.7 linear or logarithmic? After how many half-lives does the number of parent isotopes equal the number of daughter isotopes? Is the result unique to that example? Why or why not?

33. The destruction of the parent isotope is an example of exponential decay (Figure 8.7). Is the growth of the daughter isotope an example of exponential growth? How can you tell?

34. Compare Figures 8.18 and 8.23. Which region is older? How do you know?

35. Assume that Earth and Mars are perfect spheres with radii of 6,371 km and 3,390 km, respectively.
    a. Calculate the surface area of Earth.
    b. Calculate the surface area of Mars.
    c. If 0.72 (72 percent) of Earth's surface is covered with water, compare the amount of Earth's land area with the total surface area of Mars.

36. Compare the kinetic energy ($= 1/2\, mv^2$) of a 1-gram piece of ice (about half the mass of a dime) entering Earth's atmosphere at a speed of 50 km/s with that of a 2-metric-ton SUV (mass $= 2 \times 10^3$ kg) speeding down the highway at 90 km/h.

37. The object that created Arizona's Meteor Crater was estimated to have a radius of 25 meters and a mass of 300 million kg. Calculate the density of the impacting object, and explain what that may tell you about its composition.

38. Using the information in Table 8.1 and Working It Out 8.2, determine the relative rates of internal energy loss experienced by Earth and the Moon.

39. Earth's mean radius is 6,371 km, and its mass is 6.0×10^{24} kg. The Moon's mean radius is 1,738 km, and its mass is 7.2×10^{22} kg.
    a. Calculate Earth's average density. Show your work; do not look the value up.
    b. The average density of Earth's crust is 2,600 kg/m^3. What does that tell you about the composition of Earth's interior?
    c. Compute the Moon's average density. Show your work.
    d. Compare the average densities of the Moon, Earth, and Earth's crust. What do those values tell you about the Moon's composition in comparison with that of Earth and of Earth's crust?

40. The glaze on a piece of ancient pottery contains radium, a radioactive element that decays to radon and has a half-life of 1,620 years. Radon could not have been in the glaze when the pottery was being fired, but now it contains three atoms of radon for each atom of radium. How old is the pottery?

41. Archaeological samples are often dated by using radiocarbon dating. The half-life of carbon-14 is 5,700 years.
    a. After how many half-lives will the sample have only 1/64 as much carbon-14 as it originally contained?
    b. How much time will have passed?
    c. If the daughter product of carbon-14 is present in the sample when it forms (even before any radioactive decay happens), you cannot assume that every daughter product you see is the result of carbon-14 decay. If you did make that assumption, would you overestimate or underestimate the age of a sample?

42. Different radioisotopes have different half-lives. For example, the half-life of carbon-14 is 5,700 years, the half-life of uranium-235 is 704 million years, the half-life of potassium-40 is 1.3 billion years, and the half-life of rubidium-87 is 49 billion years.
    a. Why would an isotope with a half-life like that of carbon-14 be a poor choice to get the age of the Solar System?
    b. The age of the universe is approximately 14 billion years. Does that mean that no rubidium-87 has decayed yet?

43. Assume that the east coast of South America and the west coast of Africa are separated by an average distance of 4,500 km. Assume also that GPS measurements indicate that those continents are now moving apart at 3.75 cm/yr. If that rate has been constant over geological time, how long ago were the continents joined as part of a supercontinent?

44. Shield volcanoes are shaped like flattened cones. The volume of a cone is equal to the area of its base multiplied by one-third of its height. The largest volcano on Mars, Olympus Mons, is 27 km high and has a base diameter of 550 km. Compare its volume with that of Earth's largest volcano, Mauna Loa, which is 9 km high and has a base diameter of 120 km.

45. Using the data in Table 8.1, compare the surface gravity on Mars with that on Earth. How does that difference help explain why volcanoes on Mars can grow so high?

USING THE WEB

46. Go to the U.S. Geological Survey's "Earthquake" website (https://earthquake.usgs.gov/earthquakes/map). Set "Zoom to . . ." to "World," set the "Settings" icon in the upper right to "7 Days, Magnitude 2.5+," and look at the earthquakes for the past week. Did any really large ones occur? Compare the map of recent earthquakes with Figure 8.17 in the text. Are any of the quakes in surprising locations? Where was the most recent one? Now change the "Zoom to . . ." to the United States and change the settings to "30 Days, Magnitude 2.5+." Has seismic activity occurred, and if so, where?

47. Citizen science:
    a. Go to http://planetfour.org, a Zooniverse Citizen Science Project in which people examine images of the surface of Mars. Choose a project that has data to analyze. Log in or create a Zooniverse account if you don't have one. Read through what is available of "About," "Guide," "FAQs," "Learn More," and "Classify." Where did those data come from? What are the goals of this project? Why is having many people look at the data useful? Now classify some images.
    b. Go to the website for Cosmoquest (http://cosmoquest.org) and click on "Map Mercury." You will need to create an account for the Cosmoquest projects. Click on the circled question mark under the blue check box, and read the FAQ and watch the tutorial. What is the goal of this project? Where did the data come from? Classify some images.
    c. Go to the website for Cosmoquest (http://cosmoquest.org) and click on "Map the Moon." As in part (b), you will need an account. Click on the circled question mark under the blue check box and read the FAQ and watch the four tutorials. What are some of the basic features? How does the angle of the sunlight and the direction of illumination affect what you see? Now classify a few craters.

48. Space missions: Mars
    a. Go to the website for the *Mars Science Laboratory Curiosity* (https://mars.nasa.gov/msl/), which landed in 2012. What are the latest science results?
    b. Go to the website for the Mars InSight mission (https://mars.nasa.gov/insight/). Did it successfully land on Mars? Are any science results available?
    c. Go to the website for the Mars 2020 mission (https://mars.nasa.gov/mars2020/). What are the goals for that mission? What is its status? If it is on Mars, are any science results available yet?

49. Space Missions: The Moon
    a. India's *Chandrayaan-2* mission is planned to go to a southern region of the Moon. Are any results available from that mission?
    b. NASA, SpaceX, and China are planning new missions to send people to the Moon. Search the Internet to see the status of those missions. Are any happening soon?

50. Video:
    a. Watch one of the available documentaries about the *Apollo* missions to the Moon (such as *In the Shadow of the Moon*, 2008). Why did the United States decide to send astronauts to the Moon? Why did the *Apollo* program end?
    b. The first science fiction film was the short *A Trip to the Moon* (Georges Méliès, 1902). A version with an English narration can be viewed at https://archive.org/details/Levoyagedanslalune. A restored digitized and colorized version was released in 2011 and can be found at http://vimeo.com/39275260. Where do the "Selenians" live on the Moon? In that first cinematic depiction of contact with life from outside Earth, what do the human astronomers do to the Selenians? Compare what the astronomers in the film find on the Moon with what the *Apollo* astronauts actually saw.

EXPLORATION

Exponential Behavior

digital.wwnorton.com/astro6

Knowledge of the exponential function is critical to understanding life in modern times. That function shows up in contexts as diverse as economics, population studies, and climate change.

Radioactive decay exhibits exponential decay, where the amount of a radioactive material that remains after an elapsed time is proportional to the amount present at the beginning of the period. In this Exploration, you will use a small (1.69-ounce) bag of plain M&M's to investigate that behavior.

That size bag of M&M's contains approximately 56 pieces of candy. Each piece has an M stamped on one side. Thus, each piece can fall in two ways when dropped: M side up or M side down. That is much like what happens to a radioactive nucleus: for any given time, either it decays or it doesn't.

Instead of acquiring 10 bags of M&M's, with all the mess that would make, you will use the same bag 10 times and then add your results together. That approach makes the sample large enough that the exponential behavior becomes apparent.

Step 1 For the first trial, shake all the M&M's into your hands and then pour them onto the table.

Step 2 Count how many land M side up, and record that number in the provided table under Trial 1, Time Step 1.

Step 3 Set the M-side-up candies to one side.

Step 4 Repeat steps 1–3 for time steps 2, 3, 4, and so on, until the last candy lands M side up.

Step 5 Repeat steps 1–4 for nine more trials (for a total of 10 trials).

Step 6 Add the results of the trials for each time step, and record them in the Sum column of the table.

Step 7 Making sense of numbers in tabular form like this is hard, so plot the results on a graph with the sum on the y-axis and the time step on the x-axis.

Time Step	Trial 1	Trial 2	Trial 3	Trial 4	Trial 5	Trial 6	Trial 7	Trial 8	Trial 9	Trial 10	Sum
1											
2											
3											
4											
5											
6											
7											
8											
9											
10											
11											
12											

1 Study your graph. At first, does the number of M&M's landing M side up decrease slowly or quickly? In later time steps (such as 8 or 9), does the number decrease slowly or quickly?

_____

2 Generalize your answer to question 1 to the behavior of radioactive sources. Does the radioactivity fall off slowly or quickly at the beginning of an observation? How about at later times?

_____

3 As time goes by, what happens to the number of M&M's that remain? What happens to the number of M&M's in the pile set aside?

_____

4 Generalize your answer to question 3 to the behavior of radioactive sources. What happens to the number of radioactive isotopes over time? What happens to the number of daughter products?

_____

5 Imagine that you walk into a room and observe another student performing this experiment. She pours the M&M's onto the table and counts 10 candies that landed M side up. About how many time steps have passed since she started the experiment?

_____

6 Apply your answer to question 5 to the behavior of radioactive sources. When scientists study radioactive sources, they generally study the ratio of the number of radioactive isotopes to the number of daughter products. Explain how that method, though different from what you did in step 5, contains the same information about the amount of time that has passed.

_____

9

Atmospheres of the Terrestrial Planets

Earth's atmosphere surrounds its inhabitants like an ocean of air. It is evident in the blueness of the sky and in the breezes in the air. Without Earth's atmosphere, neither clouds nor oceans would exist. Without an atmosphere, Earth would look something like the Moon, and life would not exist on our planet. Among the five terrestrial bodies that we discussed in Chapter 8, only Venus and Earth have dense atmospheres. Mars has a very low-density atmosphere, and the atmospheres of Mercury and the Moon are so sparse that they can hardly be detected. To understand the origins of the atmospheres of Earth, Venus, and Mars—how they have changed, how they compare, and how they are likely to evolve—requires us to look back nearly 5 billion years to a time when the planets were just completing their growth.

LEARNING GOALS

Here in Chapter 9, we compare the atmospheres of the terrestrial planets. By the end of this chapter, you should be able to:

LG 1 Identify the processes that cause primary and secondary atmospheres to be formed, retained, and lost.

LG 2 Compare the strength of the greenhouse effect on Earth, Venus, and Mars and how it contributes to differences among the atmospheres of those planets.

LG 3 Describe the layers of the atmospheres on Earth, Venus, and Mars.

LG 4 Explain how Earth's atmosphere has been reshaped by the presence of life.

LG 5 Describe the evidence that shows Earth's climate is changing and how comparative planetology contributes to a better understanding of those changes.

Earth from space. ▶▶▶

Why can you breathe only on Earth?

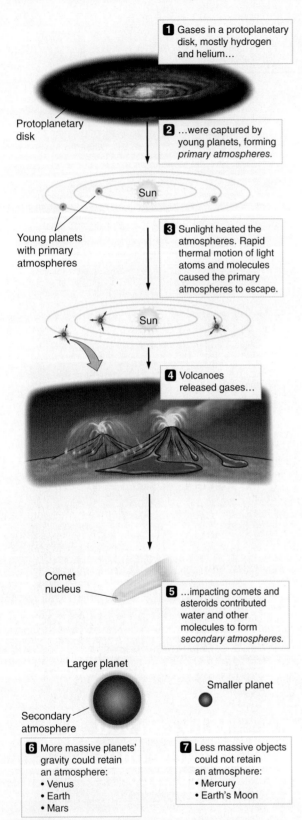

Figure 9.1 Planetary atmospheres form and evolve in phases.

9.1 Atmospheres Change over Time

An **atmosphere** is a layer of gas that sits above the surface of a body such as a terrestrial planet. A blanket of atmosphere warms and sustains Earth's temperate climate. On Venus, a thick atmosphere of carbon dioxide pushes the planet's surface temperature very high. A thin atmosphere leaves the surface of Mars unprotected and frozen. Mercury and the Moon have essentially no atmosphere. Why do some terrestrial planets have dense atmospheres, whereas others have little or none? In this section, we look at the formation of planetary atmospheres.

Formation and Loss of Primary Atmospheres

Planetary atmospheres formed in phases (**Figure 9.1**). When they formed, the young planets were initially enveloped by the remaining hydrogen and helium that filled the protoplanetary disk surrounding the Sun, and they captured some of that surrounding gas. Gas capture continued until soon after the planets formed, when the gaseous disk dissipated and the supply of gas ran out. The gaseous atmosphere collected by a newly formed planet is called its primary atmosphere. The terrestrial planets lost that primary atmosphere as the lightweight atoms and molecules escaped from the planet's gravity. To understand that, we must consider how particles move within a planetary atmosphere.

How can gas molecules escape from a planet? Any object—from a molecule to a spacecraft—can escape a planet if the object exceeds the escape velocity (Chapter 4) and is pointed in the right direction. The escape velocity depends on the mass of the planet. In Chapter 7 you saw that the terrestrial planets are less massive than the giant planets. Therefore, the terrestrial planets have weaker gravitational attraction and could not hold light gases such as hydrogen and helium. When the supply of gas in the protoplanetary disk ran out, the primary atmospheres of the terrestrial planets began leaking back into space. At the same time, intense radiation from the Sun—the primary source of thermal (kinetic) energy in the atmospheres of the terrestrial planets—raised the temperature of the atmosphere and increased the speed of atmospheric atoms and molecules enough for some to escape. In addition, giant impacts by large planetesimals early in the history of the Solar System may have blasted away some of their primary atmospheres.

How do molecules move within a planetary atmosphere? Imagine a large box that contains air. In thermal equilibrium, each type of molecule or atom in the box, from the lightest to the most massive, will have the same average kinetic energy. That average kinetic energy of the gas atoms or molecules is directly proportional to the temperature of the gas. The kinetic energy of a molecule (or any object) depends on its mass and its speed. If each type has the same average energy, the lightest molecules must be moving faster than the more massive ones. So for a gas at a given temperature, the less massive molecules must be moving faster than the more massive ones if each type of molecule has the same average kinetic energy. For example, in a mixture of hydrogen and oxygen at room temperature, hydrogen molecules will be rushing around the box at about 2,000 meters per second (m/s) on average, whereas the much more massive oxygen molecules will be moving at only 500 m/s. Remember, though, that those are average speeds. A few molecules will always be moving much faster or much slower than average.

Deep within a planet's atmosphere, fast molecules near the ground will almost certainly collide with other molecules before the fast molecules can escape. The upper levels of the atmosphere, though, contain fewer molecules.

Therefore, fast molecules there are less likely to collide with other molecules and have a better chance of escaping (as long as they are heading more or less upward). Because the giant planets were farther from the Sun, they were much more massive and were cooler: stronger gravity and lower temperatures enabled them to retain nearly all their massive primary atmospheres.

AstroTour: Atmospheres: Formation and Escape

The Formation of Secondary Atmospheres

Although their primary atmospheres were lost, some of the terrestrial planets do have an atmosphere today, known as a secondary atmosphere. Where did that secondary atmosphere come from? Accretion, volcanism, and impacts are responsible for the atmospheres of Earth, Venus, and Mars today. During the planetary accretion process, minerals containing water, carbon dioxide, and other volatile matter collected in the planetary interiors. Later, as the interior heated up, those gases were released from the minerals that had held them. Volcanism then brought the gases to the surface, where they accumulated and created a secondary atmosphere (see step 4 of Figure 9.1).

Impacts by comets and asteroids were another important source of gases. Huge numbers of comets formed in the outer Solar System and were therefore rich in volatiles. As the giant planets there grew to maturity or migrated their orbits, their gravitational perturbations stirred up the comets and asteroids that orbited relatively nearby. Many of those icy bodies were flung outward by the giant planets to join other planetesimals in the Kuiper Belt, 30–50 astronomical units (AU) from the Sun. Other bodies joined the Oort Cloud, a spherical cloud of icy planetesimals that surrounds the Sun at a distance ranging from the Kuiper Belt to about 50,000 AU from the Sun—nearly one-quarter of the way to the nearest star. Other comets were scattered into the inner Solar System. Upon impact with the terrestrial planets, those objects brought ices such as water, carbon monoxide, methane, and ammonia. On the terrestrial planets, cometary water mixed with the water that had been released into the atmosphere by volcanism (see step 5 of Figure 9.1). On Earth, and perhaps Mars and Venus, most of the water vapor then condensed as rain and flowed into the lower areas to form the earliest oceans.

Sunlight also influenced the composition of secondary atmospheres. Ultraviolet (UV) light from the Sun easily breaks down molecules such as ammonia (NH_3) and methane (CH_4). Ammonia, for example, breaks down into hydrogen and nitrogen. When that happens, the lighter hydrogen atoms quickly escape to space, leaving behind the much heavier nitrogen atoms. Nitrogen atoms then combine in pairs to form more massive nitrogen molecules (N_2), and those molecules are even less likely to escape into space. Decomposition of ammonia by sunlight became the primary source of molecular nitrogen in the atmospheres of the terrestrial planets. Molecular nitrogen makes up ~78% of Earth's atmosphere.

Among the terrestrial planets, today only Venus, Earth, and Mars have significant secondary atmospheres. What happened to Mercury and the Moon? Even if those two bodies experienced less volcanism than the other terrestrial planets (see Chapter 8), they would have had the same early bombardment of comet nuclei from the outer Solar System. Some carbon dioxide and water must have accumulated during volcanic eruptions and comet impacts. A body can lose a secondary atmosphere through the same processes that cause the loss of a primary atmosphere. Large impacts were less frequent as the Solar System aged, but atmospheric escape continued over the past 4 billion years. In addition, decreases in the magnetic field as Mercury and the Moon cooled might have contributed to atmospheric escape. With a weaker magnetic field, the planet became less protected from the **solar wind**—a constant stream of charged particles from the Sun. The solar wind can accelerate atmospheric particles to escape velocity.

Both the Moon and Mercury have virtually no atmosphere today. Mercury lost nearly its entire secondary atmosphere to space, just as it had previously lost its primary atmosphere. Even molecules as massive as carbon dioxide can escape from a small planet if the temperature is high enough, as it is on Mercury's sunlit side. Furthermore, intense UV radiation from the Sun can break molecules into less massive fragments, which are lost to space even more quickly. Because the distance from the Sun to the Moon is much farther than the distance from the Sun to Mercury, the Moon is much cooler than Mercury, but its mass is so small that molecules easily escaped even at relatively low temperatures. The ability of planets to hold on to their atmospheres is explored further in **Working It Out 9.1**.

CHECK YOUR UNDERSTANDING 9.1

Which are reasons Mercury has so little gas in its atmosphere? (Choose all that apply.) (a) Its mass is small. (b) It has a high temperature. (c) It is close to the Sun. (d) Its escape velocity is low. (e) It has no moons.

9.2 Secondary Atmospheres Evolve

Although Venus, Earth, and Mars most likely started out with atmospheres of similar composition, they ended up being very different from one another. Earth is volcanically active, Venus might still be volcanically active, and Mars has been volcanically active in the recent past. All three planets must have shared the intense cometary showers of the early Solar System. Their similar geological histories suggest that their early secondary atmospheres might also have been similar. The development of life, however, increased the amount of oxygen in Earth's secondary atmosphere. Earth's atmosphere is now made up primarily of nitrogen and oxygen, with only a trace of carbon dioxide; the compositions of the atmospheres of Venus and Mars today are nearly identical—mostly carbon dioxide, with much smaller amounts of nitrogen. The atmospheres of those planets differ for two main reasons—planetary mass and the greenhouse effect.

Planetary Mass Affects a Planet's Atmosphere

Table 9.1 shows that the *compositions* of the atmospheres of Venus and Mars today are nearly identical. They are both composed mostly of carbon dioxide, with much smaller amounts of nitrogen. Carbon dioxide and water vapor came from volcanic gases, and nitrogen came from decomposed cometary ammonia. The total *amount* of atmosphere, however, differs widely among the three planets. The atmospheric pressure on Venus is nearly 100 times greater than that on Earth. By contrast, the average surface pressure on Mars is less than a hundredth of that on Earth. Venus is nearly 8 times as massive as Mars, so Venus probably had about 8 times as much carbon within its interior to produce carbon dioxide, the principal secondary-atmosphere component of both planets.

The large difference in atmospheric mass comes from the relative strengths of each planet's surface gravity, which involves both the mass and the radius of a planet. Venus has the gravitational pull necessary to hang on to its atmosphere; Mars has less gravitational attraction to keep its atmosphere (see Working It Out 9.1). Furthermore, when a planet such as Mars began to lose its atmosphere to space, the process began to take on a *runaway* behavior. With a thinner

TABLE 9.1 Atmospheres of the Terrestrial Planets

Physical Properties and Composition

Property	PLANET		
	Venus	Earth	Mars
Surface Pressure (bars)	92	1.0	0.006
Atmospheric Mass (kg)	4.8×10^{20}	5.1×10^{18}	2.5×10^{16}
Surface Temperature (K)	740	288	210
Carbon Dioxide (%)	96.5	0.041	95.3
Nitrogen (%)	3.5	78.1	2.7
Oxygen (%)	0.00	20.9	0.13
Water (%)	0.002	0.1 to 3	0.02
Argon (%)	0.007	0.93	1.6
Sulfur Dioxide (%)	0.015	0.02	0.00

9.1 Working It Out Atmosphere Retention

To estimate a planet's ability to retain its atmosphere, we compare the escape velocity from the planet (which depends on the planet's gravity, determined from its mass and radius) with the average speed of the molecules in a gas (which depends on the temperature of the gas and the mass of the molecules that make up the gas). The escape velocity is

$$v_{esc} = \sqrt{\frac{2GM}{R}}$$

and the values of v_{esc}, in kilometers per second (km/s), are given for the inner planets in Table 8.1.

The temperature, T, of a gas is proportional to the kinetic energy of the particles, $1/2\,mv^2$ (Chapter 4). We can rearrange that relationship to solve for v and insert the constants of proportionality to get the average speed of a molecule in a gas:

$$v_{molecule} = \sqrt{\frac{3kT}{m}}$$

where T is the temperature of the gas in kelvins, m is the mass of the molecule in kilograms (kg), and k is the Boltzmann constant. The atomic mass of a molecule is found by adding the atomic masses of its composite atoms as specified in the periodic table. (Atomic masses of atoms come from the total number of neutrons and protons; the electron weighs little in comparison.) An oxygen molecule (O_2), for example, has an atomic weight of 32 (16 for each O atom), whereas a hydrogen molecule (H_2) has an atomic weight of 2 (1 for each H atom). Thus, O_2 has 16 times the mass of H_2.

If we put in the value of the Boltzmann constant ($k = 1.38 \times 10^{-23}$ joule per kelvin, or J/K) and the mass of the hydrogen atom ($m = 1.67 \times 10^{-27}$ kg), then $v_{molecule}$, in kilometers per second, is given by

$$v_{molecule} = 0.157 \text{ km/s} \times \sqrt{\frac{\text{Temperature of gas}}{\text{Atomic weight of molecule}}}$$

The higher the temperature, the higher the average kinetic energy of the individual molecules and the faster the average speed of the particles. That difference explains why Earth can hold on to the oxygen in its atmosphere but loses hydrogen to space.

In our example of hydrogen and oxygen molecules, H_2 has an atomic weight of 2 (1 for each hydrogen atom), and O_2 has an atomic weight of 32 (16 for each oxygen atom, from 8 protons and 8 neutrons). Earth has an average temperature of 288 K. The average speeds of H_2 and O_2 are thus

$$\text{For } H_2: v_{molecule} = 0.157 \times \sqrt{\frac{288}{2}} = 1.88 \text{ km/s}$$

$$\text{For } O_2: v_{molecule} = 0.157 \times \sqrt{\frac{288}{32}} = 0.47 \text{ km/s}$$

Thus, in a gas containing both H_2 and O_2, the hydrogen molecule will, on average, be moving 4 times faster than the average oxygen molecule. At any given temperature, the lighter molecules will be moving faster. Not all molecules in a gas are moving at the average speed: some move faster and some move slower (**Figure 9.2**). The general rule is that over the age of the Solar System, a planet can keep its atmosphere if, for that type of gas molecule,

$$v_{molecule} \lesssim 1/6\, v_{esc}$$

The escape velocity from Earth is 11.2 km/s, and one-sixth of that is 1.87 km/s (that is, slightly less than the average speed of H_2). Those values explain why Earth has kept its O_2 but not its H_2. A similar analysis shows that on the Moon, with its lower v_{esc} value, both H_2 and O_2 escape. Jupiter, with its much colder temperatures and higher v_{esc} value, keeps both H_2 and O_2.

Figure 9.2 The distribution of the speeds of molecules in a gas. The shape of the curve and the exact numbers depend on the temperature of the gas and the masses of the molecules. In all cases, some of the speedier molecules may escape if they exceed the escape velocity.

atmosphere, fewer slow molecules were present to keep fast molecules from escaping, and the rate of escape increased. That process led, in turn, to even less atmosphere and still greater escape rates.

Another possibility is that Mars lost atmosphere in a giant impact. In addition, scientists debate how much atmospheric loss arises from the solar wind's interaction with planetary atmospheres, especially in the absence of a planetary magnetic field. Even though Earth has a magnetic field and Venus and Mars do not, all three planets are losing some atmosphere to space. The extent to which the lack of a magnetic field on Mars played a role in its atmospheric loss is being studied by NASA's *Mars Atmosphere and Volatile EvolutioN* (*MAVEN*) mission, which arrived at Mars in September 2014.

The Atmospheric Greenhouse Effect

Astronomy in Action: Changing Equilibrium

Differences in the present-day masses of the atmospheres of Venus, Earth, and Mars have a large effect on their surface temperatures. In Chapter 5, we calculated the temperature of a planet by finding the equilibrium between the amount of energy it absorbs from sunlight and the amount of energy it radiates back into space. Recall that we found that this calculation gives a good result for planets without atmospheres. But Earth is warmer than expected, and Venus is very much hotter than that simple model predicted. When the predictions of a model fail, the implication is that something was left out of the model. Here, that something was the *atmospheric greenhouse effect*, which traps solar radiation.

The atmospheric greenhouse effect in planetary atmospheres and the conventional **greenhouse effect** operate differently, although the results are much the same. Planetary atmospheres and the interiors of greenhouses are both heated by trapping the Sun's energy, but there the similarities end. The conventional greenhouse effect is the rise in temperature in a car on a sunny day when you leave the windows closed, or what allows plants to grow in the winter in a greenhouse. Sunlight pours through the glass, heating the interior and raising the internal air temperature. With the windows closed, hot air is trapped, and temperatures can climb as high as 80°C (about 180°F). Heating by solar radiation is most efficient when an enclosure is transparent, which is why the walls and roofs of real greenhouses are made mostly of glass.

The atmospheric greenhouse effect is illustrated in **Figure 9.3**. Atmospheric gases freely transmit visible light (Figure 6.20), allowing the Sun to warm the planet's surface. The warmed surface radiates the energy in the infrared region of the spectrum. Some of the atmospheric gases strongly absorb that infrared radiation and convert it to thermal energy, which is released in random directions. Some of the thermal energy continues into space, but much of it returns to the ground, raising the planet's surface temperature. As a result of that radiation, the planet's surface receives thermal energy from both the Sun and the atmosphere. Gases that transmit visible radiation but absorb infrared radiation are known as **greenhouse gases**. Examples include water vapor, carbon dioxide, methane, and nitrous oxide, as well as industrial chemicals such as halogens. The presence of greenhouse gases in a planet's atmosphere will cause its surface temperature to rise.

That rise in temperature continues until the surface becomes hot enough—and therefore radiates enough energy—that the fraction of infrared radiation leaking out through the atmosphere balances the absorbed sunlight, and equilibrium is reached. Convection

Figure 9.3 In the atmospheric greenhouse effect, greenhouse gases such as water vapor and carbon dioxide trap infrared radiation, increasing a planet's temperature.

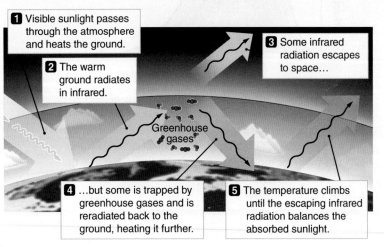

also helps maintain equilibrium by transporting thermal energy to the top of the atmosphere, where it can be more easily radiated to space. If the amount of greenhouse gases increases in the atmosphere, the trapping effect increases, and the temperature at which energy input and output balances also increases. Even though the mechanisms are different, the conventional greenhouse effect and the atmospheric greenhouse effect produce the same result: the local environment is heated by trapped solar radiation.

AstroTour: Greenhouse Effect

Similarities and Differences among the Terrestrial Planets

How does the atmospheric greenhouse effect operate on Mars, Earth, and Venus? What really matters is the actual number of greenhouse molecules in a planet's atmosphere, not the fraction they represent. For example, even though the atmosphere of Mars is composed almost entirely of carbon dioxide (see Table 9.1)—an effective greenhouse molecule—the atmosphere is very thin and contains fewer greenhouse molecules than the atmosphere of Venus or Earth. As a result, the atmospheric greenhouse effect is relatively weak on Mars and raises the average surface temperature by only about 5 K (9°F). At the other extreme, Venus's massive atmosphere of carbon dioxide and sulfur compounds raises its average surface temperature by more than 400 K, to about 740 K (467°C, 870°F). At such high temperatures, any remaining water and most carbon dioxide locked up in surface rocks are driven into the atmosphere, further enhancing the atmospheric greenhouse effect.

The atmospheric greenhouse effect on Earth is not as severe as it is on Venus—the average global temperature near Earth's surface is about 288 K (15°C, 59°F). Temperatures on Earth's surface are about 35 K (35°C, 63°F) warmer than they would be without an atmospheric greenhouse effect, mainly because of water vapor and carbon dioxide. Without that comparatively small greenhouse effect, though, Earth's average global temperature would be 255 K (−18°C, 0°F)—well below the freezing point of water, leaving us with a world of frozen oceans and ice-covered continents.

How has the atmospheric greenhouse effect made the composition of Earth's atmosphere so different from the high–carbon dioxide atmospheres of Venus and Mars? The answer lies in Earth's location in the Solar System. Earth and Venus have about the same mass, but Venus orbits the Sun closer than Earth, at 0.7 AU. Volcanism and cometary impacts produced large amounts of carbon dioxide and water vapor to form early secondary atmospheres on both planets. Most of Earth's water quickly rained out of the atmosphere to fill vast ocean basins. But because Venus was closer to the Sun, its surface temperatures were higher than those of Earth. As the Sun itself aged and brightened, Venus got warmer, and most of the rainwater on Venus immediately reevaporated, much as water does in Earth's desert regions. Venus was left with a surface that contained very little liquid water and an atmosphere filled with water vapor. The water vapor caused even higher temperatures, which led to the release of more carbon dioxide from the rocks to the atmosphere. The continuing buildup of both water vapor and carbon dioxide in the atmosphere of Venus led to a runaway atmospheric greenhouse effect that drove up the surface temperature of the planet even more. Ultimately, the surface of Venus became so hot that no liquid water could exist on it.

That early difference between a watery Earth and an arid Venus forever changed how their atmospheres and surfaces evolved. On Earth, water erosion caused by rain and rivers continually exposed fresh minerals, which then reacted chemically with atmospheric carbon dioxide to form solid carbonates. That reaction removed some of the atmospheric carbon dioxide, burying it within

Earth's crust as a component of a rock called limestone. Later, the development of life in Earth's oceans accelerated the removal of atmospheric carbon dioxide. Tiny sea creatures built their protective shells of carbonates, and as they died they built up massive beds of limestone on the ocean floors. Water erosion and the chemistry of life tied up all but a trace of Earth's total inventory of carbon dioxide in limestone beds. Earth's particular location in the Solar System seems to have spared it from the runaway atmospheric greenhouse effect. If those reactions had not locked up all the carbon dioxide now in limestone beds, Earth's atmosphere would be composed of about 98 percent carbon dioxide, similar to that of Venus or Mars. The atmospheric greenhouse effect would be much stronger, and Earth's temperature would be much higher.

Scientists study the details of the differences in the amount of water on Venus, Earth, and Mars. Geological evidence indicates that liquid water was once plentiful on the surface of Mars. Several of the spacecraft orbiting Mars have found evidence that significant amounts of water still exist on Mars as subsurface ice—far more than the atmospheric abundance indicated in Table 9.1. Earth's liquid and solid water supply is even greater, about 0.02 percent of its total mass. More than 97 percent of Earth's water is in the oceans, which have an average depth of about 4 km. Earth today has 100,000 times more water than Venus.

Some scientists think that Venus once had as much water as Earth—as liquid oceans or as more water vapor than is measured today. As the Sun aged and became brighter, and the planets received more solar energy, water molecules high in the atmosphere of Venus were broken apart into hydrogen and oxygen by solar UV radiation. The low-mass hydrogen atoms were quickly lost to space. Oxygen escaped more slowly, so some eventually migrated downward to the planet's surface, where it was removed from the atmosphere by bonding with minerals on the surface. In support of that theory, the *Venus Express* spacecraft has measured hydrogen and some oxygen escaping from the upper levels of Venus's atmosphere.

CHECK YOUR UNDERSTANDING 9.2

The main greenhouse gases in the atmospheres of the terrestrial planets are: (a) oxygen and nitrogen; (b) methane and ammonia; (c) carbon dioxide and water vapor; (d) hydrogen and helium.

9.3 Earth's Atmosphere Has Detailed Structure

Now that we have considered some of the overall processes that have influenced how the terrestrial planet atmospheres evolved, we look in depth at each of them. We begin with the composition and structure of Earth's atmosphere, not only because we know it best but also because it helps us better understand the atmospheres of other worlds.

Life and the Composition of Earth's Atmosphere

Earth's atmosphere is about four-fifths nitrogen (N_2) and one-fifth oxygen (O_2) (see Table 9.1). Many important minor constituents such as water vapor and carbon dioxide (CO_2) are also present, the amounts of which vary depending on global location and season. The composition of Earth's atmosphere is relatively uniform on a global scale, but temperatures can vary widely. Atmospheric temperatures near Earth's surface can range from as high as 60°C (140°F) in the deserts to as low as −90°C (−130°F) in the polar regions. The mean global temperature is about 15°C (59°F).

OXYGEN Table 9.1 shows that Earth's atmosphere contains abundant amounts of oxygen (O_2), whereas the atmospheres of other planets do not. Oxygen is a highly reactive gas: it chemically combines with, or oxidizes, almost any material it touches. The rust (iron oxide) that forms on steel is an example. The reddish surface of Mars is coated with oxidized iron-bearing minerals—one reason the martian atmosphere is almost completely free of oxygen. A planet with significant amounts of oxygen in its atmosphere requires a way to replace oxygen lost through oxidation. On Earth, plants perform that role.

The oxygen concentration in Earth's atmosphere has changed over the history of the planet (**Figure 9.4**). When Earth's secondary atmosphere first appeared about 4 billion years ago, it had very little oxygen because O_2 is not found in volcanic gases or comets. Studies of ancient sediments show that about 2.8 billion years ago, an ancestral form of cyanobacteria—single-celled organisms that contain chlorophyll, which enables them to obtain energy from sunlight—began releasing oxygen into Earth's atmosphere as a waste product of their metabolism. At first, that biologically generated oxygen combined with exposed metals and minerals in surface rocks and soils, so it was removed from the atmosphere as quickly as it formed. Ultimately, the explosive growth of cyanobacteria and then plant life accelerated the production of oxygen, building up atmospheric concentrations that approached today's levels only about 250 million years ago.

All plants, from tiny green algae to giant redwoods, use the energy of sunlight to build carbon compounds out of carbon dioxide and produce oxygen as a metabolic waste product in a process called photosynthesis. In that way, emerging life dramatically changed the very composition and appearance of Earth's surface—the first of many such widespread modifications imposed on Earth by living organisms. Earth's atmospheric oxygen content is held in a delicate balance primarily by plants. If plant life on the planet were to disappear, so, too, would nearly all Earth's atmospheric oxygen, and therefore all animal life—including us.

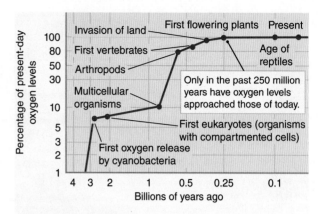

Figure 9.4 The amount of oxygen in Earth's atmosphere has built up as a result of photosynthesizing plant life.

OZONE Ozone (O_3) is another constituent in Earth's atmosphere. Ozone is formed when UV light from the Sun breaks molecular oxygen (O_2) into its individual atoms. Those oxygen atoms can then recombine with other oxygen molecules to form ozone (the net reaction is $O_2 + O \rightarrow O_3$). Most of Earth's natural ozone is concentrated in the upper atmosphere at altitudes between 10 and 50 km (the **stratosphere**). There it absorbs UV sunlight very strongly. Without the ozone layer, that radiation would reach Earth's surface, where it would be lethal to nearly all forms of life. Ozone exists in the lower atmosphere, too, where it is primarily a by-product of power plants, factories, and automobiles. That ozone is a human-made pollutant and a health hazard that raises the risk of respiratory and heart problems.

In the mid-1980s, scientists began noticing that the measured amount of ozone in Earth's upper atmosphere had been decreasing seasonally since the 1970s, primarily over the polar latitudes during springtime in both the Northern and Southern hemispheres. Scientists called those depleted regions "ozone holes": reduced concentrations of ozone and not an actual hole in the ozone layer. Ozone depletion is caused by a seasonal buildup of atmospheric halogens—mostly chlorine, fluorine, and bromine—such as those found in industrial refrigerants, especially chlorofluorocarbons (CFCs). Halogens diffuse upward into the stratosphere, where they destroy ozone without themselves being consumed. Such agents are called catalysts—materials that participate in and accelerate chemical reactions but are not themselves modified in the process. Because they are not modified or used up, halogens may remain in Earth's upper atmosphere for decades or even centuries. Even though more of the chemicals originated in the north, the

Figure 9.5 Polar stratospheric clouds form in the polar springtime and serve as the surface upon which ozone destruction takes place.

depletions are greater in the Southern Hemisphere because the colder temperatures in the southern polar regions produce a type of cloud that provides a surface on which the ozone-destroying chemical reactions can take place (**Figure 9.5**).

Scientists predicted that the continuing removal of ozone from the high atmosphere could cause trouble for terrestrial life as more and more UV radiation reached the ground. Measured increases in the levels of UV radiation appear to be related to increases in skin cancer in humans, and the mutating effects it may have on other life-forms are not completely understood. By the late 1980s, international agreements on phasing out the production of CFCs and other ozone-depleting chemicals were signed (the Montreal Protocol), and world consumption has steadily declined. The largest Southern Hemisphere ozone hole occurred over Antarctica in 2006. The southern polar ozone holes have mostly stabilized; the hole was large in 2015 but smaller in 2016 and 2017. The ozone hole in the north has been smaller than in the south but would affect more populated areas when it grows, as it did in 2011. Ozone seems to be still decreasing slowly in the lower stratosphere in midlatitudes over more populated regions, perhaps from the release of other chemicals or from changes in the climate. Full recovery to 1980 levels is not expected before the late 21st century.

CARBON DIOXIDE Carbon dioxide (CO_2) is another variable component of Earth's atmosphere. How much carbon dioxide is in the atmosphere at any one time depends on a complex pattern of carbon dioxide *sources* (places where it originates) and *sinks* (places where it goes). Plants consume carbon dioxide in great quantities as part of their metabolic process. Coral reefs are colonies of tiny ocean organisms that build their protective shells with carbonates produced from dissolved carbon dioxide. Fires, decaying vegetation, and the human burning of fossil fuels all release carbon dioxide back into the atmosphere. That balance between carbon dioxide sources and sinks changes. As we describe later, the amount of carbon dioxide in the atmosphere has varied historically but has been increasing more rapidly since the industrial revolution (since 1840 or so). That recent increase in carbon dioxide, in turn, has directly affected global temperature because carbon dioxide is a powerful greenhouse gas.

WATER VAPOR Water vapor (H_2O) in Earth's atmosphere also affects daily life and is a powerful greenhouse gas. Over the range of temperatures on Earth, the amount of water in the atmosphere varies from time to time and from place to place. In warm, moist climates, water vapor may account for as much as 3 percent of the total atmospheric composition. In cold, arid climates, it may be less than 0.1 percent. The continual process of condensation and evaporation of water involves the exchange of thermal and other forms of energy, making water vapor a major contributor to Earth's weather.

The Layers of Earth's Atmosphere

Earth's atmosphere is a blanket of gas several hundred kilometers thick. It has a total mass of approximately 5×10^{18} kg, less than one-millionth of Earth's total mass. The weight of Earth's atmosphere creates a force of approximately 100,000 newtons (N) acting on each square meter of the planet's surface, equivalent to about 14.7 pounds pressing on every square inch. That amount of pressure is called a **bar** (from the Greek *baros*, meaning "weight" or "heavy"). Earth's average atmospheric pressure at sea level is approximately 1 bar. A millibar (mb) is one-thousandth of 1 bar and is more commonly used in meteorology and in weather reports. Underwater, each depth of 10 meters adds 1 bar of pressure. We are largely unaware of Earth's atmospheric pressure because the same pressure exists both inside and outside our bodies, so the force pushing out precisely balances the force pushing in.

Recall from Chapter 8 that the pressure at any point within a planet's interior must be great enough to balance the weight of the overlying layers. The same principle holds true in a planetary atmosphere. The atmospheric pressure on a planet's surface must be great enough to support the weight of the overlying atmosphere. Different forms of matter provide the pressure within a planet's interior and in its atmosphere. In the interior of a solid planet, solid materials exert pressure as they resist being compressed. In a planetary atmosphere, the motions of gas molecules exert enough pressure to support the atmosphere.

Earth's atmosphere consists of several distinct layers (**Figure 9.6**). Those layers are distinguished by the changes in temperature and pressure through the atmosphere. The lowest layer, the one in which humans live and breathe, is called the **troposphere**. It contains 90 percent of Earth's atmospheric mass and is the source of all our weather. At Earth's surface, usually called sea level, the troposphere has an average temperature of 15°C (288 K). Within the troposphere, atmospheric pressure, density, and temperature all decrease as altitude increases. For example, at an altitude of 5.5 km (18,000 feet, a few thousand feet below the

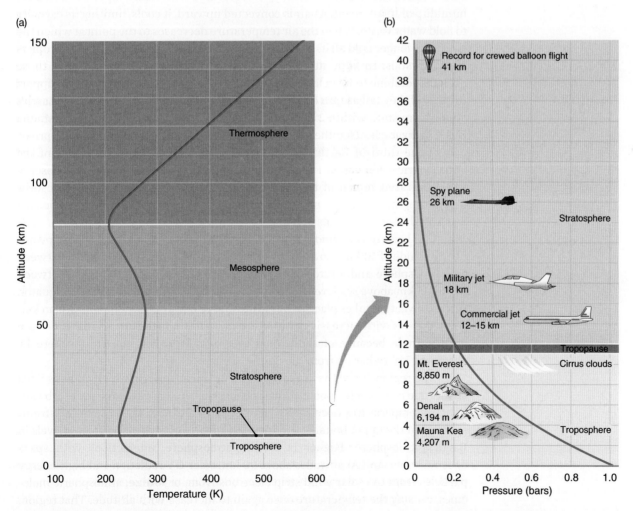

Figure 9.6 These graphs show (a) temperature and (b) pressure plotted for Earth's atmospheric layers as a function of altitude. Most human activities are confined to the bottom layers of Earth's atmosphere.

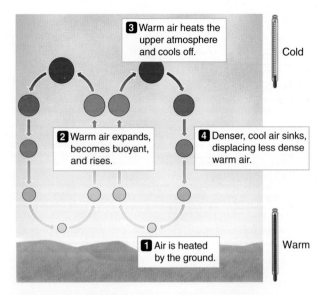

Figure 9.7 Atmospheric convection carries thermal energy from the Sun-heated surface of Earth upward through the atmosphere.

summit of Denali in Alaska), the atmospheric pressure and density are only 50 percent of their sea-level values, and the average temperature has dropped to −20°C (253 K). In the lower stratosphere, at an altitude of 12–15 km, where commercial jets cruise, the temperature is −60°C (213 K), and the density and pressure are less than one-fifth what they are at sea level.

The atmosphere is warmer near Earth's surface because the air is closer to the sunlight-heated ground, which warms the air by infrared radiation. The atmosphere is cooler at very high altitudes because there the atmosphere freely radiates its thermal energy into space. In fact, it would get colder with increasing altitude even faster if not for convection. **Figure 9.7** illustrates how convection carries thermal energy upward through Earth's atmosphere. At a given pressure, cold air is denser than warm air. So, when cold air encounters warm air, the denser cold air slips under the less dense warm air, pushing the warm air upward. That convection sets up air circulation between the lower and upper levels of the atmosphere and tends to diminish the temperature extremes caused by heating at the bottom and cooling at the top.

Convection also affects the vertical distribution of atmospheric water vapor. Air's ability to hold water in the form of vapor depends very strongly on the air temperature: the warmer the air, the more water vapor it can hold. The amount of water vapor in the air relative to what the air could hold at a particular temperature is called the relative humidity. Air saturated with water vapor has a relative humidity of 100 percent. As air is convected upward, it cools, limiting its capacity to hold water vapor. When the air temperature decreases to the point at which the air can no longer hold all its water vapor, water begins to condense to tiny droplets or ice crystals. In large numbers those become visible as clouds. When those droplets combine to form large drops, convective updrafts can no longer support them, and they fall as rain or snow. Therefore, most of the water vapor in Earth's atmosphere stays within 2 km of the surface. At an altitude of 4 km, the Mauna Kea Observatories (see the Chapter 6 opening figure) are higher than approximately one-third of Earth's atmosphere, but they lie above nine-tenths of the atmospheric water vapor. That altitude is important for astronomers who observe in the infrared region of the spectrum because water vapor strongly absorbs infrared light. The water in the atmosphere is more often visible as condensed water in the form of clouds and ice.

The layer of the atmosphere above the troposphere and extending upward to an altitude of 50 km above sea level is the stratosphere. The boundary between the troposphere and stratosphere is called the **tropopause**. It varies between 10 and 15 km above sea level, depending on latitude, and is highest at the equator. Little convection takes place in the stratosphere because the temperature no longer decreases with increasing altitude. In fact, the temperature begins to *increase* with altitude because of the ozone layer, which warms the stratosphere by absorbing UV radiation from the Sun.

The region above the stratosphere is the **mesosphere**, which extends from an altitude of 50 km to about 90 km. The mesosphere has no ozone to absorb sunlight, so temperatures once again decrease with altitude. The base of the stratosphere and the upper boundary of the mesosphere are two of the coldest levels in Earth's atmosphere. Higher in Earth's atmosphere, interactions with space become important. At altitudes above 90 km, solar UV radiation and high-energy particles from the solar wind strip electrons from, or **ionize**, atmospheric molecules, causing the temperature once again to increase with altitude. That region, called the **thermosphere**, is the hottest part of the atmosphere. The temperature can reach 1000 K near the top of the thermosphere, at an altitude of 600 km.

The atoms and molecules in the gases within and beyond the thermosphere are ionized by UV photons and high-energy particles from the Sun. That region of ionized atmosphere is called the **ionosphere**, and it not only overlaps the thermosphere but also extends farther into space. The ionosphere reflects certain frequencies of radio waves back to the ground. For example, the frequencies used by AM radio bounce back and forth between the ionosphere and the surface, enabling radio receivers to pick up stations at great distances from the transmitters. Amateur radio operators can communicate with one another around the world by bouncing their signals off the ionosphere.

CHECK YOUR UNDERSTANDING 9.3A

List the layers of the atmosphere in order from nearest to the surface to farthest from the surface: (a) stratosphere; (b) thermosphere; (c) troposphere; (d) ionosphere; (e) mesosphere.

Earth's Magnetosphere

Even farther out than the ionosphere is Earth's magnetosphere, which surrounds Earth and its atmosphere. The magnetosphere is a large region filled with electrons, protons, and other charged particles from the Sun that have been captured by the planet's magnetic field. That region has a radius approximately 10 times that of Earth and fills a volume more than 1,000 times as large as the volume of the planet itself. Magnetic fields affect only moving charges. Charged particles move freely along the direction of the magnetic field but cannot cross magnetic field lines. If they try to move across the direction of the field, they experience a force that causes them to loop around the direction of the magnetic field, as illustrated in **Figure 9.8a**. That force is perpendicular both to the motion of the particle and to the direction of the magnetic field.

If the magnetic field is pinched together at some point, particles moving into the pinch will experience a magnetic force that reflects them back along the direction they came from, creating a sort of "magnetic bottle" that contains the charged particles. If charged particles are located in a region where the field is pinched on both ends, as shown in **Figure 9.8b**, they may bounce back and forth many times. Earth's magnetic field is pinched together at the two magnetic poles and spreads out around the planet.

Earth and its magnetic field are immersed in the solar wind. When the charged particles of the solar wind first encounter Earth's magnetic field, the smooth flow is interrupted and their speed suddenly drops—they are diverted by Earth's magnetic field like a river is diverted around a boulder. As they flow past, some of those charged particles become trapped by Earth's magnetic field, where they bounce back and forth between Earth's magnetic poles, as illustrated in **Figure 9.8c**.

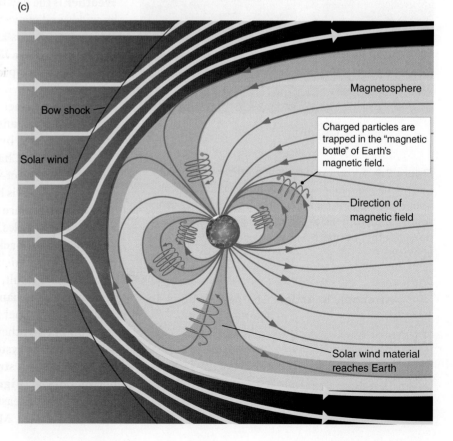

Figure 9.8 (a) Charged particles, here electrons, spiral in a uniform magnetic field. (b) When the field is pinched, charged particles can be trapped in a "magnetic bottle." (c) Earth's magnetic field acts like a bundle of magnetic bottles, trapping particles in Earth's magnetosphere. The bow shock is where the solar wind slows as it meets the magnetosphere. In all these images, the radius of the helix that the charged particle follows is greatly exaggerated.

Regions in the magnetosphere that contain especially strong concentrations of energetic charged particles, called **radiation belts**, can be very damaging to both electronic equipment and astronauts. However, leaving the surface of the planet is not necessary to witness the dramatic effects of the magnetosphere. Disturbances in Earth's magnetosphere caused by changes in the solar wind can lead to changes in Earth's magnetic field that are large enough to trip power grids, cause blackouts, and disrupt communications.

Earth's magnetic field also funnels energetic charged particles down into the ionosphere in two rings located around the magnetic poles. Those charged particles (mostly electrons) collide with atoms and molecules such as oxygen, nitrogen, and hydrogen in the upper atmosphere, causing them to glow like the gas in a neon sign. Interactions with different atoms cause different colors. Those glowing rings, called **auroras**, can be seen from space (**Figure 9.9a**). When viewed from the ground (**Figure 9.9b**), auroras appear as eerie, shifting curtains of multicolored light. People living far from the equator are often treated to spectacular displays of the aurora borealis (the "northern lights") in the Northern Hemisphere or the aurora australis in the Southern Hemisphere. When the solar wind is particularly strong, auroras can be seen at lower latitudes, far from their usual zone. Auroras also have been seen on Venus, Mars, all the giant planets, and some moons.

The general structure we have described here is not limited to Earth's atmosphere. The major vertical structural components—troposphere, tropopause, stratosphere, and ionosphere—also exist in the atmospheres of Venus and Mars, as well as in the atmospheres of Titan and the giant planets. The magnetospheres of the giant planets are among the largest structures in the Solar System.

Wind and Weather

Weather is the local day-to-day state of the atmosphere. Local weather is caused by winds and convection. Recall from Chapter 5 that heating a gas increases its pressure, which causes it to push into its surroundings. Those pressure differences cause winds. Winds are the natural movement of air, both locally and on a global scale, in response to variations in temperature from place to place. The air is usually warmer in the daytime than at night, warmer in the summer than in winter, and warmer at the equator than in the polar regions. Large bodies of water, such as oceans, also affect atmospheric temperatures. The strength of the winds is governed by the size of the temperature difference from place to place.

Recall from Chapter 2 that the effect of Earth's rotation on winds—and on the motion of any object—is called the Coriolis effect (see Figure 2.12). As air in Earth's equatorial regions is heated by the warm surface, convection causes that air to rise. The warmed surface air displaces the air above it, which then has nowhere to go but toward the poles. That air becomes cooler and denser as it moves toward the poles, so it sinks back down through the atmosphere. In the process, it displaces the surface polar air, which is forced back toward the equator, completing the circulation. As a result, the equatorial regions remain cooler and the polar regions remain warmer than they otherwise would be. Air moves between the equator and poles of a planet in a pattern known as **Hadley circulation** (**Figure 9.10a**).

On Earth, other factors break up the planetwide flow into smaller Hadley cells. Most Solar System planets and their atmospheres rotate rapidly enough that the Coriolis effect strongly interferes with Hadley circulation by redirecting the horizontal flow (**Figure 9.10b**). The Coriolis effect creates winds that blow predominantly in an east–west direction and are often confined to relatively narrow bands of latitude. Meteorologists call those **zonal winds**. Planets that rotate

Figure 9.9 Auroras result when particles trapped in Earth's magnetosphere collide with molecules in the upper atmosphere. (a) An auroral ring around Earth's south magnetic pole, as seen from space. (b) Aurora borealis—the "northern lights"—viewed from the ground in Alaska.

Astronomy in Action: Charged Particles and Magnetic Fields

Figure 9.10 (a) Hadley circulation covers an entire hemisphere. (b) On Earth, Hadley circulation breaks up into smaller circulation cells because of the Coriolis effect, which diverts the north–south flow into east–west zonal flow.

faster have a stronger Coriolis effect and stronger zonal winds. Between the equator and the poles in most planetary atmospheres, the zonal winds alternate between winds blowing from the east toward the west (easterlies) and winds blowing from the west toward the east (westerlies).

In Earth's atmosphere, several bands of alternating zonal winds lie between the equator and each hemisphere's pole. That zonal pattern is called Earth's **global circulation** because its extent is planetwide. The best-known zonal currents are the subtropical trade winds—more or less easterly winds that once carried sailing ships from Europe westward to the Americas—and the midlatitude prevailing westerlies that carried them home again.

Embedded within Earth's global circulation pattern are systems of winds associated with large high-pressure and low-pressure regions. A combination of a low-pressure region and the Coriolis effect produces a circulating pattern called **cyclonic motion** (**Figure 9.11**). Cyclonic motion is associated with stormy weather, including hurricanes. Similarly, high-pressure systems are localized regions where the air pressure is higher than average. Owing to the Coriolis effect, high-pressure regions rotate in a direction opposite to that of low-pressure regions. Those high-pressure circulating systems experience **anticyclonic motion** and are generally associated with fair weather.

Earth has a water cycle in which water from the surface enters the air and later returns to the oceans. When liquid water in Earth's oceans, lakes, and rivers absorbs enough thermal energy from sunlight, it turns to water vapor. The water vapor carries that thermal energy as it circulates throughout the atmosphere, releasing the energy to its surroundings when the water vapor condenses back

Figure 9.11 As a result of the Coriolis effect, air circulates around regions of low pressure on the rotating Earth.

into rain or snow. That process powers rainstorms, thunderstorms, hurricanes, and other dramatic weather.

For example, Coriolis forces acting on air rushing into regions of low atmospheric pressure create huge circulating systems that result in hurricanes. The conditions must be just right: warm tropical seawater, light winds, and a region of low pressure in which air spirals inward. Sustained winds near the center of the storm can exceed 300 kilometers per hour (km/h), causing widespread damage and fatalities. Tornadoes are small but violent circulations of air associated with storm systems. Dust devils are similar in structure to tornadoes, but generally smaller and less intense and usually occur in fair weather in places such as the deserts of the American Southwest. Diameters of dust devils range from a few meters to a few dozen meters, with average heights of several hundred meters. The lifetime of a typical tornado or dust devil is brief—usually a dozen or so minutes.

CHECK YOUR UNDERSTANDING 9.3B

All weather and wind on Earth are a result of convection in the: (a) troposphere; (b) stratosphere; (c) mesosphere; (d) ionosphere; (e) thermosphere.

9.4 The Atmospheres of Venus and Mars Differ from Earth's

The atmospheres of Venus, Earth, and Mars are very different (see Table 9.1). The lower atmosphere of Venus is very hot and dense compared with Earth's, whereas the atmosphere of Mars is very cold and thin. The greenhouse effect has turned Venus hellish, with extremely high temperatures and choking amounts of sulfurous gases. Compared with Venus, the surface of Mars is almost hospitable. Understanding why and how those atmospheres are so different helps us understand how Earth's atmosphere may evolve.

Venus

Venus and Earth are similar enough in size and mass that they were once thought of as sister planets. Indeed, when we used the laws of radiation in Chapter 5 to predict temperatures for the two planets, we concluded that they should be very similar. However, spacecraft visits to Venus in the 1960s revealed that the temperature, density, and pressure of Venus's atmosphere were all much higher than for Earth's atmosphere. Ninety-six percent of Venus's massive atmosphere is carbon dioxide, with only 3.5 percent nitrogen and lesser amounts of other gases. Those atmospheric properties are due to the greenhouse effect and the role of carbon dioxide in retaining the infrared radiation typically emitted by a planetary surface. That thick blanket of carbon dioxide traps the infrared radiation, raising the temperature at the surface of the planet to a sizzling 740 K (872°F) (**Figure 9.12**), hot enough to melt lead. The atmospheric pressure at the surface of Venus is 92 times greater than that at Earth's surface: that is equal to the water pressure at an ocean depth of 900 meters—more than enough to crush the hull of a submarine.

The graph in Figure 9.12 shows that the atmospheric temperature of Venus decreases continuously with altitude throughout the planet's troposphere—similar to Earth, dropping to a low of about 160 K at the tropopause. At an altitude of approximately 50 km, Venus's atmosphere has an average temperature and

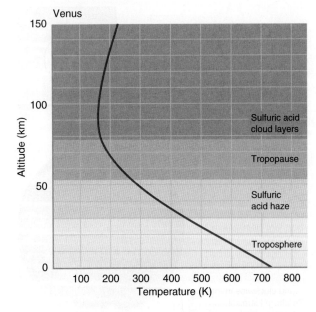

Figure 9.12 The temperature of the atmosphere on Venus primarily decreases as altitude increases, unlike temperatures in Earth's atmosphere, which fall and rise and fall again through the troposphere, stratosphere, and mesosphere (compare Figure 9.6a).

pressure similar to those of Earth's atmosphere at sea level. At altitudes between 50 and 80 km, the atmosphere is cool enough for sulfurous oxide vapors to react with water vapor to form dense clouds of concentrated droplets of sulfuric acid (H_2SO_4). Those dense clouds block the view of the surface of Venus (**Figure 9.13**). Large variations in the observed amounts of sulfurous compounds in the high atmosphere of Venus suggest that the sulfur arises from sporadic episodes of volcanic activity. That finding, along with some hot spots seen near a large shield volcano, strengthens the possibility that Venus is volcanically active.

In the 1960s, radio telescopes and spacecraft with cloud-penetrating radar provided low-resolution views of the surface of Venus. Not until 1975, when the Soviet Union landed cameras there, did scientists get a clear picture of the surface. Those images showed fields of rocks 30–40 centimeters (cm) across and basalt-like slabs surrounded by weathered material. Soviet landers in the 1980s revealed similar landscapes (**Figure 9.14**). Radar images taken by the *Magellan* spacecraft in the early 1990s (see Figure 8.20) produced a global map of the surface of Venus. The high atmospheric temperatures on Venus also mean that neither liquid water nor liquid sulfurous compounds can exist on its surface, leaving an extremely dry lower atmosphere with only 0.01 percent water and sulfur dioxide vapor.

Imagine standing on the surface of Venus. Because sunlight cannot easily penetrate the dense clouds above you, noontime is no brighter than a very cloudy day on Earth. High temperatures and very light winds keep the lower atmosphere free of clouds and hazes. The local horizon can be seen clearly, but strong scattering of light by molecules in the dense atmosphere would greatly soften any view you might have of distant mountains.

Unlike the other Solar System planets, Venus rotates on its axis in a direction opposite to its motion around the Sun. Astronomers call that *retrograde rotation*. Relative to the stars, Venus rotates on its axis once every 243 Earth days. However, a solar day on Venus—the time it takes for the Sun to return to the same place in the sky—is only 117 Earth days. The slow rotation means that Coriolis effects on the atmosphere are small. Global Hadley circulation seldom occurs in planetary atmospheres because other factors, such as planet rotation, break up the planetwide flow into smaller Hadley cells. Because of its slow rotation, however, Venus is an exception, making it the only planet with global circulation close to a classic Hadley pattern (see Figure 9.10a).

The massive atmosphere on Venus transports thermal energy around the planet very efficiently, so the polar regions are only a few degrees cooler than the equatorial regions, with almost no temperature difference between day and night. Because Venus's equator is nearly in the plane of the planet's orbit, seasonal effects are small, producing only negligible changes in surface temperature. Such small temperature variations also mean that wind speeds near the surface of Venus are low, typically about a meter per second, so wind erosion is weaker than that on Earth and Mars. High in the atmosphere, 70 km up, temperature differences are larger, contributing to superhurricane-force winds that reach speeds of 110 m/s (400 km/h), circling the planet in only 4 days. The variation of that high-altitude wind speed with latitude can be seen in the chevron, or V-shaped, cloud patterns.

When the *Pioneer Venus* spacecraft was orbiting Venus during the 1980s, its radio receiver picked up many bursts of lightning static—so many that Venus appears to have a rate of lightning activity comparable to that of Earth. On Venus, as on Earth, lightning is created in the clouds; but Venus's clouds are so high—typically 55 km above the surface of the planet—that the lightning bolts never hit

Figure 9.13 This image from *Akatsuki* shows the thick clouds obscuring the view of the surface of Venus.

Figure 9.14 This image of Venus is from the 1982 Soviet *Venera 14* mission. The spacecraft is in the foreground. Note the rocky ground and the orange sky.

Figure 9.15 This true-color image of the surface of Mars was taken by the rover *Spirit*. Without dust, the sky's thin atmosphere would appear deep blue. In this image, windblown dust turns the sky pinkish.

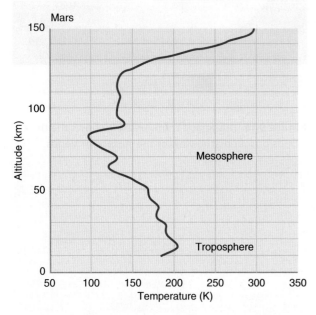

Figure 9.16 The temperature profile of the atmosphere of Mars. Note the differences in temperature and structure between this profile and the profile of the atmosphere of Venus in Figure 9.12.

Figure 9.17 Patches of early-morning water vapor fog forming in canyons on Mars.

the ground. The Japanese Space Agency mission *Akatsuki* has been orbiting Venus since late 2015 to study its weather, lightning, and signs of volcanic activity.

Mars

Mars has a stark landscape, colored reddish by the oxidation of iron-bearing surface minerals. The sky is sometimes dark blue but more often has a pinkish color due to windblown dust (**Figure 9.15**). The lower density of the martian atmosphere makes it more responsive than Earth's atmosphere to heating and cooling, so Mars has greater temperature extremes. The surface near the equator at noontime is a comfortable 20°C—a cool room temperature (68°F) on Earth. However, nighttime temperatures typically drop to a frigid −100°C, and during the polar night the air temperature can reach −150°C—cold enough to freeze carbon dioxide out of the air in the form of a dry-ice frost. The temperature profile of the atmosphere of Mars (**Figure 9.16**) has a range of only 100 degrees up to about 125 km. Above that the temperature rises because sunlight is absorbed in the upper atmosphere. The temperature profile of Mars is more similar to that of Earth than that of Venus.

The average atmospheric surface pressure of Mars is equivalent to the pressure at 35 km above sea level on Earth (well into the stratosphere; see Figure 9.6). Mars has no "sea level" because it has no oceans. Surface pressure varies from 11.5 mb in the lowest impact basins of Mars to 0.3 mb at the summit of Olympus Mons. Earth's pressure at sea level is about 1 bar, so the highest pressure on Mars is only 1.1 percent of that. Like Earth, Mars has some water vapor in its atmosphere, but its low temperatures condense much of the water vapor out as clouds of ice crystals. Mars can have early-morning ice fog in the lowlands (**Figure 9.17**) and clouds hanging over the mountains.

In the absence of plants, Mars has only a tiny trace of oxygen, which is crucial to life on Earth. Like Venus, the atmosphere of Mars is composed almost entirely of carbon dioxide (95 percent) and a lesser amount of nitrogen (2.7 percent). The near absence of oxygen means that Mars has very little ozone. Without ozone, solar UV radiation reaches the surface. Those UV rays could be lethal to any surface life-forms, so any life would either need to develop protective layers or be located away from direct exposure on the surface of the planet, such as in caves or below the surface.

The tilt of the rotation axis of Mars is similar to Earth's at present, so both planets have similar seasons. Seasonal effects on Mars, though, are larger for two reasons. First, the elliptical orbit of Mars has a higher eccentricity, so the annual orbital distance of Mars from the Sun varies more than Earth's does. Second, the low density of the martian atmosphere makes it more responsive to seasonal change. The large daily, seasonal, and latitudinal surface temperature differences on Mars often create locally strong winds—some estimated to be greater than 100 m/s (360 km/h). High winds can stir up huge quantities of dust and distribute it around the planet's surface. For more than a century, astronomers have watched the seasonal development of springtime dust storms on Mars. The stronger storms spread quickly and can envelop the planet in a shroud of dust within a few weeks (**Figure 9.18**). Such large amounts of windblown dust can take many months to settle out of the atmosphere. Seasonal movement of dust from one area to another alternately exposes and covers large areas of dark, rocky surface. That phenomenon led some astronomers of the late 19th and early 20th centuries to believe erroneously that they were witnessing the seasonal growth and decay of vegetation on Mars.

The *Viking* landers first noticed dust devils on Mars in 1976. More recently, *Mars Reconnaissance Orbiter* spotted many dust devils that were visible because of the shadows they cast on the martian surface. **Figure 9.19** shows a 20-km-high, 70-meter-wide dust devil. Most martian dust devils leave dark meandering trails behind them where they have lifted bright surface dust, revealing the dark surface rock that lies beneath. Dust devils on Mars—typically higher, wider, and stronger than those on Earth—reach heights of up to 20 km and have diameters ranging from a few dozen to a few hundred meters.

Mars probably had a more massive secondary atmosphere in the distant past. Geological evidence strongly suggests that liquid water once flowed across its surface (Chapter 8), but the low incidence (or possibly cessation) of volcanism and the planet's low gravity, and perhaps the decrease of its magnetic field, was responsible for the loss of much of that earlier atmosphere. Scientists have not yet reached a consensus on how massive the martian atmosphere was in the past.

Recent analysis of five years of data from instruments on the Mars *Curiosity* Rover have indicated the presence of organic (carbon-based) compounds. These could have formed on the surface with or without life, or come by space on comets or meteors. One of these compounds, methane gas, has been detected in Mars' atmosphere, and varies with the seasons. The methane is thought to have originated on or below the surface of Mars.

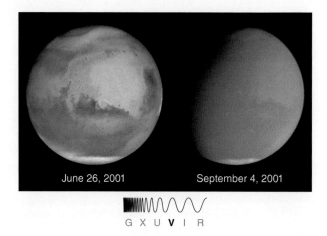

Figure 9.18 Hubble Space Telescope images show the development of a global dust storm that enshrouded Mars in September 2001. The same region of the planet is shown in both images; surface features are obscured by the thick layer of dust.

Mercury and the Moon

The ultrathin atmospheres of Mercury and the Moon are known as **exospheres**, and they are less than a million-billionth (10^{-15}) as dense as Earth's atmosphere. The recent NASA *Lunar Atmosphere and Dust Environment Explorer* (*LADEE*) mission found helium, argon, and dust in the Moon's exosphere. Other atoms, such as sodium, calcium, and even water-related ions, were seen in Mercury's exosphere by the *Messenger* spacecraft, and they may have been blasted loose from Mercury's surface by the solar wind or micrometeoroids. The exospheres of Mercury and the Moon probably vary with the strength of the solar wind and the atoms of hydrogen and helium they capture from it. Exospheres do not affect local surface temperatures, but astronomers study how exospheres interact with the solar wind.

CHECK YOUR UNDERSTANDING 9.4

Rank, from greatest to smallest, the seasonal variations on (a) Mercury, (b) Venus, (c) Earth, and (d) Mars.

9.5 Greenhouse Gases Affect Global Climates

Climate is the *average* state of an atmosphere, including its temperature, humidity, winds, and so on. Climate describes the planet as a whole, over timescales of years or decades. That is an important distinction from weather, which is the state of an atmosphere at any given time and place. The study of climate change on Earth and Mars is not new to the 21st century. Nineteenth-century scientists found evidence of past ice ages and knew that Earth's climate had been very different earlier in its history. Observations of changes in the martian ice caps led to speculation about whether Mars also had ice ages. In this section, we look at the natural factors that can cause climates to change on planets, and we then examine the additional factors that affect Earth.

Figure 9.19 This dust devil on Mars was imaged by the *Mars Reconnaissance Orbiter*.

Factors That Can Cause Climate Change on a Planet

Scientists study the astronomical, geological, and (on Earth) biological mechanisms controlling climate on the planets. Astronomical mechanisms that influence changes in planetary temperature include changes in the Sun's energy output, which has increased slowly as the Sun ages, and possibly changes in the galactic environment as the Sun travels in its orbit around the center of the Milky Way. Scientists have suggested that sporadic bursts of gamma rays or cosmic rays (fast-moving protons) from distant exploding stars could interact with planetary atmospheres. Those mechanisms would affect *all* planets in the Solar System at the same time.

Other astronomical mechanisms relevant to climate change and specific to each planet are the Milankovitch cycles, named for geophysicist Milutin Milanković (1879–1958). Milankovitch cycles may arise because a planet's energy balance can be related to a planet's motion and affected by periodic changes in its orbital eccentricity, the tilt of its rotational axis, and its precession. Recall from Working It Out 5.4 that many factors affect a planet's energy balance and therefore its temperature. If a planet's orbit becomes more eccentric, the amount of energy it receives from the Sun will vary more over its year. If the tilt of a planet increases, its seasons will become more extreme, and its temperature variation during the year will increase. The precession cycle affects which hemisphere is pointed toward the Sun at different times of the elliptical orbit, so that one hemisphere may have longer winters and the other, longer summers.

The tilt (obliquity) of Earth's axis varies from 22.1° to 24.5°, and Earth's relatively large Moon keeps that tilt from changing more than that. In contrast, the moons of Mars are small, and the gravitational influence of Jupiter is a greater factor on Mars. The tilt of Mars is thought to vary from 13° to 40° or possibly more. Given the precession of Mars and its more eccentric orbit, which creates differences in season length between its northern and southern hemispheres, Mars may have had very large swings in its climate throughout its history as the obliquity changed.

A second set of factors that can affect climate is the geological activity of a planet. Volcanic eruptions can produce dust, aerosol particles, clouds, or hazes that block sunlight over the entire globe and lower the temperature. Impacts by large objects can kick up sunlight-blocking particles. Tectonic activity also can affect climate. On Earth, for example, the shifting of the plates has led to different configurations of the oceans and the shifting continents, thus affecting global atmospheric and oceanic circulatory patterns. The albedo of a planet can increase if it has more clouds and ice or can decrease if ice melts or is covered by volcanic ash. Changes also may arise from variations in carbon cycles. On Earth, long-term interactions of the oceans, land, and atmosphere affect the levels of greenhouse gases such as carbon dioxide and water vapor.

A third set of mechanisms that trigger climate changes is biological. Over billions of years on Earth, photosynthesis by bacteria and later by plants removed carbon dioxide from the atmosphere and replaced it with oxygen (Section 9.3). Biological (and geological) activity on Earth can produce methane, a strong greenhouse gas. Certain microorganisms produce methane as a metabolic by-product. For example, bubbles rising to the surface of a stagnant pond—swamp gas—contain biologically produced methane. Methane also is emitted from the guts of grain-fed livestock (and, in the past, from some large dinosaurs) as well as from termites. Another biological effect on climate could come from phytoplankton. If the oceans get more solar energy and warm up because of one of the astronomical mechanisms, the phytoplankton in the ocean may grow faster, leading to the release of more aerosols from sea spray and the formation of more clouds, which increases albedo. Finally, human activities are triggering some major changes, as discussed in the next subsection.

In short, many factors affect the temperature of a planet. Earth's climate is the most complicated of those of the terrestrial planets because Earth is the most geologically and biologically active. How do scientists sort out all those factors? They use the scientific method. Scientists create mathematical models to simulate the general circulation and energy balance of a planet, incorporating all the appropriate factors. The goal is to create a global climate model that reproduces the empirical data from observations of a planet. Once the model correctly predicts past and present climate, it can be used to predict future climate. The first simple climate models for Earth were run on the earliest computers in the 1950s and 1960s. (Fifty years later, scientists are impressed with how well a climate model from 1967 quantified the factors affecting Earth's climate.) One set of models developed at NASA's Goddard Institute for Space Studies in the 1970s was a spinoff of a program originally designed to study Venus. The insights from comparative planetology are important for producing better models that will aid in scientific predictions of climate change on Earth.

Climate Change on Earth

Paleoclimatology is the study of changes in Earth's climate throughout history. Scientists use evidence from geology and paleontology, such as sediments, ice sheets, rocks, tree rings, coral, shells, and fossils, to get data on Earth's past climate. Researchers have found that Earth's climate has lengthy temperature cycles, some lasting hundreds of thousands of years and some tens of thousands of years. As you can see in the middle plot in **Figure 9.20**, Earth has had periods

Figure 9.20 Global variations in carbon dioxide (CO_2), temperature, and methane (CH_4) concentrations over the past 800,000 years of Earth's history. Notice the multiple y-axes on this graph. The axis on the right relates to the temperature data (black); the axes on the left relate to the CO_2 (blue) and CH_4 (red) data. These data sets have been plotted on the same graph to make the similarities and differences easier to see. The low points correspond to ice ages. ppb = parts per billion; ppm = parts per million. 2017 values for CH_4 (red) and CO_2 (blue), indicated by the arrows, are larger than at any time in the past 800,000 years.

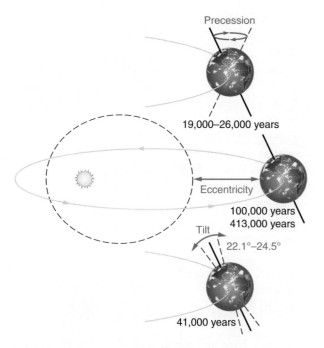

Figure 9.21 The Milankovitch cycles for Earth. The precession of Earth and the rotation of its elliptical orbit combine to yield a cyclic variation of about 21,000 years, its eccentricity varies with two cycles of about 100,000 and 413,000 years, and its tilt varies in cycles of 41,000 years. (Not to scale.)

of colder temperatures, known as ice ages. Earth's atmosphere is so sensitive to even small temperature changes that a drop of only a few degrees in the mean global temperature can plunge the planet's climate into an ice age. Those oscillations in the mean global temperature are far smaller than typical geographic or seasonal temperature variations.

Periodic Milankovitch cycle changes in Earth's orbit correspond to some temperature cycles. As shown in **Figure 9.21**, Earth's axial tilt varies in cycles of 41,000 years, the eccentricity of Earth's orbit varies with two cycles of about 100,000 and 413,000 years, and the time of the year when Earth is closest to the Sun varies in cycles of about 21,000 years. Global climate models using those Milankovitch cycles have replicated much of the observed paleoclimatology data. Temperature changes that are not periodic may have been triggered internally by volcanic eruptions or long-term interactions between Earth's oceans and its atmosphere or by other factors already mentioned.

Some scientists think that at least four of the five major extinction events in Earth's history arose from changes in the level of greenhouse gases. Those mass extinctions occurred 439 million, 364 million, 250 million, and 200 million years ago. The increase in carbon dioxide emitted by active volcanoes, along with methane released from the ground, led to ocean acidification and acid rain and perhaps reduced oxygen in the ocean. Most species died off. An increase in CO_2 from volcanic activity may also have been a factor, along with the impacting asteroid discussed in Chapter 8, for the fifth major extinction, which led to the death of the dinosaurs 65 million years ago.

If Earth's climate has been changing naturally for most of its history, why are scientists especially concerned about the current trend in global climate? Figure 9.20 shows the carbon dioxide levels (top), methane levels (bottom), and temperature (middle) of Earth's atmosphere over the past 800,000 years, obtained from measuring deep ice cores in Antarctica. Notice that those three factors are correlated: when one rises, so do the others. Those data show the naturally occurring ranges since before the first humans existed. The temperature difference between ice ages and interglacial periods is only 10°C–15°C, and those changes are gradual, occurring over tens of thousands of years.

Two major changes have taken place on Earth during the past 150 years. First, the industrial revolution led to an increase in the production of greenhouse gases, especially from the burning of fossil fuels, which releases carbon dioxide into the atmosphere. In 1896, Svante Arrhenius (1859–1927), a Nobel Prize–winning chemist from Sweden, calculated that CO_2 released from burning fossil fuels could increase the greenhouse effect and raise Earth's surface temperature. The second change has been the rapid growth in human population. When populations increase, people burn more forests to clear land for agriculture and industry, reducing the amount of CO_2 absorbed by photosynthesizing plants, increasing CO_2 emissions, and locally changing Earth's albedo. A larger population means more agricultural soil that releases nitrous oxide, more livestock that releases methane, and more people who release carbon dioxide by burning fossil fuels. The data graphed in **Figure 9.22** show that concentrations of those greenhouse gases in the atmosphere have been increasing since the industrial revolution. The CO_2 level has risen and is higher than any of the levels seen in Figure 9.20. Zooming in to more recent times, **Figure 9.23** shows the rise in the level of CO_2 plotted with the average global temperature on Earth to show that they rise together. **Figure 9.24** shows that levels of carbon dioxide, nitrous oxide, and methane have steadily increased over the past decades.

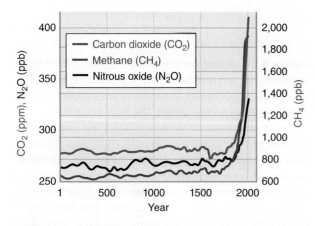

Figure 9.22 Concentrations of greenhouse gases from the year 1 to 2005, showing the increases beginning at the time of industrialization.

9.5 Greenhouse Gases Affect Global Climates 253

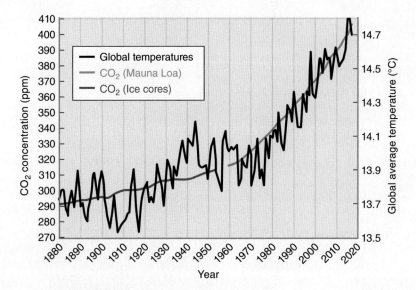

Figure 9.23 Earth's global average temperature and CO_2 concentrations since 1880. This graph shows that global temperatures are climbing along with concentrations of carbon dioxide. Annual variations in atmospheric CO_2 arises from seasonal variations in plant life and fossil fuel use, whereas the overall steady climb is due to human activities.

Nearly all climatologists accept the computer models indicating that this trend represents the beginning of a long-term change in temperature caused by the buildup of human-produced greenhouse gases. That anthropogenic (human-caused) change is happening much faster than the changes seen in Figure 9.20. Even an average increase of a few degrees can change the climate from an ice age to an interglacial period. Earth's atmosphere is a delicately balanced mechanism, and its climate is a complex system within which tiny changes can produce enormous and often unexpected results. To add to the complexity, Earth's climate is intimately tied to ocean temperatures and currents. Ocean currents are critical in transporting energy from one part of Earth to another, and how increased temperatures may affect those systems is uncertain. Warmer oceans evaporate more, leading to wetter air, which can mean more intense summer and winter storms (including more snow). We see examples of that connection in the periodic El Niño and La Niña conditions, in which small shifts in ocean temperature cause much larger global changes in air temperature and rainfall. **Figure 9.25** shows that in the past several decades, however, natural factors have been overwhelmed by the contribution from **anthropogenic climate change**: the release of greenhouse gases into the atmosphere from the burning of fossil fuels and other human activities. Models including only natural causes fail to fit the data as well as models that include both human activities and natural factors. The **Process of Science Figure** discusses how scientists think about such complex issues.

Changes in climate affect where plants and animals can live, and scientists have observed that the dates and locations of breeding, migration, hibernation, and so on, have changed. Agricultural growing seasons and pollination also are affected, as is the availability of freshwater. The melting of mountain glaciers and polar sea ice from the increase in temperature is already being observed. The levels of the oceans are predicted to rise not only from melted ice in Greenland and Antarctica but also from the thermal expansion of the water as the ocean

Figure 9.24 National Oceanic and Atmospheric Administration (NOAA) plots of global average amounts of the major greenhouse gases.

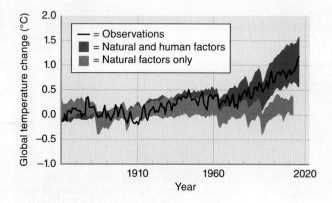

Figure 9.25 Global temperature variations on Earth since 1900, overlaid with model results that include only natural causes of temperature change (pink) and those that include both natural and human factors (purple).

PROCESS OF SCIENCE

Thinking about Complexity

Climate change is a complex scientific issue. When confronted with complex science, there are several questions you should ask.

What is the quality of the evidence?
- How many studies have been done?
- How many different kinds of studies have been done?

Does the basic physics make sense?
- Is it logical?
- Is the explanation natural or supernatural?
- If this claim were not true, what else would that imply?
- If this claim were true, what else would that imply?

Is the claim falsifiable?
- If I hold this in my head as if it were true, what would convince me it was false?
- If I hold this in my head as if it were false, what would convince me it was true?

Scientific issues become remarkably complex when they relate to policy decisions with a broad reach. Keeping an open mind in such cases means thinking carefully about the quality of the evidence, as well as what data would make you change your mind. If your mind can*not* be changed, then you are not making decisions based on evidence.

temperature increases. That is a serious problem for the many people who live in coastal or low-lying regions, and it is already an issue in Miami. A warmer Arctic can lead to a greater release of methane from the permafrost. Less ice and snow can decrease Earth's albedo, allowing more sunlight to reach the surface, although albedo may rise because of an increase in cloud cover, caused by more water in the atmosphere.

The processes are so complex that scientists still cannot predict accurately all the long-term outcomes of small changes that humans are now making to the composition of Earth's atmosphere and to Earth's albedo. In a real sense, we are *experimenting* with Earth. We are asking the question, What happens to Earth's climate if we steadily increase the number of greenhouse molecules in its atmosphere? We do not yet know the full answer, but we are already seeing some of the consequences.

CHECK YOUR UNDERSTANDING 9.5

Over the past 800,000 years, ice-core data indicate which of the following are correlated? (Choose all that apply.) (a) temperature; (b) CO_2 levels; (c) methane levels; (d) the size of the ozone hole

Origins Our Special Planet

Why are we here on Earth instead of elsewhere in the Solar System? The habitable zone of a planetary system (Chapter 7) is the range of distances from its star within which a planet could have a surface temperature such that large amounts of water could exist in liquid form. The temperature of a planet, though, can depend on more than just distance from the star. Temperature can change because of many factors, such as orbital variations, the atmospheric greenhouse effect, and the planet's ability to hold on to its atmosphere. Mercury and the Moon were too small to hold on to either their primary or their secondary atmospheres, leaving them as essentially airless rocks. Venus, Earth, and Mars all had some liquid water at one time, so early in the history of the Solar System all three might have been classified as habitable. Their atmospheres were probably similar then, too, before the atmosphere of Mars escaped, the atmosphere of Venus was heated by the greenhouse effect, and the atmosphere of Earth was oxygenated. Primitive life may even have developed on Venus and Mars at the same time it developed on Earth, about a billion years after the Solar System formed.

Ultimately, the three planets evolved differently, and only Earth now has the liquid water vital for our biology. Venus receives more energy from the Sun and was slightly warmer than the young Earth. The young Venus might have been too hot for liquid water to form oceans, or any liquid water might have quickly evaporated. Because Venus had a lot of water vapor in its atmosphere and lacked a liquid ocean to store carbon dioxide, the greenhouse effect created a thicker atmosphere. That thicker atmosphere fed back to a stronger greenhouse effect, which created an even thicker atmosphere. With that resulting runaway greenhouse effect, Venus became just too hot: the evaporated water molecules broke apart, the hydrogen escaped to space, and any water cycle was destroyed. Primitive bacteria may have lingered in water vapor in the clouds on Venus, but life did not evolve into anything more complex.

Mars is smaller and less massive than Venus or Earth, so it has a weaker gravity (**Figure 9.26**). The red planet has a larger orbit than Venus and Earth and so receives less energy from the Sun. Over time, much of its atmosphere escaped and was not replaced by volcanic emissions, and the atmospheric pressure became too low to maintain liquid water. Whereas Venus was too hot, the young Mars was too cold. Back then it had a thicker atmosphere and liquid rain. Images of the surface of Mars show flood basins, indicating huge rivers in the past. The process that prevented Earth from becoming a Venus-like hothouse continued further on Mars, and the temperature fell until the water froze. Hints of subsurface water exist on Mars, and perhaps some form of martian bacteria will be found beneath the ground.

Thus, our Solar System contains astronomical evidence of how the greenhouse effect can influence planetary atmospheres, including Earth's. Planetary scientists view those other planets as a cautionary tale, showing the results of varied "doses" of greenhouse gases. Only Earth stayed "just right" and could retain the liquid oceans in which more complex life evolved. We owe our lives to the blanket of atmosphere that covers the planet. The study of the Solar System reveals that Earth is maintained by the most delicate of balances. Over billions of years, life has shaped Earth's atmosphere, and today, through the activities of humans, life is reshaping the planet's atmosphere once again. Human civilization is much younger than Earth and has been brief in comparison with the cycles of climate on the planet. The past 10,000 years has been a relatively stable period of Earth's climate: many people argue that this stability enabled agriculture and civilization to develop. Scientists are uncertain what will happen to agriculture, civilization, or the planet itself if the climate undergoes fast, intense changes. The other planets in the Solar System are not places where billions of people from our planet can live. Earth is the only planet suitable for human life.

Figure 9.26 Physical and orbital differences among Venus, Earth, and Mars cause the planets to have very different atmospheres.

READING ASTRONOMY NEWS

ARTICLES | QUESTIONS

The Red Planet was probably a pretty nice place before the Sun's solar winds stripped the atmosphere away and left it to freeze.

NASA's MAVEN Spacecraft Unlocks the Mystery of Mars's Lost Atmosphere

By JAY BENNETT
https://popularmechanics.com/space/moon-mars/a25898/nasa-maven-mars-lost-atmosphere/

Mars is about 4.5 billion years old, and after it formed out of the dust and debris of the early Solar System, it likely had a nice thick atmosphere with clouds that looked a bit like Earth. Temperatures would have been much higher than the current average of about negative 80 degrees Fahrenheit, and liquid water could have flowed over the surface of our planetary neighbor. It is very possible that basic life existed on the surface of Mars sometime in the planet's early history.

But that's all gone now. All that remains is a desolate freezing wasteland with an atmosphere too thin to insulate the planet, and any remaining liquid or life has been driven deep underground. For the last 4 billion years, Mars is thought to have had an ambient surface temperature below freezing, as determined by studying Martian meteorites that make it to Earth. The question is, what happened?

"We've determined that most of the gas ever present in the Mars atmosphere has been lost to space," said Bruce Jakosky in a press release, principal investigator for the Mars Atmosphere and Volatile Evolution Mission (MAVEN) spacecraft run by the Laboratory for Atmospheric and Space Physics (LASP) at the University of Colorado Boulder. A paper detailing the team's latest findings was published in *Science* today, and the research shows that a whopping 65 percent of the argon that was ever in Mars's atmosphere has been blown out into space.

Argon was chosen for the study because it is a noble gas, meaning it virtually never reacts chemically. This is important because solar wind is only one way that a planet's atmosphere can dissipate. Rocks and minerals on the surface can chemically react with the gases of a planet's atmosphere and siphon them away, trapping the gas inside the surface of the planet, for example. Because argon won't react, it can only be stripped away from the planet by high-energy particles blowing from the sun, a process called "sputtering."

After the CU team determined the amount of argon that had been lost due to sputtering, they were able to measure the amount of other gases that were lost to solar winds as well, including CO_2 which makes up 95 percent of the Martian atmosphere. In the early ages of the Solar System, the sun emitted much more volatile ultraviolet radiation and solar winds, and much of the atmosphere was likely blasted away during this time.

There are some wild ideas out there to artificially allow Mars to re-form a thick atmosphere, such as launching a magnetic shield in front of the Red Planet that would block and divert solar particles, an idea recently put forth by NASA Planetary Science Director Jim Green. It might just be possible in the centuries to come, but the first astronauts to walk on Mars are going to have to make sure they keep their spacesuits on tight.

Figure 9.27

ARTICLES | **QUESTIONS**

1. Why did the scientists specifically study argon on Mars?
2. What factors led to Mars losing its atmosphere?
3. What is meant by *volatile ultraviolet radiation*—is that the same as volatile gases?
4. Why would astronauts on Mars need to "keep their spacesuits on tight"?
5. Do a news search for "Mars atmosphere." What are the most recent discoveries?

Source: Jay Bennett, "NASA's MAVEN Spacecraft Unlocks the Mystery of Mars's Lost Atmosphere," PopularMechanics.com, March 31, 2017. Reprinted with permission.

Summary

Earth, Venus, and Mars are warmer than they would be from solar illumination alone. Earth's atmosphere is thick enough to warm the surface to life-sustaining temperatures, but not so thick that Earth becomes overheated. Earth, Venus, and Mars all have significant atmospheres that are different from the original atmospheres they captured when they formed. Their atmospheres are complex, both in chemical composition and in physical characteristics such as temperature and pressure. The climates of Earth, Venus, and Mars are all determined by their individual atmospheres. The atmospheres of Earth, Venus, and Mars have different chemical compositions. They, in turn, led to dramatic differences in temperature and pressure. Life has altered Earth's atmosphere several times, most notably in the distant past from an increase in the amount of oxygen in the atmosphere and in modern times from an increase in greenhouse gases. Mars and Venus might have been habitable early in the history of the Solar System, but now only Earth has substantial liquid water on its surface.

LG 1 **Identify the processes that cause primary and secondary atmospheres to be formed, retained, and lost.** Planetary atmospheres evolve. The primary atmospheres consisted mainly of hydrogen and helium captured from the protoplanetary disk. The terrestrial planets lost their primary atmospheres soon after the planets formed. Secondary atmospheres were created by volcanic gases and from volatiles brought in by impacting comets and asteroids. Planetary bodies must have enough mass to hold on to their atmospheres.

LG 2 **Compare the strength of the greenhouse effect on Earth, Venus, and Mars and how it contributes to differences among the atmospheres of those planets.** Earth, Venus, and Mars have naturally occurring greenhouse gases, which increase the average surface temperature of each planet. The amount by which those greenhouse gases raise the temperature of a planet depends on the number of greenhouse gas molecules in the atmosphere. The differences in global temperatures among those planets can be explained in part by their distances from the Sun. However, their having different compositions and atmospheric densities is a highly significant factor in determining their global temperatures. The atmospheric greenhouse effect keeps Earth from freezing, but it turns Venus into an inferno.

LG 3 **Describe the layers of the atmospheres on Earth, Venus, and Mars.** The atmospheres of Earth, Venus, and Mars have different temperatures, pressures, and compositions. Earth's atmosphere, in particular, has many layers. The layers are determined by the vertical variations in temperature and absorption of solar radiation throughout the atmosphere. Temperature and pressure decrease with altitude in the tropospheres of Earth, Venus, and Mars. Earth's magnetosphere shields the planet from the solar wind. Venus has a massive, hot atmosphere of carbon dioxide and sulfur compounds. Venus has surprisingly fast winds for a slowly rotating planet. Mars has a thin, cold, carbon dioxide atmosphere that may have been much thicker in the past. Earth's atmosphere is thicker than that of Mars but thinner than that of Venus and is composed primarily of nitrogen.

LG 4 **Explain how Earth's atmosphere has been reshaped by the presence of life.** The oxygen levels in Earth's atmosphere have been enhanced through photosynthesis by bacteria and then by plants. Increased oxygen levels led to the formation of the ozone layer and to the development of more advanced forms of life.

LG 5 **Describe the evidence that shows Earth's climate is changing and how comparative planetology contributes to a better understanding of those changes.** Astronomical, geological, and biological processes can lead to large changes in the climate of planets. The study of climate on the terrestrial planets expands scientists' knowledge of Earth's past, present, and future conditions. Large variations in global temperature over the past 800,000 years correlate strongly with the number of greenhouse molecules in the atmosphere. The current level of greenhouse gases in Earth's atmosphere is higher than any seen during that period and correlates with a later increase in temperature.

Unanswered Questions

- Was Venus once hospitable to life? Earlier in the history of the Solar System, the young Sun was fainter and Venus was cooler. One group is using three-dimensional climate simulations to model conditions on a young Venus and concluded Venus could have had moderate temperatures if it was rotating slowly but in a prograde direction. Another group modeled the conditions on early Venus and concluded that Venus may have had an ocean of liquid water early in its history, if it had the right balance of cloud cover and the same amount of carbon dioxide as today. Understanding the differences in the evolution of Venus and Earth is of great interest to astronomers as they detect exoplanets of that size near the edge of their stars' habitable zone.

- Will humans find a way to slow or stop the rise in greenhouse gases on Earth? Climate scientists worry that current changes are abrupt compared with the natural cycles of changes in climate that take place gradually over thousands of years. Will nations reach an agreement to reduce the production of those gases, as they did to reduce the use of chemicals that created the ozone hole?

Questions and Problems

Test Your Understanding

1. Place in chronological order the following steps in the formation and evolution of Earth's atmosphere.
    a. Plant life converts carbon dioxide (CO_2) to oxygen.
    b. Hydrogen and helium are lost from the atmosphere.
    c. Volcanoes, comets, and asteroids increase the inventory of volatile matter.
    d. Hydrogen and helium are captured from the protoplanetary disk.
    e. Oxygen enables the growth of new life-forms.
    f. Life releases CO_2 from the subsurface into the atmosphere.

2. On which of the following planets is the atmospheric greenhouse effect strongest?
    a. Venus
    b. Earth
    c. Mars
    d. Mercury

3. The oxygen molecules in Earth's atmosphere
    a. were part of the primary atmosphere.
    b. arose when the secondary atmosphere formed.
    c. are the result of life.
    d. are being rapidly depleted by the burning of fossil fuels.

4. The differences in the climates of Venus, Earth, and Mars are caused primarily by
    a. the composition of their atmospheres.
    b. their relative distances from the Sun.
    c. the thickness of their atmospheres.
    d. the time at which their atmospheres formed.

5. The words *weather* and *climate*
    a. mean essentially the same thing.
    b. refer to very different timescales.
    c. refer to very different-sized scales.
    d. both b and c

6. Less massive molecules tend to escape from an atmosphere more often than more massive molecules because
    a. the gravitational force on them is less.
    b. they are moving faster.
    c. they are more buoyant.
    d. they are smaller, so they experience fewer collisions on their way out.

7. Venus is hot and Mars is cold primarily because
    a. Venus is closer to the Sun.
    b. Venus has a much thicker atmosphere.
    c. the atmosphere of Venus is dominated by CO_2, but the atmosphere of Mars is not.
    d. Venus has stronger winds.

8. Studying climate on other planets is important to understanding climate on Earth because (select all that apply)
    a. the underlying physical processes are the same on every planet.
    b. other planets offer a variety of extremes to which Earth can be compared.
    c. comparing climates on other planets helps scientists understand which factors are important.
    d. other planets can be used to test atmospheric models.

9. The atmosphere of Mars is often pink-orange because
    a. it is dominated by carbon dioxide.
    b. the Sun is at a low angle in the sky.
    c. Mars has no oceans to reflect blue light to the sky.
    d. winds lift dust into the atmosphere.

10. Auroras are the result of
    a. the interaction of particles from the Sun and Earth's atmosphere and magnetic field.
    b. upper-atmosphere lightning strikes.
    c. the destruction of stratospheric ozone, which leaves a hole.
    d. the interaction of Earth's magnetic field with Earth's atmosphere.

11. The stratospheric ozone layer protects life on Earth from
    a. high-energy particles from the solar wind.
    b. micrometeorites.
    c. ultraviolet radiation.
    d. charged particles trapped in Earth's magnetic field.

12. Hadley circulation is broken into zonal winds by
    a. convection from solar heating.
    b. hurricanes and other storms.
    c. interactions with the solar wind.
    d. the planet's rapid rotation.

13. The _____ of greenhouse gas molecules affects the temperature of an atmosphere.
    a. percentage
    b. fraction
    c. number
    d. mass

14. Over the past 800,000 years, Earth's temperature has closely tracked with
    a. solar luminosity.
    b. oxygen levels in the atmosphere.
    c. the size of the ozone hole.
    d. carbon dioxide levels in the atmosphere.

15. Convection in the _____ causes weather on Earth.
    a. stratosphere
    b. mesosphere
    c. troposphere
    d. ionosphere

Thinking about the Concepts

16. Primary atmospheres of the terrestrial planets were composed almost entirely of hydrogen and helium. Explain why they contained only those gases and not others.

17. How were the secondary atmospheres of the terrestrial planets created?

18. Nitrogen, the principal gas in Earth's atmosphere, was not a significant component of the protostellar disk from which the Sun and planets formed. Where did Earth's nitrogen come from?

19. What are the likely sources of Earth's water?

20. In what way is the atmospheric greenhouse effect beneficial to terrestrial life?

21. In what ways does plant life affect the composition of Earth's atmosphere?

22. What is the difference between ozone in the stratosphere and ozone in the troposphere? Which is a pollutant, and which protects terrestrial life?

23. What is the principal cause of winds in the atmospheres of the terrestrial planets?

24. Global warming appears to be responsible for increased melting of the ice in Earth's polar regions.
    a. Why does the melting of Arctic ice, which floats on the Arctic Ocean, *not* affect the level of the oceans?
    b. How is the melting of glaciers in Greenland and Antarctica affecting the level of the oceans?

25. Why can we not get a clear view of the surface of Venus, as we have so successfully done with the surface of Mars?

26. What is the evidence that the greenhouse effect exists on Earth, Venus, and Mars?

27. Explain why surface temperatures on Venus hardly vary between day and night and between the equator and the poles.

28. Why do scientists think that Mars and Venus were once more habitable but no longer are?

29. The last step in the Process of Science Figure is one that anyone can carry out about any complex issue. Write down your current take on the issue of anthropogenic climate change: do you accept the evidence? Then write down a piece of scientific evidence that would persuade you to change your mind. This exercise may help you to think critically about any issue.

30. Given the current conditions on Venus and Mars, which planet might be easier to engineer to make it habitable to humans? Explain.

Applying the Concepts

31. Study Figure 9.20.
    a. Are the axes linear or logarithmic?
    b. Compare the CO_2 levels (top) with the temperature relative to present (middle). How would you describe the relationship between the graphs?
    c. How much higher is the current CO_2 level than the previous highest value?

32. Study Figure 9.22.
    a. According to the figure, when (approximately) did greenhouse gases begin rising exponentially?
    b. These graphs show that several greenhouse gases have behaved similarly in recent times. Was that true in the past (in general)?
    c. Speculate on possible causes for the common behavior of greenhouse gases in modern times.

33. Why do commercial jet planes fly at the altitudes shown in Figure 9.6b?

34. Commercial jets are pressurized. If you take an unopened bag of chips on a commercial jet airplane, the bag puffs up as you travel to the cruising altitude of 14 km.
    a. Is the pressure in the cabin higher or lower than the pressure on the ground?
    b. If a second bag of chips were attached to the outside of the plane, which bag would puff up more? About how much more (assume an unbreakable bag)? (See Figure 9.6b.)

35. Atmospheric pressure is caused by the weight of a column of air above you pushing down. At sea level on Earth, that pressure is equal to 10^5 newtons per square meter (N/m^2).
    a. Estimate the total force on the top of your head from that pressure. (Note: Force = pressure × area.)
    b. Recall that the acceleration due to gravity is 9.8 m/s^2. If the force in part (a) were caused by a kangaroo sitting on your head, what would the mass of the kangaroo be?
    c. Assume that a typical kangaroo has a mass of 60 kg. How many kangaroos would have to be sitting on your head to equal the mass of the extremely massive kangaroo in part (b)?
    d. Why are you *not* crushed by that astonishing force on your head?

36. Repeat the calculations in question 35 for Venus.

37. Repeat the calculations in question 35 for Mars.

38. Increasing the temperature of a gas inside a closed, rigid container increases the pressure. (That is why you should not put an unopened can of soup directly on the stove!) Explain how Figure 9.6b shows that phenomenon at work in Earth's atmosphere.

39. The total mass of Earth's atmosphere is 5×10^{18} kg. Carbon dioxide (CO_2) makes up about 0.06 percent of Earth's atmospheric mass.
    a. What is the mass of CO_2 (in kilograms) in Earth's atmosphere?
    b. The annual global production of CO_2 is now estimated to be 3×10^{13} kg. What annual fractional increase does that value represent?
    c. The mass of a molecule of CO_2 is 7.31×10^{-26} kg. How many molecules of CO_2 are added to the atmosphere each year?
    d. Why does an increase in CO_2 have such a big effect, even though it represents a small fraction of the atmosphere?

40. Wind's ability to erode the surface of a planet is related in part to the wind's kinetic energy.
    a. Compare the kinetic energy of a cubic meter of air at sea level on Earth (mass, 1.23 kg) moving at a speed of 10 m/s with a cubic meter of air at the surface of Venus (mass, 64.8 kg) moving at 1 m/s.
    b. Compare the kinetic-energy value you determined for Earth in part (a) with that of a cubic meter of air at the surface of Mars (mass, 0.015 kg) moving at a speed of 50 m/s.
    c. Why do you think more evidence of wind erosion on Earth is not apparent?

41. Suppose you seal a rigid container that has been open to air at sea level when the temperature is 0°C (273 K). The pressure inside the sealed container is now equal to the outside air pressure: 10^5 N/m^2.
    a. What would the pressure inside the container be if it were left sitting in the desert shade, where the surrounding air temperature was 50°C (323 K)?
    b. What would the pressure inside the container be if it were left sitting out in an Antarctic night, where the surrounding air temperature was −70°C (203 K)?
    c. What would you observe in each case if the walls of the container were not rigid?

42. Oxygen molecules (O_2) are 16 times as massive as hydrogen molecules (H_2). Carbon dioxide molecules (CO_2) are 22 times as massive as H_2.
    a. Compare the average speed of O_2 and CO_2 molecules in a volume of air.
    b. Does the ratio of the speeds in part (a) depend on air temperature?

43. Calculate the average speed of a carbon dioxide molecule in the atmospheres of Earth and Mars. Compare those speeds with their respective escape velocities. What does that tell you about each planet's hold on its atmosphere?

44. The average surface pressure on Mars is 6.4 mb. Using Figure 9.6, estimate how high you would have to go in Earth's atmosphere to experience the same atmospheric pressure that you would experience if you were standing on Mars.

45. Water pressure in Earth's oceans increases by 1 bar for every 10 meters of depth. Compute how deep you would have to go to experience pressure equal to the atmospheric surface pressure on Venus.

USING THE WEB

46. Look up the data on this year's ozone hole. NASA's "Ozone Watch" website (https://ozonewatch.gsfc.nasa.gov) shows a daily image of southern ozone, as well as animations for current and previous years and some comparative plots. Other comparative plots are available on NOAA's "Meteorological Conditions & Ozone in the Polar Stratosphere" Web page; click on "ozone hole size" (http://cpc.ncep.noaa.gov/products/stratosphere/polar/polar.shtml). At what time of year is the hole the largest, and why? How do the most recent ozone holes compare with previous ones in size and minima? Do they seem to be getting smaller?

47. **Mars:**
    a. Go to https://planetfour.org, a Zooniverse Citizen Science Project in which people examine images of the surface of Mars. Log in or create a Zooniverse account if you don't have one. Read through "About": Where did these data come from? What are the goals of the project? Why is having many people look at the data useful? Read through the sections in "About," and then return to the top page to "Start Exploring." Classify some images.
    b. Go to the website for the *MAVEN* mission, which entered the orbit of Mars in 2014 (http://lasp.colorado.edu/home/maven). What are the scientific goals of the mission? Is that mission a lander, an orbiter, or a flyby? What instruments are on the mission? How will the mission contribute to understanding climate change on Mars? Go to the NASA Web page for *MAVEN* (https://nasa.gov/mission_pages/maven/main/). What are some results of that mission?

48. **Earth:**
    a. Go to the National Snow & Ice Data Center (NSIDC) websites (https://nsidc.org/data/seaice_index/ and http://nsidc.org/arcticseaicenews/). What are the current status and the trend of the Arctic sea ice? How do those compare with previous years? Is anything new reported about Antarctic ice? Qualitatively, how might a change in the amount of ice at Earth's poles affect the albedo of Earth, and how does the albedo affect Earth's temperature?
    b. Go to the website for NASA's Goddard Institute for Space Studies (https://www.giss.nasa.gov), click on "Datasets & Images," and select "Surface Temperature." Click on "Graphs" in the upper right and scroll to "Global Annual Mean Surface Air Temperature Change." The graphs are updated every year. Note that the temperature is compared to a baseline of the average temperature in the period 1951–1980. What has happened with the temperature in the past few years? If the annual mean decreased, does that change the trend? What does the 5-year running mean show? How much warmer is it on average now than in 1880? Click on "News and Feature" (usually from January) to see a video map and report of the previous year's temperatures—how does the map compare with previous years'?
    c. Go to NOAA's "Trend in Atmospheric Carbon Dioxide" Web page on carbon dioxide levels at the observatory on Mauna Loa (https://esrl.noaa.gov/gmd/ccgg/trends/mlo.html). What is the current level of CO_2? How does that value compare with the level from 1 year ago? Since 1960? Why is this a good site for measuring CO_2? What exactly is measured?

49. **Climate change:**
    a. Go to the timeline on the "Discovery of Global Warming" Web page of the American Institute of Physics (https://history.aip.org/history/climate/timeline.htm). When did scientists first suspect that CO_2 produced by humans might affect Earth's temperature? When were other anthropogenic greenhouse gases identified? When did scientific opinions about global warming start to converge? Click on "Venus & Mars" (under 1971): How did observations of those planets add to an understanding of global climate change? Click on "Aerosols" (under 1970): How do those contribute to "global dimming"?
    b. The Fifth Assessment report from the Intergovernmental Panel on Climate Change (IPCC) was released in 2014. Go to the IPCC website section on the 2014 Synthesis report (http://ipcc.ch/report/ar5/syr/) and click through the "Presentation." What are some causes of the increase in warming? What are some effects of warming seen in the polar regions? How are measurements from the past and present used to predict the climate in the future? A report on the effects of 1.5-degree warming was published in 2018 (http://www.ipcc.ch/report/sr15/). What are some of the conclusions of the 2018 report?
    c. Advanced: Go to the website for "Educational Global Climate Modeling," or EdGCM (http://edgcm.columbia.edu). That is a version of the NASA GISS modeling software that will enable you to run a functional three-dimensional global climate model on your computers. Download the trial version and install it on your computer. What can you study with the program? What factors that contribute to global warming or to global cooling on Earth can you adjust in the model? Your instructor may give you an assignment to use this program and the Earth Exploration Toolbook (http://serc.carleton.edu/eet/envisioningclimatechange).

50. **Mars movies:**
    a. Watch a science fiction film about people going to Mars. How does the film handle the science? Can people breathe the atmosphere? Are the low surface gravity and atmospheric pressure correctly portrayed? Do the astronauts have access to water?
    b. At the end of the film *Total Recall* (1990), Arnold Schwarzenegger's character presses an alien button. The martian volcanoes start spewing, and within a few minutes the martian sky is blue, the atmospheric pressure is Earth-like, and the atmosphere is totally breathable. (You may be able to find the scene online.) What, scientifically, is wrong with that scene? That is, why would volcanic gases *not* quickly create a breathable atmosphere on Mars?

EXPLORATION

Climate Change

digital.wwnorton.com/astro6

One prediction about climate change is that as the planet warms, ice in the polar caps and in glaciers will melt. Such melting certainly seems to be occurring in almost all glaciers and ice sheets around the planet. Does that actually matter, and if so, why? In this Exploration, we explore several consequences of the melting ice on Earth.

Experiment 1: Floating Ice

For this experiment, you will need a permanent marker, a translucent plastic cup, water, and ice cubes. Place a few ice cubes in the cup and add water until the cubes float (so they don't touch the bottom). Mark the water level on the outside of the cup with the marker, and label that mark "initial water level."

1 As the ice melts, what do you expect to happen to the water level in the cup?

_____

_____

Wait for the ice to melt completely, and then mark the cup again.

2 What happened to the water level in the cup when the ice melted?

_____

_____

3 From the results of your experiment, what do you predict will happen to global sea levels when the Arctic ice sheet, which floats on the ocean, melts?

_____

_____

Experiment 2: Ice on Land

For this experiment, you will need the same materials as in experiment 1, plus a paper or plastic bowl. Fill the cup about halfway with water and then mark the water level, labeling the mark "initial water level." Poke a hole in the bottom of the bowl and set the bowl over the cup. Add some ice cubes to the bowl.

4 As the ice melts, what do you expect will happen to the water level in the cup?

_____

_____

Wait for the ice to melt completely, and then mark the cup again.

5 What happened to the water level in the cup when the ice melted?

_____

_____

6 In this experiment, the water in the cup is analogous to the ocean, and the ice in the bowl is analogous to ice on land. From the results of your experiment, what do you predict will happen to global sea levels when the Antarctic ice sheet, which sits on land, melts?

_____

_____

Experiment 3: Why Does It Matter?

Search online for the phrase "Earth at night" to find a satellite picture of Earth taken at night. The bright spots on the image trace out population centers. In general, the brighter a spot, the more populous the area (although a confounding factor relates to technological advancement).

7 Where do humans tend to live—near coasts or inland? Coastal regions are, by definition, near sea level. If both the Greenland and Antarctic ice sheets melted completely, sea levels would rise by as much as 80 meters. How would a sea-level rise of a few meters (in the range of reasonable predictions) over the next few decades affect the global population? (Note that each story of a building is about 3 meters.)

_____

_____

_____

10

Worlds of Gas and Liquid— The Giant Planets

Unlike the solid planets of the inner Solar System, the four worlds in the outer Solar System captured and retained gases and volatile materials from the Sun's protoplanetary disk and grew to enormous size and mass. Those planets have dense cores, have very large atmospheres, and rotate faster than Earth.

LEARNING GOALS

By the end of this chapter, you should be able to:

LG 1 Differentiate the giant planets from one another and from the terrestrial planets.

LG 2 Describe the atmosphere of each giant planet.

LG 3 Explain the extreme conditions deep inside the giant planets.

LG 4 Describe the magnetosphere of each giant planet.

LG 5 Compare the planets of our Solar System with those in exoplanetary systems.

This *Juno* image of Jupiter was taken in 2017. The color has been enhanced. The white ovals are part of eight large massive storms in the southern hemisphere. ▶▶▶

What causes the atmospheric features on Jupiter?

TABLE 10.1 Physical Properties of the Giant Planets

Property	Jupiter	Saturn	Uranus	Neptune
Orbital semimajor axis (AU)	5.20	9.6	19.2	30
Orbital period (Earth years)	11.9	29.5	84.0	164.8
Orbital velocity (km/s)	13.1	9.7	6.8	5.4
Mass ($M_{Earth} = 1$)	317.8	95	14.5	17.2
Equatorial radius (km)	71,490	60,270	25,560	24,300
Equatorial radius ($R_{Earth} = 1$)	11.2	9.4	4.0	3.8
Oblateness	0.065	0.098	0.023	0.017
Density (water = 1)	1.33	0.69	1.27	1.64
Rotation period (hours)	9.9	10.7	17.2	16.1
Tilt (degrees)	3.13	26.7	97.8	28.3
Surface gravity ($g_{Earth} = 1$)	2.53	1.07	0.91	1.14
Escape speed (km/s)	59.5	35.5	21.3	23.5

10.1 The Giant Planets Are Large, Cold, and Massive

Collectively, Jupiter, Saturn, Uranus, and Neptune are known as the *giant planets*. They are sometimes called the *Jovian planets*, after Jupiter, the largest of the giant planets. (*Jove* is another name for Jupiter, the highest-ranking deity of ancient Roman mythology.) As with the terrestrial planets, we learn much about the giant planets by comparing them with one another. We begin our discussion of the giant planets by comparing their physical properties and their compositions (**Table 10.1**). Comparative planetology is useful both within planetary groups and between groups. Throughout most of the chapter, we compare giant planets with giant planets, but comparing at least one giant planet with Earth also is useful. For example, you can better understand and appreciate the size of the giant planets if you know that Jupiter is 11 times larger than Earth, with a mass 318 times greater than Earth's. In this section, we examine the physical properties of the giant planets.

Characteristics of the Giant Planets

The giant planets orbit the Sun far beyond the orbits of Earth and Mars. Jupiter, the closest giant planet, is more than 5 astronomical units (AU) from the Sun. At that distance, the Sun is very faint and provides little warmth. From Jupiter, the Sun appears to be a tiny disk, 1/27 as bright as it appears from Earth. At the distance of Neptune, the Sun no longer looks like a disk at all: it appears as a brilliant star about 500 times brighter than the full Moon is in Earth's sky. Daytime on Neptune is only as bright as twilight on Earth. With so little sunlight available for warmth, daytime temperatures hover around 123 kelvins (K) at the cloud tops on Jupiter, and they can dip to just 58 K on Neptune.

Jupiter, Saturn, Uranus, and Neptune are enormous in comparison with the rocky terrestrial planets. Jupiter is the largest of the eight planets and is more than 1/10 the diameter of the Sun. Saturn is slightly smaller than Jupiter, with a diameter of 9.5 Earths. Uranus and Neptune are each about 4 Earth diameters across. **Figure 10.1** shows the actual relative diameters of the giant planets (a) and their apparent relative diameters as seen from Earth (b).

Figure 10.1 (a) Relative sizes of the giant planets, shown to the same physical scale. (b, next page) The same images, scaled according to the relative sizes as they appear from Earth. The full moon is 1,800 arcseconds across in our sky.

10.1 The Giant Planets Are Large, Cold, and Massive

The most accurate method of finding the diameter of a giant planet is to observe a **stellar occultation**, which occurs when the planet eclipses a star in the sky. As shown in **Figure 10.2**, the star disappears behind the planet and then reappears a short while later. Because we know the relative orbital speeds of Earth and the giant planets, we can calculate the size of the eclipsing giant planet from the length of time the star is eclipsed. Occultations of the radio signals transmitted from orbiting spacecraft and images taken by spacecraft cameras also provide accurate measurements of the diameters and shapes of planets and their moons.

Although far away, the giant planets are so large that all but Neptune can be seen with the unaided eye. Jupiter and Saturn were known to the ancients, but Uranus was not discovered until 1781, when William Herschel accidentally noticed a tiny disk in the eyepiece of his 6-inch telescope. At first he thought he had found a comet, but the object's slow nightly motion soon convinced him that it was a planet beyond the orbit of Saturn. During the decades that followed Herschel's discovery, astronomers found that Uranus's position differed from the path predicted from Newton's laws of motion and suggested that the gravitational pull of an unknown planet caused Uranus's surprising behavior. Using the astronomers' measured positions of Uranus, two young mathematicians—Urbain-Jean-Joseph Le Verrier (1811–1877) in France and John Couch Adams (1819–1892) in England—independently predicted the location of the hypothetical planet. The German astronomer Johann Gottfried Galle (1812–1910) found the planet on his first observing night, just where Le Verrier and Adams had predicted it would be. Thus, Galle's discovery of Neptune in 1846 became a triumph for mathematical prediction based on physical law—and for the later confirmation of theory by observation (see the **Process of Science Figure**). Neptune remains the outermost known planet in the Solar System and cannot be seen without binoculars.

Figure 10.2 (a) Occultations occur when a planet, moon, or ring passes in front of a star. (b) As the planet moves (from right to left as seen from Earth), the starlight is blocked. (c) The amount of time that the star is hidden, combined with information about how fast the planet is moving, gives the size of the planet.

PROCESS OF SCIENCE

Scientific Laws Make Testable Predictions

How did astronomers predict that Neptune existed, before it was discovered? Newton's laws of motion and gravity do more than describe what we see. They also enable us to predict the existence of things yet unseen.

- **Uranus is discovered in 1781.**
- *That's odd: Uranus's orbit does not match predictions.*
- *Could another planet's gravitational pull be acting on Uranus?*
- Mathematicians used observations of Uranus and Newton's laws to predict the location of an unknown planet.

Neptune is discovered in 1846.

Newton's laws pass another test!

Laws make predictions that can be tested in order to verify or falsify them. Each test that does not falsify a scientific law strengthens scientists' confidence in its predictions.

The giant planets contain 99.5 percent of all the nonsolar mass in the Solar System. All other Solar System objects—terrestrial planets, dwarf planets, moons, asteroids, and comets—make up the remaining 0.5 percent. Even though Jupiter is only about a thousandth as massive as the Sun, it contains more than twice the mass of all the other planets combined. Jupiter is 318 times as massive as Earth, 3.3 times as massive as Saturn, and about 20 times as massive as either Uranus or Neptune.

The mass of a planet can be calculated by observing the motions and orbital size of a planet's moon. Newton's law of gravitation and Kepler's third law (Chapter 4) together show a relation between the motion of an orbiting object and the mass of the body it is orbiting. Planetary spacecraft now make it possible to measure the masses of planets even more accurately. As a spacecraft flies by a planet, the planet's gravity deflects it. By using several antennae on Earth to track and compare the spacecraft's radio signals, astronomers can detect tiny changes in the spacecraft's path and accurately measure the planet's mass.

Composition of the Giant Planets

The giant planets are made up primarily of gases and liquids. Jupiter and Saturn are composed of hydrogen and helium and are therefore known as **gas giants**. Uranus and Neptune are known as **ice giants** because they both contain much larger amounts of water and other ices than Jupiter and Saturn. On a giant planet, a relatively shallow atmosphere merges seamlessly into a deep liquid ocean, which merges smoothly, in turn, into a denser liquid or solid core. No abrupt transition exists from atmosphere to solid ground, as is found on the terrestrial planets. Although shallower than the liquid layers below, the atmospheres of giant planets are still much thicker than the atmospheres of the terrestrial planets—thousands of kilometers rather than hundreds. As with Venus, only the very highest levels of the atmospheres of the gas giants are visible to us. For Jupiter and Saturn, we see the top of a layer of thick clouds that obscures deeper layers (see the chapter-opening figure and **Figure 10.3a**). Only a few thin clouds are visible on Uranus, although atmospheric models suggest that thick cloud layers must lie below. Neptune displays a few high clouds with a deep, clear atmosphere showing between them (**Figure 10.3b**).

The terrestrial planets (Chapter 8) are composed mostly of rocky minerals, such as silicates, along with various amounts of iron and other metals. Although the atmospheres of the terrestrial planets contain lighter materials, the masses of those atmospheres—and even of Earth's oceans—are insignificant compared with the total planetary masses. The terrestrial planets are the densest objects in the Solar System, with densities ranging from 3.9 times that of water for Mars and 5.5 times for Earth, respectively. In contrast, the giant planets have lower densities because they are composed almost entirely of lighter materials, such as hydrogen, helium, and water. Among the giant planets, Neptune has the highest density, about 1.6 times that of water. Saturn has the lowest density, only 0.7 times that of water. Therefore, Saturn would float—if you had an immobile and deep enough body of water—with 70 percent of its volume submerged. The densities of Jupiter and Uranus are between those of Neptune and Saturn.

Jupiter's chemical composition is similar to that of the Sun—mostly hydrogen and helium. (Recall Figure 5.15, the astronomer's periodic table.) Only a few percent of Jupiter's mass is made up of **heavy elements**, which astronomers

Figure 10.3 (a) Saturn, imaged in visible light by *Cassini*. (b) Neptune, imaged in visible light by *Voyager*.

Figure 10.4 This Hubble Space Telescope image of Saturn was taken in 1999. The oblateness of the planet is apparent when compared with the white circle. The large orange moon Titan appears near the top of the disk of Saturn, along with its black shadow.

define as all elements more massive than helium. Many of those heavy elements combine chemically with hydrogen (H). For example, atoms of oxygen (O), carbon (C), nitrogen (N), and sulfur (S) combine with hydrogen to form molecules of water (H_2O), methane (CH_4), ammonia (NH_3), and hydrogen sulfide (H_2S). More complex combinations produce materials such as ammonium hydrosulfide (NH_4HS). Jupiter's core, which contains most of the planet's iron and silicates, is left over from the original rocky planetesimals around which Jupiter grew. Computer models of the density are required to understand the compositions deep in the cores of the giant planets.

The principal compositional differences among the four giant planets lie in the amounts of hydrogen and helium that each one contains. Because of its larger mass, Jupiter accumulated more hydrogen and helium when it formed than the other planets did. Saturn contains more heavy elements and less hydrogen and helium than Jupiter. Heavy elements are significant components of Uranus and Neptune. Methane is a particularly important molecule in the atmospheres of those two planets, giving them their characteristic blue-green color. Those differences in composition are important clues to understanding how the giant planets formed.

Days and Seasons on the Giant Planets

The giant planets rotate rapidly (Table 10.1), so their days are short (ranging from 10 to 17 hours) and their shapes are distorted from perfectly spherical to oblate—they bulge at their equators and have an overall flattened appearance. Saturn's appearance is noticeably oblate (**Figure 10.4**): its equatorial diameter is almost 10 percent greater than its polar diameter. By contrast, Earth's polar and equatorial diameters differ by only 0.3 percent.

The intensity of a planet's seasons is determined by the tilt of its axis (Chapter 2). With a tilt of only 3°, for example, Jupiter has almost no seasons at all. The tilts of Saturn (27°) and Neptune (28°), which are slightly larger than those of Earth (23.5°) and Mars (25°), cause moderate but well-defined seasons. Uranus spins on an axis, which lies nearly in the plane of its orbit—its tilt is about 98°. Uranus's high tilt causes its seasons to be extreme, with each polar region alternately experiencing 42 years of continuous sunshine followed by 42 years of total darkness. Averaged over an entire orbit, the poles receive more sunlight than the equator—a situation markedly different from that of any other Solar System planet.

Viewed from Earth, Uranus appears to be either spinning face-on or rolling along on its side (or something between), depending on where Uranus happens to be in its orbit. A tilt greater than 90° indicates that the planet rotates clockwise when seen from above its orbital plane. Why is Uranus tilted so differently from most other planets? One possible explanation is that Uranus was "knocked over" by the impact of one huge or several large planetesimals near the end of its accretion phase. Venus, Pluto, Pluto's moon Charon, and Neptune's moon Triton also have tilts greater than 90°.

CHECK YOUR UNDERSTANDING 10.1

How are Uranus and Neptune different from Jupiter and Saturn? (a) Uranus and Neptune have a higher percentage of ices in their interiors; (b) Uranus and Neptune have more hydrogen; (c) Uranus and Neptune have no storms; (d) Uranus and Neptune are closer to the Sun.

10.2 The Giant Planets Have Clouds and Weather

When we observe the giant planets through a telescope or in visible images from a spacecraft, we are seeing only the top layers of the atmosphere. Sometimes we can see a bit deeper into the clouds, but in essence, we are seeing a two-dimensional view of the cloud tops. The existence of deeper cloud layers on those giant planets is inferred from physical models of temperature as a function of depth. In this section, we explore the atmospheres of the giant planets.

Viewing the Cloud Tops

Even when viewed through small telescopes, Jupiter is very colorful. Parallel bands—ranging in hue from bluish gray to various shades of orange, reddish brown, and pink—stretch out across its large, pale yellow disk. Astronomers call the darker bands *belts* and the lighter bands *zones*. Many clouds—some dark and some bright, some circular and others more oval—appear along the edges of, or within, the belts. The most prominent cloud structure is a large, red, oval feature in Jupiter's southern hemisphere known as the **Great Red Spot** (**Figure 10.5**).

Figure 10.5 A digitally enhanced image of Jupiter taken by *Juno*. The Great Red Spot is a hurricane about 1.3 times the diameter of Earth.

The Great Red Spot was first observed more than three centuries ago, shortly after the telescope was invented. Since then, it has varied unpredictably in size, shape, color, and motion as it drifts among Jupiter's clouds. In the 1800s, the Great Red Spot was greater than twice the diameter of Earth, but now it has shrunk to 1.3 Earth diameters. Observations of small clouds circling the perimeter of the Great Red Spot show that it is an enormous atmospheric whirlpool, swirling counterclockwise with a period of about a week. Its cloud pattern looks a lot like that of a terrestrial hurricane, but it rotates in the opposite direction—exhibiting *anticyclonic* rather than cyclonic flow, indicating a high-pressure system. Data from Juno suggest that the roots of the Great Red Spot go at least 300 kilometers (km) into the atmosphere, about 20 times more than the largest storms on Earth.

Comparable whirlpool-like behavior is observed in many of the smaller oval-shaped clouds found elsewhere in Jupiter's atmosphere and in similar clouds observed in the atmospheres of Saturn and Neptune. Whirlpool-like, swirling features are known as vortices (the singular is "vortex"). Those vortices are familiar to us on Earth as high- and low-pressure systems, hurricanes, and supercell thunderstorms.

Jupiter's vortices are so complex that scientists still do not fully understand how they interact with one another. *Voyager 2* observed several Alaska-sized clouds being swept into the Great Red Spot. Some of those clouds were carried around the vortex a few times and then ejected, whereas others were swallowed up and never seen again. Other smaller clouds with structure and behavior similar to that of the Great Red Spot are seen in Jupiter's middle latitudes.

The *Juno* mission entered an elongated orbit around Jupiter in 2016. The spacecraft uses infrared and microwave instruments to further analyze the atmosphere. When closest to Jupiter, *Juno* is only a few thousand kilometers above the cloud tops. *Juno* has shown that the clouds are present even in Jupiter's polar regions and that the depths of the clouds vary greatly. For example, *Juno* observed a 4,000-km-wide cyclone near the north pole of Jupiter, with a ring of eight smaller cyclones surrounding it (**Figure 10.6**). In the southern polar region, five smaller ones circle the large cyclone.

Figure 10.6 *Juno* image of cyclones at the north pole of Jupiter. The colors represent heat: the thinner yellow clouds are about −13°C, and the dark red thicker clouds are about −118°C.

Figure 10.7 Saturn images from *Cassini*. (a) A northern hemisphere 2010 storm in Saturn's atmosphere is shown in true color. (b) The eye of this storm in Saturn's north polar region is about 2,000 km across and has wind speeds up to 150 meters per second. The false colors represent three near-infrared wavelengths; red shows low clouds and green indicates higher clouds.

Figure 10.8 The ground-based Keck telescope image of Uranus (a) and a 2016 Hubble Space Telescope (HST) image of Neptune (b) were taken at wavelengths of light that are strongly absorbed by methane. The visible clouds are high in the atmosphere. (c) The Great Dark Spot on Neptune disappeared between the time *Voyager 2* flew by Neptune in 1989 and the time earlier HST images were obtained in 1994, but a new spot has now appeared.

Because Saturn is farther away than Jupiter and smaller in radius, from Earth Saturn appears less than half as large as Jupiter (see Figure 10.1b). Saturn also displays atmospheric bands, but they tend to be wider than those on Jupiter, with much more subdued colors and contrasts. The largest atmospheric features on Saturn are about the size of the continental United States, but many are smaller than terrestrial hurricanes. Close-up views from the *Cassini* spacecraft show immense lightning-producing storms in a region of Saturn's southern hemisphere known as "storm alley." In December 2010 a large storm appeared (**Figure 10.7a**) that was visible in even small amateur telescopes. Individual clouds are not seen often on Saturn, so that was an unusual event. That large storm eventually wrapped itself around the planet. *Cassini* also discovered a spinning vortex at Saturn's northern polar region (**Figure 10.7b**), but it is not known how long it has been active.

Through most telescopes on Earth, Uranus and Neptune look like tiny, featureless, pale bluish green disks. With the largest ground-based telescopes or the telescopes in space, however, optical and infrared imaging reveals several individual clouds and belts. Images (**Figure 10.8a**) show atmospheric bands and small clouds suggestive of those seen on Jupiter and Saturn, but more subdued. Methane's strong absorption of reflected sunlight causes the atmospheres of Uranus and Neptune to appear dark in the near infrared, allowing the highest clouds and bands to stand out against the dark background.

Several bright cloud bands appear in the Hubble Space Telescope (HST) image of Neptune's atmosphere (**Figure 10.8b**). Located near the planet's tropopause, those cloud bands cast their shadows downward through the clear upper atmosphere onto a dense cloud layer 50 km below. A large, dark, oval feature seen in the southern hemisphere first observed in images taken by *Voyager 2* in 1989 reminded astronomers of Jupiter's Great Red Spot, so they called it the Great Dark Spot (**Figure 10.8c**). However, the Neptune feature was gray rather than red, and it changed in length and shape more rapidly than the Great Red Spot. When HST observed Neptune in 1994, the Great Dark Spot had disappeared. A different spot of comparable size first seen in 2016 has continued to grow (Figure 10.8b), indicating that such storms come and go.

The Structure below the Cloud Tops

As we saw in discussing the terrestrial planets, atmospheric temperature, density, pressure, and even chemical composition vary with height and over horizontal distances. As a rule, atmospheric temperature, density, and pressure all decrease with increasing altitude, although temperature is sometimes higher at very high altitudes, as in Earth's thermosphere. The stratospheres above the cloud tops of the giant planets appear relatively clear, but closer inspection shows that they contain layers of thin haze that show up best when seen in profile above the edges of the planets. The composition of the haze particles remains unknown, but they may be smoglike products created when ultraviolet sunlight acts on hydrocarbon gases such as methane.

Water is the only substance in Earth's lower atmosphere that can condense into clouds, but the atmospheres of the giant planets have a much larger range of temperatures and pressures, so more kinds of volatiles (materials that become gases at moderate temperatures) can condense and form clouds. **Figure 10.9** shows how the ice layers are stacked in the tropospheres of the giant planets. Because each kind of volatile, such as water or ammonia, condenses at a particular temperature and pressure, each forms clouds at a different altitude. Convection carries volatile materials upward along with other atmospheric gases, and when each particular volatile reaches an altitude with its condensation temperature, most of that volatile condenses and separates from the other gases, so very little of it is carried higher aloft. Those volatiles form dense layers of cloud separated by regions of relatively clear atmosphere.

The farther the planet is from the Sun, the colder its troposphere will be. Therefore, the distance from the Sun determines the altitude at which a particular volatile, such as ammonia or water, will condense to form a cloud layer on each planet. If temperatures are too high, some volatiles may not condense at all. The highest clouds in the frigid atmospheres of Uranus and Neptune are crystals of methane ice. The highest clouds on Jupiter and Saturn are made up of ammonia ice. Methane never freezes to ice in the warmer atmospheres of Jupiter and Saturn.

In 1995, an atmospheric probe on the *Galileo* spacecraft descended slowly via parachute into the atmosphere of Jupiter. Near the top of Jupiter's troposphere at a temperature of about 130 K (about −140°C), the probe found that ammonia had condensed. Next it found a layer of ammonium hydrosulfide clouds at a temperature of about 190 K (about −80°C). Soon after descending to an atmospheric pressure of 22 bars and a temperature of about 373 K (100°C), the *Galileo* probe failed, presumably because its transmitter got too hot. In 2017, *Juno* detected a deep plume of ammonia arising from the depths near Jupiter's equator.

Why are some clouds so colorful, especially Jupiter's? In their purest form, the ices that make up the clouds of the giant planets are all white, similar to snow on Earth. The colorful tints and hues must come from impurities in the ice crystals. Those impurities are elemental sulfur and phosphorus, as well as various organic materials produced when ultraviolet sunlight breaks up hydrocarbons such as methane, acetylene, and ethane. The molecular fragments can then recombine to form complex organic compounds that condense into solid particles, many of which are colorful. Such reactions also occur in Earth's atmosphere. Some of the photochemical products produced close to the ground on Earth are called *smog*.

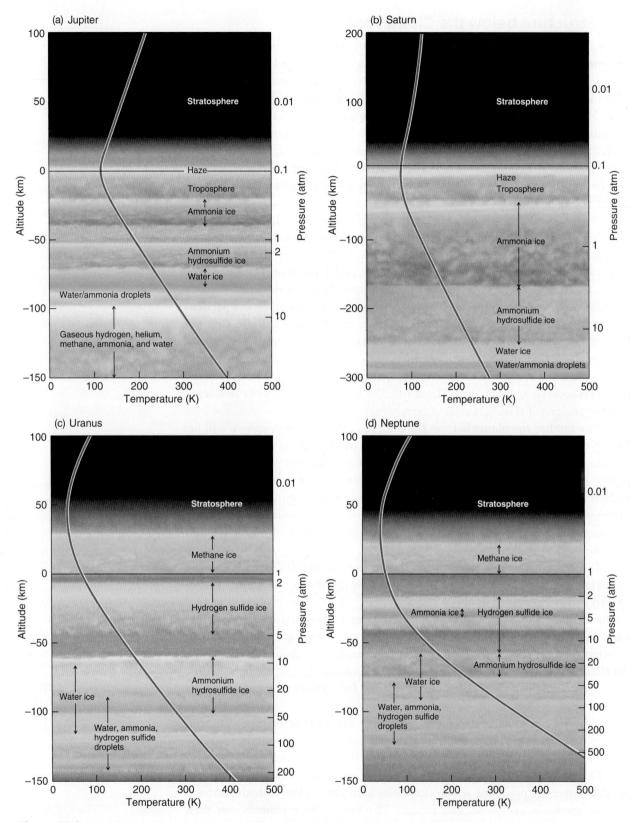

Figure 10.9 Volatile materials condense at different levels in the atmospheres of the giant planets, leading to chemically different types of clouds at different depths in the atmospheres. The red line in each diagram shows how atmospheric temperature changes with height. Because those planets have no solid surface, the zero point of altitude is arbitrary. Here, the arbitrary zero points of altitude are at 0.1 atmosphere (atm) for Jupiter (a) and Saturn (b) and at 1.0 atm for Uranus (c) and Neptune (d). The value of 1.0 atm corresponds to the atmospheric pressure at sea level on Earth. Saturn's altitude scale is compressed to better show the layered structure.

The atmospheric composition of Uranus and Neptune gives those planets their bluish green color. The upper tropospheres of Uranus and Neptune are relatively clear, with only a few white clouds that are probably composed of methane ice crystals. Methane gas is much more abundant in the atmospheres of Uranus and Neptune than in those of Jupiter and Saturn. Like water, methane gas tends to selectively absorb the longer wavelengths of light—yellow, orange, and red. Absorption of the longer wavelengths leaves only the shorter wavelengths—green and blue—to be scattered from the atmospheres of Uranus and Neptune.

Winds and Weather

On the giant planets, the thermal energy that drives convection comes both from the Sun and from the hot interiors of the planets themselves. Convection results from vertical temperature differences (Chapter 9). As heating drives air up and down, the Coriolis effect shapes that convection into atmospheric vortices, visible as isolated circular or oval cloud structures, such as the Great Red Spot on Jupiter and the Great Dark Spot on Neptune. As the atmosphere rises near the center of a vortex, it expands and cools. Cooling condenses certain volatile materials into liquid droplets, which then fall as rain. As they fall, the raindrops collide with surrounding molecules, stripping electrons from the molecules and thereby developing tiny electric charges in the air. The cumulative effect of countless falling raindrops can be an electric field so great that it ionizes the molecules in the atmosphere and creates a surge of current and a flash of lightning. A single observation of Jupiter's night side by *Voyager 1* revealed several dozen lightning bolts within an interval of 3 minutes. *Cassini* also has imaged lightning flashes in Saturn's atmosphere, and radio receivers on *Voyager 2* picked up lightning static in the atmospheres of both Uranus and Neptune.

The giant planets have much stronger zonal winds than those of the terrestrial planets. Because the giant planets are farther from the Sun, less thermal energy is available. They rotate rapidly, however, making the Coriolis effect very strong. In fact, the Coriolis effect is more important than atmospheric temperature patterns in determining the structure of the global winds. If we know the radius of the planet, we can find out how fast the features are moving, as shown in **Working It Out 10.1**, and find rotation speeds and wind speeds. **Figure 10.10** shows the wind speeds at various latitudes on the different planets.

JUPITER On Jupiter (Figure 10.10a), the strongest winds are equatorial, blowing from the west, at speeds up to 550 kilometers per hour (km/h). At higher latitudes, the winds alternate between blowing from the west or east in a pattern that might be related to Jupiter's banded structure. Near a latitude of 20° south, the Great Red Spot vortex appears to be caught between a pair of winds blowing from the west and east with opposing speeds of more than 200 km/h—indicating a relationship between zonal flow and vortices. Data from *Juno* suggest that the jet stream goes 3,000 km down into the atmosphere, much deeper than expected. The jet stream layer rotates differently according to latitude and contains about 1 percent of the mass of the planet.

SATURN The equatorial winds on Saturn (Figure 10.10b) also blow from the west but are stronger than those on Jupiter. The maximum wind speeds at any given time vary between 990 and 1,650 km/h. Saturn's winds appear to decrease with height in the atmosphere, so the apparent time variability of Saturn's equatorial winds may be nothing more than changes in the height of the cloud tops.

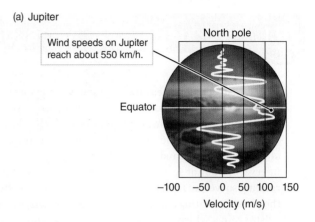

(a) Jupiter

Wind speeds on Jupiter reach about 550 km/h.

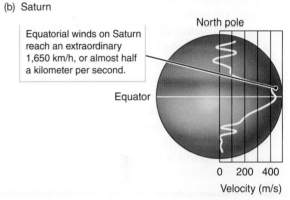

(b) Saturn

Equatorial winds on Saturn reach an extraordinary 1,650 km/h, or almost half a kilometer per second.

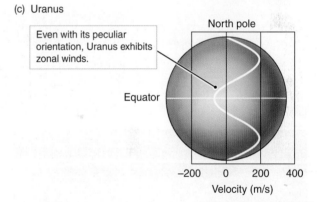

(c) Uranus

Even with its peculiar orientation, Uranus exhibits zonal winds.

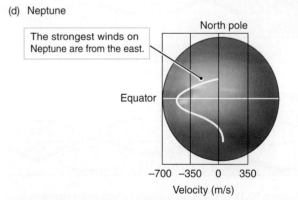

(d) Neptune

The strongest winds on Neptune are from the east.

Figure 10.10 The speed of the wind at various latitudes on the giant planets is shown by the white line. When the speed is positive (to the right of zero), the wind is from the west. When the speed is negative (to the left of zero), the wind is from the east.

10.1 Working It Out Measuring Wind Speeds on Different Planets

How do astronomers measure wind speeds on planets so far away? If we can see individual clouds in their atmospheres, we can measure their winds. As on Earth, clouds are carried by local winds. The local wind speed can be calculated by measuring the positions of individual clouds and noting how much they move during an interval of a day or so. We also need to know how fast the planet is rotating so we can measure the speed of the winds with respect to the planet's rotating surface. For the giant planets, no solid surface exists against which to measure the winds. Scientists must instead assume a hypothetical surface—one that rotates as though it were somehow "connected" to the planet's deep interior. Periodic bursts of radio energy caused by the rotation of a planet's magnetic field tell us how fast the interior of the planet is rotating.

If we know the radius of the planet, we can find the speed of the features. Consider a small white cloud in Neptune's atmosphere. The cloud, on Neptune's equator, is observed to be at longitude 73.0° west on a given day. (That longitude system is anchored in the planet's deep interior and already takes into account Neptune's rotation.) The spot is then seen at longitude 153.0° west exactly 24 hours later.

Neptune's equatorial winds have carried the white spot 80.0° in longitude in 24 hours.

The circumference (C) of a planet is given by $2\pi r$, where r is the equatorial radius. The equatorial radius of Neptune is 24,760 km, so Neptune's circumference is

$$C = 2\pi r = 2\pi \times 24{,}760 \text{ km} = 155{,}600 \text{ km}$$

The full circle represented by the circumference has 360° of longitude. The spot has moved 80°/360° of the circumference and thus has traveled

$$\frac{80}{360} \times 155{,}600 \text{ km} = 34{,}580 \text{ km}$$

in 24 hours. Therefore, the wind speed is given by

$$\text{Speed} = \frac{\text{Distance}}{\text{Time}} = \frac{34{,}580 \text{ km}}{24 \text{ h}} = 1{,}440 \text{ km/hr} = 400 \text{ m/s}$$

The equatorial winds are very strong and are blowing in a direction opposite to the planet's rotation. On Earth, the much slower equivalents of those winds are called *trade winds* (see Figure 9.10).

Figure 10.11 (a) This jet stream in Saturn's northern hemisphere, imaged by *Voyager*, is similar to jet streams in Earth's atmosphere. (b) The jet stream dips toward the equator below regions of low pressure and is forced toward the pole above regions of high pressure.

Alternating winds blowing from the east or west also occur at higher latitudes; but unlike on Jupiter, that alternation seems to bear no clear association with Saturn's atmospheric bands. That example is just one of the many unexplained differences among the giant planets.

Saturn's jet stream at latitude 45° north (**Figure 10.11a**) is a narrow meandering river of atmosphere with alternating crests and troughs. **Figure 10.11b** shows how the jet stream curves around regions of high and low pressure to create a wavelike structure. That behavior is similar to Earth's jet streams, in which high-speed winds blow generally from west to east but wander toward and away from the poles. Nested within the crests and troughs of Saturn's jet stream are anticyclonic and cyclonic vortices. They appear similar in both form and size to terrestrial high- and low-pressure systems, which bring alternating periods of fair and stormy weather.

URANUS Less is known about global winds on Uranus (Figure 10.10c) than about those on the other giant planets. When *Voyager 2* flew by Uranus in 1986, the few visible clouds were in the southern hemisphere because the northern hemisphere was in complete darkness at the time. The strongest winds observed were 650 km/h from the west in the middle to high southern latitudes, and no winds from the east were seen. Because Uranus's peculiar orientation makes its poles warmer than its equator, some astronomers had predicted that the global wind system of Uranus might differ greatly from that of the other giant planets. But *Voyager 2* observed that the Coriolis forces dominate on Uranus as they do on other planets, so the dominant winds on Uranus are zonal, just as on the other giant planets.

As Uranus moved along in its orbit, regions previously unseen by modern telescopes have become visible (**Figure 10.12**). Observations by HST and ground-based telescopes showed bright cloud bands in the far north extending more than 18,000 km and revealed wind speeds of up to 900 km/h. As Uranus approaches northern summer solstice in the year 2027, much more about its northern hemisphere will be learned.

NEPTUNE On Neptune (Figure 10.10d), the southern hemisphere's summer solstice occurred in 2005, so much of the north is still in darkness. Observers will have to wait until Neptune's equinox in 2045 to see its northern hemisphere. The strongest winds on Neptune occur in the tropics, similar to winds on Jupiter and Saturn. On Neptune, however, the winds are from the east rather than from the west, with speeds in excess of 2,000 km/h. Winds from the west with speeds higher than 900 km/h have been seen in Neptune's south polar regions. With wind speeds 5 times greater than those of the fiercest hurricanes on Earth, Neptune and Saturn are the windiest planets known.

CHECK YOUR UNDERSTANDING 10.2

Why does Jupiter appear reddish in color? (a) It is very hot; (b) its atmosphere consists of reddish chemical compounds; (c) it is moving very quickly; (d) it is rusty, like Mars.

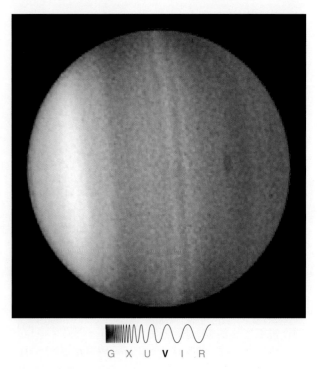

Figure 10.12 Uranus is approaching equinox in this 2006 Hubble Space Telescope image. Much of its northern hemisphere is becoming visible. The dark spot in the northern hemisphere (to the right) is similar to but smaller than the Great Dark Spot seen on Neptune in 1989.

10.3 The Interiors of the Giant Planets Are Hot and Dense

At the center of each giant planet is a dense, liquid core consisting of a very hot mixture of heavier materials such as water, molten rock, and metals. **Figure 10.13** illustrates the interior structure of the giant planets. As you can see, the gas giants differ from the ice giants in their amounts of hydrogen and ices. We look at each in turn.

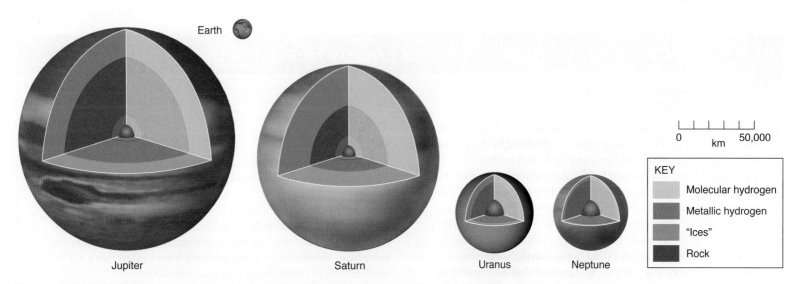

Figure 10.13 The interiors of the giant planets have central cores and outer liquid shells. Only Jupiter and Saturn have significant amounts of the molecular and metallic forms of liquid hydrogen surrounding their cores.

The Cores of Jupiter and Saturn

The overlying layers of Jupiter and Saturn press down on the liquid core, raising its temperature. For example, the pressure at Jupiter's core is about 45 million bars, and a pressure that high heats the fluid to 35,000 K. Central temperatures and pressures of the other, less massive giant planets are correspondingly lower than those of Jupiter. Water is still liquid at those enormously high temperatures because the extremely high pressures at the centers of the giant planets prevent water from turning to steam.

The internal energy that lies deep within the giant planets is left over from their formation. The giant planets are still contracting and converting their gravitational potential energy into thermal energy today as they did when they first formed, but they are doing it more slowly. The annual amount of contraction necessary to sustain their internal temperature is only a tiny fraction of their radius. Jupiter, for example, would need to contract by a bit less than 2 centimeters per year. The thermal energy from the core drives convection in the atmosphere and eventually escapes to space as radiation (**Working It Out 10.2**). For Saturn, moreover, and perhaps for Jupiter, liquid helium separates from a hydrogen-helium mixture under the right conditions and rains downward toward the core. As the droplets of liquid helium sink, they release their gravitational potential energy as thermal energy. Planetary physicists think that most of Saturn's internal energy and perhaps some of Jupiter's internal energy come from that separation of liquid helium. The continuous production of thermal energy is enough to replace the energy escaping from their interiors.

The pressure within the atmospheres of Jupiter and Saturn increases with depth because overlying layers of atmosphere press down on lower layers. At depths of a few thousand kilometers, the atmospheric gases of Jupiter and Saturn are so compressed by the weight of the overlying atmosphere that they turn to liquid. That depth roughly marks the lower boundary of the atmosphere. The difference between a liquid and a highly compressed, very dense gas is subtle, so Jupiter and Saturn have no clear boundary between the atmosphere and the ocean of liquid that lies below. Jupiter's atmosphere is about 20,000 km deep, and Saturn's atmosphere is about 30,000 km deep; at those depths, the pressure

---

10.2 Working It Out Internal Thermal Energy Heats the Giant Planets

The equilibrium between the absorption of sunlight and the radiation of infrared light into space was introduced in Chapter 5. How the resulting equilibrium temperature is modified by the greenhouse effect on Venus, Earth, and Mars was explained in Chapter 9. When that equilibrium is calculated for the giant planets, however, it doesn't match the measurements. According to those calculations, the equilibrium temperature for Jupiter should be 109 K, but the average temperature turns out to be 124 K. A difference of 15 K might not seem like much, but the energy radiated by an object depends on its temperature raised to the fourth power (the Stefan-Boltzmann law; Chapter 5). Applying that relationship to Jupiter, we get:

$$\left(\frac{T_{\text{actual}}}{T_{\text{expected}}}\right)^4 = \left(\frac{124\,\text{K}}{109\,\text{K}}\right)^4 = 1.67$$

That result implies that Jupiter is radiating roughly two-thirds more energy into space than it absorbs from sunlight. Similarly, the internal energy escaping from Saturn is observed to be about 1.8 times greater than the sunlight that it absorbs. Neptune emits 2.6 times more energy than it absorbs from the Sun. Those planets, then, are *not* in equilibrium; they are slowly contracting and thus generating more heat. The internal energy escaping from Uranus turns out to be smaller than the solar energy the planet absorbs.

climbs to 2 million bars and the temperature reaches 10,000 K (hotter than the surface of the Sun). Under those conditions, hydrogen molecules are battered so violently that their electrons are stripped free, and the hydrogen acts like a liquid metal. In that state, it is called metallic hydrogen. Those oceans of hydrogen and helium are tens of thousands of kilometers deep.

Differentiation has occurred and is still occurring in Saturn and perhaps in Jupiter. On Saturn, helium condenses out of the hydrogen-helium oceans. Helium also can be compressed to a metal, but it does not reach that metallic state under the physical conditions of the interiors of the giant planets. Because those droplets of helium are more dense than the hydrogen-helium liquid in which they condense, they sink toward the center of the planet, converting gravitational energy to thermal energy. That process heats the planet and enriches the helium concentration in the core while depleting it in the upper layers. In Jupiter's hotter interior, by contrast, the liquid helium is mostly dissolved together with the liquid hydrogen.

The heavy-element components of the cores of Jupiter and Saturn have masses of about 5–20 Earth masses. Jupiter and Saturn have total masses of 318 and 95 Earth masses, respectively. The heavy materials in their cores contribute little to their average chemical composition. Therefore, Jupiter and Saturn have approximately the same composition as the Sun and the rest of the universe: about 98 percent hydrogen and helium, leaving only 2 percent for everything else.

The Cores of Uranus and Neptune

Uranus and Neptune are less massive than Jupiter and Saturn, have lower interior pressures, and contain smaller fractions of hydrogen—their interiors probably contain only a small amount of liquid hydrogen, with little or none in a metallic state. Uranus and Neptune are made of denser material than Saturn and Jupiter. Neptune, the most dense of the giant planets, is about 1.6 times denser than water and only about half as dense as rock. Uranus is less dense than Neptune. Those observations tell us that water and other low-density ices, such as ammonia and methane, must be the major compositional components of Uranus and Neptune, along with lesser amounts of silicates and metals. The total amount of hydrogen and helium in those planets is probably limited to no more than 1 or 2 Earth masses, and most of those gases reside in the relatively shallow atmospheres of the planets. Computer simulations suggest that under their conditions of high pressure and temperature, the water that makes up so much of Uranus and Neptune might be *super-ionic*—a state between a liquid and a solid.

Why do Jupiter and Saturn have so much more hydrogen and helium than Uranus and Neptune? Although each giant planet formed around cores of rock and metal, they turned out differently. Those differences are an important clue to their origins. The variation may be due to the time those planets took to form and to the distribution of material from which they formed. The cores of Uranus and Neptune were smaller and formed much later than those of Jupiter and Saturn, when most of the gas in the protoplanetary disk had been blown away by the emerging Sun. The icy planetesimals from which they formed were more widely dispersed at their greater distances from the Sun. With more space between planetesimals, their cores would have taken longer to build up. Saturn may have captured less gas than Jupiter both because its core formed later and because less gas was available at its greater distance from the Sun. Some astronomers hypothesize that Uranus and Neptune might have formed in a location different from where they are now (see the Origins section).

Figure 10.14 The magnetic fields of the giant planets can be approximated by the fields from bar magnets offset and tilted with respect to the planets' axes. Compare those magnetic fields with Earth's, shown in Figure 8.12.

(a) Jupiter

(b) Saturn

(c) Uranus

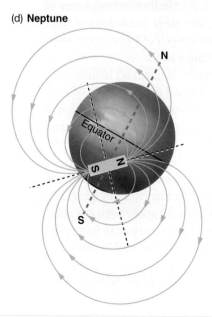

(d) Neptune

CHECK YOUR UNDERSTANDING 10.3

The interiors of the giant planets are heated by gravitational contraction. We know that because: (a) the cores are very hot; (b) most of the giant planets radiate more energy than they receive from the Sun; (c) the giant planets have strong winds; (d) the giant planets are mostly atmosphere.

10.4 The Giant Planets Are Magnetic Powerhouses

All the giant planets have magnetic fields much stronger than Earth's: their field strengths range from 50 to 20,000 times stronger. Field strength falls off with distance, however, so the fields at the cloud tops of Saturn, Uranus, and Neptune have comparable strength to that of Earth's surface field. Even for Jupiter's exceptionally strong field, the field strength at the cloud tops is only about 15 times that of Earth's surface field. In Jupiter and Saturn, circulating currents within deep layers of metallic hydrogen generate magnetic fields. In Uranus and Neptune, magnetic fields arise within deep oceans of water and ammonia made electrically conductive by dissolved salts or ionized molecules. The magnetospheres of the giant planets are very large and interact with not only the solar wind (as Earth's does) but also the rings and moons that orbit the giant planets.

The Size and Shape of the Magnetospheres

Just as Earth's magnetic field traps energetic charged particles to form Earth's magnetosphere, the magnetic fields of the giant planets also trap energetic particles to form magnetospheres of their own. The most colossal of those by far is Jupiter's magnetosphere. Its radius is 100 times that of the planet itself, roughly 10 times the radius of the Sun. Even the relatively weak magnetic fields of Uranus and Neptune form magnetospheres that are comparable in size to the Sun. Evidence of the giant planets' magnetospheres comes from spacecraft in the outer Solar System, from telescopes orbiting Earth, and from radio emissions received on Earth.

Figure 10.14 illustrates the geometry of the magnetic fields of the giant planets as though they came from bar magnets. The differences in the orientations of the magnetic field axes are not well understood. Jupiter's magnetic axis is inclined 10° with respect to its rotation axis—an orientation similar to Earth's—but it is offset about a tenth of a radius from the planet's center. Saturn's magnetic axis is located almost precisely at the planet's center and is almost perfectly aligned with the rotation axis. The magnetic axis of Uranus is inclined nearly 60° to its rotation axis and is offset by a third of a radius from the planet's center. The orientation of Neptune's rotation axis is similar to that of Earth, Mars, and Saturn. Neptune's magnetic-field axis is inclined 47° to its rotation axis, however, and the center of that magnetic field is displaced from the planet's center by more than half the radius—an offset even greater than that of Uranus. The field is displaced primarily toward Neptune's southern hemisphere, thereby creating a field 20 times stronger at the southern cloud tops than at the northern cloud tops. The reason for the unusual geometries of the magnetic fields of Uranus and Neptune remains uncertain, but it is not related to the orientations of their rotation axes.

The magnetosphere also is influenced by the solar wind, which supplies some of the particles (Chapter 9). In addition, the pressure of the solar wind pushes on and compresses a magnetosphere, so the size and shape of a planet's magnetosphere depend on how the solar wind is blowing at any particular time. Planetary magnetic fields also divert the solar wind, which flows around magnetospheres the way a stream flows around boulders. Just as a rock in a river creates a wake that extends downstream (**Figure 10.15a**), the magnetosphere of a planet produces a wake that can extend for great distances.

Figure 10.15b shows that the wake of Jupiter's magnetosphere extends well past the orbit of Saturn. Jupiter's magnetosphere is the largest permanent structure in the Solar System, surpassed in size only by the tail of an occasional comet. If your eyes were sensitive to radio waves, the second-brightest object in the sky would be Jupiter's magnetosphere. The Sun would still be brighter, but even at a distance from Earth of 4.2–6.2 AU, Jupiter's magnetosphere would appear roughly twice as large as the Sun in the sky.

Saturn's magnetosphere also would be large enough to see, if we could see radio waves, but it would be much fainter than Jupiter's. Even though Saturn has a strong magnetic field, pieces of rock, ice, and dust in Saturn's spectacular rings act like sponges, soaking up magnetospheric particles soon after they enter the magnetosphere. With far fewer magnetospheric electrons, Saturn has much less radio emission. The magnetic tails of Uranus and Neptune have a curious structure. For both Uranus and Neptune, the tilt and the large displacement of the magnetic field from the center of each planet cause the magnetosphere to wobble as the planet rotates. That wobble causes the tail of the magnetosphere to twist like a corkscrew as it stretches away.

Rapidly moving electrons in planetary magnetospheres spiral around the magnetic field lines, and as they do so they emit a type of radiation, known as **synchrotron radiation**, concentrated in the low-energy radio part of the spectrum. Precise measurement of periodic variations in the radio signals "broadcast" by the giant planets indicates the planets' true rotation periods. The magnetic field of each planet is locked to the conducting liquid layers deep within the planet's interior, so the magnetic field rotates with the same period as the deep interior of the planet. With the fast and highly variable winds that

Figure 10.15 (a) Water flowing past a rock sweeps the algae against the rock and into a "tail" pointing in the direction of the water's flow. (b) The solar wind compresses Jupiter's (or any other planet's) magnetosphere on the side toward the Sun and draws it out into a magnetic tail away from the Sun. Jupiter's tail stretches beyond the orbit of Saturn. (Not to scale.)

push around clouds in the atmospheres of the giant planets, measuring the radio emission is the only way to determine their true rotation periods.

Radiation Belts and Auroras of the Giant Planets

As a planet rotates on its axis, it drags its magnetosphere around with it, and charged particles are swept around at high speeds. Those fast-moving charged particles slam into neutral atoms, and the energy released in the resulting high-speed collisions heats the **plasma** to extreme temperatures. (A plasma is a gas consisting of electrically charged particles.) In 1979, while passing through Jupiter's magnetosphere, *Voyager 1* encountered a region of plasma with a temperature of more than 300 million K—20 times the temperature at the center of the Sun. *Voyager 1* did not melt when passing through that region because the plasma was so tenuous that the plasma's particles were very far apart in space. Each particle was extraordinarily energetic; too few of them were present to harm the probe.

Charged particles trapped in planetary magnetospheres are concentrated in *radiation belts*. Although Earth's radiation belts are severe enough to worry astronauts, the radiation belts that surround Jupiter are searing by comparison. In 1974, the *Pioneer 11* spacecraft passed through the radiation belts of Jupiter. Several of the instruments onboard were permanently damaged as a result, and the spacecraft itself barely survived to continue its journey to Saturn.

In addition to protons and electrons from the solar wind, the magnetospheres of the giant planets contain large amounts of various elements (some ionized), including sodium, sulfur, oxygen, nitrogen, and carbon. Those elements come from several sources, including the planets' extended atmospheres and the moons that orbit within them. The most intense radiation belt in the Solar System is a doughnut-shaped ring, or *torus*, of plasma associated with Io, the innermost of Jupiter's four Galilean moons. As we discuss in more detail in Chapter 11, Io has low surface gravity and violent volcanic activity. Some of the gases erupting from Io's interior escape and become part of Jupiter's radiation belt. As charged particles are slammed into Io by the rotation of Jupiter's magnetosphere, even more material is knocked free of its surface and ejected into space. Images of the region around Jupiter, taken in the light of emission lines from atoms of sulfur or sodium, show a faintly glowing ring of plasma supplied by Io. Other moons also influence the magnetospheres of the planets they orbit. *Cassini* found that Saturn's moon Enceladus leaks ionized molecules (including nitrogen), water vapor, and ice grains from icy geysers and provides most of the torus of plasma in Saturn's magnetosphere.

Figure 10.16 The Hubble Space Telescope took images of auroral rings around the poles of Jupiter (a) and Saturn (b). The auroral images (bright rings near the north pole) were taken in ultraviolet light and then superimposed on visible-light images.

Charged particles spiral along the magnetic-field lines of the giant planets, bouncing back and forth between each planet's two magnetic poles, just as they do around Earth. As with Earth, those energetic particles collide with atoms and molecules in a planet's atmosphere, knocking them into excited energy states that decay and emit light. The results are bright auroral rings (**Figure 10.16**). Those auroral rings surround the magnetic poles of the giant planets, just as the aurora borealis and aurora australis ring the north and south magnetic poles, respectively, of Earth (see Figure 9.9a).

Jupiter's auroras are the most powerful in the Solar System, and they have an added twist not seen on Earth. As Jupiter's magnetic field sweeps past Io, electrons spiral along Jupiter's magnetic-field lines. The result is a magnetic channel, called a **flux tube**, that connects Io with Jupiter's atmosphere near the planet's magnetic poles (**Figure 10.17**). Io's flux tube carries power roughly equivalent to the total power produced by all electrical generating stations on Earth. Much of the power generated within the flux tube is radiated away as radio energy. Those radio signals are received on Earth as intense bursts. However, a substantial fraction of the energy of the particles in the flux tube also is deposited into Jupiter's atmosphere. At the very location where Io's flux tube intercepts Jupiter's atmosphere is a spot of intense auroral activity. As Jupiter rotates, that spot leaves behind an auroral trail in Jupiter's atmosphere.

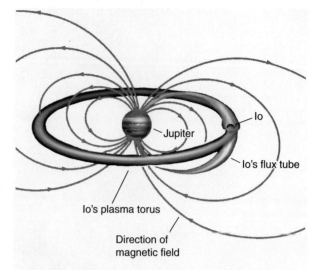

Figure 10.17 The geometry of Io's plasma torus and flux tube.

CHECK YOUR UNDERSTANDING 10.4

Why are the radiation belts around Jupiter much stronger than those around Earth? (a) Jupiter has larger storms than Earth; (b) Jupiter is colder than Earth; (c) Jupiter rotates faster than Earth; (d) Jupiter has a stronger magnetic field than Earth.

10.5 The Planets of Our Solar System Might Not Be Typical

In Chapters 8–10, we discussed the planets of our Solar System in detail. The categories of inner terrestrial rocky planets versus outer gas and ice giants were based on our eight planets—but how typical are those categories when compared with the many exoplanets that have been detected? Do other multiplanet systems resemble our own? As the number of confirmed exoplanets increases, astronomers can compare them statistically with those of our own system. Planetary scientists were surprised to find that exoplanets differ from the planets of our Sun. In the Kepler-20 system, for example, the large planets alternate with the smaller ones (**Figure 10.18**). In this section, we examine some of those differences.

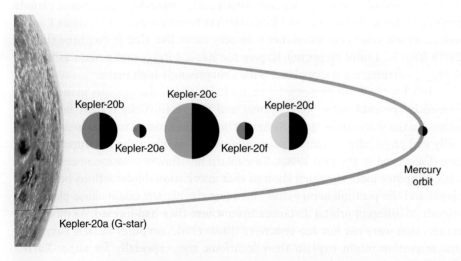

Figure 10.18 Illustration of the Kepler-20 system. Large and small planets alternate, but they are all closer to their star than Mercury is to the Sun.

Figure 10.19 Artist's depiction of the super-Jupiter planet Kappa Andromedae b.

Different Types of "Jupiters"

The first exoplanets were discovered using the radial velocity method (Chapter 7), which measures the wobble of a star caused by the gravity of its planet. That method is most successful at finding a massive planet located close to its star, where the planet's gravitational tug on the star is stronger than if the planet were of smaller mass or farther away. Those *hot Jupiters* are gas giants, within a few percent of an AU to half an AU from their respective stars, with correspondingly short and sometimes highly elliptical orbits. Current models of planetary formation suggest that not enough excess hydrogen would have existed that close to their stars for those planets to form there. Some of those planets may be *puffy Jupiters*, which have a larger radius and a lower density than Jupiter. Their densities are closer to that of Saturn. The larger radius is thought to come from a heated, and thus expanding, gaseous atmosphere.

Astronomers have identified several hundred exoplanets known as *super-Jupiters* because their masses are 2–13 times that of Jupiter. Some are also hot Jupiters, but most are not. Their higher mass gives them stronger gravitational contraction than that in planets with Jupiter's mass, so they shrink. As a result, most have a higher density than Jupiter. The gases can be compressed by self-gravity so much that the super-Jupiter could be denser than Earth. The stronger gravitational contraction would create hotter cores, so they might have more intense winds and weather than Jupiter has. An artist's depiction of the super-Jupiter planet Kappa Andromedae b, with about 13 times the mass of Jupiter, is shown in **Figure 10.19**.

Super-Earths to Mini-Neptunes

Currently, observations suggest that the most common exoplanet size is one with a radius between that of Earth and that of Neptune (4 times larger than Earth). However, no planet in our Solar System falls in that range. Planets with about 1.5–10 M_{Earth} are called *super-Earths*. Up to about 2 R_{Earth}, those planets get denser as they get larger, as expected for rocky planets. Above 2 R_{Earth}, however, most planets are puffier—a gaseous envelope surrounds the rocky core. **Figure 10.20** shows a plot of the size of planets versus their frequency. The gaseous planets at the higher end of that range are sometimes called *mini-Neptunes*. Some planets don't fit those rules; for example, KOI-314c has Earth's mass but 1.6 times Earth's radius, which gives that exoplanet a density more like that of Neptune than of Earth (that is, a mini-Neptune). Kepler-10c has 2.3 R_{Earth} and a mass as high as 17 M_{Earth}—making it a rocky planet with a surprisingly high mass.

In Chapter 7, we presented a scenario in which the gaseous giant planets formed in the cold outer Solar System and the small rocky terrestrial planets formed in the warm inner Solar System. That model seemed to make sense chemically and physically, so astronomers were shocked when the hot Jupiters were first discovered in the mid-1990s. To explain hot Jupiters, astronomers worked with computer models, which showed that gravitational interactions between a planet and the protoplanetary disk or among the planets could cause planets to migrate to different orbital distances from where they had formed. As more exoplanets that were *not* hot Jupiters were discovered, computer models suggested that migration might explain their locations, too, especially for super-Earths close in to their respective stars.

To date, the mix of planet types in our Solar System—outer giant gaseous planets and inner small rocky planets, all with nearly circular orbits—has not

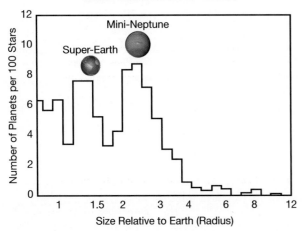

Figure 10.20 Smaller planets tend to be super-Earths or mini-Neptunes. Data on nearby super-Earths suggest that at about 2 R_{Earth}, planet density decreases with size because the planet has more gas accumulated on its rocky core.

been seen in several hundred extrasolar multiplanet systems. Observations and computer models show that many combinations of planetary sizes, masses, and compositions can exist within planetary systems. It is not yet known whether planetary systems like ours are rare or if they just haven't yet been discovered.

CHECK YOUR UNDERSTANDING 10.5

Place in order of increasing diameter the following types of exoplanets: (a) super-Earths; (b) puffy Jupiters; (c) super-Jupiters; (d) mini-Neptunes.

Origins Giant Planet Migration and the Inner Solar System

As long as only two objects are involved, the motions that result from Newton's law of gravitation are simple. Kepler's laws describe the regular, repeating elliptical orbits of planets around the Sun. When more than two objects are involved, however, the resulting motions may be anything but simple and regular. Each planet in the Solar System moves under the gravitational influence of the Sun combined with that of all the other planets. Although those extra influences are small, they are not negligible. Over millions of years, they lead to significant differences in the locations of planets in their orbits. In many possible exoplanetary systems, such interactions among planets might cause planets to dramatically change their orbits or even be ejected from the system entirely.

Computer models developed to understand the exoplanet systems have been applied to the Solar System as well, and the results are intriguing. Computer models of the formation of the Solar System show that the giant planets may not have formed in their current locations and could have migrated substantially in the early Solar System. A key point seems to be the gravitational influence of Saturn on Jupiter, especially the ratio of their orbital periods. **Figure 10.21** shows one type of migration model. When Saturn's orbital period became twice that of Jupiter, their respective orbits became more elongated. The result was an outward migration of Uranus and Neptune, whose orbits grew larger. Uranus and Neptune may actually have switched places. That shuffling cleared away nearby planetesimals, sending some to the inner Solar System, and made the planetary orbits more stable. In another set of models, Jupiter migrated inward to 1.5 AU—the current orbital distance of Mars. Then Saturn migrated inward even faster to a point at which its orbital period was 1.5 times that of Jupiter, and finally they both migrated outward, pushing Uranus and Neptune into larger orbits.

In some exoplanetary systems, gas giants are observed at the right distance from their respective stars to be in the habitable zone (see Chapter 7), but it is not known whether gas giants can support life. Super-Earths also are sometimes found in the habitable zone of their stars. In the Solar System, migrations of the giant planets could have helped confine and stabilize the orbits of the inner planets, so that they reside in or near the habitable zone. Migration of Jupiter may have destabilized the orbits of close-in hot planets, so that they crashed into the Sun. Planetesimals closer to 1.5 AU may have scattered, keeping Mars from growing so that it stayed small. A small Mars could not hold on to its atmosphere and its liquid water, making it a less likely starting point for life. In addition, the shuffling about of the outer planets may be responsible for the period of late heavy bombardment, which brought at least some of the atmospheric gases, water, and possibly organic molecules needed for life to form on Earth.

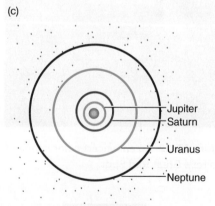

Figure 10.21 Snapshots from computer simulations of planet migration show how Jupiter and Saturn affect the outer Solar System. The four circles are the orbits of the Jovian planets, and the black dots are planetesimals. (a) The orbital period of Saturn becomes twice that of Jupiter. (b) Planetesimals are scattered as the giant planets change orbits. (c) In some models, Neptune and Uranus change places. The inner Solar System is left more stable.

READING ASTRONOMY NEWS

The newest mission to Jupiter observes its clouds and atmosphere.

Juno Spacecraft Spots Jupiter's Great Red Spot

jpl.nasa.gov

Images of Jupiter's Great Red Spot (**Figure 10.22**) reveal a tangle of dark, veinous clouds weaving their way through a massive crimson oval. The JunoCam imager aboard NASA's *Juno* mission snapped pics of the most iconic feature of the solar system's largest planetary inhabitant during its Monday (July 10) flyby. The images of the Great Red Spot were downlinked from the spacecraft's memory on Tuesday and placed on the mission's JunoCam website Wednesday morning.

"For hundreds of years scientists have been observing, wondering and theorizing about Jupiter's Great Red Spot," said Scott Bolton, *Juno* principal investigator from the Southwest Research Institute in San Antonio. "Now we have the best pictures ever of this iconic storm. It will take us some time to analyze all the data from not only Juno-Cam, but Juno's eight science instruments, to shed some new light on the past, present and future of the Great Red Spot."

As planned by the *Juno* team, citizen scientists took the raw images of the flyby from the JunoCam site and processed them, providing a higher level of detail than available in their raw form. The citizen-scientist images, as well as the raw images they used for image processing, can be found at: https://www.missionjuno.swri.edu/junocam/processing.

"I have been following the *Juno* mission since it launched," said Jason Major, a Juno-Cam citizen scientist and a graphic designer from Warwick, Rhode Island. "It is always exciting to see these new raw images of Jupiter as they arrive. But it is even more thrilling to take the raw images and turn them into something that people can appreciate. That is what I live for."

Measuring in at 10,159 miles (16,350 kilometers) in width (as of April 3, 2017) Jupiter's Great Red Spot is 1.3 times as wide as Earth. The storm has been monitored since 1830 and has possibly existed for more than 350 years. In modern times, the Great Red Spot has appeared to be shrinking.

All of *Juno*'s science instruments and the spacecraft's JunoCam were operating during the flyby, collecting data that are now being returned to Earth. *Juno*'s next close flyby of Jupiter will occur on Sept. 1.

Juno reached perijove (the point at which an orbit comes closest to Jupiter's center) on July 10 at 6:55 p.m. PDT (9:55 p.m. EDT). At the time of perijove, *Juno* was about 2,200 miles (3,500 kilometers) above the planet's cloud tops. Eleven minutes and 33 seconds later, *Juno* had covered another 24,713 miles (39,771 kilometers), and was passing directly above the coiling, crimson cloud tops of the Great Red Spot. The spacecraft passed about 5,600 miles (9,000 kilometers) above the clouds of this iconic feature.

Juno launched on Aug. 5, 2011, from Cape Canaveral, Florida. During its mission of exploration, *Juno* soars low over the planet's cloud tops—as close as about 2,100 miles (3,400 kilometers). During these flybys, *Juno* is probing beneath the obscuring cloud cover of Jupiter and studying its auroras to learn more about the planet's origins, structure, atmosphere and magnetosphere.

Early science results from NASA's *Juno* mission portray the largest planet in our solar system as a turbulent world, with an intriguingly complex interior structure, energetic polar aurora, and huge polar cyclones.

"These highly-anticipated images of Jupiter's Great Red Spot are the 'perfect storm' of art and science. With data from *Voyager, Galileo, New Horizons,* Hubble and now *Juno*, we have a better understanding of the composition and evolution of this iconic feature," said Jim Green, NASA's director of planetary science. "We are pleased to share the beauty and excitement of space science with everyone."

Figure 10.22 An enhanced-color image of Jupiter's Great Red Spot.

QUESTIONS

1. Why has the Great Red Spot been seen for only ~350 years?
2. Why is it difficult for astronomers to understand what is happening below the cloud tops?
3. Do a search to see whether the Great Red Spot is still shrinking.
4. Go to the website mentioned in the article. Why is *Juno* using "citizen scientists" to process the images?
5. What new observational capabilities does the *Juno* mission bring to understanding the Jovian system?

Summary

The giant planets are much larger and less dense than the terrestrial planets. They consist primarily of light elements rather than rock, and their outer atmospheres are much colder. Because of their rapid rotation and the Coriolis effect, zonal winds are very strong on those planets. Storms, such as those on Saturn, tend to be larger and longer-lived than storms on Earth. Volatiles become ices at various heights in those atmospheres, leading to a layered cloud structure. Jupiter, Saturn, and Neptune are still shrinking, and their gravitational energy is being converted to thermal energy, heating both the cores and the atmospheres from the inside. Uranus does not seem to have as large a heat source inside. Temperatures and pressures in the cores of the giant planets are very high, leading to unusual states of matter, such as metallic hydrogen and super-ionic water. The current locations of those planets might be different from where they formed: models suggest that their positions may have migrated.

LG 1 **Differentiate the giant planets from one another and from the terrestrial planets.** Uranus and Neptune were discovered by telescope, unlike all the other planets, which have been known since ancient times. Jupiter and Saturn are gas giants, whereas Uranus and Neptune are ice giants. Jupiter and Saturn are made up mostly of hydrogen and helium (a composition similar to that of the Sun), whereas Uranus and Neptune contain larger amounts of ices, such as water, ammonia, and methane, than are found in Jupiter and Saturn. Those compositions set them apart from the terrestrial planets. In addition, all four giant planets are much larger than Earth.

LG 2 **Describe the atmosphere of each giant planet.** We see only atmospheres on the giant planets because solid or liquid surfaces, if they exist, are deep below the cloud layers. Clouds on Jupiter and Saturn are composed of various kinds of ice crystals colored by impurities. Uranus and Neptune have relatively few clouds, and so their atmospheres appear more uniform. The most prominent atmospheric feature is the Great Red Spot in Jupiter's southern hemisphere.

LG 3 **Explain the extreme conditions deep inside the giant planets.** The ongoing collapse of the giant planets converts gravitational energy to thermal energy. That process heats most of the giant planets from within, producing convection. Powerful convection and the Coriolis effect drive high-speed winds in the upper atmospheres of all the giant planets. The interiors of the giant planets are very hot and very dense because of the high pressures exerted by their overlying atmospheres.

LG 4 **Describe the magnetosphere of each giant planet.** The giant planets have enormous magnetospheres that emit synchrotron radiation and interact with their moons. The rotation speed of a gas giant is found from periodic bursts of radio waves generated as the planet's magnetic field rotates.

LG 5 **Compare the planets of our Solar System with those in exoplanetary systems.** Systems of exoplanets found to date do not contain our Solar System's distribution of small, rocky inner planets and large, giant outer planets. Most planets found so far fall between Earth and Neptune in size. Our understanding of how solar systems form is incomplete.

? Unanswered Questions

- Did the Solar System start with more planets? In the same types of computer models of the early Solar System that we discussed in Origins, astronomers can run simulations with different initial configurations of planets to see which configurations evolve and then compare those with what is observed today. In one set of models, astronomers found that starting with five giant planets best reproduced the current outer Solar System. The fifth planet, which would have been kicked out of the Solar System after a close encounter with Jupiter, may still be wandering through the Milky Way.

- What are the mass and size of the core of each giant planet? Is a rocky core underneath the thick atmosphere? Is the core the size of a terrestrial planet or larger? The NASA *Juno* mission is measuring Jupiter's gravitational and magnetic fields to map the amount and distribution of mass in its core and atmosphere. Preliminary results from *Juno* suggest that Jupiter's core is larger and "fuzzier" than previously thought, so that question might be answered for Jupiter in a few years.

Questions and Problems

Test Your Understanding

1. The following steps lead to convection in the atmospheres of giant planets. After (a), place (b)–(f) in order.
   a. Gravity pulls particles toward the center.
   b. Warm material rises and expands.
   c. Particles fall toward the center, converting gravitational energy to kinetic energy.
   d. Expanding material cools.
   e. Thermal energy heats the material.
   f. Friction converts kinetic energy to thermal energy.

2. Deep inside the giant planets, water is still a liquid even though the temperatures are tens of thousands of degrees above the boiling point of water. That can happen because
   a. the density inside the giant planets is so high.
   b. the pressure inside the giant planets is so high.
   c. the outer Solar System is so cold.
   d. space has very low pressure.

3. Assume you want to deduce the radius of a Solar System planet as it occults a background star when the relative velocity between the planet and Earth is 30 km/s. If the star crosses through the middle of the planet and disappears for 26 minutes, what is the planet's radius?
   a. 3,000 km
   b. 23,000 km
   c. 15,000 km
   d. 5,000 km

4. Neptune's existence was predicted because
   a. Uranus did not seem to obey Newton's laws of motion.
   b. Uranus wobbled on its axis.
   c. Uranus became brighter and fainter in an unusual way.
   d. some of the solar nebula's mass was unaccounted for.

5. Which giant planet has the most extreme seasons?
   a. Jupiter
   b. Saturn
   c. Uranus
   d. Neptune

6. The magnetic fields of the giant planets
   a. align closely with the planet's rotation axis.
   b. extend far into space.
   c. are thousands of times stronger at the cloud tops than at Earth's surface field.
   d. have an axis that passes through the planet's center.

7. An occultation occurs when
   a. a star passes between Earth and a planet.
   b. a planet passes between Earth and a star.
   c. a planet passes between Earth and the Sun.
   d. Earth passes between the Sun and a planet.

8. Occultations directly determine a planet's
   a. diameter.
   b. mass.
   c. density.
   d. orbital speed.

9. The chemical compositions of Jupiter and Saturn are most similar to those of
   a. Uranus and Neptune.
   b. the terrestrial planets.
   c. their moons.
   d. the Sun.

10. Individual cloud layers in the giant planets have different compositions because
    a. the winds are all in the outermost layer.
    b. the Coriolis effect occurs only close to the "surface" of the inner core.
    c. convection does not occur on the giant planets.
    d. different volatiles freeze out at different temperatures.

11. The Great Red Spot on Jupiter is
    a. a surface feature.
    b. a storm that has been raging for more than 300 years.
    c. caused by the interaction between the magnetosphere and Io.
    d. currently about the size of North America.

12. How are Uranus and Neptune different from Jupiter and Saturn?
    a. Uranus and Neptune have a higher percentage of ices in their interiors.
    b. Uranus and Neptune have no rings.
    c. Uranus and Neptune have no magnetic field.
    d. Uranus and Neptune are closer to the Sun.

13. What could have caused the planets to migrate through the Solar System?
    a. gravitational pull from the Sun
    b. interaction with the solar wind
    c. accreting gas from the solar nebula
    d. gravitational pull from other planets

14. Zonal winds on the giant planets are stronger than those on the terrestrial planets because
    a. they have more thermal energy.
    b. the giant planets rotate faster.
    c. the moons of giant planets provide additional pull.
    d. the moons feed energy to the planet through the magnetosphere.

15. A "hot Jupiter" gets its name from the fact that
    a. its temperature has been measured to be higher than Jupiter's.
    b. it is located around a much hotter star than the Sun.
    c. it has very high density, so its temperature is high.
    d. it orbits very close to its central star.

Thinking about the Concepts

16. Describe how the giant planets differ from the terrestrial planets.

17. Jupiter's chemical composition is more like that of the Sun than Earth's is, yet both planets formed from the same protoplanetary disk. Explain why they are different today.

18. What can be learned about a Solar System object when it occults a star?

19. What drives the zonal winds in the atmospheres of the giant planets?

20. Compare the sequence of events in the Process of Science Figure in this chapter with the flowchart of the Process of Science Figure in Chapter 1. Redraw the flowchart, incorporating each event leading to the discovery of Uranus as examples in the appropriate boxes.

21. None of the giant planets are perfectly spherical. Explain why they have a flattened appearance.

22. What is the source of color in Jupiter's clouds? Uranus and Neptune, when viewed through a telescope, appear distinctly bluish green. What are the two reasons for their striking appearance?

23. Which giant planets have seasons similar to Earth's, and which one experiences extreme seasons?

24. Jupiter's core is thought to consist of rocky material and ices, all in a liquid state at a temperature of 35,000 K. How can materials such as water be liquid at such high temperatures?

25. Explain how astronomers measure wind speeds in the atmospheres of the giant planets.

26. What is the Great Red Spot? How has it changed over the centuries it has been observed?

27. Jupiter, Saturn, and Neptune radiate more energy into space than they receive from the Sun. What is the source of the additional energy?

28. When viewed by radio telescopes, Jupiter is the second-brightest object in the sky. What is the source of its radiation?

29. What creates auroras in the polar regions of Jupiter and Saturn?

30. How might migration of the outer giant planets affect the sizes and orbits of the inner planets?

Applying the Concepts

31. Figure 10.1 shows two sets of pictures of the outer planets. What is the difference between Figure 10.1a and Figure 10.1b?

32. Figure 10.10d shows the winds on Neptune. The graph, however, does not cover the full planet. Is that likely to mean that the wind speed is zero where no white line is present or that the wind speed is unknown where no white line is present? Explain your reasoning.

33. What creates metallic hydrogen in the interiors of Jupiter and Saturn, and why do we call it metallic?

34. Use Figure 10.17 to estimate the radius of Io's plasma torus in terms of the radius of Jupiter. Convert that value to kilometers, and then look up the answer on the Internet. How close did you get with your simple measurement?

35. The Sun appears 400,000 times brighter than the full Moon in Earth's sky. How far from the Sun (in astronomical units) would you have to go for the Sun to appear only as bright as the full Moon appears in Earth's nighttime sky? How does the distance you would have to travel compare with the semimajor axis of Neptune's orbit?

36. Uranus occults a star when the relative motion between Uranus and Earth is 23.0 km/s. An observer on Earth sees the star disappear for 37 minutes 2 seconds and notes that the center of Uranus passed directly in front of the star.
    a. On the basis of those observations, what value would the observer calculate for the diameter of Uranus?
    b. What could you conclude about the planet's diameter if its center did not pass directly in front of the star?

37. Jupiter's equatorial radius (R_{Jup}) is 71,500 km, and its oblateness is 0.065. What is Jupiter's polar radius (R_{polar})? Oblateness is given by $(R_{Jup} - R_{polar})/R_{Jup}$.

38. Ammonium hydrosulfide (NH_4HS) is a molecule in Jupiter's atmosphere responsible for many of its clouds. Using the periodic table in Appendix 3, calculate the molecular weight of an ammonium hydrosulfide molecule, where the atomic weight of a hydrogen atom is 1. (Recall from Working It Out 9.1 that the weight of a molecule is equal to the sum of the weights of its component atoms.)

39. Jupiter is an oblate planet with an average radius of 69,900 km. Earth's average radius is 6,370 km.
    a. Given that volume is proportional to the cube of the radius, how many Earth volumes could fit inside Jupiter?
    b. Jupiter is 318 times as massive as Earth. Show that Jupiter's average density is about one-fourth that of Earth.

40. The tilt of Uranus is 98°. From one of the planet's poles, how far from the zenith would the Sun appear on summer solstice?

41. A small cloud in Jupiter's equatorial region is observed to be at a longitude of 122.0° west in a coordinate system and to be rotating at the same rate as the deep interior of the planet. (West longitude is measured along a planet's equator toward the west.) Another observation, made exactly 10 Earth hours later, finds the cloud at a longitude of 118.0° west. Jupiter's equatorial radius is 71,500 km. What is the observed equatorial wind speed in kilometers per hour? Is that wind from the east or west?

42. The equilibrium temperature for Saturn should be 82 K, but the observed temperature is 95 K. How much more energy does Saturn radiate than it absorbs?

43. Neptune radiates 2.6 times as much energy into space as it absorbs from the Sun. Its equilibrium temperature (see Chapter 5) is 47 K. What is its true temperature?

44. Compare the graphs in Figures 10.9a and 10.9b. Does atmospheric pressure increase more rapidly with depth on Jupiter or on Saturn? Compare the graphs in Figures 10.9c and 10.9d. Does pressure increase more rapidly with depth on Uranus or on Neptune? Of the four giant planets, which has the fastest pressure rise with depth? The slowest?

45. Use Figure 10.9 to find the temperature at an altitude of 100 km on each of the four giant planets.

USING THE WEB

46. a. Go to websites for the NASA *Juno* mission (https://www.nasa.gov/mission_pages/juno/ and http://missionjuno.swri.edu), a spacecraft launched in 2011, which has been in orbit around Jupiter since 2016. Examine the mission's trajectory. Why did it loop around the Sun and pass Earth again in 2013 before heading to Jupiter? Why does the spacecraft carry a plaque dedicated to Galileo Galilei?
    b. What are the science goals of the mission? What is a recent discovery? Is the mission still active?

47. Go to the website for the *Voyager 1* and *2* missions (https://voyager.jpl.nasa.gov), which celebrated their 40th anniversary in 2017. *Voyager* visited and collected data on all four giant planets.
    a. Where are the spacecraft now? Click on "Galleries" and then "Images." Those are still the only close-up images of Uranus and Neptune. What was learned about those planets?
    b. Click on the "The Golden Record," and then on the right, look at scenes, greetings, music, and sounds from Earth. Suppose you were asked to make a new version of the Golden Record, a playlist to send on an upcoming space mission to outside the Solar System. What would you include in one or more of those categories?

48. Go to the *Cassini* website (https://saturn.jpl.nasa.gov). That mission ended in 2017, but data is still being analyzed. Why did NASA crash *Cassini* into Saturn? What are some of the main discoveries from that mission? Are new discoveries still being reported?

49. Go to the Extrasolar Planets encyclopedia (http://exoplanet.eu/catalog/).
    a. Under "Mass," look for a super-Jupiter planet with a mass significantly larger than that of Jupiter. How far is it from its star—is it a hot Jupiter? Click on the planet name—how was it discovered? If a radius is given, is it denser or less dense than Jupiter? Click twice on "Mass" to get a list in descending order—what is the most massive super-Jupiter in the catalog?
    b. Under "Mass," click on "M_{Jup}" so it changes to "M_{Earth}"; do the same under "Radius" so it shows "R_{Earth}." Look for a "Super-Earth." What is its radius? How was it detected? Is an estimated mass listed—if so, what is its density compared with that of Earth? Is it a hot or cold super-Earth?

50. Go to the website in Problem 49 or to the NASA Exoplanet Archive (https://exoplanetarchive.ipac.caltech.edu/ and click on "Multi-Planet Systems"). Look for a star that has multiple planets. Make a graph showing the distances of the planets from that star, and note the masses and sizes of the planets. Put the Solar System planets on the same axis. How does that exoplanet system differ from our Solar System?

EXPLORATION

digital.wwnorton.com/astro6

Estimating Rotation Periods of the Giant Planets

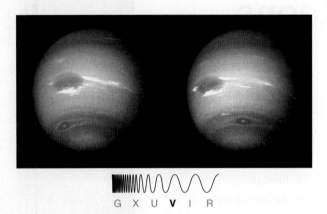

Figure 10.23 Images of Neptune, taken 17.6 hours apart by *Voyager 2*.

Study the two images of Neptune in **Figure 10.23**. The image on the left was taken first, whereas the image on the right was taken 17.6 hours later. During that time, the Great Dark Spot completed nearly one full rotation. The small storm at the bottom of the image completed slightly more than one rotation. You would be very surprised to see that result for locations on Earth.

1 What do those observations tell you about the rotation of visible cloud tops of Neptune?

You can find the rotation period of the smaller storm by equating two ratios. First, use a ruler to find the distance (in millimeters) from the left edge of the planet to the small storm in each image (**Figure 10.24a**). The right edge of the planet is not illuminated, so you will have to estimate the radius of the circles traveled by the storms. You can do that by measuring from the edge of the planet to a line through the planet's center (**Figure 10.24b**). Because the small storm travels along a line of latitude close to a pole, the distance it travels is significantly less than the circumference of the planet.

Figure 10.24 (a) How to measure the position of the storm. (b) How to measure the radius of the circle that the storm traveled.

2 Estimate the radius of the circle (in millimeters) that this small storm makes around Neptune by measuring from the edge of the disk to the line through the center of the planet.

3 Find the circumference of that circle (in millimeters).

Because the small storm rotated *more* than one time, the total distance it traveled is the circumference of the circle plus the distance between its locations in the two images.

4 Add the numbers you obtained in steps 2 and 3 to get the total distance traveled (in millimeters) between those images.

Now take a ratio and find the rotation period. The ratio of the rotation period (T) to the time elapsed (t) must be equal to the ratio of the circumference of the circle around which it travels (C, in millimeters) to the total distance traveled (D, in millimeters): $T/t = C/D$.

5 You have all the numbers you need to solve for T. What value do you calculate for the small storm's rotation period? (To check your work, note that your answer should be less than 17.6 hours. Why?)

Why does that calculation work at all? The actual distance the small storm traveled is *not* a few millimeters, nor is the circumference of the circle around which it travels. To find the actual distance or circumference, you would multiply both values by the same constant of proportionality. Because you are taking a ratio, however, that constant cancels out, so you might as well leave it out from the beginning.

6 Perform the corresponding measurements and calculations for the Great Dark Spot. What is its rotation period? Think carefully about how to find the total distance traveled because the Great Dark Spot has rotated around the planet *less* than once. (To check your work, note that your answer should be more than 17.6 hours. Why?)

7 How similar are the rotation periods for those two storms?

8 What does that comparison tell you about using that method to determine the rotation periods of the giant planets?

9 What method do astronomers use instead?

11

Planetary Moons and Rings

For centuries, Saturn's rings and the Galilean moons of Jupiter have delighted people who looked through telescopes. Since the dawn of the space age, robotic explorers traveling through the Solar System have revealed even more of the diverse collection of moons and rings orbiting other planets.

LEARNING GOALS

By the end of this chapter, you should be able to:

LG 1 Compare the orbits and formation of regular and irregular moons.

LG 2 Describe the evidence for geological activity and liquid oceans on some of the moons.

LG 3 Describe the composition, origin, and general structure of the rings of the giant planets.

LG 4 Explain the role gravity plays in the structure of the rings and the behavior of ring particles.

Saturn and its rings as viewed by the *Cassini* spacecraft. ▶▶▶

Why do some planets have rings?

11.1 Many Solar System Planets Have Moons

Most of the planets and some of the dwarf planets in our Solar System have moons orbiting them. New moons in the outer Solar System are still being discovered. As of 2018, the planets and dwarf planets of the Solar System have nearly 200 moons (some are provisional and need further confirmation). Many of those moons are unique worlds of their own, exhibiting geological processes similar to those on the terrestrial planets. Some moons have volcanic activity and atmospheres, and some are likely to contain liquid water under their icy surfaces. A few of those moons could have conditions suitable to some forms of life. Some of those planetary moons are listed in Appendix 4 (moons around asteroids are discussed in Chapter 12), and an updated list can be found through the "Using the Web" problems at the end of this chapter. In this section, we discuss the orbits and the formation of the moons.

The Distribution of the Moons

The moons of the Solar System are not distributed equally; most are among the giant planets. Only three moons are in the inner Solar System: Mars has two, Earth has one, and Mercury and Venus have none. Among the dwarf planets,

Figure 11.1 Various spacecraft have imaged the major moons of the Solar System. The images are to scale, with the planet Mercury and dwarf planet Pluto shown for comparison. At this scale, the martian moons, Phobos and Deimos, are too small to be shown.

Pluto has five known moons, Haumea has two, and Eris and Makemake each have one. The giant planets have the remaining moons. While the larger planets were forming, they had greater attracting mass and more debris around them; consequently, they have more moons.

Figure 11.1 shows the major moons in the Solar System. Some, like Earth's Moon, are made of rock. Others, especially in the outer Solar System, are mixtures of rock and water ice. A few are made almost entirely of ice. Two moons, Jupiter's Ganymede and Saturn's Titan, are larger in diameter than Mercury. The smallest known moons are only a kilometer in diameter. Although most moons have no atmosphere, Titan has an atmosphere denser than Earth's, and several have very low-density atmospheres. Moons probably accreted from smaller bodies in much the same way that planets accreted from planetesimals, although some may be the product of collisions.

The Orbits of the Moons

A moon can be classified according to its orbit as either *regular* or *irregular*. A **regular moon** lies near its planet's equatorial plane, is close to its planet, and has a nearly circular orbit (**Figure 11.2a**). A regular moon also moves in the same direction the planet rotates. About one-third of the moons in the Solar System are regular. Those moons probably formed from an accretion disk around a host planet at around the time the planet was forming. Our Moon, the Galilean moons of Jupiter, and Saturn's Titan are large, regular moons. Most regular moons are tidally locked to their parent planets. Recall from Chapter 4 that tidal locking causes a body to rotate synchronously with respect to its orbit, as Earth's Moon does. When a moon is tidally locked to its planet, the leading hemisphere permanently faces "forward," in the direction in which the moon is traveling in its orbit around the planet. The trailing hemisphere faces backward. The leading hemisphere is always flying directly into any local debris surrounding the planet, so that hemisphere may have more impact craters on its surface than the trailing hemisphere.

Compared with a regular moon, an **irregular moon** has a more elliptical and more inclined orbit and generally is farther away from its planet (**Figure 11.2b**). Most irregular moons orbit in a direction *opposite* to the rotation of their respective planets; that is, they have retrograde orbits. (Recall from Chapter 3 that apparent backward motion of a planet in the sky is called retrograde motion.) The largest irregular moons are Neptune's Triton and Saturn's Phoebe. Most of the recently discovered moons of the outer planets are irregular, and many are only a few kilometers across. Those moons are almost certainly bodies that formed elsewhere and were later captured by the planets.

Some of the regular moons have strange orbital characteristics. For example, Phobos, one of the two small moons of Mars (**Figure 11.3**), is so close to Mars that it actually orbits Mars faster than Mars rotates. As seen from Mars, Phobos rises in the west and sets in the east twice a day. Phobos is closer to its planet than any known moon. It is not known whether Phobos and the other moon of Mars, Deimos, were captured from the nearby asteroid belt or whether they evolved together with Mars, possibly after a collision early in the history of Mars.

Another strange regular moon is Pluto's Charon, which has a radius one-half that of Pluto. The two are in synchronous rotation, so each has one hemisphere that always faces the other body and another hemisphere that never faces the

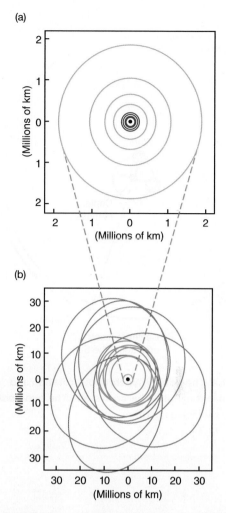

Figure 11.2 These diagrams illustrate the view from "above" the orbits of some of Jupiter's moons. (a) Closer moons, including the Galilean moons, are regular, with nearly circular orbits in Jupiter's equatorial plane. (b) Most of the more distant moons are irregular, with more elliptical, retrograde orbits not in the equatorial plane.

Figure 11.3 The *Mars Reconnaissance Orbiter* photographed the two tiny moons of Mars: (a) Phobos and (b) Deimos.

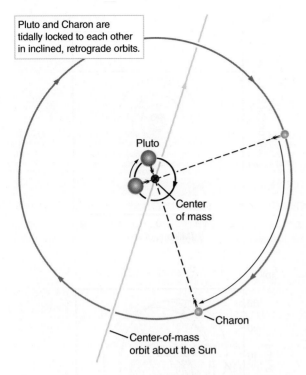

Figure 11.4 The doubly synchronous rotation and revolution in the Pluto-Charon system. The two bodies permanently face each other.

Pluto and Charon are tidally locked to each other in inclined, retrograde orbits.

other body (**Figure 11.4**). Pluto and Charon are the only known pair in the Solar System in which both objects are tidally locked to each other in that way. Pluto's highly compact moon system may have been created by a massive collision between Pluto and another planetesimal, producing a cloud of debris that coalesced to form Charon and the four smaller moons of Pluto, perhaps similar to the way Earth's Moon formed (Chapter 8).

Some sets of moons are in synchronized orbits, called **orbital resonances**, in which the orbital period of one is a multiple of the orbital period of another. Jupiter's moons Ganymede, Europa, and Io are in a resonance of 1:2:4; for every one orbit of Ganymede, Europa orbits twice and Io orbits four times. When the moons are aligned, gravitational effects elongate Io's orbit, which varies from the tidal forces of Jupiter on Io. Pluto's five moons appear to be in a 1:3:4:5:6 sequence of near resonances. Pairs of some of Saturn's moons are in resonance, and resonances also exist between its moons and gaps in its rings.

The orbits of the moons not only indicate something about their origin. They also can be used to estimate the masses of their host planets by using Kepler's laws, just as the orbital properties of the planets can be used to find the mass of the Sun. An example of that calculation is shown in **Working It Out 11.1**.

CHECK YOUR UNDERSTANDING 11.1

Which of the following are characteristics of regular moons? (Choose all that apply.)
(a) They revolve around their planets in the same direction as the planets rotate.
(b) They have orbits that lie nearly in the equatorial planes of their planets. (c) They are usually tidally locked to their parent planets. (d) They are much smaller than all the known planets.

11.1 Working It Out Using Moons to Compute the Mass of a Planet

Recall from Working It Out 4.3 that Newton's version of Kepler's law for planets orbiting the Sun can be used to estimate the mass of the Sun. In Chapter 4, we used the following equation to calculate the mass (M) of the Sun:

$$M = \frac{4\pi^2}{G} \times \frac{A^3}{P^2}$$

where A is the semimajor axis of the orbit, P is the orbital period of a planet, and G is the universal gravitational constant.

For moons orbiting a planet, the same equation applies, as long as the moon is much less massive than the planet. Thus, we can use the orbital motion of the moons to estimate the mass of the planet. For example, Jupiter's moon Io has an orbital semimajor axis of $A = 422{,}000$ kilometers (km) and an orbital period of $P = 1.77$ days. To match the units in G, we need to convert P into seconds:

$$1.77 \text{ days} = 1.77 \text{ days} \times 24\frac{\text{h}}{\text{day}} \times 60\frac{\text{min}}{\text{h}} \times 60\frac{\text{s}}{\text{min}} = 152{,}928 \text{ s}$$

$G = 6.67 \times 10^{-20}$ km³/(kg s²), so the mass of Jupiter (M_{Jup}) is given by

$$M_{\text{Jup}} = \frac{4\pi^2}{G} \times \frac{A^3}{P^2} = \frac{4\pi^2}{6.67 \times 10^{-20} \text{ km/(kg s)}^2} \times \frac{(422{,}000 \text{ km})^3}{(152{,}92 \text{ s})^2}$$

$$M_{\text{Jup}} = 1.90 \times 10^{27} \text{ kg}$$

Using any other moon of Jupiter, you would get the same answer.

Back before Newton published his law of gravity and before any measured value of the gravitational constant G was possible, Galileo and Kepler showed that P^2/A^3 was the same for each of the four Galilean moons of Jupiter. That finding demonstrated that Kepler's law applied to systems other than planets orbiting the Sun.

11.2 Some Moons Have Geological Activity and Water

The moons of the Solar System can be grouped in several ways. Some groupings are based on the sequence of the moons in their orbits around their parent planets; others are based on the sizes or compositions of the moons. In this section, we organize our discussion of the moons by considering some of the same properties we discussed for the terrestrial planets: the history of the moons' geological activity and the presence of water and an atmosphere.

Some moons in the Solar System have been frozen in time since their formation during the early history of the Solar System, whereas others are even more geologically active than Earth. As with the terrestrial planets and Earth's Moon, surface features provide critical clues to a moon's geological history. For example, water ice is a common surface material among the moons of the outer Solar System, and the freshness of that ice indicates the age of those surfaces. Meteorite dust darkens the icy surfaces of moons just as dirt darkens snow in urban areas. A bright surface often means a fresh surface. The size and number of impact craters indicate the relative timing of events such as volcanism, and that timing enables scientists to gauge whether and when a moon may have been active in the past. Older surfaces have more craters. Observations of erupting volcanoes, which are found on Io and Enceladus, for example, are direct evidence that some moons are geologically active today.

Io, the Most Geologically Active Moon

One surprise in Solar System exploration was the discovery of active volcanoes on Io, the innermost of the four large moons of Jupiter. Because Io's orbit is elliptical, Jupiter's gravitational pull flexes Io's crust and generates enough thermal energy to melt parts of it. Did you ever take a piece of metal and bend it back and forth, eventually breaking it in half? Touching the crease line can burn your fingers. Just as bending metal in your hands creates heat, the continual flexing of Io generates enough energy to melt parts of its mantle. In that way, Jupiter's gravitational energy is converted into thermal energy, powering the most active volcanism in the Solar System.

Io is just slightly larger than our Moon. Its surface is covered with volcanic features, including vast lava flows, volcanoes, and volcanic craters (**Figure 11.5a**). Lava flows and volcanic ash bury impact craters as quickly as they form, so no impact craters have been observed on the surface. The *Voyager*, *Galileo*, and *New Horizons* spacecraft and the Keck telescope have observed hundreds of volcanic vents and active volcanoes on Io. The most vigorous eruptions spray sulfurous gases, ash, and lava hundreds of kilometers above the surface. Some of that material escapes from Io. Ash and other particles rain onto the surface as far as 600 km from the vents. The moon is so active that several huge eruptions often occur at the same time. **Figure 11.5b** shows the volcanic activity on Io—the source of the material supplying Io's plasma torus and flux tube (Figure 10.17).

The surface of Io displays pale shades of red, yellow, orange, and brown. Mixtures of sulfur, sulfur dioxide frost, and sulfurous salts of sodium and potassium probably cause that variety of colors. Bright patches may be fields of sulfur dioxide snow. Liquid sulfur dioxide flows beneath Io's surface, held at

Figure 11.5 (a) Composite image of Jupiter's volcanically active moon Io constructed from pictures obtained by *Galileo*. (b) The plume from the crater Pillan Patera rises 140 km above the limb of the moon on the left, whereas the shadow of a 75-km-high plume can be seen to the right of the vent of Prometheus Patera, a volcanic crater near the moon's center.

Figure 11.6 This *Galileo* image of Io shows regions where lava has erupted within a caldera. The molten lava flow is shown in false color to make it more visible.

Figure 11.7 This high-resolution *Galileo* image of Jupiter's moon Europa shows where the icy crust has been broken into slabs that have shifted into new positions. Those areas of chaotic terrain are characteristic of a thin, brittle crust of ice floating atop a liquid or slushy ocean.

Figure 11.8 Artist's conception of Europa, showing liquid bubbling up from the liquid ocean underneath the icy surface. Jupiter and Io are visible in the Europan sky.

high pressure by the weight of overlying material. Like water from a spring, that pressurized sulfur dioxide is pushed out through fractures in the crust, producing sprays of sulfur dioxide snow crystals that travel for up to hundreds of kilometers before settling back to the moon's surface. A similar process takes place with a carbon dioxide fire extinguisher—namely, liquid carbon dioxide contained at high pressure immediately turns to "dry ice" snow as it leaves the nozzle.

Spacecraft images reveal the plains, irregular volcanic craters, and flows, all related to the eruption of mostly silicate magmas onto the surface of Io. The images also show tall mountains, some nearly twice the height of Mount Everest. Huge structures have multiple summit craters showing a long history of repeated eruptions followed by the collapse of partially emptied magma chambers. Many of the chamber floors are very hot (**Figure 11.6**) and might still contain molten material similar to the magnesium-rich lavas that erupted on Earth more than 1.5 billion years ago. Volcanoes on Io are spread much more randomly than those on Earth, implying a lack of plate tectonics.

Because of its active volcanism, Io's mantle has turned inside-out more than once, leading to chemical differentiation. Volatiles such as water and carbon dioxide probably escaped into space long ago, whereas most of the heavier materials sank to the interior to form a core. Sulfur and various sulfur compounds, as well as silicate magmas, are constantly being recycled to form the complex surface we see today.

Evidence of Liquid Oceans on Europa and Enceladus

Jupiter's moon Europa is slightly smaller than our Moon, is made of rock and ice, and has an iron core. *Voyager* observed an outer shell of water ice with surface cracks and creases. Few impact craters are present, so the surface must be young. Regions of chaotic terrain (**Figure 11.7**) are places where the icy crust has been broken into slabs that have shifted into new positions. In other areas, the crust has split apart, and the gaps have filled in with new dark material rising from the interior. The young surface implies geological activity, which varies as Europa orbits Jupiter. As on Io, that activity is probably powered by continually changing tidal forces from Jupiter. However, the forces are not as strong on Europa as on Io because Europa is farther from Jupiter (**Working It Out 11.2**).

The *Galileo* spacecraft measured Europa's magnetic field and found that it is variable, indicating an internal electrically conducting fluid. Detailed computer models of the interior of Europa suggest that Europa has a global ocean 100 km deep that contains more water than held in all of Earth's oceans. That ocean might be salty as a result of dissolved minerals. The lightly cratered and thus geologically young surface indicates that energy exchange occurs between the icy crust and liquid water, but it is not known whether the icy crust is tens or hundreds of kilometers thick. It also is not yet known whether Europa has volcanic activity at the seafloor.

The Hubble Space Telescope may have detected two transient plumes of water vapor erupting from the icy surface, but they were not seen in later observations. The surface is too cold for liquid water, but lakes might lie a few kilometers underneath the surface. Scientists have reanalyzed observations of Europa in light of what has been learned about the many subsurface lakes in Antarctica, subglacial volcanoes in Iceland, and ice sheets at both poles of Earth. Many shallow "great lakes" may exist underneath the ice, and those would be prime targets for future exploration. **Figure 11.8** is an artist's schematic of Europa showing

11.2 Working it Out Tidal Forces on the Moons

Recall from Chapter 4 that the tidal force between a planet and its moon depends on the masses of the planet and the moon, and the size of the moon, divided by the cube of the distances between them:

$$F_{tidal} = \frac{2GM_{Jup}M_{moon}R_{moon}}{d_{Jup\text{-}moon}^3}$$

We can use that equation to compare the tidal forces between Jupiter and two of its moons by taking a ratio (as we did in Working It Out 4.4 to compare tidal forces between Earth and the Sun and Moon). For Io and Europa:

$$\frac{F_{tidal\text{-}Io}}{F_{tidal\text{-}Europa}} = \frac{\dfrac{2GM_{Jup}M_{Io}R_{Io}}{d_{Io}^3}}{\dfrac{2GM_{Jup}M_{Europa}R_{Europa}}{d_{Europa}^3}} = \frac{\dfrac{M_{Io}R_{Io}}{d_{Io}^3}}{\dfrac{M_{Europa}R_{Europa}}{d_{Europa}^3}}$$

After canceling out the mass of Jupiter (M_{Jup}) and the constant $2G$, and using data on Io and Europa from Appendix 4—namely, $M_{Io} = 8.9 \times 10^{22}$ kg, $M_{Europa} = 4.8 \times 10^{22}$ kg, $R_{Io} = 1,820$ km, $R_{Europa} = 1,560$ km, $d_{Io} = 422,000$ km, and $d_{Europa} = 671,000$ km—we have

$$\frac{F_{tidal\text{-}Io}}{F_{tidal\text{-}Europa}} = \frac{\dfrac{(8.9 \times 10^{22}) \times 1,820}{(422,000)^3}}{\dfrac{(4.8 \times 10^{22}) \times 1,560}{(671,000)^3}} = 8.7$$

Thus, the tidal forces on Io are much stronger than those on Europa.

---

some of the water from the ocean leaking out at the surface, making it easier to study.

Enceladus, one of Saturn's icy moons, shows a variety of ridges, faults, and smooth plains. That evidence of tectonic processes is unexpected for a small (500 km) body. The activity on Enceladus is an example of **cryovolcanism**, which is similar to terrestrial volcanism but is driven by subsurface low-temperature liquids such as water and hydrogen rather than by molten rock. Some impact craters appear softened, perhaps by the viscous flow of ice, such as the flow that occurs in the bottom layers of glaciers on Earth. Parts of the moon have no craters, indicating recent resurfacing. Terrain near the south pole of Enceladus is cracked and twisted (**Figure 11.9a**). The cracks are warmer than their surroundings, suggesting that tidal heating and radioactive decay within the moon's rocky core heat the surrounding ice and drive it to the surface.

Enceladus has a liquid ocean buried beneath 30–40 km of ice crust and is 10 km deep (**Figure 11.9b**). Active cryovolcanic plumes expel water vapor, tiny ice crystals, and salts. Some of the crystals fall back onto the surface as an extremely fine, powdery snow. The rate of accumulation is very low—a fraction of a millimeter per year—but over time the snow builds up. *Cassini* scientists estimate that the snow may be 100 meters thick in one area near the south pole of Enceladus, indicating that the plume activity has continued on and off for at least tens of millions of years. Planetary scientists are very interested in sending a mission to Enceladus to explore this ocean.

Flexing from tidal forces is the likely source of the heat energy coming from Enceladus, as on Io and Europa. Enceladus has an orbital resonance with Saturn's moon Dione. A moon made completely of ice would be too stiff for tidal heating to be effective; tidal heating works more effectively with ice over liquid water or cracked ice with some liquid. It remains a mystery why Enceladus is so active, whereas Mimas, a neighboring moon of about the same size—but closer to Saturn and also subject to tidal heating—appears geologically dead.

Figure 11.9 Images of Enceladus taken by *Cassini*. (a) The deformed ice cracks (shown blue in false color) were found to be the sources of cryovolcanism. (b) Cryovolcanic plumes in the south polar region are seen spewing ice particles into space.

Figure 11.10 Images of Saturn's largest moon, Titan, taken by *Cassini*. (a) Titan's orange atmosphere is caused by organic, smoglike particles. (b) Infrared-light imaging penetrates Titan's smoggy atmosphere and reveals surface features.

Figure 11.11 Radar imaging (false color) near Titan's north pole shows lakes of liquid hydrocarbons covering 100,000 square kilometers (km^2) of the moon's surface. Features such as islands, bays, and inlets are visible in many of those radar images.

Titan's Atmosphere, Lakes, and Ocean

Saturn's moon Titan is slightly larger than Mercury and has a composition of about 45 percent water ice and 55 percent rocky material. What makes Titan especially remarkable is its thick atmosphere and the presence of liquid methane on its surface. Whereas Mercury's secondary atmosphere has escaped to space due to proximity to the Sun, Titan's greater mass and distance from the Sun have allowed it to retain an atmosphere that is 30 percent denser than that of Earth. Titan's atmosphere, like Earth's, is mostly nitrogen. As Titan differentiated, various ices, including methane (CH_4) and ammonia (NH_3), emerged from the interior to form an early atmosphere. Ultraviolet photons from the Sun have enough energy to break apart ammonia and methane molecules—a process called **photodissociation**. Photodissociation of ammonia is the likely source of Titan's atmospheric nitrogen. Methane breaks into fragments that recombine to form organic compounds, including complex hydrocarbons such as ethane. Those compounds tend to cluster in tiny particles, creating organic smog much like the air over Los Angeles on a bad day; that gives Titan's atmosphere its characteristic orange hue (**Figure 11.10a**).

The *Cassini* spacecraft obtained close-up views of Titan's surface. Haze-penetrating infrared imaging showed broad regions of dark and bright terrain (**Figure 11.10b**). Radar imaging of Titan revealed irregularly shaped features in its northern hemisphere that appear to be widespread lakes and seas of methane, ethane, and other hydrocarbons (**Figure 11.11**). The photodissociative process by sunlight should have destroyed all atmospheric methane within a geologically brief period of about 50 million years, so a process must be renewing the methane being destroyed by solar radiation. That, along with the near absence of impact craters on the surface of Titan, suggests recent methane-producing activity. Radar views also indicate an active surface, showing features that resemble terrestrial sand dunes and channels. Heat supplied by radioactive decay could cause cryovolcanism that releases "new" methane from underground. The evidence of active cryovolcanism on Titan is indirect—the presence of abundant atmospheric methane and of methane lakes strongly suggests that Titan has some geological activity.

Titan has terrains reminiscent of those on Earth, with networks of channels, ridges, hills, and flat areas that may be dry lake basins. Those terrains suggest a methane cycle (analogous to Earth's water cycle) in which methane rain falls to the surface, washes the ridges free of the dark hydrocarbons, and then collects into drainage systems that empty into low-lying, liquid methane pools. Stubby, dark channels appear to be springs where liquid methane emerges from the subsurface; bright, curving streaks could be water ice that has oozed to the surface to feed glaciers. An infrared camera photographed a reflection of the Sun from such a lake surface. The type of reflection observed proves that the lake contains a liquid and is not frozen or dry. Recent observations might indicate waves on one of those lakes.

Titan is the only moon (besides Earth's) that has been landed upon. In 2005, *Cassini* released the *Huygens* probe, which plunged through Titan's atmosphere, taking pictures and measuring the moon's composition, temperature, pressure, and wind speeds. *Huygens* confirmed the presence of nitrogen-bearing organic compounds in the clouds. During its descent, *Huygens* encountered 120-m/s winds and temperatures as low as 88 kelvins (K). As it reached the surface, though, winds died down to less than 1 m/s and the temperature warmed to 112 K. The pictures taken by *Huygens* showed that the surface was wet with liquid methane, which evaporated as the probe—heated during its passage through the atmosphere—landed in the frigid soil. The surface also was rich with other organic (carbon-bearing) compounds, such as cyanogen and ethane. **Figure 11.12** shows that the surface around the landing site is relatively flat and littered with rounded "rocks" of water ice. The dark "soil" is probably a mixture of water and hydrocarbon ices.

As with Saturn's moon Enceladus and Jupiter's moon Europa, gravitational mapping provides indirect evidence that Titan also has an ocean buried beneath its surface. Some of Titan's surface features move by as much as 35 km, which suggests that the crust is sliding on an underlying liquid layer. The current model of Titan (**Figure 11.13**) is that its rigid ice shell varies in thickness and surrounds an ocean 100 km below the surface. That ocean would be made of water mixed with dissolved salts—possibly saltier than Earth's Dead Sea. In that model, methane outgassing would occur in hot spots.

Titan is the only moon with a significant atmosphere and the only Solar System body other than Earth that has standing liquid on the surface and a cycle of liquid rain and evaporation. In many ways, Titan resembles a primordial Earth, albeit at much lower temperatures. The presence of liquids and of organic compounds that could be biological precursors for life in the right environment makes Titan another high-priority target for continued exploration.

Figure 11.12 The two water-ice "rocks" just below the center of this *Huygens* image are about 85 centimeters (cm) from the camera and roughly 15 and 4 cm across, respectively.

CHECK YOUR UNDERSTANDING 11.2A

Which of the following moons is *not* thought to have an ocean of water beneath its surface? (a) Io; (b) Europa; (c) Enceladus; (d) Titan

Cryovolcanism on Triton

Cryovolcanism also occurs on Triton, Neptune's largest moon. Triton is an irregular moon; its retrograde orbit suggests that Triton was captured by Neptune after the planet's formation. As Triton achieved its current circular, synchronous orbit, it experienced extreme tidal stresses from Neptune, generating large amounts of thermal energy. The interior may have melted, allowing Triton to become chemically differentiated.

Triton has a thin atmosphere and a surface composed mostly of ices and frosts of methane and nitrogen at a temperature of about 38 K. The relative absence of craters tells us the surface is geologically young. Part of Triton is covered with terrain that looks like the skin of a cantaloupe (**Figure 11.14**), with irregular pits and hills that may be caused by slushy ice emerging onto the surface from the interior. Veinlike features include grooves and ridges that could result from ice oozing out along fractures. The rest of Triton is covered with smooth volcanic plains. Irregularly shaped depressions as wide as 200 km formed when mixtures of water, methane, and nitrogen ice melted in the interior of Triton and erupted onto the surface, much as rocky magmas erupted onto the lunar surface and filled impact basins on Earth's Moon.

Colorless nitrogen ice creates a localized greenhouse effect, in which solar energy trapped beneath the ice raises the temperature at the base of the ice layer. A temperature increase of only 4 K vaporizes the nitrogen ice. As that gas is formed, the expanding vapor exerts very high pressures beneath the ice cap. Eventually, the ice ruptures and vents the gas explosively into the low-density atmosphere. *Voyager 2* found four of those active geyserlike cryovolcanoes on Triton. Each consisted of a plume of gas and dust as much as 1 km wide rising 8 km above the surface, where the plume was caught by upper atmospheric winds and carried for hundreds of kilometers downwind. Dark material, perhaps

Figure 11.13 Artist's conception of Titan's internal structure showing how Titan is differentiated, with a core of water-bearing rocks and a subsurface ocean of liquid water. A layer of high-pressure ice surrounds the core, with an outer ice shell on top of the subsurface ocean.

Figure 11.14 *Voyager 2* mosaic showing various terrains on the Neptune-facing hemisphere of Triton. The lack of impact craters in the "cantaloupe terrain," visible at the top, indicates a geologically younger age than that of the bright, cratered terrain at the bottom.

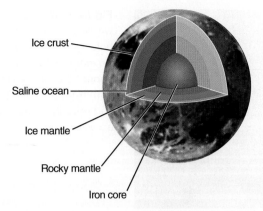

Figure 11.15 Artist's conception of the interior of Jupiter's moon Ganymede, showing that the ocean and ice may be stacked in multiple layers.

Figure 11.16 *Voyager* image showing filled impact scars on Jupiter's moon Ganymede.

silicate dust or radiation-darkened methane ice grains, is carried along with the expanding vapor into the atmosphere, from which it later settles to the surface, forming dark patches streaked out by local winds, as seen near the lower right of Figure 11.14.

Formerly Active Moons

Some moons show clear evidence of past ice volcanism and tectonic deformation, but no current geological activity. For example, Jupiter's moon Ganymede, the largest moon in the Solar System, is larger than the planet Mercury. When Jupiter was forming, low temperatures enabled grains of water ice to survive and coalesce along with dust grains into larger bodies at the distance of Ganymede's orbit. In less than half a million years, those bodies accreted to form Ganymede. Heating from accretion melted parts of Ganymede so that it is fully differentiated (**Figure 11.15**), with outer water layers, an inner silicate zone, and an iron-rich liquid core. As the moon cooled, much of the outer water layer froze, forming a dirty ice crust. Most of the denser materials sank to the central core, leaving an intermediate ice-silicate zone. Ganymede might also have a large, salty ocean underneath its icy surface, maybe 800 km deep, containing 25 times the volume of Earth's oceans.

Its surface is composed of two prominent terrains: a dark, heavily cratered (and therefore ancient) terrain, and a bright terrain characterized by ridges and grooves. The abundance of impact craters on Ganymede's dark terrain reflects the period of intense bombardment during the early history of the Solar System. The largest region of ancient dark terrain includes a semicircular area more than 3,200 km across on the leading hemisphere. Furrowlike depressions occurring in many dark areas are among Ganymede's oldest surface features. They may represent surface deformation from internal processes or may be relics of impact-cratering processes.

Impact craters on Ganymede range up to hundreds of kilometers in diameter, and the larger craters are proportionately shallower. The icy crater rims slowly slump, like a lump of soft clay. They are seen as bright, flat, circular patches found principally in the moon's dark terrain (**Figure 11.16**) and are thought to be scars left by early impacts onto a thin, icy crust overlying water or slush (**Figure 11.17**). In Chapter 8, we discussed how planetary surfaces can be fractured by faults or folded by compression resulting from movements initiated in the mantle. On Ganymede, the tectonic processes have been so intense that the fracturing and faulting have completely deformed the icy crust, destroying all

Figure 11.17 Filled impact scars form as viscous flow smooths out structures left by impacts on icy surfaces.

signs of older features, such as impact craters, and creating the bright terrain. The energy that powered Ganymede's early activity was liberated during a period of differentiation when the moon was very young. After differentiation was complete, that source of internal energy ran out, and geological activity ceased.

Many other moons show evidence that they experienced an early period of geological activity that resulted in a dazzling array of terrains. A 400-km impact crater scars Saturn's moon Tethys, covering 40 percent of its diameter, and an enormous canyonland wraps around at least three-fourths of the moon's equator. Saturn's moon Dione shows bright ice cliffs up to several hundred meters high, created by tectonic fracturing. The trailing hemisphere of Saturn's Iapetus is bright, reflecting half the light that falls on it, whereas much of the leading hemisphere is as black as tar. Those dark deposits appear *only* in the leading hemisphere of Iapetus, suggesting that they might be debris that was blasted off small retrograde moons of Saturn by micrometeoritic impacts and swept up by Iapetus as it moved along in its prograde orbit around Saturn.

Saturn's moon Mimas, no larger than the state of Ohio, is heavily cratered with deep, bowl-shaped depressions. The most striking feature on Mimas is a huge impact crater in the leading hemisphere (**Figure 11.18**). Named "Herschel" after astronomer Sir William Herschel, who discovered many of Saturn's moons, the crater is 130 km across—a third the size of Mimas itself. It is doubtful that Mimas could have survived the impact of a body much larger than the one that created Herschel. Some astronomers think that Mimas (and perhaps other small, icy moons as well) was hit many times in the past by objects so large as to fragment the moon into many small pieces. Each time that happened, the individual pieces still in Mimas's orbit would coalesce to re-form the moon, perhaps in much the same way that Earth's Moon coalesced from fragments that remained in orbit around Earth after a large planetesimal impacted Earth early in its history.

Areas on Uranus's small moon Miranda have been resurfaced by eruptions of icy slush or glacierlike flows. Other moons of Uranus—Oberon, Titania, and Ariel—are covered with faults and additional signs of early tectonism. On Ariel, in particular, very old, large craters appear to be missing, perhaps obliterated by earlier volcanism.

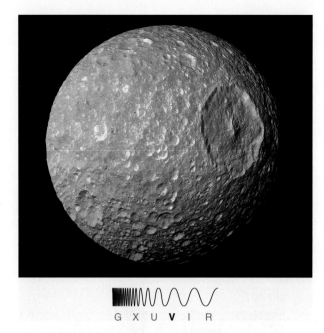

Figure 11.18 *Cassini* image showing Saturn's moon Mimas and the crater Herschel.

Geologically Dead Moons

Geologically dead moons, such as Jupiter's Callisto, Saturn's Hyperion, Uranus's Umbriel, and a large assortment of irregular moons, are moons for which little or no evidence exists of internal activity having occurred at any time since their formation. The surfaces of those moons are heavily cratered and show no modification other than the cumulative degradation caused by a long history of impacts.

Callisto is the third-largest moon in the Solar System, just slightly smaller than Mercury. Callisto also is the darkest of the Galilean moons of Jupiter, yet it is still twice as reflective as Earth's Moon. That brightness indicates that Callisto is rich in water ice, but with a mixture of dark, rocky materials. Except in areas that experienced large impact events, the surface is essentially uniform, consisting of relatively dark, heavily cratered terrain. Callisto's most prominent feature is a 2,000-km, multiringed structure of impact origin named Valhalla (the largest bright feature visible on Callisto's face in **Figure 11.19**). *Galileo* results suggest that a liquid ocean containing water or water mixed with ammonia could exist beneath the heavily cratered surface. Callisto may have partially differentiated, with rocky material separating from ices and sinking deeper into the interior.

Figure 11.19 *Galileo* image showing Jupiter's second-largest moon, Callisto. The moon's ancient surface is dominated by impact craters and shows no sign of early internal activity.

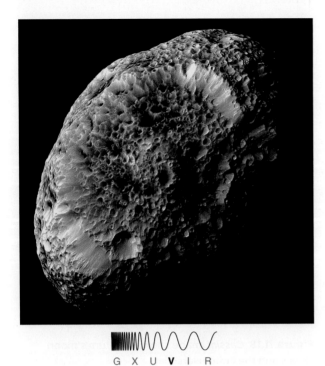

G X U **V** I R

Figure 11.20 Saturn's moon Hyperion rotates chaotically, with its rotation period and spin axis constantly changing. The 250-km moon's low density and spongelike texture, seen in this *Cassini* image, suggest that its interior houses a vast system of caverns.

Saturn's Hyperion (**Figure 11.20**) is one of the largest irregularly shaped moons and could be the remnant of an impact. The extensive craters look almost like sponges. Hyperion crosses Saturn's magnetosphere in its chaotic orbit, which seems to have left the moon with some electric charge. Umbriel, the darkest and third largest of Uranus's moons, appears uniform in color, reflectivity, and general surface features, indicating an ancient surface. The real puzzle posed by Umbriel is why it is geologically dead, whereas the surrounding large moons of Uranus have been active at least at some time in their past.

CHECK YOUR UNDERSTANDING 11.2B

Rank the following moons according to the density of impact craters (from most to least) that you would expect to observe on the surface. (a) Callisto; (b) Titan; (c) Io; (d) Ganymede

11.3 Rings Surround the Giant Planets

A planetary ring is a collection of particles—ranging in size from tiny grains to house-sized boulders—that orbit individually around a planet, forming a flat disk. Ring systems do not occur around the terrestrial planets but are found around each of the giant planets. **Figure 11.21** shows how the ring system of each giant planet varies in size and complexity: some systems extend for hundreds of thousands of kilometers, and some have detailed structure that includes many small rings. In this section, we discuss ring formation, composition, and evolution.

The Discovery of Planetary Rings

Saturn's rings have been observed for centuries. In 1610, Galileo observed two small objects next to Saturn and thought they might be similar to the four moons orbiting Jupiter. But Saturn's "moons" did not move, and 2 years later they disappeared. In 1655, Dutch instrument maker Christiaan Huygens (1629–1695) pointed a superior telescope of his own design at Saturn. Huygens observed that an apparently continuous flat ring surrounds the planet and that the ring's visibility changes with its apparent tilt as Saturn orbits the Sun. Over the next three centuries, astronomers discovered more rings around Saturn, but searches failed to detect rings around any other planet.

In 1977, a team of astronomers studying the atmosphere of Uranus during stellar occultations saw brief, minute changes in the brightness of a star as it first approached and then receded from the planet. The astronomers realized that meant that Uranus has rings. Over the next several years, stellar occultations revealed nine rings surrounding the planet. In 1986, *Voyager 2* imaged two additional rings of Uranus, and in 2005 the Hubble Space Telescope recorded two more, bringing the total to 13. In 1979, cameras on *Voyager 1* recorded a faint ring around Jupiter. The occultation technique also revealed arclike ring segments around Neptune, which were determined to be complete rings when *Voyager 2* reached Neptune in 1989.

The Orbits of Ring Particles

Ring particles follow Kepler's laws, so the speed and orbital period of each particle must vary with its distance from the planet. The closest particles move the fastest and have the shortest orbital periods (see Working It Out 11.1). The orbital

Figure 11.21 The ring systems of the four giant planets vary in size and complexity. Saturn's system, with its broad E Ring, is by far the largest and has the most complex structure in its inner rings.

periods of particles in Saturn's bright rings, for example, range from 5 hours 45 minutes at the inner edge of the innermost bright ring to 14 hours 20 minutes at the outer edge of the outermost bright ring. Ring particles have low speeds relative to one another because they are all orbiting in the same direction. A particle moving on an upward trajectory will bump into another particle on a downward trajectory, and the upward and downward motion will cancel—leaving the

Figure 11.22 Hubble Space Telescope image showing Saturn and its A Ring, B Ring, Cassini Division, and Encke Gap. The C Ring is too dim to be seen clearly.

particles moving in the same plane. A similar process occurs for particles moving inward and outward, leaving the particles moving at a constant radius.

The orbits of ring particles can also be influenced by the planet's larger moons. If the moon is massive enough, it exerts a gravitational tug on the ring particles as it passes by. If that happens over and over through many orbits, the particles are pulled out of the area, leaving a lower-density gap (see Figure 11.21). Such is the case with Saturn's moon Mimas, which causes the gap in the rings around Saturn called the **Cassini Division** (**Figure 11.22**). Mimas is in a 1:2 orbital resonance with the Cassini Division, giving a ring particle located in the Cassini Division an orbital period about Saturn that is equal to half the orbital period of Mimas. Such resonances with other moons produce some of the gaps that appear in Saturn's bright rings. One of the gaps is caused by a 4:1 resonance between the ring particles and Mimas.

Other kinds of orbital resonances also are possible. For example, most narrow rings are caught up in a periodic gravitational tug-of-war with nearby moons, known as **shepherd moons** for how they "herd" the flock of ring particles. Shepherd moons are usually small and often come in pairs, with one orbiting just inside and the other just outside a narrow ring. A shepherd moon just outside a ring robs orbital energy from any particles that drift outward beyond the edge of the ring, causing the particles to move back inward. A shepherd moon just inside a ring gives up orbital energy to a ring particle that has drifted too far in, nudging it back in line with the rest of the ring. Sometimes narrow rings are trapped between two shepherd moons in slightly different orbits.

Ring Formation and Evolution

Much of the material found in planetary rings is thought to be the result of tidal stresses. If a moon (or other planetesimal) orbits a large planet, the force of gravity will be stronger on the side of the moon close to the planet and weaker on the side farther away. That difference in gravitational force stretches out the moon, as you saw in the discussion of tidal forces in Chapter 4. If the tidal stresses are greater than the self-gravity that holds the moon together, the moon will be torn apart. The distance at which the tidal stresses equal the self-gravity is known as the Roche limit. The Roche limit applies only to objects held together by their own gravity; it does not apply to objects, such as people or cars, that are held together by other forces. If a moon or planetesimal comes within a planet's Roche limit, the object is pulled apart by tidal stresses, leaving many small pieces orbiting the planet. Those pieces gradually spread out, and their orbits are circularized and flattened out by collisions. The fragmented pieces of the disrupted body are then distributed around the planet in the form of a ring.

Planetary rings do not have the long-term stability of most Solar System objects. Ring particles are constantly colliding with one another in their tightly packed environment, either gaining or losing orbital energy. That redistribution of orbital energy can cause particles at the ring edges to leave the rings and drift away, aided by nongravitational influences such as the pressure of sunlight. Although moons may help guide the orbits of ring particles and delay the dissipation of the rings themselves, at best that condition can be only temporary. Saturn's brightest rings might be nearly as old as Saturn, but most planetary rings eventually disperse.

Even Earth may have had several short-lived rings at various times during its long history. Many comets or asteroids must have passed within Earth's Roche limit (about 25,000 km for rocky bodies and more than twice that for icy bodies) to disintegrate into a swarm of small fragments to create a temporary ring. Unlike the giant planets, however, Earth lacks shepherd moons to provide orbital stability to rings.

The Composition of Ring Material

Because much of the material in the rings of the giant planets comes from their moons, the composition of the rings is similar to that of the moons. Saturn's bright rings probably formed when a moon or planetesimal came within the Roche limit of Saturn. Those rings reflect about 60 percent of the sunlight falling on them. They are made of water ice, though a slight reddish tint indicates that they must contain small amounts of other materials, such as silicates. The icy moons around Saturn or the frozen comets of the outer Solar System could easily supply that material.

Saturn's rings are the brightest in the Solar System and are the only ones known to be composed of water ice. In stark contrast, the rings of Uranus and Neptune are among the darkest objects known in the Solar System. Only 2 percent of the sunlight falling on them is reflected back into space, making the ring particles blacker than coal or soot. No silicates or similar rocky materials are that dark, so those rings are likely to be composed of organic materials and ices that have been radiation darkened by high-energy, charged particles in the magnetospheres of those planets. (Radiation blackens organic ices such as methane by releasing carbon from the ice molecules.) Jupiter's rings are of intermediate brightness, suggesting that they may be rich in silicate materials, like the innermost of Jupiter's small moons.

The jumble of fragments that make up Saturn's rings is understood to be a product of tidal disruption of a moon or planetesimal, but moons can contribute material to rings in other ways. The brightest of Jupiter's rings is a relatively narrow strand only 6,500 km across, consisting of material from the moons Metis and Adrastea. Those two moons orbit in Jupiter's equatorial plane, and the ring they form is narrow. Beyond that main ring, however, are the very different wispy rings called gossamer rings. The gossamer rings are supplied with dust by the moons Amalthea and Thebe. The innermost ring in Jupiter's system, called the halo ring, consists mostly of material from the main ring. As the dust particles in the main ring drift slowly inward toward the planet, they pick up an electric charge and are pulled into that thick torus by electromagnetic forces associated with Jupiter's powerful magnetic field.

Finally, moons may contribute ring material through volcanism. Volcanoes on Jupiter's moon Io continually eject sulfur particles into space, many of which are pushed inward by sunlight and find their way into a ring. The particles in Saturn's E Ring are ice crystals ejected from icy geysers on the moon Enceladus, in the very densest part of the E Ring (**Working It Out 11.3**). Ice particles ejected into space replace particles continually lost from Saturn's E Ring (**Figure 11.23**). The E Ring will survive for as long as Enceladus remains geologically active.

CHECK YOUR UNDERSTANDING 11.3

If rings are observed around a planet, that means: (a) a recent source of ring material is present; (b) the planet is newly formed; (c) the rings formed with the planet; (d) the rings are made of fine dust.

Figure 11.23 Saturn's moon Enceladus (the large bright spot appearing to be on the ring) is the source of material in Saturn's E Ring. Note the distortion in the distribution of ring material around the moon. That distortion is caused by the moon's gravitational influence on the orbits of ring particles. Other bright objects in this *Cassini* image are distant stars.

11.3 Working It Out Feeding the Rings

The moons of the giant planets have a low surface gravity and a much lower escape velocity than the 11.2 km/s of Earth. Thus, volcanic emissions from some of those small moons can escape and supply material to a ring. Recall the equation from Working It Out 4.2 for the escape velocity from a spherical object of mass M and radius R:

$$v_{esc} = \sqrt{\frac{2GM}{R}}$$

Saturn's moon Enceladus has a mass of 1.08×10^{20} kg and a radius of 250 km. The escape velocity from Enceladus is given by

$$v_{esc} = \sqrt{\frac{2 \times [6.67 \times 10^{-20}\,\text{km}^3/(\text{kg s}^2)] \times (1.08 \times 10^{20}\,\text{kg})}{250\,\text{km}}}$$

$v_{esc} = 0.24$ km/s; or multiply by 3,600 s/h to get 864 km/h

That escape velocity is much lower than the speed of the volcanic plumes on Enceladus, which is nearly 2,200 km/h. The icy particles from the plumes supply particles to Saturn's E Ring.

11.4 Ring Systems Have a Complex Structure

Huygens, with his mid-17th-century understanding of physics, thought that Saturn's ring was a solid disk surrounding the planet. Not until the mid-19th century did the brilliant Scottish mathematical physicist James Clerk Maxwell show that solid rings would be unstable and would quickly break apart. In this section, we examine the details of the rings of each outer planet.

Saturn's Magnificent Rings—A Closer Look

Saturn is adorned by a magnificent and complex system of rings, unmatched by any other planet in the Solar System. Figure 11.21b shows the components of Saturn's ring system and its major divisions and gaps. Among the four giant planets, Saturn's rings are the widest and brightest. The outermost bright ring, the A Ring, is the narrowest of the three bright rings. It has a sharp outer edge and contains several narrow gaps.

In 1675, the Italian-French astronomer Jean-Dominique Cassini (1625–1712) found a gap in the planet's seemingly solid ring. Saturn appeared to have two rings rather than one, and the gap that separated them became known as the *Cassini Division*. The Cassini Division is so wide (4,700 km) that the planet Mercury would almost fit within it. Astronomers once thought that the gap was empty, but images taken by *Voyager 1* show that the Cassini Division is filled with material, albeit less dense than the material in the bright rings.

The B Ring, whose width is roughly twice Earth's diameter, is the brightest of Saturn's rings and has no internal gaps on the scale of those seen in the other bright rings. The C Ring is so much fainter than neighboring rings that it often fails to show up in normally exposed photographs. Through the eyepiece of a telescope, that ring appears like delicate gauze. No known gap exists between the C Ring and either adjacent ring; only an abrupt change in brightness marks the boundary between them. The cause of that sharp change in the amount of ring material remains unknown. Too dim to be seen next to Saturn's bright disk, the D Ring is a fourth wide ring that was unknown until it was imaged by *Voyager 1*. The D Ring shows less structure than any of the bright rings, and it does not appear to have a definable inner edge. The D Ring may extend all the way down to the top of Saturn's atmosphere, where its ring particles would burn up as meteors.

(a)

(b)

Figure 11.24 (a) *Cassini* image of the rings of Saturn showing many ringlets and minigaps. The cause of most of that structure has yet to be explained in detail. (b) Hubble Space Telescope took these images of Saturn over 6 years, showing how the rings appear to change their shape as Saturn moves through its orbit around the Sun. That effect is caused by changes in the position of Earth relative to the rings.

Saturn's bright rings are not uniform. The A and C rings contain hundreds, and the B Ring, thousands, of individual ringlets, some only a few kilometers wide (**Figure 11.24a**). Each ringlet is a narrowly confined concentration of ring particles bounded on both sides by regions of relatively little material. About every 15 years, the plane of Saturn's rings lines up with Earth, and we view them edge on (**Figure 11.24b**). The rings are so thin that they almost vanish for a day or so in even the largest telescopes. While the glare of the rings is absent, astronomers search for undiscovered moons or other faint objects close to Saturn. In 1966, an astronomer looking for moons found weak but compelling evidence for a faint ring near the orbit of Saturn's moon Enceladus. In 1980, *Voyager 1* confirmed the existence of that faint ring, now called the E Ring, and found another, closer one known as the G Ring.

The E and G rings are examples of diffuse rings. In a diffuse ring, particles are far apart, and rare collisions between them can cause their individual orbits to become eccentric, inclined, or both. Because collisions are rare, the particles tend to remain in those disturbed orbits. Diffuse rings spread out horizontally and thicken vertically, sometimes without any obvious boundaries.

Diffuse rings contain tiny particles that show up best when the viewer is looking at those rings *into* the light—that is, in the direction of the Sun. In contrast, larger objects such as pebbles and boulders are easiest to see when the light illuminating them is coming from behind the viewer. For example, dust particles on your windshield appear brightest when you are driving toward the Sun. Photographers call that effect backlighting and often place their subjects in front of a bright light to highlight hair (**Figure 11.25a**). Backlighting happens when light falls on very small objects—those with dimensions a few times to several dozen times the wavelength of light. Cat hair and human hair are near the upper end of that range. Light falling on strands of hair tends to continue in the direction away from the source of illumination. Very little light is scattered off to the side, and almost none is scattered back toward the source.

Some of the dustier planetary rings are filled with particles just a few times larger than the wavelength of visible light. To a spacecraft approaching from the direction of the Sun, such rings may be difficult or even impossible to see. Those tiny ring particles scatter very little sunlight back toward the Sun and the approaching spacecraft. However, when the spacecraft passes by the planet and looks backward toward the Sun, those same dusty rings suddenly appear as a circular blaze of light, much like a halo surrounding the nighttime hemisphere of the planet. Many planetary rings are best seen with backlighting, and some, such as Saturn's G Ring (**Figure 11.25b**), have been observed only under those conditions. In 2009, astronomers using the infrared Spitzer Space Telescope discovered another diffuse ring around Saturn (**Figure 11.26**). That dusty ring is thicker than other rings, about 20 times larger than Saturn from top to bottom, and is tilted 27° with respect to the plane of the rest of the rings.

Although Saturn's bright rings are very wide—more than 62,000 km from the inner edge of the C Ring to the outer edge of the A Ring—they are extremely thin. Saturn's bright rings are no more than 100 meters thick and probably only a few tens of meters from their lower to upper surfaces. The diameter of Saturn's bright ring system is 10 million times the thickness of the rings. If the bright rings of Saturn were the thickness of a page in a book, their diameter would equal six football fields laid end to end.

Voyager 1 images showed that Saturn's F Ring is separated into several strands that appear to be intertwined as well as displaying what appear to be several knots and kinks. Saturn's F Ring is now understood to be a dramatic example of the action of a pair of shepherd moons. The F Ring is flanked by Prometheus,

Figure 11.25 (a) Backlighting of hair creates a halo effect. (b) This backlit image of Saturn shows the G Ring.

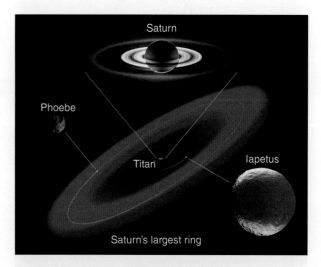

Figure 11.26 Artist's conception showing the highly inclined giant dust ring recently discovered around Saturn. That ring is so large that the rest of Saturn appears as a speck in the center (magnified in the inset).

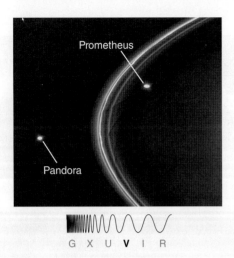

Figure 11.27 *Cassini* image showing Saturn's F Ring and its shepherd moons, Pandora and Prometheus. Also visible are some "kinks" in the inner ring.

Figure 11.28 In this high-resolution view from *Cassini*, Saturn's Encke Gap reveals a scalloped pattern, caused by the moon Pan, along its inner edge.

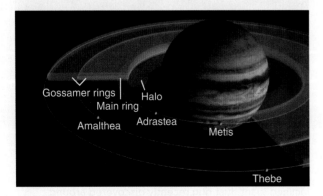

Figure 11.29 A diagram of the Jupiter ring system and the small moons that form the rings.

orbiting 860 km inside the ring, and Pandora, orbiting 1,490 km on the outside, as seen in a more recent *Cassini* image (**Figure 11.27**). Both moons are irregular in shape, with average diameters of 85 and 80 km, respectively. Because of their relatively large size and their proximity, the moons exert significant gravitational forces on nearby ring particles. The resulting tug-of-war between Prometheus pulling nearby ring particles into larger orbits and Pandora drawing its neighboring particles into smaller orbits causes the bizarre structure in the F Ring (see the **Process of Science Figure**).

The F Ring is not an isolated case. The 360-km-wide Encke Gap in the outer part of Saturn's A Ring contains two narrow rings that show bright knots and dark gaps—a structure that must be related to a 20-km moon named Pan that orbits within the gap. Small moons orbiting within ring gaps also can disturb ring particles along the edges of the gaps. **Figure 11.28** shows the scalloped pattern caused by Pan and that is found along the inner edge of the Encke Gap. Similarly, the 7-km moon Daphnis disrupts the inner and outer edges of Saturn's 35-km Keeler Gap, located near the outer edge of the A Ring.

Voyager 1 and then *Cassini* observed dozens of dark, spokelike features in the outer part of Saturn's B Ring. Those temporary features grow in a radial direction and are seasonal, lasting for less than half an orbit around Saturn, indicating that the particles in the spokes must be suspended above the ring plane, probably by electrostatic forces. One explanation is that when the charged particles interact with Saturn's magnetic field, the spokes rotate as the planet spins.

Rings around the Other Outer Planets

Ring structure among the other giant planets is not as diverse as Saturn's. Most rings other than Saturn's are narrow, although a few are diffuse. When *Voyager 1* scientists looked at Jupiter's ring system with the Sun behind the camera, they saw only a narrow, faint strand. But when *Voyager 2* looked back toward the Sun while in the shadow of the planet, Jupiter's rings suddenly blazed into prominence. **Figure 11.29** shows Jupiter's moons orbiting among that ring system. Most of the material in Jupiter's rings is made up of fine dust dislodged by meteoritic impacts on the surfaces of Jupiter's small inner moons.

Of the 13 rings of Uranus, nine are very narrow and widely spaced relative to their widths (see Figure 11.21c). Most are only a few kilometers wide, but they are many hundreds of kilometers apart. The two rings discovered by the Hubble Space Telescope in 2005 (**Figure 11.30**) are much wider and more distant than the narrow rings. The most prominent ring of Uranus, the Epsilon Ring, is eccentric and the widest of the planet's inner narrow rings, ranging between 20 and 100 km. The innermost ring is wide and diffuse, with an undefined inner edge. As with Saturn's D Ring, material in that ring may be spiraling into the top of the planetary atmosphere. When viewed under backlit conditions by *Voyager 2*, the space *between* the rings of Uranus turned out to be filled with dust, much as in Jupiter's ring system.

The rings in the Encke Gap are unusual but not unique. If shepherd moons are in eccentric or inclined orbits, they cause the confined ring also to be eccentric or inclined. That is the case for the Epsilon Ring of Uranus. Because shepherd moons can be so small, they often escape detection. According to current theories of ring dynamics, several still-unknown shepherd moons must be interspersed among the ring systems of the outer Solar System.

For a while, Neptune seemed to be the only giant planet devoid of rings. Then, in the early to mid-1980s, occultation searches by teams of astronomers began yielding confusing results. Several occultation events that appeared to be due to

PROCESS OF SCIENCE

Following Up on the Unexpected

Scientists expected to find dust particles in the rings of Saturn moving on undisturbed orbits. Instead, the F Ring particles seemed to disobey the laws of physics!

The media proclaimed:
" The laws of physics are wrong! "

Observers found more examples of deformed rings and the "shepherd moons" that cause them.

The *Voyager 1* spacecraft discovered multiple intertwining strands, knots, and kinks in Saturn's F Ring.

Further observations revealed previously unobserved moons that produced the unexpected behavior of F Ring particles.

The theory that rings can be distorted by the gravitational influence of nearby moons became widely accepted because it repeatedly passed tests of observation and simulation.

The scientists proclaimed:
" We must not have accounted for everything. "

Theorists verified gravitational effects with simulations.

Scientists are excited by apparent violations of well-supported theories because they may lead to new discoveries. One consequence is that unexpected or contradictory results often receive more attention than confirming results.

Figure 11.30 The appearance of rings depends dramatically on lighting conditions and the angle from which they are seen. Earth's view of the rings of Uranus changes over several years, as shown.

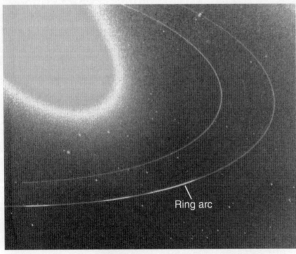

Figure 11.31 *Voyager 2* image showing the three brightest arcs in Neptune's Adams Ring. Neptune itself is greatly overexposed.

rings were seen on only one side of the planet. The astronomers concluded that Neptune was surrounded not by complete rings but rather by several arclike ring segments. When *Voyager 2* reached Neptune in 1989, it was determined that Neptune's rings were complete. The **ring arcs** are high-density segments within one of its narrow rings. All of Neptune's rings are faint and, except for the ring arcs, contain too little material to be detected through the stellar occultation technique.

Four of Neptune's six rings are very narrow, similar to the 13 narrow rings surrounding Uranus. The other two are a few thousand kilometers wide (see Figure 11.21d). Neptune's rings are named for 19th century astronomers who made major contributions to Neptune's discovery. Of those, the Adams Ring attracts the greatest attention. Much of the material in the Adams Ring is clumped into several ring arcs. Those high-density ring segments extend 4,000–10,000 km yet are only about 15 km wide. When first discovered, the ring arcs were a puzzle because mutual collisions among their particles should cause the particles to be spread more or less uniformly around their orbits. Most astronomers now attribute that clumping to orbital resonances with the moon Galatea, which orbits just inside the Adams Ring (**Figure 11.31**). Images obtained by the Hubble Space Telescope in 2004 and 2005 compared with those taken by *Voyager 2* in 1989 show that some parts of the Neptune ring arcs are unstable. Slow decay is evident in two of the arcs, Liberté and Courage, suggesting that they may disappear before the end of the century. Uranus's Lambda Ring and Saturn's G Ring also show ring arcs.

Moons and Rings around Exoplanets

Large and small planets, many in multiplanet systems, have been detected within our galaxy, demonstrating that exoplanets are common. If our Solar System is typical, we might expect that other planetary systems contain planets with rings and planets with large moons, perhaps with geological activity and water. However, identifying exoplanet moons and rings is at the limit of current astronomical instrumentation, and as of this writing no confirmed detections have been made.

The proposed methods for detecting *exomoons* are similar to those for detecting exoplanets. Recall that *Kepler* detects planets through the transit method (see Figures 7.19 and 7.20). The presence of a moon near the planet can slightly alter the depth and duration of the change in the light curve. The moon is probably in a different place in its orbit each time the planet orbits the star, so those alterations in the light curve will be different for each cycle. Over several cycles, the signature of a moon might be detected. It is also possible that astronomers could detect a large exomoon as the moon itself transits its star. Or a large moon could cause its planet's orbit to "wobble," which could be detected in the transit signal. NASA supercomputers are being used to analyze the large database from *Kepler* to look for signatures of exomoons.

Similarly, a large exoplanet with an extensive ring system, especially if the ring system has gaps, might be detectable through changes in the transit signal. The depth and length of the changes in the star's light curve could indicate the presence of such a system. **Figure 11.32** is an artist's conception of an exoplanet with rings and a large exomoon.

CHECK YOUR UNDERSTANDING 11.4

If you wanted to search for faint rings around a giant planet by sending a spacecraft on a flyby, you could make your best observations: (a) as the spacecraft approached the planet; (b) after the spacecraft passed the planet; (c) while the spacecraft is orbiting the planet; (d) during the closest flyby; (e) while the spacecraft is orbiting one of its moons.

Origins Extreme Environments

During the 1980s and 1990s, as the *Voyager* spacecraft were exploring the outer Solar System, biologists back on Earth were identifying strange forms of life. Off the coast of the Galápagos Islands, 2,500 meters beneath the ocean's surface, plates grind against each other, creating friction, high temperatures, and seafloor volcanism. Mineral-rich, superheated water pours out of hydrothermal vents. The surrounding water contains very little dissolved oxygen. No sunlight reaches those depths, yet in the total darkness of the ocean bottom, life abounds. From tiny bacteria to shrimp to giant clams and tube worms, sea life thrives in that severe environment. Being without sunlight, the small, single-celled organisms at the bottom of the local food chain get their energy from *chemosynthesis*, a process by which inorganic materials are converted into food by using chemical energy. Biologists call such life-forms *extremophiles*.

Figure 11.32 Artist's conception showing a hypothetical Earth-like moon in the foreground, orbiting around a Saturn-like exoplanet.

Similarly, robust types of bacteria are found flourishing in the scalding waters of Yellowstone's hot springs; in the bone-dry oxidizing environment of Chile's Atacama Desert; and in the Dead Sea, where salt concentrations run as high as 33 percent. Bacteria have even been found in core samples of ancient ice 3,600 meters below the East Antarctic ice sheet. When it comes to harsh habitats, life is amazingly adaptable. If life can exist under such extreme conditions on Earth, might it also exist on those moons of the giant planets that have the ingredients necessary for life on Earth: liquid water, an energy source, and the presence of organic compounds?

When scientists realized that Mars and Venus were not as Earth-like as once imagined, prospects for finding life elsewhere in the Solar System seemed dim. Now astrobiologists are turning their attention to some of the small worlds around the giant planets far from the Sun. Those moons may supply clues about the history of life in the Solar System. Their environments may be similar to some of the ecological niches on Earth that support extremophiles. The conditions necessary to create and support life on Earth—liquid water, heat, and organic material—could all be present in oceans on some of those moons.

Enceladus, Saturn's geologically active ice moon, spews salty water-ice grains, indicating liquid water below the surface. Perhaps its south polar region is habitable. The fractured ice floes on Jupiter's Europa may cover an ocean warmed and enriched by geothermal vents similar to those that dot the floors of Earth's oceans. Like Europa, Callisto also shows magnetic variability, possibly indicating a salty ocean. Saturn's Titan has an atmosphere, liquid methane lakes, and a possible subsurface ocean. Titan is similar in several ways to a much earlier Earth.

The presence of comet-borne organic material in large bodies of water on Europa, Callisto, Enceladus, or Titan cannot yet be confirmed. Methane can arise from biological processes or can come from chemical or geochemical processes, so its detection on Enceladus and Titan is tantalizing but is not definitive evidence of life. In addition to the presence of methane, spectroscopy reveals organic gases in Titan's massive atmosphere. Titan's nitrogen atmosphere contains compounds of biological interest. For example, five molecules of hydrogen cyanide (HCN) will spontaneously combine to form adenine, one of the four primary components of DNA and RNA. HCN also is a building block of amino acids, which combine, in turn, to form proteins. Photodissociation and recombination of those various gases produce complex organic molecules that then rain out onto Titan's surface as a frozen tarry sludge. Biochemists think that many of those substances are biological precursors, similar to the organic molecules that preceded the development of life on Earth.

Astronomers anticipate future exploration of those moons, which may yield fascinating clues to the origins of terrestrial life.

READING ASTRONOMY NEWS

Spacecraft's 13 years in orbit have produced suggestions of potentially habitable worlds.

NASA killed *Cassini* to avoid contaminating Saturn's moons

By **NICOLE MORTILLARO**, CBC News

NASA scientists killed the hard-working *Cassini* spacecraft to avoid contaminating Saturn's moons with Earth microbes because they may have the potential to support life.

On Friday at 7:55 a.m. ET, the world said goodbye to *Cassini* after 20 years in space and 13 years orbiting Saturn and its moons, providing incredibly detailed, high-resolution photos of what many consider the jewel of our solar system.

Though *Cassini*'s life was extended twice with new missions, the spacecraft had a limited fuel supply. So NASA announced in April that it would carry out one final mission, dubbed The Grand Finale.

Scientists and engineers altered the orbit of the spacecraft to run into Saturn, where it would safely burn up in its atmosphere.

Cassini's fate arose directly from its success in shedding a light on the possibility of life on other worlds.

On Wednesday, NASA's planetary science division director Jim Green spoke about Enceladus, a small, icy moon, spewing organic material that likely originates from a subsurface ocean.

"What we thought was an icy ball, when we observed the southern hemisphere, and geysers of water spewing out into the Saturn system, it amazed us," he told reporters.

"And it began changing the way we view the habitability or potential habitability of moons in the outer part of our solar system."

Allowing *Cassini* to run out of fuel would leave NASA no way to control the spacecraft. One day it could crash into Enceladus or even Titan, Saturn's largest moon, which is also being studied for potential habitability.

While there is no guarantee that there is microbial life on any of these moons, there is a chance of contaminating their surface with Earth microbes.

"Now because of the importance of Enceladus that *Cassini* has shown us, and of Titan, another potential world that could be habitable for life, perhaps not like we know it, but perhaps completely different than ours, we had to make decisions on how to dispose of the spacecraft," Green said.

"And that led us, inevitably, to the plan of taking *Cassini* and plunging into Saturn."

On the plus side, *Cassini*'s Grande Finale has given us unprecedented views of the planet's north pole as well as its intricate ring system. For the first time in its mission, the spacecraft flew between the planet and the rings.

And, with its success and valuable insight into the potential habitability of other worlds, it's likely there could be new missions.

Missions have been suggested to explore Enceladus, including its interior and the subsurface liquid body, Larry Soderblom, an interdisciplinary scientist on the Cassini mission, told CBC News.

"We might find in fact a higher probability than I'm willing to admit right now that life exists in the interior of Enceladus."

While astronomers and planetary enthusiasts may miss the data and photos gathered by Saturn, NASA's *Juno* spacecraft is also orbiting Jupiter. And there are plans for the *Europa Clipper*, which will conduct flybys of another icy, potentially habitable moon, Europa.

If there is a chance for life to be found in the Saturn system, NASA is eager to protect it.

"Because of planetary protection and our desire to go back to Enceladus and go back to Titan and go back to the Saturn system, we must protect those bodies for future exploration," Green said.

1. How could *Cassini* "contaminate" a moon?
2. Do a search on YouTube to see the end of *Cassini*. What did the final dives reveal about Saturn's rings?
3. Do a search to see whether new missions to Saturn's moons have been approved. Which moons will be explored?
4. Go to the Web page for the NASA Office of Planetary Protection (https://planetaryprotection.nasa.gov/overview). Which moons are affected by that policy? How might that policy affect future plans for a lander on Europa? What was done to protect Mars from contamination during the NASA Mars lander missions?

Source: Nicole Mortillaro, "NASA killed Cassini to avoid contaminating Saturn's moons," CBC News, September 15, 2017. Reprinted by permission of CBC Licensing.

Summary

The moons of the outer Solar System are composed of rock and ice. A few moons are geologically active, but most are dead. All four giant planets have ring systems, which are temporary, created from and maintained by moons also in orbit around those planets. Scientists are excited about the evidence for oceans on several of the moons and the possibility that some form of life may exist in those oceans.

LG 1 **Compare the orbits and formation of regular and irregular moons.** Most of the regular moons were formed along with their parent planets and have short and nearly circular orbits. Irregular moons were captured later, have more elongated orbits, and often orbit in the opposite direction of the planet's rotation. Observations of the orbits of moons can be used to find the masses of their host planets.

LG 2 **Describe the evidence for geological activity and liquid oceans on some of the moons.** Observations indicate that Jupiter's Io is the most volcanically active body in the Solar System. Jupiter's moon Europa contains an enormous subsurface ocean and probably has some geological activity. Saturn's moon Titan has lakes of liquid methane and perhaps a deep, salty ocean. Saturn's moon Enceladus and Neptune's moon Triton have cryovolcanoes. The large Galilean moons of Jupiter, Ganymede and Callisto, also may have subsurface oceans. Some moons were geologically active in the past, as indicated by crater scars and areas smoothed by flowing fluids. Moons that have always been geologically dead show nothing but impact craters on their surfaces.

LG 3 **Describe the composition, origin, and general structure of the rings of the giant planets.** Rings are formed of countless numbers of particles all in the same plane, held to the host planet by gravity. Some rings form when moons cross a planet's Roche limit. The composition of those moons determines the composition of the rings that form from them. Shepherd moons often maintain and shape rings by gravitationally pulling and pushing those particles as they pass by. Ring particles also interact gravitationally with one another. Some rings may be temporary features held in place by moons. Saturn's bright rings and its E Ring are made primarily of water ice: the rings of the other planets are composed of darker materials.

LG 4 **Explain the role gravity plays in the structure of the rings and the behavior of ring particles.** Saturn's ring system is the most complex, and it is the best laboratory for understanding gravitational interactions between moons and rings, interactions between rings, and ring formation and dissipation. Gravity holds the ring particles in orbit around the planet, and gravitational interactions with moons determine the size and shape of rings.

? Unanswered Questions

- What is the source of Titan's nitrogen atmosphere? One group of experimenters studied that by blasting a laser at water-ammonia (H_2O-NH_3) ice to simulate cometary impacts and see whether nitrogen gas (N_2) forms. They concluded that the observed amount of N_2 in Titan's atmosphere could have been created from ammonia ice in that way. Another idea is that those gases were accreted during the formation of the moon. Using data from the *Huygens* probe, other researchers conclude that if Titan had differentiated like Ganymede, hydrothermal activity released the gases from a hot core. Astronomers want to understand why Titan has an atmosphere and the other larger moons do not.

- What is the origin of Saturn's brightest rings? One early hypothesis is that the rings come from a moon that approached too close to the planet, but Saturn's moons are composed of rock and ice, and the rings are solely ice. Some recent computer models start with a differentiated moon the size of Titan that had a rock and iron core and a large, icy mantle. As the moon in the model slowly migrates toward Saturn and crosses the Roche limit, tidal forces rip away its water ice, but not the rocky core. According to that model, the core might have continued migrating inward until it fell into Saturn, with the ice forming Saturn's ring. As time went on, the ring spread, and as material crossed the Roche limit outward, Saturn's small moonlets formed. Computers are only now getting fast enough to test those models, which suggest a unified origin for many of the moons and rings.

Questions and Problems

Test Your Understanding

1. Categorizing moons by geological activity is helpful because
   a. comparing them reveals underlying physical processes.
   b. geological activity levels drop with distance from the Sun.
   c. the size and composition of the moons depend on geological activity.
   d. most moons are very similar to one another.

2. Why are Ganymede and Callisto geologically dead, whereas the other two Galilean moons of Jupiter are active?
   a. They are larger.
   b. They are farther from Jupiter.
   c. They are more massive.
   d. They have retrograde orbits.

3. Moons of outer planets may serve as a home for life because
   a. some have liquid water.
   b. some have organic molecules.
   c. some have an interior source of energy.
   d. all of the above

4. Io has the most volcanic activity in the Solar System because
   a. it is continually being bombarded with material in Saturn's E Ring.
   b. it is one of the largest moons and its interior is heated by radioactive decays.
   c. of gravitational friction caused by the moon Enceladus.
   d. its interior is tidally heated as it orbits around Jupiter.
   e. the ice on the surface creates a large pressure on the water below.

5. Gravitational interactions with moons produce
   a. fine structure within rings.
   b. short-lived rings.
   c. smoothed-out rings.
   d. rings with spokes.

6. Saturn's bright rings are located within the planet's Roche limit, supporting the theory that those rings (select all that apply)
   a. formed of moons torn apart by tidal stresses.
   b. formed when Saturn formed.
   c. are relatively recent.
   d. are temporary.

7. The story of the F Ring of Saturn is an example of
   a. an unexplained phenomenon.
   b. media bias.
   c. the self-correcting nature of science.
   d. a violation of causality.

8. If a moon revolves opposite to its planet's rotation, it probably
   a. was captured after the planet formed.
   b. had its orbit altered by a collision.
   c. has a different composition from that of other moons.
   d. formed very recently in the Solar System's history.

9. Under what lighting conditions are the tiny dust particles found in some planetary rings best observed?
   a. viewed from the shadowed side of the planet, looking toward deep space
   b. viewed from the shadowed side of the planet, looking toward the Sun
   c. viewed from the near side of the planet, looking toward deep space
   d. viewed from the near side of the planet, looking toward the Sun

10. Planets in the outer Solar System have more moons than those in the inner Solar System because
    a. the solar wind was weaker there.
    b. more debris was present around the outer planets when they were forming.
    c. the outer planets captured most of their moons.
    d. more planetesimal collisions occurred far from the Sun.

11. The difference between a moon and a planet is that
    a. moons orbit planets, whereas planets orbit stars.
    b. moons are smaller.
    c. moons and planets have different compositions.
    d. moons and planets formed differently.

12. Scientists determine the geological history of the moons of the outer planets from
    a. seismic probing.
    b. radioactive dating.
    c. surface features.
    d. time-lapse photography.

13. The energy that keeps Io's core molten comes from
    a. the Sun.
    b. radioactivity in the core.
    c. residual heat from the collapse.
    d. Jupiter's gravity.

14. We classify moons as formerly active if they
    a. are covered in craters.
    b. have no young craters.
    c. have regions with few craters.
    d. have regions with no craters.

15. Volcanoes on Enceladus affect the E Ring of Saturn by
    a. pushing the ring around.
    b. stirring the ring particles.
    c. supplying ring particles.
    d. dissipating the ring.

Thinking about the Concepts

16. Explain the process that drives volcanism on Jupiter's moon Io.

17. Describe cryovolcanism and explain its similarities and differences with respect to terrestrial volcanism. Which moons show evidence of cryovolcanism?

18. Discuss evidence supporting the idea that Europa has a subsurface ocean of liquid water.

19. Titan contains abundant amounts of methane. What process destroys methane in that moon's atmosphere?

20. In certain ways, Titan resembles a frigid version of the early Earth. Explain the similarities.

21. Some moons display signs of geological activity in the past. Identify some of the evidence for past activity.

22. Why do the outer planets but not the inner planets have rings? Describe a ground-based technique that led to the discovery of rings around the outer planets.

23. What are ring arcs, and where are they found?

24. Identify and explain two possible mechanisms that can produce planetary ring material.

25. Explain two mechanisms that create gaps in Saturn's bright-ring system.

26. Describe ways in which diffuse rings differ from other planetary rings.

27. In Chapter 1, we stated that "all scientific theories are provisional." Explain how the discovery of the detailed structure of Saturn's F Ring (as described in the Process of Science Figure) challenged a scientific theory and how that apparent conflict was ultimately resolved.

28. Astronomers think that most planetary rings eventually dissipate. Explain why the rings do not last forever. Describe and explain a mechanism that keeps planetary rings from dissipating.

29. Name one ring that might continue to exist indefinitely, and explain why it could survive when others might not.

30. Make a case for sending a space mission to one of the moons. Which moon would you choose, and what observations would you try to obtain?

Applying the Concepts

31. Io has a mass of 8.9×10^{22} kg and a radius of 1,820 km.
    a. Using the formula provided in Working It Out 11.3, calculate Io's escape velocity.
    b. How does Io's escape velocity compare with the vent velocities of 1 km/s from its volcanoes?

32. Use the value of P^2/A^3 for Europa, as in Working It Out 11.1, to estimate the mass of Jupiter.

33. Follow Working It Out 11.1 to use one of Saturn's moons to estimate the mass of that planet

34. Study Figure 11.2.
    a. Are the scales on Figure 11.2a and 11.2b linear or logarithmic?
    b. About how much larger is the space shown in Figure 11.2b than in Figure 11.2a?

35. Planetary scientists have estimated that Io's extensive volcanism could be covering the moon's surface with lava and ash to an average depth of up to 3 millimeters (mm) per year.
    a. If Io is a sphere with a radius of 1,820 km, what are its surface area and volume?
    b. What is the volume of volcanic material deposited on Io's surface each year?
    c. How many years would it take for volcanism to perform the equivalent of depositing Io's entire volume on its surface?
    d. How many times might Io have "turned inside-out" over the age of the Solar System?

36. Use the formula for tidal forces in Working It Out 11.2 to answer the following questions:
    a. If the radius of the moon increases but its mass stays the same, what happens to the tidal force?
    b. If the radius of the moon's orbit decreases, what happens to the tidal force?
    c. If the mass of the central planet increases, what happens to the tidal force?

37. Use Working It Out 11.2 to compare the tidal force between Jupiter and Io with the tidal force between Earth and its Moon.

38. A 60-kg astronaut is spacewalking outside the International Space Station, 380 km above Earth. Use Working It Out 11.2 to find the tidal force on the astronaut. Assume that her feet point toward the center of Earth.

39. Assuming that all other numbers are held constant, make a graph of the tidal force versus the distance between a planet and its moon. On the same graph, plot the gravitational force (which falls off as $1/d^2$). Compare the two graphs to determine the relative importance of tidal forces and gravitational forces at various distances.

40. Particles at the very outer edge of Saturn's A Ring are in a 7:6 orbital resonance with the moon Janus. If the orbital period of Janus is 16 hours 41 minutes ($16^h 41^m$), what is the orbital period of the outer edge of Ring A?

41. Use Working It Out 11.3 to find the escape velocity from Saturn's moon Janus.

42. The inner and outer diameters of Saturn's B Ring are 184,000 and 235,000 km, respectively. If the average thickness of the ring is 10 meters and the average density is 150 kilograms per cubic meter (kg/m³), what is the mass of Saturn's B Ring?

43. The mass of Saturn's small, icy moon Mimas is 3.8×10^{19} kg. How does that mass compare with the mass of Saturn's B Ring, as calculated in Question 42? Why is that comparison meaningful?

44. The inner and outer diameters of Saturn's B Ring are 184,000 and 235,000 km, respectively. Use that information to find the ratio of the periods of particles at those two diameters. Does the B Ring orbit like a solid disk or like a collection of separate particles?

45. Use the escape velocity equation in Working It Out 11.3 to answer the following questions
    a. For more massive planets, is the escape velocity higher or lower?
    b. For larger planets, is the escape velocity higher or lower?
    c. If you know the escape velocity of a planet, what other piece of information do you need to find the planet's mass?

USING THE WEB

46. Go to *Sky & Telescope*'s "Jupiter's Moons" Web page (http://www.skyandtelescope.com/wp-content/observing-tools/jupiter_moons/jupiter.html). Enter your date and time. Where are the four Galilean moons? Keep clicking on "+1 hour" to see how their positions change. Which moon passes in front of (transits) Jupiter? If possible, observe those moons for a couple of nights through a small telescope, binoculars, or telephoto camera lens. Sketch the positions of the moons.

47. Look at the updated lists of giant planet moons on NASA's "Our Solar System" website (https://solarsystem.nasa.gov/planets; click on "Moons") or on the Carnegie Institution of Washington Department of Terrestrial Magnetism's (DTM) "Jupiter Satellite Page" (http://home.dtm.ciw.edu/users/sheppard/satellites). What are two of the more recently discovered moons of one of the planets? Are the orbits retrograde? What are the eccentricities and inclinations of the orbits? Where would those new moons fit on the graph of orbits on the DTM's website? Where did the names come from? Why are some of the more recent moons labeled "provisional"?

48. Go to the website for the former *Cassini* mission (https://saturn.jpl.nasa.gov). Watch the video at https://saturn.jpl.nasa.gov/resources/732/ to listen to the "hiss" from the aurora. What moon causes Saturn's aurora?

49. Missions to the moons:
    a. Go to the website for NASA's *Clipper* mission to Europa (https://www.nasa.gov/europa). What is the science plan for the mission? Why will the spacecraft orbit Jupiter and not Europa? What is its status? When will it get to Europa? Will it include a lander?
    b. Go to the website for the European Space Agency's *JUICE* (*JUpiter ICy moons Explorer*) mission (http://sci.esa.int/juice/). Which moons will that mission study? What are the goals of the mission? What is the status of the project? Why will *JUICE* take longer to get to Jupiter than the *Clipper* mission?
    c. Do a search to see whether a new mission to the one of the moons of Saturn has been approved.

50. Do a search to see whether moons or rings have been confirmed on exoplanets. Why is that topic of interest to astronomers?

EXPLORATION

Measuring Features on Io

digital.wwnorton.com/astro6

Part A: Finding the Image Scale

Finding the scale of an image is like finding the scale on a map. On a map, each inch or centimeter represents miles or kilometers of actual space. The same thing is true in an image. If you take a picture of a meter stick and then measure the meter stick in the picture to be 10 cm long, you know that 10 cm in the picture represents 1 meter of actual space, and 1 cm in the picture represents 10 cm of actual space.

To find the scale, you must compare the size of something in the image with its actual size in space. In **Figure 11.33**, the moon Io is the known object.

1 Use a ruler to measure the diameter of that image of Io in millimeters.

2 Estimate the error in your measurement. (How far off could your measurement be?)

Figure 11.33 A *Galileo* image of Jupiter's moon Io, a very active moon. This image, which appeared previously in the chapter as Figure 11.5b, offers an opportunity to make some measurements.

3 Find the radius from that diameter.

4 Look up the actual radius of Io (in kilometers) in Appendix 4 or online.

5 Find the image scale (s) as follows:

$$s = \frac{\text{Actual size of object}}{\text{Size of object on the image}}$$

6 What are the units of that image scale?

Part B: Finding the Sizes of Features on Io

7 Near the center of Io is a geyser, surrounded by a black circle and a white ring. That is Prometheus Patera. What is the diameter of the black ring around Prometheus in that image (in millimeters)? (Do not use the inset image.)

8 Multiply the measured diameter by the image scale to find the actual size of the circle around Prometheus. Identify something on Earth that is about the same size.

9 On the limb of the moon is a purple plume from an erupting sulfur geyser. Follow steps 7 and 8 to find the height of that plume. Identify something on Earth that is about the same height.

That measurement has at least two sources of error. One is the error of measurement—how well you use a ruler and how accurately you determined the top and the bottom of the plume. The other source of error is uncertainty about whether the plume is *exactly* at the limb of the moon. If the eruption occurred on the far side of Io, would your calculation be an overestimate or underestimate of the plume height?

12 Dwarf Planets and Small Solar System Bodies

Here in Chapter 12, we explore the small bodies remaining from the formation of the Solar System (see Chapter 7), including dwarf planets, irregular moons, asteroids, and comets. Those remaining planetesimals, and the fragments that some of them continually create, have revealed much about the physical and chemical conditions of the earliest moments in the history of the Solar System and how the Solar System formed and evolved. In addition, those planetesimals are important because some fraction of the water, gases, and organic material found on Earth and in the inner Solar System came from comets and asteroids.

LEARNING GOALS

By the end of this chapter, you should be able to:

LG 1 List the categories of small bodies and identify their locations in the Solar System.

LG 2 Describe the defining characteristics of the dwarf planets in the Solar System.

LG 3 Describe the origin of the types of asteroids, comets, and meteorites.

LG 4 Explain how asteroids, comets, and meteoroids provide important clues about the history and formation of the Solar System.

LG 5 Describe what has been learned from observations of recent impacts in the Solar System.

The European Space Agency landed a probe on Comet 67P/Churyumov-Gerasimenko in 2014. ▶▶▶

Why land a spacecraft on a comet?

12.1 Dwarf Planets May Outnumber Planets

Recall from Chapter 7 that very early in the history of the Solar System—when the Sun was becoming a star—tiny grains of primitive material stuck together to produce swarms of small bodies called *planetesimals*. Those that formed in the hotter, inner part of the Solar System were composed mostly of rock and metal, whereas those in the colder, outer part were composed of ice, organic compounds, and rock. Some of the objects collided to become planets and moons. Many are still present, however, and they remain a scientifically important component of the present-day Solar System.

Dwarf planets, asteroids, Kuiper Belt objects, comets, and meteoroids are smaller than planets and orbit the Sun. Dwarf planets are found in the asteroid belt and in the **Kuiper Belt**. The asteroid belt in the region between the orbits of Mars and Jupiter contains most of the asteroids in the Solar System. The Kuiper Belt is a disk-shaped population of comet nuclei extending from Neptune's orbit (30 astronomical units [AU]) to about 50 AU.

The dwarf planets orbit the Sun and have round shapes, but because they have relatively small mass, they have not cleared the area around their orbits. As of this writing, the Solar System has five officially recognized dwarf planets: Pluto, Eris, Haumea, Makemake, and Ceres (their properties are tabulated in Appendix 4). Ceres is a large object in the main asteroid belt, whereas the others are found in the Kuiper Belt. Hundreds of dwarf planet candidates exist, but their shapes have not yet been measured well enough to classify them definitively.

Pluto

Throughout the 19th century, discrepancies were observed between the observed and predicted orbital positions of Uranus and Neptune. Early in the 20th century, astronomers hypothesized that an unseen body was perturbing the orbits of those planets. Astronomers called that body Planet X and estimated that it had 6 times Earth's mass and was located beyond Neptune's orbit. Astronomer Clyde W. Tombaugh (1906–1997) discovered Planet X in 1930, not far from its predicted position. It became the Solar System's ninth planet and was named Pluto for the Roman god of the underworld. However, observational evidence soon indicated that the mass of Pluto was far too small to have produced the perturbations in the orbits of Uranus and Neptune. When astronomers reanalyzed the 19th century observations, they found that the orbital "discrepancies" were a mistake. Pluto's discovery thus turned out to be a coincidence.

Pluto's orbital period is 248 Earth years. Its orbit is elliptical and is tilted with respect to the plane of the Solar System. Pluto's orbit periodically crosses inside Neptune's nearly circular orbit—from 1979 to 1999, Pluto was closer to the Sun than Neptune. Pluto has only two-thirds the diameter of our Moon. It has five known moons, the largest of which is Charon, about half the size of Pluto. The total mass of the Pluto-Charon system is 1/400 that of Earth, or 1/5 the mass of the Moon. Similar to Uranus, Pluto rotates nearly on its side—that is, its equatorial plane is nearly perpendicular to its orbital plane. Pluto and Charon are a tidally locked pair: each has one hemisphere that always faces the other.

As they discovered more about Pluto and other objects beyond Neptune's orbit, some astronomers questioned Pluto's classification as a planet, and a debate ensued. In 2005, astronomers identified an object more distant than Pluto, later named Eris, and then Eris's moon, Dysnomia. Observations of Dysnomia's orbit yielded a mass for Eris, which turned out to be about 28 percent greater than

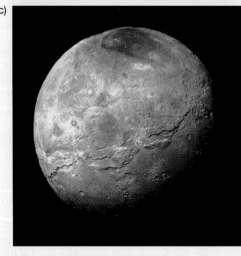

Figure 12.1 All images are from the *New Horizons* spacecraft flyby in 2015. (a) Enhanced color image of Pluto. (b) Dark highlands shown in the lower right border a section of icy plains. (c) Pluto's largest moon, Charon.

Pluto's mass. Pluto and Eris have similar nitrogen and methane abundances and a relatively large moon. Then the inevitable question emerged: Should astronomers consider Eris the Solar System's tenth planet? Or should neither Pluto nor Eris be called a planet? The members of the International Astronomical Union (IAU) decided in August 2006: Pluto is round like the classical planets, but it cannot clear its neighborhood, so it was reclassified as a dwarf planet. (see the **Process of Science Figure**).

Pluto and Charon were not on *Voyager*'s route through the Solar System, so until recently only limited information was available about their surface properties or geological history. That changed when NASA's *New Horizons* flyby spacecraft passed within 12,500 km of Pluto in July 2015. Astronomers were surprised to find that Pluto and Charon had geological activity. *New Horizons* images of Pluto showed varied surface features, including a large bright region whose western half is a basin containing nitrogen, methane, and carbon monoxide ices, possibly from a large impact (**Figure 12.1a**). **Figure 12.1b** shows highlands and icy plains. Pluto's surface contains an icy mixture of frozen water, carbon dioxide, methane, and carbon monoxide, with flowing nitrogen ice. Pluto has a thin atmosphere of nitrogen, methane, ethane, and carbon monoxide: those gases freeze out of the atmosphere when Pluto is more distant from the Sun and therefore colder. Astronomers suspect, but have not confirmed, that Pluto has cryovolcanoes (Section 2 in Chapter 11) and a subsurface ocean of liquid water. Charon has no atmosphere. Its surface has deep canyons, which might have formed as an ancient ocean froze and pushed the surface outward (**Figure 12.1c**).

Ceres

In 1801, Sicilian astronomer Giuseppe Piazzi found a bright object between the orbits of Mars and Jupiter. He named the new object Ceres. Piazzi thought he might have found a hypothetical "missing planet," but as more objects were discovered orbiting the region between Mars and Jupiter, astronomers classified Ceres as belonging to a new category of Solar System objects called asteroids. Ceres is the largest body in the main asteroid belt. It also is now called a dwarf planet because although it is round (**Figure 12.2a**), it has not cleared its surroundings. With a diameter of about 940 km, Ceres is larger than most moons but smaller than any planet. It contains about a third of the total mass in the asteroid belt but only about 1.3 percent of the mass of Earth's Moon. Ceres rotates on its axis with a period of about 9 hours, typical of many asteroids.

Ceres is being explored by the *Dawn* mission, which went into orbit around Ceres in 2015 and is expected to remain until its fuel runs out in 2018. The spacecraft found many features similar to those seen on the inner planets. For example, the 34-km-wide Haulani Crater (**Figure 12.2b**) has landslides on the crater's limb and blue streaks radiating outward from the central crater peak. The color in that image is enhanced, and blue indicates young surfaces on Ceres. Those color variations indicate that the surface of Ceres has a different composition from that of the subsurface layer. Ceres has a large mountain, Ahuna Mons, that is 4–5 km (13,000–16,000 feet) high. It seems to be a salty-mud cryovolcano that may have been active in the last few hundred million years. Evidence also indicates that Ceres may have had cryovolcanoes that were flattened out by gravity over hundreds of millions of years. Because those volcanoes are made of ice rather than rock, they flow downward slowly, much like a blob of honey gradually spreads out in the bottom of a mug.

Figure 12.2 (a) Dwarf planet Ceres photographed by the *Dawn* mission. (b) The 34-km-wide Haulani Crater shows landslides on the crater's limb, as well as smooth material and a central peak on the crater floor. Blue indicates a younger surface on Ceres. The color in this image is enhanced. (c) Diagram of the interior of Ceres.

PROCESS OF SCIENCE

How to Classify Pluto

Pluto's reclassification from planet to dwarf planet in 2006 received a lot of publicity. The reclassification of Pluto is a clear example of the scientific method in practice.

"Planet" → "Dwarf Planet"

1930
Pluto was discovered. It was classified as a planet because it orbited the Sun.

1978
Pluto's moon Charon was discovered. As information about Pluto accumulated, it became clear that Pluto was different from the other planets:
- Pluto's composition is not like the other planets.
- Pluto crosses Neptune's orbit.
- Pluto has a moon nearly as large as itself.

2006
In light of recent discoveries of new Pluto-like objects, the International Astronomical Union redefined the word *planet*. Pluto fell outside the new definition and was classified as a "dwarf planet."

Scientific decision making must follow the weight of the evidence, not historical precedent.

Through gravity mapping, astronomers have confirmed that Ceres is much less dense than the other rocky bodies in the solar system. Ceres is at least partly differentiated, with a rocky core and a lower-density, 40-km-thick crust of ice and salt (**Figure 12.2c**). Water ice has been detected at about a dozen sites, including within craters. Because hydrogen has been detected on the surface, scientists think that water probably exists only 1–2 meters beneath. That water is thought to have been with Ceres since it formed instead of having come in from comets and asteroids. Minerals that contain water are seen on the surface, and the icy crust might be the remains of an ancient ocean. As Ceres has moved in its orbit around the Sun, *Dawn* has seen changes, as some crater ice releases water vapor, condenses, and causes more landslides.

Eris, Haumea, and Makemake

Eris, Haumea, and Makemake were discovered in the 21st century. Eris is about the same size as Pluto but has more mass. The highly eccentric orbit of Eris carries it from 37.8 AU out to 97.6 AU away from the Sun, with an orbital period of 557 years (**Working It Out 12.1**). Eris is near the most distant point in its orbit, making it the most remote known object in the Solar System. At that distance it is about 100 times fainter than Pluto. The eccentric orbits of other Solar System bodies will eventually carry them farther away than Eris, however, so Eris will not always be the most distant object known.

When astronomers combine the observed brightness of Eris with its diameter, they find it has a surprisingly high albedo of 0.96. (Recall from Chapter 5 that albedo is the percentage of light an object reflects.) The surface of Eris is more reflective than that of any other major Solar System body except Enceladus, so Eris too must have a coating of pristine ice. The surface of Enceladus is water ice, whereas Eris is covered with methane ice. At its present location, the average surface temperature on Eris is cold enough to freeze out any atmospheric methane, but it will probably develop a methane atmosphere when it comes closest to the Sun in the year 2257.

Haumea and Makemake are both smaller than Pluto (**Figure 12.3**) and have slightly larger orbits. Haumea has two moons—Hi'iaka and Namaka—enabling astronomers to calculate the system's mass. Although Haumea has enough mass to pull itself into a spherical shape, it spins so rapidly on its axis that its shape is flattened, with an equatorial radius approximately twice its polar radius. That difference between the equatorial and polar radii gives Haumea an oblateness (a measure of how far an object is from being perfectly round) of 0.5, giving it the most distorted shape of any of the planets or dwarf planets. Hubble Space Telescope (HST) infrared imaging indicates that Haumea and its two moons are covered in water ice. Astronomers think that those three objects and some smaller debris were left over after a larger body broke up after a collision. Makemake is about 2/3 the diameter of Pluto and has a surface covered in nitrogen, ethane, and methane ices. In 2016, HST discovered a moon orbiting Makemake.

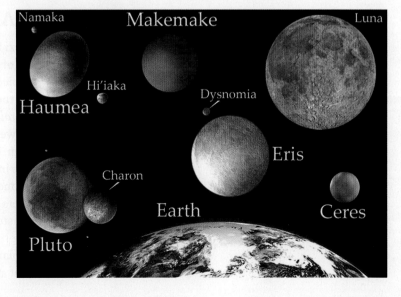

Figure 12.3 This NASA illustration shows the five dwarf planets compared with our Moon (Luna) and Earth.

CHECK YOUR UNDERSTANDING 12.1

Why are these objects called dwarf *planets* even though they are actually smaller than some moons?

12.1 Working It Out Eccentric Orbits

Many of the objects discussed in this chapter are far from the Sun. Their complete orbits take many years, but observing a complete orbit is not necessary to determine an object's semimajor axis and eccentricity: those values can be obtained from watching how the object moves in just a fraction of its orbit. Astronomers can calculate the orbits of distant objects as they approach the Sun in a highly elliptical orbit and determine whether they will come near Earth.

Kepler's third law for objects orbiting the Sun is $P^2 = A^3$, where P is the period of the orbit and A is the semimajor axis (Chapter 3). **Figure 12.4** shows how eccentricity (e) is defined mathematically as the distance from the center of the orbit to one focus (the Sun) divided by the semimajor axis (A). (The eccentricities of the orbits of the planets in the Solar System range from 0.007 for Venus, which is nearly circular, to 0.2 for Mercury.) The types of objects discussed here in Chapter 12 generally have higher eccentricities.

The eccentricity can be related to the closest approach and the farthest distance in the orbit (Figure 12.4). That is, the object's closest approach to the Sun, its **perihelion**, equals $A \times (1 - e)$, and the object's farthest distance from the Sun, its **aphelion**, equals $A \times (1 + e)$. So, if we know the semimajor axis and eccentricity of an orbit, we can calculate how close to and how far away from the Sun an object's orbit takes it.

What are perihelion and aphelion for the dwarf planet Eris? The eccentricity of Eris's orbit is 0.44, and the semimajor axis of its orbit is 67.7 AU. Therefore,

$$\text{Perihelion} = A \times (1 - e) = 67.7 \times (1 - 0.44) = 67.7 \times 0.56 = 37.9 \text{ AU}$$

$$\text{Aphelion} = A \times (1 + e) = 67.7 \times (1 + 0.44) = 67.7 \times 1.44 = 97.5 \text{ AU}$$

Eris is now close to aphelion. When approaching perihelion, though, Eris will cross the orbit of Pluto, whose distance varies from 29.7 to 48.9 AU.

What are perihelion and aphelion for Apollo asteroid 2005 YU55, which has a semimajor axis of 1.14 AU and an orbital eccentricity of 0.43?

$$\text{Perihelion} = A \times (1 - e) = 1.14 \times (1 - 0.43) = 1.14 \times 0.57 = 0.65 \text{ AU}$$

$$\text{Aphelion} = A \times (1 + e) = 1.14 \times (1 + 0.43) = 1.14 \times 1.43 = 1.63 \text{ AU}$$

Those results indicate that the orbit of 2005 YU55 crosses the orbits of Earth and Mars. In November 2011, that asteroid passed 324,900 km from Earth—about 85 percent of the distance to the Moon.

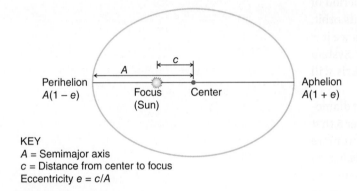

Figure 12.4 The relationships between eccentricity, aphelion, and perihelion in an elliptical orbit.

12.2 Asteroids Are Pieces of the Past

After Piazzi found Ceres in 1801, several similar objects were discovered between the orbits of Mars and Jupiter. Because those new objects appeared in astronomers' eyepieces as nothing more than faint points of light, William and Caroline Herschel (the brother-sister pair of astronomers who discovered Uranus) named them *asteroids*, a Greek word meaning "starlike." As the years went by, more asteroids were discovered. Now, an estimated 1 million to 2 million asteroids larger than 1 km exist, as well as many more that are smaller. Because those objects are in the Solar System, many of them move quickly enough across the sky that their motion is noticeable over a few hours. Both professional and amateur astronomers have discovered asteroids.

An asteroid is a primitive planetesimal that did not become part of the accretion process that formed planets. The planetesimals that formed our Solar System's planets and moons have been so severely modified by planetary processes that nearly all information about their original physical condition and chemical composition has been lost. By contrast, asteroids and comet nuclei constitute an ancient and far more pristine record of what the early Solar System was like. Asteroids are composed of the same types of rocky and metallic materials that became the inner planets, and comets are composed of the same types of icy materials that became the outer planets. Thus, planetary scientists study asteroids to learn about the inner planets and their formation. In this section, we study the orbits and composition of the asteroids.

The Distribution of Asteroids

Asteroids are found throughout the Solar System. Most orbit the Sun in several distinct zones, with many residing between the orbits of Mars and Jupiter in the **main asteroid belt**. The main belt contains at least 1,000 objects larger than 30 km, of which about 200 are larger than 100 km. Although a great many asteroids exist, they account for only a tiny fraction of the matter in the Solar System. Some of the asteroids are bound to another asteroid in a double system, and more than 200 asteroids have moons, some similar in size to the asteroids themselves. At least one asteroid has a ring.

Asteroids are not distributed randomly throughout the main asteroid belt: several empty regions exist. **Figure 12.5** shows that very few asteroids orbit at specific distances from the Sun. Those "gaps" in the asteroid belt are called **Kirkwood gaps**, after Daniel Kirkwood (1814–1895), the astronomer who first recognized them. The orbital periods of some moons around their planets are numerically related—that idea of orbital resonances was introduced in Chapter 11. Similarly, all the Kirkwood gaps in the asteroid belt correspond to resonances: asteroid orbits related to the orbital period of Jupiter by the ratio of two small integers. The boundaries of the asteroid belt are set by some of those resonances. The inner boundary of the asteroid belt, at 1.8 AU, corresponds to the 5:1 orbital resonance of Jupiter, wherein the asteroids make five orbits for every orbit of Jupiter. The outer boundary, at 3.3 AU, corresponds to the 2:1 orbital resonance of Jupiter.

Those gaps are caused by gravitational interactions with Jupiter. Imagine an asteroid orbiting the Sun. While crossing the line between Jupiter and the Sun, that asteroid passes closest to Jupiter; the asteroid will receive an outward tug from Jupiter's gravity. The gravitational force of the Sun on the asteroid is more than 360 times stronger than the gravitational force that Jupiter exerts on the asteroid there, so a single close pass between Jupiter and the asteroid does very little to the asteroid's orbit. For an asteroid *not* in orbital resonance with Jupiter, the tiny gravitational tugs from Jupiter come at a different place in its orbit each time. The effects of those random tugs average out, and as a result even multiple passes close to Jupiter have little overall effect.

Now consider an asteroid starting with an orbital period exactly half that of Jupiter, a 2:1 orbital resonance. After 11.86 years—two complete asteroid orbits and one Jupiter orbit—the asteroid, Jupiter, and the Sun line up in the same location. The repeated tugs from Jupiter at the same location in the asteroid's orbit add together and move the asteroid out of that orbit, so a gap in the distribution of asteroids forms there. That is why no asteroids are in a 2:1 resonance with Jupiter. Asteroids in other orbital resonances, such as a 3:1 resonance, are similarly moved from their orbits. Asteroids are less likely to be found in the Kirkwood gaps because their gravitational interaction with Jupiter prevents them from staying there.

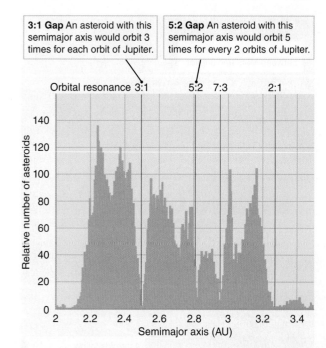

Figure 12.5 The relative number of asteroids in the main belt with a given orbital semimajor axis. The gaps in the distribution of asteroids, called *Kirkwood gaps*, are caused by orbital resonances with Jupiter.

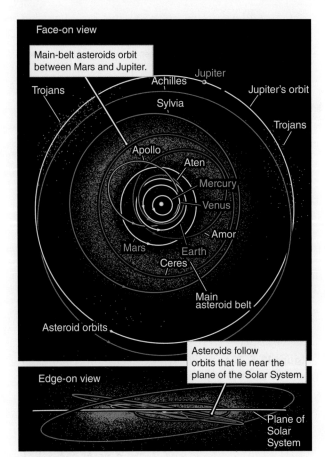

Figure 12.6 Face-on and edge-on views of asteroid orbits. Blue dots show the locations of known asteroids at a single point in time. The orbits of Aten, Amor, and Apollo (prototype members of some groups of asteroids) are shown. Most asteroids, such as Sylvia, are main-belt asteroids. Achilles was the first Trojan asteroid discovered.

Several groups of asteroids exist outside the main asteroid belt. They are divided according to their orbital characteristics (**Figure 12.6**). **Trojan** asteroids share Jupiter's orbit and are held in place by interactions with Jupiter's gravitational field. Three other groups are defined by their relationship to the orbits of Earth and Mars: **Apollo** asteroids cross the orbits of Earth and Mars, **Aten** asteroids cross Earth's orbit but not that of Mars, and **Amor** asteroids cross the orbit of Mars but not Earth's. All three of those groups are named for a prototype asteroid that represents the group.

Asteroids whose orbits bring them within 1.3 AU of the Sun are called **near-Earth asteroids** because their orbits bring them close to Earth's orbit, at 1 AU. Those asteroids, along with a few comet nuclei, are known collectively as **near-Earth objects** (**NEOs**). NEOs occasionally collide with Earth or the Moon. Astronomers estimate that 500–1,000 NEOs have diameters larger than a kilometer. Collisions with NEOs are geologically important and have dramatically altered the history of Earth and life on Earth (see Chapter 8).

Part of NASA's mission is to identify and track NEOs. NASA's Wide-field Infrared Survey Explorer (WISE), an infrared telescope in space, surveyed the entire sky during 2010. The data suggest that about 20,000 mid-sized asteroids (100 meters to 1 km) exist near Earth. WISE also observed more than 150,000 asteroids in the main belt, including 33,000 new ones, as well as 2,000 Jovian Trojans. The spacecraft was reactivated in late 2013 as NEOWISE; it has discovered asteroids and comets and now is searching for NEOs.

The Composition and Classification of Asteroids

Most asteroids are relics of rocky or metallic planetesimals that originated between the orbits of Mars and Jupiter. Although early collisions between those planetesimals created several bodies large enough to differentiate, Jupiter's tidal disruption and possible orbital migration prevented them from forming a single Moon-sized planet. As they orbit the Sun, asteroids continue to collide with one another, producing small fragments of rock and metal. Most meteorites are pieces of those asteroidal fragments that have found their way to Earth and crashed to its surface.

Studies of meteorites led to a scheme for classifying asteroids by composition. Meteorites found on Earth come from asteroids, which come from planetesimals, as shown in **Figure 12.7**. As larger planetesimals accreted smaller objects, thermal energy from impacts and the decay of radioactive elements heated them. Despite that heating, some planetesimals never reached the high temperatures needed to melt their interiors: they simply cooled. The planetesimals look like rubble piles, pretty much as they were when they formed. Those planetesimals, the most common type of asteroid in the main belt, are called **C-type** (carbon-type) asteroids. They are composed of primitive material that has largely been unmodified since the origin of the Solar System almost 4.6 billion years ago.

In contrast, some planetesimals were heated enough by impacts and radioactive decay to cause them to melt and differentiate, with denser matter such as iron sinking to their centers. Lower-density material—such as compounds of calcium, silicon, and oxygen—floated toward the surfaces of those planetesimals and formed mantles and crusts of silicate rocks. **S-type** (stony) asteroids may be pieces of the mantles and crusts of such differentiated planetesimals and are chemically similar to volcanic rocks found on Earth. S-type asteroids were hot enough at some point to lose their carbon compounds and other volatile materials

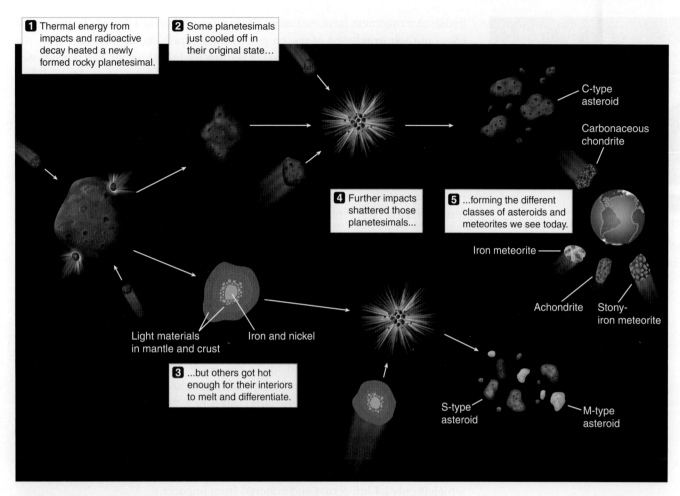

Figure 12.7 The fate of a rocky planetesimal in the young Solar System depends on whether it gets large and hot enough to melt and differentiate, as well as on the impacts it experiences. Different histories led to the varieties of asteroids and meteorites found today. (Images not to scale.)

to space. **M-type** (metal) asteroids are fragments of the iron- and nickel-rich cores of one or more differentiated planetesimals that shattered into small pieces during collisions with other planetesimals.

Recently, some asteroids have been shown to have ice on their surface. Using a ground-based infrared telescope, astronomers found ice on 24 Themis, one of the largest main-belt asteroids (diameter 200 km) that orbits the Sun at the outer edge of the asteroid belt. Water ice covers its surface, and organic molecules also were found there. Hydrated minerals have been found on meteorites thought to have come from outer-main-belt asteroids, but that was the first direct detection of water ice on an asteroid. The discovery may indicate that a continuum rather than a strict boundary exists between icy comets and rocky asteroids. The observations support the idea that both asteroids and comets brought water and organic material to the early Earth.

With a few exceptions, the mass of an asteroid is too small for self-gravity to have pulled it into a spherical shape. Some asteroids have highly elongated irregular shapes, like potatoes, suggesting objects that either are fragments of larger bodies or were created haphazardly from collisions between smaller

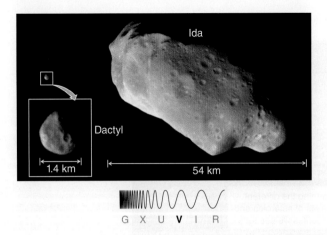

Figure 12.8 This *Galileo* spacecraft image shows the asteroid Ida with its tiny moon, Dactyl (shown enlarged in the inset).

bodies. Astronomers have measured the masses of several asteroids by noting the effect of their gravity on Mars, on passing spacecraft, on one another, or by the orbits of their moons. The total mass of the asteroids in the main belt is estimated to be about 3 times the mass of Ceres, or 4 percent of the mass of the Moon. Their densities can be found from their mass and size and range between 1.3 and 3.5 times the density of water. The lower-density asteroids are shattered heaps of rubble, with large voids between the fragments.

Asteroids rotate just as planets and moons do, although irregularly shaped asteroids can wobble a lot as they spin. Their rotation periods range from 2 hours to longer than 40 Earth days. Rotation periods for asteroids are measured by watching changes in their brightness as they alternately present their broad and narrow faces to Earth. Different groups of asteroids have different average rotations.

Asteroids Viewed Up Close

Spacecraft have visited several asteroids. In 1991, the *Galileo* mission passed by two S-type asteroids while on its way to Jupiter. The small asteroid Gaspra is cratered and irregular in shape, about 18 × 11 × 9 km. Faint groovelike patterns may be fractures from the impact that chipped Gaspra from a larger planetesimal. Distinctive colors imply that Gaspra is covered with a variety of rock types. Later, *Galileo* passed close to asteroid Ida in the outer part of the main asteroid belt (**Figure 12.8**). *Galileo* flew so close to Ida that the probe's cameras could see details as small as 10 meters across. Ida is 60 × 19 × 25 km, and its surface is about a billion years old, twice the age estimated for Gaspra. Like Gaspra, Ida contains fractures, indicating that those asteroids must be made of relatively solid rock. That finding supports the idea that some asteroids are chips from larger, solid objects. The *Galileo* images also revealed a tiny moon orbiting Ida, called Dactyl, which is only 1.4 km across and cratered from impacts.

The first spacecraft to land on an asteroid was *NEAR Shoemaker*, which was gently crash-landed onto asteroid Eros in 2002 after a year of taking observations. Chemical analyses confirmed that the composition of Eros is like that of primitive meteorites. In November 2005, the Japanese spacecraft *Hayabusa* made contact with the small (less than 0.5 km) S-type asteroid Itokawa. *Hayabusa* collected small samples of dust that were returned to Earth in 2010—the first sample-return mission from an asteroid. Chemical analysis showed that such S-type asteroids are the parents of a type of meteorite found on Earth. The findings also suggested that Itokawa had been much larger, more than 20 km, when it formed.

In 2011, NASA's *Dawn* spacecraft went into orbit around Vesta (**Figure 12.9**), the second-most-massive body in the asteroid belt (after Ceres). Vesta is smaller (diameter, 525 km) than the terrestrial planets but larger than the other visited asteroids. The data from *Dawn* indicate that Vesta is a leftover intact protoplanet that formed within the first 2 million years of the aggregation of the first solid bodies in the Solar System. Vesta has an iron core and is differentiated, so it is more like the planets than it is like other asteroids.

Vesta's spectrum matches the reflection spectrum of a peculiar group of meteorites that look like rocks taken from iron-rich lava flows on Earth and the Moon. A collision—or two—that created the two large impact basins in the south polar region of Vesta (**Figure 12.10a**) blasted material into space that then landed on Earth as those meteorites. Those basins are only 1 billion to 2 billion years old. The younger basin is 500 km across and 19 km deep—a depth greater than the height of Mauna Kea in Hawaii (more than 10 km, measured

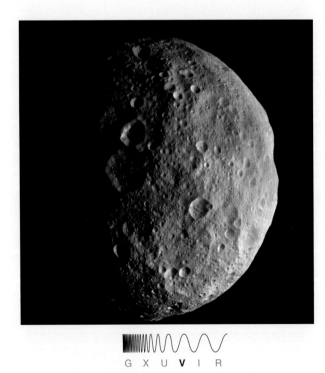

Figure 12.9 This image of Vesta was taken by *Dawn* in 2012. Its north pole is in the middle of the image.

from the ocean floor). Smaller adjacent impact craters in the northern hemisphere (**Figure 12.10b**) were nicknamed "Snowman."

Two spacecraft are visiting asteroids, with the intent to return samples to Earth for analysis. The Japanese space agency mission *Hayabusa 2* has landed on asteroid 162173 Ryugu. After studying the asteroid for 1.5 years, *Hayabusa 2* will return a sample to Earth in late 2020. NASA's *OSIRIS-REx* (Origins, Spectral Interpretation, Resource Identification, Security-Regolith Explorer) mission is in transit to near-Earth asteroid 101955 Bennu. The probe will arrive in late 2018 and bring a sample to Earth in 2023. For both missions, the goal is to study those relics from the early Solar System to look for water and organic compounds.

CHECK YOUR UNDERSTANDING 12.2

Remnants of volcanic activity on the asteroid Vesta indicate that members of the asteroid belt: (a) were once part of a single protoplanet that was shattered by collisions; (b) have all undergone significant chemical evolution since formation; (c) occasionally grow large enough to become differentiated and geologically active; (d) used to be volcanic moons orbiting other planets.

12.3 Comets Are Clumps of Ice

Comets are icy planetesimals that formed from primordial material. They spend most of their time adrift in the frigid outer reaches of the Solar System. Comet nuclei put on a show only when their orbit brings them deep enough into the inner Solar System to undergo destructive heating from the Sun—they emit streams of dust and gas. In this section, we examine the orbits and composition of comets.

Early cultures viewed the sudden and unexpected appearance of a bright comet as an omen. Comets were often seen as dire warnings of disease, destruction, and death, or sometimes as portents of victory in battle or as heavenly messengers announcing the impending birth of a great leader. The earliest records of comets date from as long ago as the 23rd century BCE. Until the end of the Middle Ages, comets were regarded as mysterious temporary atmospheric phenomena rather than as astronomical objects. In the 16th century, Tycho Brahe reasoned that if comets were atmospheric phenomena like clouds, their appearance and location in the sky should be very different to observers located many miles apart. But when Tycho compared sightings of comets made by observers at several sites, he found no evidence of such differences and concluded that comets must be at least as far away as the Moon.

The Homes of the Comets

A *comet* is a complex object consisting of a small, solid, icy nucleus; an atmospheric halo; and a tail of dust and gas: a comet nucleus is the "heart" of the comet and contains most of the comet's mass. When very distant from the Sun, the comet consists entirely of frozen nucleus. As a comet comes near enough to the Sun to heat up, we can see visible changes. When a comet is near enough to the Sun to show the effects of solar heating, it is called an **active comet**, or often simply *comet*. Most comet nuclei are much too small and far away to be seen, so no one really knows how many exist. Estimates for our Solar System range as high as a trillion (10^{12}) comet nuclei—more than the number of stars in the Milky Way Galaxy—but astronomers have seen only several thousand.

Figure 12.10 (a) Impact basins at the south pole of Vesta. By counting the craters on top of it, astronomers estimate Rheasilvia to be 1 billion years old. Veneneia is partly beneath Rheasilvia and is estimated to be 2 billion years old. Red indicates higher elevation. (b) Three craters in the northern hemisphere—60, 50, and 22 km across, respectively. The feature was nicknamed "Snowman."

We can extrapolate where comets come from by observing their orbits as they pass through the inner Solar System. Comets fall into two distinct groups named for scientists Gerard Kuiper (1905–1973) and Jan Oort (1900–1992).

KUIPER BELT The Kuiper Belt is a disk-shaped population of comet nuclei that begins about 30 AU from the Sun, near the orbit of Neptune, and extends outward to about 55 AU (**Figure 12.11**). Comets from the Kuiper Belt orbit the Sun in a disk-shaped region aligned with the Solar System. The innermost part of the Kuiper Belt contains tens of thousands of icy planetesimals known as Kuiper Belt objects (KBOs) or sometimes as *trans-Neptunian objects* (*TNOs*). The largest KBOs are similar in size to Pluto and Eris. With a few exceptions, the sizes of KBOs are difficult to determine because although brightness and approximate distance are known, their albedos are uncertain. Reasonable limits for the albedos can set maximum and minimum values for their size. Like asteroids, some KBOs have moons, and at least one has three moons. We know very little of the chemical and physical properties of KBOs because of their great distance. After encountering Pluto in 2015, the *New Horizons* spacecraft continued outward into the Kuiper Belt, where it will fly close to the 40-km object 2014 MU69 in early 2019.

One of the larger known KBOs, called Quaoar (pronounced "kwa-whar"), is also one of the few whose size astronomers have independently measured—about 900 km. From its apparent brightness, distance, and size, astronomers calculate Quaoar's albedo to be 0.20, making it more reflective than the nuclei of those comets that have entered the inner Solar System but far less reflective than Pluto. Quaoar's remote location and pristine condition have allowed some volatile ices to survive on its surface, including crystalline water ice, methane, and ethane. Quaoar has a nearly circular orbit about the Sun and has a small moon, which enables astronomers to estimate the larger body's mass.

The icy planetesimals in the Kuiper Belt are packed closely enough to interact gravitationally from time to time. When that happens, one object gains energy, whereas the other loses it. The "winner" may gain enough energy to be sent into an orbit that reaches far beyond the boundary of the Kuiper Belt. The "loser" may fall inward toward the Sun.

OORT CLOUD Unlike the flat disk of the Kuiper Belt, the **Oort Cloud** is a spherical distribution of planetesimals (Figure 12.11) much too distant to be seen by even the most powerful telescopes. Astronomers determine the size and shape of the Oort Cloud from the orbits of its comets, which approach the Sun from all directions and from as far away as 100,000 AU—nearly halfway to the nearest stars.

Sedna is an object in the inner Oort Cloud whose highly elliptical orbit around the Sun takes it from 76 AU out to 937 AU. With such an extended orbit, Sedna requires more than 11,000 years to make a single trip around the Sun. When discovered in 2003, Sedna was about 90 AU from the Sun and getting closer. It will reach its perihelion in 2076. Herschel Space Observatory data suggest an albedo of 0.30 and a size of 1,000 km. Sedna has no known moon, so its mass is hard to estimate. Water and methane ices have been detected in its spectrum. Like dwarf planet Eris, Sedna has a highly eccentric orbit. A second object in the inner Oort Cloud, 2012 VP113, was recently detected. Its distance ranges from 80 to 452 AU from the Sun, and it is thought to be about half the size of Sedna.

Inner Solar System objects are close enough to the Sun that disturbances external to the Solar System never exert more than a tiny fraction of the gravitational force of the Sun on them. In the distant Oort Cloud, however, comet nuclei are so far from the Sun, and the Sun's gravitational force on them is so weak, that

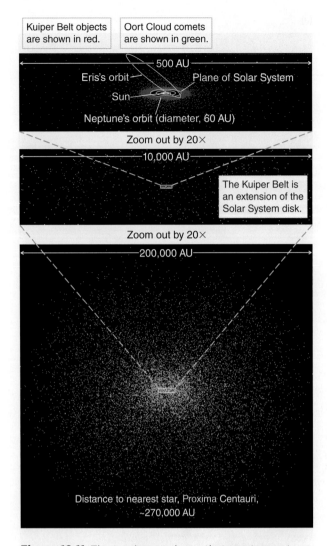

Figure 12.11 The top image shows that most comets near the inner Solar System populate an extension to the disk of the Solar System called the Kuiper Belt (red). The orbit of dwarf planet Eris is highly inclined with respect to the disk. The middle image zooms out to show the expanded Kuiper Belt. The bottom image zooms out to illustrate the spherical Oort Cloud, which is far larger and contains many more comet nuclei (green).

they are barely bound to the Sun at all. The tug of a slowly passing star or interstellar cloud can compete with the Sun's gravity, significantly stirring up the Oort Cloud and changing the orbits of its objects. If the interaction adds to the orbital energy of a comet nucleus, the comet may move outward to an even more distant orbit or perhaps escape from the Sun. A comet nucleus that loses orbital energy as a result of that type of interaction will fall inward. Some of those comet nuclei come all the way into the inner Solar System, where they may appear briefly in Earth's skies before returning once again to the Oort Cloud.

The Orbits of Comets

The lifetime of a comet nucleus depends on how often it passes by the Sun and how close it gets. About 400 **short-period comets** are known, which by definition have periods less than 200 years. In addition, each year astronomers discover about six new **long-period comets**, whose orbital periods are longer than 200 years. The total number of long-period comets observed to date is about 3,000.

Figure 12.12 shows the orbits of several comets, nearly all highly elliptical, with one end of the orbit close to the Sun and the other in the distant parts of the

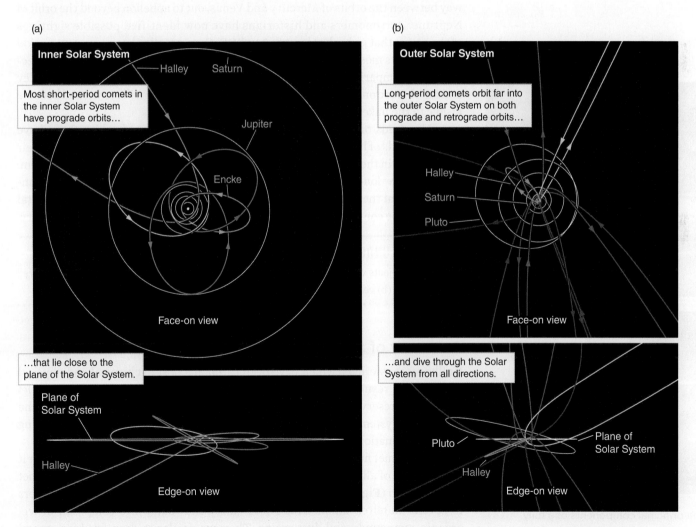

Figure 12.12 The orbits of several comets in face-on and edge-on views of the Solar System. Populations of (a) short-period comets and (b) long-period comets have very different orbital properties. Comet Halley, which appears in both diagrams for comparison, is a short-period comet.

Solar System. Most comets passing through the inner Solar System have long orbital periods that carry them back to the Oort Cloud or the Kuiper Belt. Long-period comets were scattered to the outer Solar System by gravitational interactions, so they come into the inner Solar System from all directions. Some orbit the Sun in the same direction that the planets orbit (prograde), whereas others orbit in the opposite direction (retrograde).

Conversely, short-period comets tend to be prograde and to have orbits in the ecliptic plane, often passing close enough to a planet for its gravity to change the comet's orbit about the Sun. Short-period comets likely originated in the Kuiper Belt, but as they fell in toward the Sun, gravitational encounters with Jupiter forced them into their current short-period orbits relatively close to the Sun.

Comet Halley is the most famous short-period comet. In 1705, Edmund Halley, using the gravitational laws of his colleague Isaac Newton, noted that a bright comet from 1682 had an orbit remarkably similar to those of comets seen in 1531 and 1607. He concluded that all three were the same comet and predicted that it would return in 1758. When it reappeared, astronomers quickly named it Halley's Comet and heralded it as a triumph for the genius of both Newton and Halley. Comet Halley's highly elongated orbit takes it from perihelion, about halfway between the orbits of Mercury and Venus, out to aphelion beyond the orbit of Neptune. Astronomers and historians have now identified possible sightings of the comet that go back at least to 240 BCE. Comet Halley has an average period of 76 years. Its most recent appearance was in 1986, and it was not especially spectacular in comparison with its appearance in 1910 because in 1986, Comet Halley and Earth were on opposite sides of the Sun. Comet Halley will return and become visible to the naked eye once again in summer 2061.

Hundreds of long-period comets have well-determined orbits. Some have orbital periods of hundreds of thousands or even millions of years. Almost all their time is spent in the Oort Cloud in the frigid, outermost regions of the Solar System. Orbits of a few long-period comets are shown in Figure 12.12b. Those are the comets that reveal the existence of the Oort Cloud. Because of their very long orbital periods, those comets have been observed only once throughout recorded history.

CHECK YOUR UNDERSTANDING 12.3A

Oort Cloud objects will pass close to the Sun, and become comets, only if their orbits are (a) very large; (b) very small; (c) highly elliptical; (d) highly inclined.

Anatomy of an Active Comet

Unlike most asteroids, which have been through a host of chemical and physical changes as a result of collisions, heating, and differentiation, most comet nuclei have been preserved over the past 4.6 billion years by the "deep freeze" of the outer Solar System. Comet nuclei are made of nearly pristine material remaining from the formation of the Solar System.

The comet nucleus at the center is the smallest component of a comet, but it is the source of all the material that we see stretched across the sky as the comet nears the Sun (**Figure 12.13**). Comet nuclei range in size from a few dozen meters to several hundred kilometers. Those "dirty snowballs" are composed of ice, organic compounds, and dust grains. They are similar to deep-fried ice cream, with a soft porous interior surrounded by a crunchy crust of hardened water-ice crystals, topped off with sooty dust and organic molecules.

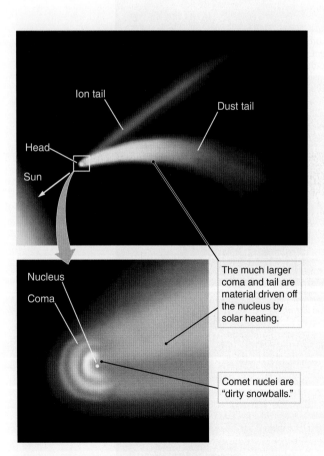

Figure 12.13 The principal components of a fully developed active comet are the nucleus, the coma, and two types of tails, called the dust tail and the ion tail. Together, the nucleus and the coma are called the head.

As a comet nucleus nears the Sun, sunlight heats its surface, vaporizing ices that stream away, carrying dust particles along. That change from solid to gas is called **sublimation**. For example, dry ice (frozen carbon dioxide) does not melt like water ice but instead sublimates directly into carbon dioxide gas—that is why it is called "dry." Set a piece of dry ice out in the Sun on a summer day, and you will get a pretty good idea of what happens to a comet. The gases and dust driven from the nucleus of an active comet form a nearly spherical atmospheric cloud around the nucleus called the **coma**. The nucleus and the inner part of the coma form the comet's **head**. Pointing from the head of the comet in a direction more or less away from the Sun are long streamers of dust, gas, and ions called **tails**.

The tails are the largest and most spectacular part of a comet. The tails also are the "hair" for which comets are named. (*Comet* comes from the Greek word *kometes*, which means "hairy one.") Active comets have two types of tails, called the **ion tail** and the **dust tail** (Figure 12.13). Many of the atoms and molecules that make up a comet's coma are ions. Being electrically charged, ions in the coma experience the effect of the solar wind—the stream of charged particles that blows continually away from the Sun. The solar wind pushes on those ions, rapidly accelerating them to speeds of more than 100 kilometers per second (km/s)—far greater than the orbital velocity of the comet itself—and sweeps them out into a long wispy structure. Because the particles that make up the ion tail are so quickly picked up by the solar wind, an ion tail is usually very straight: beginning at the head of the comet, an ion tail points directly away from the Sun.

Dust particles in the coma can also have a net electric charge and are subject to the force of the solar wind. Sunlight also exerts a force on cometary dust. But dust particles are much more massive than individual ions, so they are accelerated more gently and do not reach such high relative speeds as those of the ions. As a result, the dust particles cannot keep up with the comet, and the dust tail often curves away from the head of the comet as the dust particles are gradually pushed from the comet's orbit in the direction away from the Sun (Figure 12.13).

Figure 12.14 shows the tails of a comet at various points in its orbit. Both types of tails always point *away* from the Sun, regardless of which direction the comet is moving. As the comet approaches the Sun, then, its two tails trail *behind* its nucleus, but the tails extend *ahead* of the nucleus as the comet moves away from the Sun. Tails vary greatly from one comet to another. Some comets display both types of tails simultaneously; others, for reasons not yet understood, produce no tails at all. A tail often forms as a comet crosses the orbit of Mars, where the increase in solar heating drives gas and dust away from the nucleus.

The gas in a comet's tail is even more tenuous than the gas in its coma, with densities of no more than a few hundred particles per cubic centimeter. That is much, much less than the density of Earth's atmosphere, which at sea level contains more than 10^{19} molecules per cubic centimeter. Dust particles in the tail are typically about 1 micron (μm) in diameter, roughly the size of smoke particles.

The nuclei of short-period comets have been badly worn out by repeated exposure to heating by the Sun. As the volatile ices are driven from a nucleus, some of the dust and organics are left behind on the surface. (Similarly, as a pile of dirty snow melts, the dirt left behind is concentrated on the surface of the snow.) The buildup of that covering slows down cometary activity. In contrast, long-period comets are usually relatively pristine. More of their supply of volatile ices still remains near the surface of the nucleus, and they can produce a truly magnificent show.

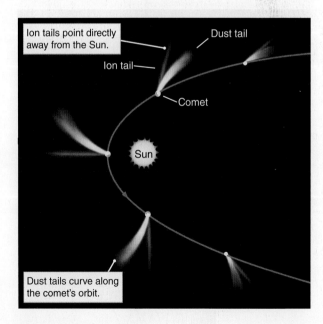

Figure 12.14 The orientation of the dust and ion tails at several points in a comet's orbit. The ion tail points directly away from the Sun, whereas the dust tail curves along the comet's orbit.

Figure 12.15 Comet McNaught in 2007 was the brightest comet to appear in decades, but its true splendor was visible only to observers in the Southern Hemisphere.

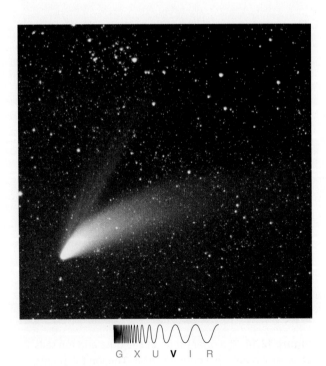

Figure 12.16 Comet Hale-Bopp was a great comet in 1997. The ion tail is blue in this image, and the dust tail is white.

Most naked-eye comets first develop a coma and then an extended tail as they approach the inner Solar System. Comet McNaught in 2007 was such a comet, the brightest to appear in nearly 50 years. The comet's nucleus and coma were visible in broad daylight as its orbit carried it within 25 million km of the Sun. When Comet McNaught had passed behind the Sun and next appeared in the evening skies to observers in the Southern Hemisphere, its tail had grown to more than 160 million km and stretched 35° across the sky (**Figure 12.15**). Comet McNaught came into the inner Solar System from the Oort Cloud, but it left on a path that will carry it out of the Solar System.

Comet Hale-Bopp in 1997 was a spectacular long-period comet with a long, beautiful tail (**Figure 12.16**). Hale-Bopp was a large comet, with a nucleus estimated at 60 km in diameter. It was discovered far from the Sun, near Jupiter's orbit, 2 years before its perihelion passage. That early discovery extended the total time available to study its development and plan observations as it approached the Sun. Warmed by the Sun, the nucleus produced large quantities of gas and dust and as much as 300 tons of water per second, with lesser amounts of carbon monoxide, sulfur dioxide, cyanogen, and other gases. Comet Hale-Bopp will continue its outward journey for more than 1,000 years, and it will not return to the inner Solar System until sometime around the year 4530.

Comet Ikeya-Seki is a member of a family of comets called **sungrazers**, comets whose perihelia are located very close to the surface of the Sun. Many sungrazers fail to survive even a single orbit of the Sun. Ikeya-Seki became so bright as it neared perihelion in 1965 that it was visible in broad daylight, close to the Sun in the sky. Sungrazers generally come in groups, with successive comets following in nearly identical orbits. Each member of such a group started as part of a single larger nucleus that broke into pieces during an earlier perihelion passage.

A half dozen or so long-period comets arrive each year. Most pass through the inner Solar System at relatively large distances from Earth or the Sun and never become bright enough to attract much public attention. On average, a spectacular comet appears about once per decade.

Visits to Comets

Comets pose an engineering challenge for spacecraft designers. Researchers seldom have enough advance knowledge of a comet's visit or its orbit to mount a successful mission to intercept it. The relative speed between an Earth-launched spacecraft and a comet can be extremely high. Observations must be made very quickly, and danger exists of high-speed collisions with debris from the nucleus. About a dozen spacecraft have been sent to rendezvous with comets, including five spacecraft sent to Comet Halley by the Soviet, European, and Japanese space agencies in 1986. Much of what we know about comet nuclei and the innermost parts of the coma comes from data sent back by those missions. The spacecraft observed gas and dust jets, impact craters, and ice and dust on the comet nuclei.

The Soviet *Vega 1* and *Vega 2* and the European *Giotto* spacecraft entered the coma of Comet Halley when they were still nearly 300,000 km from its nucleus. The dust from Comet Halley was a mixture of light organic substances and heavier rocky material, and the gas was about 80 percent water and 10 percent carbon monoxide, with smaller amounts of other organic molecules. The surface of Comet Halley's nucleus is among the darkest known objects in the Solar System. That means it is rich in complex organic matter that must have been present as dust in the disk around the young Sun—perhaps even in the interstellar cloud from which the Solar System formed. As the three spacecraft passed near Halley's nucleus, they observed jets of gas and dust moving away from its surface at speeds of up to 1 km/s, far above the escape velocity. By observing the jets of material streaming away from the nucleus of Halley, planetary scientists estimated that Comet Halley must have lost one-tenth of 1 percent of its mass as it went around the Sun.

Several space missions have visited short-period comets. In 2004, NASA's *Stardust* spacecraft flew within 235 km of the nucleus of Comet Wild 2 (**Figure 12.17**). Comet Wild 2 had previously resided between the orbits of Jupiter and Uranus, but a close encounter with Jupiter in 1974 had perturbed its orbit, bringing the relatively pristine body closer to the Sun as it traveled between the orbits of Jupiter and Earth. At the time of *Stardust*'s encounter with Wild 2, the comet had made only five trips around the Sun in its new orbit. Wild 2's nearly spherical nucleus is about 5 km across. At least 10 gas jets were active, some of which carried large chunks of surface material. The surface of Wild 2 is covered with features that may be impact craters modified by ice sublimation, small landslides, and erosion by jetting gas. Some craters show flat floors, suggesting a relatively solid interior beneath a porous surface layer.

The *Stardust* mission collected dust samples from Wild 2, which were returned to Earth in 2006. It found new kinds of organic materials unlike any seen before in materials from space. They are more primitive than those observed in meteorites and may have formed before the Solar System itself. Those grains can be used to investigate the conditions under which the Sun and planets formed. Minerals that form at high temperature also have been found, supporting the idea that the solar wind blew material out of the inner Solar System very early in the system's history. Scientists will be studying the particles from that mission for many years.

In 2005, NASA's *Deep Impact* spacecraft launched a 370-kilogram (kg) impacting projectile into the nucleus of Comet Tempel 1 at a speed of more than 10 km/s (**Figure 12.18**). The impact sent 10,000 tons of water and dust flying off into space at speeds of 50 m/s. A camera mounted on the projectile snapped

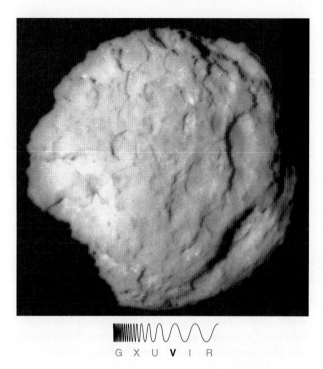

Figure 12.17 The nucleus of Comet Wild 2 was imaged by the *Stardust* spacecraft, which also sampled the comet's tail.

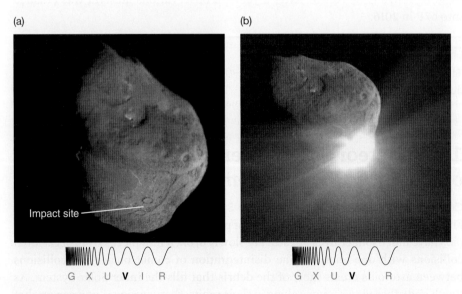

Figure 12.18 (a) The surface of the nucleus of Comet Tempel 1 is shown just before impact by the *Deep Impact* projectile. The impact occurred between the two 370-meter craters located near the bottom of the image. The smallest features appearing in this image are about 5 meters across. (b) Sixteen seconds after the impactor struck the comet, the parent spacecraft took this image of the initial ejecta.

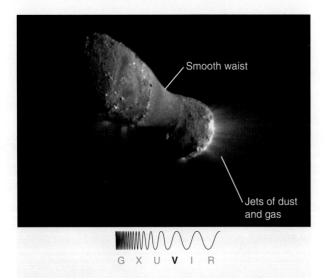

Figure 12.19 This image of Comet Hartley 2 taken by the *EPOXI* spacecraft reveals two distinct surface types. Water seeps through the dust at the smooth "waist" of the comet's nucleus, whereas carbon dioxide jets shoot gas, dust, and chunks of ice from the rough areas.

photos of its target until the projectile was vaporized by the impact. Observations of the event were made both locally by *Deep Impact* and back on Earth by orbiting and ground-based telescopes. Water, carbon dioxide, hydrogen cyanide, iron-bearing minerals, and a host of complex organic molecules were identified in the Comet Tempel 1 impact. The comet's outer layer is composed of fine dust with the consistency of talcum powder. Beneath the dust are layers made up of water ice and organic materials. Well-formed impact craters, which had been absent in previous images of Comet Wild 2, also were seen.

In 2010, the *EPOXI* spacecraft flew past Comet Hartley 2 (**Figure 12.19**), imaging not only jets of dust and gas, indicating a remarkably active surface, but also an unusual separation of rough and smooth areas. Carbon dioxide jets shoot out from the rough areas. The narrow part at the middle of Figure 12.19 is a smooth, inactive area where ejected material has fallen back onto the cometary nucleus. Further observations by the Herschel Space Observatory showed that the water on Hartley 2 has the same ratio of hydrogen isotopes as that of the water in Earth's oceans. That finding suggests that some of Earth's water could have originated in the Kuiper Belt. Measurements of water on comets from the Oort Cloud have a different ratio, and so they have been ruled out as the source of Earth's water.

The European Space Agency spacecraft *Rosetta* visited Comet 67P/Churyumov-Gerasimenko in 2014 (see the chapter-opening figure). The comet is "binary"—it consists of two pieces that bumped together billions of years ago. Organic compounds such as glycine (an amino acid), ethanol, and complex carbons were identified in the comet's dust. A small, separate spacecraft landed on the comet and sent back data for 2.5 days before running out of power. The landing site was dust-covered solid ice, too thick to drill. The coma is composed of water vapor, carbon monoxide (CO), carbon dioxide (CO_2), hydrogen sulfide, and ammonia. *Rosetta* orbited the comet as it approached the Sun and observed changes as the comet heated up. Dust and gas were released from the comet, including a bright jet. About half that dust is made up of organic molecules. The water on that comet did not have the same isotope ratio as Earth's water (see Reading Astronomy News). After 2 years of observations, *Rosetta* was crashed into 67P in 2016.

CHECK YOUR UNDERSTANDING 12.3B

The nucleus of a comet is mostly: (a) solid ice; (b) solid rock; (c) a porous mix of ice and dust; (d) frozen carbon dioxide.

12.4 Meteorites Are Remnants of the Early Solar System

Comet nuclei that enter the inner Solar System generally disintegrate within a few hundred thousand years as a result of passing repeatedly near the Sun. Asteroids have much longer lives but still are slowly broken into pieces from occasional collisions with one another. The disintegration of comet nuclei and collisions between asteroids create most of the debris that fills the inner Solar System. As Earth and other planets move along in their orbits, they continually sweep up that fine debris. Most of the meteoroids Earth encounters come from the disintegration of comets or asteroids. Meteoroids are small solid bodies ranging in size from 10 millimeters (mm) to 100 meters. When a meteoroid enters Earth's atmosphere, frictional heat causes the air to glow, producing an atmospheric phenomenon called a *meteor*. If a meteoroid survives to reach the planet's surface, we call it a *meteorite*. Earth sweeps up about 100,000 kg of meteoritic debris every day, and particles smaller than 100 mm eventually settle to the ground as fine dust. In this section, we look more closely at meteorites and what can be learned about the early Solar System from them.

Observations of Meteors

If you stand outside for a few minutes on a moonless, starry night, away from bright city lights, you will almost certainly see a meteor, commonly known as a *shooting star*. The larger pieces that survive the plunge through Earth's atmosphere are usually fragments of asteroids. Most of the smaller pieces that burn up in the atmosphere before reaching the ground are cometary fragments typically less than a centimeter across and having about the same density as cigarette ash.

A 1-gram meteoroid (about half the mass of a dime) entering Earth's atmosphere at 50 km/s has a kinetic energy comparable to that of an automobile cruising along at the fastest highway speeds. Scientists measuring meteor altitudes with radar find that the altitudes of meteors are between 50 and 150 km. Most meteoroids are so small and fragile that they burn up before reaching Earth's surface. A meteor may streak across 100 km of Earth's atmosphere and last at most a few seconds. Meteorites probably litter the surfaces of all solid planets and moons.

Nearly all ancient cultures were fascinated by those rocks from the sky. Iron from meteorites was used to make the earliest tools. Early Egyptians preserved meteorites along with the remains of their pharaohs, Japanese placed them in Shinto shrines, and ancient Greeks worshipped them. Despite many eyewitness accounts of meteorite falls, however, many people were slow to accept that those peculiar rocks actually come from far beyond Earth. By the early 1800s, scientists had documented so many meteorite falls that their true origin was indisputable. Today, hardly a year passes without a recorded meteorite fall, including some that have caused damage.

Fragments of asteroids are much denser than cometary meteoroids. If an asteroid fragment is large enough—about the size of your fist—it can survive all the way to the ground to become a meteorite. The fall of a 10-kg meteoroid can produce a fireball so bright that it lights up the night sky more brilliantly than the full Moon. Such a large meteoroid, traveling many times faster than the speed of sound, may create a sonic boom heard hundreds of kilometers away. The meteoroid may even explode into multiple fragments as it nears the end of its flight. Some fireballs glow with a brilliant green color, caused by elements in the meteoroid that created them.

Meteor showers occur when Earth's orbit crosses the orbit of a comet or asteroid and passes through a concentration of cometary or asteroid debris. During a shower, many meteors can be observed in just a few hours. More than a dozen comets and at least two asteroids have orbits that come close enough to Earth's orbit to produce annual meteor showers—see the list in **Table 12.1**. Because the meteoroids in a shower are in similar orbits, they all enter Earth's atmosphere moving in the same direction—the paths through the

TABLE 12.1 Selected Meteor Showers

Shower	Approximate Date	Parent Object
Quadrantids	January 3–4	Asteroid 2003 EH1
Lyrids	April 21–22	Comet Thatcher
Eta Aquariids	May 5–6	Comet Halley
Perseids	August 12–13	Comet Swift-Tuttle
Draconids	October 8–9	Comet Giacobini-Zinner
Orionids	October 21–22	Comet Halley
Taurids	November 5–6	Comet Encke
Leonids	November 17–18	Comet Tempel-Tuttle
Geminids	December 13–14	Asteroid Phaethon
Ursids	December 22–23	Comet Tuttle

sky are parallel to one another. Therefore, all the meteors appear to originate from the same point in the sky (**Figure 12.20a**), just as the parallel rails of a railroad track appear to vanish to a single point in the distance (**Figure 12.20b**). That point is called the shower's **radiant**.

The Perseid shower occurs in August because that's when Earth crosses the orbit of Comet Swift-Tuttle. Although spread out along the comet's orbit, the debris is more concentrated near the comet itself. In 1992, Comet Swift-Tuttle returned to the inner Solar System for the first time since its discovery in 1862, resulting in an exceptional Perseid meteor shower with counts of up to 500 meteors per hour.

In mid-November of each year, Earth passes almost directly through the orbit of Comet Tempel-Tuttle, a short-period comet with an orbital period of 33.2 years. That crossing produces the Leonid meteor shower, which is usually weak because Comet Tempel-Tuttle distributes little of its debris around its orbit in most years. In 1833 and 1866, however, Tempel-Tuttle was not far away when Earth passed through its orbit, and the Leonid showers were so intense that meteors filled the sky with as many as 100,000 meteors per hour. Further perturbations of the comet's orbit caused a spectacular Leonid shower in 1966 that may have produced as many as a half-million meteors per hour. The Leonid shower put on less spectacular but still impressive shows between 1999 and 2003, when several thousand meteors per hour were seen.

Types of Meteoroids

As asteroids orbit the Sun, they occasionally collide with one another, chipping off smaller rocks and bits of dust. Sometimes, one of those fragments is captured by Earth's gravity and survives its fiery descent through Earth's atmosphere as a meteor. Thousands of meteorites reach the surface of Earth every day, but only a tiny fraction of those are ever found and identified. Antarctica offers the best meteorite hunting in the world because in many places, the only stones to be found on the ice are meteorites. Because Antarctica has little precipitation, Antarctic meteorites also tend to show little weathering or contamination from terrestrial dust or organic compounds, making them excellent specimens for study.

Meteorites are categorized as stony, iron, or stony-iron on the basis of their materials and the degree of differentiation they experienced within their parent bodies (**Figure 12.21**). More than 90 percent of meteorites are **stony meteorites**, which are similar to terrestrial silicate rocks. A stony meteorite is characterized by the thin coating of melted rock that forms as it passes through the atmosphere. Many stony meteorites contain small, round spherules called **chondrules**, once-molten droplets that rapidly cooled to form crystallized spheres ranging in size from that of sand grains to that of marbles. Stony meteorites containing chondrules are called **chondrites** (Figure 12.21a); those without chondrules are known as **achondrites** (Figure 12.21b). The most primitive meteorites are called **carbonaceous chondrites** because they are chondrites rich in carbon. Indirect measurements suggest that those meteorites are about 4.56 billion years old—consistent with all other measurements of the time that has passed since the Solar System formed.

Iron meteorites (Figure 12.21c) come from M-type asteroids. Iron meteorites can be recognized by their melted and pitted appearance generated by frictional heating as they streaked through the atmosphere. Many iron meteorites are never found, either because they land in water or because they are not recognized as meteorites. The Mars *Opportunity* rover discovered a few iron meteorites on the

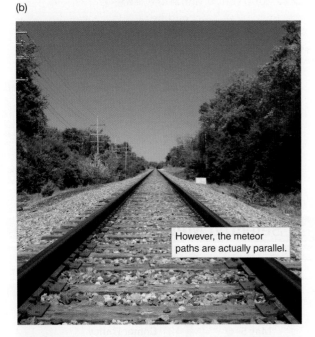

Figure 12.20 (a) Meteors appear to stream away from the radiant of the Leonid meteor shower. (b) Such streaks are actually parallel paths that appear to emerge from a vanishing point, as in our view of these railroad tracks.

martian surface (**Figure 12.22**). Both their appearance—typical of iron meteorites found on Earth—and their position on the smooth, featureless plains made them instantly recognizable.

Stony-iron meteorites consist of a mixture of rocky material and iron-nickel alloys (Figure 12.21d). They are relatively rare.

Meteorites and the History of the Solar System

Meteorites are extremely valuable because they are samples of the same relatively pristine material that makes up asteroids. Astronomers can take meteorites into the laboratory and thus study them more easily than sending spacecraft to comets or asteroids. Scientists compare meteorites to rocks found on Earth and the Moon and contrast their structure and chemical makeup with those of rocks studied by spacecraft that have landed on Mars and Venus. Comparing the spectra of meteorites with those of asteroids and planets reveals their origin.

Figure 12.21 Cross sections of several kinds of meteorites: (a) a chondrite (a stony meteorite with chondrules); (b) an achondrite (a stony meteorite without chondrules); (c) an iron meteorite; (d) a stony-iron meteorite.

Meteorites come from asteroids, which originate from stony-iron planetesimals. A few planetesimals between the orbits of Mars and Jupiter evolved toward becoming tiny planets before being shattered by collisions. Some became volcanically active, with lava erupting onto their surfaces. But instead of forming planets, those planetesimals broke into pieces in collisions with other planetesimals.

Some types of meteorites fail to follow the patterns discussed so far. Whereas most achondrites have ages of 4.5 billion to 4.6 billion years, some are less than 1.3 billion years old. Other achondrites are chemically and physically similar to the soil and the atmospheric gases that NASA's lander instruments have measured on Mars. The similarities are so strong that most planetary scientists think those meteorites are pieces of Mars knocked into space by large asteroidal impacts—as a result, researchers can study pieces of Mars in laboratories here on Earth. In 1996, a NASA research team announced that the meteorite ALH84001, found in Antarctica, showed possible physical and chemical evidence of past life on Mars, but the claim is still debated (see Chapter 24).

Another group of meteorites bear striking similarities to samples returned from the Moon. Like the meteorites from Mars, they are chunks of the Moon blasted into space by impacts and later fell to Earth. Therefore, meteorites from Earth may have fallen on the Moon. If they are ever collected from the unchanging Moon, they could tell us about what conditions were like on the early Earth.

Zodiacal Dust

Like meteoroids, **zodiacal dust** is a mixture of cometary debris and ground-up asteroid material. Just as you can "see" sunlight streaming through an open window by observing its reflection from dust drifting in the air, you can see the sunlight reflected off tiny zodiacal dust particles that fill the inner Solar System close to the plane of the ecliptic. On a clear, moonless night, not long after the western sky has grown dark, that dust is visible as a faint column of light slanting upward from the western horizon along the path of the ecliptic. That band, called the

Figure 12.22 This 2 meter long iron meteorite lying on the surface of Mars was imaged by the Mars exploration rover *Curiosity*.

Figure 12.23 Zodiacal light shines in the western sky after sunset as seen from the La Silla Observatory in Chile.

zodiacal light, also can be seen in the eastern sky just before dawn (**Figure 12.23**). With good eyes and an especially dark night, you may be able to follow the zodiacal dust band all the way across the sky. In its brightest parts, the zodiacal light can be several times brighter than the Milky Way, for which it is sometimes mistaken.

The dust grains are roughly a millionth of a meter in diameter—the size of smoke particles. Near Earth, each cubic kilometer of space contains only a few particles of zodiacal dust. The total amount of zodiacal dust in the entire Solar System is estimated to be 10^{16} kg, equivalent to a solid body about 25 km across, or roughly the size of a large comet nucleus. Grains of zodiacal dust are constantly being lost as they are swept up by planets or pushed out of the Solar System by the pressure of sunlight. Such interplanetary dust grains have been recovered from Earth's upper atmosphere by aircraft flying very high. If not replaced by new dust from comets, all zodiacal dust would be gone within the brief span of 50,000 years.

In the infrared region of the spectrum, thermal emission from the band of warm zodiacal dust makes it one of the brightest features in the sky. It is so bright that astronomers wanting to observe faint infrared sources are often hindered by its foreground glow.

CHECK YOUR UNDERSTANDING 12.4

Meteorites contain clues to which of the following? (Choose all that apply.) (a) the age of the Solar System; (b) the temperature in the early solar nebula; (c) changes in the composition of the primitive Solar System; (d) changes in the rate of cratering in the early Solar System; (e) the physical processes that controlled the formation of the Solar System.

12.5 Comet and Asteroid Collisions Still Happen Today

Almost all hard-surfaced objects in the Solar System still bear the scars of a time when tremendous impact events were common. In previous chapters, we explained how impacts from long ago may have been responsible for the formation of our Moon, the variation in hemispheric features on Mars, and the death of the dinosaurs. Although such impacts are far less frequent today than they once were, they still happen. In this section, we examine a few examples of recent collisions.

Comet Shoemaker-Levy 9 Collided with Jupiter

Early in the 20th century, the orbit of a comet nucleus called Shoemaker-Levy 9 from the Kuiper Belt was perturbed, and the comet's new orbit carried it close to Jupiter. Eventually, it was captured by Jupiter and orbited the planet. In 1992, that comet passed so close to Jupiter that tidal stresses broke it into two dozen major fragments, which then spread out along its orbit. The fragments took one more 2-year orbit around the planet, and throughout a week in 1994, the entire string of fragments crashed into Jupiter. The impacts occurred just behind the limb of the planet, so they were not visible on Earth until Jupiter's rotation put the impact points in view. Astronomers using ground-based telescopes and the HST could see immense plumes rising from the impacts to heights of more than

3,000 km above the cloud tops at the limb. The debris in those plumes then rained back onto Jupiter's stratosphere, causing ripples like pebbles thrown into a pond. **Figure 12.24** shows some HST images of the impact features. Sulfur and carbon compounds released by the impacts formed Earth-sized scars in the atmosphere that persisted for months.

Collisions with Earth

In summer 1908, a remote region of western Siberia was blasted with the energy equivalent of 2,000 times the energy of the atomic bomb dropped on Hiroshima in 1945. Eyewitness accounts detailed the destruction of dwellings, the incineration of reindeer (including one herd of 700), and the deaths of at least five people. Although trees were burned or flattened over more than 2,150 square kilometers (km²)—an area greater than metropolitan New York City—no crater was left behind. The Tunguska event (named for the nearby river) was the result of a tremendous high-altitude explosion that occurred when a small body hit Earth's atmosphere, ripped apart, and formed a fireball before reaching Earth's surface. Recent expeditions to the Tunguska area have recovered resin from the trees blasted by the event. Chemical traces in the resin suggest that the impacting object may have been a stony asteroid.

In February 2013, a known near-Earth object about half the size of an American football field passed so close to Earth that it came within the orbit of artificial satellites. That near miss was uneventful, and the object simply continued on its way. In an unrelated event on the same day, however, a previously unknown meteoroid estimated to have a radius of about 20 meters exploded over Chelyabinsk, Russia. The shock waves from that explosion damaged thousands of buildings in six cities and injured more than 1,000 people. That was likely to have been the largest impact on Earth since the Tunguska event, and many observations were recorded of the effect on that less remote location.

From car dashboard, cell phone, and security camera video and images (**Figure 12.25a**), as well as reports of the time between the brighter-than-the-Sun flash and the sonic boom that followed, scientists determined the trajectory and speed of the incoming object as it traveled through the atmosphere. They estimate a preimpact orbit of the object in the inner asteroid belt and think it originally broke off from a known 2-km asteroid. From small pieces collected

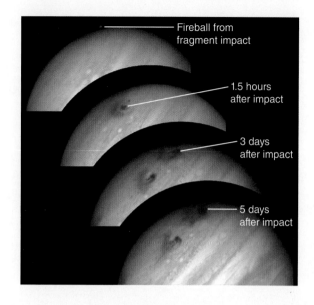

Figure 12.24 HST images of the evolution of the scar left by one fragment of Comet Shoemaker-Levy 9 when it impacted Jupiter in 1994.

Figure 12.25 (a) In February 2013, a meteoroid entered the atmosphere over Russia, creating a fireball that eyewitnesses said was brighter than the Sun. (b) A 600-kg piece of the meteorite on display.

over a wide area and from a large, 600-kg chunk found in a frozen-over lake (**Figure 12.25b**), scientists could analyze the object's composition and density. It seems to be similar in composition to that of Itokawa, the S-type asteroid whose dust was collected and returned to Earth. It is estimated that only a few hundredths of 1 percent of the original 10-million-kg mass has been found on the surface of Earth. The energy of the explosion was about 30 times the energy released by the Hiroshima atomic bomb. Most of that energy went into the atmosphere, however, heating and breaking up the meteoroid at a much higher altitude than that at which bombs are detonated, so the effects on the ground were less than those of a bomb.

Those impacts are sobering events. The distribution of relatively large asteroids in the inner Solar System indicates it is highly improbable that an asteroid will impact a populated area on Earth within your lifetime. However, many comets and smaller asteroids have unknown orbits, and several previously unknown long-period comets enter the inner Solar System each year. If on a collision course with Earth, even a large comet might not be noticed until just a few weeks or months before impact. For example, Comet Hyakutake was discovered only 2 months before it passed near Earth, and a potentially destructive asteroid that just missed Earth in 2002 was not discovered until 3 days *after* its closest approach. The Chelyabinsk meteoroid was in an orbit such that it wasn't detectable at all. Earth's geological and historical record suggests that actual impacts by large bodies are infrequent events.

As many as 10 million asteroids larger than a kilometer across may exist, but only about 130,000 have well-known orbits, and most of the unknown asteroids are too small to see until they come very close to Earth. The U.S. government—along with the governments of several other nations—is aware of the risk posed by near-Earth objects. Although the probability of a collision between a small asteroid and Earth is small, the consequences could be catastrophic (**Working It Out 12.2**), so NASA has been given a congressional mandate to catalog all NEOs and to scan the skies for those that remain undiscovered.

12.2 Working It Out Impact Energy

How much energy can be released from the impact of a comet nucleus? The kinetic energy of a moving object is given by

$$E_K = \frac{1}{2}mv^2$$

where E_K is the kinetic energy in joules (J), m is the mass in kilograms, and v is the speed in meters per second (m/s).

Suppose an asteroid or comet nucleus that is 10 km in diameter with a mass of 5×10^{14} kg hits Earth at a speed of 20 km/s = 20×10^3 m/s. Putting those values into the preceding equation gives us

$$E_K = \frac{1}{2} \times (5 \times 10^{14} \text{ kg}) \times (20 \times 10^3 \text{ m/s})^2$$

$$E_K = 1.0 \times 10^{23} \text{ J}$$

How much energy is that? A 1-megaton hydrogen bomb (67 times as energetic as the Hiroshima atomic bomb) releases 4.2×10^{15} J. If we divide the energy from our comet impact by that number, we have

$$\frac{1.0 \times 10^{23} \text{ J}}{4.2 \times 10^{15} \text{ J/megaton H-bomb}} = 2.4 \times 10^7$$

$$= 24 \text{ million 1-megaton H-bombs}$$

That's a lot of energy, and it shows why impacts have been so important in the history of the Solar System.

CHECK YOUR UNDERSTANDING 12.5

How do astronomers determine the origin of a meteorite that reaches Earth?

Origins Comets, Asteroids, Meteoroids, and Life

Water is essential to life on Earth, so where did it all come from? Scientists have thought that some of Earth's water was contributed during impacts of icy planetesimals early in the history of the Solar System. The icy planetesimals condensed from the protoplanetary disk surrounding the young Sun and grew to their current size near the orbits of the giant planets. Those planetesimals later interacted with the giant planets, which may themselves have been migrating to and from the inner Solar System. In such interactions, some of the planetesimals were flung outward to form the Kuiper Belt and Oort Cloud, and some were thrown inward toward the Sun, possibly hitting Earth. Because much of the mass in comet nuclei and some in asteroids appears to be in the form of water ice, some of Earth's current water supply may have come from that early bombardment. Spacecraft have measured the type of water in several comets and asteroids, and the water on Earth best matches that of the primitive carbonaceous chondrites and a few, but not most, comets.

Comets and asteroids can threaten life on Earth. Occasional collisions of comet nuclei and asteroids with Earth have probably resulted in widespread devastation of Earth's ecosystem and in the extinction of many species. Passing stars or the periodic passage of the Sun through giant gas clouds located in denser regions of the Milky Way Galaxy may have resulted in showers of comet nuclei into the inner Solar System, possibly changing the climate and contributing to mass extinctions. Although certainly qualifying as global disasters for the plants and animals alive at the time, such events also represent global opportunities for new life-forms to evolve and fill the niches left by species that did not survive. As noted in Chapter 8, such a collision with an asteroid or comet probably played a central role in ending the 180-million-year reign of dinosaurs and provided an opportunity for the evolution of mammals.

In studying comets, astronomers may also have found a key to the chemical origins of life on Earth. Comets are rich in complex organic material—the chemical basis for terrestrial life—and cometary impacts on the young Earth may have played a role in chemically seeding the planet. If comets are pristine samples of the material from which the Sun and planets formed, organic material must be widely distributed throughout interstellar space. Radio telescope observations of vast interstellar clouds throughout the Milky Way confirm the presence of organic material. The fact that asteroid belts and storms of comets have been observed in distant solar systems could have significant implications as astronomers consider the possibility of life elsewhere in the universe.

READING ASTRONOMY NEWS

Scientists report on the water observed on Comet 67P/Churyumov-Gerasimenko.

Rosetta Spacecraft Finds Water on Earth Didn't Come from Comets

By REBECCA JACOBSON, *PBS NewsHour*

It's a mystery that has baffled scientists for decades: Where did Earth's water come from?

Some scientists believed comets might have been the original source of the Earth's oceans. But a study published this week in the journal *Science* is sending scientists back to the drawing board. In its first published scientific data, the ROSINA mass spectrometer on board the *Rosetta* probe found that water on Comet 67P/Churyumov-Gerasimenko doesn't match the water on Earth.

The result is surprising, says Kathrin Altwegg, principal investigator for ROSINA at the University of Bern and one of the authors of the study. For decades, scientists had ruled out comets from the Oort Cloud at the very edge of our Solar System as the source of Earth's water.

But three years ago, an analysis of water on the Hartley 2 comet near Jupiter found a perfect match to the Earth's oceans. That finding led scientists to believe that Earth's water could have come from much closer comets, either near Jupiter or in the Kuiper Belt just beyond Neptune. Comet 67P/Churyumov-Gerasimenko is one of those Jupiter family comets, which scientists believe originated in the Kuiper Belt.

"That was a big surprise, but now we are back to what I expected," she said. "I think it's very nice to see the diversity we have in Kuiper Belt, to see that not everything is as simple as it seemed."

To find the origin of Earth's water, scientists analyze the water's "fingerprint," says Claudia Alexander, project scientist of the U.S. *Rosetta* Project at NASA's Jet Propulsion Laboratory. Water has a chemical isotopic signature, which works just like a fingerprint. Planets, comets, even minerals all have a fingerprint, Alexander says, and scientists are looking for a match to Earth's.

On Earth, water is mostly two parts hydrogen and one part oxygen—H_2O. But there's also "heavy" water, Alexander explained, which is made with deuterium—a hydrogen atom with a neutron. That heavy water is what the *Rosetta* spacecraft found on Comet 67P/Churyumov-Gerasimenko. It's also a closer match to the water scientists have found on other comets, ruling them out as Earth's water source, Alexander said.

"The clues don't quite all add up," she said.

Altwegg agrees, saying that it's not likely the other Kuiper Belt comets have a match to Earth's water, but further studies would be helpful.

"You would have to assume that 67P is the exception in the Kuiper Belt," she said. "We need more missions to Kuiper Belt comets, which would be fabulous."

There are several ideas to explain the origin of Earth's water, Alexander said. Some believe that water has been on Earth since its formation, that it was beaten out of other minerals as the planet formed. Others think "wet planetesimals" near Jupiter—which were like planetary Silly Putty, loose sticky blobs of rock and ice, Alexander said—collided with Earth in the early formation of the Solar System.

Alexander believes the answer could be a combination of any of these ideas. The *Dawn* mission in 2015 [is studying] the water on the asteroid Ceres, near Jupiter. If it's a match for Earth's water, it may be another clue, Alexander said. But the *Rosetta* finding is a huge step in solving the mystery, she said.

"I think this is a big deal. . . . For me, I've not always been a believer in the story that comets brought the water," Alexander said. "In some respects, I'm somewhat relieved (this finding) doesn't confirm it. It's more complicated than that. I think we need more forensic evidence to settle the score."

QUESTIONS

1. This article uses the word *fingerprint* when discussing isotope ratios. Earlier in this book, we used that word in our discussion of spectral lines. How are the two phenomena similar?
2. What might explain why the water on Comet Hartley 2 matched Earth's water, but the water on Comet 67P/Churyumov-Gerasimenko did not?
3. Is the water on other comets more like the water on Comet Hartley 2 or on Comet 67P/Churyumov-Gerasimenko?
4. What are other possible sources of Earth's water?
5. Do an Internet search to see whether any results from the *Dawn* mission indicate the type of water observed on Ceres.

Source: "Rosetta spacecraft finds water on Earth didn't come from comets," by Rebecca Jacobson. *PBS NewsHour*, December 10, 2014. © 2014 NewsHour Productions LLC. Reprinted by permission.

Summary

The story of how planetesimals, asteroids, and meteorites are related is a great success of planetary science. Scientists have assembled a wealth of information about that diverse collection of objects to piece together a picture of how planetesimals grow, differentiate, and then shatter in collisions. That story fits well with the even larger story of how most planetesimals were accreted into the planets and their moons (Chapter 7). Comets, asteroids, and meteoroids may have supplied Earth with water, volatiles, and organic material in the early history of the Solar System. Impacts with large asteroids or comet nuclei may have led to mass extinctions on Earth that eventually enabled mammals (for example) to evolve. Recent spacecraft missions to comets and asteroids have begun to reveal details about the composition of those bodies.

LG 1 **List the categories of small bodies and identify their locations in the Solar System.** Small bodies in the Solar System that orbit the Sun include dwarf planets, asteroids, comets, Kuiper Belt objects, and meteoroids. Most asteroids are located in the main asteroid belt between the orbits of Mars and Jupiter. Comets are small, icy planetesimals that reside in the frigid regions of the Kuiper Belt and the Oort Cloud, beyond the planets. The orbits of nearly all comets are highly elliptical, with one end of the orbit close to the Sun and the other in the distant parts of the Solar System.

LG 2 **Describe the defining characteristics of the dwarf planets in the Solar System.** Pluto, Eris, Haumea, Makemake, and Ceres are massive enough to have pulled themselves into spherical shapes. But they are classified as dwarf planets, not planets, because they are not massive enough to have cleared their surroundings of other bodies.

LG 3 **Describe the origin of the types of asteroids, comets, and meteorites.** Asteroids are small Solar System bodies made of rock and metal. Although early collisions between those planetesimals created several bodies large enough to differentiate, Jupiter's tidal disruption (and possible migration) prevented them from forming a single planet. Comets that venture into the inner Solar System are warmed by the Sun, producing an atmospheric coma and a tail. Meteoroids are small fragments of asteroids and comets. When a meteoroid enters Earth's atmosphere, frictional heat causes the air to glow, producing a meteor. Meteor showers occur when Earth passes through a trail of cometary debris. A meteoroid that survives to a planet's surface is called a meteorite. The various types of meteoroids that are formed depend on the differentiation of the parent body.

LG 4 **Explain how asteroids, comets, and meteoroids provide important clues about the history and formation of the Solar System.** Asteroids, comets, and meteoroids are leftover debris from the formation of the Solar System. Asteroids are composed of the same type of material that became the inner planets, and comets are composed of the same type of material that became the outer planets. They provide samples of the initial composition and properties of the Solar System and furnish samples of material from its entire history.

LG 5 **Describe what has been learned from observations of recent impacts in the Solar System.** Impacts by comets and meteoroids have been observed recently on Jupiter and Earth. The debris from those impacts helps astronomers understand the conditions in the early Solar System when impacts were much more frequent.

? Unanswered Questions

- What can asteroids and comets reveal about the migration of the giant planets in the early Solar System? It is not accidental that the main asteroid belt and the Kuiper Belt are at the boundaries of the orbits of the giant planets. As noted in Chapter 10, the giant planets may have moved around quite a bit in the early Solar System. The migration of the giant planets may explain the spread in the orbits of the objects in the main asteroid belt. Migration may also have brought icy objects—such as comet nuclei—out of the Kuiper Belt and into the main asteroid belt. The Kuiper Belt would have been closer to the Sun originally, and Jupiter and Saturn would have pushed it outward. And, finally, the migration might have sent objects from both belts into the inner Solar System, creating the heavy bombardment of 4 billion years ago.

- Does a "Planet Nine" exist? That is a hypothetical planet in the far Solar System that was proposed to explain some irregularities in the orbits of objects in the Kuiper belt and beyond. The planet would be about 10 times the mass of Earth, about 2–4 times the size of Earth—similar to Uranus and Neptune. It would have an elongated orbit, with a distance from the Sun of about 200–1,200 AU, and would take 10,000–20,000 years to orbit the Sun. Originally it might have migrated outward in the early Solar System or have been captured from another star. Scientists have extensively searched through existing data and made new observations, but nothing has been found. Because the planet would be cold, it would radiate in the infrared and might be detected by the James Webb Space Telescope.

- Does a small, dim star exist in the neighborhood of the Sun that comes close periodically and stirs up the Oort Cloud, sending a much larger than average number of comets into the inner Solar System? Some scientists think the fossil data show periodic mass extinctions on Earth and have investigated astronomical causes of the extinctions. One hypothesis is that a distant companion to the Sun has a wide orbit and periodically has come closer to the Sun as they both have traveled around the Milky Way Galaxy. That companion may have stirred up the Oort Cloud as a result. That Oort Cloud disturbance sent many comets to the inner Solar System, which could have caused many impacts on Earth, leading to a mass extinction similar to the one that wiped out the dinosaurs. But some scientists dispute that Earth's extinctions have occurred at regular intervals, and surveys with infrared telescopes have found no evidence of a small companion star.

Questions and Problems

Test Your Understanding

1. This chapter has focused on leftover planetesimals. What became of most of the other planetesimals?
   a. They evaporated.
   b. They left the Solar System.
   c. They became part of larger bodies.
   d. They fragmented into smaller pieces.

2. The three types of meteorites come from different parts of their parent bodies. Stony-iron meteorites are rare because
   a. they are hard to find.
   b. the volume of a differentiated body that has both stone and iron is small.
   c. the Solar System has very little iron.
   d. the magnetic field of the Sun attracts the iron.

3. As a comet leaves the inner Solar System, the ion tail points
   a. back along the orbit.
   b. forward along the orbit.
   c. toward the Sun.
   d. away from the Sun.

4. Congress tasked NASA with searching for near-Earth objects because
   a. they might impact Earth, as others have.
   b. they are close by and easy to study.
   c. they are moving fast.
   d. they are scientifically interesting.

5. Meteorites can provide information about all of the following except
   a. the early composition of the Solar System.
   b. the composition of asteroids.
   c. the composition of comets.
   d. the Oort Cloud.

6. Perihelion is the point in an orbit _____ the Sun; aphelion is the point in an orbit _____ the Sun.
   a. closest to; farthest from
   b. farthest from; closest to
   c. at one focus of; at the other focus of

7. Kuiper Belt objects (KBOs) are mostly comet nuclei. Why don't they display comae and tails?
   a. Most of the material has already been stripped from the objects.
   b. They are too far from the Sun.
   c. They are too close to the Sun.
   d. The comae and tails are pointing away from Earth, behind the object.

8. Asteroids are small
   a. rock and metal objects orbiting the Sun.
   b. icy objects orbiting the Sun.
   c. rock and metal objects found only between the orbits of Mars and Jupiter.
   d. icy bodies found only in the outer Solar System.

9. Aside from their periods, short-period and long-period comets differ in that
   a. short-period comets orbit prograde, whereas long-period comets orbit retrograde.
   b. short-period comets contain less ice, whereas long-period comets contain more.
   c. short-period comets do not develop ion tails, whereas long-period comets do.
   d. short-period comets come closer to the Sun at closest approach than long-period comets.

10. On average, a bright comet appears about once each decade. Statistically, that means that
    a. one will definitely be observed every tenth year.
    b. one will definitely be observed in each 10-year period.
    c. exactly 10 comets will be observed in a century.
    d. about 10 comets will be observed in a century.

11. Most asteroids are located between the orbits of
    a. Earth and Mars.
    b. Mars and Jupiter.
    c. Jupiter and Saturn.
    d. the Kuiper Belt and the Oort Cloud.

12. Comets, asteroids, and meteoroids may be responsible for delivering a significant fraction of the current supply of _____ to Earth.
    a. mass
    b. water
    c. oxygen
    d. carbon

13. An iron meteorite most likely came from
    a. an undifferentiated asteroid.
    b. a differentiated asteroid.
    c. a planet.
    d. a comet.

14. Meteor showers occur because Earth passes through the path of
    a. another planet.
    b. a planetesimal.
    c. a comet.
    d. the Moon.

15. Dwarf planets differ from the other planets in that they
    a. have no atmosphere.
    b. have no moons.
    c. are all very far from the Sun.
    d. have lower mass.
    e. are covered in ice.

Thinking about the Concepts

16. Describe ways in which Pluto differs significantly from the Solar System planets.

17. By what criteria did Pluto fail to be considered a planet under the new IAU definition? Explain how that decision demonstrates the self-correcting nature of science.

18. How does the composition of an asteroid differ from that of a comet nucleus?

19. Define *meteoroid*, *meteor*, and *meteorite*.

20. What are the differences between a comet and a meteor in size, distance, and how long they remain visible?

21. Most meteorites are 4.54 billion years old. Carbonaceous chondrites, however, are 4.56 billion years old (20 million years older). What determines the time of "birth" of those pieces of rock? What does that information tell you about the history of their parent bodies?

22. Most asteroids are found between the orbits of Mars and Jupiter, but astronomers are especially interested in the relative few whose orbits cross that of Earth. Why?

23. How could you and a friend, armed only with your cell phones and knowledge of the night sky, prove conclusively that meteors are an atmospheric phenomenon?

24. Suppose you find a rock that has all the characteristics of a meteorite. You take it to a physicist friend, who confirms that it is a meteorite but says that radioisotope dating indicates an age of only a billion years. What might be the origin of that meteorite?

25. Describe differences between the Kuiper Belt and the Oort Cloud as sources of comets. What is the ultimate fate of a comet from each of those reservoirs?

26. What are the three parts of a comet? Which part is the smallest in radius? Which is the most massive?

27. In 1910, Earth passed directly through the tail of Comet Halley. Among the various gases in the tail was hydrogen cyanide (HCN). HCN is deadly to humans, yet nobody became ill from that event. Why?

28. Comets have two types of tails. Describe them and explain why they sometimes point in different directions.

29. What is zodiacal light, and what is its source?

30. How might comets and asteroids have contributed to the origin of life on Earth?

Applying the Concepts

31. Comet 67P/Churyumov-Gerasimenko has a diameter of 4 km and a mass of 10^{13} kg.
    a. What is the density of the comet? How does that value compare with the density of water?
    b. What is the escape velocity from the surface of that comet?

32. Ceres has a diameter of 975 km and a period of about 9 hours. What is the rotational speed of a point on the surface of that dwarf planet?

33. Figure 12.11 shows the scale of the Solar System out to the Oort Cloud. Judging from that figure, what fraction of the distance between the Sun and Proxima Centauri is occupied by the Oort Cloud, a part of the Solar System?

34. What are the perihelion and aphelion distances for Pluto and Eris? (Hint: Use Working It Out 12.1 and the information in Appendix 4.)

35. Use Working It Out 12.2 to find the impact energy (in joules) of an asteroid with a mass of 4.6×10^{11} kg traveling at 40 km/s. Does that energy depend on the "target" of impact? What is the equivalent in 1-megaton hydrogen bombs?

36. Earth's Moon has a diameter of 3,474 km and orbits at an average distance of 384,400 km. At that distance, it subtends an angle just slightly larger than half a degree in Earth's sky. Pluto's moon Charon has a diameter of 1,186 km and orbits at a distance of 19,600 km from the dwarf planet.
    a. Compare the appearance of Charon in Pluto's skies with the Moon in Earth's skies.
    b. Describe where in the sky Charon would appear as seen from various locations on Pluto.

37. One recent estimate concludes that nearly 800 meteorites with mass greater than 100 grams (massive enough to cause personal injury) strike Earth each day. If you present a target of 0.25 square meter (m^2) to a falling meteorite, what is the probability that you will be struck by a meteorite during your 100-year lifetime? (The surface area of Earth is approximately 5×10^{14} m^2.)

38. Electra is a 182-km asteroid accompanied by a small moon orbiting at a distance of 1,350 km in a circular orbit with a period of 3.92 days.
    a. What is the mass of Electra?
    b. What is Electra's density?

39. Calculate the orbital radius of the Kirkwood gap that is in a 3:1 orbital resonance with Jupiter.

40. The orbital periods of Comets Encke, Halley, and Hale-Bopp are 3.3, 76, and 2,530 years, respectively. Their orbital eccentricities are 0.847, 0.967, and 0.995, respectively.
    a. What are the semimajor axes (in astronomical units) of the orbits of those comets?
    b. What are the minimum and maximum distances from the Sun (in astronomical units) reached by Comets Halley and Hale-Bopp in their respective orbits?
    c. Which region of the Solar System did each probably come from?
    d. Which would you guess is the most pristine comet among the three? Which is the least? Explain your reasoning.

41. Comet Halley has a mass of approximately 2.2×10^{14} kg. It loses about 3×10^{11} kg each time it passes the Sun.
    a. The first confirmed observation of the comet was made in 240 BCE. If you assume a constant period of 76.4 years, how many times has it reappeared since that early sighting?
    b. How much mass has the comet lost since 240 BCE?
    c. What percentage of the comet's total mass today does that amount represent?

42. If Comet Halley is approximated as a sphere 5 km in radius, what is its density if it has a mass of 2.2×10^{14} kg? How does that density compare with that of water (1,000 kg/m³)?

43. A cubic centimeter of the air you breathe contains about 10^{19} molecules. A cubic centimeter of a comet's tail may typically contain 200 molecules. Calculate the volume of comet tail material that would hold 10^{19} molecules.

44. Some near-Earth objects are in binary systems, so it is possible to estimate their mass. How much energy would be released if a near-Earth asteroid with mass $m = 4.6 \times 10^{11}$ kg hit Earth at a velocity $v = 5$ km/s?

45. The estimated amount of zodiacal dust in the Solar System remains constant at approximately 10^{16} kg, yet zodiacal dust is constantly being swept up by planets or removed by the pressure of sunlight.
    a. If all the dust disappeared (at a constant rate) over a span of 30,000 years, what would the average production rate have to be, in kilograms per second, to maintain the current content?
    b. Is that an example of static or dynamic equilibrium? Explain your answer.

USING THE WEB

46. Dwarf planets:
    a. Go to planetary astronomer Mike Brown's website of dwarf planets (http://gps.caltech.edu/~mbrown/dps.html). How many dwarf planets does he think are in the Solar System? Why is it difficult officially to certify an object as a dwarf planet?
    b. Go to the website for the *New Horizons* mission (http://pluto.jhuapl.edu), which reached Pluto in 2015 and is scheduled to pass Kuiper Belt Object (KBO) 2014 MU69 in early 2019. Click on "News Center." What has been learned from that mission?

47. Citizen Science: Go to Backyard Worlds: Planet 9 (https://www.zooniverse.org/projects/marckuchner/backyard-worlds-planet-9). In that project you help search for new objects in the far reaches of the Solar System. Read all the sections in "About," and then click on "Classify" and look at some images.

48. Go to NASA's Asteroid Watch website (https://www.jpl.nasa.gov/asteroidwatch). What is new? Has a new discovery or a recent flyby occurred? Was the asteroid studied with a spacecraft, an orbiting telescope, or a ground-based telescope? What has been learned about the object?

49. Go to the Space Weather website (http://spaceweather.com). Are any comets now visible with the naked eye? Scroll down to "Near Earth Asteroids." Are any "close encounters" coming up in the next few months? Click on a few asteroid names to access the JPL Small-Body Database. Note the values of *e* and *a* in the table under the orbit. Calculate the NEO's closest and farthest distances from the Sun. In each case, how close will the NEO be to Earth when it is at its closest? How large is the object?

50. Do a search for the Japanese *Hayabusa 2* and the NASA *OSIRIS-REx* missions, each of which is visiting an asteroid in 2018 and is scheduled to bring back a sample to Earth after making observations. Did they arrive at their asteroid? What has been discovered from the observations? Has a sample been returned to Earth—if so, what has been learned?

EXPLORATION

Asteroid Discovery

digital.wwnorton.com/astro6

Astronomers often discover asteroids and other small Solar System objects by comparing two (or more) images of the same star field and looking for bright spots that have moved between the images. The four images in **Figure 12.26** are all "negative images": every dark spot would actually be bright in the sky, and all the white space is dark sky. A negative image sometimes helps the observer pick out faint details, and it is preferable for printing and photocopying. Study those four images. Can you find an asteroid that moves across the field in those images? That's the hard way to do it. A much easier method is to use a "blink comparison," which lets you look at one image and then another very quickly. Make three photocopies of each image, cut out each one, and align them all carefully on top of one another in sequence, so that the stars overlap. You should have 12 pieces of paper, in this order: Image 1, Image 2, Image 3, Image 4; Image 1, Image 2, . . . , and so on. Staple the top edge and flip the pages with your thumb, looking carefully at the images. Can you find the asteroid now?

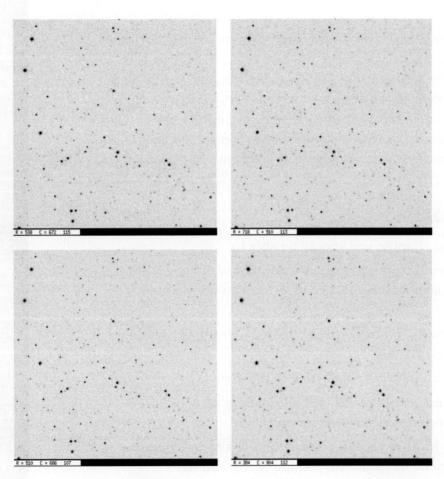

Figure 12.26

1 Circle the asteroid in each image.

The "blink comparison" method takes advantage of a feature of the human eye-brain connection. Humans are much better at noticing things that move than things that are stationary.

2 Why might that feature be a helpful evolutionary adaptation?

A third method is available, which sometimes makes things easier to see but requires high-quality digital images. In that method, one image is subtracted from another.

3 If you used the subtraction method with two of those images, what would you expect to see in the resulting image?

13 Taking the Measure of Stars

To all but the largest telescopes, even nearby stars are just points of light in the night sky. Astronomers study the stars by observing their light, by using the laws of physics discussed in earlier chapters, and by finding patterns in subgroups of stars that are then applied to other stars. Astronomers use knowledge of geometry, radiation, and orbits to begin to answer basic questions about stars, such as how they are similar to or different from the Sun, and whether they might have planets orbiting around them as the Sun does.

LEARNING GOALS

By the end of this chapter, you should be able to:

LG 1 Explain how astronomers use the brightness of nearby stars and their distances from Earth to determine how luminous they are.

LG 2 Explain how astronomers obtain the temperatures, sizes, and composition of stars.

LG 3 Describe how astronomers estimate the masses of stars.

LG 4 Classify stars and organize that information on a Hertzsprung-Russell (H-R) diagram.

LG 5 Explain how the mass and composition of a main-sequence star determine its luminosity, temperature, and size.

The constellation Orion. In addition to stars we can see numerous nebula of gas and dust. ▶▶▶

Why do stars have different colors?

13.1 Astronomers Measure the Distance, Brightness, and Luminosity of Stars

The stars in the sky differ in brightness and color, but does one star appear brighter than another because it has a higher luminosity and is emitting more light or because it is closer to us? In this section, we explain how to find the distances to nearby stars and how to use distance and apparent brightness to find the luminosity of a star.

Stereoscopic Vision

Your two eyes have different views that depend on the distance to the object you are viewing. Hold up your finger in front of you, close to your nose. View it with your right eye only and then with your left eye only. Each eye sends a slightly different image to your brain, so your finger *appears* to move back and forth relative to the background behind it. Now hold up your finger at arm's length, and blink your right eye and then your left. Your finger appears to move much less. The way your brain combines the different information from your eyes to perceive the distances to objects around you is called **stereoscopic vision**. **Figure 13.1a** shows an overhead view of the experiment you just performed with your finger. The left eye sees the blue pencil almost directly between the green balls on the bookshelf. But the right eye sees the blue pencil to the left of both balls. Similarly, the position of the pink pencil appears to vary. Because the pink pencil is closer to the observer, its position appears to change more than the position of the blue pencil—seeming to move from the right of the blue pencil to the left of the blue pencil.

Stereoscopic vision enables you to judge the distances of objects as far away as a few hundred meters, but beyond that it is of little use. Your right eye's view of a mountain several kilometers away is indistinguishable from the view seen by your left eye—all you can determine is that the mountain is too far away for you to judge its distance stereoscopically. The distance over which your stereoscopic vision works is limited by the separation between your eyes—about 6 centimeters (cm). If you could increase the distance between your eyes by several meters, the view from each eye would be different enough for you to determine the distances to objects kilometers away.

You cannot separate your eyes, but you can compare pictures taken with a camera from two widely separated locations. The greatest separation possible without leaving Earth is to let Earth's orbital motion carry us from one side of the Sun to the other. If you take a picture of the sky tonight and then wait 6 months and take another picture, the distance between the two locations is the diameter of Earth's orbit (2 astronomical units [AU]), which gives us more powerful stereoscopic vision.

Astronomy in Action: Parallax

Distances to Nearby Stars

Figure 13.1b, which depicts Earth's orbit from far above the Solar System, shows how astronomers apply the concept of stereoscopic vision to measure the distances to stars. The change in position of Earth over 6 months is like the distance between the right eye and the left eye in Figure 13.1a. The nearby (pink and blue) stars are like the pink and blue pencils, whereas the distant yellow stars are like the green balls on the bookcase. Because of the shift in perspective as Earth orbits the Sun, nearby stars appear to shift their positions. The pink star, which is closer,

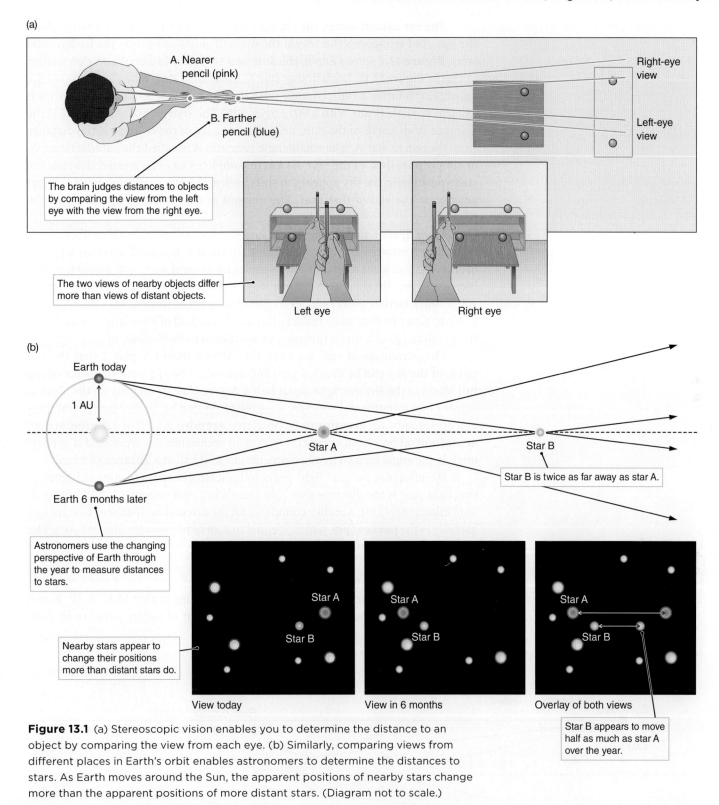

Figure 13.1 (a) Stereoscopic vision enables you to determine the distance to an object by comparing the view from each eye. (b) Similarly, comparing views from different places in Earth's orbit enables astronomers to determine the distances to stars. As Earth moves around the Sun, the apparent positions of nearby stars change more than the apparent positions of more distant stars. (Diagram not to scale.)

appears to move farther than the more distant blue star. Over 1 full year, the nearby star appears to move one way and then back again with respect to distant background stars, returning to its original position at the end of that year. We can determine the distance to the star by applying geometry to the amount of that apparent shift.

The eye cannot detect the changes in position of a nearby star throughout the year, but telescopes can reveal those small shifts relative to the background stars. **Figure 13.2** shows Earth, the Sun, and two stars in a configuration similar to that in Figure 13.1b. Look first at star A, the closest star. When at the top of the figure, Earth forms a right triangle with the Sun and star A at the other corners. (A right triangle is one with a 90° angle in it.) The short leg of the triangle is the distance from Earth to the Sun, 1 AU. The long leg of the triangle is the distance from the Sun to star A. The small angle near star A is called the *parallactic angle*, or simply **parallax**, of the star. As Earth completes an orbit around the Sun, the star's position in the sky appears to shift back and forth, returning to its original position at the end of that year. The amount of that shift is equal to twice the parallax.

More distant stars make longer and skinnier triangles with smaller parallaxes. Star B is twice as far away as star A, and so star B's parallax is only half that of star A. If you were to draw such triangles for several stars, you would find that increasing the distance to the star always reduces the star's parallax. Moving a star 3 times farther away reduces its parallax to 1/3 of its original value. Moving a star 10 times farther away reduces its parallax to 1/10 of its original value. Thus, the parallax (p) of a star is inversely proportional to its distance (d).

The parallaxes of real stars are tiny. Recall from Chapter 2 that the full circle of the sky can be divided into 360 degrees. The apparent diameter of the full Moon in the sky averages about half a degree. Just as an hour on the clock is divided into minutes and seconds, a degree of sky can be divided into arcminutes and arcseconds. An **arcminute** (abbreviated **arcmin**) is 1/60 of a degree, and an **arcsecond** (abbreviated **arcsec**) is 1/60 of an arcminute. An arcsecond is approximately the angle formed by the diameter of a golf ball at a distance of 9 km.

We often use units of light-years to indicate distances to stars (Chapter 1). One light-year is the distance that light travels in 1 year—about 9.5×10^{12} (9.5 trillion) kilometers (km). Another common unit for discussing distances to stars and galaxies is the **parsec** (**pc**), which is equal to 3.26 light-years (or 206,265 AU). The term is short for *parallax second*—a star at a distance of 1 parsec has a parallax of 1 arcsecond.

When astronomers began to measure the parallax angles of stars, they discovered that stars are very distant objects (**Working It Out 13.1**). F. W. Bessel (1784–1846) made the first successful measurement of stellar parallax in 1838,

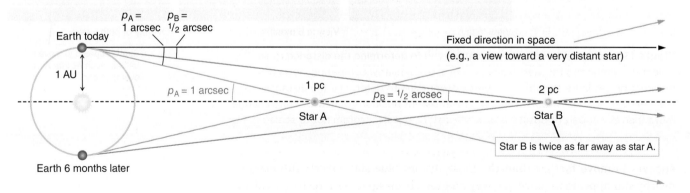

Figure 13.2 The parallax (*p*) of a star is inversely proportional to its distance. More distant stars have smaller parallaxes. (Diagram not to scale.)

13.1 Working It Out Parallax and Distance

If star B in Figure 13.2 is twice as far away from us as star A, star B will have half the parallax of star A. The parallax of a star (p) is inversely proportional to its distance (d):

$$p \propto \frac{1}{d} \quad \text{or} \quad d \propto \frac{1}{p}$$

If the angle at the apex of a triangle is 1 arcsec, and the base of the triangle is 1 AU, the length of the triangle is 1 parsec. One parsec equals 206,265 AU (see Appendix 1), or 3.09×10^{13} km, or 3.26 light-years. Astronomers use the unit of parsecs because doing so makes the relationship between distance and parallax easier to manage than when the distance is measured in light-years.

A star with a parallax of 1 arcsec is at a distance of 1 pc (Figure 13.2). The inverse proportionality between distance and parallax becomes

$$\left(\begin{array}{c}\text{Distance measured} \\ \text{in parsecs}\end{array}\right) = \frac{1}{\left(\begin{array}{c}\text{Parallax measured} \\ \text{in arcseconds}\end{array}\right)}$$

or

$$d(\text{pc}) = \frac{1}{p(\text{arcsec})}$$

Suppose that the parallax of a star is measured to be 0.5 arcsec. The distance can be found by

$$d(\text{pc}) = \frac{1}{0.5} = 2 \text{ parsecs}$$

Similarly, a star with a measured parallax of 0.01 arcsec is located at a distance of 1/0.01 = 100 pc.

After the Sun, the next-closest star to Earth is Proxima Centauri. Located at a distance of 4.24 light-years, Proxima Centauri is a faint member of a system of three stars called Alpha Centauri. What is that star's parallax? First, we convert the distance to parsecs:

$$d = 4.24 \text{ light-years} \times \frac{1 \text{ parsec}}{3.26 \text{ light-years}} = 1.30 \text{ parsecs}$$

Then,

$$p(\text{arcsec}) = \frac{1}{1.30 \text{ pc}} = 0.77 \text{ arcsec}$$

Even the closest star to the Sun has a parallax of only about 3/4 arcsec.

---

reporting a parallax of 0.314 arcsec for the star 61 Cygni. That finding implied that 61 Cygni was 3.2 pc away, or 660,000 times as far away as the Sun. With that one measurement, Bessel increased the known volume of the universe by a factor of 10,000. Today, astronomers know of at least 60 stars, some in multiple-star systems, within 5 pc (16.3 light-years) of the Sun. In the neighborhood of the Sun, each star or star system has on average a volume of about 260 cubic light-years of space to itself.

Most stars are so far away that the parallax angle is too small to measure using ground-based telescopes, which are limited by Earth's atmosphere. In the 1990s, the European Space Agency's *Hipparcos* satellite measured the positions and parallaxes of over 100,000 stars with an uncertainty of about ±0.001 arcsec. *Hipparcos*' successor is *Gaia*, a space mission that has been studying stellar parallaxes since the end of 2013. *Gaia*'s second data release in 2018 included parallaxes of more than 1 billion stars, with uncertainties of ±0.00004 to ±0.0007 arcsec, depending on the star's brightness.

Note that since the distance depends on 1 over the parallax, an uncertainty in the parallax will mean a corresponding range of distances. For example, a star with a measured parallax of 0.004 ± 0.001 arcsec really has a parallax between 0.003 and 0.005 arcsec. That range of parallaxes gives a corresponding distance range of 200–333 pc from Earth. If the measured parallax is 0.004 ± 0.0001 arcsec, then the distance range is 244–256 parsecs. Thus astronomers are very excited about this recent data from *Gaia*. Other methods of measuring distance to more remote stars are discussed later.

Luminosity, Brightness, and Distance

The stars in Earth's sky are of different brightnesses, where brightness corresponds to the amount of energy falling on a square meter of area each second in the form of electromagnetic radiation (Chapter 5). Although the brightness of a star can be measured directly, it does not immediately give much information about the star itself. As illustrated in **Figure 13.3**, a star in the night sky may in fact appear bright only because it is nearby. Conversely, a faint star may be a powerful beacon, still visible despite its tremendous distance.

Astronomers measure the brightness of stars by comparing them with one another. The system they use dates back 2,100 years, when the Greek astronomer Hipparchus classified stars according to their brightness. The details of his system, still in use today, are discussed in **Working It Out 13.2** and **Appendix 7**.

To learn about the actual properties of a star, astronomers need to know the total energy radiated by a star each second—the star's luminosity. Recall from

Figure 13.3 The brightness of a star visible in our sky depends on both its luminosity—how much light it emits—and its distance.

13.2 Working It Out The Magnitude System

The **magnitude** system of brightness for celestial objects can be traced back 2,100 years to the ancient Greek astronomer Hipparchus, who classified the brightest stars he could see as being "of the first magnitude" and the faintest as being "of the sixth magnitude." Later, astronomers defined Hipparchus's 1st magnitude stars as being exactly 100 times brighter than his 6th magnitude stars. Hipparchus must have had typical eyesight because an average person under dark skies can see stars only as faint as 6th magnitude. Today, telescopes extend our vision far into space. The Hubble Space Telescope (HST) can integrate for long exposures and detect stars as faint as 30th magnitude. The limits also have been extended to stars brighter than 1st magnitude by using zero and negative numbers. A negative magnitude signifies that an object is *brighter* than an object at zero magnitude. For example, Sirius, the brightest star in the sky, has a magnitude of −1.46. Venus can be as bright as magnitude −4.4, or about 15 times brighter than Sirius and bright enough to cast a shadow. The magnitude of the full Moon is −12.6, and that of the Sun is −26.7 (**Figure 13.4**).

Looking at that system mathematically, with five steps between the 1st and 6th magnitudes, you see that each step is equal to the fifth root of 100, or $100^{1/5}$, approximately 2.512. That system is logarithmic, but instead of the usual base 10, it is base 2.512. Thus, a 5-magnitude difference in brightness equals $(2.512)^5$, or 100, times the difference in brightness. Fifth magnitude stars are 2.512 times brighter than 6th magnitude stars, and 4th magnitude stars are $2.512 \times 2.512 = 6.310$ times brighter than 6th magnitude stars. A 2.5-magnitude difference equals $(2.512)^{2.5}$, or 10, times difference in brightness. The brightness ratio between any two stars is equal to $(2.512)^N$, where N is the magnitude difference between them.

Because the limit of the Hubble Space Telescope (HST) is 30, and $30 - 6 = 24$, HST can detect stars that are $(2.512)^{24} = 4 \times 10^9$, or 4 billion, times fainter than the magnitude 6 that the naked eye can see. Or if we compare the Sun and the Moon, the Sun is 14 magnitudes, or $(2.512)^{14} = 4 \times 10^5$ (400,000) times brighter than the full Moon. (For more detailed calculations and a table of magnitudes and brightness differences, see Appendix 7.)

The magnitude of a star, as we have discussed it so far, is called the star's **apparent magnitude** because it is the brightness of the star as it *appears* in Earth's sky. If all stars were located exactly 10 pc (32.6 light-years) away from Earth, the brightness of each star would reflect its luminosity. If the distance from Earth to a star is known, astronomers compute how bright the star would appear if it were located at 10 pc. The **absolute magnitude** of a star—its apparent magnitude at a distance of 10 pc—measures the star's luminosity.

Figure 13.4 Apparent magnitude indicates the apparent brightness of an object in our sky. The brightest objects have a negative apparent magnitude, while telescopes have extended the observable range to fainter objects with higher magnitudes.

Chapter 5 that the brightness of an object that has a known luminosity L and is located at a distance d is given by

$$\text{Brightness} = \frac{\text{Total light emitted per second}}{\text{Area of sphere of radius } d} = \frac{\text{Luminosity}}{4\pi d^2}$$

You can rearrange that equation, moving the quantities you know how to measure (distance and brightness) to the right-hand side and the quantity you would like to know (luminosity) to the left, to get

$$\text{Luminosity} = 4\pi d^2 \times \text{Brightness}$$

That equation is used to find how much total light a star must be giving off to appear as bright as it does when seen from Earth.

Different stars have different luminosities. The Sun provides a convenient comparison when measuring the properties of stars, including their luminosity. The measured luminosity of the Sun is $L_{\text{Sun}} = 3.8 \times 10^{26}$ watts (W). The most luminous stars exceed the luminosity of the Sun by a million times ($10^6\ L_{\text{Sun}}$), whereas the least luminous stars have luminosities less than 1/10,000 that of the Sun ($10^{-4}\ L_{\text{Sun}}$). The most luminous stars are therefore more than 10 billion (10^{10}) times more luminous than the least luminous stars. Only a very small fraction of stars are near the upper end of that range of luminosities. Most stars are at the faint end of that distribution, less luminous than the Sun. **Figure 13.5** shows the relative number of stars compared with their luminosities in solar units. (Distances for the nearest stars are obtained from their parallaxes; other methods—discussed later—are used for the more distant stars.)

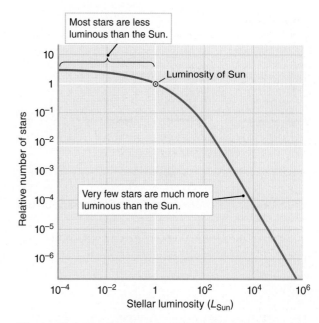

Figure 13.5 The distribution of the luminosities of stars is plotted logarithmically, with increments in powers of 10.

CHECK YOUR UNDERSTANDING 13.1

Stars A and B appear equally bright, but star A is twice as far away from us as star B. Which of the following is true? (a) Star A is twice as luminous as star B. (b) Star A is 4 times as luminous as star B. (c) Star B is twice as luminous as star A. (d) Star B is 4 times as luminous as star A. (e) Star A and star B have the same luminosity because they have the same brightness.

13.2 Astronomers Can Determine the Temperature, Size, and Composition of Stars

Two everyday concepts—stereoscopic vision and the fact that objects appear brighter when closer—have provided the tools needed to measure the distance and luminosity of the closest stars. Stars that appear to be faint points of light in the night sky are in fact luminous beacons located at great distances. The laws of radiation described in Chapter 5 reveal still more about stars.

The gas in the outer layers of stars is dense enough that the radiation from a star comes close to obeying the same laws as the radiation from solid objects, such as the heating element on an electric stove. We can therefore use what we already know about Planck blackbody radiation (Chapter 5) to understand the radiation from stars. According to the Stefan-Boltzmann law, the hotter of two same-sized objects is more luminous, and according to Wien's law, hotter objects are bluer. In this section, we use those two laws to measure the temperatures and sizes of stars. We also develop a more detailed understanding of the line emission mentioned in Chapter 5 to obtain information about the composition of stars.

AstroTour: Stellar Spectrum

Wien's Law Revisited: The Color and Surface Temperature of Stars

Wien's law (see Working It Out 5.3) shows that the temperature of an object determines the peak wavelength of its spectrum. As the temperature of the surface of an object increases, the light that it emits gets bluer. Stars with especially hot surfaces are blue, stars with especially cool surfaces are red, and yellow-white stars such as the Sun are in between. If you obtain a spectrum of a star and measure the wavelength at which the spectrum peaks, Wien's law can tell you the temperature of the star's surface. The color of a star tells you about the temperature only at the surface because that layer is giving off most of the radiation that we see. (Stellar interiors are far hotter, as we discuss in Chapter 14.)

Astronomers can determine the temperature of a star without a spectrum. They often measure the colors of stars by comparing the brightness at two specific wavelengths. The brightness of a star is usually measured through an optical **filter**—sometimes just a piece of colored glass—that lets through only a small range of wavelengths. Two of the most common filters are a blue filter, which allows light with wavelengths of about 440 nanometers (nm) to pass through, and a "visual" (yellow-green) filter, which allows light with wavelengths of about 550 nm to pass through. The ratio of brightness between the blue and visual filters is called the *color index* of the star (see Appendix 7 for more details). From a pair of pictures of a group of stars, each taken through a different filter, we can find an approximate value of the surface temperature of every star in the picture—perhaps hundreds or even thousands—all at once. That type of analysis shows that many more cool stars exist than hot stars and that most stars have surface temperatures lower than that of the Sun.

Classifying Stars by Surface Temperature

Although the hot "surface" of a star emits radiation with a spectrum very close to a smooth Planck blackbody curve, that light must then escape through the outer layers of the star's atmosphere. The atoms and molecules in the cooler layers of the star's atmosphere leave their absorption line fingerprints in the escaping light (**Figure 13.6**). Under some circumstances, the atoms and molecules in the star's atmosphere, along with any gas that might be found near the star, can produce emission lines in stellar spectra. Absorption and emission lines complicate how astronomers use the laws of Planck blackbody radiation to interpret light from stars, but spectral lines provide a wealth of information about the state and composition of the gas in a star's atmosphere.

The spectra of stars were first classified during the late 1800s, long before stars, atoms, and radiation were well understood. Stars were classified by the appearance of the dark bands (now known as absorption lines) seen in their spectra. The original ordering of that classification was arbitrarily based on the prominence of particular absorption lines known to be associated with the element hydrogen. Stars with the strongest hydrogen lines were denoted *A stars*, stars with weaker hydrogen lines were denoted *B stars*, and so on.

Annie Jump Cannon (1863–1941) led an effort at the Harvard College Observatory to examine and classify the spectra of hundreds of thousands of stars systematically. She dropped many of the earlier spectral types, keeping only seven that were later reordered on the basis of surface temperatures. Spectra of stars of several spectral types are shown in **Figure 13.7**. The hottest stars, with surface temperatures above 30,000 K, are denoted *O stars*. O stars have only weak absorption lines from hydrogen and helium. The coolest stars—*M stars*—have temperatures as low

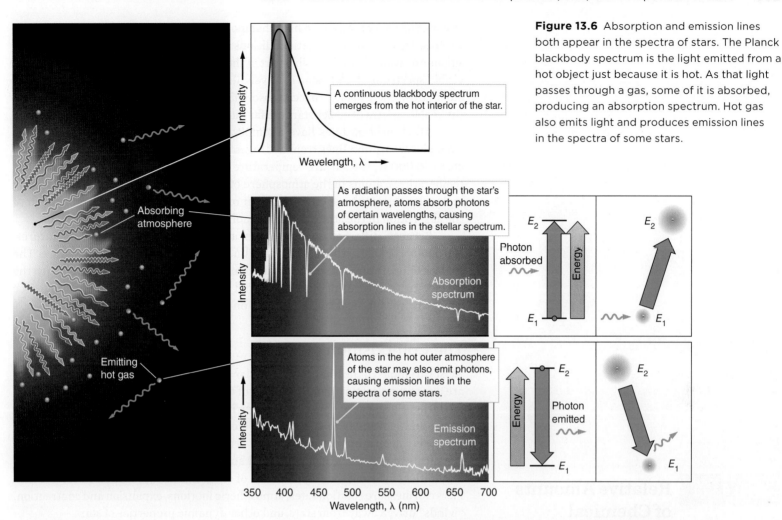

Figure 13.6 Absorption and emission lines both appear in the spectra of stars. The Planck blackbody spectrum is the light emitted from a hot object just because it is hot. As that light passes through a gas, some of it is absorbed, producing an absorption spectrum. Hot gas also emits light and produces emission lines in the spectra of some stars.

as about 2800 K. M stars have absorption lines from many types of atoms and molecules. The sequence of **spectral types** of stars, from hottest to coolest, is O, B, A, F, G, K, M. That sequence has undergone several modifications, most recently to add cooler objects known as brown dwarfs with spectral types L, T, and Y.

Astronomers divide the main spectral types into a finer sequence of subclasses by adding numbers to the letter designations. For example, the hottest B stars are B0 stars, slightly cooler B stars are B1 stars, and so on. The coolest B stars are B9 stars, only slightly hotter than A0 stars. The boundaries between spectral types are not always easy to determine. A hotter-than-average G star is very similar to a cooler-than-average F star. The Sun is a G2 star.

In Figure 13.7, notice that not only are hot stars bluer than cool stars but also the absorption lines in their spectra are different. The temperature of the gas in the atmosphere of a star affects the state of the atoms in that gas, affecting in turn the energy level transitions available to absorb radiation (see Section 5.2 to review atomic energy levels). In

Figure 13.7 Visible light spectra of stars with different spectral types, ranging from hot, blue O stars to cool, red M stars. Hotter stars are more luminous at shorter wavelengths. The dark lines are absorption lines. These spectra have been adjusted to show only the brightest part of each spectrum.

O stars, the temperature is so high that most atoms have had one or more electrons stripped from them by energetic collisions within the hot gas. As a result, few transitions are available in the visible part of the electromagnetic spectrum, making the visible spectrum of an O star relatively featureless. At lower temperatures, more atoms are in energy states that can absorb light in the visible part of the spectrum, so the visible spectra of cooler stars are far more complex than the spectra of O stars.

All absorption lines have a temperature at which they are strongest. For example, absorption lines from hydrogen are most prominent at temperatures of about 10,000 K, the surface temperature of an A star. At the very lowest stellar temperatures, atoms in the atmosphere of a star react with one another, forming molecules. Molecules such as titanium oxide (TiO) are responsible for much of the absorption in the atmospheres of cool M stars.

Because different spectral lines are formed at different temperatures, astronomers can use those absorption lines to measure a star's temperature. The surface temperatures of stars measured in that way agree extremely well with the surface temperatures of stars measured by using Wien's law, confirming once again that the physical laws that apply on Earth apply to stars as well.

The Composition of Stars

Most variations in the lines of a particular chemical element seen in stellar spectra are due to temperature, but the details of the absorption and emission lines found in starlight also carry a wealth of other information. By applying the physics of atoms and molecules to the study of stellar absorption lines, astronomers can accurately determine not only surface temperatures of stars but also pressures, chemical compositions, magnetic-field strengths, and other physical properties. In addition, by using the Doppler shift of emission and absorption lines, astronomers can measure rotation rates, atmospheric motions, expansion and contraction, "winds" driven away from stars, and other dynamic properties of stars.

The chemical composition of a star also is found from its spectra. Each type of atom has different energy levels and thus different spectral lines (Chapter 5). The patterns of spectral lines are measured in laboratories on Earth and then used to identify the atoms (or molecules) present in stars. For example, if a star has absorption lines that correspond to the energy difference between two levels in the calcium atom, we know that calcium is present in the atmosphere of the star.

The strengths of various absorption lines tell us not only what kinds of atoms are present in the gas but also the amount of each. However, we must take great care in interpreting spectra to account properly for the temperature and density of the gas in the atmosphere of a star. Recall Figure 5.15—the "astronomer's periodic table of the elements." Typically, hydrogen composes more than 90 percent of the atoms in the atmosphere of a star, whereas helium accounts for most of what remains. All the other chemical elements, collectively called **heavy elements** or **massive elements**, are present in only very small amounts.

Table 13.1 shows the chemical composition of the atmosphere of the Sun. The Sun's composition is fairly typical for stars nearby, but the percentages of specific heavy elements can vary tremendously from star to star. Some stars have smaller amounts of heavier elements than the Sun. The existence of such stars, essentially devoid of more massive elements, provides important clues about the origin of chemical elements and the chemical evolution of the universe. Many of the atoms that make up Earth and its atmosphere (such as iron, silicon, nitrogen, oxygen, and carbon) exist as only a small percentage of the Sun.

TABLE 13.1 Relative Amounts of Chemical Elements in the Atmosphere of the Sun

Element	Percentage of Atoms in the Sun	Percentage of Sun's Mass
Hydrogen	92.5	74.5
Helium	7.4	23.7
Oxygen	0.064	0.82
Carbon	0.039	0.37
Neon	0.012	0.19
Nitrogen	0.008	0.09
Silicon	0.004	0.09
Magnesium	0.003	0.06
Iron	0.003	0.16
Sulfur	0.001	0.04
Total of others	0.001	0.03

The Stefan-Boltzmann Law and Finding the Sizes of Stars

Stars are so far away that most cannot be imaged as more than point sources. Finding the size of a star, then, involves putting together other observable measurements, such as temperature and luminosity.

The temperature of a star can be found directly, either from its color through Wien's law (**Figure 13.8a**) or from the strength of its spectral lines. The temperature of the surface of a star is one factor that influences its luminosity. If a large star and a small star are the same temperature, they will emit the same energy from every patch of surface, but the large star has more patches, so it will be more luminous altogether. Conversely, if two stars are the same size, the hotter one will be more luminous than the cooler one. That relationship is an application of the Stefan-Boltzmann law (**Figure 13.8b**). A small, hot star might even be more luminous than a larger cool star.

The luminosity of a star can be found from its brightness and its distance. Because the luminosity depends on both the temperature and the size of the star, we can use the luminosity and the temperature to determine the radius of the star (**Figure 13.8c** and **Working It Out 13.3**). Astronomers have used the luminosity-temperature-radius relationship to estimate the radii of many thousands of stars.

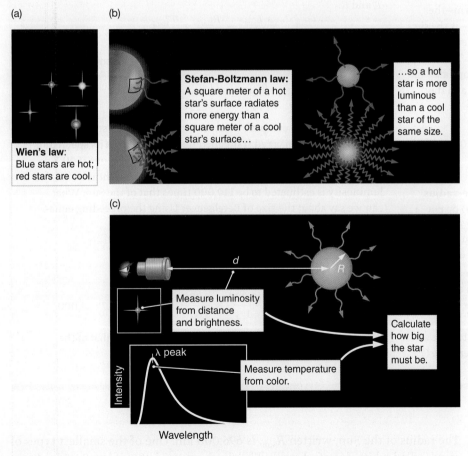

Figure 13.8 (a) The temperature of a star can be found from its color through Wien's law. (b) The luminosity depends on both the temperature and the size of the star. Blue stars are hotter; red stars are cooler. (c) Once the temperature and the luminosity are known, the size of the star can be calculated.

13.3 Working It Out Estimating the Sizes of Stars

According to the Stefan-Boltzmann law (Chapter 5), the amount of energy radiated each second by each square meter of the surface of a star is equal to the constant σ multiplied by the surface temperature of the star raised to the fourth power:

$$\left(\begin{array}{c}\text{Energy radiated each}\\\text{second by 1 m}^2\text{ of surface}\end{array}\right) = \sigma T^4$$

To find the total amount of light radiated each second by the star, we need to multiply the radiation emitted per second from each square meter by the number of square meters of the star's surface:

$$\left(\begin{array}{c}\text{Energy radiated}\\\text{each second}\end{array}\right) = \left(\begin{array}{c}\text{Energy radiated each}\\\text{second by 1 m}^2\text{ of surface}\end{array}\right) \times \left(\begin{array}{c}\text{Surface}\\\text{area}\end{array}\right)$$

The left-hand term in that equation—the total energy emitted by the star per second (in units of joules per second [J/s] = watts [W])—is the star's luminosity, L. The middle term—the energy radiated by each square meter of the star per second (in units of joules per square meter per second [J/m²/s])—can be replaced with the σT^4 factor from the Stefan-Boltzmann law. The remaining term—the number of square meters covering the surface of the star—is the surface area of a sphere, $A_{\text{sphere}} = 4\pi R^2$ (in units of square meters [m²]), where R is the radius of the star.

If we replace the words in the equation with the appropriate mathematical expressions for the Stefan-Boltzmann law and the area of a sphere, the equation for the luminosity becomes

$$L = \sigma T^4 \times 4\pi R^2$$

Combining gives

$$L = 4\pi R^2 \sigma T^4 \text{ J/s (W)}$$

That last equation is called the **luminosity-temperature-radius relationship** for stars. Because the constants (4, π, and σ) do not change, the luminosity of a star is proportional only to $R^2 T^4$. Make a star 3 times as large, and its surface area becomes $3^2 = 9$ times as large. Because 9 times as much area is present to radiate, 9 times as much radiation is emitted. Make a star twice as hot, and each square meter of the star's surface radiates $2^4 = 16$ times as much energy. Thus, larger, hotter stars are more luminous than smaller, cooler stars.

Now, how large must a star of a given temperature be to have a total luminosity of L? The star's luminosity (L) and temperature (T) are measurable quantities, and the star's radius (R) is what we want to know. We can rearrange the previous equation, moving the properties that we know how to measure (temperature and luminosity) to the right-hand side of the equation and the property that we would like to know (the radius of the star) to the left-hand side. After doing some algebra, we find that

$$R = \sqrt{\frac{L}{4\pi\sigma T^4}} = \frac{1}{T^2} \times \sqrt{\frac{L}{4\pi\sigma}}$$

Again, the right-hand side of the equation contains only things that we know or can measure. The constants 4, π, and σ are always the same. We can find L, the luminosity of the star, from the measurements of the star's brightness and parallax (although only for nearby stars with known parallax). T is the surface temperature of the star, which can be measured from its color. From the relationship of those measurements, we now know something new: the size of the star.

If we compare two stars, the constants all cancel out, leaving L, T, and R:

$$\frac{L_{\text{star 1}}}{L_{\text{star 2}}} = \frac{R^2_{\text{star 1}}}{R^2_{\text{star 2}}} \times \frac{T^4_{\text{star 1}}}{T^4_{\text{star 2}}}$$

rearranging the terms, and taking the square root on both sides:

$$\frac{R_{\text{star 1}}}{R_{\text{star 2}}} = \sqrt{\frac{L_{\text{star 1}}}{L_{\text{star 2}}} \times \frac{T^2_{\text{star 2}}}{T^2_{\text{star 1}}}}$$

How does the Sun compare with the second-brightest star in the constellation Orion, a red star called Betelgeuse? From its spectrum, we know that Betelgeuse's surface temperature T is about 3500 K. Its distance is about 200 pc, and from that and its brightness, its luminosity is estimated to be 140,000 times that of the Sun. What can we say about the size of Betelgeuse? Using the preceding equation, we can determine the following:

$$\frac{R_{\text{Betelgeuse}}}{R_{\text{Sun}}} = \sqrt{\frac{L_{\text{Betelgeuse}}}{L_{\text{Sun}}} \times \frac{T^2_{\text{Sun}}}{T^2_{\text{Betelgeuse}}}}$$

$$\frac{R_{\text{Betelgeuse}}}{R_{\text{Sun}}} = \sqrt{\frac{140,000}{1} \times \frac{5,800^2}{3,500^2}} = \sim 370 \times 2.7 \sim 1,000$$

Betelgeuse has a radius about 1,000 times larger than that of the Sun, making it a *supergiant*.

The radius of the Sun, written R_{Sun}, is 696,000 km. One of the smallest types of stars, called white dwarfs, have radii only about 1 percent of the Sun's radius—about the size of Earth. The largest stars, called red supergiants, can have radii more than 1,000 times that of the Sun. Many more stars are toward the small end of that range—smaller than the Sun—than giant stars.

CHECK YOUR UNDERSTANDING 13.2

If star A has twice the surface temperature of the Sun but has the same luminosity, the radius of star A must be _____ the radius of the Sun. (a) 16 times; (b) 4 times; (c) 1/2; (d) 1/4

Figure 13.9 The center of mass of two objects is the "balance" point on a line joining the centers of two masses.

13.3 Measuring the Masses of Stars in Binary Systems

Determining the mass of a star is difficult. Astronomers cannot use the amount of light from a star or the star's size as a measure of its mass. Stars can be large or small, faint or luminous. However, more massive stars *always* have stronger gravity. When astronomers are trying to determine the masses of astronomical objects, they almost always wind up looking for the effects of gravity. In Chapter 4, you learned that Kepler's laws of planetary motion are the result of gravity and that the properties of the orbit of a planet can be used to measure the mass of the Sun. Similarly, astronomers can study two stars that orbit each other to determine their masses.

About half of the higher-mass stars in the sky are actually systems consisting of several stars moving about under the influence of their mutual gravity. Most of those are **binary stars**, in which two stars orbit each other in elliptical orbits as predicted by Newton's version of Kepler's laws. That version of Kepler's laws can be used to find the mass of a star, as we will demonstrate. However, most low-mass stars are single, and low-mass stars far outnumber higher-mass stars, so most stars are single and their mass cannot be found that way. In this section, we look at how astronomers measure the masses of stars in binary systems.

Binary Star Orbits

The **center of mass** is the balance point of a system. If the two objects were sitting on a seesaw in a gravitational field, the support of the seesaw would have to be directly under the center of mass for the objects to balance (**Figure 13.9**). When Newton applied his laws of motion to the problem of orbits, he found that two objects must move in elliptical orbits around each other and that their common center of mass lies at one focus shared by both ellipses (**Figure 13.10**). The center of mass, which lies along the line between the two objects, remains stationary. The two objects will always be found on exactly opposite sides of the center of mass.

Imagine that you are watching a binary star, as shown in **Figure 13.11a**. As seen from above, two stars orbit the common center of mass. Star 1, which is less massive, must complete its orbit in the same time as star 2, which is more massive. Because the less massive star has farther to go around the center of mass, it must be moving *faster* than the more massive star. In that face-on view, no determination of the Doppler shift (Chapter 5) can be made because all the motion is in the plane of the sky, and none is toward or away from the observer.

When a system is edge-on to the observer, however, the observer can take advantage of the Doppler shift to find out about the motion. **Figure 13.11b**

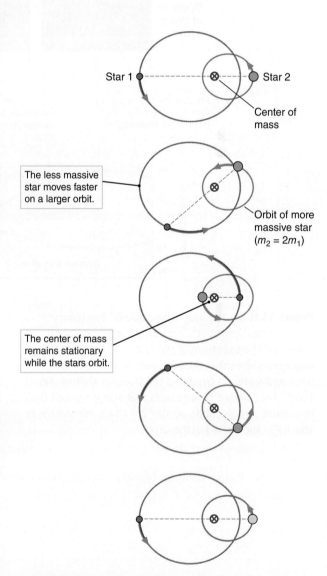

Figure 13.10 In a binary star system, the two stars orbit on elliptical paths about their common center of mass. Here, the blue star has twice the mass of the red one. The eccentricity of the orbits is 0.5. The time steps between the frames are equal.

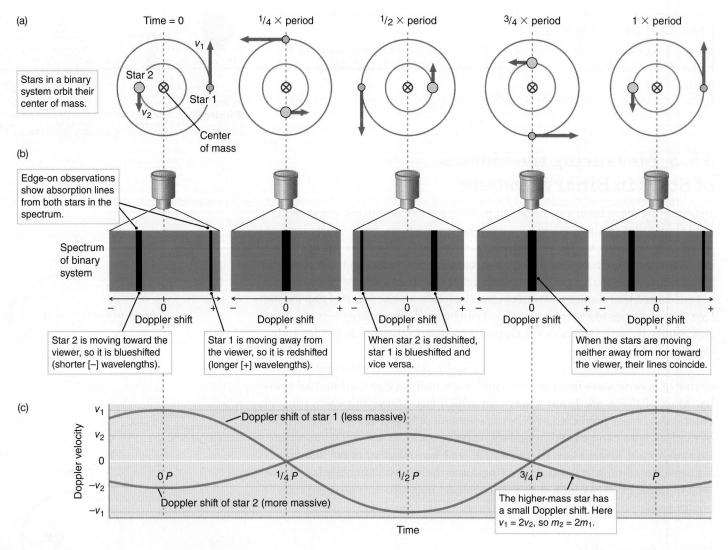

Figure 13.11 (a) The view from "above" the binary system shows that both stars orbit a common center of mass. (b) The spectrum of the combined system (seen edge-on) shows that the spectral lines of each star shift back and forth. (c) Graphing the Doppler shift of star 1 with star 2 versus time reveals that star 2 has half the maximum Doppler shift, so star 2 is twice as massive as star 1. P is the period of the orbit.

shows observations of the spectrum of the combined system associated with each position in Figure 13.11a, if it were edge-on instead of face-on. The spectral lines of the stars shift back and forth as they move toward and away from the observer. Because the two stars are always on exactly opposite sides of their center of mass, they always move in opposite directions. Thus, when star 2 approaches, star 1 recedes. The light coming from star 2 will be shifted to *shorter* wavelengths because of the Doppler effect as it approaches, so the light will be *blue*shifted, and the light coming from star 1 will be shifted to *longer* wavelengths as it recedes, so the light will be *red*shifted. Half an orbital period later, the situation is reversed: lines from star 1 are blueshifted, and lines from star 2 are redshifted.

The less massive star has a larger orbit—and consequently moves more quickly—than the more massive star. **Figure 13.11c** compares the velocity obtained from the maximum Doppler shift for star 1 with the velocity obtained from the maximum Doppler shift for star 2. That comparison gives the ratio of the masses of the two stars:

$$\frac{v_1}{v_2} = \frac{m_2}{m_1}$$

By observing the spectrum of the system, we can find from that equation the relative masses of the two stars—star 2 is 2 times as massive as star 1—but we can't find the actual mass of either star from those observations alone.

The Masses of Binary Stars

In Chapter 4, we ignored the complexity of the motion of two objects around their common center of mass. Now, however, that very complexity enables us to measure the masses of the two stars in a binary system. If we can measure the period of the binary system and the average separation between the two stars, Kepler's third law gives us the total mass in the system: the sum of the two masses. Because the analysis in the previous subsection gives us the ratio of the two masses, we now have two relationships between two unknowns. We have all we need to determine the mass of each star separately. In other words, if we know that star 2 is 2 times as massive as star 1, and we know that star 1 and star 2 together are 3 times as massive as the Sun, we can calculate separate values for the masses of star 1 and star 2.

Depending on the type of system, several ways are available to measure the average separation and the orbital period. In a **visual binary** system, the system is close enough to Earth, and the stars are far enough from each other, that we can take pictures that show the two stars separately (**Figure 13.12**). Then, astronomers can directly measure the shapes and periods of the orbits of the two stars just by watching them as they orbit each other. Those values can be used with Doppler measurements of the radial (line-of-sight) velocities of the stars to solve for the ratio of the two masses.

In most binary systems, however, the two stars are so close together and so far away from us that we cannot see the stars separately. The identification of those stars as binary systems is indirect and comes from observing periodic variations in the *light* from the star or from observing periodic changes in the *spectrum* of the star. If we view a binary system nearly edge-on, so that one star passes in front of the other, it is called an **eclipsing binary**. An observer will see a repeating dip in brightness as one star passes in front of (eclipses) the other (**Figure 13.13**). If the stars have different temperatures, a repeating pattern will occur of a smaller dip in brightness when the hotter star eclipses the cooler one, followed by a larger dip in brightness when the cooler star eclipses the hotter one. The pattern of those dips also gives an estimate of the relative sizes (radii) of the two stars. That procedure for identifying binary systems—similar to the transit method for finding extrasolar planets discussed in Chapter 7—works only when the system is viewed nearly edge-on. The Kepler space telescope has discovered thousands of eclipsing binaries in addition to finding new extrasolar planets.

In a **spectroscopic binary** system, the spectral lines of the two stars show periodic changes as they are Doppler-shifted away from each other, first in one direction and then in the other (Figure 13.11). The period of the orbit is determined from the time for a set of spectral lines to go from approaching to receding and back again. The orbital velocities of the stars and the period of the orbit give the size of the orbit because distance equals velocity multiplied by time. Consequently, astronomers can estimate the combined masses of the two stars. To calculate the individual masses, an estimate of the tilt of the orbit is needed. Thus, spectroscopic binary masses are more approximate than those in eclipsing binary systems.

Figure 13.12 The two stars of a visual binary are resolved. These stars are two components of the Alpha Centauri system, the nearest star system to the Solar System.

Figure 13.13 In an eclipsing binary system, the system is viewed nearly edge-on, so that the stars repeatedly pass behind each other, blocking some of the light. When the blue star passes in front of the larger, cooler star, less light is blocked than when the smaller star passes behind the red star. The shape of the dips in the light curve of an eclipsing binary can reveal information about the relative size and surface brightness of the two stars.

13.4 Working It Out Measuring the Mass of an Eclipsing Binary Pair

In Working It Out 4.3, we used Newton's version of Kepler's third law to calculate the mass of the Sun by observing the orbital period of one of its planets. In that special case, the Sun's mass is so much greater than the mass of a planet that the planet's mass is negligible. For two stars, however, the masses are similar, so neither is negligible, and we need to keep both in the equations. Newton showed that if two objects with masses m_1 and m_2 are in orbit about each other, the period of the orbit, P, is related to the average distance between the two masses, the semimajor axis A, by the equation

$$P^2 = \frac{4\pi^2 A^3}{G(m_1 + m_2)}$$

Rearranging that equation turns it into an expression for the sum of the masses of the two objects:

$$m_1 + m_2 = \frac{4\pi^2}{G} \times \frac{A^3}{P^2}$$

We could use the equation that way, with the masses of the two stars in kilograms (kg), the distance between them in kilometers (km), the period of their orbit in seconds (s), and the gravitational constant G in $km^3/(kg\ s)$. However, astronomers often think about stellar masses in units of the Sun's mass. If we divide that equation by the "mass of the Sun" equation in Working It Out 4.3,

$$M_{Sun} = \frac{4\pi^2}{G} \times \frac{A^3}{P^2}$$

(where M_{Sun} = mass of the Sun, $A = 1$ AU, and $P = 1$ year), the constants cancel out and the equation simplifies to

$$\frac{m_1}{M_{Sun}} + \frac{m_2}{M_{Sun}} = \frac{A^3_{AU}}{P^2_{years}}$$

Therefore, if we know both m_1/m_2 from measuring velocities by Doppler shifts and $m_1 + m_2$ from the observed orbital properties, we can solve for the separate values of m_1 and m_2.

Suppose you are an astronomer studying a binary star system. After observing the star for several years, you accumulate the following information about the system:

1. The star is an eclipsing binary.
2. The period of the orbit is 2.63 years.
3. Star 1 has a Doppler velocity that varies between +20.4 and −20.4 km/s.
4. Star 2 has a Doppler velocity that varies between +6.8 and −6.8 km/s.
5. The stars are in circular orbits. You know that because the Doppler velocities about the star are symmetric; the approach and recession speeds of the star are equal.

Those data are summarized in **Figure 13.14**. You begin your analysis by noting that the star is an eclipsing binary, which tells you that the orbit of the star is edge-on to your line of sight. The Doppler velocities tell you the total orbital velocity of each star, and you determine the size of the orbits by using the relationship

$$\text{Distance} = \text{Speed} \times \text{Time}$$

In one orbital period, star 1 travels around a circle—a distance of

$$d = (20.4 \text{ km/s}) \times (2.63 \text{ yr}) = 53.7 \text{ km} \times \text{yr/s}$$

A binary system can fall into more than one of those three categories, regardless of how it was originally discovered. If a spectroscopic binary system is also a visual or eclipsing binary, the orbit and masses of the stars can be completely solved (**Working It Out 13.4**). Historically, most stellar masses were measured for stars in eclipsing binary systems rather than for those in visual or spectroscopic binaries. New observational capabilities, however, have increased the number of known visual binaries by greatly improving the ability to see the stars in a binary directly. Accurate measurements of masses have been obtained for several hundred binary stars, about half of which are eclipsing binaries. The range of stellar masses found in that way is not nearly as great as the range of stellar luminosities. The least massive stars have masses of about 0.08 M_{Sun}, whereas the most massive stars appear to have masses up to about 200 M_{Sun}.

CHECK YOUR UNDERSTANDING 13.3

Which of the following properties must be measured to determine the masses of stars in a typical binary system? (Choose all that apply.) (a) the period of the orbits of the two stars; (b) the average separation between the two stars; (c) the radii of the two stars; (d) the velocities of the two stars.

Multiply by the number of seconds in a year:

$$d = 53.7 \frac{\text{km} \times \text{yr}}{\text{s}} \times \frac{3.16 \times 10^7 \, \text{s}}{\text{yr}} = 1.70 \times 10^9 \, \text{km}$$

That distance is the circumference of the star's orbit, or 2π times the radius of the star's orbit, A_1. Thus, star 1 is following an orbit with a radius of

$$A_1 = \frac{d}{2\pi} = \frac{1.70 \times 10^9 \, \text{km}}{2\pi} = 2.7 \times 10^8 \, \text{km}$$

To convert that result to astronomical units, use the relation $1 \, \text{AU} = 1.50 \times 10^8 \, \text{km}$:

$$A_1 = 2.7 \times 10^8 \, \text{km} \times \frac{1 \, \text{AU}}{1.50 \times 10^8 \, \text{km}} = 1.8 \, \text{AU}$$

A similar analysis of star 2 shows that its orbit has a radius of $A_2 = 0.6$ AU.

Next, apply Newton's version of Kepler's third law. Because the stars are always on opposite sides of the center of mass, the semi-major axis $A_{\text{AU}} = 1.8 \, \text{AU} + 0.6 \, \text{AU} = 2.4 \, \text{AU}$. Because you know A and the period P (measured as 2.63 years), you can calculate the total mass of the two stars:

$$\frac{m_1}{M_{\text{Sun}}} + \frac{m_2}{M_{\text{Sun}}} = \frac{(A_{\text{AU}})^3}{(P_{\text{years}})^2} = \frac{(2.4)^3}{(2.63)^2} = 2.0$$

Thus, the combined mass of the two stars is twice the mass of the Sun. To sort out the individual masses of the stars, use the measured velocities and the fact that the mass and velocity are inversely proportional:

$$\frac{m_2}{m_1} = \frac{v_1}{v_2} = \frac{20.4 \, \text{km/s}}{6.8 \, \text{km/s}} = 3.0$$

Star 2 is 3 times as massive as star 1. In mathematical terms, $m_2 = 3 \times m_1$. Substituting into the equation

$$m_1 + m_2 = 2.0 \, M_{\text{Sun}}$$

gives

$$m_1 + 3m_1 = 2.0 \, M_{\text{Sun}}$$

or $4m_1 = 2.0 \, M_{\text{Sun}}$, so $m_1 = 0.5 \, M_{\text{Sun}}$. Because $m_2 = 3 \times m_1$, $m_2 = 1.5 \, M_{\text{Sun}}$. Star 1 has a mass of $0.5 \, M_{\text{Sun}}$, and star 2 has a mass of $1.5 \, M_{\text{Sun}}$. You have just found the masses of two distant stars.

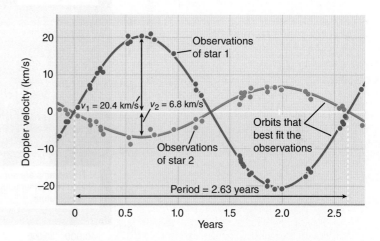

Figure 13.14 Doppler velocities of the stars in an eclipsing binary are used to measure the masses of the stars.

13.4 The Hertzsprung-Russell Diagram Is the Key to Understanding Stars

Measuring basic properties of stars is only the first step in understanding; the next step is to look for patterns in their properties. In the early 20th century, Ejnar Hertzsprung (1873–1967) and Henry Norris Russell (1877–1957) independently studied the properties of stars. Both astronomers plotted the luminosities of stars versus their surface temperatures—a diagram that came to be known as the *Hertzsprung-Russell diagram*, or simply **H-R diagram**. We use that diagram often for the study of stars. In this section, we take a first look at that important diagram and how stars are organized within it.

The H-R Diagram

The H-R diagram (**Figure 13.15**) is a graph of luminosity versus temperature that can be used to study how stars change over time. The spectral type is plotted on the horizontal axis (the *x*-axis), along with the surface temperature plotted backward: temperature is higher on the left and lower on the right. Hot, blue stars are on the

Figure 13.15 The Hertzsprung-Russell (H-R) diagram is used to plot the properties of stars. More luminous stars are at the top of the diagram. Hotter stars are on the left. Stars of the same radius (R) lie along the dotted lines moving from upper left to lower right. Absolute magnitudes are discussed in Working It Out 13.2 and Appendix 7.

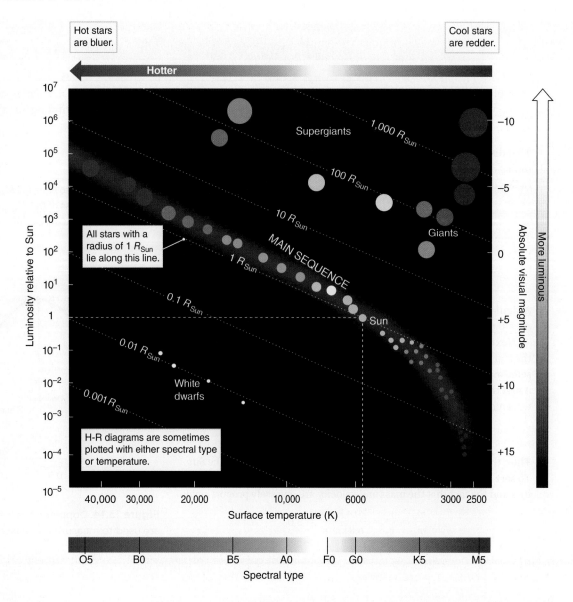

AstroTour: H-R Diagram

Simulation: H-R Diagram

left side of the H-R diagram; cool, red stars are on the right. Temperature is plotted logarithmically. For example, consider two points on the H-R diagram. One represents a star with a surface temperature of 40,000 K, and the other a star with a surface temperature of 20,000 K—a temperature change of a factor of 2. The interval along the axis between those points is therefore the same size as the interval between points representing stars with surface temperatures of 10,000 and 5000 K—with that same change by a factor of 2. The horizontal axis is sometimes labeled with another characteristic that corresponds to temperature, such as the color.

The luminosity of stars is plotted along the vertical axis (the y-axis). More luminous stars are toward the top of the diagram; less luminous stars are toward the bottom. Sometimes the luminosity axis is labeled with the absolute visual magnitude instead of luminosity, as shown on the right-hand y-axis. As with the temperature axis, luminosities are plotted logarithmically. Here, each step along the left-hand y-axis corresponds to a multiplicative factor of 10 in the luminosity. To understand why the plotting is done that way, recall that the most luminous stars are 10 billion times more luminous than the least luminous stars, yet all those stars must fit on the same plot.

Each point on the H-R diagram is specified by a surface temperature and luminosity. Therefore, we can use the luminosity-temperature-radius relationship described earlier in the chapter to find the radius of a star at that point as well. A star in the upper right corner of the H-R diagram is very cool, so each square meter of its surface radiates only a small amount of energy. That star also is extremely luminous. It must be huge, then, to account for its high luminosity, despite the feeble radiation coming from each square meter of its surface. Conversely, a star in the lower left corner of the H-R diagram is very hot, which means that a large amount of energy is coming from each square meter of its surface. That star has a very low overall luminosity, however, so it must be very small. Moving up and to the right takes you to larger and larger stars; moving down and to the left takes you to smaller and smaller stars. All stars of the same radius lie along slanted lines across the H-R diagram. Astronomers can note the properties of a star—its temperature, color, size, and luminosity—from a glance at its position on the H-R diagram. The discovery and study of those patterns led to an understanding of the astrophysics of stars (see the **Process of Science Figure**).

The Main Sequence

Figure 13.16 shows 4 million stars plotted on an H-R diagram. The data are based on observations of stars within 5,000 light-years of the Sun, near enough for parallax measurements obtained by the *Gaia* satellite. A quick look at the diagram immediately shows what was first discovered in the original diagrams of Hertzsprung and Russell. About 90 percent of the stars in the sky lie in a well-defined region running across the H-R diagram from lower right to upper left, known as the **main sequence**. On the left end of the main sequence are the O stars: hotter, larger, and more luminous than the Sun. On the right end of the main sequence are the M stars: cooler, smaller, and fainter than the Sun. If you know where a star lies on the main sequence, you know its approximate luminosity, surface temperature, and size.

The H-R diagram offers a useful method for finding the distance to main-sequence stars. Astronomers can determine whether a star is on the main sequence by looking at the absorption lines in its spectrum. The spectral type also is determined from the spectral lines, and that spectral type indicates the star's temperature. Once that value on the *x*-axis is known, we can then read up to the main sequence and then across to the *y*-axis to find the star's luminosity. Recall that the luminosity, brightness, and distance are all connected. We can think about how far away a star of a particular luminosity must be to have the observed brightness. So we can find the star's distance by comparing a star's luminosity, obtained from the H-R diagram, with its apparent brightness. That method of determining distances to main-sequence stars from the spectra, luminosity, and brightness of stars is called **spectroscopic parallax**. Details of that method are discussed in Appendix 7. Despite the similarity between the names, the method is very different from the parallax method using trigonometry. Spectroscopic parallax is useful to much larger distances than trigonometric parallax, although it is less precise.

From a combination of observations of binary star masses, parallax, luminosity measurements, and mathematical models, astronomers have determined that stars of different masses lie on different parts of the main sequence. Stellar mass increases smoothly from the lower right to the upper left along the main sequence (**Figure 13.17**). If a main-sequence star is *less* massive than the Sun, it is also smaller, cooler, redder, and less luminous than the Sun and is located to the lower right of the Sun on the main sequence. Conversely, if a main-sequence star is *more* massive than the Sun, it is also larger, hotter, bluer, and more luminous

Figure 13.16 An H-R diagram for 4 million stars within 5,000 light-years of the Sun plotted from data obtained by the *Gaia* satellite clearly shows the main sequence. Most of the stars lie along this band running from the lower right of the diagram toward the upper left.

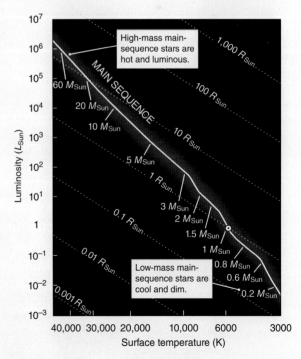

Figure 13.17 Mass determines the location of a star along the main sequence.

PROCESS OF SCIENCE

Science Is Collaborative

It took decades, and the contributions of dozens of people, all working toward a common goal, to understand the meaning behind stellar data.

Henry Norris Russell

Cecilia Payne-Gaposchkin

Annie Jump Cannon

Ejnar Hertzsprung

Meghnad Saha

1800s
The Observations

About 500,000 photographs of stellar spectra were obtained by many astronomers at many telescopes.

1900s
The Classification

Annie Jump Cannon led a team that classified all the spectra according to the strengths of particular absorption lines at particular wavelengths.

1910s
The Graph

Hertzsprung and Russell independently developed what will later be called the H-R diagram. They did not understand why the *x*-axis, ordered O-B-A-F-G-K-M, revealed a nice band across the middle of the diagram. Russell hypothesized this result must be caused by a single stellar characteristic.

1920s
The Understanding

Meghnad Saha showed that the stellar characteristic in question was temperature. Cecilia Payne-Gaposchkin showed that stars were mostly composed of hydrogen and helium. Modern astrophysics was born; others went on to develop the understanding of stellar atmospheres.

Scientific discoveries sometimes seem to occur suddenly. Instead, new scientific knowledge usually results from the effort of many people working for many years to solve a problem.

than the Sun and is located to the upper left of the Sun on the main sequence. The mass of a star determines where on the main sequence the star will lie.

Table 13.2 summarizes the properties of the spectral classes of main-sequence stars. *All* main-sequence stars with a mass of 1 M_{Sun} are G2 stars like the Sun and have the *same* surface temperature, size, and luminosity as the Sun. Similarly, if a main-sequence star is classified as B0, it has a surface temperature of about 30,000 K, a luminosity about 32,500 times that of the Sun, a mass of about 17.5 M_{Sun}, and a radius of about 6.7 R_{Sun}. If a different main-sequence star is classified as M5, it has a surface temperature of 3170 K, a luminosity of about 0.008 L_{Sun}, a mass of about 0.21 M_{Sun}, and a radius of about 0.29 R_{Sun}.

The relationship between the mass and the luminosity of stars is very sensitive. Relatively small differences in the masses of stars result in large differences in their main-sequence luminosities. From determining the luminosities of binary stars with measured mass, a relationship between the mass and luminosity was found. That **mass-luminosity relationship**, usually expressed as $L \propto M^{3.5}$, is shown in **Figure 13.18a**. The exact exponent varies from about 3 to 4 for different ranges of stellar masses, but the method is useful for estimating masses of single stars. The mass also correlates to the size of a star (**Figure 13.18b**) and to the temperature of a star (**Figure 13.18c**).

The mass and chemical composition of a main-sequence star determine how large it is, what its surface temperature is, how luminous it is, what its internal structure is, how long it will live, how it will evolve, and what its final fate will be. A star must have a balance between gravity trying to hold the star together and the energy released by nuclear reactions within the star trying to blow it apart. The mass of the star determines the strength of its gravity, which in turn determines how much energy must be generated in its interior to prevent it from collapsing under its own weight. The mass of a star determines where that balance is struck.

TABLE 13.2 Properties of Main-Sequence Stars

Spectral Type	Temperature (K)	Mass (M_{Sun})	Radius (R_{Sun})	Luminosity (L_{Sun})
O5	42,000	60	13	500,000
B0	30,000	17.5	6.7	32,500
B5	15,200	5.9	3.2	480
A0	9800	2.9	2.0	39
A5	8200	2.0	1.8	12.3
F0	7300	1.6	1.4	5.2
F5	6650	1.4	1.2	2.6
G0	5940	1.05	1.06	1.25
G2 (Sun)	5780	1.00	1.00	1.0
G5	5560	0.92	0.93	0.8
K0	5150	0.79	0.93	0.55
K5	4410	0.67	0.80	0.32
M0	3840	0.51	0.63	0.08
M5	3170	0.21	0.29	0.008

Stars Not on the Main Sequence

Although 90 percent of stars are main-sequence stars, some stars are found in the upper right portion of the H-R diagram, well above the main sequence (see Figure 13.15). Those stars, which must be luminous, cool, and large, with radii hundreds or thousands of times the radius of the Sun, are called giants. At the other extreme are stars found in the far lower left corner of the H-R diagram. Those stars are the tiny white dwarfs, comparable to the size of Earth. Their small surface areas explain why they have such low luminosities, despite having high temperatures.

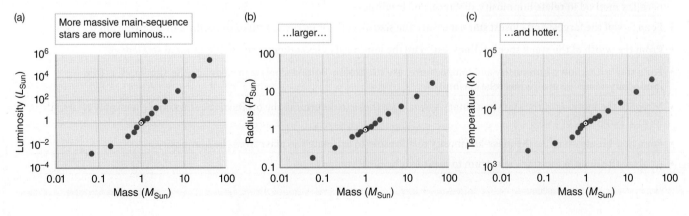

Figure 13.18 These graphs plot luminosity (a), radius (b), and temperature (c) versus mass for stars along the main sequence. The mass of a main-sequence star determines its other properties.

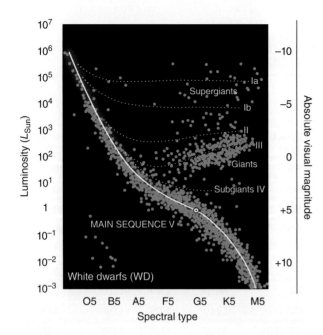

Figure 13.19 Stellar luminosity classes indicate the size (radius) of a star at each spectral type.

Stars that lie off the main sequence on the H-R diagram can be identified by their luminosities (determined by their distance) or by slight differences in their spectral lines. The width of a star's spectral lines indicates the density and surface pressure of gas in the star's atmosphere. In general, denser stars have broader lines. Puffed-up stars above the main sequence have lower densities and lower surface pressure and narrower absorption lines than main-sequence stars.

When using the H-R diagram to estimate the distance to a star by the spectroscopic parallax method, astronomers must know whether the star is on, above, or below the main sequence to find the star's luminosity. The spectral line widths of stars both on and off the main sequence indicate **luminosity class**, which tells us the relative *size* of the star within each spectral class. Supergiant stars, the largest stars that we see, are luminosity class I, bright giants are class II, giants are class III, subgiants are class IV, main-sequence stars are class V, and white dwarfs are class WD. Luminosity classes I–IV lie above the main sequence, whereas class WD falls below and to the left of the main sequence (**Figure 13.19**). Thus, the complete spectral classification of a star includes both its spectral type (indicating temperature and color) and its luminosity class (indicating relative size).

The existence of the main sequence, together with the fact that the mass of a main-sequence star determines where on the sequence it will lie, creates a grand pattern that makes it possible to understand stars in fundamental ways. The existence of stars that do *not* follow that pattern raises yet more questions. In the coming chapters, you will learn that the main sequence tells us what stars are and how they work, and that stars off the main sequence reveal how stars form, evolve, and die. **Table 13.3** summarizes the techniques that astronomers use to determine some of the basic properties of stars. Of the properties listed in the table,

TABLE 13.3 Taking the Measure of Stars

Property	Methods
Luminosity	• For a star with a known distance, measure the brightness and then apply the inverse square law of radiation: $$\text{Luminosity} = 4\pi \times \text{Distance}^2 \times \text{Brightness}$$ • For a star without a known distance, take a spectrum of the star to determine its spectral and luminosity classes, plot them on an H-R diagram, and read the luminosity from the diagram.
Temperature	• Measure the color index of the star by using blue and visual filters. Use Wien's law to relate the color to a temperature. • Take a spectrum of the star, and estimate the temperature from its spectral class by noting which spectral lines are present.
Distance	• For a relatively nearby star (within a few hundred parsecs), measure the parallax shift of the star over the year. • For a more distant star, find the luminosity by using the H-R diagram as noted earlier, and then use the spectroscopic parallax method to relate luminosity, distance, and brightness.
Size	• For a few of the largest and closest stars, measure the size directly or by the length of the eclipse in eclipsing binary stars. • From the width of the star's spectral lines, estimate the luminosity class (supergiant, giant, or main sequence). • For a star with known luminosity and temperature, use the Stefan-Boltzmann law to calculate the star's radius (the luminosity-temperature-radius relationship).
Mass	• Measure the motions of the stars in a binary system, use those to determine the orbits of the stars, and then apply Newton's form of Kepler's third law. • For a non-binary star, use the mass-luminosity relationship to estimate the mass from the luminosity.
Composition	• Analyze the lines in the star's spectrum to measure chemical composition.

only temperature, distance, and composition can be *measured*. Luminosity must be *inferred* from the H-R diagram or calculated from distance and brightness, and size and mass must be *calculated*. Other measurable properties include brightness, color, spectral type, and parallax shift.

CHECK YOUR UNDERSTANDING 13.4

Choose the two qualities that describe a star located in the lower right of the H-R diagram: (a) hot; (b) cool; (c) high luminosity; (d) low luminosity.

Origins Habitable Zones

How might a basic property of a star, such as its luminosity, color, mass, or surface temperature, affect the chance of a planet with life being in orbit around that star? The only known life is that on planet Earth, where liquid water was essential for life to form and evolve. Whether liquid water is an absolute requirement for life elsewhere is not known, but the presence of water is a good starting point for determining where to look. So astronomers look for planets at the right distance from their stars to have a planetary temperature that permits water to exist in a liquid state on their surfaces—a range of distances known as the **habitable zone**. On planets whose orbits are closer to their star than the habitable zone, water would exist only as a vapor—if at all. On planets that have orbits beyond the habitable zone, water would be permanently frozen as ice.

An important factor for estimating the temperature of a planet is the brightness of the sunlight that falls on that planet (see the Chapter 5 "Origins" section and Working It Out 5.4). That factor depends on the luminosity of the star and the planet's distance from the star. In the Solar System, the habitable zone ranges from ~0.9 to ~1.4 AU, which includes Earth but just misses Venus and Mars. Main-sequence stars less luminous than the Sun are cooler and have narrower habitable zones, minimizing the chance that a planet will form within that slender zone. Main-sequence stars more massive than the Sun are hotter and have larger habitable zones. **Figure 13.20** illustrates those zones around Sun-like, hotter, and cooler stars.

Astronomers are now finding planets in the habitable zones of their respective stars. Methods of planet detection, as discussed in Chapter 7, work best when the planet is close to its star. As of this writing, the Kepler Mission has identified and confirmed a few dozen planets in habitable zones by using the transit method and has found more candidate planets that need to be confirmed.

The distance from a star at a certain temperature is not the only consideration for whether a planet has water. The presence of a planetary atmosphere also is a factor. More massive planets can retain their atmospheres, which can trap heat and raise the planet's temperature, as we saw in Chapter 5 for Venus and Earth. Smaller planets have a lower gravitational pull and may not be able to retain an atmosphere. In addition, some habitable zones may be near planets, not stars. Some of the giant planets in the cold, outer part of our own Solar System have moons with liquid water (Chapters 11 and 24). The heat keeping the water liquid comes from the nearby planet, not from the Sun.

Finally, *habitable* does not mean *inhabited*; it means only that the planet is at the right distance from its star that it could have liquid water. Identifying planets in their habitable zone is a first step to selecting which planets are most interesting for further study.

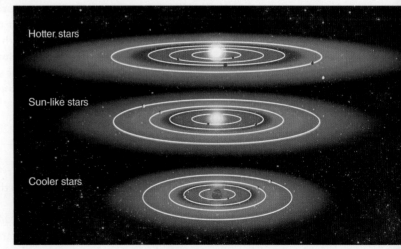

Figure 13.20 The distance and extent of a habitable zone (green) surrounding a star depends on the star's temperature. Regions too close to the star are too hot (red) and those too far away are too cold (blue) to permit the existence of liquid water. The orbits of Mercury, Venus, Earth, and Mars have been drawn around these stars for scale.

READING ASTRONOMY NEWS

ARTICLES | QUESTIONS

NASA reports on a new technique for using the Hubble Space Telescope to measure parallax angles.

NASA's Hubble Extends Stellar Tape Measure 10 Times Farther into Space

NASA Press Release

Using NASA's Hubble Space Telescope, astronomers now can precisely measure the distance of stars up to 10,000 light-years away—10 times farther than previously possible.

Astronomers have developed yet another novel way to use the 24-year-old space telescope by employing a technique called spatial scanning, which dramatically improves Hubble's accuracy for making angular measurements. The technique, when applied to the age-old method for gauging distances called astronomical parallax, extends Hubble's tape measure 10 times farther into space (**Figure 13.21**).

"This new capability is expected to yield new insight into the nature of dark energy, a mysterious component of space that is pushing the universe apart at an ever-faster rate," said Nobel laureate Adam Riess of the Space Telescope Science Institute (STScI) in Baltimore, Maryland.

Parallax, a trigonometric technique, is the most reliable method for making astronomical distance measurements, and a practice long employed by land surveyors here on Earth. The diameter of Earth's orbit is the base of a triangle, and the star is the apex where the triangle's sides meet. The lengths of the sides are calculated by accurately measuring the three angles of the resulting triangle.

Astronomical parallax works reliably well for stars within a few hundred light-years of Earth. For example, measurements of the distance to Alpha Centauri, the star system closest to our Sun, vary only by 1 arcsec. This variance in distance is equal to the apparent width of a dime seen from 2 miles away.

Stars farther out have much smaller angles of apparent back-and-forth motion that are extremely difficult to measure. Astronomers have pushed to extend the parallax yardstick ever deeper into our galaxy by measuring smaller angles more accurately.

This new long-range precision was proven when scientists successfully used Hubble to measure the distance of a special class of bright stars called Cepheid variables, approximately 7,500 light-years away in the northern constellation Auriga. The technique worked so well, they are now using Hubble to measure the distances of other far-flung Cepheids.

Such measurements will be used to provide firmer footing for the so-called cosmic "distance ladder." This ladder's "bottom rung" is built on measurements to Cepheid variable stars that, because of their known brightness, have been used for more than a century to gauge the size of the observable universe. They are the first step in calibrating far more distant extra-galactic milepost markers such as Type Ia supernovae.

Riess and the Johns Hopkins University in Baltimore, Maryland, in collaboration with Stefano Casertano of STScI, developed a technique to use Hubble to make measurements as small as five-billionths of a degree.

To make a distance measurement, two exposures of the target Cepheid star were taken 6 months apart, when Earth was on opposite sides of the Sun. A very subtle shift in the star's position was measured to an accuracy of 1/1,000 the width of a single image pixel in Hubble's Wide Field Camera 3, which has 16.8 megapixels total. A third exposure was taken after another 6 months to allow for the team to subtract the effects of the subtle space motion of stars, with additional exposures used to remove other sources of error.

Riess shares the 2011 Nobel Prize in Physics with another team for his leadership in the 1998 discovery that the expansion rate of the universe is accelerating—a phenomenon widely attributed to a mysterious, unexplained dark energy filling the universe. This new high-precision distance measurement technique is enabling Riess to gauge just how much the universe is stretching. His goal is to refine estimates of the universe's expansion rate to the point where dark energy can be better characterized.

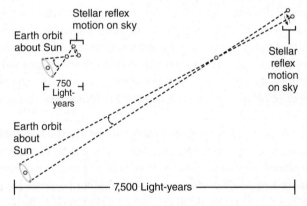

Figure 13.21 By applying a technique called spatial scanning to an age-old method for gauging distances called astronomical parallax, scientists now can use NASA's Hubble Space Telescope to make precision distance measurements 10 times farther into our galaxy than was previously possible.

ARTICLES | **QUESTIONS**

1. How many parsecs are in 7,500 light-years? In 10,000 light-years?
2. What is the parallax angle of a star 7,500 light-years away? Ten thousand light-years away?
3. Why were the exposures taken 6 months apart?
4. Why is making parallax measurements from space better than doing so from the ground on Earth?
5. How does improving the accuracy of the distances to nearby stars from trigonometric parallax affect astronomers' estimates of farther stars' distances, measured using spectroscopic parallax?

Summary

Finding the distances to stars is a difficult but important task for astronomers. Parallax and spectroscopic parallax are two methods that astronomers use to determine distances to stars. Brightness and distance can be used to obtain the luminosity. Careful study of the light from a star, including its spectral lines, gives the temperature, size, and composition of the star. Study of binary systems gives the masses of stars of various spectral types, which we can extend to all stars of the same spectral type. The H-R diagram shows the relationship among the various physical properties of stars. The mass of a star is the major determining factor in the changes it undergoes. The habitable zone is the distance from a star in which a planet could have the right temperature for liquid water to exist on its surface. Stars of different luminosities and temperatures have habitable zones of different widths at different distances from the star.

LG 1 **Explain how astronomers use the brightness of nearby stars and their distances from Earth to determine how luminous they are.** The distance to a nearby star is measured by finding its parallax—by measuring how the star's apparent position changes in the sky over a year. The nearest star (other than the Sun) is about 4 light-years (1.3 parsecs) away. The brightness of a star in the sky can be measured directly, and brightness and distance can be used to obtain the star's luminosity—how much light the star emits.

LG 2 **Explain how astronomers obtain the temperatures, sizes, and composition of stars.** The color of a star depends on its temperature; blue stars are hotter and red stars are cooler. The radius can be computed from the temperature and luminosity of the star. Small, cool stars greatly outnumber large, hot stars. Spectral lines carry a great deal of information about a star, including what chemical elements and molecules are present in the star.

LG 3 **Describe how astronomers estimate the masses of stars.** Masses of stars are measured in binary star systems by observing the effects of the gravitational pull between the stars. Newton's universal law of gravitation and Kepler's laws connect the motion of the star to the forces they experience and thus to their masses.

LG 4 **Classify stars and organize that information on a Hertzsprung-Russell (H-R) diagram.** The H-R diagram shows the relationship among the various physical properties of stars. Temperature increases to the left, so that hotter stars lie on the left side of the diagram, whereas cooler stars lie on the right. Luminosity increases vertically, so that the most luminous stars lie near the top of the diagram. A star's luminosity class and temperature indicate its size. The mass and composition of a main-sequence star determine its luminosity, temperature, and size. Ninety percent of stars lie along the main sequence.

LG 5 **Explain how the mass and composition of a main-sequence star determine its luminosity, temperature, and size.** The mass and composition of a main-sequence star determine its original position on the H-R diagram. The main sequence on the H-R diagram is actually a sequence of masses. That position connects its other properties such as its luminosity, temperature, and size.

? Unanswered Questions

- What is the *upper* limit for stellar mass? Both theory and observation have shown that the *lower* limit for stellar mass is approximately 0.08 M_{Sun} with temperature about 2000 K. However, neither theory nor observation has yielded a definitive value for the upper limit. Many astronomers believe that the upper limit lies somewhere around 150–200 M_{Sun}. The very first stars that formed in the universe might have been even larger.

- Are planets with life likely to be orbiting around main-sequence stars of type M? Those low-luminosity stars are the most common type in the Milky Way, and the Kepler telescope has detected many planets orbiting such stars. However, the habitable zone of an M star is very close to the star, so that the planet may be strongly affected by streams of charged particles blowing off the star. Unless the planet has a strong protective magnetic field, that radiation may decrease or completely strip a planet of its atmosphere. Another complication is that a close-in planet is likely to be tidally locked to the star, so that one hemisphere of the planet receives light and the other hemisphere is permanently dark. That imbalance of light and heat might make the planet uninhabitable.

Questions and Problems

Test Your Understanding

1. Star A and star B are nearly the same distance from Earth. If star A is half as bright as star B, which of the following statements must be true?
   a. Star B is farther away than star A.
   b. Star B is twice as luminous as star A.
   c. Star B is hotter than star A.
   d. Star B is larger than star A.

2. Star A and star B are two nearby stars. If star A is blue and star B is red, which of the following statements must be true?
   a. Star A is hotter than star B.
   b. Star A is cooler than star B.
   c. Star A is farther away than star B.
   d. Star A is more luminous than star B.

3. Star A and star B are two stars nearly the same distance from Earth. If star A is blue and star B is red, but they have equal brightness, which of the following statements is true?
   a. Star A is more luminous than star B.
   b. Star A is larger than star B.
   c. Star A is smaller than star B.
   d. Star A is less luminous than star B.

4. What does it most likely mean when a star has very weak hydrogen lines and is blue?
   a. The star is too hot for hydrogen lines to form.
   b. The star has no hydrogen.
   c. The star is too cold for hydrogen lines to form.
   d. The star is moving too fast to measure the lines.

5. Star A and star B are a binary system. If the Doppler shift of star A's absorption lines is 3 times the Doppler shift of star B's absorption lines, which of the following statements is true?
   a. Star A is 3 times as massive as star B.
   b. Star A is one-third as massive as star B.
   c. Star A is closer than star B.
   d. The binary pair is moving toward Earth, but star A is farther away.

6. Star A and star B are two red stars at nearly the same distance from Earth. If star A is many times brighter than star B, which of the following statements is true?
   a. Star A is a main-sequence star, whereas star B is a red giant.
   b. Star A is a red giant, whereas star B is a main-sequence star.
   c. Star A is hotter than star B.
   d. Star A is a white dwarf, whereas star B is a red giant.

7. Star A and star B are two blue stars at nearly the same distance from Earth. If star A is many times brighter than star B, which of the following statements is true?
   a. Star A is a main-sequence star, whereas star B is a red giant.
   b. Star A is a main-sequence star, whereas star B is a blue giant.
   c. Star A is a white dwarf, whereas star B is a blue giant.
   d. Star A is a blue giant, whereas star B is a white dwarf.

8. In which region of an H-R diagram would you find the main-sequence stars with the widest habitable zones?
   a. upper left
   b. upper right
   c. center
   d. lower left
   e. lower right

9. If star A is more massive than star B, and both are main-sequence stars, star A is _____ than star B. (Choose all that apply.)
   a. more luminous
   b. less luminous
   c. hotter
   d. colder
   e. larger
   f. smaller

10. A telescope on Mars could measure the distances to more stars than can be measured from Earth because
    a. the resolution of the telescope would be better.
    b. Mars has a thin atmosphere.
    c. it would be closer to the stars.
    d. the parallax "baseline" would be longer.

11. Star A and star B are two nearby stars. If star A has a parallactic angle 4 times as large as star B's, which of the following statements is true?
    a. Star A is one-quarter as far away as star B.
    b. Star A is 4 times as far away as star B.
    c. Star A has moved through space one-quarter as far as star B.
    d. Star A has moved through space 4 times as far as star B.

12. If star A appears twice as bright as star B but is also twice as far away, star A is _____ as luminous as star B.
    a. 8 times
    b. 4 times
    c. twice
    d. half

13. In Table 13.1, the percentage of hydrogen in the Sun decreases when changing from percentage by number of atoms to percentage by mass, but the percentage of helium increases. Why?
    a. Hydrogen is more massive than helium.
    b. Helium is more massive than hydrogen.
    c. Hydrogen is located in a different part of the Sun.
    d. The mass of hydrogen is hard to measure.

14. Capella (in the constellation Auriga) is the sixth-brightest star in the sky. When viewed with a high-power telescope, it turns out that Capella is actually two pairs of binary stars: the first pair are G-type giants, whereas the second pair are M-type main-sequence stars. What color does Capella appear to be?
    a. red
    b. yellow
    c. blue
    d. The color cannot be determined from that information.

15. An eclipsing binary system has a primary eclipse (star A is eclipsed by star B) that is deeper (more light is removed from the light curve) than the secondary eclipse (star B is eclipsed by star A). What does that information tell you about stars A and B?
    a. Star A is hotter than star B.
    b. Star B is hotter than star A.
    c. Star B is larger than star A.
    d. Star B is moving faster than star A.

Thinking about the Concepts

16. The distances of nearby stars are determined by their parallaxes. Why are the distances of stars farther from Earth more uncertain?

17. To know certain properties of a star, you must first determine the star's distance. For other properties, knowing its distance is unnecessary. Explain why an astronomer does or does not need to know a star's distance to determine each of the following properties: size, mass, temperature, color, spectral type, and chemical composition.

18. Albireo, in the constellation Cygnus, is a visual binary system whose two components can easily be seen with even a small, amateur telescope. Viewers describe the brighter star as "golden" and the fainter one as "sapphire blue."
    a. What does that description tell you about the relative temperatures of the two stars?
    b. What does that description tell you about their respective sizes?

19. Very cool stars have temperatures around 2500 K and emit Planck spectra with peak wavelengths in the red part of the spectrum. Do those stars emit any blue light? Explain your answer.

20. The stars Betelgeuse and Rigel are both in the constellation Orion. Betelgeuse appears red, whereas Rigel is bluish white. To the eye, the two stars seem equally bright. If you can compare the temperature, luminosity, or size from just that information, do so. If not, explain why.

21. Explain why the stellar spectral types (O, B, A, F, G, K, M) are not in alphabetical order. What sequence of temperatures is defined by those spectral types?

22. Other than the Sun, the only stars whose mass astronomers can measure *directly* are those in eclipsing or visual binary systems. Why? How do astronomers estimate the masses of stars that are not in eclipsing or visual binary systems?

23. Once the mass of a certain spectral type of star located in a binary system has been determined, it can be assumed that all other stars of the same spectral type and luminosity class have the same mass. Why is that a reasonable assumption?

24. Explain why the Kepler Mission is finding eclipsing binary stars while it is searching for extrasolar planets by using the transit method.

25. Scientific advances often require the participation of scientists from all over the world, working on the same problem over many decades, even centuries. Compare that mode of "collaboration" with collaborations in your courses (perhaps on final projects or papers). What mechanisms must be in place to allow scientists to collaborate across space and time in that way?

26. What would happen to our ability to measure stellar parallax if we were on Mars? What if we were on Venus or Jupiter?

27. Figure 13.7 has an absorption line at about 410 nm that is weak for O stars and weak for G stars but very strong in A stars. That particular line comes from the transition from the second excited state of hydrogen up to the sixth excited state. Why is that line weak in O stars? Why is it weak in G stars? Why is it strongest in the middle of the range of spectral types?

28. Which kinds of binary systems are best observed edge-on? Which kinds are best observed face-on?

29. In Figure 13.10, two stars orbit a common center of mass.
    a. Explain why star 2 has a smaller orbit than star 1.
    b. Re-sketch that picture for the case in which star 1 has a very low mass, perhaps close to that of a planet.
    c. Re-sketch that picture for the case in which star 1 and star 2 have the same mass.

30. If our Sun were a blue main-sequence star, and Earth was still 1 AU from the Sun, would you expect Earth to be in the habitable zone? What about if our Sun were a red main-sequence star?

Applying the Concepts

31. Suppose that Figure 13.1b included a third star, located 4 times as far away as star A. How much less than star A would the third star appear to move each year? How much less than star B?

32. Suppose you see an object jump from side to side by half a degree as you blink back and forth between your eyes. How much farther away is an object that moves only one-third of a degree?

33. Logarithmic (log) plots show major steps along an axis scaled to represent equal factors, most often factors of 10. Why do astronomers sometimes use a log plot instead of the more conventional linear plot? Is the horizontal axis of the H-R diagram in Figure 13.15 logarithmic or linear?

34. Figure 13.5 is plotted logarithmically on both axes. The luminosities are in units of solar luminosities.
    a. How much more luminous than the Sun is a star on the far right side of the plot?
    b. How much less luminous than the Sun is a star on the far left side of the plot?

35. Compared with the Sun, how luminous, large, and hot is a star that has 10 times the mass of the Sun? Use Figure 13.17 to answer that question.

36. Sirius, the brightest star in the sky, has a parallax of 0.379 arcsec. What is its distance in parsecs? In light-years? How long does the light take to reach Earth?

37. Sirius is actually a binary pair of two A-type stars. The brighter star is called the "Dog Star" and the fainter is called the "Pup Star" because Sirius is in the constellation Canis Major (meaning "big dog"). The Dog Star appears about 6,800 times brighter than the Pup Star, even though both stars are at the same distance from Earth. Compare the temperatures, luminosities, and sizes of those two stars.

38. Sirius and its companion orbit around a common center of mass with a period of 50 years. The mass of Sirius is 2 times the mass of the Sun.
    a. If the orbital velocity of the companion is 2.35 times greater than that of Sirius, what is the mass of the companion?
    b. What is the semimajor axis of the orbit?

39. Sirius is 25 times more luminous than the Sun, and Polaris (the "North Pole Star") is 2,500 times more luminous than the Sun. Sirius appears 24 times brighter than Polaris. How much farther away is Polaris than Sirius? Use your answer from Problem 36 to find the distance of Polaris in light-years.

40. Betelgeuse (in Orion) has a parallax of 0.00451 ± 0.00080 arcsec, as measured by the Hipparcos satellite. What is the distance to Betelgeuse, and what is the uncertainty in that measurement?

41. The star Achernar has a Hipparchus parallax of 0.02339 arcsec and appears about as bright as Betelgeuese (previous problem) in the sky. Which star is actually more luminous? Knowing that Betelgeuse appears reddish, whereas Achenar appears bluish, which star would you say is hotter, and why?

42. The Sun is about 16 trillion (1.6×10^{13}) times brighter than the faintest stars visible to the naked eye.
    a. How far away (in astronomical units) would an identical solar-type star be if it were just barely visible to the naked eye?
    b. What would be its distance in light-years?

43. If $m_1 = m_2$ in Figure 13.9, where would the center of mass be located? If $m_1 = 2m_2$, where would the center of mass be located?

44. Find the peak wavelength of blackbody emission for a star with a temperature of about 10,000 K. In what region of the spectrum does that wavelength fall? What color is that star?

45. About 1,470 watts (W) of solar energy hits each square meter of Earth's surface. Use that value and the distance to the Sun to calculate the Sun's luminosity.

USING THE WEB

46. Go to the European Space Agency's *Gaia* mission website (https://www.esa.int/Our_Activities/Space_Science/Gaia). How has it helped astronomers determine the distances to more stars? Why is making parallax measurements from space better than measuring from the ground on Earth? Do a search for *Gaia* news (for example, at http://sci.esa.int/gaia/)—what has been learned from the released data?

47. Go to the Kepler Eclipsing Binary Catalog (http://keplerebs.villanova.edu) to see what new observations look like. Pick a few stars to study. Look at the last two columns (figures, labeled "LC Figs" and "SC Figs"). The "raw" and "dtr" figures are rough, but the "pf" figure shows a familiar light curve. How deep is the eclipse; that is, how much lower is the "normalized flux" during maximum eclipse?

48. Do a search for a photograph of your favorite constellation (or go outside and take a picture yourself). Can you see different colors in the stars? What do the colors tell you about the surface temperatures of the stars? From your photograph, can you tell which are the three brightest stars in the constellation? Those stars will be named "alpha" (α), "beta" (β), and "gamma" (γ) for that constellation. Look up the constellation online and see whether you chose the right stars. What are their temperatures and luminosities? What are their distances?

49. Citizen science: Go to the website for the Stellar Classification Online Public Exploration (SCOPE) project (http://scope.pari.edu/takepart.php). The project uses crowdsourcing to classify stars seen on old photographic plates of pictures taken in the Southern Hemisphere. Create an account, review the science and the FAQs, and then click on "To Take Part" to see some practice examples. Then go to "Classify," choose a photographic plate, and classify a few stars.

50. Watch the short NASA video about planets in binary star systems: https://nasaviz.gsfc.nasa.gov/11106. Why do astronomers think these systems are not very common? Do a search on "circumbinary planets"—have many more been found?

EXPLORATION

The H-R Diagram

digital.wwnorton.com/astro6

Visit the Digital Resources Page and on the Student Site open the "H-R Diagram" Interactive Simulation for Chapter 13. This simulation enables you to compare stars on the H-R diagram in two ways. You can compare an individual star (marked by a red X) to the Sun by varying its properties in the box in the left half of the window. Or you can compare groups of the nearest and brightest stars. Play around with the controls for a few minutes to familiarize yourself with the simulation.

Begin by exploring how changes to the properties of the individual star change its location on the H-R diagram. First, press the "Reset" button at the top right of the window.

Decrease the temperature of the star by dragging the temperature slider to the left. Notice that the luminosity remains the same. Because the temperature has decreased, each square meter of star surface must be emitting less light. What other property of the star changes to keep the total luminosity of the star constant?

Predict what will happen when you move the temperature slider all the way to the right. Now do it. Did the star behave as you expected?

1 As you move to the left across the H-R diagram, what happens to the radius?

_____

_____

_____

2 What happens as you move to the right?

_____

_____

Press "Reset" and experiment with the luminosity slider.

3 As you move up on the H-R diagram, what happens to the radius?

_____

_____

4 What happens as you move down?

_____

_____

_____

Press "Reset" again and predict how you would have to move the slider bars to move your star into the red giant portion of the H-R diagram (upper right). Adjust the slider bars until the star is in that area. Were you correct?

5 How would you adjust the slider bars to move the star into the white dwarf area of the H-R diagram?

_____

_____

Press the "Reset" button and explore the right-hand side of the window. Add the nearest stars to the graph by clicking their radio button under "Plotted Stars." Using what you learned above, compare the temperatures and luminosities of those stars with the Sun (marked by the X).

6 Are the nearest stars generally hotter or cooler than the Sun?

_____

7 Are the nearest stars generally more or less luminous than the Sun?

_____

Press the radio button for the brightest stars. That action will remove the nearest stars and add the brightest stars in the sky to the plot. Compare those stars with the Sun.

8 Are the brightest stars generally hotter or cooler than the Sun?

_____

9 Are the brightest stars generally more or less luminous than the Sun?

_____

10 How do the temperatures and luminosities of the brightest stars in the sky compare with the temperatures and luminosities of the nearest stars? Does that information support the claim in the chapter that more low-luminosity stars than high-luminosity stars exist? Explain.

_____

_____

_____

_____

14

Our Star—The Sun

Because the Sun is the only star close to Earth, much of our detailed knowledge about stars has come from studying the Sun. In Chapter 13, we looked at the physical properties of distant stars, including their mass, luminosity, size, temperature, and chemical composition. Here in Chapter 14, we ask fundamental questions about Earth's local star: How does the Sun work? Where does it get its energy? How has its luminosity changed over the billions of years since the Solar System formed?

LEARNING GOALS

By the end of this chapter, you should be able to:

LG 1 Describe the balance between the forces that determine the structure of the Sun.

LG 2 Explain how mass is converted to energy in the Sun's core and how long the Sun will take to use up its fuel.

LG 3 Sketch a physical model of the Sun's interior, and list the ways that energy moves outward from the Sun's core toward its surface.

LG 4 Describe how observations of solar neutrinos and seismic vibrations on the surface of the Sun test astronomers' models of the Sun.

LG 5 Describe the solar activity cycles of 11 and 22 years, and explain how those cycles are related to the Sun's changing magnetic field.

This image is a combination of several extreme ultraviolet images of the Sun from the Solar Dynamics Observatory. ▶▶▶

What makes the Sun shine?

14.1 The Sun Is Powered by Nuclear Fusion

Energy from the Sun is responsible for daylight, for Earth's weather and seasons, and for terrestrial life itself. At a luminosity of 3.85×10^{26} watts (W), the Sun produces more energy in a second than all the electric power plants on Earth could generate in a half-million years. In this section, we look at energy production in the Sun.

Hydrostatic Equilibrium

Geologists learn about the interior of Earth by using a combination of physics, detailed computer models, and experiments that test the predictions of those models. The task of exploring the interior of the Sun is much the same. Like Earth's structure, the structure of the Sun is governed by several physical processes and relationships. Using physics, chemistry, and the properties of matter and radiation, astronomers can express those processes and relationships as mathematical equations. They then solve those equations and create a model of the Sun. One of the great successes of 20th century astronomy is the construction of a theoretical model of the Sun that agrees with observations of the mass, composition, size, temperature, and luminosity of the real thing.

The structure of the Sun results from a balance between the outward force due to the pressure of radiation produced inside and the inward force of gravity. That balance is known as **hydrostatic equilibrium**. The outward pressure is a result of local energy in the process of finding its way to the surface of the Sun from deep in its interior. To understand how hydrostatic equilibrium affects the Sun, we need to know how those forces are produced and how they continually change to balance each other.

The balance between the forces due to radiation pressure and gravity is illustrated in **Figure 14.1**. The Sun is a huge ball of hot gas. Deep in the Sun's interior, the outer layers press downward because of gravity, producing a large inward force. To maintain balance, the outward force due to pressure must be equally large. If gravity were not balanced by radiation pressure, the Sun would collapse. If radiation pressure were greater than gravity, the Sun would blow itself apart. At every point within the Sun's interior, the radiation pressure must be just enough to hold up the weight of all the layers above that point. If the Sun were not in a stable hydrostatic equilibrium, forces within it would not be in balance, and the size of the Sun would change.

Hydrostatic equilibrium becomes an even more powerful concept when combined with the way gases behave. Deeper in the interior of the Sun, the weight of the material above becomes greater, and hence the radiation pressure increases. In a gas, higher pressure means higher density, higher temperature, or both. **Figure 14.2** shows how conditions vary inside the Sun. The graphs in Figure 14.2b, which are based on calculations, show that as the radiation pressure increases toward the center of the Sun, the density and temperature of the gas increase as well.

Figure 14.1 The structure of the Sun is determined by the balance between the forces due to pressure and gravity and the balance between the energy generated in its core and the energy radiated from its surface.

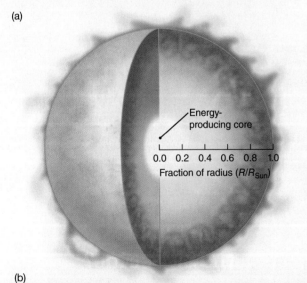

Figure 14.2 (a) This cutaway figure shows how the fraction of radius given in the x-axis of the graphs in (b) is measured. The energy produced by the Sun is generated in the Sun's core. (b) Pressure, density, and temperature all increase toward the center of the Sun.

CHECK YOUR UNDERSTANDING 14.1A

Hydrostatic equilibrium in the Sun means that: (a) the Sun does not change; (b) the Sun absorbs and emits equal amounts of energy; (c) the outward force from radiation pressure balances the weight of overlying layers; (d) energy produced in the core per unit time equals energy emitted at the surface per unit time.

Nuclear Fusion

A second fundamental balance within the Sun is the balance of energy (see Figure 14.1). Stars like the Sun are remarkably stable objects. To remain in balance, the Sun must produce just enough energy in its interior to replace the energy radiated away from its surface. That energy balance tells us how much energy must be produced in the interior of the Sun and how that energy finds its way from the interior to the Sun's surface, where it is radiated away. Theoretical models of stellar evolution indicate that the luminosity of the Sun is increasing, but very, very slowly. From geological records and the models, astronomers estimate that the Sun's luminosity 4.5 billion years ago was about 70 percent of its current luminosity.

One of the most basic questions facing the pioneers of stellar astrophysics was how the Sun and other stars get their energy. In the 19th century, physicists proposed that the Sun was slowly shrinking and that the core was heating up as a result of that gravitational contraction. Calculations soon showed, however, that gravitational contraction could power the Sun for only millions of years. Geological and biological evidence available at the time suggested that Earth was tens of millions or hundreds of millions of years old. By the early 20th century, radiometric dating suggested that Earth was more than a billion years old, and therefore gravitational contraction could not be the source of the Sun's energy. In the 1930s, using theoretical and laboratory physics, nuclear physicists concluded that the Sun's energy comes from nuclear reactions at its core, capable of powering the star for billions of years.

The nucleus of most hydrogen atoms consists of a single proton (Chapter 5). Nuclei of all other atoms are built from a mixture of protons and neutrons. Most helium nuclei, for example, consist of two protons and two neutrons. Protons have a positive electric charge, whereas neutrons have no electric charge. Because like charges repel, and the closer they are the stronger the force, all the protons in an atomic nucleus are continually repelling each other with a tremendous force. The nuclei of atoms should fly apart because of electric repulsion. They don't, though, because they are held together by the **strong nuclear force**, which overcomes that repulsion. However, the strong nuclear force acts only over very short distances, on the order of 10^{-15} meters—about the size of the atomic nucleus, or about a hundred-thousandth the size of an atom.

Compared with the energy required to free an electron from an atom, the amount of energy required to tear a nucleus apart is enormous. Conversely, when an atomic nucleus (with a mass up to the nucleus of iron) is formed from component parts, energy is released. **Nuclear fusion**—the process of combining two less massive atomic nuclei into a single, more massive atomic nucleus—occurs when atomic nuclei are brought close enough together for the strong nuclear force to overcome the force of electric repulsion (**Figure 14.3**). Many kinds of nuclear fusion can occur in stars. In main-sequence stars like the Sun, the primary energy generation process is the fusion of hydrogen into helium. That process is sometimes called "hydrogen burning" even though it has nothing to do with fire or other chemical combustion. The fusion of hydrogen into helium always takes several steps, but the net result is that four hydrogen nuclei become one helium nucleus plus energy.

The energy produced in nuclear reactions comes from converting mass to energy. The exchange rate between mass and energy is given by Einstein's equation, $E = mc^2$, in which E is energy, m is mass, and c^2 is the speed of light squared. For any nuclear reaction, we can determine the mass turned into energy by calculating the mass lost. To find that lost mass, we subtract the mass of the outputs from the mass of the inputs. In hydrogen fusion, the inputs are four hydrogen nuclei, and the output is a helium nucleus plus energy. The mass of four separate

Figure 14.3 (a) Atomic nuclei are positively charged and electrically repel each other. (b) The faster that two nuclei are moving toward each other, the closer they will get before veering away. (c) At the temperatures and densities found in the centers of stars, nuclei can overcome that electric repulsion, so fusion takes place.

hydrogen nuclei is 1.007 times greater than the mass of a single helium nucleus (with two protons and two neutrons); so when hydrogen fuses to make helium, 0.7 percent of the mass of the hydrogen is converted to energy.

Although each fusion reaction produces a small amount of energy, the total mass of the Sun is very large, so a great deal of hydrogen is available to fuse. When the amount of energy produced by nuclear fusion is compared with the luminosity of the Sun, we see that those reactions can power the Sun for 10 billion years—a time frame longer than the 4.6-billion-year age of the Solar System (measured from radioactive dating of meteorites from space and the Moon). Details of that calculation are provided in **Working It Out 14.1**.

Energy is produced in the Sun's innermost region, the **core**, where the conditions are the most extreme. The density of matter in the core is about 150 times the density of water, and the temperature is about 15 million kelvins (K). Under those conditions, the atomic nuclei have tens of thousands of times more kinetic energy than that of atoms at room temperature and can slam into each other hard enough to overcome the electric repulsion, allowing the strong nuclear force to act (Figure 14.3c). In hotter and denser gases, such collisions happen more often. For that reason, the rate of nuclear fusion reactions is extremely sensitive to the

14.1 Working It Out The Source of the Sun's Energy

Like all stars, the Sun has a lifetime limited by the amount of fuel available. We can calculate how long the Sun will live by comparing the mass involved in nuclear fusion with the amount of mass available. Converting four hydrogen nuclei (protons) into a single helium nucleus results in a loss of mass. The mass of a single hydrogen nucleus is 1.6726×10^{-27} kilogram (kg). So, four hydrogen nuclei have a mass of 4 times that, or 6.6904×10^{-27} kg. The mass of a helium nucleus is 6.6447×10^{-27} kg, which is less than the mass of the four hydrogen nuclei. The amount of mass lost, m, is

$$m = 6.6904 \times 10^{-27} \text{ kg} - 6.6447 \times 10^{-27} \text{ kg} = 0.0457 \times 10^{-27} \text{ kg}$$

We can write that result as 4.57×10^{-29} kg—a mass loss of about 0.7 percent. Conversion of 0.7 percent of the mass of the hydrogen into energy might not seem very efficient—until we compare it with other sources of energy and discover that it is millions of times more efficient than even the most efficient chemical reactions.

Using Einstein's equation $E = mc^2$, where c is the speed of light (3×10^8 m/s), along with the definition of a joule (1 J = 1 kg m^2/s^2), we can calculate the energy released by that mass-to-energy conversion:

$$E = mc^2 = (4.57 \times 10^{-29} \text{ kg}) \times (3.00 \times 10^8 \text{ m/s})^2 = 4.11 \times 10^{-12} \text{ J}$$

Each reaction that takes four hydrogen nuclei and turns them into a helium nucleus releases 4.11×10^{-12} J of energy, which doesn't seem like very much. But atoms are very small. Fusing a single kilogram of hydrogen into helium releases about 6.3×10^{14} J of energy—about the equivalent of the chemical energy released in burning 100,000 barrels of oil. To see how much the Sun must be fusing per second to produce its current luminosity, we divide the luminosity of the Sun by that amount of energy per kilogram:

$$\frac{\text{Luminosity of Sun}}{\text{Energy per kilogram}} = \frac{3.9 \times 10^{26} \text{ J/s}}{6.3 \times 10^{14} \text{ J/kg}} = 6.2 \times 10^{11} \text{ kg/s}$$

For the Sun to produce as much energy as it does, it must convert roughly 620 billion kg of hydrogen into helium every second (and about 4 billion kg of matter—0.7 percent—is converted to energy in the process). The Sun has been fusing hydrogen at that rate for at least the age of Earth and the Solar System—4.6 billion years. How much longer will the Sun last?

Astronomers estimate that only 10 percent of the Sun's total mass will ever be involved in fusion because the other 90 percent will never get hot enough or dense enough for the strong nuclear force to make fusion happen. Ten percent of the mass of the Sun is $(0.1) \times (2 \times 10^{30})$ kg, or 2×10^{29} kg. That is the amount of "fuel" the Sun has available. The Sun consumes hydrogen at a rate of 620 billion kg/s, so each year the Sun consumes about:

$$M_{\text{year}} = (6.2 \times 10^{11} \text{ kg/s}) \times (3.16 \times 10^7 \text{ s/yr}) \sim 2 \times 10^{19} \text{ kg/yr}$$

If we know how much fuel the Sun has (2×10^{29} kg), and we know how much the Sun fuses each year (2×10^{19} kg/yr), we can divide the amount by the rate to find the lifetime of the Sun:

$$\text{Lifetime} = \frac{M_{\text{fuel}}}{M_{\text{year}}} = \frac{2 \times 10^{29} \text{ kg}}{2 \times 10^{19} \text{ kg/yr}} = 10^{10} \text{ yr}$$

When the Sun was formed, it had enough fuel to power it for about 10 billion years. The Sun is nearly halfway through its lifetime of hydrogen fusion.

temperature and the density of the gas, which is why those energy-producing collisions are concentrated in the Sun's core. Half the energy produced by the Sun is generated within the inner part of the core: the inner 9 percent of the Sun's radius, or less than 0.1 percent of the volume of the Sun.

The conversion of four hydrogen nuclei to one helium nucleus is the most significant source of energy in main-sequence stars. Hydrogen is the most abundant element in the universe, so it is the most abundant source of nuclear fuel at the beginning of a star's lifetime. Hydrogen also is the easiest atom to fuse. Hydrogen nuclei—protons—have an electric charge of +1. The electric barrier that must be overcome to fuse protons is the repulsion of one proton against another. To fuse two carbon nuclei, for example, the repulsion of the six protons in one carbon nucleus pushing against the six protons in another carbon nucleus must be overcome. The repulsion between two carbon nuclei is 36 times stronger than that between two hydrogen nuclei. Therefore, hydrogen fusion occurs at a much lower temperature than any other type of nuclear fusion.

AstroTour: The Solar Core

The Proton-Proton Chain

To test the theory that the Sun shines because of nuclear fusion, astronomers can analyze the predicted by-products of the nuclear reactions. In the core of the Sun and in other low-mass stars, hydrogen fusion takes place in a series of nuclear reactions called the **proton-proton chain**, which has three branches. The most important branch, responsible for about 85 percent of the energy generated in the Sun, consists of the three steps illustrated in **Figure 14.4**. Each step produces particles and/or energy in the form of light. We begin by following the creation of the helium nucleus and then go back to find out what happens to the other products of the reaction.

Simulation: Proton-Proton Chain

Follow along in Figure 14.4 as we step through the proton-proton chain. The nucleus of hydrogen consists of one proton. In the first step, two protons fuse. During that process, one proton is transformed into a neutron. To conserve spin and charge, two particles are emitted: a positively charged particle called a **positron** and a neutral particle called a **neutrino**. Energy also is emitted in the form of gamma ray photons. The new atomic nucleus formed by the first step in the chain consists of a proton and a neutron. Recall from Chapter 5 that an *isotope* of an element has the same number of protons but a different number of neutrons. Thus, the new atomic nucleus is still hydrogen because it has only one proton. That particular isotope of hydrogen is common enough that it has its own name—deuterium, written as ^{2}H (H is the element symbol for hydrogen, which always has one proton, and 2 is the *atomic mass number*—the total number of protons and neutrons in the nucleus).

Figure 14.4 The Sun and all other main-sequence stars get their energy by fusing the nuclei of four hydrogen atoms to make a single helium atom. In the Sun, about 85 percent of the energy produced comes from the branch of the proton-proton chain illustrated here.

In the second step of the proton-proton chain, another proton slams into the deuterium nucleus, forming the nucleus of an isotope of helium, ^{3}He, consisting of two protons and a neutron. The energy released in that step is carried away as a highly energetic gamma-ray photon. Those first two steps are shown twice in Figure 14.4a because they must occur twice to produce a single ^{4}He nucleus.

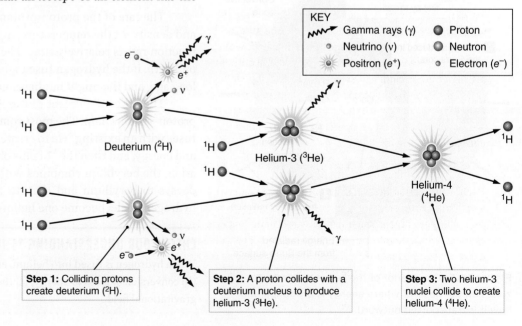

Step 1: Colliding protons create deuterium (^{2}H).

Step 2: A proton collides with a deuterium nucleus to produce helium-3 (^{3}He).

Step 3: Two helium-3 nuclei collide to create helium-4 (^{4}He).

In the third and final step of the proton-proton chain, two ^{3}He nuclei collide and fuse, producing an ordinary ^{4}He nucleus and ejecting two protons in the process. The energy released in that step is the kinetic energy of the helium nucleus and two ejected protons. Overall, four hydrogen nuclei have combined to form one helium nucleus.

Now let's go back and look at what happens to the other products of the reaction. In step 1, a positron—a particle of antimatter—is produced. **Antimatter** particles have the same mass as a corresponding matter particle but have opposite values of other properties, such as charge. The positron ($e+$) is the antimatter counterpart of an electron (e^-). When matter (electrons) and antimatter (positrons) meet, they annihilate each other, and their total mass is converted to energy in the form of gamma-ray photons (γ). That happens to the emitted positrons inside the Sun, and the emitted photons from the annihilation carry away part of the energy released when the two protons fused. Those photons heat the surrounding gas. The gamma rays emitted in step 2 similarly heat the gas. The thermal energy produced in the core of the Sun takes one hundred thousand years or more to find its way to the Sun's surface, and so the light we see from the Sun indicates what the Sun's core was doing a very long time ago.

The neutrino emitted in step 1 has a very different fate. Neutrinos are particles that have no charge, have very little mass, and travel at nearly the speed of light. They interact so weakly with ordinary matter that the neutrino escapes from the Sun without further interactions with any other particles. The core of the Sun is buried beneath 700,000 kilometers (km) of dense, hot matter, yet the Sun is transparent to neutrinos—essentially all of them travel into space as though the outer layers of the Sun were not there. Because they travel at nearly the speed of light, neutrinos from the center of the Sun arrive at Earth after 8.3 minutes. Therefore, we can use them to probe what the Sun is doing today.

The dominant branch of the proton-proton chain can be written symbolically as follows:

Step 1: $^1H + {}^1H \to {}^2H + e^+ + \nu$ and then $e^+ + e^- \to \gamma + \gamma$

Step 2: $^2H + {}^1H \to {}^3He + \gamma$

Step 3: $^3He + {}^3He \to {}^4He + {}^1H + {}^1H$

The rate of the proton-proton chain reaction depends on both temperature and density. At the temperature and pressure that exist within the Sun's core, the reaction rate is relatively slow. The Sun's slow nuclear fusion is fortunate for life on Earth: if the hydrogen fused quickly, the Sun would have exhausted its supply long ago, and life might not have had time to evolve.

The other 15 percent of the Sun's energy is generated by variations of the proton-proton chain. The most common variation happens in step 3, in which ^{3}He fuses with an existing ^{4}He to create beryllium (^{7}Be), which decays to lithium (^{7}Li) and energy, and then the ^{7}Li plus one ^{1}H become two ^{4}He. In a less common variation, the beryllium combines with hydrogen to become boron (^{8}B), which then decays to beryllium and then to two ^{4}He. In both variations, ultimately four hydrogen nuclei become one helium nucleus.

Figure 14.5 The interior of the Sun is divided into zones on the basis of where energy is produced and how it is transported outward.

1. Energy produced in the Sun's core…
2. …is carried outward first by radiation…
3. …then by convection…
4. …then away from the Sun by radiation.

Energy radiated from the Sun's surface

Core · Radiative zone · Convective zone

CHECK YOUR UNDERSTANDING 14.1B

When hydrogen is fused into helium, energy is released from: (a) gravitational collapse; (b) conversion of mass to energy; (c) the increase in pressure; (d) the decrease in the gravitational field.

14.2 Energy Is Transferred from the Interior of the Sun

Although geologists cannot travel deep inside Earth to find out how it is structured, they can build a model of its interior by using data on how seismic waves travel during earthquakes. Similarly, astronomers can create a model of the Sun's interior by using their knowledge of the balance of forces and energy within the Sun and an understanding of how energy moves from one place to another. Those models can be tested by observations of waves traveling through the Sun and by studying neutrinos from the Sun.

Energy Transport

Some of the energy released by hydrogen fusion in the core of the Sun escapes directly into space in the form of neutrinos. However, most of the energy heats the solar interior and then moves outward through the Sun to the surface, a process known as **energy transport**. Energy transport, a key determinant of the Sun's structure, can occur by *conduction, convection,* or *radiation.*

Conduction is important primarily in solids. When you pick up a hot object, for example, your fingers are heated by conduction. That happens because energetic thermal vibrations of atoms and molecules cause neighboring atoms and molecules to vibrate more rapidly as well. Conduction is typically ineffective in a gas because the atoms and molecules are too far apart to transmit vibrations to one another efficiently. Conduction does not play a key role in transporting energy from the core of the Sun to its surface, but conduction will be relevant later when we discuss dying stars.

In the Sun, energy is transported by convection and radiation through zones, as shown in **Figure 14.5**. The mechanism of energy transport from the center of the Sun outward depends on the decreasing temperature and density as the radius increases. First, energy moves outward through the inner layers of the Sun as radiation in the form of photons. Next, energy moves by convection in parcels of gas. Finally, energy radiates from the Sun's surface as light. We look at each process in turn.

Near the core, **radiation** transfers energy from hotter to cooler regions via photons (**Figure 14.6**), which carry the energy with them. Recall from your study of radiation in Chapter 5 that the hotter region contains more (and more energetic) photons than the cooler region. More photons move from the hotter, very crowded region to the cooler, less crowded region than in the reverse direction. A net transfer of photons and photon energy occurs from the hotter region to the cooler region, and radiation carries energy outward from the Sun's core.

The transfer of energy from one point to another by radiation also depends on how freely radiation can move from one point to another within a star. The degree to which matter blocks the flow of photons through it is called **opacity**. The opacity of a material depends on many things, including the density of the material, its composition, its temperature, and the wavelength of the photons moving through it.

Energy transfer by radiation is most efficient in regions with low opacity. The radiative zone (see Figure 14.5) is the region in the inner part of the Sun in which the opacity is relatively low, and radiation carries the energy produced in the

Astronomy in Action: Random Walk

Figure 14.6 Higher-temperature regions deep within the Sun produce more radiation than do lower-temperature regions farther out. Although radiation flows in both directions, more radiation flows from the hotter regions to the cooler regions than from the cooler regions to the hotter regions. Therefore, radiation carries energy outward from the inner parts of the Sun. For simplicity, this illustration includes only the most common photons (those with wavelengths at the peak of the blackbody curve). Photons of wavelengths representing all colors are present in all regions, with more of all kinds in the hotter regions and fewer of all kinds in the cooler regions.

Figure 14.7 (a) When a crowd of people plays with a beach ball, the ball never travels very far before someone hits it, turning it in another direction. A ball often takes a long time to make its way from one edge of the crowd to the other. (b) Similarly, a photon traveling through the Sun takes a long time to make its way out of the Sun.

Figure 14.8 The top of the Sun's convective zone shows the bubbling of the surface caused by rising and falling packets of gas.

core outward through the star. That radiative zone extends about 70 percent of the way out toward the surface of the Sun. Even though the region's opacity is low enough for radiation to dominate convection as an energy transport mechanism, photons still travel only a short distance within the region before being absorbed and then reemitted, or scattered by matter, much like a beach ball being batted about by a crowd of people (**Figure 14.7**). Each interaction sends the photon in an unpredictable direction—not necessarily toward the surface of the star. The distances between interactions are so short that, on average, the energy of a gamma-ray photon produced in the interior of the Sun takes about one hundred thousand or more years to find its way to the outer layers of the Sun. Opacity holds energy inside the Sun and lets it seep away only slowly. As it travels, the gamma-ray photon gradually is converted to lower-energy photons, emerging mostly as optical and infrared radiation from the surface.

From a peak of 15 million K at the center of the Sun, the temperature falls to about 2 million to 3 million K at the outer margin of the radiative zone. At that cooler temperature, the opacity is higher, so radiation is less efficient at carrying energy from one place to another. The energy flowing outward through the Sun "piles up" against that edge of the radiative zone.

Nearer the surface of the Sun, transfer by radiation becomes inefficient and the temperature changes quickly. Instead, **convection** takes over. Convection carries energy from inside a planet to its surface or from the Sun-heated surface of Earth upward through Earth's atmosphere. Convection also plays an important role in transporting energy outward from the interior of the Sun. Convection transports energy by moving packets of hot gas, like hot-air balloons, which become buoyant and rise up through the lower-temperature gas above them, carrying energy with them. The solar convective zone (see Figure 14.5) extends from the outer boundary of the radiative zone outward to just below the visible surface of the Sun, where evidence of convection can be seen in the bubbling surface (**Figure 14.8**).

In the outermost layers of the Sun, radiation again takes over as the primary mode of energy transport, and it is radiation that transports energy from the Sun's outermost layers off into space.

Observing Neutrinos from the Core of the Sun

The model of energy production and energy transport in the Sun just presented correctly matches observed global properties of the Sun such as its size, temperature, and luminosity. The nuclear fusion model of the Sun predicts exactly which nuclear reactions should be occurring in the core of the Sun and at what rate. The nuclear reactions that make up the proton-proton chain produce a vast number of neutrinos. Because neutrinos barely interact with other ordinary matter, almost all the neutrinos produced in the heart of the Sun travel freely through the outer parts of the Sun and on into space as though the outer layers of the Sun were not there. Solar neutrinos produced in the core of the Sun, traveling at nearly the speed of light, take only 8.3 minutes to reach Earth—much quicker than the 100,000-year journey of photons leaving the core.

Neutrinos interact so weakly with matter that they are extremely difficult to observe. Nevertheless, an extremely large number of nuclear reactions take place in the Sun, so the Sun produces an enormous number of neutrinos. As you read this sentence, about 400 trillion solar neutrinos are passing through your body. That happens even at night because neutrinos easily pass through Earth, too. With that many neutrinos moving about, a neutrino detector does not have to detect a very large percentage of them to be effective.

A neutrino telescope looks very different from other telescopes. The first apparatus designed to detect solar neutrinos was built 1,500 meters underground, within the Homestake Mine in Lead, South Dakota. Astronomers filled a tank with 100,000 gallons of a dry-cleaning fluid called perchloroethylene (C_2Cl_4). Astronomers predicted that, over 2 days, roughly 10^{22} solar neutrinos passed through the Homestake detector. Of those, on average only one neutrino interacted with a chlorine atom within the fluid to form a radioactive isotope of argon. Over time, a measurable amount of argon was produced.

The Homestake experiment operated from the late 1960s to the early 1990s and detected that argon isotope—evidence of neutrinos from the Sun, confirming that nuclear fusion powers the Sun. Astronomers noticed, however, that they were measuring only about one-third as many solar neutrinos as predicted by solar models. The difference between the predicted and measured number of solar neutrinos was called the **solar neutrino problem**.

One possible explanation of the solar neutrino problem was that the working model of the structure of the Sun was somehow wrong. That possibility seemed unlikely, however, because of the many other successful predictions of the solar model. A second possibility was that an understanding of the neutrino itself was incomplete. The neutrino was long thought to have zero mass, like photons, and to travel at the speed of light. But if neutrinos actually do have a tiny amount of mass, particle physics suggests that solar neutrinos should oscillate—alternate back and forth—among three kinds of neutrinos: the electron, muon, and tau neutrinos. According to that explanation, early neutrino experiments could detect only the electron neutrino and, consequently, observed only about a third of the expected number of neutrinos. Since then, many other neutrino detectors have been built, each using different reactions to detect neutrinos of different energies or different types. Experiments at high-energy physics labs, nuclear reactors, and neutrino telescopes around the world have shown that neutrinos do have a nonzero mass and do oscillate among neutrino types.

Solving the solar neutrino problem is a good example of how science works—how a better model of the neutrino showed that the solar neutrino problem was real and not merely an experimental mistake, and how a single set of anomalous observations was later confirmed by other, more sophisticated experiments. All that effort led to a better understanding of basic physics (see the **Process of Science Figure**).

Probing the Sun's Interior

Models of Earth's interior predict how density and temperature change from place to place. Those differences affect the seismic waves traveling through Earth, bending the paths that they travel. Geologists test models of Earth's interior by comparing measurements of seismic waves from earthquakes with model predictions of how seismic waves should travel through the planet.

Just as geologists use seismic waves from earthquakes to probe the interior of Earth, solar physicists use the surface oscillations of the Sun to probe the solar interior. The science that uses solar oscillations to study the Sun is called **helioseismology**. Detailed observations of motions of material from place to place across the surface of the Sun show that the Sun vibrates or rings, something like a bell that has been struck. Unlike a well-tuned bell—which vibrates primarily at one frequency—the vibrations of the Sun are very complex. In the Sun, many frequencies of vibrations occur simultaneously, causing some parts of the Sun to bulge outward and some to draw inward. Those motions help

PROCESS OF SCIENCE

Learning from Failure

The first detection of solar neutrinos raised more questions than it answered.

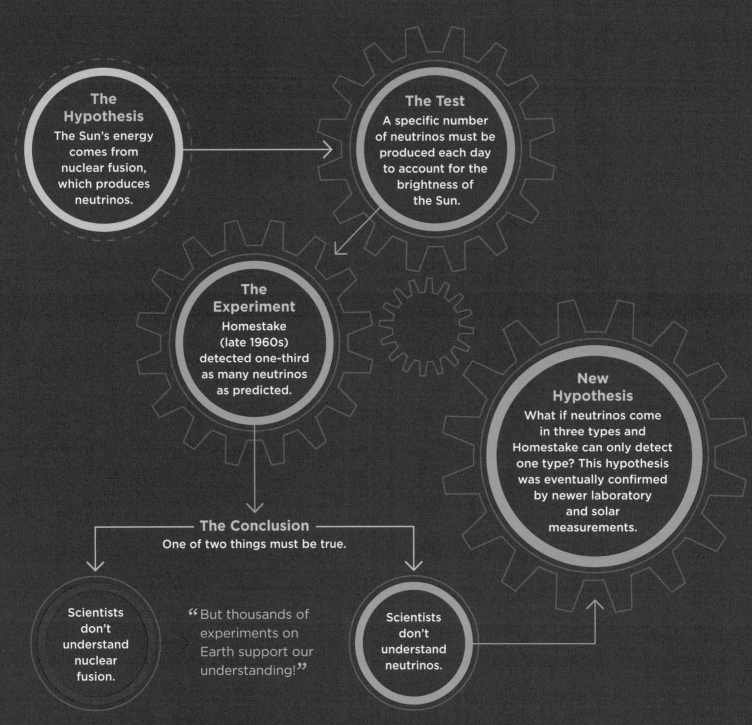

The Hypothesis
The Sun's energy comes from nuclear fusion, which produces neutrinos.

The Test
A specific number of neutrinos must be produced each day to account for the brightness of the Sun.

The Experiment
Homestake (late 1960s) detected one-third as many neutrinos as predicted.

The Conclusion
One of two things must be true.

Scientists don't understand nuclear fusion.

"But thousands of experiments on Earth support our understanding!"

Scientists don't understand neutrinos.

New Hypothesis
What if neutrinos come in three types and Homestake can only detect one type? This hypothesis was eventually confirmed by newer laboratory and solar measurements.

Part of the "scientific attitude" is to find failure exciting. When experiments do not turn out as expected, good scientists get excited because there is something new to understand!

researchers probe what lies below. **Figure 14.9** illustrates the motions of the different parts of the Sun, with red and blue areas moving in opposite directions. Some waves are amplified and some are suppressed, depending on how they overlap as they travel through the Sun. Astronomers study those waves by using the Doppler effect (see Chapter 5), which distinguishes between parts of the Sun that move toward the observer and those that move away.

To detect the disturbances of helioseismic waves on the surface of the Sun, astronomers must measure Doppler shifts of less than 0.1 m/s while detecting changes in brightness of only a few parts per million at any given location on the Sun. Tens of millions of wave motions are possible within the Sun. Some waves travel around the circumference of the Sun, yielding information about the density of the upper convection zone. Other waves travel through the interior of the Sun, revealing the density structure of the Sun close to its core. Still others travel inward toward the center of the Sun, until they are bent by the changing solar density and return to the surface.

All those wave motions are going on at the same time, so sorting them out requires computer analysis of long, unbroken strings of solar observations from several sources. The Global Oscillation Network Group (GONG) is a network of six solar observation stations spread around the world that enables astronomers to observe the surface of the Sun approximately 90 percent of the time.

To interpret helioseismology data, scientists compare the measurements of the strength, frequency, and wavelengths of the waves against predicted vibrations calculated from models of the solar interior. That technique serves as a powerful test of models of the solar interior and has led both to some surprises and to improvements in the models. For example, some scientists proposed that the solar neutrino problem might be solved if the models had overestimated the amount of helium in the Sun. That explanation was ruled out by analyzing the waves that penetrate to the core of the Sun. Helioseismology showed that the value for opacity used in early solar models was too low. That realization led astronomers to recalculate the location of the bottom of the convective zone. Both theory and observation now put the base of the convective zone at 70 percent of the way out from the center of the Sun, with an uncertainty in that number of less than 0.5 percent.

Working back and forth between observation and theory has enabled astronomers to probe the otherwise inaccessible interior of the Sun. We now know that the energy is produced by nuclear fusion deep in the core and that it moves outward by radiation to a point about 70 percent of the radius of the Sun. Then it travels outward by convection to the surface. We also know how the temperature, density, and pressure change with radius and how those factors change the opacity at different distances from the center. Even though researchers usually cannot sample directly or set up controlled experiments, collaboration between theory and observation is essential to observational sciences such as astronomy.

Figure 14.9 The interior of the Sun rings like a bell as helioseismic waves move through it. This figure shows one particular mode of the Sun's vibration. Red indicates regions where gas is traveling inward, whereas blue indicates regions where gas is traveling outward. Astronomers observe those motions via Doppler shifts

CHECK YOUR UNDERSTANDING 14.2

How do neutrinos help us understand what is going on in the core of the Sun?
(a) Neutrinos from distant objects pass through the Sun, probing the interior.
(b) Neutrinos from the Sun pass easily through Earth. (c) Neutrinos created in fusion reactions at the core of the Sun easily escape.

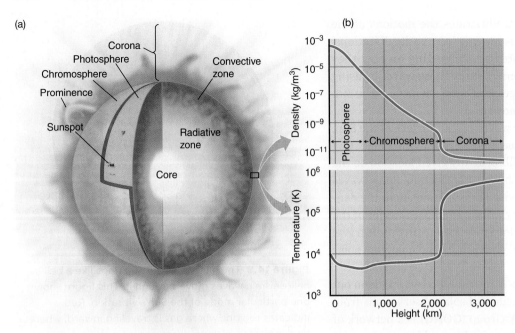

Figure 14.10 (a) The components of the Sun's atmosphere are located above the convective zone. (b) The density and temperature of the Sun's atmosphere change abruptly at the boundary between the chromosphere and corona. The y-axes are logarithmic.

Figure 14.11 (a) When viewed in visible light, the Sun appears to have a sharp outline, even though it has no true surface. The center of the Sun appears brighter, whereas the limb of the Sun is darker—an effect known as limb darkening. (b) Looking at the center of the Sun allows us to see deeper into the Sun's interior than we do when looking at the edge of the Sun. Because higher temperature means more luminous radiation, the center of the Sun appears brighter than its limb.

14.3 The Atmosphere of the Sun

Beyond the convective zone lie the outer layers of the Sun, collectively known as the Sun's atmosphere. Those layers, shown in **Figure 14.10a**, include the *photosphere*, the *chromosphere*, and the *corona*. We can observe those layers of the Sun directly by using telescopes and satellites. Observations of the Sun's atmosphere are important because activity in the Sun's atmosphere has consequences for human infrastructure such as power grids and satellites in orbit around Earth.

The Sun is a large ball of gas and so, unlike Earth, has no solid surface. Its apparent surface is like a fog bank on Earth. After people walk into a fog bank and disappear from view, you would say they were inside the fog bank, even though they never passed through a definite boundary. The apparent surface of the Sun is similar. Light from the Sun's surface can escape directly into space, so we can see it. Light from below the Sun's surface cannot escape directly into space, so we cannot see it.

At the base of the atmosphere is the **photosphere**: the apparent surface of the Sun or its visible boundary. That is where features such as sunspots can be seen. Above that photosphere is the **chromosphere**, a region of strong emission lines. The top layer is the **corona**, which can be viewed during a solar eclipse as a halo around the Sun. In the Sun's atmosphere, the density of the gas drops very rapidly with increasing altitude. The graphs in **Figure 14.10b** show how density and temperature change across the atmosphere of the Sun. The temperature increases sharply across the boundary between the chromosphere and the corona, whereas the density falls sharply across the same boundary. In this section, we explore each layer, beginning at the bottom with the photosphere.

The Photosphere

The **effective temperature** of the photosphere is calculated from the Sun's luminosity and radius by using the Stefan-Boltzmann law (see Chapter 5). The photosphere has an effective temperature of 5780 K, ranging from 6600 K to 4500 K over a 500-km-thick zone. The Sun appears to have a well-defined surface and a sharp outline when viewed from Earth because 500 km does not look very thick when viewed from a distance of 150 million km.

In **Figure 14.11a**, the Sun appears fainter near its edges than near its center, an effect called **limb darkening**. That effect is an artifact of the structure of the Sun's photosphere. When looking near the edge of the Sun, you are looking through the photosphere at a steep angle. As a result, you do not see as deeply into the Sun as when you are looking directly down through the photosphere near the center of the Sun's disk. The light from the limb of the Sun comes from a shallower layer that is cooler and fainter, as shown in **Figure 14.11b**.

In the Sun's atmosphere, the density of the gas drops very rapidly with increasing altitude. All visible solar phenomena take place in the Sun's atmosphere. Most of the radiation from below the Sun's photosphere is absorbed by matter and reemitted at the photosphere as a blackbody spectrum.

As we examine the structure of the Sun in more detail, however, that simple description of the spectra of stars turns out to be incomplete. Light from the solar photosphere must escape through the upper layers of the Sun's atmosphere, which affects the spectrum we observe. In Chapter 13, we discussed the presence of absorption lines in the spectra of stars. Now we can take a closer look at how those absorption lines form. As photospheric light travels upward, atoms in the solar atmosphere absorb the light at discrete wavelengths, forming absorption lines. Because the Sun appears so much brighter than any other star, its spectrum can be studied in far more detail, so specially designed telescopes and high-resolution spectrometers have been built specifically to study the Sun's light. The solar spectrum is shown in **Figure 14.12**. Absorption lines from more than 70 elements have been identified. Analysis of those lines forms the basis for much of astronomers' knowledge of the solar atmosphere, including the composition of the Sun. That is also the starting point for understanding the atmospheres and spectra of other stars.

The Chromosphere and Corona

The temperature falls from 6600 K at the photosphere's bottom to 4400 K at its top. At that point, the trend reverses and the temperature slowly begins to climb, rising to about 6000 K at a height of 1,500 km above the top of the photosphere (see Figure 14.10b). This region of increasing temperature is called the chromosphere (**Figure 14.13a**). The reason for the chromosphere's temperature reversal with increasing height is not well understood, but it may be caused by magnetic waves propagating through the region and depositing their energy at the top of the chromosphere.

The chromosphere was discovered in the 19th century during observations of total solar eclipses (**Figure 14.13b**). The chromosphere is seen most strongly at the solar limb as a source of emission lines, especially a particular hydrogen line produced when the electron falls from the third energy state to the second energy state. That line is known as the Hα line (the "hydrogen alpha line"). The deep red color of the Hα line is what gives the chromosphere its name; that word means "the place where color comes from." The element helium was discovered in 1868 from a spectrum of the chromosphere of the Sun nearly 30 years before it was found on Earth. Helium is named after *helios*, the Greek word for "Sun."

At the top of the chromosphere, across a transition region only about 100 km thick, the temperature suddenly soars, whereas the density abruptly drops (see Figure 14.10b). Above that transition lies the outermost region of the Sun's atmosphere, the corona, where temperatures reach 1 million to 2 million K. The corona is thought to be heated by magnetic fields and micro solar flares.

The Sun's corona has been known since ancient times: it is visible during total solar eclipses as an eerie outer glow stretching several solar radii beyond the Sun's surface (**Figure 14.13c**). Because it is so hot, the solar corona also is a strong source of X-rays. Those X-ray photons have so much energy that many

Figure 14.12 This high-resolution spectrum of the Sun stretches from 400 nanometers (nm) in the lower left corner to 700 nm in the upper right corner and shows dark absorption lines. The spectrum was produced by passing the Sun's light through a prism-like device and then cutting and folding the single long spectrum (from blue to red) into rows so that it will fit in a single image taken by a camera.

Figure 14.13 (a) This spacecraft image of the Sun shows fine structure in the chromosphere extending outward from the photosphere. (b) The chromosphere is visible during a total eclipse. (c) This eclipse image shows the Sun's corona, consisting of million-kelvin gas that extends for millions of kilometers beyond the surface of the Sun.

Figure 14.14 This close-up image of the Sun shows the tangled structure of coronal loops.

electrons are stripped away from nuclei, leaving atoms in the corona highly ionized.

CHECK YOUR UNDERSTANDING 14.3

The surface of the Sun appears sharp in visible light because: (a) the photosphere is cooler than the layers below it; (b) the photosphere is thinner than the other layers in the Sun; (c) the photosphere is less dense than the convection zone; (d) the Sun has a distinct surface.

14.4 The Atmosphere of the Sun Is Very Active

The atmosphere of the Sun is a very turbulent place. The best-known features on the surface of the Sun are relatively dark blemishes in the solar photosphere, called **sunspots**. Sunspots come and go over time, though they remain long enough for us to determine the rotation rate of the Sun. Those spots are associated with active regions: loops of material and explosions that fling particles far out into the Solar System. Long-term patterns have been observed in the variations of sunspots and active regions, revealing that the magnetic field of the Sun is constantly changing.

Solar Activity Is Caused by Magnetic Effects

The magnetic field (see Chapter 5) of the Sun causes virtually all the structure seen in the Sun's atmosphere. High-resolution images of the Sun show *coronal loops* that make up much of the Sun's lower corona (**Figure 14.14**). That texture is the result of magnetic structures called flux tubes. Magnetic fields are responsible for much of the structure of the corona as well. The corona is far too hot to be held in by the Sun's gravity, but over most of the surface of the Sun, coronal gas is confined by magnetic loops with both ends firmly anchored deep within the Sun. The magnetic field in the corona acts almost like a network of rubber bands that coronal gas is free to slide along but cannot cross. In contrast, about 20 percent of the surface of the Sun is covered by an ever-shifting pattern of **coronal holes**, large regions where the magnetic field points outward, away from the Sun, and where coronal material is free to stream away into interplanetary space as the solar wind. In extreme ultraviolet images of the Sun we see coronal holes as dark regions, which indicates that they are cooler and less dense than their surroundings (see the chapter-opening photograph).

The relatively steady part of the solar wind consists of lower-speed flows with velocities of about 350 km/s and higher-speed flows with velocities up to about 700 km/s. The higher-speed flows originate in coronal holes. Depending on their speed, particles in the solar wind take 2–5 days to reach Earth. Often, 2–5 days after a coronal hole passes across the center of the face of the Sun, the speed and density of the solar wind reaching Earth increases. The solar wind drags the Sun's magnetic field along with it (**Figure 14.15**), so the magnetic field in the solar wind gets "wound up" by the Sun's rotation. Consequently, the solar wind has a spiral structure resembling the stream of water from a rotating lawn sprinkler.

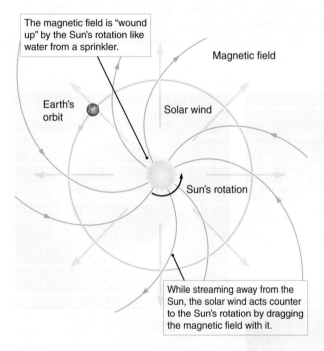

Figure 14.15 The solar wind (yellow arrows) streams away from active areas and coronal holes on the Sun. As the Sun rotates, the solar wind takes on a spiral structure, much like the spiral of water that streams away from a rotating lawn sprinkler.

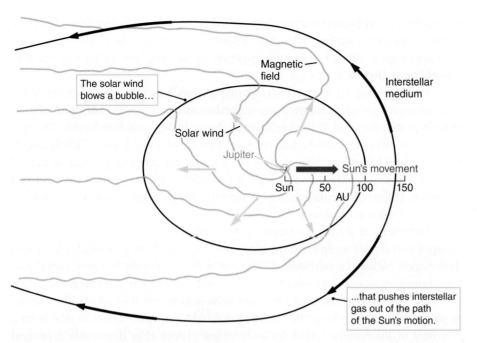

Figure 14.16 The solar wind (yellow arrows) streams away from the Sun for about 100 AU until it finally piles up against the pressure of the interstellar medium through which the Sun is traveling. The *Voyager 1* spacecraft crossed that boundary in 2012.

The effects of the solar wind are felt throughout the Solar System. The solar wind blows the tails of comets away from the Sun, shapes the magnetospheres of the planets, and supplies the energetic particles that power Earth's spectacular auroral displays. Using space probes, astronomers have observed the solar wind extending out to 100 astronomical units (AU) from the Sun. But the solar wind does not go on forever. The farther it gets from the Sun, the more it spreads out. Just like radiation, the density of the solar wind follows an inverse square law. At a distance of about 100 AU from the Sun, the solar wind stops abruptly. Here it piles up against the pressure of the **interstellar medium**, the gas and dust that lie between stars in a galaxy. **Figure 14.16** shows the region of space over which the solar wind is measured. The *Voyager 1* spacecraft has crossed the outer edge of that boundary and sent back the first direct measurements of true interstellar space. The *Interstellar Boundary Explorer* spacecraft, launched in 2008, also is exploring the region.

Sunspots and Changes in the Sun

Sunspots have been noted since antiquity. Telescopic observations of sunspots date back almost 400 years, and records exist of naked-eye observations by Chinese, Greek, and medieval astronomers centuries before that. *But remember: never look directly at the Sun!* Direct viewing through a commercial solar filter is safe, as is projecting the image through a telescope or binoculars onto a surface such as paper and looking only at the projection. Many websites have live images of the Sun viewed through ground and space telescopes (see the "Using the Web" problems at the end of the chapter). Sunspots are places where material is trapped at the surface of the Sun by magnetic-field lines. When that material

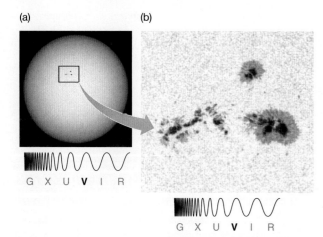

Figure 14.17 (a) This image from the Solar Dynamics Observatory (SDO), taken in 2010, shows a large sunspot group. Sunspots are magnetically active regions cooler than the surrounding surface of the Sun. (b) This high-resolution view shows the sunspots in this group.

cools, convection cannot carry it downward, so it makes a cooler (and therefore darker) spot on the surface of the Sun. **Figure 14.17** shows a large sunspot group. Sunspots appear dark, but only in contrast to the brighter surface of the Sun (**Working It Out 14.2**).

Early telescopic observations of sunspots made during the 17th century led to the discovery of the Sun's rotation, which has an average period of about 27 days as seen from Earth and 25 days relative to the stars. Because Earth orbits the Sun in the same direction that the Sun rotates, observers on Earth see a slightly longer rotation period. Observations of sunspots also show that the Sun's rotation period is shorter at its equator than at higher latitudes, an effect called **differential rotation**. Differential rotation is possible only because the Sun is a large ball of gas rather than a solid object.

Figure 14.18 shows the structure of a sunspot on the surface of the Sun. A sunspot consists of an inner dark core called the **umbra**, surrounded by a less dark region called the **penumbra**. The penumbra shows an intricate radial pattern, reminiscent of the petals of a flower. Sunspots are caused by magnetic fields thousands of times greater than the magnetic field at Earth's surface. They occur in pairs connected by loops in the magnetic field. Sunspots range in size from a few tens of kilometers across up to complex groups that may contain several dozen individual spots and span as much as 150,000 km. The largest sunspot groups can be seen without a telescope.

Although sunspots occasionally last 100 days or longer, half of all sunspots come and go in about 2 days, and 90 percent are gone within 11 days. The number and distribution of sunspots change in a pattern averaging 11 years called the **sunspot cycle**. **Figure 14.19a** shows data for several recent cycles. At the beginning of a cycle, sunspots appear at solar latitudes of about 30° north and south of

14.2 Working It Out Sunspots and Temperature

Sunspots are about 1500 K cooler than their surroundings. What does that lower temperature tell us about their luminosity? Think back to the Stefan-Boltzmann law in Chapter 5. The flux, $\mathcal{F}$, from a blackbody is proportional to the fourth power of the temperature, T:

$$\mathcal{F} = \sigma T^4$$

The constant of proportionality is the Stefan-Boltzmann constant, σ, which has a value of 5.67×10^{-8} W/(m² K⁴).

The flux is the amount of energy coming from a square meter of surface every second. How much less energy comes out of a sunspot than out of the rest of the Sun? Using 4500 and 6000 K for the temperatures of a typical sunspot and the surrounding photosphere, respectively, we can set up two equations:

$$\mathcal{F}_{\text{spot}} = \sigma T^4_{\text{spot}} \quad \text{and} \quad \mathcal{F}_{\text{surface}} = \sigma T^4_{\text{surface}}$$

We could solve each of those separately and then divide the value of $\mathcal{F}_{\text{spot}}$ by $\mathcal{F}_{\text{surface}}$ to find out how much fainter the sunspot is, but solving for the ratio of the fluxes eliminates σ:

$$\frac{\mathcal{F}_{\text{spot}}}{\mathcal{F}_{\text{surface}}} = \frac{\sigma T^4_{\text{spot}}}{\sigma T^4_{\text{surface}}} = \frac{T^4_{\text{spot}}}{T^4_{\text{surface}}} = \left(\frac{T_{\text{spot}}}{T_{\text{surface}}}\right)^4$$

Plugging in our values for T_{spot} and T_{surface} gives

$$\frac{\mathcal{F}_{\text{spot}}}{\mathcal{F}_{\text{surface}}} = \left(\frac{4500 \text{ K}}{6000 \text{ K}}\right)^4 = 0.32$$

and multiplying both sides by $\mathcal{F}_{\text{surface}}$ gives

$$\mathcal{F}_{\text{spot}} = 0.32 \mathcal{F}_{\text{surface}}$$

So, the amount of energy coming from a square meter of sunspot every second is about one-third as much as the amount of energy coming from a square meter of surrounding surface every second. In other words, the sunspot is about one-third as bright as the surrounding photosphere. If you could cut out the sunspot and place it elsewhere in the sky, it would be brighter than the full Moon.

the solar equator. Over the following years, sunspots are found closer to the equator as their number increases to a maximum and then declines. As the last few sunspots approach the equator, new sunspots again begin appearing at middle latitudes, and the next cycle begins. **Figure 14.19b** shows the number of sunspots at a given latitude plotted against time: that diagram of opposing diagonal bands is called the sunspot "butterfly diagram."

In the early 20th century, solar astronomer George Ellery Hale (1868–1938) was the first to show that the 11-year sunspot cycle is actually half of a 22-year magnetic cycle during which the direction of the Sun's magnetic field reverses after each 11-year sunspot cycle. **Figure 14.19c** shows how the average strength of the magnetic field at every latitude has changed over more than 35 years. The direction of the Sun's magnetic field flips at the maximum of each sunspot cycle. Sunspots come in pairs, with one spot (the leading sunspot) in front of the other with respect to the Sun's rotation. In one 11-year sunspot cycle, the leading sunspot in each pair tends to be a north magnetic pole, whereas the trailing sunspot tends to be a south magnetic pole. In the next 11-year sunspot cycle, that polarity is reversed: the leading sunspot in each pair is a south magnetic pole, whereas the trailing sunspot tends to be a north magnetic pole. The

Figure 14.18 This very high-resolution view of a sunspot shows the dark umbra surrounded by the lighter penumbra. The solar surface around the sunspot bubbles with separate cells of hot gas called *granules*. The smallest features are about 100 km across.

Figure 14.19 (a) The number of sunspots varies, as shown in this graph of the past few solar cycles. (b) The "solar butterfly" diagram shows the fraction of the Sun covered by sunspots at each latitude. The data are color coded to show the percentage of the strip at that latitude that is covered in sunspots at that time: black, 0 to less than 0.1 percent; red, 0.1–1.0 percent; yellow, greater than 1.0 percent. (c) The Sun's magnetic poles flip every 11 years. Yellow indicates magnetic north, whereas blue indicates magnetic south.

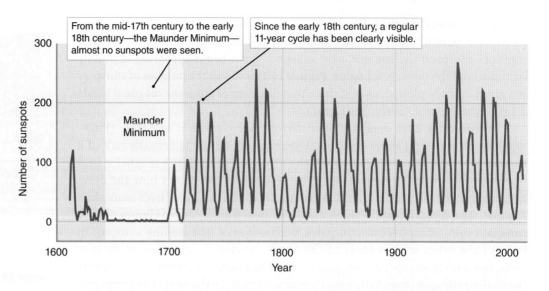

Figure 14.20 Sunspots have been observed for hundreds of years. In this plot, the 11-year cycle in the number of sunspots (half of the 22-year solar magnetic cycle) is clearly visible. Sunspot activity varies greatly. The period from the middle of the 17th century to the early 18th century, when almost no sunspots were seen, is called the *Maunder minimum*.

transition between those two magnetic polarities occurs near the peak of each sunspot cycle. Magnetic activity on the Sun affects the photosphere, chromosphere, and corona.

The graph in **Figure 14.20** shows 400 years of sunspot observations. The 11-year cycle is neither perfectly periodic nor especially reliable. The time between peaks in the number of sunspots actually varies between about 9.7 and 11.8 years. The number of spots seen during a given cycle fluctuates as well, and in some periods sunspot activity has disappeared almost entirely. An extended lull in solar activity, called the **Maunder minimum**, lasted from 1645 to 1715. Typically, about six peaks of solar activity occur in 70 years, but virtually no sunspots were seen during the Maunder minimum, and auroral displays were less frequent than usual.

Sunspots are only one of several phenomena that follow the Sun's 22-year cycle of magnetic activity. The peaks of the cycle, called **solar maxima**, are times of intense activity. Sunspots are often accompanied by a brightening of the solar chromosphere that is seen most clearly in emission lines such as Hα. Those bright regions are known as solar active regions. The magnificent loops arching through the solar corona, shown in **Figure 14.21**, are solar **prominences**, magnetic flux tubes of relatively cool (5000–10,000 K) but dense gas extending through the million-kelvin gas of the corona. Those prominences are anchored in the active regions. Although most prominences are relatively quiet, others can erupt out through the corona, towering a million kilometers or more over the surface of the Sun and ejecting material into the corona at velocities of 1,000 km/s.

Solar flares are the most energetic form of solar activity, violent eruptions in which enormous amounts of magnetic energy are released over a few minutes to a few hours. **Figure 14.22** shows solar flares erupting from two sunspot groups. The images in Figures 14.22a and b, taken in ultraviolet light, show material at very high temperatures. The spots in the visible-light image (Figure 14.22c) are at the base of the activity seen in Figures 14.22a and b. Solar flares can heat gas to temperatures of 20 million K, and they are the source of intense X-rays and gamma rays. Hot **plasma** (consisting of atoms stripped of some or all of their electrons) moves outward from flares at speeds that can reach 1,500 km/s. Magnetic effects can then accelerate subatomic particles to almost the speed of light. Such

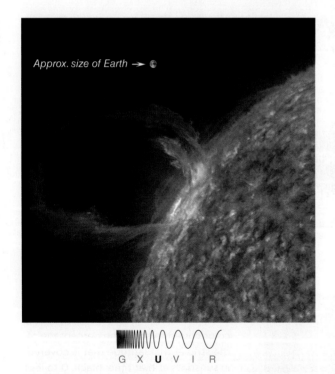

Figure 14.21 Solar prominences are magnetically supported arches of hot gas that rise high above active regions on the Sun. Here, you can see a close-up view at the base of a large prominence. An image of Earth is included for scale (it is not actually that close to the Sun).

Figure 14.22 The Solar Dynamics Observatory (SDO) observed these active regions of the Sun that produced solar flares in August 2011. (a) Activity near the surface at 60,000 K is visible in extreme ultraviolet light (along with a prominence rising up from the Sun's edge). (b) Viewed at other ultraviolet wavelengths, many looping arcs and plasma heated to about 1 million K become visible. (c) The dark spots in this image are the magnetically intense sunspots that are the sources of all the activity.

events, called **coronal mass ejections** (**CMEs**) (**Figure 14.23**), send powerful bursts of energetic particles outward through the Solar System. CMEs occur about once per week during the minimum of the sunspot cycle and as often as several times per day near the maximum of the cycle.

Solar Activity Affects Earth

The amount of solar radiation received at the distance of Earth from the Sun has been measured to be, on average, 1,361 watts per square meter (W/m^2). Satellite measurements of the amount of radiation coming from the Sun (**Figure 14.24**) show that this value varies by as much as 0.2 percent over periods of a few weeks, as dark sunspots in the photosphere and bright spots in the chromosphere move across the disk. Overall, however, the increased radiation from active regions on the Sun more than makes up for the reduction in radiation from sunspots. On average, the Sun seems to be about 0.1 percent brighter during the peak of a solar cycle than it is at its minimum.

Solar activity affects Earth in many ways. Solar active regions are the source of most of the Sun's extreme ultraviolet and X-ray emissions—energetic radiation that heats Earth's upper atmosphere and, during periods of increased solar activity, causes Earth's upper atmosphere to expand. When that happens, the swollen upper atmosphere can significantly increase the atmospheric drag on spacecraft

Figure 14.23 This Solar and Heliospheric Observatory (*SOHO*) image shows a coronal mass ejection (upper right); a simultaneously recorded ultraviolet image of the solar disk is superimposed.

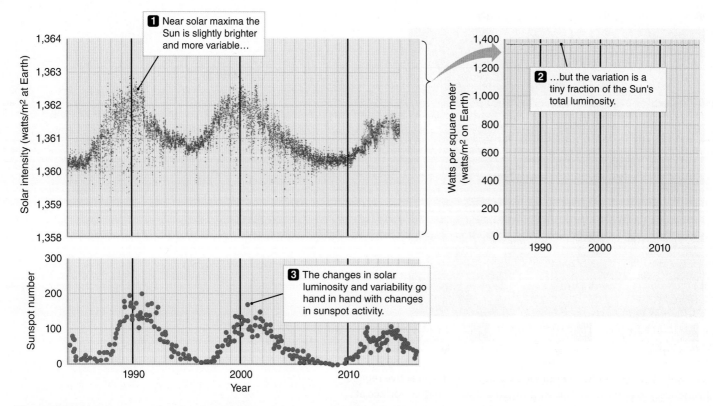

Figure 14.24 Measurements taken by satellites show that the amount of light from the Sun changes slightly.

orbiting at relatively low altitudes, such as that of the Hubble Space Telescope, causing their orbits to decay. Periodic boosts have been necessary to keep the Hubble Space Telescope in its orbit.

Earth's magnetosphere is the result of the interaction between Earth's magnetic field and the solar wind. Increases in the solar wind accompanying solar activity, especially CMEs directed at Earth, can disrupt Earth's magnetosphere. Spectacular auroras can accompany such events, as can magnetic storms that have disrupted electric power grids and caused blackouts across large regions. CMEs emitted in the direction of Earth also hinder radio communication and navigation, and they can damage sensitive satellite electronics, including communication satellites. In addition, energetic particles accelerated in solar flares pose one of the greatest dangers to human exploration of space.

Detailed observations from the ground and from space help astronomers understand the complex nature of the solar atmosphere. The Solar and Heliospheric Observatory (*SOHO*) spacecraft is a joint mission between NASA and the European Space Agency (ESA). *SOHO* moves in lockstep with Earth at a location approximately 1,500,000 km from Earth that is almost directly in line between Earth and the Sun. *SOHO* carries 12 scientific instruments that monitor the Sun and measure the solar wind upstream of Earth. In addition, NASA's Solar Dynamics Observatory (SDO) studies the solar magnetic field to predict when major solar events will occur, instead of simply responding after they happen.

CHECK YOUR UNDERSTANDING 14.4

Sunspots appear dark because: (a) they have very low density; (b) magnetic fields absorb most of the light that falls on them; (c) they are regions of very high pressure; (d) they are cooler than their surroundings.

Origins The Solar Wind and Life

Energetic particles accelerated in solar flares need to be considered when astronauts are orbiting Earth in a space station or, someday, traveling to the Moon or farther. Earth's magnetic field protects life on the surface from those energetic particles, which travel along the magnetic-field lines to Earth's poles, creating the auroras. But the Moon does not have that protection because its magnetic field is very weak. Astronauts on the lunar surface would be exposed to as much radiation as astronauts traveling in space. The strength of the solar wind varies with the solar cycle, as noted in Section 14.4, so the exposure danger varies as well.

The Solar System is surrounded by the **heliosphere** (**Figure 14.25**), in which the solar wind blows against the interstellar medium and clears out an area like the inside of a bubble. As the Sun and Solar System move through the Milky Way Galaxy, passing in and out of interstellar clouds, that heliosphere protects the entire Solar System from galactic high-energy particles known as cosmic rays that originate primarily in high-energy explosions of massive dying stars. When the Sun is in its lower-activity state, the heliosphere is weaker, so more galactic cosmic rays enter the Solar System. In addition, the intensity of those cosmic rays depends on where the Sun and Solar System are located in their orbit about the center of the Milky Way Galaxy.

Some scientists have theorized that at times when the Sun was quiet and the heliosphere was weaker than average, and the Solar System was passing through a particular part of the galaxy, the cosmic-ray flux in the Solar System—and on Earth—increased. That increased flux possibly led to a disruption in Earth's ozone layer and possibly contributed to a mass extinction in which many species died out on Earth.

Thus, in addition to heat and light on Earth, the extension of the Sun through the solar wind may have affected the evolution of life on Earth—and may affect the ability of humans to live and work in space.

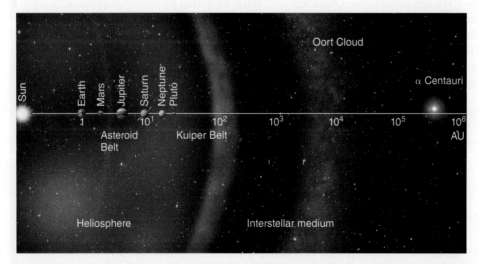

Figure 14.25 The heliosphere of the Sun is a bubble of charged particles covering the Solar System. The heliosphere is formed by the solar wind blowing against the interstellar medium. Both the *Voyager* spacecraft are now past 100 AU. The scale, indicating how far objects are from the Sun, is logarithmic.

READING ASTRONOMY NEWS

Scientists discuss major solar storms that affect Earth.

Carrington-Class CME Narrowly Misses Earth

By DR. TONY PHILLIPS, Science@NASA

Last month (April 8–11, 2014), scientists, government officials, emergency planners, and others converged on Boulder, Colorado, for NOAA's Space Weather Workshop—an annual gathering to discuss the perils and probabilities of solar storms.

The current solar cycle is weaker than usual, so you might expect a correspondingly low-key meeting. On the contrary, the halls and meeting rooms were abuzz with excitement about an intense solar storm that narrowly missed Earth.

"If it had hit, we would still be picking up the pieces," says Daniel Baker of the University of Colorado, who presented a talk entitled *The Major Solar Eruptive Event in July 2012: Defining Extreme Space Weather Scenarios*.

The close shave happened almost two years ago. On July 23, 2012, a plasma cloud or "CME" rocketed away from the Sun as fast as 3,000 km/s, more than 4 times faster than a typical eruption. The storm tore through the Earth's orbit, but fortunately Earth wasn't there. Instead it hit the *STEREO-A* spacecraft. Researchers have been analyzing the data ever since, and they have concluded that the storm was one of the strongest in recorded history. "It might have been stronger than the Carrington Event itself," says Baker.

The Carrington Event of September 1859 was a series of powerful CMEs that hit Earth head-on, sparking Northern Lights as far south as Tahiti. Intense geomagnetic storms caused global telegraph lines to spark, setting fire to some telegraph offices and disabling the "Victorian Internet." A similar storm today could have a catastrophic effect on modern power grids and telecommunication networks. According to a study by the National Academy of Sciences, the total economic impact could exceed $2 trillion or 20 times greater than the costs of a Hurricane Katrina. Multi-ton transformers fried by such a storm could take years to repair and impact national security.

A recent paper in *Nature Communications* authored by UC Berkeley space physicist Janet G. Luhmann and former postdoc Ying D. Liu describes what gave the July 2012 storm Carrington-like potency. For one thing, the CME was actually *two* CMEs separated by only 10 to 15 minutes. This double storm cloud traveled through a region of space that had been cleared out by another CME only four days earlier. As a result, the CMEs were not decelerated as much as usual by their transit through the interplanetary medium.

Had the eruption occurred just one week earlier, the blast site would have been facing Earth, rather than off to the side, so it was a relatively narrow escape.

When the Carrington Event enveloped Earth in the 19th century, technologies of the day were hardly sensitive to electromagnetic disturbances. Modern society, on the other hand, is deeply dependent on Sun-sensitive technologies such as GPS, satellite communications, and the Internet.

"The effect of such a storm on our modern technologies would be tremendous," says Luhmann.

During informal discussions at the workshop, Nat Gopalswamy of the Goddard Space Flight Center noted that "without NASA's *STEREO* probes, we might never have known the severity of the 2012 superstorm. This shows the value of having 'space weather buoys' located all around the Sun."

It also highlights the potency of the Sun even during so-called "quiet times." Many observers have noted that the current solar cycle is weak, perhaps the weakest in 100 years. Clearly, even a weak solar cycle can produce a very strong storm. Says Baker, "We need to be prepared."

1. What is a coronal mass ejection (CME)?
2. Why would a CME cause disruptions on Earth?
3. Explain how the 2012 storm missed Earth by 1 week.
4. More sensationalistic headlines for this story claimed that Earth almost "was sent back to the Dark Ages." What did they mean by that exaggeration?
5. Go to the NASA press release for this event (https://science.nasa.gov/science-news/science-at-nasa/2014/23jul_superstorm/), and click to watch the 4-minute "ScienceCast" video. What happened during the Carrington CME in 1859? Is that video effective at communicating the science information to the nonspecialist?

Summary

The forces due to pressure and gravity balance each other in hydrostatic equilibrium, maintaining the Sun's structure. Nuclear reactions converting hydrogen to helium are the source of the Sun's energy. Energy created in the Sun's core moves outward to the surface, first by radiation and then by convection. The solar wind may adversely affect astronauts located in space or on planets that lack a protective magnetic field, but it also has protected the Solar System from galactic, high-energy, cosmic-ray particles.

LG 1 **Describe the balance between the forces that determine the structure of the Sun.** The outward radiation pressure of the hot gas inside the Sun balances the inward pull of gravity at every point. That balance is dynamically maintained. An energy balance is also maintained, with the energy produced in the core of the Sun balancing the energy lost from the surface.

LG 2 **Explain how mass is converted to energy in the Sun's core and how long the Sun will take to use up its fuel.** In the core of the Sun, mass is converted to energy via the proton-proton chain. When four hydrogen atoms fix to one helium atom, some mass is lost. That mass is released as energy, nearly all of which leaves the Sun either as photons or as neutrinos. Neutrinos are elusive, almost massless particles that interact only very weakly with other matter. Observations of neutrinos confirm that nuclear fusion is the Sun's primary energy source.

LG 3 **Sketch a physical model of the Sun's interior, and list the ways that energy moves outward from the Sun's core toward its surface.** The interior of the Sun is divided into zones defined by how energy is transported in that region. Energy moves outward through the Sun by radiation and by convection.

LG 4 **Describe how observations of solar neutrinos and seismic vibrations on the surface of the Sun test astronomers' models of the Sun.** The Sun has multiple layers, each with a characteristic pressure, density, and temperature. Neutrinos directly probe the interior of the Sun. That model of the interior of the Sun has been tested by helioseismology, in much the same way that the model of Earth's interior has been tested by seismology.

LG 5 **Describe the solar activity cycles of 11 and 22 years, and explain how those cycles are related to the Sun's changing magnetic field.** Activity on the Sun follows a cycle that peaks every 11 years but takes 22 full years for the magnetic field to reverse. Sunspots are photospheric regions cooler than their surroundings, and they reveal the cycles in solar activity. Material streaming away from the Sun's corona creates the solar wind, which moves outward through the Solar System until it meets the interstellar medium. Solar storms, including ejections of mass from the corona, produce auroras and can disrupt power grids and damage satellites.

? Unanswered Questions

- Will nuclear fusion become a major source of energy production on Earth? Scientists have been working on controlled nuclear fusion for more than 60 years, since the first hydrogen bombs were developed. So far, however, replicating the conditions inside the Sun has posed too many difficulties. Nuclear fusion requires that we have hydrogen isotopes at very high temperature, density, and pressure, just as when a hydrogen bomb explodes. However, controlled nuclear fusion requires that we confine that material long enough to get more energy out than we put in. Several major experiments have attempted to fuse isotopes of hydrogen. An alternative approach is to fuse an isotope of helium, ^{3}He, which has only three particles in the nucleus (two protons and one neutron). On Earth, ^{3}He is found in very limited supply. But ^{3}He is in much greater abundance on the Moon, so some people propose setting up mining colonies on the Moon to extract ^{3}He for use in fusion reactions on Earth or possibly even on the Moon (see question 49b at the end of the chapter).

- Are large timescale variations in Earth's climate—ice ages—related to solar activity? Solar activity affects Earth's upper atmosphere and may affect weather patterns as well. Variations in the amount of radiation from the Sun might be responsible for past variations in Earth's climate. Current models indicate that observed variations in the Sun's luminosity could account for only about 0.1-K differences in Earth's average temperature—much less than the effects due to the ongoing buildup of carbon dioxide in Earth's atmosphere. Triggering the onset of an ice age may require a sustained drop in global temperatures of only 0.2–0.5 K, so astronomers are continuing to investigate a possible link between solar variability and long-timescale changes in Earth's climate.

Questions and Problems

Test Your Understanding

1. The physical model of the Sun's interior has been confirmed by observations of
   a. neutrinos and seismic vibrations.
   b. sunspots and solar flares.
   c. neutrinos and positrons.
   d. sample returns from spacecraft.
   e. sunspots and seismic vibrations.

2. Place in order the following steps in the fusion of hydrogen into helium. If two or more steps happen simultaneously, use an equals sign (=).
   a. A positron is emitted.
   b. One gamma ray is emitted.
   c. Two hydrogen nuclei are emitted.
   d. Two ^{3}He collide and become ^{4}He.
   e. Two hydrogen nuclei collide and become ^{2}H.
   f. Two gamma rays are emitted.
   g. A neutrino is emitted.
   h. One deuterium nucleus and one hydrogen nucleus collide and become ^{3}He.

3. Sunspots, flares, prominences, and coronal mass ejections are all caused by
   a. magnetic activity on the Sun.
   b. electrical activity on the Sun.
   c. the interaction of the Sun's magnetic field and the interstellar medium.
   d. the interaction of the solar wind and Earth's magnetic field.
   e. the interaction of the solar wind and the Sun's magnetic field.

4. The structure of the Sun is determined by not only the balance between the forces due to _____ and gravity but also the balance between energy generation and energy _____.
   a. pressure; production
   b. pressure; loss
   c. ions; loss
   d. solar wind; production

5. In the proton-proton chain, four hydrogen nuclei are converted to a helium nucleus. That does not happen spontaneously on Earth because the process requires
   a. vast amounts of hydrogen.
   b. very high temperatures and densities.
   c. hydrostatic equilibrium.
   d. very strong magnetic fields.

6. The solar neutrino problem pointed to a fundamental gap in our knowledge of
   a. nuclear fusion.
   b. neutrinos.
   c. hydrostatic equilibrium.
   d. magnetic fields.

7. Sunspots change in number and location during the solar cycle. That phenomenon is connected to
   a. the rotation rate of the Sun.
   b. the temperature of the Sun.
   c. the magnetic field of the Sun.
   d. the tilt of the axis of the Sun.

8. Suppose an abnormally large amount of hydrogen suddenly fused in the core of the Sun. Which of the following would be observed first?
   a. The Sun would become brighter.
   b. The Sun would swell and become larger.
   c. The Sun would become bluer.
   d. The Sun would emit more neutrinos.

9. The solar corona has a temperature of 1 million to 2 million K; the photosphere has a temperature of only about 6000 K. Why isn't the corona much, much brighter than the photosphere?
   a. The magnetic field traps the light.
   b. The corona emits only X-rays.
   c. The photosphere is closer to us.
   d. The corona has a much lower density.

10. The Sun rotates once every 25 days relative to the stars. The Sun rotates once every 27 days as seen from Earth. Why are those two numbers different?
    a. The stars are farther away.
    b. Earth is smaller.
    c. Earth moves in its orbit as the Sun rotates.
    d. The Sun moves relative to the stars.

11. Place the following regions of the Sun in order of increasing radius.
    a. corona
    b. core
    c. radiative zone
    d. convective zone
    e. chromosphere
    f. photosphere
    g. a sunspot

12. Coronal mass ejections
    a. carry away 1 percent of the mass of the Sun each year.
    b. are caused by breaking magnetic fields.
    c. are always emitted in the direction of Earth.
    d. are unimportant to life on Earth.

13. As energy moves out from the Sun's core toward its surface, it first travels by _____, then by _____, and then by _____.
    a. radiation; conduction; radiation
    b. conduction; radiation; convection
    c. radiation; convection; radiation
    d. radiation; convection; conduction

14. Energy is produced primarily in the center of the Sun because
    a. the strong nuclear force is too weak elsewhere.
    b. that's where neutrinos are created.
    c. that's where most of the helium is.
    d. the temperature and density are high enough in the core.

15. The solar wind pushes on the magnetosphere of Earth, changing its shape, because
    a. the solar wind is so dense.
    b. the magnetosphere is so weak.
    c. the solar wind contains charged particles.
    d. the solar wind is so fast.

Thinking about the Concepts

16. Explain how hydrostatic equilibrium acts as a safety valve to keep the Sun at its constant size, temperature, and luminosity.

17. Two of the three atoms in a molecule of water (H_2O) are hydrogen. Why are Earth's oceans not fusing hydrogen into helium and setting Earth ablaze?

18. Why are neutrinos so difficult to detect?

19. Explain the proton-proton chain through which the Sun generates energy by converting hydrogen to helium.

20. On Earth, nuclear power plants use *fission* to generate electricity. In fission, a heavy element such as uranium is broken into many atoms, where the total mass of the fragments is less than that of the original atom. Explain why fission could not be powering the Sun today.

21. If an abnormally large amount of hydrogen suddenly fused in the core of the Sun, what would happen to the rest of the Sun? Would the Sun change as seen from Earth?

22. Study the Process of Science Figure. If the follow-up experiments did not detect the other types of neutrinos, what would have been the next step for scientists at that point?

23. What is the solar neutrino problem, and how was it solved?

24. The Sun's visible "surface" is not a true surface, but a feature called the photosphere. Explain why the photosphere is not a true surface.

25. How are orbiting satellites and telescopes affected by the Sun?

26. Describe the solar corona. Under what circumstances can it be seen without special instruments?

27. In the proton-proton chain, the mass of four protons is slightly greater than the mass of a helium nucleus. Explain what happens to that "lost" mass.

28. What have sunspots revealed about the Sun's rotation?

29. Why are different parts of the Sun best studied at different wavelengths? Which parts are best studied from space?

30. Why is studying the interaction of the solar wind with the interstellar medium important?

Applying the Concepts

31. In Figure 14.10, density and temperature are both graphed versus height.
    a. Is the height axis linear or logarithmic? How do you know?
    b. Is the density axis linear or logarithmic? How do you know?
    c. Is the temperature axis linear or logarithmic? How do you know?

32. Using the data in Figures 14.19b and c, present an argument that sunspots occur in regions of strong magnetic field.

33. Study Figure 14.17a and Figure 14.20.
    a. Estimate the fraction of the Sun's surface covered by the large sunspot group in Figure 14.17a. (Remember that you are seeing only one hemisphere of the Sun.)
    b. From the graph in Figure 14.20, estimate the average number of sunspots that occurs at solar maximum.
    c. On average, what fraction of the Sun could be covered by sunspots at solar maximum? Is that a large fraction?
    d. Compare your conclusion with the graph of intensity in Figure 14.24. Does that graph make sense to you?

34. The Sun has a radius equal to about 2.3 light-seconds. Explain why a gamma ray produced in the Sun's core does not emerge from the Sun's surface 2.3 seconds later.

35. Assume that the Sun's mass is about 300,000 Earth masses and that its radius is about 100 times that of Earth. The density of Earth is about 5,500 kg/m^3.
    a. What is the average density of the Sun?
    b. How does that value compare with the density of Earth? With the density of water?

36. The Sun shines by converting mass into energy according to $E = mc^2$. Show that if the Sun produces 3.85×10^{26} J of energy per second, it must convert 4.3 million metric tons (4.3×10^9 kg) of mass per second into energy.

37. Assume that the Sun has been producing energy at a constant rate over its lifetime of 4.6 billion years (1.4×10^{17} seconds).
    a. How much mass has it lost creating energy over its lifetime?
    b. The current mass of the Sun is 2×10^{30} kg. What fraction of its current mass has been converted into energy over the lifetime of the Sun?

38. Suppose our Sun was an A5 main-sequence star, with twice the mass and 12 times the luminosity of the Sun, a G2 star. How long would that A5 star fuse hydrogen to helium? What would that mean for Earth?

39. Imagine that the source of energy inside the Sun changed abruptly.
    a. How long would it take before a neutrino telescope detected the event?
    b. When would a visible-light telescope see evidence of the change?

40. On average, how long do particles in the solar wind take to reach Earth from the Sun if they are traveling at an average speed of 400 km/s?

41. If a sunspot appears only 70 percent as bright as the surrounding photosphere, and the photosphere has a temperature of approximately 5780 K, what is the temperature of the sunspot?

42. The hydrogen bomb represents an effort to create a process similar to what takes place in the core of the Sun. The energy released by a 5-megaton hydrogen bomb is 2×10^{16} J.
    a. This textbook, *21st Century Astronomy*, has a mass of about 1.6 kg. If all its mass were converted to energy, how many 5-megaton bombs would it take to equal that energy?
    b. How much mass did Earth lose each time a 5-megaton hydrogen bomb was exploded?

43. Verify the claim made at the start of this chapter that the Sun produces more energy per second than all the electric power plants on Earth could generate in a half-million years. Estimate or look up how many power plants are on the planet and how much energy an average power plant produces. Be sure to account for different kinds of power, such as coal, nuclear, and wind.

44. Let's examine the reason that the Sun cannot power itself by chemical reactions. Using Working It Out 14.1 and the fact that an average chemical reaction between two atoms releases 1.6×10^{-19} J, estimate how long the Sun could emit energy at its current luminosity. Compare that estimate with the known age of Earth.

45. The Sun could get energy from gravitational contraction for a time period of $(GM^2_{Sun}/R_{Sun}L_{Sun})$. How long would the Sun last at its current luminosity? (Be careful with units!)

USING THE WEB

46. a. Go to the SDO website (http://sdo.gsfc.nasa.gov). Under "Data," select "The Sun Now" and view the Sun at many wavelengths. What activity do you observe in the images at the location of any sunspots seen in the "HMI Intensitygram" images? (You can download a free SDO app by Astra to get real-time images on your mobile device.) Look at a recent news story from the SDO website. What was observed, and why is it newsworthy?
    b. Go to the *STEREO* mission's website (https://stereo.gsfc.nasa.gov). What is *STEREO*? Where are the spacecraft located? How does that configuration enable observations of the entire Sun at once? (You can download the app "3-D Sun" to get the latest images on your mobile device.)

47. a. What are the science goals of NASA's *Interface Region Imaging Spectrograph* (*IRIS*) mission (https://www.nasa.gov/mission_pages/iris/)? What has it discovered?
    b. An older space mission, *SOHO* (Solar and Heliospheric Observatory; https://sohowww.nascom.nasa.gov), was launched in 1995 by NASA and ESA. Click on "The Sun Now" to see today's images. The Extreme Ultraviolet imaging Telescope (EIT) images are in the far ultraviolet and show violent activity. How do those images differ from the ones of SDO in question 46b?
    c. Go to the Daniel K. Inouye Solar Telescope (DKIST) website (https://atst.nso.edu). That adaptive-optics telescope under construction on Haleakala, Maui, will be the largest solar telescope. Why is studying the magnetic field of the Sun important? What are some advantages of studying the Sun from a ground-based telescope instead of a space-based telescope? What wavelengths does the DKIST observe? Why is Maui a good location? When is the telescope scheduled to be completed?

48. a. Go to the Space Weather website (http://spaceweather.com). Are any solar flares happening today? What is the sunspot number? Is it about what you would expect for this year? (In the column at left, click on "What is the sunspot number?" to see a current graph.) Are any coronal holes present today?
    b. Citizen science: Go to the Solar Stormwatch II website (https://www.zooniverse.org/projects/shannon-/solar-stormwatch-ii), a Zooniverse project from the Royal Observatory in Greenwich, England. Create an account for Zooniverse if you don't already have one. Log in and click on "Get Started" and "Project Tutorial." Mark up two sets of images (and save a screen shot for your homework).

49. a. Go to the National Ignition Facility (NIF) & Photon Science website (https://lasers.llnl.gov/about/). Under "Science," click on "How to Make a Star." How are lasers used in experiments to develop controlled nuclear fusion on Earth? How does the fusion reaction here differ from that in the Sun?
    b. An alternative approach is to fuse ^{3}He + ^{3}He instead of the hydrogen isotopes. On Earth, however, ^{3}He is in limited supply. Helium-3 is in much greater abundance on the Moon, so some people propose setting up mining colonies on the Moon to extract ^{3}He for fusion reactions on Earth. Do an Internet search on "helium 3 Moon." Which countries are talking about going to the Moon for that purpose? What is the timeline for when that might happen? What are the difficulties?

50. a. Go to https://voyager.jpl.nasa.gov/mission/status/. Where are the *Voyager* spacecraft now? Has *Voyager 2* crossed into interstellar space?
    b. Go to the website for the Interstellar Boundary Explorer (*IBEX*) (https://www.nasa.gov/mission_pages/ibex/). What has *IBEX* learned about the solar wind and the interstellar medium?

EXPLORATION

The Proton-Proton Chain

digital.wwnorton.com/astro6

The proton-proton chain powers the Sun by fusing hydrogen into helium. That fusion process produces several types of particles as by-products, as well as energy. In this Exploration, we study the steps of the proton-proton chain in detail, with the intent of helping you keep them straight.

Visit the Digital Resources Page on the Student Site and open the "Proton-Proton Chain" Interactive Simulation in Chapter 14.

Watch the animation all the way through once.

Play the animation again, pausing after the first collision. Two hydrogen nuclei (both positively charged) have collided to produce a new nucleus with only one positive charge.

1 Which particle carried away the other positive charge?

_____
_____
_____

2 What is a neutrino? Did the neutrino enter the reaction, or was the neutrino produced in the reaction?

_____
_____
_____

Compare the interaction on the top with the interaction on the bottom.

3 Did the same reaction occur in each instance?

_____
_____
_____

Resume playing the animation, pausing it after the second collision.

4 What two types of nuclei entered the collision? What type of nucleus resulted?

_____
_____
_____

5 Was charge conserved in that reaction, or did a particle have to carry charge away?

_____
_____
_____

6 What is a gamma ray? Did the gamma ray enter the reaction, or was it produced by the reaction?

_____
_____
_____

Resume the animation again, and allow it to run to the end.

7 What nuclei enter the final collision? What nuclei are produced?

_____
_____
_____

8 In chemistry, a catalyst is a species that facilitates a reaction but is not used up in the process. Do any nuclei act like catalysts in the proton-proton chain?

_____
_____
_____

Make a table of inputs and outputs. Which particles in the final frame of the animation were inputs to the reaction? Which were outputs? Fill in your table with those inputs and outputs.

9 Which outputs are converted into energy that leaves the Sun as light?

_____
_____
_____

10 Which outputs could become involved in another reaction immediately?

_____
_____
_____

11 Which output is likely to stay in that form for a very long time?

_____
_____
_____

24 Life

Throughout history, many people have wondered whether we are alone in the universe. But only in recent times has science has been able to address the underlying questions that must be answered, such as: How common are planets? What is the range of conditions under which life can thrive? How does life begin and evolve? Even answering the question "What is life?" is surprisingly complicated. Answering those questions is part of a science called **astrobiology**—the study of the origin, evolution, distribution, and future of life in the universe. Throughout this book, the "Origins" section in each chapter has discussed how astronomers think about those questions. In this chapter, we expand on some of those topics and provide a more systematic overview of how astronomers think about life in the universe and how they search for signs of it.

LEARNING GOALS

By the end of this chapter, you should be able to:

LG 1 Explain our current understanding of how and when life began on Earth and how it has evolved.

LG 2 Explain how life is a structure that has evolved through the action of the physical and chemical processes that shape the universe.

LG 3 Describe the locations in our Solar System and around other stars where astronomers think life might be possible.

LG 4 Describe some methods used to search for intelligent extraterrestrial life.

The Allen Telescope Array searches for signs of intelligent life. ▶▶▶

Are we alone in the universe?

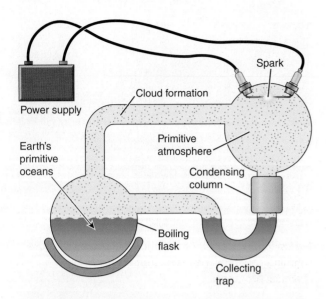

Figure 24.1 The Urey-Miller experiment was designed to simulate conditions in an early-Earth atmosphere.

Figure 24.2 (a) Life on Earth may have arisen near ocean hydrothermal vents such as this one. Similar environments might exist elsewhere in the Solar System. (b) Living organisms around hydrothermal vents, such as the giant tube worms shown here, rely on hydrothermal rather than solar energy for their survival.

24.1 Life Evolves on Earth

What is **life**? Many scientists suggest that no single definition of life would encompass all the forms of life that may exist in the universe. To date, we have discovered only a single example of life: that found here on Earth. Life on Earth may be very different from life found in other places in the universe. Indeed, even comparing varied organisms on Earth leads to complications in the definition of life. Viruses, for example, meet some criteria for life but not others. A complete definition of life may also have to take into account life-forms not yet discovered. In general, life draws energy from the environment to survive and reproduce. From studies of known life, we conclude that like planets, stars, and galaxies, life is a structure that has evolved in the universe. On Earth, all life involves carbon-based chemistry and uses liquid water as its biochemical solvent, whereas specific biological molecules such as ribonucleic acid (RNA) and deoxyribonucleic acid (DNA) enable life to reproduce and evolve. In this section, we will briefly review what is known about the origin and evolution of life on Earth.

The Origin of Life on Earth

How did life begin on Earth? Recall from Chapter 7 that Earth's secondary atmosphere was formed in part by carbon dioxide and water vapor emitted by volcanoes. Comets and asteroids probably added large quantities of water, methane, and ammonia to the mix. Liquid water is considered essential for any terrestrial-type life to get its start and evolve because water is an effective solvent that can move other atoms and molecules around, making them more accessible to cells. Early Earth had abundant sources of energy, such as lightning and ultraviolet solar radiation, that fragmented those simple molecules. Those fragments later reassembled into molecules of greater mass and complexity. Some of those were organic molecules—that is, molecules that contained hydrogen and carbon. Rain carried the heavier molecules out of the atmosphere into Earth's oceans, forming a primordial soup.

In 1952, chemists Harold Urey (1893–1981) and Stanley Miller (1930–2007) attempted to create conditions similar to those early-Earth conditions. Using equipment illustrated in **Figure 24.1**, they placed water in a sterilized laboratory jar to represent the ocean and then added methane, ammonia, and hydrogen as a primitive atmosphere; electric sparks simulated lightning as a source of energy. Within a week, the Urey-Miller experiment yielded molecules associated with life: amino acids and components of nucleic acids. Proteins, the structural molecules of life, are made of 20 amino acids. Eleven of those acids were synthesized in the Urey-Miller experiment. Nucleic acids are the precursors of RNA and DNA.

Additional sealed samples from that old experiment were examined 50 years later. In those samples, hydrogen sulfide had been added to the "primitive atmosphere." When the samples were analyzed, 23 amino acids were found. That finding suggests that hydrogen sulfide, which would have come from volcanic plumes in the early Earth, was important. More recent experiments with carbon dioxide and nitrogen as the primitive atmosphere have produced results similar to those of Urey and Miller. A plausible atmospheric composition with an energy source can produce significant quantities of amino acids and other substances important to life.

From laboratory experiments such as those, scientists have developed various models to explain how life might have begun in an early-Earth environment. However, the details of how those precursor molecules evolved into the molecules of life are not yet clear. Some biologists think life began in the ocean depths, where volcanic vents provided the localized heat needed to create the highly organized molecules responsible for biochemistry (**Figure 24.2**). Other researchers think that life originated in tide pools, where lightning and ultraviolet radiation supplied the energy (**Figure 24.3**). In either case, short strands of molecules that could replicate themselves may have formed first, later evolving into RNA, and finally into DNA, the huge molecule that serves as the biological "blueprint" for self-replicating organisms.

A few scientists have suggested that life on Earth may have been "seeded" from space in the form of microorganisms brought here by meteoroids or comets. However, no scientific evidence exists to support the seeding hypothesis. In addition, although that hypothesis might explain how life came to Earth, it does not explain how life itself began.

When Life Began

If life did indeed get its start in Earth's oceans, when did it happen? Recall from Chapter 7 that Earth was bombarded by Solar System debris for several hundred million years after it formed roughly 4.6 billion years ago. Those conditions might have been too harsh for life to form and evolve on Earth. Once the bombardment abated and oceans formed, the opportunities for life to begin greatly improved. Terrestrial life seems to have quickly taken advantage of that more favorable environment. In the Nuvvuagittuq belt in Quebec, Canada, scientists recently found what they think are fossilized microorganisms that are 3.77 billion to 4.28 billion years old and seem similar to microorganisms from modern hydrothermal vents. Scientists debate whether carbonized material in Greenland rocks dating back 3.65 billion to 3.85 billion years provides indirect evidence of early life. Stronger and more direct evidence for early life appears in the form of fossilized **stromatolites** (masses of simple microorganisms) that date back about 3.5 billion years. Fossilized stromatolites have been found in western Australia and southern Africa, and living examples still exist today (**Figure 24.4**). An analysis of eleven 3.465-billion-year-old microfossils from western Australia found five species—two of which carried out primitive photosynthesis and the other three produced or consumed methane. That evidence suggests that the earliest life formed less than a billion years after the Solar System formed and within a few hundred million years of the end of the late heavy-bombardment period.

All life on Earth shares a similar genetic code that originated from a common ancestor. Close comparison of DNA of different species enables biologists to trace backward to the time when different types of life first appeared on Earth and to identify the species from which those life-forms evolved. The earliest organisms were **extremophiles**—life-forms that not just survive but actually thrive under extreme environmental conditions. Extremophiles include organisms such as thermophiles, which flourish in water temperatures as high as 120°C that occur near deep-ocean hydrothermal vents, such as the one shown in Figure 24.2a. Other extremophiles thrive in conditions of extraordinary cold, low oxygen, salinity, pressure, dryness, acidity, or alkalinity. Scientists today study extremophiles in boiling-hot sulfur springs in Yellowstone National Park, in salt crystals beneath the Atacama Desert in Chile, at the bottoms of glaciers, in ice fields in the Arctic, and in other extreme environments (**Figure 24.5**).

Among the early life-forms was an ancestral form of **cyanobacteria**, single-celled organisms otherwise known as *blue-green algae*. Those microorganisms

Figure 24.3 Life may have begun in tide pools, where lightning and ultraviolet light supply the energy for chemical processes.

Figure 24.4 These modern-day stromatolites are growing in colonies along an Australian shore.

Figure 24.5 These thermophiles in the Grand Prismatic Spring in Yellowstone National Park live in temperatures of 70°C. The colors result from different amounts of chlorophyll.

(a)

(b)

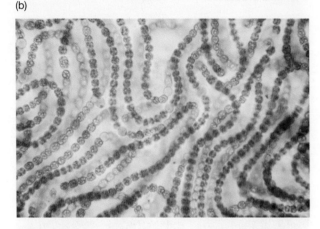

Figure 24.6 (a) Cyanobacteria today form sheets on lakes and other bodies of water. (b) Under a microscope, the individual microorganisms are visible.

Figure 24.7 The amount of oxygen in Earth's atmosphere has built up as a result of cyanobacteria and plant life on the planet.

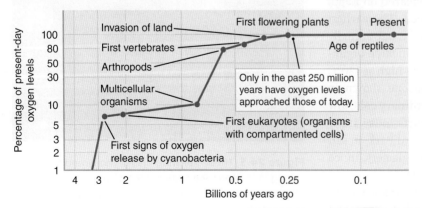

form extensive sheets on the surface of bodies of water, as shown in **Figure 24.6a**. Under a microscope, it becomes clear that those sheets are colonies of individual microorganisms, as seen in **Figure 24.6b**. Cyanobacteria **photosynthesize**, using sunlight and carbon dioxide as food and generating oxygen as a waste product. Initially, the highly reactive oxygen that cyanobacteria produced was quickly removed from Earth's atmosphere by oxidation, or rusting, of surface minerals. Once most of the exposed minerals were oxidized and could no longer absorb oxygen, atmospheric levels of oxygen began to rise. Oxygenation of Earth's atmosphere and oceans began about 2 billion years ago, and the current level was reached only about 250 million years ago, as shown in **Figure 24.7**. Without cyanobacteria and other photosynthesizing organisms, Earth's atmosphere would be as oxygen-free as the atmospheres of Venus and Mars.

Biologists comparing DNA sequences find that terrestrial life is divided into two types: prokaryotes and eukaryotes. Prokaryotes, which include Bacteria and Archaea, are simple organisms that consist of free-floating DNA inside a cell wall; as shown in **Figure 24.8a**, they lack both cell structure and a nucleus. Eukaryotes, which form the cells in animals, plants, and fungi, have a more complex form of DNA contained within the cell's membrane-enclosed nucleus, illustrated in **Figure 24.8b**. The first eukaryote fossils date from about 2 billion years ago, coincident with the rise of free oxygen in the oceans and atmosphere, although the first multicellular eukaryotes did not appear until a billion years later.

Increasing Complexity

Scientists have used DNA sequencing to establish what is known as the "phylogenetic tree of life," shown in **Figure 24.9**. That complex tree describes the evolutionary interconnectivity of all species of Bacteria, Archaea, and Eukarya and has revealed some interesting relationships. For example, Archaea were initially thought to be the same as Bacteria, but genetic studies show they diverged long ago, and the Archaea have genes and metabolic pathways more similar to those of Eukarya than to those of Bacteria. On the macroscopic scale, the phylogenetic tree places animals closest to fungi, which branched off the evolutionary tree after slime molds and plants.

Living creatures in Earth's oceans remained much the same—a mixture of single-celled and relatively primitive multicellular organisms—for more than 3 billion years after terrestrial life appeared. Between 540 million and 500 million years ago, the number and diversity of biological species increased spectacularly. Biologists call that event the **Cambrian explosion**. The trigger of that sudden surge in biodiversity remains unknown, but possibilities include rising oxygen levels, an increase in genetic complexity, major climate change, or some combination of those factors. The "Snowball Earth" hypothesis suggests that before the Cambrian explosion, Earth was in a period of extreme cold between about 750 million and 550 million years ago and was covered almost entirely by ice. During that period of extreme cold, predatory animals died out, making it easier for new species to adapt and thrive. Another possibility is that the marked increase in atmospheric oxygen (O_2) would have been accompanied by a corresponding increase in stratospheric ozone (O_3), which shields Earth's surface from deadly solar ultraviolet radiation. With that protective ozone layer in place, life was free to leave the oceans and move to land. *Tiktaalik*, a fish with limblike fins and ribs, was an animal in a midevolutionary step of leaving the water for dry land, as shown in the artist's illustration in **Figure 24.10**.

The first plants appeared on land about 475 million years ago. Large forests and insects go back 360 million years. The age of dinosaurs began 230 million years ago and ended abruptly 65 million years ago, when an asteroid or comet collided with Earth. The collision threw so much dust into the atmosphere that the sunlight was dimmed for months, causing the extinction of more than 70 percent of all existing plant and animal species. Mammals were the big winners in the aftermath. Primates evolved from the last ancestor common with other mammals about 70 million years ago. The great apes (gorillas, chimpanzees, bonobos, and orangutans) split off from the lesser apes about 20 million years ago (Figure 24.9, inset). DNA tests show that humans and chimpanzees share about 98 percent of their DNA, indicating that they evolved from a common ancestor about 6 million years ago. By comparison, all humans share 99.9 percent of their DNA. The earliest human ancestors appeared a few million years ago, and the first civilizations occurred a mere 10,000 years ago. Present-day industrial society, barely more than two centuries old, is but a moment in the history of life on Earth.

Humans are here today because of a series of events that occurred throughout the history of the universe. Some of those events are common in the universe, such as the formation of heavy elements in earlier generations of stars and the formation of planets. Other events in Earth's history may have been less likely to happen elsewhere, such as the formation of a planet with life-supporting conditions like Earth or the development of self-replicating molecules that led to Earth's earliest life. A few events stand out, such as major extinctions that allowed the evolution of mammalian life and, ultimately, human beings.

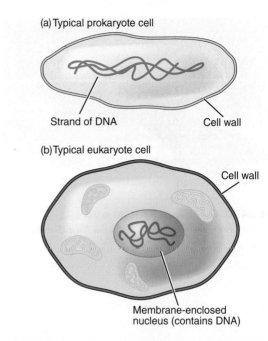

Figure 24.8 (a) A simple prokaryote cell contains little more than the cell's genetic material. (b) A eukaryote cell contains several membrane-enclosed structures, including a nucleus that houses the cell's genetic material.

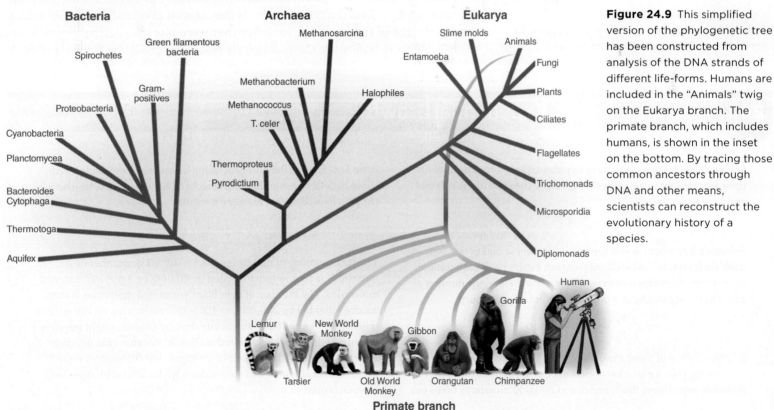

Figure 24.9 This simplified version of the phylogenetic tree has been constructed from analysis of the DNA strands of different life-forms. Humans are included in the "Animals" twig on the Eukarya branch. The primate branch, which includes humans, is shown in the inset on the bottom. By tracing those common ancestors through DNA and other means, scientists can reconstruct the evolutionary history of a species.

Figure 24.10 This illustration is an artist's reconstruction from a fossil of *Tiktaalik* found in the Canadian Arctic.

Evolution as a Mechanism of Change

Imagine that just once during the first few hundred million years after Earth formed, a single molecule formed somewhere in Earth's oceans. That molecule had a very special property: chemical reactions between that molecule and other molecules in the surrounding water caused the molecule to make a copy of itself. The molecule became *self-replicating*. Chemical reactions then produced copies of each of those two molecules, making four molecules. Four molecules became eight, eight became 16, 16 became 32, and so on. By the time the original molecule had copied itself just 100 times, more than a million trillion trillion (10^{30}) of those molecules existed. That is about 100 million times more of those molecules than the number of stars in the observable universe. The molecules of DNA that make up the chromosomes in the nuclei of the cells of all advanced life today are direct descendants of those early self-replicating molecules that flourished in the oceans of the young Earth.

Over time, not all replications are exact. The likelihood that a copying variation will occur while a molecule is replicating increases significantly with the number of copies being made. For DNA, which contains the genetic code for an entire organism, a change in the genetic code is called a **mutation**. Sometimes a mutation has no effect. Other times, a mutation can prevent an organism from flourishing. In still other cases, a mutation can make an organism better suited to its environment. Organisms with those advantageous mutations will survive to reproduce successfully. Even if mutations are rare and only a small fraction turn out to be beneficial, after just 100 generations trillions of mutations occur that, by luck, might improve on the original (**Working It Out 24.1**). **Heredity**—the ability of one generation to pass on its genetic code to future generations—allow beneficial mutations to persist and be incorporated into a species' genetic code.

As the organisms of the early Earth continued to interact with their surroundings and make copies of themselves, mutations caused them to diversify into many species. Sometimes the resources they needed to reproduce became scarce. In the face of that scarcity, varieties that were more successful reproducers became more numerous. Competition for resources, predation by one species on

24.1 Working It Out Exponential Growth

Self-replication is an example of exponential growth. The doubling time, *n*, for exponential growth is given by the ratio of the original and final amounts:

$$\frac{P_F}{P_O} = 2^n$$

Assume a hypothetical self-replicating molecule makes one copy of itself each minute, and each copy in turn copies itself each minute. How many molecules will exist after an hour? Here, the number of generations is given by $n = 60$, for 60 minutes in an hour:

$$\frac{P_F}{P_O} = 2^{60} = 1.2 \times 10^{18}$$

A billion billion of those molecules will exist after 1 hour.

Now suppose a mutation occurs once every 50,000 times that a molecule reproduces itself, and one of 200,000 mutations turns out to be beneficial. After 100 generations, how many molecules with those beneficial mutations might exist? That equation is similar to the previous equation, but here $n = 100$:

$$\frac{P_F}{P_O} = 2^{100} = 1.3 \times 10^{30}$$

The total number of molecules is 1.3×10^{30}. The number of mutations is that number divided by 50,000, or 2.6×10^{25} mutated molecules. The number of beneficially mutated molecules is that number divided by 200,000, or 1.3×10^{20} molecules. So 100 million trillion (10^{20}) mutations will occur that, by chance, might improve the survivability of the original molecule. Because that number does not count earlier beneficial changes that themselves replicated, the total number of molecules with beneficial changes will be even larger!

another, and cooperation between organisms became important to the survival of different varieties. Some varieties were more successful and reproduced to become more numerous, whereas less successful varieties became less and less common. That process, in which better-adapted organisms reproduce and thrive, whereas less well-adapted organisms become extinct, is called **natural selection**.

Life has existed on Earth for about 4 billion years, which is a very long time—long enough for the combined effects of heredity and natural selection to shape the descendants of that early self-copying molecule into a huge variety of complex, competitive, successful structures. Geological processes on Earth have preserved a fossil record of the history of some of those structures (**Figure 24.11**). Among those descendants are human "structures" that can think about their own existence and unravel the mysteries of the stars.

CHECK YOUR UNDERSTANDING 24.1

Extremophiles are organisms that: (a) are extremely reactive; (b) are extremely rare; (c) have an extreme quality, such as mass or size; (d) live in extreme conditions.

Figure 24.11 Fossils, such as this *Parasaurolophus* ("near crested lizard"), record the history of the evolution of life on Earth. This 10 meter long plant-eating dinosaur lived in North America about 75 million years ago.

24.2 Life Involves Complex Chemical Processes

The evolution of life on Earth cannot be separated from the narrative of astronomy: it is one of many examples of the emergence of structure in an evolving universe (see the **Process of Science Figure**). That emergence leads naturally to a profound question: Has life arisen elsewhere? Unlike the study of planets, stars, and galaxies, only one known case exists for the study of life—Earth—and scientists do not know how much can be generalized to other places. To explore that question, we need to take a closer look at the processes that have led to life on Earth. In this section, we explore the chemical and physical properties of life on Earth.

The infant universe was composed basically of hydrogen and helium and very little else. After 9 billion years of stellar nucleosynthesis, all the heavier chemical elements essential to life were present and available in the molecular cloud that gave birth to the Solar System. Those heavier elements were formed by nuclear fusion in the cores of earlier generations of stars and were then dispersed into space. At times, that dispersal was passive. For example, low-mass stars such as the Sun, when they become puffed-up, dying red giants, may shed their extended atmospheres, sending some newly created carbon into space. Other dispersals were more violent. High-mass stars produce even heavier elements through nucleosynthesis in their cores—up to and including iron. But some of the trace elements essential to biology on Earth are even more massive than iron. They are produced within a matter of minutes during the violent supernova explosions that mark the death of high-mass stars and then are thrown into the chemical mix found in molecular clouds. Or they are produced in the collisions of neutron stars, the dense remains of prior supernova explosions (**Figure 24.12**).

All known living organisms on Earth are composed of a more or less common suite of complex chemicals. Approximately two-thirds of the atoms in the

Figure 24.12 This version of the periodic table shows the astronomical origin of the naturally occurring elements (up to uranium) in the Solar System.

Big bang fusion	Cosmic ray fission
Merging neutron stars	Exploding massive stars
Dying low mass stars	Exploding white dwarfs

PROCESS OF SCIENCE

All of Science Is Interconnected

More than most other subjects, astrobiology relies on concepts from a number of scientific fields—showing that all of science is interconnected.

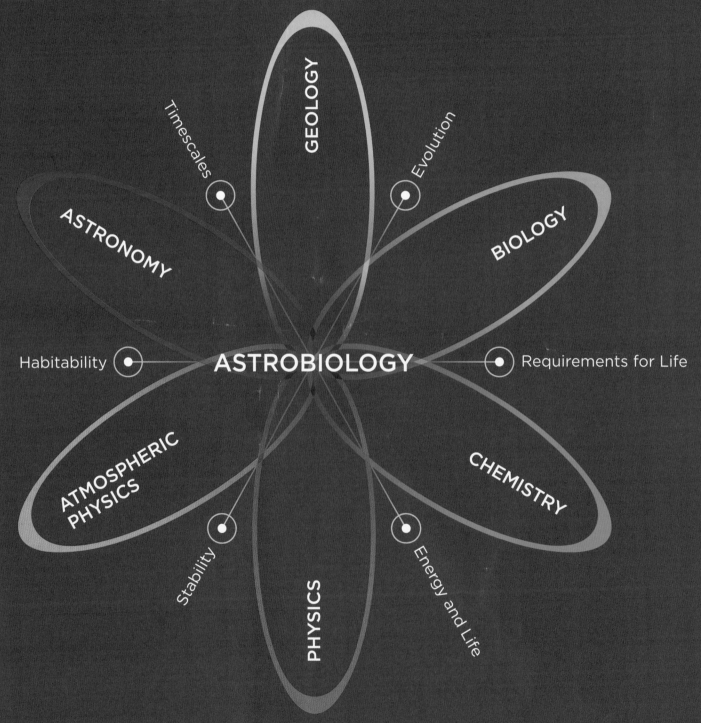

No science stands alone. All are connected. Interdisciplinary fields of study like astrobiology provide opportunities for new tests for theories of many fields.

human body are hydrogen (H), about one-fourth are oxygen (O), a tenth are carbon (C), and a few hundredths are nitrogen (N). Carbon, nitrogen, and oxygen are the three most abundant products of stellar nucleosynthesis after helium (see the "astronomer's periodic table" in Figure 5.15). The several dozen remaining atomic elements in the human body make up only 0.2 percent of the total. All known living creatures are assemblages of molecules composed almost entirely of those four elements, sometimes called CHON (carbon, hydrogen, oxygen, nitrogen), along with small amounts of phosphorus and sulfur. Some of those molecules are enormous. Consider DNA, which is responsible for genetic codes, illustrated in **Figure 24.13**. DNA is made up entirely of only five atomic elements: CHON and phosphorus. But the DNA in each cell of the human body is composed of combinations of *tens of billions* of atoms of those same five elements. Proteins, the huge molecules responsible for the structure and function of living organisms, are long chains of smaller molecules called amino acids. Terrestrial life uses 20 specific amino acids, which also consist of no more than five atomic elements—CHON plus sulfur.

Although life primarily uses the half-dozen atoms of CHON plus sulphur and phosphorus, some other elements, present in smaller amounts, are essential to the chemical processes that living organisms carry out. Those elements include sodium, chlorine, potassium, calcium, magnesium, iron, manganese, and iodine. Trace elements such as copper, zinc, selenium, and cobalt also play a crucial role in biochemistry but are needed in only tiny amounts.

Notice that carbon, which can bond to four other atoms or molecules, forms the backbone of the DNA molecule shown in Figure 24.13. That is why carbon is so important to life on Earth. Forms of extraterrestrial life could exist that are also carbon-based but have chemistries different from that of life on Earth. For example, countless varieties of amino acids exist in addition to the 20 used by terrestrial life. Most other atoms are more limited than carbon in the number of bonds they can make, but silicon, like carbon, can bond to four other atoms, so that many combinations are possible. As a potential life-enabling atom, silicon has both advantages and disadvantages in comparison with carbon. Silicon-based molecules remain stable at much higher temperatures than carbon-based molecules, perhaps enabling possible silicon-based life to thrive in high-temperature environments, such as on planets that orbit close to their parent star. But silicon is also a larger and more massive atom than carbon: it cannot form molecules as complex as those based on carbon. Any silicon-based life probably would be simpler than life-forms here on Earth, but it might exist in high-temperature niches somewhere within the universe. Although carbon's unique properties make it readily adaptable to the chemistry of life on Earth, other types of life might be found elsewhere.

Figure 24.13 DNA, the heritable molecule that forms the basis for life on Earth, contains the atoms of only five elements. Even so, those atoms are combined in billions of ways, giving rise to the diversity of life on Earth.

CHECK YOUR UNDERSTANDING 24.2

Carbon is a favorable base for life because: (a) it can bond to many other atoms in long chains; (b) it is nonreactive; (c) it forms weak bonds that can be readily reorganized as needed; (d) it is organic.

24.3 Where Do Astronomers Look for Life?

One approach to the scientific search for extraterrestrial life is to use robotic spacecraft to explore the planets and moons of the Solar System (see Chapter 6). Spacecraft have visited all the planets and some moons and sent back at least some information about the conditions on those worlds. Another approach is to use telescopes to detect planets outside our Solar System (see Chapter 7). Telescopes on the ground and in space have detected several thousand planets orbiting other stars. In this section, we will survey the locations where life might be found, both in our own Solar System and around other stars.

Life within Our Solar System

Scientists start the search for evidence of extraterrestrial life here in our own Solar System. Early conjectures about life in our Solar System seem naïve, considering what we now know. Two centuries ago, the eminent astronomer Sir William Herschel, discoverer of Uranus, proclaimed, "We need not hesitate to admit that the Sun is richly stored with inhabitants." In 1877, astronomer Giovanni Schiaparelli (1835–1910) observed what appeared to be linear features on Mars and dubbed them *canali* ("channels" in Italian). The famous observer of Mars, Percival Lowell (1855–1916), misinterpreted Schiaparelli's *canali* as "canals," suggesting that they were constructed by intelligent beings.

Because Mercury and the Moon lacked atmospheres, astronomers determined that those worlds were not conducive to life. The giant planets and their moons were thought to be too remote and too cold to sustain life. By the 1960s it was understood that the surface of Venus is too hot to permit life. At the same time, astronomers using ground-based telescopes had discovered that Mars has a thin atmosphere, water ice, and carbon dioxide ice. During the 1960s, the United States and the Soviet Union sent reconnaissance spacecraft to the Moon, Venus, and Mars, but the instruments on those spacecraft probed the physical and geological properties of those astronomical bodies instead of searching for life. Serious efforts to look for signs of life—past or present—require more advanced spacecraft with specialized instrumentation.

In the mid-1970s, two American *Viking* spacecraft were sent to Mars with detachable landers containing a suite of instruments designed to find evidence of a terrestrial type of life. When the *Viking* landers failed to find convincing evidence of life on Mars, hopes faded for finding life on any other body orbiting the Sun. Since then, however, further exploration of the Solar System has generated renewed optimism. A better understanding of the history of Mars indicates the planet's climate has changed. Mars was once wetter and warmer than it is today.

In the 1990s, Mars missions began to map the planet's surface from the ground and from space. In 2008, NASA's *Phoenix* spacecraft landed at a far-northern latitude, inside the planet's arctic circle, where specialized instruments dug into and analyzed the martian water-ice permafrost. *Phoenix* found that the martian arctic soil has a chemistry similar to that of the Antarctic dry valleys on Earth, where life exists deep below the surface at the ice-soil boundary. Minerals that form in water, such as calcium carbonate, have been detected. That finding suggests that oceans existed in the past on Mars. However, *Phoenix* found no evidence of life.

The *Curiosity* rover (originally known as the *Mars Science Laboratory*) landed in Gale Crater on Mars in 2012. That rover studies the rocks and soil of Mars to acquire data for a better understanding of the history of the planet's climate and geology. Shortly after landing, *Curiosity* found evidence that a stream of liquid water had once flowed in the crater. The rover observed rounded, gravelly pebbles stuck together, which have been interpreted as coming from a stream that varied at times from ankle-deep to hip-deep, and moved at about 1 meter per second (**Figure 24.14**). The rover also found sedimentary rocks containing clay, which suggested the presence of a freshwater lake bed at one time. Later observations found that the surface soil contained up to 2 percent water by weight—or about 1 quart per cubic foot of

Figure 24.14 The Mars *Curiosity* rover detected evidence that Mars had a watery past. The rounded gravel surrounding the bedrock suggests that an ancient, flowing stream once existed.

martian dirt. Those conditions are too dry to support Earthlike plant life, which permanently wilts in soil that is less than about 10% water by weight.

Curiosity also detected evidence of organic molecules in 3-billion-year-old rocks located in that crater (**Figure 24.15**). Those organic molecules could have come from space on asteroids and comets, from geological processes on Mars, or from ancient life. *Curiosity* also detected seasonal variation in the amount of methane gas within the crater; methane can come from water-rock chemistry or from life. Thus, a water lake inside Gale Crater could have been hospitable to life billions of years ago, but astrobiologists do not know whether any life existed.

The *Mars Atmosphere and Volatile EvolutioN* (*MAVEN*) mission, which arrived at Mars in September 2014, is studying the upper atmosphere to learn more about the escape of carbon dioxide, hydrogen, and nitrogen from the planet's atmosphere and how losing those gases affected surface pressure and the existence of liquid water. Recently the *Mars Reconnaissance Orbiter* (*MRO*) found water ice on eight steep, eroded slopes located at midlatitudes on Mars (**Figure 24.16**). *Mars Express* may have found evidence of a lake of liquid water under the south polar ice cap. Future experiments will look for liquid water—and fossil or living microorganisms—below the martian surface. Eventually other missions will return samples of Martian rock and soil to Earth for more advanced analysis.

NASA's instrumented robotic spacecraft reached the outer Solar System starting in the 1970s, and many astrobiologists were surprised by the findings. Although the outer planets themselves did not appear to be habitats for life, some of their moons became objects of special interest. Jupiter's moon Europa is covered with a layer of water ice that appears to overlie a great ocean of briny liquid water (**Figure 24.17**). The water remains liquid because of high pressure and tidal heating by Jupiter. Impacts by comet nuclei may have added a mix of organic material, another essential ingredient for life. Once thought to be a frozen, inhospitable world, Europa is now a candidate for biological exploration. Recently, scientists using the Hubble Space Telescope observed water geysers taller than Mount Everest erupting from the icy surface of Europa. Ejected material from those geysers may make searching for life on Europa possible without drilling through the ice. NASA's *Europa Clipper* mission, scheduled for launch in the 2020s, will fly by Europa a few dozen times. Jupiter's moons Ganymede and Calisto also have icy surfaces and possibly subsurface salty oceans.

Figure 24.15 A "selfie" of the *Curiosity* Mars rover at the site where it drilled into some rocks and identified organic compounds.

Figure 24.16 This image from the *MRO* shows 80 meters of exposed water ice on a steep slope. The color is exaggerated. Similar steep slopes with ice at other midlatitude locations also were seen.

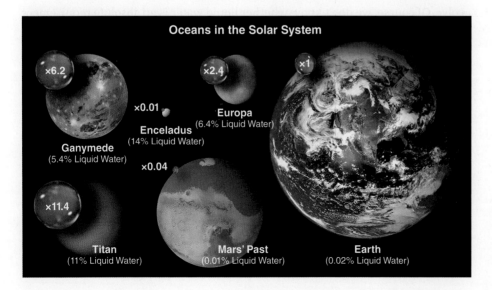

Figure 24.17 The total amount of liquid water (blue balls) on Jupiter's moons and Saturn's moon Titan compared with the amount on Earth. All figures are drawn to scale and assume average ocean depths of 4 km (Earth), 100 km (Europa), and 200 km (Titan).

Saturn's moon Titan has an atmosphere rich in organic chemicals, many of which are thought to be precursor molecules of a type that existed on Earth before life appeared here. A probe from the *Cassini* spacecraft in orbit around Saturn descended through Titan's atmosphere and found additional evidence for a variety of molecules that might be necessary for life, as well as liquid lakes of methane on the surface and probably a liquid-water ocean under the surface. The *Cassini* spacecraft also detected water-ice crystals spouting from cryovolcanoes (which erupt ice crystals instead of rocks) near the south pole of Saturn's tiny moon Enceladus. Liquid water must lie beneath its icy surface, and Enceladus therefore is also a possible habitat of extremophile life, perhaps life similar to that found near hydrothermal vents deep within Earth's oceans.

The discovery of life on even one Solar System body beyond Earth would be exciting. If scientists discover that life arose independently *twice* in the same planetary system, that finding could suggest that the spontaneous appearance of life is not rare at all.

CHECK YOUR UNDERSTANDING 24.3A

Which of the following Solar System objects is *not* a good candidate for future searches for life? (a) Mars; (b) Jupiter's moon Europa; (c) Saturn's moon Titan; (d) Uranus

Habitable Zones

Recall from Chapter 7 that thousands of confirmed and thousands of candidate exoplanets exist within the Milky Way Galaxy. To decide which planets to focus on for further study, astronomers are narrowing the possibilities by searching for planets with environments conducive to the formation and evolution of life, as we understand it, while eliminating clearly unsuitable planets. Astronomers consider issues such as each planet's orbit, inferred temperature, distance from its star, and location in the galaxy. One criterion astrobiologists look for is stability of planetary systems. As noted in Chapters 2, 3, and 4, astronomers think about the effects of a planet's rotation and orbit. Planets in stable systems have nearly circular orbits that preserve relatively uniform climatological environments. Planets in very elliptical orbits or planets with a large axial tilt can experience more intense temperature swings that could be detrimental to the survival of life. A stable temperature that maintains the existence of water in a liquid state might be important. We know that liquid water was essential for life on Earth to form and evolve. Of course, we don't know whether liquid water is an absolute requirement for life elsewhere, but it's a reasonable starting assumption.

In the "Origins" sections of Chapters 7 and 13, we discussed the idea of the habitable zone, the location of a planet relative to its parent star that provides a range of temperatures in which liquid water can exist on the surface. On planets too close to their parent stars, water would exist only as a vapor—if at all. On planets too far from their stars, water would be permanently frozen as ice. Planet size is another consideration: Large gas giants retain most of their light gases during formation and have no solid surface. Small planets may be rocky or a mix of water, rock, and ice: measuring the mass and radius enables scientists to estimate the density. Very small planets may not have enough surface gravity to retain their atmospheric gases and so end up like our Moon. Calculating whether any particular planet is in the habitable zone is complicated. Recall from Chapter 7 that even if a planet is located in the habitable zone, that means only that liquid water *could* exist on the surface: it does not mean that astronomers have confirmed the presence of liquid water or that the planet has inhabitants.

In our own Solar System, Venus, which orbits at 0.7 times Earth's distance from the Sun, has become an inferno because of its runaway greenhouse effect. Any liquid water that might once have existed on Venus has long since evaporated and been lost to space. Mars orbits about 1.5 times farther from the Sun than the orbit of Earth, and the water that we see on Mars today is nearly always frozen. But the orbit of Mars is more elliptical and variable than Earth's, giving the planet a greater variety of climate, including long-term cycles that might occasionally permit liquid water to exist. Most astrobiologists put the habitable zone of our Solar System at about 0.9–1.4 astronomical units (AU), which includes Earth but just misses Venus and Mars. However, that range ignores the possibility of liquid water under ice, as occurs on the moons of the outer Solar System.

Astronomers also must think about the type of star they are observing in their search for planets that could have liquid water. Stars less massive than the Sun and thus cooler will have narrower habitable zones. A planet in the habitable zone of a cool star is close in to its star. As a result, the planet is more likely to be tidally locked to the star so that no day/night cycle exists. Stars more massive than the Sun are hotter and will have a larger and more distant habitable zone (**Figure 24.18**). However, massive stars have shorter main-sequence lifetimes and might not last long enough for evolution to take place. For example, a star of 3 M_{Sun} has a lifetime of only a few hundred million years. Here on Earth, a billion years was long enough for bacterial life to form and cover the planet, but insufficient for anything more advanced to evolve. It took 3.5 billion years of evolution on Earth to reach the period of the Cambrian explosion. Even though evolution might happen at a different pace elsewhere, stellar lifetime is still a sufficiently strong consideration that astronomers focus their efforts on stars with longer lifetimes—specifically, stars of 0.6–1.4 M_{Sun}, which corresponds to spectral types F, G, K, and M. Most stars in our galaxy are low-mass M main-sequence stars, and many of those seem to have small, rocky planets. One example is the Trappist-1 system of seven planets in close orbits to their cool M dwarf star (**Figure 24.19**). Some of those planets could be tidally locked (see the Chapter 4 "Origins") to their star. Two or three of the planets are in the star's habitable zone.

Another factor to consider is a planet's atmosphere. A planet's ability to keep an atmosphere depends on the planet's mass and radius (and therefore its escape velocity) and its temperature. Very small planets may not have enough surface gravity to retain their atmospheric gases. In the inner Solar System, the Moon and Mercury were too small to keep any atmosphere. Mars lost most of its atmosphere, but the larger Earth and Venus kept thick atmospheres. Another important consideration is the greenhouse effect, which traps heat underneath an atmosphere and raises the temperature

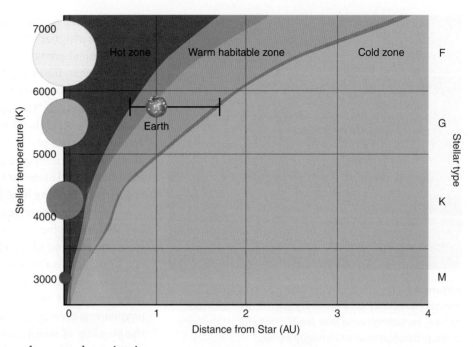

Figure 24.18 The habitable zone changes with the mass and temperature of a star. Habitable zones around hot, high-mass stars are larger and more distant than the zones around cooler, lower-mass stars.

Figure 24.19 The Trappist-1 system. The Trappist planets are very close to their star: their distances are more comparable to those of Jupiter and its large moons than to our Solar System planets. (The star is not to scale.)

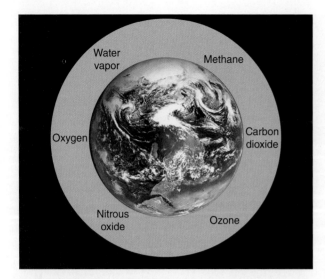

Figure 24.20 Spectroscopy of rocky exoplanet atmospheres may pick up many chemicals important for life, including some that are biosignatures, thought to be produced overwhelmingly by life.

on a planetary surface. That has happened on Venus, Earth, and Mars, each of which has a higher surface temperature because of its atmosphere. The thickness and chemical content of the atmosphere affect the strength of the greenhouse effect, so that, for example, Venus is much hotter than its distance from the Sun would suggest because of its thick atmosphere of carbon dioxide (see Figure 5.24 and the Chapter 5 "Origins"). Some of those exoplanets might be more like Venus than like Earth if they have atmospheres filled with greenhouse gases. The total amount of atmosphere affects the atmospheric pressure at the surface, which, along with the temperature, determines whether water (or other molecules) can exist in a liquid state on the surface. The current thin atmosphere of Mars does not permit standing liquid water on its surface.

In the next few years, with new telescopes, many more observations will identify atmospheres on exoplanets. Water vapor, hydrogen, and carbon monoxide have been found on a few exoplanets already. In particular, discovering oxygen in the atmosphere of an exoplanetary atmosphere would be exciting, but not definitive. The oxygen in Earth's atmosphere makes it stand out from the rest of the planets and moons in the Solar System, and we know that most terrestrial oxygen was created by photosynthetic life. However, oxygen also can come from the breakup of water molecules, so the presence of oxygen alone doesn't necessarily mean life exists there. Aside from oxygen, astrobiologists are thinking about other possible **biosignatures**—chemical compounds that would show up in the spectra of an exoplanet atmosphere that would suggest life may be present (**Figure 24.20**).

Astrobiologists continue to develop ways to classify planets more clearly as they try to narrow the possibilities about where life might exist. One such classification uses the currently available data on an exoplanet to estimate how much it is like Earth. Factors include the radius, density, escape velocity, and surface temperature. That is an Earth-centric approach based on the experience of life on Earth. Another measure aims to be less Earth-centric and to broaden the options for habitability, but it depends on factors not yet measured or measurable for most exoplanets. That measure depends on whether the planet has a surface on which organisms can grow, as well as the right kind of chemistry, a source of energy, and the ability to hold a liquid solvent. Saturn's moon Titan or Jupiter's moon Europa might satisfy those conditions, and Mars might have done so in the past. Over the next decade, improvements in observations will probably lead to enough information that at least some exoplanets can be classified in that way. As observations of exoplanets become more complete, astrobiologists will undoubtedly develop new classification schemes that are more accurate and informative.

Astronomers also consider the *galactic habitable zone*—the idea that the Milky Way Galaxy may have some locations where planets might have a higher probability of hosting life. Stars situated too far from the galactic center may be without enough heavy elements—such as oxygen, silicon (silicates), iron, and nickel—in their protoplanetary disks to form rocky planets like Earth. Conversely, regions too close to the galactic center experience less star formation and therefore fewer opportunities to gather heavy elements into planetary environments. Stars too close to the galactic center may be affected by the high-energy radiation environment (X-rays and gamma rays from supermassive black holes or gamma-ray bursts), which can damage RNA and DNA. Stars that migrate within the galaxy and change their distance from the galactic center may move in and out of any galactic habitable zone.

CHECK YOUR UNDERSTANDING 24.3B

The habitable zone around a star depends most on the star's: (a) mass and age; (b) radius and distance; (c) age and radius; (d) color and distance; (e) luminosity and velocity.

24.4 Scientists Are Searching for Signs of Intelligent Life

Are we alone? Scientists approach that question from many directions. Biologists consider the origin and evolution of life and the definition of intelligence. Astronomers send messages and search for alien signals in the vast array of astronomical data. In this section, you will learn about the search for intelligent life and how scientists think about the probability of finding it.

Sending Messages

During the 1970s, messages were sent from Earth to space. The *Pioneer 11* spacecraft, which will probably spend eternity drifting through interstellar space, carries the plaque shown in **Figure 24.21**. It pictures humans and the location of Earth for any future interstellar traveler who might happen to find it and understand its content. Another message to the cosmos accompanied the two *Voyager* spacecraft on identical phonograph records that contained greetings from planet Earth in 60 languages, samples of music, animal sounds, and a message from then-President Jimmy Carter. Some politicians were concerned that scientists were dangerously advertising our location in the galaxy, even though radio signals had already been broadcast into space for nearly 80 years. Some philosophers also worried that those messages contained anthropomorphic assumptions about aliens being enough like us to decode the messages. However, sending messages on spacecraft is not an efficient way to make contact with extraterrestrial life. The probability that an alien species will actually find any of those messages is very, very small.

In 1974, astronomers used the 300-meter-wide dish of the Arecibo radio telescope to beam a message in binary code (**Figure 24.22**) toward the star cluster M13, located 25,000 light-years away. That distance is far enough that by the time the message arrives, the core of M13 will have moved, and the radio signal will not actually arrive there. However, the intention of the experiment was to show that such a message could be sent, not to make contact, because 50,000 years would pass before any reply could come back. In 2008, a radio telescope in Ukraine sent a message to the exoplanet Gliese 581c. The message was composed of 501 digitized images and text messages selected by users on a social networking site and will arrive at Gliese 581c in 2029.

The Drake Equation

The first serious effort to quantify the probability of the existence of intelligent extraterrestrial life was made by astronomer Frank Drake in 1960. He developed the **Drake equation**, which estimates the likely number (N) of intelligent civilizations currently existing in the Milky Way Galaxy. The Drake equation is different from the other equations in this book because the values for many of the variables are uncertain. However, it is a useful way to categorize some of the

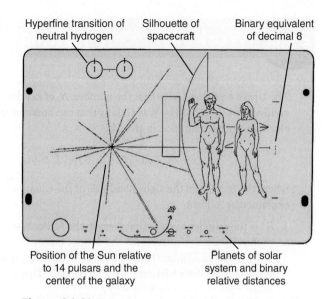

Figure 24.21 This plaque is carried by the *Pioneer 11* probe, which launched in 1973 and will eventually leave the Solar System to travel in interstellar space.

Figure 24.22 This message was beamed toward the star cluster M13 in 1974. Reading right to left, this binary-encoded message contains the numbers 1–10, hydrogen and carbon atoms, some interesting molecules, DNA, a human figure and its size, the basics of the Solar System, and a depiction of the Arecibo telescope.

24.2 Working It Out | Putting Numbers into the Drake Equation

The Drake equation states that the number, N, of extraterrestrial civilizations in the Milky Way Galaxy that can communicate by electromagnetic radiation is given by

$$N = R^* \times f_p \times n_e \times f_l \times f_i \times f_c \times L$$

where the factors on the right-hand side of the equation are explained as follows:

1. R^* is the number of stars that form in the Milky Way Galaxy each year that are suitable for the development of intelligent life. Astronomers consider those to be F, G, K, or M spectral-type stars because their lifetimes are long enough. That is about five to seven stars per year, so $R^* = 5$ to 7.

2. f_p is the fraction of stars that form planetary systems. The discoveries of exoplanets over the past two decades have shown that planets form as a natural by-product of star formation and that many—perhaps most—stars have planets. For that factor, astronomers assume that f_p is between 0.5 and 1.

3. n_e is the number of planets and moons in each planetary system with an environment suitable for life. In the Solar System, that number is at least 1 (for Earth), but it could be more if Mars or an outer-planet moon or two has suitability for life. Only recently have stars with multiple planets been discovered, so astronomers are just starting to get data on that factor. Generally they estimate that $n_e = 0$ to 3.

4. f_l is the fraction of suitable planets and moons on which life actually arises. Remember that just a single self-replicating molecule may be enough to start the process. Some biochemists think that life *will* develop if the right chemical and environmental conditions are present, but others disagree. Values of f_l range from 100 percent (life always develops) to 1 percent (life is more rare). Astronomers use a range of 0.01–1 for f_l.

5. f_i is the fraction of those planets harboring life that eventually develop intelligent life. Intelligence is certainly the kind of survival trait that might often be strongly favored by natural selection. Yet on Earth, tool-building intelligence took about 4 billion years—nearly half the expected lifetime of the Sun—to evolve. The correct value for f_i might be close to 0.01 or it might be closer to 1. No one knows.

6. f_c is the fraction of intelligent life-forms that develop technologically advanced civilizations—that is, civilizations that send communications into space. With only one example of a technological civilization to work with, f_c also is unknown. Astronomers estimate f_c to be between 0.1 and 1.

7. L is the number of years that technologically advanced civilizations exist. That factor is certainly difficult to estimate because it depends on the long-term stability of those civilizations. On Earth, the longest-lived civilizations have existed for, at most, thousands of years. Those civilizations, however, were not at the level of technology that allows interstellar communication—thus far, the first "technologically advanced civilization" on Earth is less than 100 years old. We do not know whether life intelligent enough to manipulate its environment technologically can maintain its planetary resources for any extended length of time. Astronomers usually put a value of L between 1,000 years and 1 million years in their estimates.

factors that relate to the conditions that must be met for a civilization to exist. The equation is discussed further in **Working It Out 24.2**.

As illustrated in **Figure 24.23**, the conclusions we draw using the Drake equation depend a great deal on the assumptions we make and therefore on the numbers used in the equation. For the most pessimistic estimates, the Drake equation sets the number of technological civilizations in our galaxy at about 1, in which case we are the *only* technological civilization in the Milky Way at this time. Such a universe could still be full of intelligent life. With 100 billion galaxies in the observable universe, even those pessimistic assumptions mean that 100 billion technological civilizations could be out there somewhere. However, if the nearest neighbors are in another galaxy, they are *very* far away—millions of parsecs, on average.

At the other extreme—with the most optimistic numbers, which assume that intelligent life arises and survives everywhere it gets the chance—the Milky Way alone could have tens of millions of technological civilizations! If so, the nearest neighbors may be "only" 40 or 50 light-years away.

If humans did meet a technologically advanced civilization, what would it be like? The Drake equation suggests that neighbors nearby are highly unlikely, unless civilizations typically live for many thousands or even millions of years (see Working It Out 24.2). If so, any civilization we encountered would probably have been around for much longer than we have.

Technologically Advanced Civilizations

During lunch with colleagues, the physicist Enrico Fermi (1901–1954), a firm believer in extraterrestrial life, is reported to have asked, "If the universe is teeming with aliens... where is everybody?" Fermi's question—first posed in 1950 and sometimes called the *Fermi paradox*—remains unanswered. If intelligent life-forms are common but interstellar travel is difficult or impossible, would the aliens send out messages—perhaps by electromagnetic waves instead? And if they did, why haven't astronomers detected their signals?

Drake used what was then astronomy's most powerful radio telescope to listen for signals from intelligent life around two nearby stars, but he found nothing unusual. His original project has grown over the years into a much more elaborate program called the Search for Extraterrestrial Intelligence, or **SETI**. Scientists from around the world have thought carefully about what strategies might be useful for finding life in the universe. Most of those endeavors use radio telescopes to listen for signals from space that bear an unambiguous signature of an intelligent source. Some have focused on significant parts of the spectrum, assuming that a civilization will broadcast on a channel that astronomers throughout the galaxy should find interesting—for example, the 21-centimeter (21-cm) line from hydrogen gas. More recent searches have used advances in technology to record as broad a range of radio signals from space as possible. Analysts search those databases for regular signals that might be intelligent in origin.

The SETI Institute's Allen Telescope Array (ATA) received much of its initial financing from Microsoft cofounder Paul Allen. The ATA consists of a "farm" of small, inexpensive radio dishes like those used to capture signals from orbiting television-broadcasting satellites (see the chapter-opening photograph). One key project of the ATA is to observe the planets discovered by the Kepler Mission. Each dish has a diameter of 6.1 meters, but all the telescopes working together have a total signal-receiving area greater than that of a 100-meter radio telescope. Just as your brain can sort out sounds coming from different directions, that array of radio telescopes can determine the direction a signal is coming from, allowing it to listen to many stars at the same time. Over several years' time, astronomers using the ATA are expected to survey as many as a million stars, hoping to find a civilization that has sent a signal toward Earth.

As stated earlier, finding even one nearby civilization in the Milky Way Galaxy—that is, a *second* technological civilization in Earth's small corner of the universe—will make scientists optimistic that the universe as a whole is teeming with intelligent life. The likelihood of SETI's success is difficult to predict, but its potential payoff is enormous. Few discoveries would have a more profound impact than the certain knowledge that we on Earth are not alone.

Science fiction is filled with tales of humans who leave Earth to "seek out new life and new civilizations." Unfortunately, those scenarios are not scientifically realistic. The distances to the stars and their planets are enormous: to explore a significant sample of stars would require extending the physical search over tens or hundreds of light-years. Special relativity limits how fast one can travel. The speed of light is the limit, and even at that rate reaching the *nearest*

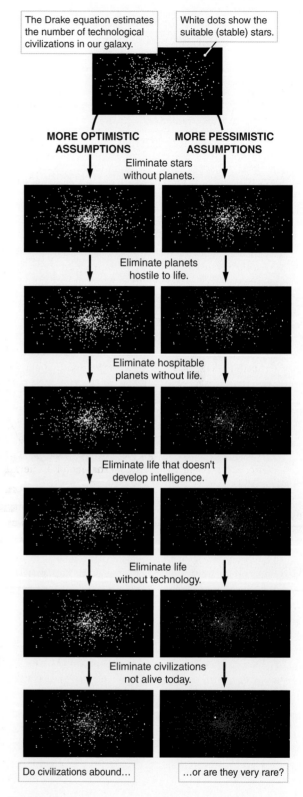

Figure 24.23 The two columns show estimates of the existence in the Milky Way Galaxy of intelligent, communicative civilizations as made using the Drake equation. White dots represent stars with possible civilizations. Notice how widely the estimates vary, given optimistic and pessimistic assumptions about the seven factors in the equation.

star would take more than 4 years. The relativistic effect of time dilation means that time would pass slower for astronauts traveling at very high speeds, and they would return to Earth younger than if they had stayed at home. For example, suppose astronauts visited a star 15 light-years distant. Even if they traveled at speeds close to the speed of light, by the time they returned to Earth, 30 years would have passed at home.

Some science fiction writers get around that problem by invoking "warp speed" or "hyperdrive," which enables travel faster than the speed of light, or by using wormholes as shortcuts across the galaxy—but absolutely no evidence exists that any of those options are possible. And most of those imaginative stories ignore the vast number of other complications that accompany human space travel. Humans are just beginning to learn to live in space even for short periods. Much work remains to be done before we can realistically contemplate even voyages within the Solar System.

Some people claim that aliens have already visited Earth: tabloid newspapers, books, and websites are filled with tales of sightings of unidentified flying objects (UFOs), government conspiracies and cover-ups, alleged alien abductions, and UFO religious cults. However, none of those reports meet the basic standards of science. They are not falsifiable—they lack verifiable evidence and repeatability—and we must conclude that no scientific evidence exists for any alien visitations.

CHECK YOUR UNDERSTANDING 24.4

The Drake equation enables astronomers to: (a) calculate precisely the number of alien civilizations; (b) organize their thoughts about probabilities for life; (c) locate the stars they should study to find life; (d) find new kinds of life.

Origins The Fate of Life on Earth

Astronomers have used their understanding of physics and cosmology to look back through time and watch as structure formed throughout the universe and to look forward to the future of our Sun, our galaxy, and the ultimate fate of the universe. About 5 billion years from now, the Sun will end its long period of relative stability. The Sun will expand to become a red giant star, swelling to hundreds of times larger than it is now and thousands of times more luminous. The giant planets, orbiting outside the extended red giant atmosphere, will probably survive. But at least some of the planets of the inner Solar System will not. Just as an artificial satellite is slowed by drag in Earth's tenuous outer atmosphere and eventually falls to the ground, so, too, will a planet caught in the Sun's atmosphere be engulfed by the expanding Sun. If that happens to Earth, no trace of this planet will remain other than a slight increase in the amount of massive elements in the Sun's atmosphere.

Another possibility is that the red giant Sun will lose mass in a powerful wind, its gravitational pull on the planets will weaken, and the orbits of both the inner and outer planets will spiral outward. If Earth moves out far enough, it may survive as a seared cinder, orbiting the small, hot, white dwarf star that the Sun will become. Barely larger than Earth and with its nuclear fuel exhausted, the white dwarf Sun will slowly cool, eventually becoming a cold sphere of densely packed carbon, orbited by what remains of its planets. The ultimate outcome for Earth—consumed in the Sun or left behind as a cold, burned rock orbiting a long-dead white dwarf—is not yet known.

Or the threat to the future of Earth can come from within, as humanity itself changes conditions on Earth that are necessary for our habitability (for example, by releasing chemicals into the atmosphere that alter the ozone layer, or are greenhouse gases that alter the trapping of radiation).

In either case, however, Earth's status as a garden spot in the habitable zone will be at an end. If the Sun does not expand too far or Mars also migrates outward, Mars could become the habitable planet in the Solar System, at least for a while. As the dying Sun loses more and more of its atmosphere in a stellar wind, Earth's atoms might be expelled back into the reaches of interstellar space from which they came, perhaps to be recycled into new generations of stars, planets, and even life itself.

But even before the Sun's change into a red giant star, the Sun's luminosity will begin to rise. As solar luminosity increases, so will temperatures on all the planets. The inner edge of the Sun's habitable zone will slowly move out past the orbit of Earth. Eventually, Earth's temperatures will climb so high that all animal and plant life will perish. Even the extremophiles that inhabit the oceanic depths will die as the oceans themselves boil away. Models of the Sun's evolution are still not precise enough for astronomers to predict with certainty when that fatal event will occur, but the end of all terrestrial life may be 1 billion to 4 billion years away. In addition, the Milky Way Galaxy will collide with the Andromeda Galaxy in 4 billion to 5 billion years' time. Galaxies are mostly empty space, so the Sun is not likely to collide with another star, but one effect of the collision is that our Solar System may be gravitationally flung to a different part of our galaxy.

However, that the descendants of today's humanity will even be around a billion years from now is far from certain. Some threats to life come from beyond Earth. For the rest of the Sun's life, the terrestrial planets, including Earth, will continue to be bombarded by asteroids and comets. Perhaps a hundred or more of those impacts will involve kilometer-sized objects, capable of causing the kind of devastation that eradicated the dinosaurs. Those impacts may create new surface scars but they will have little effect on the integrity of Earth itself. Earth's geological record is filled with such events, and each time they happen, life manages to recover and reorganize. But we do not know if human life or civilization could survive.

Humans might protect themselves from the fate of the dinosaurs, but in the long run humanity will either leave this world or die out. Planetary systems are common to other stars, and many other Earth-like planets may well exist throughout the Milky Way Galaxy. Colonizing other planets is currently the stuff of science fiction, but if the descendants of modern-day humans are ultimately to survive the death of the home planet, off-Earth colonization must at some point become science fact.

By studying astronomy, you have learned where you come from. You have learned about the self-correcting nature of science—how it continually adapts to new information to give us the ability to make better and better predictions about the behavior of the physical world. That predictive ability makes science an extremely powerful tool. No other species in the history of Earth has been able to understand its position, predict what will happen next, and therefore adapt its behavior to seek the best possible future for its members.

Figure 24.24 shows Earth as seen by the *Cassini* spacecraft, looking near Saturn's rings. That tiny dot, which is Earth, is the only place in the entire universe where we know that life exists. Compare the size of that dot with the size of the universe. Compare the history of life on Earth with the history of the universe. Compare Earth's future with the fate of the universe. Astronomy is humbling. We occupy a tiny part of space and time. Yet we are unique, as far as we know. Think for a moment about what that means to you. That may be the most important lesson the universe has to offer.

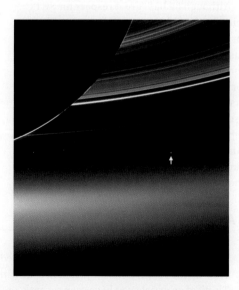

Figure 24.24 This image from the *Cassini* spacecraft shows Earth as seen from Saturn. The pale dot in the lower right (arrow) is Earth.

READING ASTRONOMY NEWS

ARTICLES | QUESTIONS

Astronomers struggle with how to handle announcements (or nonannouncements) about extraterrestrial contact.

When Reporting News about Aliens, Caution Is Advised

By ELIZABETH HOWELL, airspacemag.com

Speculation about extraterrestrials seems to be everywhere these days. Last week it was "Tabby's Star" (more officially known as KIC 8462852), whose mysterious dimming and brightening, according to the latest analysis, is likely due to dust blocking different wavelengths of light rather than "alien megastructures." Before that came reports of an interstellar asteroid—not a spacecraft—entering our solar system and a UFO monitoring program conducted by the Department of Defense.

The attention given to such stories has some scientists worried, especially as social media amplifies claims of alien contact over other, more prosaic explanations.

"Currently, most SETI-related news seems to be interfering with conventional scientific discoveries, stealing the limelight—without following basic rules of science," wrote Dutch exoplanet researcher Ignas Snellen of Leiden Observatory, on a Facebook exoplanets discussion group for professional astronomers.

Although he has "great respect for SETI scientists," Leiden wrote, "there is no place for alien civilizations in a scientific discussion on new astrophysical phenomena, in the same way as there is no place for divine intervention as a possible solution. One may view it as harmless fun, but I see parallels in athletes taking banned substances. It may lead to short-term fame and medals, but in the long run it harms the sport. Same for astronomy: we should be very careful not to be ridiculed. I really hope we can stop mentioning SETI for every unexplained phenomenon."

Such worries aren't new. Nearly 20 years ago researchers in the field came up with the Rio Scale to guide them in reporting the significance of any candidate SETI signal. Modeled after the Richter Scale for earthquakes, it assigns a value between 0—no significance—and 10—extraordinary significance—to any detected signal. The method was first proposed by Ivan Almar and Jill Tarter at the 51st International Astronautical Congress in Rio de Janeiro, in 2000. Although the International Academy of Astronautics' SETI Permanent Committee adopted the scale two years later, it hasn't seen widespread use.

Meanwhile, the news business has changed drastically since 2000. Today much of the public doesn't get its information from TV and newspapers, but from Facebook updates and Twitter posts, which move—and change—at a much faster pace. In recognition of the altered media landscape, a new paper submitted to the *International Journal of Astrobiology* proposes streamlining the Rio Scale and having scientists take a short quiz to answer a few questions about their discovery. Journalists and other news providers could use the answers in their stories.

"You can go through the steps and ask yourself, do I believe the instrumentation is working properly? Do I believe I am looking at it objectively? Do I have enough people who looked at the signal with different instruments? Has there been a lot of scientific discussion? Have there been alternative explanations? Is it a hoax?" says the paper's lead author, Duncan Forgan, a postdoctoral scholar at the University of St. Andrews' School of Physics and Astronomy in Scotland, whose work includes problems related to astrobiology and SETI.

Forgan says that although it can be difficult to convey uncertainty about a particular result to the public, the revised Rio Scale would make that job easier. Some scientific speculation can be "a bit wibbly-wobbly," he says, "and those wibbly-wobbly bits end up in the press." He agrees that spurious SETI claims can sometimes distract from more legitimate science.

In recognition of the sensitivity around alien signal detection, SETI has a voluntary list of protocols to follow when something interesting is found. The first principle urges researchers to "verify that the most plausible explanation for the evidence is the existence of extraterrestrial intelligence, rather than some other natural phenomenon or anthropogenic phenomenon, before making any public announcement."

Morris Jones, an Australian space observer with both scientific and journalistic training, says we should label "fringe SETI" claims for what they are. "The media is under pressure to deliver attention-grabbing news, but it's hard to expect them to judge fringe SETI as spurious when it comes from reputable institutions and qualified researchers. The best way to reduce these reports is to stop the production of questionable scientific papers in the first place."

ARTICLES | QUESTIONS

1. Why are astrobiologists concerned about media coverage of SETI?
2. How do the suggestions for journalists fit with what you have learned about the process of science?
3. Think of an article you have seen recently on evidence for alien life. Was it overly speculative?
4. How do you think people on Earth would react to an actual discovery of alien life?

Source: Elizabeth Howell, "When Reporting News About Aliens, Caution Is Advised," AirSpaceMag.com, January 9, 2018. Reprinted with permission.

Summary

Astrobiology seeks to answer the question, Are we alone? All the sciences have a part to play in answering that question, from astronomy to zoology. Theories of how life began and evolved on Earth help limit the number of targets astronomers study in their search for life elsewhere. Even before the Sun expands and evolves to a red giant star, its luminosity will increase enough to alter the location of its habitable zone. When that happens, Earth may no longer be the planet at the right location for maintaining liquid water.

LG 1 **Explain our current understanding of how and when life began on Earth and how it has evolved.** Life on Earth is a form of complex carbon-based chemistry, made possible by self-replicating molecules. Life probably formed in Earth's oceans and then evolved chemically from simple molecules into self-replicating organisms through a combination of mutation and heredity.

LG 2 **Explain how life is a structure that has evolved through the action of the physical and chemical processes that shape the universe.** All terrestrial life is composed primarily of only six elements: carbon, hydrogen, oxygen, nitrogen, sulfur, and phosphorus. However, life-forms very different from those on Earth, including those based on silicon chemistry, cannot be ruled out.

LG 3 **Describe the locations in our Solar System and around other stars where astronomers think life might be possible.** Within the Solar System, Mars and some moons of Jupiter and Saturn are the most promising candidates for life. A habitable zone around a star is a location within which the temperature of a planet will support liquid water on its surface. Thus, astronomers look for exoplanets that orbit in habitable zones surrounding solar-type stars. That concept of habitable zone can be extended by special circumstances, such as tidal heating from a giant planet or atmospheres containing greenhouse gases.

LG 4 **Describe some methods used to search for intelligent extraterrestrial life.** The Drake equation includes factors that astronomers consider when thinking about the possibility of life in the universe. Astronomers use radio telescopes to search for signals from extraterrestrial life, particularly in astronomically important regions of the spectrum. In recent times, that effort has been expanded to a search through as broad a region of the radio spectrum as possible. No intelligent extraterrestrial life has yet been detected.

? Unanswered Questions

- Will humans spread life into space? Some scientists have suggested that seeding from Earth may have already happened as Earth microorganisms scattered into space after giant impacts. Humans also may have unintentionally sent microorganisms to space aboard our spacecraft. More intentional methods of seeding include sending microorganisms from Earth to other planets or moons to try to jump-start evolution, "terraforming" Mars or a moon to change conditions on it to make it more habitable for humanity, or sending humans in spaceships to colonize the galaxy.

- Will humans themselves spread into space? At some point, humans must leave planet Earth if the species is to survive. But space is a dangerous place, and many of the problems that humans encounter in space have not yet been solved. Those problems are as diverse as purely physical issues, such as the loss of bone density in low gravity, and the societal problems that occur when a few people are confined together for long periods. We do not yet know whether humanity can overcome those problems and journey to other planets.

Questions and Problems

Test Your Understanding

1. The study of life and the study of astronomy are connected because (select all that apply)
   a. life may be commonplace in the universe.
   b. studying other planets may help explain why life exists on Earth.
   c. explorations of extreme environments on Earth suggest where to look for life elsewhere.
   d. life is a structure that evolved through physical processes, and life on Earth may not be unique.
   e. life elsewhere is most likely to be found by astronomers.

2. Scientists look for water to indicate places where life might exist because
   a. water is a common molecule in interstellar space.
   b. life on Earth depends on it.
   c. no other molecules are solvents.
   d. the spectrum of water is very complex.

3. The Urey-Miller experiment produced _____ in a laboratory jar.
   a. life
   b. RNA and DNA
   c. amino acids
   d. proteins

4. A mutation is
   a. always a deadly change to DNA.
   b. always a beneficial change to DNA.
   c. a change to DNA that is sometimes beneficial and sometimes not.
   d. a change to DNA that cannot be inherited.

5. Scientists think that terrestrial life probably originated in Earth's oceans, rather than on land, because (select all that apply)
   a. all the chemical pieces were in the ocean.
   b. energy was available in the ocean.
   c. the earliest evidence for life on Earth is from fossils of ocean-dwelling organisms.
   d. the deepest parts of the ocean have hydrothermal vents.

6. Any system with the processes of heredity, mutation, and natural selection will, over time (choose all that apply),
   a. change.
   b. become larger.
   c. become more complex.
   d. develop intelligence.

7. The fact that no alien civilizations have yet been detected indicates that
   a. they are not there.
   b. they are rare.
   c. Earth is in a "blackout," and they are not talking to us.
   d. we don't know enough yet to draw any conclusions.

8. During the Cambrian explosion
   a. the dinosaurs were killed.
   b. all the carbon that is now here on Earth was produced.
   c. biodiversity increased significantly.
   d. a lot of carbon dioxide was released into the atmosphere.

9. The habitable zone is the place around a star where
   a. life has been found.
   b. atmospheres can contain oxygen.
   c. liquid water has been confirmed.
   d. liquid water can exist on the surface of a planet.

10. The difference between a prokaryote and a eukaryote is that prokaryotes
    a. have no DNA.
    b. have no cell wall.
    c. have no nucleus.
    d. do not exist today.

11. A thermophile is an organism that lives in extremely _____ environments.
    a. salty
    b. hot
    c. cold
    d. dry

12. The search for life elsewhere in the Solar System is carried out primarily by
    a. astronauts.
    b. robotic spacecraft.
    c. astronomers using optical telescopes.
    d. astronomers using radio telescopes.

13. Astronomers think that intelligent life is more likely to be found around stars of types F, G, K, and M because
    a. those stars are hot enough to have planets and moons with liquid water.
    b. those stars are cool enough to have planets and moons with liquid water.
    c. those stars live long enough for life to begin and evolve.
    d. those stars produce no ultraviolet radiation or X-rays.

14. Life first appeared on Earth
    a. billions of years ago.
    b. millions of years ago.
    c. hundreds of thousands of years ago.
    d. thousands of years ago.

15. In the phrase "theory of evolution," the word *theory* means that evolution
    a. is an idea that can't be tested scientifically.
    b. is an educated guess to explain natural phenomena.
    c. probably doesn't happen anymore.
    d. is a tested and corroborated scientific explanation of natural phenomena.

Thinking about the Concepts

16. How does the evolution of life on Earth depend on RNA and DNA?

17. How do scientists think that amino acids first formed on Earth?

18. Today, most organisms on Earth enjoy relatively moderate climates and temperatures. Compare that environment with some of the conditions in which early life developed.

19. The Process of Science Figure represents many areas of science that all inform the science of astrobiology. Choose any two of those, and explain how one depends on the other.

20. How was Earth's carbon dioxide atmosphere changed into today's oxygen-rich atmosphere? How long did that transformation take?

21. What was the Cambrian explosion, and what might have caused it?

22. Why did plants and forests appear in large numbers before large animals did?

23. What is a habitable zone? What defines its boundaries?

24. Which general conditions were needed on Earth for life to arise?

25. Is evolution under way on Earth today? If so, how might humans continue to evolve?

26. Where did all the atoms in your body come from?

27. The two *Viking* spacecraft found no convincing evidence of life on Mars when they visited that planet in the late 1970s, nor did the *Phoenix* lander when it examined the martian soil in 2009. Do those results imply that life never existed on the planet? Why or why not?

28. Some scientists believe that humans may be the only advanced life in the galaxy today. If so, which factors in the Drake equation must be extremely small?

29. The second law of thermodynamics says that the entropy (a measure of disorder) of the universe is always increasing. Yet complex living organisms exist. Why does that fact not violate the second law of thermodynamics?

30. Why is life on Earth as we know it likely to end long before the Sun runs out of nuclear fuel?

Applying the Concepts

31. Several missions are searching for planets in the habitable zones of stars. Explain which factors in the Drake equation are affected by that search and how the final number N will be affected if these missions find that most stars have planets in their habitable zones.

32. Study Figure 24.22. The white blocks at the top represent the numbers 1–10. The bottom row of the set of four is a placeholder, and the top three rows of the four represent the numbers, in order from right to left. Explain the "rule" for the kind of counting shown here. (For example, how do three white blocks represent the number 7?)

33. Suppose that an organism replicates itself each second. If you start with a single specimen, what will the final population be after 10 seconds?

34. Suppose that an organism replicates itself each second. After how many seconds will the population increase by a factor of 1,024?

35. In general, you can find the doubling time for exponential growth by dividing 70 by the rate of increase. So, if the population increases by 7 percent per year, the doubling time is 10 years (70/7 = 10). Suppose Earth's human population continues to grow by 1 percent annually. What is the doubling time? How much time will pass before Earth has 4 times as many humans?

36. As discussed in Working It Out 24.1, the doubling time for exponential growth is given by $P_F/P^n = 2n$. Assume a self-replicating molecule that makes one copy of itself each second. Make a graph of the number of molecules versus time for the first 60 seconds after the molecule begins replicating.

37. The doubling time for *Escherichia coli* is 20 minutes, and you start getting sick when just 10 bacteria enter your system. How many bacteria are in your body after 12 hours?

38. If the chance that a given molecule will mutate is one in 100,000, how many generations are needed before, on average, at least one mutation has occurred?

39. Study the Drake equation in Working It Out 24.2. Make your own most optimistic and most pessimistic assumptions for each variable in the equation. What values do you find for N?

40. Study the Drake equation in Working It Out 24.2. Keep the other variables as is, but put in different values for the lifetime of a technological civilization. How does doing so affect your value of N? How does that affect the possibility of making contact with extraterrestrials?

41. Look back at Figure 5.24 in Chapter 5. Trace (or photocopy) that graph, and then add horizontal lines for the temperatures at which water freezes and boils. Which planets in the Solar System have measured temperatures that fall within those lines? Which planets have predicted temperatures (based on the equilibrium model) that fall within those lines? What does that tell you about assumptions about the habitable zone?

42. Figure 5.24 shows that Mercury's measured range of temperatures overlaps the temperatures at which water is a liquid. Is Mercury in the habitable zone? Why or why not?

43. Suppose astronomers announce the discovery of a new planet around a star with a mass equal to the Sun's. The planet has an orbital period of 87 days. Is that planet in the habitable zone for a Sun-like star?

44. Why do you think astronomers sent a coded radio signal to the globular cluster M13 in 1974—rather than, say, to a nearby star?

45. As noted in Section 24.1, some scientists suspect the early Earth was "seeded" with primitive life stored in comets and meteoroids. Now that you know when and how our Solar System, galaxy, and universe formed, what time line is required for such seeding to be possible?

USING THE WEB

46. a. Go to the online *Astrobiology Magazine* (https://www.astrobio.net), which covers many topics included in this chapter. Under "Origins," click on "Extreme Life." What are some new findings about extremophiles? Why is that important for astrobiologists? Under "Deep Space" click on "New Planets." What is a recent discovery?

    b. Go to the "Life, Unbounded" blog (https://blogs.scientificamerican.com/life-unbounded). What is a recent topic of interest? Is the discussion based on some new data? A new theory?

47. Solar System space missions:
    a. Go to the website for the *Curiosity Mars Science Laboratory* (https://mars.nasa.gov/msl/mission). Are there any results from the *Mars Insight* mission (https://mars.nasa.gov/insight/)? What is the status of the *Mars 2020* lander, scheduled for launch in 2020?

    b. The *MAVEN* Mars mission (https://mars.nasa.gov/maven/) arrived at Mars in 2014. What are the mission's science goals? Why are astrobiologists interested in the history of the climate and atmosphere of Mars? Are any results available?

48. Go to the "Habitable Exoplanets Catalog" website (http://phl.upr.edu/projects/habitable-exoplanets-catalog). Click on "Methods" to read about the criteria for evaluating a planet's habitability. How many confirmed habitable exoplanets exist? How many candidates? How might the number of confirmed exoplanets affect the terms in the Drake equation?

49. a. Go to the website for the European Space Agency's *Gaia* satellite (http://sci.esa.int/gaia). What are the mission's science objectives? Under "Mission Science" on the left, click on "Exoplanets." How does *Gaia* identify new planets? What properties of the planet will it be able to measure? What are some recent results?

    b. Go to the website for the NASA TESS (Transiting Exoplanet Survey Satellite; https://tess.gsfc.nasa.gov/). What is the mission's goal? What are some of its discoveries?

    c. Go to the website for the NASA JWST (James Web Space Telescope; https://www.jwst.nasa.gov/). On the left, click on "Science" and then "Planets & Life." How will the telescope study exoplanet atmospheres? Click on "News"—are any results available?

50. Go to the website for the Breakthrough Initiative (https://breakthroughinitiatives.org/), a privately funded series projects started in 2015 to address the question, Are we alone? What are those projects and what is their status?

EXPLORATION Fermi Problems and the Drake Equation

digital.wwnorton.com/astro5

The Drake equation is a way to organize ideas about other intelligent, communicating civilizations in the galaxy. That type of thinking is useful for estimating a value, particularly when analyzing systems for which counting is not possible. The types of problems that can be solved in that way are often called Fermi problems after Enrico Fermi, who was mentioned in this chapter. For example, we might ask, "What is the circumference of Earth?"

You could Google that question, or you could already "know" the answer, or you might look it up in this textbook. Alternatively, you could very carefully measure the shadow of a stick in two locations at the same time on the same day. Or you could drive around the planet.

Or you could reason this way:

How many time zones are between New York and Los Angeles?
Three time zones. You know that from traveling or from television.

How many miles is it from New York to Los Angeles?
Three thousand miles. You know that from traveling.

So, how many miles per time zone?
3,000/3 = 1,000

How many time zones are in the world?
Twenty-four, because 24 hours are in a day, and each time zone marks an hour.

So, what is the circumference of Earth?
Twenty-four thousand miles, because Earth has 24 time zones, each 1,000 miles wide.

The measured circumference is 24,900 miles, which agrees with our estimate to within 4 percent.

Here we list several Fermi problems. Time yourself for an hour, and work as many of them as possible. (You don't have to do them in order!)

1 How much has the mass of the human population on Earth increased in the past year?

2 How much energy does a horse consume in its lifetime?

3 How many pounds of potatoes are consumed in the United States annually?

4 How many cells are in the human body?

5 If your lifetime earnings were given to you by the hour, how much is your time worth per hour?

6 What is the weight of solid garbage that U.S. families throw out each year?

7 How fast does human hair grow (in feet per hour)?

8 If all the people on Earth were crowded together, how much area would be covered?

9 How many people could fit on Earth if every person occupied 1 square meter of land?

10 How much carbon dioxide (CO_2) does an automobile emit each year?

11 What is the mass of Earth?

12 What is the average annual cost of an automobile, including overhead (maintenance, looking for parking, insurance, cleaning, and so on)?

APPENDIX 1:
Mathematical Tools

Mathematics helps scientists understand the patterns they see and then communicate that understanding to others. Appendix 1 presents some tools that will be useful in our study of astronomy.

Powers of 10 and Scientific Notation

Astronomy is a science of both the very large and the very small. The mass of an electron, for example, is

$$0.0000000000000000000000000000009109 \text{ kilograms (kg)}$$

whereas the distance to a galaxy far, far away might be about

$$100{,}000{,}000{,}000{,}000{,}000{,}000{,}000{,}000 \text{ meters}$$

A quick glance at those two numbers shows why astronomers need a more convenient way to express numbers.

Our number system is based on **powers of 10**. Going to the left of the decimal place,

$$10 = 10 \times 1$$
$$100 = 10 \times 10 \times 1$$
$$1{,}000 = 10 \times 10 \times 10 \times 1$$

and so on. Going to the right of the decimal place,

$$0.1 = \frac{1}{10} \times 1$$
$$0.01 = \frac{1}{10} \times \frac{1}{10} \times 1$$
$$0.001 = \frac{1}{10} \times \frac{1}{10} \times \frac{1}{10} \times 1$$

and so on. In other words, each place to the right or left of the decimal place in a number represents a power of 10. For example, 1 million can be written as

$$1 \text{ million} = 1{,}000{,}000 = 1 \times 10 \times 10 \times 10 \times 10 \times 10 \times 10$$

That is, 1 million is "1 multiplied by six factors of 10." **Scientific notation** combines those factors of 10 in convenient shorthand.

Rather than all being written out, the six factors of 10 are expressed using an exponent:

$$1 \text{ million} = 1 \times 10^6$$

which also means "1 multiplied by six factors of 10."

Moving to the right of the decimal place, each step *removes* a power of 10 from the number. One-millionth can be written as

$$1 \text{ millionth} = 1 \times \frac{1}{10} \times \frac{1}{10} \times \frac{1}{10} \times \frac{1}{10} \times \frac{1}{10} \times \frac{1}{10}$$

That divides by powers of 10, so that number can be written by use of a negative exponent

$$\frac{1}{10} = 10^{-1}$$

as

$$1 \text{ millionth} = 1 \times 10^{-1} \times 10^{-1} \times 10^{-1} \times 10^{-1} \times 10^{-1} \times 10^{-1}$$
$$= 1 \times 10^{-6}$$

Let's return to our earlier examples. The mass of an electron is 9.109×10^{-31} kg, and the distant galaxy is located 1×10^{26} meters away. Scientific notation is a much more convenient ways of writing those values. Notice that *the exponent in scientific notation gives you a feel for the size of a number at a glance*. The exponent of 10 in the electron mass is -31, which quickly indicates that it is a very small number. The exponent of 10 in the distance to a remote galaxy, $+26$, quickly indicates that it is a very large number. The exponent is often called the *order of magnitude* of a number. When you see a number written in scientific notation while reading *21st Century Astronomy* (or elsewhere), just remember to look at the exponent to understand the size of the number.

Scientific notation also is convenient because it makes multiplying and dividing numbers easier. For example, 2 billion multiplied by eight-thousandths can be written as

$$2{,}000{,}000{,}000 \times 0.008$$

but writing those two numbers in scientific notation is:

$$(2 \times 10^9) \times (8 \times 10^{-3})$$

We can regroup those expressions in the following form:

$$(2 \times 8) \times (10^9 \times 10^{-3})$$

The first part of the problem is just $2 \times 8 = 16$. The more interesting part of the problem is the multiplication in the right-hand parentheses. The first number, 10^9, is just shorthand for $10 \times 10 \times 10 \ldots$ nine times. That is, it represents nine factors of 10. The second number stands for three factors of 1/10—or removing three factors of 10 if you prefer to think of it that way. Altogether, that makes $9 - 3 = 6$ factors of 10. In other words,

$$10^9 \times 10^{-3} = 10^{(9-3)} = 10^6$$

Putting the problem together, we get

$$(2 \times 10^9) \times (8 \times 10^{-3}) = (2 \times 8) \times (10^9 \times 10^{-3})$$
$$= 16 \times 10^6$$

By convention, when a number is written in scientific notation, only one digit is placed to the left of the decimal point. Here, though, we have two. However, 16 is 1.6×10, so we can add that additional factor of 10 to the exponent at right, making the final answer

$$1.6 \times 10^7$$

Dividing is just the inverse of multiplication. Dividing by 10^3 means removing three factors of 10 from a number. Using the previous number,

$$(1.6 \times 10^7) \div (2 \times 10^3) = (1.6 \div 2) \times (10^7 \div 10^3)$$
$$= 0.8 \times 10^{(7-3)}$$
$$= 0.8 \times 10^4$$

This time we have only a zero to the left of the decimal point. To get the number into proper form, we can substitute 8×10^{-1} for 0.8, giving

$$0.8 \times 10^4 = (8 \times 10^{-1}) \times 10^4 = 8 \times 10^3$$

Adding and subtracting numbers in scientific notation is harder because all numbers must be written as values multiplied by the *same* power of 10 before they can be added or subtracted. Therefore, you will need to use a calculator that has scientific notation. Most scientific calculators have a button that says EXP or EE. Those abbreviations mean "times 10 to the ____." So for 4×10^{12}, you would type [4][EXP][1][2] or [4][EE][1][2] into your calculator. Usually, that number shows up in the window on your calculator either just as you see it written in this book or as a 4 with a smaller 12 all the way over in the right side of the window.

Significant Figures

In the previous example, we actually broke some rules in the interest of explaining how powers of 10 are treated in scientific notation. The rules we broke involve the *precision* of the numbers. When expressing quantities in science, it is extremely important to know not only the value of a number but also how precise that value is.

The most complete way to keep track of the precision of numbers is to actually write down the uncertainty in the number. For example, suppose you know that the distance to a store (call it d) is between 0.8 and 1.2 kilometers (km); you can then write

$$d = 1.0 \pm 0.2 \text{ km}$$

where the symbol "$\pm$" is pronounced "plus or minus." Here, d is between $1.0 - 0.2 = 0.8$ km and $1.0 + 0.2 = 1.2$ km. That is an unambiguous statement about the limitations on knowing the value of d, but carrying along the formal errors with every number written would be cumbersome at best. Instead, you keep track of the approximate precision of a number by using *significant figures*.

The convention for significant figures is this: Assume the written number has been rounded from a number that had one additional digit to the right of the decimal point. If a quantity d, which might represent the distance to the store, is "1.", then d is close to 1. It is probably not as small as "0.", and it is probably not as large as "2.". If instead it is written as

$$d = 1.0$$

then d is probably not 0.9 and is probably not 1.1. It is roughly 1.0 to the nearest tenth. The greater the number of significant figures, the more precisely the number is being specified. For example, 1.00000 is not the same number as 1.00. The first number, 1.00000, represents a value that is probably not as small as 0.99999 and probably not as large as 1.00001. The second number, 1.00, represents a value that is probably not as small as 0.99 nor as large as 1.01. The number 1.00000 is much more precise than the number 1.00.

In mathematical operations, significant figures are important. For example, consider $2.0 \times 1.6 = 3.2$. It does *not* equal 3.20000000000. *The product of two numbers cannot be known to any greater precision than the numbers themselves*. In general, when you multiply and divide, the answer should have the same number of significant figures as the less precise of the numbers being multiplied or divided. In other words, $2.0 \times 1.602583475 = 3.2$. Because all you know is that the first factor is probably closer to 2.0 than to 1.9 or 2.1, all you know about the product is that it is between about 3.0 and 3.4. It is 3.2. It is not 3.205166950 (*even if that is the answer on your calculator*). The rest of the digits to the right of 3.2 just do not mean anything.

When two numbers are added or subtracted, if one number has a significant figure with a particular place value but another number does not, their sum or difference cannot have a significant figure in that place value. For example,

$$\begin{array}{r}1{,}045.\\ +1.34567\\ \hline 1{,}046.\end{array}$$

The answer is "1,046." *not* "1,046.34567". Again, the extra digits to the right of the decimal place have no meaning because "1,045." is not known to that precision.

What is the precision of the number 1,000,000? As it is written, the answer is unclear. Are all those zeros really significant, or are they merely placeholders? If the number is written in scientific notation, however, you never have to wonder. Instead of 1,000,000, you write 1.0×10^6 for a number that is known to the nearest hundred thousand or so, or you write 1.00000×10^6 for a number that is known to the nearest 10.

So the earlier example would have been more correct if written as

$$(2.0 \times 10^9) \times (8.0 \times 10^{-3}) = 1.6 \times 10^7$$

Algebra

Mathematics has many branches. The branch that focuses on the relationships between quantities is called **algebra**. Basically, algebra begins by using symbols to represent quantities.

For example, you could write the distance you travel in a day as d. As it stands, d has no value. It might be 10,000 miles. It might be 30 feet. It does, however, have **units**—here, the units of distance. The average speed at which you travel is equal to the distance you travel divided by the time you take. By using the symbol v to represent your average speed and the symbol t to represent the time you take, instead of writing out "Your average speed is equal to the distance you travel divided by the time you take" you can write

$$v = \frac{d}{t}$$

The meaning of that algebraic expression is the same as the sentence quoted before it, but it is much more concise. As it stands, v, d, and t still have no specific values. No numbers are assigned to them yet. However, the expression indicates what the relationship between those numbers will be when you look at a specific example. For example, if you go 500 km ($d = 500$ km) in 10 hours ($t = 10$ hours), that expression tells you that your average speed is

$$v = \frac{d}{t} = \frac{500\,\text{km}}{10\,\text{h}} = 50\,\text{km/h}$$

Notice that the units in this expression act exactly like the numerical values. Dividing the two shows that the units of v are kilometers divided by hours, or km/h (pronounced "kilometers per hour").

We introduced algebra as shorthand for expressing relations between quantities, but it is far more powerful than that. Algebra provides rules for manipulating the symbols used to represent quantities. We begin with a bit of notation for *powers* and *roots*. Similar to powers of 10, raising a quantity to a power means multiplying the quantity by itself some number of times. For example, if S is a symbol for something (anything), S^2 (pronounced "S-squared" or "S to the second power") means $S \times S$, and S^3 (pronounced "S-cubed" or "S to the third power") means $S \times S \times S$. Suppose S represents the length of the side of a square. The area of the square is given by

$$\text{Area} = S \times S = S^2$$

If $S = 3$ meters (m), the area of the square is

$$S^2 = 3\,\text{m} \times 3\,\text{m} = 9\,\text{m}^2$$

(pronounced "9 square meters"). That is why raising a quantity to the second power is called *squaring* the quantity. We could have done the same thing for the sides of a cube and found that the volume of the cube is

$$\text{Volume} = S \times S \times S = S^3$$

If $S = 3$ meters, the volume of the cube is

$$S^3 = 3\,\text{m} \times 3\,\text{m} \times 3\,\text{m} = 27\,\text{m}^3$$

(pronounced "27 cubic meters"). That is why raising a quantity to the third power is called *cubing* the quantity.

Roots are the reverse of that process. The square root of a number is the value that, when squared, gives the original quantity. The square root of 4 is 2, which means that $2 \times 2 = 4$. The square root of 9 is 3, which means that $3 \times 3 = 9$. Similarly, the cube root of a quantity is the value that, when cubed, gives the original quantity. The cube root of 8 is 2, which means that $2 \times 2 \times 2 = 8$. Roots are written with the symbol $\sqrt{}$. For example, the square root of 9 is written as

$$\sqrt{9} = 3$$

and the cube root of 8 is written as

$$\sqrt[3]{8} = 2$$

If the volume of a cube is $V = S^3$, then

$$S = \sqrt[3]{V} = \sqrt[3]{S^3}$$

Roots can also be written as powers. Powers and roots behave like the exponents of 10 in our discussion of scientific notation. (The exponents used in scientific notation are just powers of 10.) For example, if a, n, and m are all algebraic quantities, then

$$a^n \times a^m = a^{n+m} \quad \text{and} \quad \frac{a^n}{a^m} = a^{n-m}$$

(The square root of a can also be written $a^{1/2}$ and the cube root of a can be written $a^{1/3}$.)

The rules of arithmetic can be applied to the symbolic quantities of algebra. As long as the rules of algebra are applied properly, the relationships among symbols arrived at through algebraic manipulation remain true for the physical quantities that those symbols represent.

Here we summarize a few algebraic rules and relationships. Here, a, b, c, m, n, r, x, and y are all algebraic quantities.

Associative rule:
$$a \times b \times c = (a \times b) \times c = a \times (b \times c)$$

Commutative rule:
$$a \times b = b \times a$$

Distributive rule:
$$a \times (b + c) = (a \times b) + (a \times c)$$

Cross-multiplication:
$$\text{If } \frac{a}{b} = \frac{c}{d}, \text{ then } ad = bc$$

Working with exponents:
$$\frac{1}{a^n} = a^{-n} \qquad a^n a^m = a^{n+m}$$

$$\frac{a^n}{a^m} = a^{n-m} \qquad (a^n)^m = a^{n \times m} \qquad \left(\frac{a}{b}\right)^n = \frac{a^n}{b^n}$$

Equation of a line with slope m and y-intercept b:
$$y = mx + b$$

Equation of a circle with radius r centered at $x = 0, y = 0$:
$$x^2 + y^2 = r^2$$

Angles and Distances

The farther away something is, the smaller it appears. That is common sense and everyday experience. Because astronomers cannot walk up to the object they are studying and measure it with a meterstick, their knowledge about the sizes of things usually depends on relating the size of an object, its distance, and the angle it covers in the sky.

One way to measure angles is to use a unit called the **radian**. As shown in **Figure A1.1a**, the size of an angle in radians is the length of the arc subtending the angle, divided by the radius of the circle. In the figure, the angle $x = S/r$ radians.

Because the circumference of a circle is 2π multiplied by the radius ($C = 2\pi r$), a complete circle has an angular measure of $(2\pi r)/r = 2\pi$ radians. In more conventional angular measure, a complete circle is 360°, so

$$360° = 2\pi \text{ radians}$$

or

$$1 \text{ radian} = \frac{360°}{2\pi} = 57.2958°$$

Often, seconds of arc (**arcseconds**) are used to measure angles for stars and galaxies. A degree is divided into 60 minutes of arc (**arcminutes**), each of which is divided into 60 seconds of arc—so 3,600 seconds of arc are in a degree. Therefore,

$$3{,}600 \frac{\text{arcseconds}}{\text{degree}} \times 57.2958 \frac{\text{degree}}{\text{radian}} = 206{,}265 \frac{\text{arcseconds}}{\text{radian}}$$

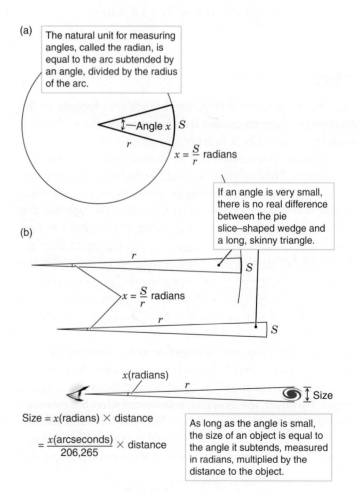

Figure A1.1 Measuring angles.

If the angle is small enough (which it usually is in astronomy), very little difference exists between the pie slice just described and a long skinny triangle with a short side of length S, as Figure A1.1b illustrates. So, if you know the distance d to an object and you can measure the angular size x of the object, the size of the object is given by

$$S = x \text{ (in radians)} \times d = \frac{x \text{ (in degrees)}}{57.2958 \text{ degrees/radian}} \times d$$

$$S = \frac{x \text{ (in arcseconds)}}{206{,}265 \text{ arcseconds/radian}} \times d$$

That is how astronomers relate an object's angular size, distance, and physical size.

Circles and Spheres

To round out those mathematical tools, here are a few useful formulas for circles and spheres. The circle or sphere in each case has a radius r.

$$\text{Circumference}_{\text{circle}} = 2\pi r$$
$$\text{Area}_{\text{circle}} = \pi r^2$$
$$\text{Surface area}_{\text{sphere}} = 4\pi r^2$$
$$\text{Volume}_{\text{sphere}} = \frac{4}{3}\pi r^3$$

Working with Proportionalities

Most of the mathematics in *21st Century Astronomy* involves **proportionalities**—statements about how one physical quantity changes when another quantity changes. We began a discussion of proportionality in Working It Out 1.1; here, we offer a few examples of working with proportionalities.

To use a statement of proportionality to compare two objects, begin by turning the proportionality into a ratio. For example, the price of a bag of apples is **proportional** to the weight of the bag:

$$\text{Price} \propto \text{Weight}$$

Here, the symbol $\propto$ is pronounced "is proportional to." That means the ratio of the prices of two bags of apples is equal to the ratio of the weights of the two bags:

$$\text{Price} \propto \text{Weight means}$$

$$\frac{\text{Price of A}}{\text{Price of B}} = \frac{\text{Weight of A}}{\text{Weight of B}}$$

Let's work a specific example. Suppose bag A weighs 2 pounds and bag B weighs 1 pound. That means bag A will cost twice as much as bag B. We can turn that proportionality into the following equation:

$$\frac{\text{Price of A}}{\text{Price of B}} = \frac{\text{Weight of A}}{\text{Weight of B}} = \frac{2 \text{ lb}}{1 \text{ lb}} = 2$$

In other words, the price of bag A is 2 times the price of bag B. The price per pound is an example of a **constant of proportionality**.

Now let's work another, more complicated example. In Chapter 13, we discuss how the luminosity, brightness, and distance of stars are related. The luminosity of a star—the total energy that the star radiates each second—is proportional to the star's brightness multiplied by the square of its distance:

$$\text{Luminosity} \propto \text{Brightness} \times \text{Distance}^2$$

We can turn that proportionality into a ratio for two stars, A and B:

$$\frac{\text{Luminosity of A}}{\text{Luminosity of B}} = \frac{\text{Brightness of A}}{\text{Brightness of B}} \times \left(\frac{\text{Distance of A}}{\text{Distance of B}}\right)^2$$

If we use the symbols L, b, and d to represent luminosity, brightness, and distance, respectively, that equation becomes

$$\frac{L_A}{L_B} = \frac{b_A}{b_B} \times \left(\frac{d_A}{d_B}\right)^2$$

As an example, suppose that star A appears twice as bright in the sky as star B, but star A is located 10 times as far away as star B. How do the luminosities of the two stars compare? We know that

$$\text{Luminosity} \propto \text{Brightness} \times \text{Distance}^2$$

we write

$$\frac{\text{Luminosity of A}}{\text{Luminosity of B}} = \frac{\text{Brightness of A}}{\text{Brightness of B}} \times \left(\frac{\text{Distance of A}}{\text{Distance of B}}\right)^2$$

$$\frac{\text{Luminosity of A}}{\text{Luminosity of B}} = \frac{2}{1} \times \left(\frac{10}{1}\right)^2 = 200$$

In other words, star A is 200 times as luminous as star B.

Two quantities may also be inversely proportional, such that making one of them smaller makes the other larger. For example, when you are driving to another town, if you drive twice as fast, getting there takes half the time. The travel time is inversely proportional to the travel speed. We write that relationship as

$$\text{Time} \propto \frac{1}{\text{Speed}}$$

Proportionalities are used to compare one object with another. Constants of proportionality are used to calculate actual values. In *21st Century Astronomy*, the proportionality is usually what is important.

APPENDIX 2:
Physical Constants and Units

Fundamental Physical Constants

Constant	Symbol	Value
Speed of light in a vacuum	c	2.99792×10^8 m/s
Universal gravitational constant	G	6.6741×10^{-11} m^3/(kg s^2) 6.6741×10^{-20} km^3/(kg s^2)
Planck constant	h	6.62607×10^{-34} J-s
Boltzmann constant	k	1.38065×10^{-23} J/K
Stefan-Boltzmann constant	σ	5.67037×10^{-8} W/(m^2 K^4)
Mass of electron	m_e	9.10938×10^{-31} kg
Mass of proton	m_p	1.67262×10^{-27} kg
Mass of neutron	m_n	1.67493×10^{-27} kg
Electric charge of electron or proton	e	1.60218×10^{-19} C

SOURCE: Data from the Particle Data Group (http://pdg.lbl.gov).

Unit Prefixes

Prefix*	Name	Factor†
n	nano-	10^{-9}
μ	micro-	10^{-6}
m	milli-	10^{-3}
k	kilo-	10^{3}
M	mega-	10^{6}
G	giga-	10^{9}
T	tera-	10^{12}

*When appended to a unit, these prefixes change the size of the unit by the factor (†) given. For example, 1 km is 10^3 meters.

Units and Values

Quantity	Fundamental Unit	Values
Length	meters (m)	radius of Sun (R_{Sun}) = 6.957×10^8 m astronomical unit (AU) = 1.49598×10^{11} m 1 AU = 149,598,000 km light-year (ly) = 9.4605×10^{15} m 1 ly = 6.324×10^4 AU 1 parsec (pc) = 3.262 *ly* = 3.0857×10^{16} m 1 m = 3.281 feet
Volume	cubic meters (m^3)	1 m^3 = 1,000 liters = 264.2 gallons
Mass	kilograms (kg)	1 kg = 1,000 grams mass of Earth (M_{Earth}) = 5.9724×10^{24} kg mass of Sun (M_{Sun}) = 1.9885×10^{30} kg
Time	seconds (s)	1 hour (h) = 60 minutes (min) = 3,600 s solar day (noon to noon) = 86,400 s sidereal day (Earth rotation period) = 86,164.1 s tropical year (equinox to equinox) = 365.24219 days = 3.15569×10^7 s sidereal year (Earth orbital period) = 365.25636 days = 3.15581×10^7 s
Speed	meters/second (m/s)	1 m/s = 2.237 miles/h 1 km/s = 1,000 m/s = 3,600 km/h c = 2.99792×10^8 m/s = 299,792 km/s
Acceleration	meters/second^2 (m/s^2)	g = gravitational acceleration on Earth = 9.81 m/s^2
Energy	joules (J)	1 J = 1 kg m^2/s^2 1 megaton = 4.18×10^{15} J
Power	watts (W)	1 W = 1 J/s solar luminosity (L_{Sun}) = 3.828×10^{26} W
Force	newtons (N)	1 N = 1 kg m/s^2 1 pound (lb) = 4.448 N 1 N = 0.22481 lb
Pressure	newtons/meter^2 (N/m^2)	atmospheric pressure at sea level = 1.013×10^5 N/m^2 = 1.013 bar
Temperature	kelvins (K)	absolute zero = 0 K = $-273.15°C$ = $-459.67°F$

SOURCE: Data from the Particle Data Group (http://pdg.lbl.gov).

APPENDIX 3:
Periodic Table of the Elements

1 1A																	18 8A
1 **H** Hydrogen 1.00794	2 2A											13 3A	14 4A	15 5A	16 6A	17 7A	2 **He** Helium 4.002602
3 **Li** Lithium 6.941	4 **Be** Beryllium 9.012182											5 **B** Boron 10.811	6 **C** Carbon 12.0107	7 **N** Nitrogen 14.0067	8 **O** Oxygen 15.9994	9 **F** Fluorine 18.9984032	10 **Ne** Neon 20.1797
11 **Na** Sodium 22.98976928	12 **Mg** Magnesium 24.3050	3 3B	4 4B	5 5B	6 6B	7 7B	8	9 8B	10	11 1B	12 2B	13 **Al** Aluminum 26.9815386	14 **Si** Silicon 28.0855	15 **P** Phosphorus 30.973762	16 **S** Sulfur 32.065	17 **Cl** Chlorine 35.453	18 **Ar** Argon 39.948
19 **K** Potassium 39.0983	20 **Ca** Calcium 40.078	21 **Sc** Scandium 44.955912	22 **Ti** Titanium 47.867	23 **V** Vanadium 50.9415	24 **Cr** Chromium 51.9961	25 **Mn** Manganese 54.938045	26 **Fe** Iron 55.845	27 **Co** Cobalt 58.933195	28 **Ni** Nickel 58.6934	29 **Cu** Copper 63.546	30 **Zn** Zinc 65.38	31 **Ga** Gallium 69.723	32 **Ge** Germanium 72.64	33 **As** Arsenic 74.92160	34 **Se** Selenium 78.96	35 **Br** Bromine 79.904	36 **Kr** Krypton 83.798
37 **Rb** Rubidium 85.4678	38 **Sr** Strontium 87.62	39 **Y** Yttrium 88.90585	40 **Zr** Zirconium 91.224	41 **Nb** Niobium 92.90638	42 **Mo** Molybdenum 95.96	43 **Tc** Technetium [98]	44 **Ru** Ruthenium 101.07	45 **Rh** Rhodium 102.90550	46 **Pd** Palladium 106.42	47 **Ag** Silver 107.8682	48 **Cd** Cadmium 112.411	49 **In** Indium 114.818	50 **Sn** Tin 118.710	51 **Sb** Antimony 121.760	52 **Te** Tellurium 127.60	53 **I** Iodine 126.90447	54 **Xe** Xenon 131.293
55 **Cs** Cesium 132.9054519	56 **Ba** Barium 137.327	57 **La** Lanthanum 138.90547	72 **Hf** Hafnium 178.49	73 **Ta** Tantalum 180.94788	74 **W** Tungsten 183.84	75 **Re** Rhenium 186.207	76 **Os** Osmium 190.23	77 **Ir** Iridium 192.217	78 **Pt** Platinum 195.084	79 **Au** Gold 196.966569	80 **Hg** Mercury 200.59	81 **Tl** Thallium 204.3833	82 **Pb** Lead 207.2	83 **Bi** Bismuth 208.98040	84 **Po** Polonium [209]	85 **At** Astatine [210]	86 **Rn** Radon [222]
87 **Fr** Francium [223]	88 **Ra** Radium [226]	89 **Ac** Actinium [227]	104 **Rf** Rutherfordium [261]	105 **Db** Dubnium [262]	106 **Sg** Seaborgium [266]	107 **Bh** Bohrium [264]	108 **Hs** Hassium [277]	109 **Mt** Meitnerium [268]	110 **Ds** Darmstadtium [271]	111 **Rg** Roentgenium [272]	112 **Cn** Copernicium [285]	113 **Uut** Ununtrium [284]	114 **Fl** Flerovium [289]	115 **Uup** Ununpentium [288]	116 **Lv** Livermorium [292]	117 **Uus** Ununseptium [294]	118 **Uuo** Ununoctium [294]

Key:
- 1 — Atomic number
- H — Symbol
- Hydrogen — Name
- 1.00794 — Average atomic mass
- Metals
- Metalloids
- Nonmetals

6 Lanthanides

58 **Ce** Cerium 140.116	59 **Pr** Praseodymium 140.90765	60 **Nd** Neodymium 144.242	61 **Pm** Promethium [145]	62 **Sm** Samarium 150.36	63 **Eu** Europium 151.964	64 **Gd** Gadolinium 157.25	65 **Tb** Terbium 158.92535	66 **Dy** Dysprosium 162.500	67 **Ho** Holmium 164.93032	68 **Er** Erbium 167.259	69 **Tm** Thulium 168.93421	70 **Yb** Ytterbium 173.05	71 **Lu** Lutetium 174.967

7 Actinides

90 **Th** Thorium 232.03806	91 **Pa** Protactinium 231.03588	92 **U** Uranium 238.02891	93 **Np** Neptunium [237]	94 **Pu** Plutonium [244]	95 **Am** Americium [243]	96 **Cm** Curium [247]	97 **Bk** Berkelium [247]	98 **Cf** Californium [251]	99 **Es** Einsteinium [252]	100 **Fm** Fermium [257]	101 **Md** Mendelevium [258]	102 **No** Nobelium [259]	103 **Lr** Lawrencium [262]

We have used the U.S. system as well as the system recommended by the International Union of Pure and Applied Chemistry (IUPAC) to label the groups in this periodic table. The system used in the United States includes a letter and a number (1A, 2A, 3B, 4B, etc.), which is close to the system developed by Mendeleev. The IUPAC system uses numbers 1–18 and has been recommended by the American Chemical Society (ACS). While we show both numbering systems here, we use the IUPAC system exclusively in the book. Elements with atomic numbers higher than 112 have been reported but not yet fully authenticated.

APPENDIX 4:
Properties of Planets, Dwarf Planets, and Moons

Physical Data for Planets and Dwarf Planets

Planet	Equatorial Radius (km)	Equatorial Radius (R/R_{Earth})	Mass (kg)	Mass (M/M_{Earth})	Average Density (relative to water*)	Rotation Period (days)	Tilt of Rotation Axis (degrees, relative to orbit)	Equatorial Surface Gravity (relative to Earth†)	Escape Velocity (km/s)	Average Surface Temperature (K)§
Mercury	2,440	0.383	3.30×10^{23}	0.055	5.427	58.65	0.03	0.378	4.3	340 (100, 700)
Venus	6,052	0.950	4.87×10^{24}	0.815	5.243	243.02‡	177.4	0.905	10.36	737
Earth	6,378	1.000	5.97×10^{24}	1.000	5.514	1.000	23.44	1.000	11.19	288 (185, 331)
Mars	3,396	0.532	6.42×10^{23}	0.107	3.934	1.026	25.19	0.377	5.03	210 (120, 293)
Ceres	473	0.074	9.39×10^{20}	0.0002	2.09	0.378	4	0.26	0.5	168
Jupiter	71,492	11.209	1.90×10^{27}	317.8	1.326	0.4135	3.13	2.530	59.5	165
Saturn	60,268	9.449	5.68×10^{26}	95.16	0.687	0.444	26.73	1.065	35.5	134
Uranus	25,559	4.007	8.68×10^{25}	14.54	1.271	0.7183‡	97.77	0.905	21.3	76
Neptune	24,764	3.883	1.02×10^{26}	17.15	1.638	0.6713	28.32	1.14	23.5	72
Pluto	1,187	0.186	1.30×10^{22}	0.0022	1.860	6.387‡	122.53	0.063	1.21	50
Haumea	~800	0.13	4.0×10^{21}	0.0007	1.9–2.6	0.163	?	0.045	0.81	<50
Makemake	715	0.11	4.0×10^{21}	0.0007	1.4–3.2	0.324	3	~0.5	~0.8	~30
Eris	1,163	0.182	1.67×10^{22}	0.0028	2.5	1.08	?	0.083	1.38	43

*The density of water is 1,000 kg/m³.
†The surface gravity of Earth is 9.81 m/s².
‡Venus, Uranus, and Pluto rotate opposite to the directions of their orbits. Their north poles are south of their orbital planes.
§Values in parentheses give extremes of recorded temperatures.

Orbital Data for Planets and Dwarf Planets

Planet	Mean Distance from Sun (A*) (10^6 km)	(AU)	Orbital Period (P) (sidereal years)	Eccentricity	Inclination (degrees, relative to ecliptic)	Average Speed (km/s)
Mercury	57.9	0.387	0.241	0.2056	7.0	47.36
Venus	108.2	0.723	0.615	0.0067	3.39	35.02
Earth	149.6	1.000	1.000	0.0167	0.000	29.78
Mars	227.9	1.524	1.881	0.0935	1.85	24.07
Ceres	414.0	2.768	4.603	0.079	10.59	17.90
Jupiter	778.6	5.204	11.862	0.0484	1.30	13.06
Saturn	1,433.53	9.582	29.457	0.0565	2.49	9.68
Uranus	2,872.5	19.20	84.011	0.0457	0.77	6.80
Neptune	4,495.1	30.047	164.79	0.0113	1.769	5.43
Pluto	5,906.4	39.48	247.94	0.2488	17.16	4.67
Haumea	6,432.0	43.2	284.12	0.198	28.19	4.53
Makemake	6,783.3	45.7	309.1	0.164	29.00	4.42
Eris	10,180	67.8	558	0.434	43.8	3.43

*A is the semimajor axis of the planet's elliptical orbit.

Properties of Selected Moons*

Planet	Moon	ORBITAL PROPERTIES		PHYSICAL PROPERTIES		Relative Density (g/cm³) (water = 1.00)
		P (days)	A (10^3 km)	R (km)	M (10^{20} kg)	
Earth (1 moon)	Moon	27.32	384.4	1,738.1	735	3.34
Mars (2 moons)	Phobos	0.32	9.38	13.0 × 11.4 × 9.1	0.0001	1.9
	Deimos	1.26	23.46	7.8 × 6.0 × 5.1	0.00002	1.75
Jupiter (79 known moons)	Metis	0.30	128	30 × 20 × 17	0.001	3
	Amalthea	0.50	181.4	125 × 73 × 64	0.075	3.1
	Io	1.77	421.8	1,822	893	3.53
	Europa	3.55	671.1	1,561	480	3.01
	Ganymede	7.15	1,070	2,631	1,482	1.94
	Callisto	16.69	1,883	2,410	1,076	1.83
	Himalia	250.6	11,460	85	0.095	2.6
	Pasiphae	744[†]	23,620	18	0.0030	2.6
	Callirrhoe	759[†]	24,100	4.3	0.00001	2.6
Saturn (62 known moons)	Pan	0.58	133.58	17 × 16 × 10	0.00005	0.42
	Prometheus	0.61	139.35	68 × 40 × 30	0.0016	0.48
	Pandora	0.63	141.70	52 × 41 × 32	0.0014	0.49
	Mimas	0.94	185.52	198	0.38	1.15
	Enceladus	1.37	238.02	252	1.08	1.6
	Tethys	1.89	294.66	531	6.18	0.99
	Dione	2.73	377.40	561	11.0	1.48
	Rhea	4.52	527.04	763	23.1	1.24
	Titan	15.95	1,222	2,575	1,346	1.88
	Hyperion	21.28	1,481	180 × 133 × 103	0.056	0.55
	Iapetus	79.33	3,561	735	18.1	1.09
	Phoebe	548[†]	12,944	107	0.08	1.64
	Paaliaq	687.5	15,200	11	0.0007	2.3
Uranus (27 known moons)	Cordelia	0.34	49.8	20	0.0004	1.3
	Miranda	1.41	129.9	236	0.66	1.20
	Ariel	2.52	199.9	579	12.9	1.59
	Umbriel	4.14	266.0	585	12.2	1.46
	Titania	8.71	436.3	789	34.2	1.66
	Oberon	13.46	583.5	761	28.8	1.56
	Setebos	2,235	17,418	24	0.0008	1.5

(continued)

Properties of Selected Moons*

(continued)

Planet	Moon	ORBITAL PROPERTIES		PHYSICAL PROPERTIES		Relative Density (g/cm³) (water = 1.00)
		P (days)	A (10^3 km)	R (km)	M (10^{20} kg)	
Neptune (14 known moons)	Naiad	0.29	48.2	$48 \times 30 \times 26$	0.002	1.3
	Larissa	0.55	73.5	$108 \times 102 \times 84$	0.05	1.3
	Proteus	1.12	117.6	$220 \times 208 \times 202$	0.5	1.3
	Triton	5.88†	354.8	1,353	214	2.05
	Nereid	360.14	5,513.4	170	0.3	1.5
Pluto (5 moons)	Charon	6.39	19.60	606	15.9	1.70
Haumea (2 moons)	Namaka	18	25.66	85	0.018	~1
	Hi'iaka	49	49.88	170	0.179	~1
Eris	Dysnomia	15.8	37.4	350	?	?

*Innermost, outermost, largest, and/or a few other moons for each planet.
†Irregular moon (has retrograde orbit).

APPENDIX 5:
Space Missions

Selected Recent and Current Solar System Missions

Spacecraft	Sponsoring Nation(s)*	Destination	Launch Year	Type	Status (mid-2018)
Voyager 1 and 2	USA	Jupiter, Saturn, Uranus (2), Neptune (2)	1977	Flyby	Actively exploring outer edge of Solar System
Galileo	USA	Jupiter	1989	Orbiter/probe	Ended 2003
Ulysses	USA, Europe	Sun	1990	Solar polar orbiter	Ended 2008
SOHO	USA, Europe	Sun	1995	Orbiter	Active
Mars Global Surveyor	USA	Mars	1996	Orbiter	Ended 2006
Cassini-Huygens	USA, Europe, Italy	Saturn, Titan	1997	Saturn orbiter, Titan probe/lander	Ended 2017
Stardust	USA	Comets	1999	Sample return/flyby	Ended 2011
Mars Odyssey	USA	Mars	2001	Orbiter	Active
Mars Exploration Rover	USA	Mars	2003	Two landers	One rover active
Hayabusa	Japan	Asteroid	2003	Sample return	Ended 2010
Mars Express	Europe	Mars	2003	Orbiter	Active
Messenger	USA	Mercury (2011)	2004	Orbiter	Ended 2015
Rosetta	Europe	Comet 67P/Churyumov–Gerasimenko (2014)		Orbiter and lander	Ended 2016
Venus Express	Europe	Venus	2005	Orbiter	Ended 2014
Mars Reconnaissance Orbiter (MRO)	USA	Mars	2005	Orbiter	Active
Deep Impact/EPOXI	USA	Comet Hartley (2010)	2005	Impactor/flyby	Ended 2010
STEREO	USA	Sun	2006	Two orbiters	Active
New Horizons	USA	Pluto (2015)	2006	Flyby	Active
Chang'e 1	China	Moon	2007	Orbiter	Ended 2009
Kayuga	Japan	Moon	2007	Orbiter	Ended 2009
Artemis	USA	Moon, solar wind	2007	Orbiter	Active
Dawn	USA	Vesta (2011), Ceres (2015)	2007	Orbiter	Ended 2018
Chandrayaan	India	Moon	2008	Orbiter/impactor	Ended 2009
Lunar Reconnaissance Orbiter (LRO)	USA	Moon	2009	Orbiter	Active
Lunar Crater Observation and Sensing Satellite (LCROSS)	USA	Moon	2009	Impactor	Ended 2009

(continued)

Selected Recent and Current Solar System Missions

(continued)

Spacecraft	Sponsoring Nation(s)*	Destination	Launch Year	Type	Status (mid-2018)
Chang'e 2	China	Moon	2010	Orbiter	Ended 2011
Akatsuki	Japan	Venus (2015)	2010	Orbiter	Active
Juno	USA	Jupiter (2016)	2011	Orbiter	Active
Gravity Recovery and Interior Laboratory (*GRAIL*)	USA	Moon	2011	Two orbiters	Ended 2012
Mars Science Laboratory (*Curiosity* rover)	USA	Mars	2011	Lander	Active
Mars Atmosphere and Volatile EvolutioN (*MAVEN*) mission	USA	Mars	2013	Orbiter	Active
Chang'e 3	China	Moon	2013	Lander	On lunar surface
Mangalyaan	India	Mars	2013	Orbiter	Active
Hayabusa 2	Japan	Asteroid 162173 Ryugu	2014	Orbiter, scheduled sample return	Active
Osiris-REx (Origins, Spectral Interpretation, Resource Identification, Security-Regolith Explorer)	USA	Asteroid 101955 Bennu	2016	Orbiter, scheduled sample return	En route
ExoMars Trace Gas Orbiter (*TGO*)	Europe, Russia	Mars (2016)	2016	Orbiter (lander crashed)	Active
InSight	USA	Mars	2018	Lander	En route
Parker Solar Probe	USA	Sun	2018	Orbiter	En route

*Countries are represented by the following agencies: China = CNSA (China National Space Administration); Europe = ESA (European Space Agency); India = ISRO (Indian Space Research Organisation); Italy = Italian Space Agency; Japan = JAXA (Japan Aerospace Exploration Agency); Russia = ROSCOSMOS (State Corporation for Space Activities); USA = NASA (National Aeronautics and Space Administration).

APPENDIX 6:
Brightest Stars

The 25 Brightest Stars in the Sky

Name	Common Name	Distance (ly)	Spectral Type	Relative Visual Luminosity* (Sun = 1.000)	Apparent Visual Magnitude	Absolute Visual Magnitude
Sun	Sun	1.58×10^{-5}	G2V	1.000	−26.74	4.83
Alpha Canis Majoris	Sirius	8.60	A1V	22.9	−1.46	1.43
Alpha Carinae	Canopus	309	F0II	14,900	−0.72	−5.60
Alpha^1 Centauri	Rigil Kentaurus A	4.36	G2V	1.51	−0.01	4.38
Alpha^2 Centauri	Rigil Kentaurus B	4.36	K1V	0.44	1.33	5.71
Alpha Bootis	Arcturus	37	K1.5III	113	−0.04	−0.30
Alpha Lyrae	Vega	25	A0Va	49.2	0.03	0.60
Alpha Aurigae	Capella	43	G5IIIe+G0III	137	0.08	−0.51
Beta Orionis	Rigel	860	B8Iab	54,000	0.12	−7.0
Alpha Canis Minoris	Procyon	11.5	F5IV-V	7.73	0.34	2.61
Alpha Eridani	Achernar	140	B3Vpe	1,030	0.46	−2.70
Beta Centauri	Hadar	392	B1III	7,180	0.61	−4.81
Alpha Orionis	Betelgeuse	570	M2Iab	13,600	0.7	−5.5
Alpha Aquilae	Altair	16.7	A7V	11.1	0.77	2.22
Alpha Crucis	Acrux	325	B0.5IV+B1V	3,100	1.3	−3.9
Alpha Tauri	Aldebaran	67	K5III	163	0.85	−0.70
Alpha Scorpii	Antares	550	M1.5Ib	16,300	0.96	−5.7
Alpha Virginis	Spica	250	B1IV+B4V	1,920	1.04	−3.38
Beta Geminorum	Pollux	34	K0III	32.2	1.14	1.06
Alpha Piscis	Fomalhaut	25	A3V	17.4	1.16	1.73
Beta Crucis	Mimosa	280	B0.5III	1,980	1.25	−3.41
Alpha Cygni	Deneb	1,425	A2Ia	58,600	1.25	−7.09
Alpha Leonis	Regulus	79	B7V	146	1.35	−0.58
Epsilon Canis Majoris	Adhara	405	B2II	3,400	1.50	−4.0
Alpha Gemini	Castor	51	A1V+A5Vm	49	1.58	0.61
Gamma Crucis	Gacrux	88	M3.5III	138	1.63	−0.52

SOURCES: Data from Jim Kaler's *STARS* page (http://stars.astro.illinois.edu/sow/bright.html); SIMBAD Astronomical Database (http://simbad.u-strasbg.fr/simbad).
*Luminosity in this table refers only to radiation in "visual" light.

APPENDIX 7:
Observing the Sky

This appendix gives you enough information to make sense of a star chart or list of astronomical objects and find a few objects in the sky.

Celestial Coordinates

In Chapter 2, we discuss the **celestial sphere**—the imaginary sphere with Earth at its center on which celestial objects appear to lie. Several coordinate systems are used to specify the positions of objects on the celestial sphere. The simplest of these is the *altitude-azimuth coordinate system*. The altitude-azimuth coordinate system is based on the "map" direction to an object (the object's azimuth, with north = 0°, east = 90°, south = 180°, and west = 270°) combined with how high the object is above the horizon (the object's altitude, with the horizon at 0° and the zenith at 90°). For example, an object 10° above the eastern horizon has an altitude of 10° and an azimuth of 90°. An object 45° above the horizon in the southwest is at altitude 45°, azimuth 225°.

The altitude-azimuth coordinate system is the simplest way to tell someone where in the sky to look at the moment, but it is not a good coordinate system for cataloging the positions of objects. The altitude and azimuth of an object are different for each observer, depending on the observer's position on Earth, and they are constantly changing as Earth rotates on its axis. To specify the direction to an object in a way that is the same for everyone requires a coordinate system that is fixed relative to the celestial sphere. The most common such coordinates are called *celestial coordinates*.

Celestial coordinates are illustrated in **Figure A7.1**. Celestial coordinates are much like the traditional system of latitude and longitude used on the surface of Earth. On Earth, latitude specifies how far you are from Earth's equator, as discussed in Chapter 2. If you are on Earth's equator, your latitude is 0°. If you are at Earth's North Pole, your latitude is 90° north. If you are at Earth's South Pole, your latitude is 90° south.

The latitude-like coordinate on the celestial sphere is called **declination**, often signified with the lowercase Greek letter δ (delta). The celestial equator has $\delta = 0°$. The north celestial pole has $\delta = +90°$. The south celestial pole has $\delta = -90°$. (See Chapter 2 if you need to refresh your memory about the celestial equator or celestial poles.) Declination is usually expressed in degrees, minutes of arc, and seconds of arc. For example, Sirius, the brightest star in the sky, has $\delta = -16°42'58''$ meaning that it is located not quite 17° south of the celestial equator.

On Earth, east–west position is specified by longitude. Lines of constant longitude run north–south from one pole to the other. Unlike latitude, for which the equator provides a natural place to call "zero," longitude has no natural starting point, so one was invented. By arbitrary convention, the Royal Observatory in Greenwich, England, is defined to lie at a longitude of 0°. On the celestial sphere, the longitude-like coordinate is called **right ascension**, often signified with the lowercase Greek letter α (alpha). Unlike longitude, right ascension *does* have a natural starting point on the celestial sphere: the vernal equinox, or the point at which the ecliptic crosses the celestial equator with the Sun moving from the southern sky into the northern sky. The (Northern Hemisphere) vernal equinox defines the line of right ascension at which $\delta = 0°$. The (Northern Hemisphere) autumnal equinox, located on the opposite side of the sky, is at $\delta = 180°$.

Normally, right ascension is measured in units of time rather than degrees. Earth takes 24 hours (of sidereal time) to rotate on its axis, so the celestial sphere is divided into 24 hours of right ascension, with each hour of right ascension corresponding to 15°. Hours of right ascension are then subdivided into minutes and seconds of time. Right ascension increases going to the east. The right ascension of Sirius, for example, is $\alpha = 06^h45^m08.9^s$, meaning that Sirius is about 101° (that is, 06^h45^m) east of the vernal equinox. Time is a natural unit for measuring right ascension because time naturally tracks the motion of objects due to Earth's rotation on its axis. If stars on the meridian at a certain time have $\alpha = 06^h$, then an hour later the stars on the meridian will have $\alpha = 07^h$, and an hour after that they will have $\alpha = 08^h$. The *local sidereal time*, or *star time*, at your location right now is equal to the right ascension of the stars on your meridian at the moment. Because of Earth's motion around the Sun, a sidereal day is about 4 minutes shorter than a solar day, so local sidereal time constantly gains on solar time. At midnight on September 22, the local sidereal time is 0^h. By midnight on December 21, local sidereal time has advanced to 06^h. On March 20, local sidereal time at midnight is 12^h. And at midnight on June 20, local sidereal time is 18^h.

Putting all that together, right ascension and declination provide a convenient way to specify the location of any object on the celestial sphere. Sirius is located at $\alpha = 06^h45^m08.9^s$, $\delta = -16°42'58''$ which means that at midnight on December 21

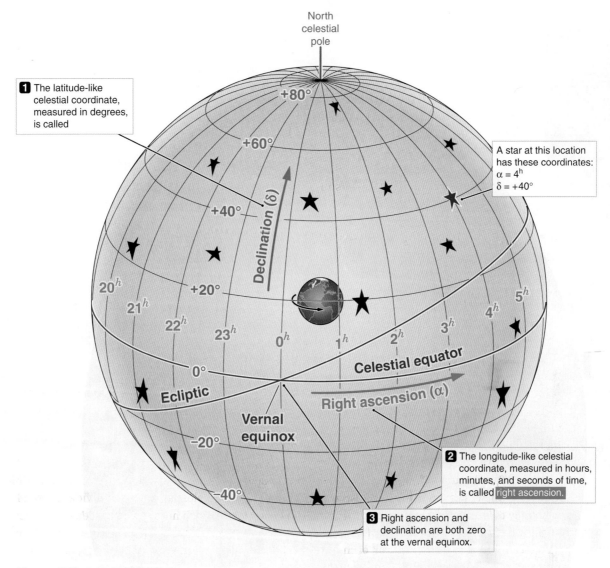

Figure A7.1 Celestial coordinates.

(local sidereal time = 06^h), you will find Sirius about 45^m east of the meridian, not quite 17° south of the celestial equator.

Just one final caveat remains. As we discussed in Chapter 2, the directions of the celestial equator, celestial poles, and vernal equinox are constantly changing as Earth's axis wobbles like the axis of a spinning top. In Chapter 2, we called that 26,000-year wobble the **precession of the equinoxes**, meaning that the location of the equinoxes is slowly advancing along the ecliptic. So when we specify the celestial coordinates of an object, we need to specify the date at which the positions of the vernal equinox and celestial poles were measured. By convention, coordinates are usually referred to with the position of the vernal equinox on January 1, 2000. A complete, formal specification of the coordinates of Sirius would then be $\alpha(2000) = 06^h45^m08.9^s$, $\delta(2000) = -16°42'58''$, where the "2000" in parentheses refers to the equinox of the coordinates.

Constellations and Names

Although it is certainly possible to specify any location on the surface of Earth exactly by giving its latitude and longitude, using a more descriptive address is usually more convenient. We might say, for example, that one of the coauthors of this book works near latitude 37° north, longitude 122° west; but it would probably mean a lot more to you if we said that George Blumenthal works in Santa Cruz, California.

Just as the surface of Earth is divided into nations and states, the celestial sphere is divided into 88 **constellations**, the names of which are often used to refer to objects within their boundaries (see the star charts in **Figure A7.2**). The brightest stars within the boundaries of a constellation are named using a Greek letter combined with the name of the constellation. For example, the star Sirius is the brightest star in the constellation Canis Major

Figure A7.2A The sky from right ascension 20ʰ to 04ʰ and declination −60° to +60°.

(b)

Figure A7.2B The sky from right ascension 02^h to 10^h and declination −60° to +60°.

Figure A7.2C The sky from right ascension 08^h to 16^h and declination −60° to +60°.

Figure A7.2D The sky from right ascension 14ʰ to 22ʰ and declination −60° to +60°.

Figure A7.2E The regions of the sky north of declination +40° and south of declination −40°. NCP = north celestial pole; SCP = south celestial pole.

(literally, the "great dog"), so it is called "Alpha Canis Majoris." The bright red star in the northeastern corner of the constellation Orion is called "Alpha Orionis," also known as Betelgeuse. Rigel, the bright blue star in the southwest corner of Orion, is also called "Beta Orionis."

Astronomical objects can take on a bewildering range of names. For example, the bright southern star Canopus, also known as "Alpha Carinae" (the brightest star in the constellation Carina), has no fewer than 34 names, most of which are about as memorable as "SAO 234480" (number 234,480 in the Smithsonian Astrophysical Observatory catalog of stars).

A constellation is spelled slightly differently when it becomes part of a star's name. For example, Sirius is called "Alpha Canis Majoris," not "Alpha Canis Major"; Rigel is referred to as "Beta Orionis," not "Beta Orion"; and Canopus becomes "Alpha Carinae," not "Alpha Carina." That is because the Latin genitive, or possessive, case is used with star names; for example, *Orionis* means "of Orion."

Astronomical Magnitudes

Apparent Magnitudes

We first introduced magnitudes in Working It Out 13.2; here, we provide more information. You are most likely to see that system if you take a lab course in astronomy or if you use a star catalog. Astronomers use the logarithmic system of **apparent magnitudes** to compare the apparent brightness of objects in the sky. Other common systems of logarithmic measurements that you may have encountered include decibels for measuring sound levels and the Richter scale for measuring the strength of earthquakes. For example, an earthquake of magnitude 6 is not just a little stronger than an earthquake of magnitude 5; it is, in fact, 10 times stronger.

As discussed in Working It Out 13.2, a difference of five magnitudes between the apparent brightness of two stars (say, a star with $m = 6$ and a star with $m = 1$) corresponds to 100 times difference in brightness, and *the greater the magnitude, the fainter the object*. If five steps in magnitude corresponds to a factor of 100 in brightness, one step in magnitude must correspond to the fifth root of 100; that is, a factor of $100^{1/5}$ = approximately 2.512 in brightness ($100^{1/5} \times 100^{1/5} \times 100^{1/5} \times 100^{1/5} \times 100^{1/5} = 100$).

If star 1 has a brightness of b_1 and star 2 has a brightness of b_2, the ratio of the brightness of the stars is given by

$$\frac{b_1}{b_2} = (2.512)^{m_2 - m_1} = 100^{\frac{(m_2 - m_1)}{5}}$$

We can put that into the more common base 10 by noting that $100 = 10^2$, so the expression becomes

$$\frac{b_1}{b_2} = 10^{2 \times \frac{(m_2 - m_1)}{5}} = 10^{0.4(m_2 - m_1)}$$

After we take the log of both sides and divide by 0.4, the difference in magnitude ($m_2 - m_1$) between the two stars is given by

$$m_2 - m_1 = 2.5 \log_{10} \frac{b_1}{b_2}$$

The following table shows some examples using the preceding equations.

Apparent Magnitude Difference ($m_2 - m_1$)	Ratio of Apparent Brightness (b_1/b_2)
1	2.512
2	$2.512^2 = 6.3$
3	$2.512^3 = 15.8$
4	$2.512^4 = 39.8$
5	$2.512^5 = 100$
10	$2.512^{10} = 100^2 = 10,000$
15	$2.512^{15} = 100^3 = 1,000,000$
20	$2.512^{20} = 100^4 = 10^8$
25	$2.512^{25} = 100^5 = 10^{10}$

Absolute Magnitudes

Recall that stars differ in their brightness for two reasons: the amount of light they are actually emitting and their distance from Earth. The magnitude system also is used for **luminosity**, with the same scale as for brightness: a difference of five magnitudes corresponds to 100 times difference in luminosity. Astronomers call those **absolute magnitudes** (M_{abs}), and the idea is to imagine how bright the star would be if it were at a distance of 10 parsecs (pc). Absolute magnitudes enable comparison of how luminous two stars really are, without the factor of distance. The Sun is very bright because it is so close (apparent visual magnitude = −27), but if the Sun were at a distance of 10 pc, its magnitude would be only about 5.[1] Thus, the absolute magnitude of the Sun is $M_{abs} = 5$. Recall that the luminosity of a star is usually expressed by comparing it with the luminosity of the Sun. As with apparent magnitudes, higher magnitude numbers correspond to lower luminosity. Thus, a star 1/100 as luminous as the Sun will be 5 absolute magnitudes fainter, or $M_{abs} = 10$. A

---

[1] The apparent and absolute magnitudes of the Sun are −26.74 and +4.83, respectively. We use +5 for the Sun's absolute magnitude as an approximation.

star 10,000 times more luminous than the Sun will be 10 absolute magnitudes brighter, or $M_{abs} = -5$.

Absolute magnitudes and luminosities follow the same equations as those we provided already, using L instead of b and M_{abs} instead of m:

$$\frac{L_1}{L_2} = 10^{2 \times \frac{M_{abs(2)} - M_{abs(1)}}{5}} = 10^{0.4(M_{abs(2)} - M_{abs(1)})}$$

and

$$M_{abs(2)} - M_{abs(1)} = 2.5 \log_{10} \frac{L_1}{L_2}$$

Most often, astronomers think about the luminosity of a star in comparison with the luminosity of the Sun. Here $L_1 = L_{star}$ and $L_2 = L_{Sun}$. The following table compares luminosity (where $L_{Sun} = 1$) with absolute magnitude of a star.

L_{Star}/L_{Sun}	M_{abs}
1,000,000	−10
10,000	−5
100	0
1	5
1/100	10
1/10,000	15

Distance Modulus

The difference between the apparent magnitude and the absolute magnitude depends on the star's distance. By definition, a star at a distance of exactly 10 pc will have an apparent magnitude equal to its absolute magnitude. Astronomers can always measure the brightness of a star and thus its apparent magnitude and can estimate the luminosity of a star and thus its absolute magnitude by using the Hertzsprung-Russell (H-R) diagram. That is how the distances to most stars are found.

Using the preceding equations and the definition of absolute magnitude, we can get to the following relatively simple expression:

$$m - M_{abs} = 5 \log_{10} d - 5$$

where distance d is in parsecs.

We can rewrite that equation to solve for distance as follows:

$$d = 10^{\frac{(m - M_{abs} + 5)}{5}}$$

The following table shows how the difference between an object's apparent and absolute magnitudes leads to its distance in parsecs.

$m - M_{abs}$	Distance (pc)
−3	2.5
−2	4.0
−1	6.3
0	10
1	16
2	25
3	40
4	63
5	100
10	1,000
15	10,000
20	100,000

Although the system of astronomical magnitudes is convenient in many ways—which is why astronomers continue to use it—it can also be confusing to new students. Just remember three things and you will probably get by:

1. The greater the magnitude, the fainter the object.
2. One magnitude *smaller* means about two and a half times *brighter*.
3. The brightest stars in the sky have magnitudes of less than 1, and the faintest stars visible to the naked eye on a dark night have magnitudes of about 6.

A final note: Astronomers sometimes use "colors" based on the ratio of the brightness of a star as seen in two parts of the spectrum. The "b_B/b_V color," for example, is the ratio of the brightness of a star seen through a blue filter, divided by the brightness of a star seen through a yellow-green (visual) filter. Normally, astronomers instead discuss the "$B - V$ color" of a star, which is equal to the difference between a star's blue magnitude and its visual magnitude. We can use the previous expression for a magnitude difference to write

$$B - V \text{ color} = m_B - m_V = -2.5 \log_{10}(b_B/b_V)$$

Thus, a star with a b_B/b_V color of 1.0 has a $B - V$ color of 0.0, and a star with a b_B/b_V color of 1.4 has a $B - V$ color of -0.37. Notice that, as with magnitudes, $B - V$ colors are "backward": the bluer a star, the greater its b_B/b_V color but the less its $B - V$ color.

APPENDIX 8:
Uniform Circular Motion and Circular Orbits

Uniform Circular Motion

In Chapter 4 (see Section 4.2 and Figure 4.7), we discuss the motion of an object moving in a circle at a constant speed. That motion, called **uniform circular motion**, is the result of the fact that centripetal force always acts toward the center of the circle. The key question when thinking about uniform circular motion is, How hard does something have to pull to keep the object moving in a circle? Part of the answer to that question is obvious: the more massive an object is, the harder it will be to keep it moving on its circular path. According to Newton's second law of motion, $F = ma$, or here, the centripetal force equals the mass multiplied by the centripetal acceleration. The larger the mass, the greater the force required to keep it moving in its circle.

The centripetal force needed to keep an object moving in constant circular motion also depends on two other quantities: the speed of the object and the size of the circle. The faster an object is moving, the more rapidly it has to change direction to stay on a circle of a given size. The second quantity that influences the needed acceleration is the radius of the circle. The smaller the circle, the greater the pull needed to keep it on track. You can understand that relationship by looking at the motion. A small circle requires a continuous "hard" turn, whereas a larger circle requires a more gentle change in direction. It takes more force to keep an object moving faster in a smaller circle than it does to keep the same object moving more slowly in a larger circle. (To get a better feel for how that works, think about the difference between riding in a car taking a tight curve at high speed and a car moving slowly around a gentle curve.)

To arrive at the circular velocity and other results discussed in Chapter 4, those intuitive ideas about uniform circular motion are turned into a quantitative expression of exactly how much centripetal acceleration is needed to keep an object moving in a circle with radius r at speed v. **Figure A8.1** shows a ball moving around a circle of radius r at a constant speed v at two times. The centripetal acceleration keeping the ball on the circle is a. Remember that the acceleration is always directed toward the center of the circle, whereas the velocity of the ball is always perpendicular to the acceleration. The ball's velocity and its acceleration are always at right angles to each other. As the object moves around the circle, the direction of motion and the direction of the acceleration change together in lockstep.

Figure A8.1 contains two triangles. Triangle 1 shows the velocity (speed and direction) at both times. The arrow labeled "Δv" connecting the heads of the two velocity arrows shows how much the velocity changed between time 1 (t_1) and time 2 (t_2). That change is the effect of the centripetal acceleration. If you imagine that points 1 and 2 are very close together—so close that the direction of the centripetal acceleration does not change by much between the two—the centripetal acceleration equals the change in the velocity divided by the time between the two, $\Delta t = t_2 - t_1$. So, $\Delta v = a\Delta t$.

Triangle 2 shows something similar. Here, the arrow labeled "Δr" indicates the change in the position of the ball between time 1 and time 2. Again, if you imagine that the time between the two points is very short, Δr is equal to the velocity multiplied by the time, or $\Delta r = v\Delta t$.

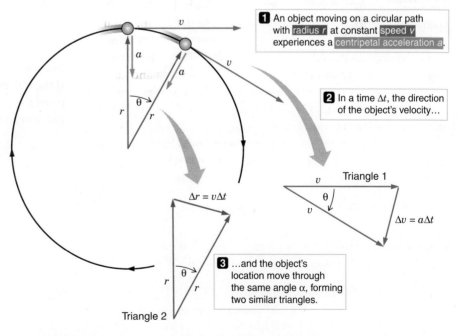

Figure A8.1 Similar triangles are used to find the centripetal force needed to keep an object moving at a constant speed on a circular path.

The line between the center of the circle and the ball is always perpendicular to the velocity of the ball. So if the direction of the ball's velocity changes by an angle θ, the direction of the line between the ball and the center of the circle must also change by the same angle α. In other words, triangles 1 and 2 are "similar triangles." They have the same *shape*. If the triangles are the same shape, the ratio of two sides of triangle 1 must equal the ratio of the two corresponding sides of triangle 2. Then,

$$\frac{a\Delta t}{v} = \frac{v\Delta t}{r}$$

If we divide by Δt on both sides of the equation and then cross-multiply, we obtain

$$ar = v^2$$

which, after we divide both sides of the equation by r, becomes

$$a_{centripetal} = \frac{v^2}{r}$$

The subscript "centripetal" is added to a to signify that this is the centripetal acceleration needed to keep the object moving in a circle of radius r at speed v. The centripetal force required to keep an object of mass m moving on such a circle is then

$$F_{centripetal} = ma_{centripetal} = \frac{mv^2}{r}$$

Circular Orbits

For an object moving in a circular orbit, no string is available to hold the ball on its circular path. Instead, that force is provided by **gravity**.

Think about an object with mass m in orbit about a much larger object with mass M. The orbit is circular, and the distance between the two objects is given by r. The force needed to keep the smaller object moving at speed v in a circle with radius r is given by the previous expression for $F_{centripetal}$. The force actually provided by gravity (see Chapter 4) is

$$F_{grav} = G\frac{Mm}{r^2}$$

If gravity is responsible for holding the mass in its circular motion, it should be true that $F_{grav} = F_{centripetal}$. That is, if mass m is moving in a circle under the force of gravity, the force provided by gravity must equal the centripetal force needed to explain that circular motion. Setting the two expressions for $F_{centripetal}$ and F_{grav} equal to each other gives

$$\frac{mv^2}{r} = G\frac{Mm}{r^2}$$

All that remains is a bit of algebra. Dividing by m on both sides of the equation and multiplying both sides by r gives

$$v^2 = G\frac{M}{r}$$

Taking the square root of both sides then yields the desired result:

$$v_{circular} = \sqrt{\frac{GM}{r}}$$

That is the **circular velocity** we presented in Chapter 4. It is the velocity at which an object in a circular orbit must be moving. If the object were not moving at that velocity, gravity would not be providing the needed centripetal force, and the object would not move in a circle.

Glossary

A

aberration of starlight A star's apparent displacement in position due to the finite speed of light and Earth's orbital motion around the Sun. (Ch. 18)

absolute magnitude A measure of the intrinsic brightness, or luminosity, of a celestial object, generally a star. Specifically, the apparent magnitude an object would have if it were located at a standard distance of 10 parsecs (pc). Compare *apparent magnitude*. (Ch. 13)

absolute zero The temperature at which thermal motions cease. The lowest possible temperature. Zero on the Kelvin temperature scale. (Ch. 5)

absorption The process by which an atom captures energy from a passing photon. Compare *emission*. (Ch. 5)

absorption line A minimum in the intensity of a spectrum that is due to the absorption of electromagnetic radiation at a specific wavelength determined by the energy levels of an atom or molecule. Compare *emission line*. (Ch. 5)

acceleration The rate at which the speed and/or direction of an object's motion is changing. (Ch. 3)

accretion disk A flat, rotating disk of gas and dust surrounding an object, such as a young stellar object, a forming planet, a collapsed star in a binary system, or a black hole. (Ch. 7)

achondrite A stony meteorite that does not contain chondrules. Compare *chondrite*. (Ch. 12)

active comet A comet nucleus that approaches close enough to the Sun to show signs of activity, such as the production of a coma and tail. (Ch. 12)

active galactic nucleus (AGN) A highly luminous, compact galactic nucleus whose luminosity may exceed that of the rest of the galaxy. (Ch. 19)

active region An area of the Sun's chromosphere anchoring bursts of intense magnetic activity. (Ch. 14)

adaptive optics Electro-optical systems that largely compensate for image distortion caused by Earth's atmosphere. (Ch. 6)

AGB See *asymptotic giant branch*. (Ch. 16)

AGN See *active galactic nucleus*. (Ch. 19)

albedo The fraction of electromagnetic radiation striking a surface that is reflected by that surface. (Ch. 5)

algebra A branch of mathematics in which letters represent numeric variables. (Appendix 1)

alpha particle A ^{4}He nucleus, consisting of two protons and two neutrons. Alpha particles are given off in the type of radioactive decay referred to as *alpha decay*. (Ch. 16)

altitude The location of an object above the horizon, measured by the angle formed between an imaginary line from an observer to the object and a second line from the observer to the point on the horizon directly below the object. (Ch. 2)

Amors A group of asteroids whose orbits cross the orbit of Mars but not the orbit of Earth. Compare *Apollos* and *Atens*. (Ch. 12)

amplitude In a wave, the maximum deviation from its undisturbed or relaxed position. For example, in a water wave the amplitude is the vertical distance from the wave's crest to the undisturbed water level. (Ch. 5)

angular momentum In a rotating or revolving system, a conserved property whose value depends on the velocity and distribution of the system's mass. (Ch. 7)

angular resolution The ability of an imaging device such as a telescope (or the eye) to separate two objects that appear close together. (Ch. 6)

annular solar eclipse The type of solar eclipse that occurs when the apparent diameter of the Moon is less than that of the Sun, leaving a visible ring of light ("annulus") surrounding the dark disk of the Moon. Compare *partial solar eclipse* and *total solar eclipse*. (Ch. 2)

Antarctic Circle The circle on Earth with latitude 66.5° south, marking the northern limit where at least one day per year is in 24-hour daylight. Compare *Arctic Circle*. (Ch. 2)

anthropic principle The idea that this universe (or this bubble in the universe) must have physical properties that allow intelligent life to develop. (Ch. 22)

anthropogenic climate change The release of greenhouse gases into the atmosphere from human activities, such as the burning of fossil fuels. (Ch. 9)

anticyclonic motion The rotation of a weather system resulting from the Coriolis effect as air moves outward from a region of high atmospheric pressure. Compare *cyclonic motion*. (Ch. 9)

antimatter Matter made up of antiparticles. (Ch. 14, 22)

antiparticle An elementary particle of antimatter identical in mass but opposite in charge and all other properties to its corresponding ordinary matter particle. (Ch. 22)

aperture The clear diameter of a telescope's objective lens or primary mirror. (Ch. 6)

aphelion (pl. aphelia) The point in a solar orbit that is farthest from the Sun. Compare *perihelion*. (Ch. 12)

Apollos A group of asteroids whose orbits cross the orbits of both Earth and Mars. Compare *Amors* and *Atens*. (Ch. 12)

apparent daily motion As seen from Earth's surface, the path along which each object seems to move across the sky. (Ch. 2)

apparent magnitude A measure of the apparent brightness of a celestial object, generally a star. Compare *absolute magnitude*. (Ch. 13)

arcminute (arcmin) A minute of arc ('), a unit used to measure angles. An arcminute is 1/60 of a degree of arc. (Ch. 6, 13)

arcsecond (arcsec) A second of arc ("), a unit used to measure very small angles. An arcsecond is 1/60 of an arcminute, or 1/3,600 of a degree of arc. (Ch. 6, 13)

Arctic Circle The circle on Earth with latitude 66.5° north, marking the southern limit where at least one day per year is in 24-hour daylight. Compare *Antarctic Circle*. (Ch. 2)

asteroid Also called *minor planet*. A primitive rocky or metallic body (planetesimal) that has survived planetary accretion. Asteroids are parent bodies of meteoroids. (Ch. 12)

asteroid belt The region between the orbits of Mars and Jupiter that contains most of the asteroids in our Solar System. (Ch. 12)

astrobiology An interdisciplinary science combining astronomy, biology, chemistry, geology, and physics to study life in the cosmos. (Ch. 1, 24)

astrology The belief that the positions and aspects of stars and planets influence human affairs and characteristics, as well as terrestrial events. (Ch. 1)

astronomical seeing A measurement of the degree to which Earth's atmosphere degrades the resolution of a telescope's view of astronomical objects. (Ch. 6)

astronomical unit (AU) The average distance from the Sun to Earth: approximately 150 million kilometers (km). (Ch. 1, 3)

astronomy The scientific study of planets, stars, galaxies, and the universe as a whole. (Ch. 1)

astrophysics The application of physical laws to the understanding of planets, stars, galaxies, and the universe as a whole. (Ch. 13)

asymptotic giant branch (AGB) The path on the H-R diagram that goes from the horizontal branch toward higher luminosities and lower temperatures, asymptotically approaching and then rising above the red giant branch. (Ch. 16)

Atens A group of asteroids whose orbits cross the orbit of Earth but not the orbit of Mars. Compare *Amors* and *Apollos*. (Ch. 12)

atmosphere The gravitationally bound outer gaseous envelope surrounding a planet, moon, or star. (Ch. 9)

atmospheric greenhouse effect A warming of planetary surfaces produced by atmospheric gases that transmit optical solar radiation but partially trap infrared radiation. Compare *greenhouse effect*. (Ch. 9)

atmospheric probe An instrumented package designed to provide on-site measurements of the chemical and/or physical properties of a planetary atmosphere. (Ch. 6)

atmospheric window A region of the electromagnetic spectrum in which radiation can penetrate a planet's atmosphere. (Ch. 6)

atom The smallest unit of a chemical element that retains the properties of that element. Each atom is composed of a nucleus (neutrons and protons) surrounded by a cloud of electrons.

AU See *astronomical unit*. (Ch. 1, 3)

aurora Emission in the upper atmosphere of a planet from atoms that have been excited by collisions with energetic particles from the planet's magnetosphere. (Ch. 9)

autumnal equinox 1. One of two points where the Sun crosses the celestial equator. 2. The day on which the Sun appears at that location, marking the first day of autumn (about September 22 in the Northern Hemisphere and March 20 in the Southern Hemisphere). Compare *vernal equinox*. See also *summer solstice* and *winter solstice*. (Ch. 2)

axion A hypothetical elementary particle first proposed to explain certain properties of the neutron and now considered a candidate for cold dark matter. (Ch. 23)

B

backlighting Illumination from behind a subject as seen by an observer. Fine material such as human hair and dust in planetary rings stands out best when viewed under backlighting conditions. (Ch. 11)

bar A unit of pressure. One bar is equivalent to 10^5 newtons per square meter—approximately equal to Earth's atmospheric pressure at sea level. (Ch. 9)

barred spiral galaxy A spiral galaxy with a bulge having an elongated, barlike shape. Compare *elliptical galaxy*, *irregular galaxy*, *S0 galaxy*, and *spiral galaxy*. (Ch. 19)

basalt Gray to black volcanic rock, rich in iron and magnesium. (Ch. 8)

beta decay 1. The decay of a neutron into a proton by emission of an electron (beta ray) and an antineutrino. 2. The decay of a proton into a neutron by emission of a positron and a neutrino. (Ch. 22)

Big Bang The event that occurred 13.8 billion years ago that marks the beginning of time and the universe. (Ch. 1, 21)

Big Bang nucleosynthesis The formation of low-mass nuclei (H, He, Li, Be) during the first few minutes after the Big Bang. (Ch. 21)

Big Crunch A hypothetical cosmic future in which the expansion of the universe reverses and the universe collapses onto itself. (Ch. 22)

Big Rip A hypothetical cosmic future in which all matter in the universe, from stars to subatomic particles, is progressively torn apart by expansion of the universe. (Ch. 22)

binary star A system in which two stars are in gravitationally bound orbits about their common center of mass. (Ch. 13)

binding energy The minimum energy required to separate an atomic nucleus into its component protons and neutrons. (Ch. 17)

biosignature Chemical compounds detectable in the spectra of an exoplanet's atmosphere that might indicate the presence of life. (Ch. 24)

biosphere The global sum of all living organisms on Earth (or any planet or moon). Compare *hydrosphere* and *lithosphere*. (Ch. 18)

bipolar outflow Material streaming away in opposite directions from both sides of the accretion disk of a young star. (Ch. 15)

black hole An object so dense that its escape velocity exceeds the speed of light; a *singularity* in spacetime. (Ch. 18)

blackbody An object that absorbs and can reemit all electromagnetic energy it receives. (Ch. 5)

blackbody spectrum See *Planck spectrum*. (Ch. 5)

blue-green algae See *cyanobacteria*. (Ch. 24)

blueshift The Doppler shift toward shorter (bluer) wavelengths of light from an approaching object. Compare *redshift*. (Ch. 5)

Bohr model A model of the atom, proposed by Niels Bohr in 1913, in which a small positively charged nucleus is surrounded by orbiting electrons, similar to a miniature solar system. (Ch. 5)

bound orbit An orbit in which an object is gravitationally bound to the body it is orbiting. A bound orbit's velocity is less than the escape velocity. Compare *unbound orbit*. (Ch. 4)

bow shock The boundary at which the speed of the solar wind abruptly drops from supersonic to subsonic in its approach to a planet's magnetosphere; the boundary between the region dominated by the solar wind and the region dominated by a planet's magnetosphere. (Ch. 9)

brightness The apparent intensity of light from a luminous object. Brightness depends on both the *luminosity* of a source and its distance. Units at the detector: watts per square meter (W/m²). (Ch. 5)

brown dwarf A "failed star" without enough mass to fuse hydrogen in its core. An object whose mass is intermediate between that of the least massive stars and that of supermassive planets. (Ch. 7, 15)

bulge The central region of a spiral galaxy that is similar in appearance to a small elliptical galaxy. (Ch. 19)

C

C See *Celsius*. (Ch. 5)

C-type asteroid An asteroid made of material that has remained mostly unmodified since the formation of the Solar System; the most primitive type of asteroid. Compare *M-type asteroid* and *S-type asteroid*. (Ch. 12)

caldera The summit crater of a volcano. (Ch. 11)

Cambrian explosion The spectacular rise in the number and diversity of biological species that occurred between 540 million and 500 million years ago. (Ch. 24)

carbon-nitrogen-oxygen (CNO) cycle One way in which hydrogen is converted to helium (hydrogen fusion in the interiors of main-sequence stars. Compare *proton-proton chain*. (Ch. 17)

carbon star A cool red giant or asymptotic giant branch star that has an excess of carbon in its atmosphere. (Ch. 16)

carbonaceous chondrite A primitive stony meteorite that contains chondrules and is rich in carbon and volatile materials. (Ch. 12)

Cassini Division The largest gap in Saturn's rings, discovered by Jean-Dominique Cassini in 1675. (Ch. 11)

catalyst An atomic and molecular structure that permits or encourages chemical and nuclear reactions but does not change its own chemical or nuclear properties. (Ch. 9, 14)

CCD See *charge-coupled device*. (Ch. 6)

celestial equator The imaginary great circle that is the projection of Earth's equator onto the celestial sphere. (Ch. 2)

celestial sphere An imaginary sphere with celestial objects on its inner surface and Earth at its center. The celestial sphere has no physical existence but is a convenient tool for picturing the directions in which celestial objects are seen from Earth's surface. (Ch. 2)

Celsius (C) Also called *centigrade scale*. The arbitrary temperature scale, formulated by Anders Celsius (1701–1744), that defines 0°C as the freezing point of water and 100°C as the boiling point of water at sea level. Unit: degrees Celsius (°C). Compare *Fahrenheit* and *Kelvin temperature scale*. (Ch. 5)

center of mass 1. The weighted average location of all the mass in a system of objects. The point in any isolated system that moves according to Newton's first law of motion. 2. In a binary star system, the point between the two stars that is the focus of both their elliptical orbits. (Ch. 4, 13)

centigrade scale See *Celsius*. (Ch. 5)

centripetal force A force directed toward the center of curvature of an object's curved path. (Ch. 4)

Cepheid variable An evolved high-mass star with an atmosphere that is pulsating, leading to variability in the star's luminosity and color. (Ch. 17)

Chandrasekhar limit The upper limit on the mass of an object supported by electron degeneracy pressure; approximately 1.4 solar masses (M_{Sun}). (Ch. 16)

chaotic Behavior in complex systems in which a small change in the initial state of a system can lead to a large change in the final state of the system. (Ch. 7)

charge-coupled device (CCD) A common type of solid-state detector of electromagnetic radiation that transforms the intensity of light directly into electric signals. (Ch. 6)

charge destruction Electrons are absorbed by protons in atomic nuclei, forming neutrons and releasing neutrinos. (Ch. 17)

chondrite A stony meteorite that contains chondrules. Compare *achondrite*. (Ch. 12)

chondrule A small, crystallized, spherical inclusion of rapidly cooled molten droplets found inside some meteorites. (Ch. 12)

chromatic aberration A detrimental property of a lens in which rays of different wavelengths are brought to different focal distances from the lens. (Ch. 6)

chromosphere The region of the Sun's atmosphere located between the *photosphere* and the *corona*. (Ch. 14)

circular velocity The orbital velocity needed to keep an object moving in a circular orbit. (Ch. 4)

circumpolar Describing the part of the sky, near either celestial pole, that can always be seen above the horizon from a specific location on Earth. (Ch. 2)

circumstellar disk See *protoplanetary disk*. (Ch. 7)

climate The state of an atmosphere averaged over an extended time. Compare *weather*. (Ch. 9)

closed universe A finite universe with a curved spatial structure such that the sum of the angles of a triangle always exceeds 180 degrees. Compare *flat universe* and *open universe*. (Ch. 22)

CMB See *cosmic microwave background radiation*. (Ch. 6, 21)

CNO cycle See *carbon-nitrogen-oxygen cycle*. (Ch. 17)

cold dark matter Particles of dark matter that move slowly enough to be gravitationally bound even in the smallest galaxies. Compare *hot dark matter*. (Ch. 23)

color index The color of a celestial object, generally a star, based on the ratio of its brightness in blue light to its brightness in "visual" (yellow-green) light. The difference between an object's blue (B) magnitude and visual (V) magnitude, $B - V$. (Ch. 13, Appendix 7)

coma (pl. **comae**) The nearly spherical cloud of gas and dust surrounding the nucleus of an active comet. (Ch. 12)

comet A complex object consisting of a small, solid, icy nucleus; an atmospheric halo; and a tail of gas and dust. (Ch. 7)

comet nucleus A primitive planetesimal composed of ices and refractory materials that has survived planetary accretion. The "heart" of a comet, containing nearly the entire mass of the comet. A "dirty snowball." (Ch. 7)

comparative planetology The study of planets by comparing their chemical and physical properties. (Ch. 8)

composite volcano A large, cone-shaped volcano formed by viscous, pasty lava flows alternating with pyroclastic (explosively generated) rock deposits. Compare *shield volcano*. (Ch. 8)

compound lens A lens made up of two or more elements with different refractive indices, the purpose of which is to minimize chromatic aberration. (Ch. 6)

concave mirror A telescope mirror with a surface that curves inward toward the incoming light. (Ch. 6)

conduction The transfer of energy in which the thermal energy of particles is transferred to adjacent particles by collisions or other interactions. Conduction is the most important way that thermal energy is transported in solid matter. Compare *convection*. (Ch. 14)

conservation law A physical law stating that the amount of a particular physical quantity (such as energy or angular momentum) of an isolated system does not change. (Ch. 7)

conservation of angular momentum The physical law stating that the amount of angular momentum of an isolated system does not change. (Ch. 7)

conservation of energy The physical law stating that the amount of energy of an isolated, closed system does not change. (Ch. 7)

constant of proportionality The number by which one quantity is multiplied to get another number. (Ch. 1, 4)

constellation An imaginary image formed by patterns of stars; any of 88 defined areas on the celestial sphere that astronomers use to locate celestial objects. (Ch. 2)

continental drift The slow motion (centimeters per year) of Earth's continents relative to one another and to Earth's mantle. See also *plate tectonics*. (Ch. 8)

continuous radiation Electromagnetic radiation with intensity that varies smoothly over a wide range of wavelengths. (Ch. 5)

continuous spectrum A spectrum containing all wavelengths, without specific spectral lines. (Ch. 5)

convection The transport of thermal energy from the lower (hotter) to the higher (cooler) layers of a fluid by motions within the fluid driven by variations in buoyancy. Compare *conduction*. (Ch. 8, 14)

convective zone A region within a star where energy is transported outward by convection. Compare *radiative zone*. (Ch. 13)

core 1. The innermost region of a planetary interior. Compare *crust* and *mantle*. 2. The innermost part of a star. (Ch. 8, 14)

core accretion–gas capture A process for forming giant planets, in which large amounts of surrounding hydrogen and helium gas are gravitationally captured onto a massive rocky core. (Ch. 7)

Coriolis effect The apparent displacement of objects in a direction perpendicular to their true motion as viewed from a rotating frame of reference. On a rotating planet, that effect arises from different latitudes' rotating at different speeds. (Ch. 2)

corona The hot, outermost part of the Sun's atmosphere. Compare *chromosphere* and *photosphere*. (Ch. 14)

coronal hole A low-density region in the solar corona containing "open" magnetic-field lines along which coronal material is free to stream into interplanetary space. (Ch. 14)

coronal mass ejection (CME) An eruption on the Sun that ejects hot gas and energetic particles at much higher speeds than are typical in the solar wind. (Ch. 14)

cosmic microwave background radiation (CMB) Also called simply *cosmic background radiation*. Isotropic microwave radiation from every direction in the sky having a 2.73-kelvin (K) blackbody spectrum. The CMB is residual radiation from the Big Bang. (Ch. 6, 21)

cosmic ray A very fast-moving particle (usually protons or another atomic nucleus) that originated in outer space; cosmic rays fill the disk of the Milky Way. (Ch. 18, 20)

cosmological constant A constant, introduced into general relativity by Einstein, that characterizes an extra, repulsive force in the universe due to the vacuum of space itself. (Ch. 22)

cosmological principle The (testable) assumption that the same physical laws that apply here and now also apply everywhere and at all times and that the universe has no special locations or directions. (Ch. 1, 21)

cosmological redshift The redshift that results from the expansion of the universe rather than from the motions of galaxies or gravity. Compare *gravitational redshift*. (Ch. 21)

cosmology The study of the large-scale structure and evolution of the universe as a whole. (Ch. 21)

crescent Any phase of the Moon, Mercury, or Venus in which the object appears less than half illuminated by the Sun. Compare *gibbous*. (Ch. 2)

Cretaceous-Paleogene (K-Pg) boundary The boundary between the Cretaceous and Paleogene periods in Earth's history. That boundary corresponds to the time of the impact of an asteroid or comet and the extinction of the dinosaurs. (Ch. 8)

critical density The value of mass density of the universe that, ignoring any cosmological constant, can just barely halt expansion of the universe. (Ch. 22)

crust The relatively thin, outermost, hard layer of a planet, which is chemically distinct from the interior. Compare *core* and *mantle*. (Ch. 8)

cryovolcanism Low-temperature volcanism in which the magmas are composed of molten ices rather than rocky material. (Ch. 11)

cyanobacteria Also called *blue-green algae*. Single-celled organisms that created oxygen in Earth's atmosphere by photosynthesizing carbon dioxide and releasing oxygen as a waste product. (Ch. 24)

cyclonic motion The rotation of a weather system resulting from the Coriolis effect as air moves toward a region of low atmospheric pressure. Compare *anticyclonic motion*. (Ch. 9)

D

Dark Ages The epoch in the history of the universe during which no visible "light" came from astronomical objects. (Ch. 23)

dark energy A form of energy that permeates all space (including the vacuum), producing a repulsive force that accelerates the expansion of the universe. (Ch. 22)

dark matter Matter in galaxies that does not emit or absorb electromagnetic radiation. Dark matter is thought to constitute most of the mass in the universe. Compare *luminous matter*. (Ch. 19)

dark matter halo The centrally condensed, greatly extended dark matter component of a galaxy that accounts for up to 95 percent of the galaxy's mass. (Ch. 19)

daughter product An element resulting from radioactive decay of a more massive *parent element*. (Ch. 8)

decay 1. The process of a radioactive nucleus changing into its daughter product. 2. The process of an atom or molecule dropping from a higher energy state to a lower energy state. 3. The process of a satellite's orbit losing energy. (Ch. 5)

declination A measure, analogous to *latitude*, that tells you the angular distance of a celestial body north or south of the celestial equator (from 0° to ±90°). Compare *right ascension*. (Ch. 2)

degeneracy pressure Pressure exerted by closely packed electrons in the collapsing core of a star. Such pressure pushes outward against gravity and keeps the star from collapsing further. (Ch. 16)

density The measure of an object's mass per unit of volume. Possible units include kilograms per cubic meter (kg/m^3). (Ch. 4, 7)

differential rotation Rotation of different parts of a system at different rates. (Ch. 14)

differentiation The process by which materials of higher density sink toward the center of a molten or fluid planetary interior. (Ch. 8)

diffraction The spreading of a wave after it passes through an opening or beyond the edge of an object. (Ch. 6)

diffraction grating An optical component with many narrow parallel lines that separate the wavelengths of light to produce a spectrum. (Ch. 6)

diffraction limit The limit of a telescope's angular resolution caused by diffraction. (Ch. 6)

diffuse ring A sparsely populated planetary ring spread out both horizontally and vertically.

dispersion The separation of rays of light into their component wavelengths. (Ch. 6)

distance ladder A sequence of techniques for measuring cosmic distances: each method is calibrated using the results from other methods that have been applied to closer objects. (Ch. 19)

Doppler effect The change in wavelength of sound or light as a result of the relative motion of the source toward or away from the observer. (Ch. 5)

Doppler shift The amount by which the Doppler effect shifts the wavelength of light. (Ch. 5)

Drake equation A prescription for estimating the number of intelligent civilizations existing in the Milky Way Galaxy. (Ch. 24)

dust devil A small tornado-like column of air containing dust or sand. (Ch. 9)

dust tail A type of comet tail consisting of dust particles pushed away from the comet's head by radiation pressure from the Sun. Compare *ion tail*. (Ch. 12)

dwarf galaxy A small galaxy with a luminosity ranging from 1 million to 1 billion solar luminosities (L_{Sun}). Compare *giant galaxy*. (Ch. 19)

dwarf planet A body with characteristics similar to those of a planet except that it has not cleared smaller bodies from the neighboring regions around its orbit. Compare *planet* (definition 2). (Ch. 12)

dynamic equilibrium A state in which a system is constantly changing but its configuration remains the same because one source of change is exactly balanced by another source of change. Compare *static equilibrium*. (Ch. 5)

dynamo theory A theory postulating that Earth's magnetic field (and those of other planets) is generated from a rotating and electrically conducting liquid core. (Ch. 8)

E

eccentricity (*e*) The ratio of the distance between the two foci of an ellipse to the length of its major axis, which measures how noncircular the ellipse is. (Ch. 3)

eclipse 1. The total or partial obscuration of one celestial body by another. 2. The total or partial obscuration of light from one celestial body as it passes through the shadow of another celestial body. (Ch. 2)

eclipse season Any time during the year when the Moon's line of nodes points towards the Sun and eclipses can occur. (Ch. 2)

eclipsing binary A binary system in which the orbital plane is oriented such that the two stars appear to pass in front of each other as seen from Earth. Compare *spectroscopic binary* and *visual binary*. (Ch. 13)

ecliptic 1. The apparent annual path of the Sun against the background of stars. 2. The projection of Earth's orbital plane onto the celestial sphere. (Ch. 2)

ecliptic plane The plane of Earth's orbit around the Sun. The ecliptic is the projection of that plane onto the celestial sphere. (Ch. 2)

effective temperature The temperature at which a blackbody, such as a star, appears to radiate. (Ch. 14)

Einstein ring Light bent by gravitational lensing into a ring. (Ch. 18)

ejecta 1. Material thrown outward by the impact of an asteroid or comet on a planetary surface, leaving a crater behind. 2. Material thrown outward by a stellar explosion. (Ch. 8)

electric field A field that can exert a force on a charged object, whether at rest or moving. Compare *magnetic field*. (Ch. 5)

electric force The force exerted on electrically charged particles such as protons and electrons, arising from their electric charges. Compare *magnetic force*. See also *electromagnetic force*. (Ch. 5)

electromagnetic force The force, including both electric and magnetic forces, that acts on electrically charged particles. One of four fundamental forces of nature, along with the *strong nuclear force*, *weak nuclear force*, and *gravity* (definition 1). The force is mediated by the exchange of photons. (Ch. 22)

electromagnetic radiation A traveling disturbance in the electric and magnetic fields caused by accelerating electric charges. In quantum mechanics, a stream of photons. Light. (Ch. 5)

electromagnetic spectrum The spectrum made up of all possible frequencies or wavelengths of electromagnetic radiation, ranging from gamma rays through radio waves and including the portion our eyes can use. (Ch. 5)

electromagnetic wave A wave consisting of oscillations in the electric-field strength and the magnetic-field strength. (Ch. 5)

electron (e^-) A subatomic particle having a negative electric charge of 1.6×10^{-19} coulomb (C), a rest mass of 9.1×10^{-31} kilogram (kg), and a rest energy of 8×10^{-14} joule (J). The antiparticle of the *positron*. Compare *proton* and *neutron*. (Ch. 5)

electron-degenerate Describing matter, compressed to the point at which electron density reaches the limit imposed by the rules of quantum mechanics. (Ch. 16)

electroweak theory The quantum theory that combines descriptions of both the electromagnetic force and the weak nuclear force. (Ch. 22)

element One of 92 naturally occurring substances (such as hydrogen, oxygen, and uranium) and more than 20 human-made ones (such as plutonium). Each element is chemically defined by the specific number of protons in the nuclei of its atoms. (Ch. 5)

elementary particle One of the basic building blocks of nature that is not known to have substructure, such as the *electron* or the *quark*. (Ch. 23)

ellipse A conic section produced by the intersection of a plane with a cone when the plane is passed through the cone at an angle to the axis other than 0° or 90°. The shape that results when you attach the two ends of a piece of string to a piece of paper, stretch the string tight with the tip of a pencil, and then draw around those two points while keeping the string taut. (Ch. 3)

elliptical galaxy A galaxy of Hubble type "E" class, with a circular to elliptical outline on the sky, and containing almost no disk and a population of old stars. Compare *barred spiral galaxy*, *irregular galaxy*, *S0 galaxy*, and *spiral galaxy*. (Ch. 19)

emission The production of a photon when an atom decays to a lower energy state. Compare *absorption*. (Ch. 5)

emission line A peak in the intensity of a spectrum that is due to the emission of electromagnetic radiation at a specific wavelength determined by the energy levels of an atom or molecule. Compare *absorption line*. (Ch. 5)

empirical science Descriptive scientific investigation based primarily on observations and experimental data rather than on theoretical inference. (Ch. 3)

energy The conserved quantity that gives objects and systems the ability to do work. Possible units include joules (J). (Ch. 5)

energy transport The transfer of energy from one location to another. In stars, energy transport is carried out mainly by radiation or convection. (Ch. 14)

entropy A measure of the disorder of a system related to the number of ways a system can be rearranged without its appearance being affected. (Ch. 24)

equator The imaginary great circle on the surface of a body midway between its poles that divides the body into northern and southern hemispheres. The equatorial plane passes through the center of the body and is perpendicular to its rotation axis. Compare *meridian*. (Ch. 2)

equilibrium The state of an object in which physical processes balance each other so that its properties or conditions remain constant. (Ch. 5)

equinox Literally, "equal night." 1. One of two positions on the ecliptic where it intersects the celestial equator. 2. Either of the two times of year (the *autumnal equinox* and *vernal equinox*) when the Sun is at one of these two positions. At this time, night and day are of the same length everywhere on Earth. Compare *solstice*. (Ch. 2)

equivalence principle The principle stating that no difference exists between a frame of reference freely floating through space and one freely falling within a gravitational field. (Ch. 18)

erosion The degradation of a planet's surface topography by the mechanical action of wind, water, or living organisms. (Ch. 8)

escape velocity The minimum velocity needed for an object to achieve a parabolic trajectory and thus permanently leave the gravitational grasp of another mass. (Ch. 4)

eternal inflation The idea that a universe might inflate forever. In such a universe, quantum effects could randomly cause regions to slow their expansion, eventually stop inflating, and experience an explosion resembling the Big Bang. (Ch. 22)

event Something that happens at a particular location in spacetime. (Ch. 18)

event horizon The effective "surface" of a black hole. Nothing inside that surface—not even light—can escape from a black hole. (Ch. 18)

evolutionary track The path that a star follows across the H-R diagram as it evolves through its lifetime. (Ch. 15)

excited state Any energy level of a system or part of a system, such as an atom, molecule, or particle, that is higher than its ground state. Compare *ground state*. (Ch. 5)

exoplanet See *extrasolar planet*. (Ch. 7)

exosphere A very thin atmosphere or layer of atmosphere, where the molecules are bound by gravity to the moon or planet but their density is too low to behave like a gas of colliding particles. (Ch. 9)

extrasolar planet Also called *exoplanet*. A planet orbiting a star other than the Sun. (Ch. 7)

extremophile A life-form that thrives under extreme environmental conditions. (Ch. 24)

eyepiece A lens that is closest to the eye in a telescope. Changing the eyepiece will change the magnification of the image in the telescope. (Ch. 6)

F

F See *Fahrenheit*. (Ch. 5)

Fahrenheit (F) The arbitrary temperature scale, formulated by Daniel Gabriel Fahrenheit (1686–1736), that defines 32°F as the melting point of water and 212°F as the boiling point of water at sea level. Unit: degrees Fahrenheit (°F). Compare *Celsius* and *Kelvin temperature scale*. (Ch. 5)

falsified A hypothesis shown to be false. (Ch. 1)

fault A fracture in the crust of a planet or moon along which blocks of material can slide. (Ch. 8)

filter An instrument element that transmits a limited wavelength range of electromagnetic radiation. For the optical range, such elements are typically made of different kinds of glass and take on the hue of the light they transmit. (Ch. 13)

first quarter Moon The phase of the Moon in which only the western half of the Moon, as viewed from Earth, is illuminated by the Sun. It occurs about a week after a new Moon. Compare *third quarter Moon*. See also *full Moon* and *new Moon*. (Ch. 2)

fissure A fracture in the planetary lithosphere from which magma emerges. (Ch. 8)

flat rotation curve A rotation curve of a spiral galaxy in which rotation rates do not decline in the outer part of the galaxy but remain relatively constant to the outermost points. (Ch. 19)

flat universe An infinite universe whose spatial structure obeys Euclidean geometry, such that the sum of the angles of a triangle always equals 180 degrees. Compare *closed universe* and *open universe*. (Ch. 22)

flatness problem The surprising result that the sum of Ω_m plus Ω_Λ is extremely close to 1 in the present-day universe; equivalent to saying that it is surprising the universe is so close to being exactly flat. (Ch. 22)

flux The total amount of energy passing through each square meter of a surface each second. Unit: watts per square meter (W/m^2). (Ch. 5)

flux tube A strong magnetic field contained within a tubelike structure. Flux tubes are found in the solar atmosphere and connecting the space between Jupiter and its moon Io. (Ch. 10)

flyby A spacecraft that first approaches and then continues flying past a planet or moon. Flybys can visit multiple objects, but they remain near their targets only briefly. Compare *orbiter*. (Ch. 6)

focal length The optical distance between a telescope's objective lens or primary mirror and the plane (called the focal plane) on which the light from a distant object is focused. (Ch. 6)

focal plane The plane, perpendicular to the optical axis of a lens or mirror, on which an image is formed. (Ch. 6)

focus (pl. **foci**) 1. One of two points that define an ellipse. 2. A point in the focal plane of a telescope. (Ch. 3)

force A push or a pull on an object. (Ch. 3)

frame of reference A coordinate system within which an observer measures positions and motions. (Ch. 2)

free fall The motion of an object when the only force acting on it is gravity. (Ch. 4)

frequency The number of times per second that a periodic process occurs. Unit: hertz (Hz), or cycles per second (1/s). (Ch. 5)

full Moon The phase of the Moon in which the near side of the Moon, as viewed from Earth, is fully illuminated by the Sun. It occurs about two weeks after a new Moon. See also *first quarter Moon* and *third quarter Moon*. (Ch. 2)

G

galaxy A gravitationally bound system that consists of stars and star clusters, gas, dust, and dark matter; typically greater than 1,000 light-years across and recognizable as a discrete, single object. (Ch. 19)

galaxy cluster A large, gravitationally bound collection of galaxies containing hundreds to thousands of members; typically 3–5 megaparsecs (Mpc) across. Compare *galaxy group* and *supercluster*. (Ch. 23)

galaxy group A small, gravitationally bound collection of galaxies containing from several to a hundred members; typically 1–2 megaparsecs (Mpc) across. Compare *galaxy cluster* and *supercluster*. (Ch. 20)

gamma ray Also called *gamma radiation*. Electromagnetic radiation with higher frequency, higher photon energy, and shorter wavelength than all other types of electromagnetic radiation. (Ch. 5)

gamma-ray burst (GRB) A brief, intense burst of gamma rays from a distant energetic explosion. (Ch. 18)

gas giant A giant planet formed mostly of hydrogen and helium. In the Solar System, Jupiter and Saturn are the gas giants. Compare *ice giant*. (Ch. 10)

general relativistic time dilation The verified prediction that time passes more slowly in a gravitational field than in the absence of a gravitational field. Compare *time dilation*. (Ch. 18)

general relativity See *general theory of relativity*. (Ch. 18)

general theory of relativity Sometimes referred to as simply *general relativity*. Einstein's theory explaining gravity as the distortion of spacetime by massive objects, such that particles travel on the shortest path between two events in spacetime. This theory deals with all types of motion. Compare *special theory of relativity*. (Ch. 18)

geocentric model A historical cosmological model with Earth at its center, and all the other objects in the universe in orbit around Earth. Compare *heliocentric model*. (Ch. 3)

geodesic The path an object will follow through spacetime in the absence of external forces. (Ch. 18)

giant galaxy A galaxy with luminosity greater than about 1 billion solar luminosities (L_{Sun}). Compare *dwarf galaxy*. (Ch. 19)

giant molecular cloud An interstellar cloud composed primarily of molecular gas and dust and having hundreds of thousands of solar masses. (Ch. 15)

giant planet Also called *Jovian planet*. One of the largest planets in the Solar System (Saturn, Jupiter, Uranus, or Neptune), typically 10 times the size and many times the mass of any *terrestrial planet* and lacking a solid surface. (Ch. 7)

gibbous Any phase of the Moon, Mercury, or Venus in which the object appears more than half illuminated by the Sun. Compare *crescent*. (Ch. 2)

global circulation The overall, planetwide circulation pattern of a planet's atmosphere. (Ch. 9)

globular cluster A spherically symmetric, highly condensed group of stars, containing tens of thousands to a million members. Compare *open cluster*. (Ch. 17)

gluon The particle that carries (or, equivalently, mediates) interactions due to the strong nuclear force. (Ch. 22)

grand unified theory (GUT) A unified quantum theory that combines the strong nuclear, weak nuclear, and electromagnetic forces but does not include gravity. (Ch. 22)

granite Rock that is cooled from magma and is relatively rich in silicon and oxygen. (Ch. 8)

grating An optical surface containing many narrow, closely and equally spaced parallel grooves or slits that spectrally disperse reflected or transmitted light. (Ch. 6)

gravitational lens A massive object that gravitationally focuses the light of a more distant object to produce multiple brighter, magnified, possibly distorted images. (Ch. 7, 18)

gravitational lensing The bending of light by gravity. (Ch. 18)

gravitational potential energy The stored energy in an object that is due solely to its position within a gravitational field. (Ch. 7)

gravitational redshift The shifting to longer wavelengths of radiation from an object deep within a gravitational well. Compare *cosmological redshift*. (Ch. 18)

gravitational wave A wave in the fabric of spacetime emitted by accelerating masses. (Ch. 6, 18)

gravity 1. The mutually attractive force between massive objects. One of four fundamental forces of nature, along with the *electromagnetic force*, the *strong nuclear force*, and the *weak nuclear force*. 2. An effect arising from the bending of spacetime by massive objects. (Ch. 4)

GRB See *gamma-ray burst*. (Ch. 18)

Great Red Spot The giant, oval, brick red anticyclone seen in Jupiter's southern hemisphere. (Ch. 10)

greenhouse effect The solar heating of air in an enclosed space, such as a closed building or car, resulting primarily from the inability of the hot air to escape. Compare *atmospheric greenhouse effect*. (Ch. 9)

greenhouse gas One of a group of atmospheric gases such as carbon dioxide that are transparent to visible radiation but absorb infrared radiation. (Ch. 9)

greenhouse molecule A molecule such as water vapor or carbon dioxide that transmits visible radiation but absorbs infrared radiation. (Ch. 9)

Gregorian calendar The modern calendar. A modification of the Julian calendar decreed by Pope Gregory XIII in 1582. By then, the less accurate Julian calendar had developed an error of 10 days over the 13 centuries since its inception. (Ch. 2)

ground state The lowest possible energy state for a system or part of a system, such as an atom, molecule, or particle. Compare *excited state*. (Ch. 5)

GUT See *grand unified theory*. (Ch. 22)

H

H II region A region of interstellar gas that has been ionized by ultraviolet radiation from nearby hot, massive stars. (Ch. 15)

H-R diagram The Hertzsprung-Russell diagram, a plot of the luminosities versus the surface temperatures of stars. The evolving properties of stars are plotted as tracks across the H-R diagram. (Ch. 13)

habitable zone The distance from its star at which a planet must be located to have a temperature suitable for water to exist in a liquid state. (Ch. 7, 13)

Hadley circulation A simplified, and therefore uncommon, atmospheric global circulation that carries thermal energy directly from the equator to the polar regions of a planet. (Ch. 9)

half-life The time that half a sample of a particular radioactive parent element takes to decay into a daughter product. (Ch. 8)

halo The spherically symmetric, low-density distribution of stars and dark matter that defines the outermost regions of a galaxy. (Ch. 19)

harmonic law See *Kepler's third law*. (Ch. 3)

Hawking radiation Radiation from a black hole. (Ch. 18)

Hayashi track The path that a protostar follows on the H-R diagram as it contracts toward the main sequence. (Ch. 15)

head The part of a comet that includes both the nucleus and the inner part of the coma. (Ch. 12)

heavy element Also called *massive element*. Any element more massive than helium. (Ch. 10, 13)

Heisenberg uncertainty principle The physical limitation that the product of the position and the momentum of a particle cannot be smaller than a well-defined value, Planck's constant (h). (Ch. 22)

heliocentric model A model of the Solar System, with the Sun at its center, and the planets, including Earth, in orbit around the Sun. Compare *geocentric model*. (Ch. 3)

helioseismology The use of solar oscillations to study the interior of the Sun. (Ch. 14)

heliosphere A region surrounding the Solar System in which the solar wind blows against the interstellar medium and clears out an area like the inside of a bubble. The heliosphere protects the Solar System from cosmic rays. (Ch. 14)

helium capture A helium nucleus is captured by another nucleus during nucleosynthesis (Ch. 17)

helium flash The runaway explosive fusion of helium in the degenerate helium core of a red giant star. (Ch. 16)

Herbig-Haro (HH) object A glowing, rapidly moving knot of gas and dust that is excited by bipolar outflows in very young stars. (Ch. 15)

heredity The process by which one generation passes on its characteristics to future generations. (Ch. 24)

hertz (Hz) A unit of frequency equivalent to cycles per second. (Ch. 5)

Hertzsprung-Russell diagram See *H-R diagram*. (Ch. 13)

HH object See *Herbig-Haro object*. (Ch. 15)

hierarchical clustering The "bottom up" process of forming large-scale structure. Small-scale structure first produces groups of galaxies, which in turn form clusters, which then form superclusters. (Ch. 23)

high-mass star A star with a main-sequence mass of greater than about 8 solar masses (M_{Sun}). Compare *low-mass star* and *medium-mass star*. (Ch. 16)

homogeneous In cosmology, describing a universe in which observers in any location would observe the same properties. Compare *isotropic*. (Ch. 21)

horizon The boundary that separates the sky from the ground. (Ch. 2)

horizon problem The puzzling observation that the cosmic background radiation is so uniform in all directions, even though widely separated regions should have been "over the horizon" from each other in the early universe. (Ch. 22)

horizontal branch A region on the H-R diagram defined by stars fusing helium to carbon in a stable core. (Ch. 16)

hot dark matter Particles of dark matter that move so fast that gravity cannot confine them to the volume occupied by a galaxy's normal luminous matter. Compare *cold dark matter*. (Ch. 23)

hot Jupiter A large, Jupiter-type extrasolar planet located very close to its parent star. (Ch. 7)

hot spot A place where hot plumes of mantle material rise near the surface of a planet. (Ch. 8)

Hubble constant (H_0) The constant of proportionality relating the recession velocities of galaxies to their distances. Compare *Hubble time*. (Ch. 19)

Hubble flow The motion of galaxies as a result of the expanding universe. (Ch. 21)

Hubble's law The law stating that the speed at which a galaxy is moving away from Earth is proportional to the distance of that galaxy. (Ch. 19)

Hubble time An estimate of the age of the universe from the inverse of the *Hubble constant*, $1/H_0$. (Ch. 21)

hurricane A large tropical cyclonic system circulating counterclockwise in the Northern Hemisphere and clockwise in the Southern Hemisphere. Hurricanes can extend outward from their center to more than 600 kilometers (km) and generate winds in excess of 300 kilometers per hour (km/h). (Ch. 8)

hydrogen fusion The release of energy from the nuclear fusion of four hydrogen atoms into a single helium atom. (Ch. 14)

hydrogen shell fusion The fusion of hydrogen in a shell surrounding a stellar core that may be either degenerate or fusing more massive elements. (Ch. 16)

hydrosphere The portion of Earth that is largely liquid water. Compare *biosphere* and *lithosphere*. (Ch. 8)

hydrostatic equilibrium The condition in which the weight bearing down at a particular point within an object is balanced by the pressure within the object. (Ch. 8, 14)

hypernova (pl. **hypernovae**) A very energetic supernova from a very high-mass star. (Ch. 18)

hypothesis A well-considered idea, based on scientific principles and knowledge, that leads to testable predictions. Compare *theory*. (Ch. 1)

Hz See *hertz*. (Ch. 5)

I

ice The solid form of a volatile material; sometimes the *volatile material* itself, in any form. (Ch. 7)

ice giant A giant planet formed mostly of the liquid form of volatile substances (ices). In the Solar System, Uranus and Neptune are the ice giants. Compare *gas giant*. (Ch. 10)

impact crater The scar of the impact left on a solid planetary or moon surface by collision with another object. Compare *secondary crater*. (Ch. 8)

impact cratering The process in which solid planetary objects collide with each other, leaving distinctive scars. (Ch. 8)

index of refraction (n) The ratio of the speed of light in a vacuum (c) to the speed of light in an optical medium (v). (Ch. 6)

inert gas A gaseous element that combines with other elements only under conditions of extreme temperature and pressure. Examples include helium, neon, and argon. (Ch. 16)

inertia The tendency for objects to retain their state of motion. (Ch. 3)

inertial frame of reference 1. A frame of reference moving in a straight line at constant speed, that is, not accelerating. 2. In general relativity, a frame of reference falling freely in a gravitational field. (Ch. 3, 18)

inertial reference frame See *inertial frame of reference*. (Ch. 18)

inferior planet A Solar System planet that orbits closer to the Sun than Earth does. Compare *superior planet*. (Ch. 3)

inflation An extremely brief phase of ultra-rapid expansion of the very early universe. After inflation, the standard Big Bang models of expansion apply. (Ch. 22)

infrared (IR) radiation Electromagnetic radiation with frequencies, photon energies, and wavelengths between those of visible light and microwaves. (Ch. 5)

instability strip A region of the H-R diagram containing stars that pulsate with a periodic variation in luminosity. (Ch. 17)

integration time The time interval during which photons are collected and added up in a detecting device. (Ch. 6)

intensity Of light, the amount of radiant energy emitted per second per unit area. Units for electromagnetic radiation: watts per square meter (W/m^2). (Ch. 5)

intercloud gas A low-density region of the interstellar medium that fills the space between interstellar clouds. (Ch. 15)

interference The interaction of two sets of waves producing high and low intensity, depending on whether their amplitudes reinforce (*constructive interference*) or cancel (*destructive interference*). (Ch. 6)

interferometer Linked optical or radio telescopes whose overall separation determines the angular resolution of the system. (Ch. 6)

interferometric array An interferometer made up of several telescopes arranged in an array. (Ch. 6)

interstellar cloud A discrete, high-density region of the interstellar medium made up mostly of atomic or molecular hydrogen and dust. (Ch. 15)

interstellar dust Small particles or grains (0.01–10 microns [μm] in diameter) of matter, primarily carbon and silicates, distributed throughout interstellar space. (Ch. 15)

interstellar extinction The dimming of visible and ultraviolet light by interstellar dust. (Ch. 15)

interstellar gas The tenuous gas, far less dense than air, composing 99 percent of the matter in the interstellar medium. (Ch. 7, 15)

interstellar medium The gas and dust that fill the space between the stars within a galaxy. (Ch. 14, 15)

inverse square law The rule stating that a quantity or effect diminishes with the square of the distance from the source. (Ch. 4)

ion An atom or molecule that has lost or gained one or more electrons. (Ch. 9, 15)

ionize see *ionization*. (Ch. 9, 15)

ionization The process by which electrons are stripped free from an atom or molecule, resulting in free electrons and a positively charged atom or molecule. (Ch. 9, 15)

ionosphere A layer high in Earth's atmosphere in which most atoms are ionized by solar radiation. (Ch. 9)

ion tail A type of comet tail consisting of ionized gas. Particles in the ion tail are pushed directly away from the comet's head in the antisolar direction at high speeds by the solar wind. Compare *dust tail*. (Ch. 12)

IR Infrared. See *infrared radiation*. (Ch. 5)

iron meteorite A metallic meteorite composed mostly of iron-nickel alloys. Compare *stony-iron meteorite* and *stony meteorite*. (Ch. 12)

irregular galaxy A galaxy without regular or symmetric appearance. Compare *barred spiral galaxy*, *elliptical galaxy*, *S0 galaxy*, and *spiral galaxy*. (Ch. 19)

irregular moon A moon that has been captured by a planet instead of having formed along with that planet. Some irregular moons revolve in a direction opposite to the rotation of the planet, and many are in distant, unstable orbits. Compare *regular moon*. (Ch. 11)

isotope A forms of the same chemical element that has the same number of protons but a different number of neutrons. (Ch. 5)

isotropic In cosmology, having the same appearance to an observer in all directions. Compare *homogeneous*. (Ch. 21)

J

J See *joule*. (Ch. 5)

jansky (Jy) The basic unit of flux density. Unit: watts per square meter per hertz (W/m^2/Hz). (Ch. 6)

jet 1. A stream of gas and dust ejected from a comet nucleus by solar heating. 2. A stream of material that moves away from a protostar or active galactic nucleus at hundreds of kilometers per second. (Ch. 15)

joule (J) A unit of energy or work. 1 J = 1 newton meter. (Ch. 5)

Jovian planet See *giant planet*. (Ch. 7)

Jy See *jansky*. (Ch. 6)

K

K See *kelvin*. (Ch. 5)

K-Pg boundary See *Cretaceous-Paleogene boundary*. (Ch. 8)

KBO See *Kuiper Belt object*. (Ch. 12)

kelvin (K) The basic unit of the Kelvin scale of temperature. (Ch. 5)

Kelvin temperature scale The temperature scale, formulated by William Thomson, better known as Lord Kelvin (1824–1907), that uses Celsius-sized degrees but defines 0 K as absolute zero instead of as the melting point of water. Unit: kelvins (K). Compare *Celsius* and *Fahrenheit*. (Ch. 5)

Kepler's first law A rule of planetary motion that Johannes Kepler inferred, stating that planets move in elliptical orbits with the Sun at one focus. (Ch. 3)

Kepler's laws The three rules of planetary motion that Johannes Kepler inferred from data collected by Tycho Brahe. (Ch. 3)

Kepler's second law Also called *law of equal areas*. A rule of planetary motion that Johannes Kepler inferred, stating that a line drawn from the Sun to a planet sweeps out equal areas in equal times as the planet orbits the Sun. (Ch. 3)

Kepler's third law Also called *harmonic law*. A rule of planetary motion that Johannes Kepler inferred, describing the relationship between the period of a planet's orbit and its distance from the Sun. The law states that the square of the period of a planet's orbit, measured in years, is equal to the cube of the semimajor axis of the planet's orbit, measured in astronomical units: $(P_{years})^2 = (A_{AU})^3$. (Ch. 3)

kiloparsec A unit of distance equal to 1,000 parsecs, or 3,260 light-years. (Ch. 19)

kinetic energy (E_K) The energy of an object resulting from its motions. $E_K = \frac{1}{2}mv^2$. Possible units include joules (J). (Ch. 5, 7)

Kirkwood gap A gap in the main asteroid belt related to orbital resonances with Jupiter. (Ch. 12)

Kuiper Belt A disk-shaped population of comet nuclei extending from Neptune's orbit to perhaps several thousand astronomical units (AU) from the Sun. The highly populated innermost part of the Kuiper Belt has an outer edge approximately 50 AU from the Sun. (Ch. 12)

Kuiper Belt object (KBO) Also called *trans-Neptunian object*. An icy planetesimal (comet nucleus) that orbits within the Kuiper Belt beyond the orbit of Neptune. (Ch. 12)

L

Lambda-CDM The standard model of the Big Bang universe in which most of the energy density of the universe is dark energy (similar to Einstein's cosmological constant), and most of the mass in the universe is cold dark matter. (Ch. 23)

lander An instrumented spacecraft designed to land on a planet or moon. Compare *rover*. (Ch. 6)

large-scale structure Observable aggregates on the largest scales in the universe, including galaxy groups, clusters, and superclusters. (Ch. 23)

latitude The angular distance north (+) or south (−) from the equatorial plane of a nearly spherical body. Compare *longitude*. (Ch. 2)

lava Molten rock flowing out of a volcano during an eruption; also the rock that solidifies and cools from that liquid. (Ch. 8)

law of equal areas See *Kepler's second law*. (Ch. 3)

law of gravitation See *universal law of gravitation*. (Ch. 4)

leap year A year that contains 366 days. Leap years occur every 4 years when the year is divisible by 4, correcting for the accumulated excess time in a normal year, which is approximately 365 1/4 days long. (Ch. 2)

length contraction The relativistic compression of moving objects in the direction of their motion. (Ch. 18)

Leonids A November meteor shower associated with the dust debris left by comet Tempel-Tuttle. (Ch. 12)

life A biochemical process in which living organisms can reproduce, evolve, and sustain themselves by drawing energy from their environment. All terrestrial life involves carbon-based chemistry, assisted by the self-replicating molecules ribonucleic acid (RNA) and deoxyribonucleic acid (DNA). (Ch. 24)

light All electromagnetic radiation, which composes the entire electromagnetic spectrum. (Ch. 5)

light-year (ly) The distance that light travels in 1 year—about 9.5 trillion kilometers (km). (Ch. 1)

limb The outer edge of the visible disk of a planet, moon, or the Sun. (Ch. 11,14)

limb darkening The darker appearance caused by increased atmospheric absorption near the limb of a planet or star. (Ch. 14)

limestone A common sedimentary rock composed of calcium carbonate. (Ch. 9)

line of nodes 1. A line defined by the intersection of two orbital planes. 2. The line defined by the intersection of Earth's equatorial plane and the plane of the ecliptic. (Ch. 2)

lithosphere The solid, brittle part of Earth (or any planet or moon), including the crust and the upper part of the mantle. Compare *biosphere* and *hydrosphere*. (Ch. 8)

lithospheric plate A separate piece of Earth's lithosphere that can move independently. See also *continental drift* and *plate tectonics*. (Ch. 8)

Local Group The group of galaxies that includes the Milky Way and Andromeda galaxies. (Ch. 1)

long-period comet A comet with an orbital period of greater than 200 years. Compare *short-period comet*. (Ch. 12)

longitude The angular distance east (+) or west (−) from the prime meridian at Greenwich, England. Compare *latitude*. (Ch. 2)

longitudinal wave A wave that oscillates parallel to the direction of the wave's propagation. Compare *transverse wave*. (Ch. 8)

look-back time The amount of time that the light from an astronomical object has taken to reach Earth. (Ch. 21)

low-mass star A star with a main-sequence mass of less than about 3 solar masses (M_{Sun}). Compare *high-mass star* and *medium-mass star*. (Ch. 16)

luminosity The total amount of light emitted by an object. Unit: watts (W). Compare *brightness*. (Ch. 5)

luminosity class A spectral classification based on stellar size, ranging from supergiants at the large end to white dwarfs at the small end. (Ch. 13)

luminosity-temperature-radius relationship A relationship among those three properties of stars indicating that if any two are known, the third can be calculated. (Ch. 13)

luminous matter Also called *normal matter*. Matter in galaxies—including stars, gas, and dust—that emits electromagnetic radiation. Compare *dark matter*. (Ch. 19)

lunar eclipse An eclipse that occurs when the Moon is partially or entirely in Earth's shadow. Compare *solar eclipse*. (Ch. 2)

lunar tide A tide on Earth caused by the differential gravitational pull of the Moon. Compare *solar tide*. (Ch. 4)

lunisolar calendar Calendar created by the Babylonians in which a month began with the first sighting of the lunar crescent, and a 13th month was added when needed to catch up to the solar year. (Ch. 2)

ly See *light-year*. (Ch. 1)

M

μm See *micron*. (Ch. 5)

M-type asteroid An asteroid made of material that was once part of the metallic core of a larger, differentiated body that has since broken into pieces; made mostly of iron and nickel. Compare *C-type asteroid* and *S-type asteroid*. (Ch. 12)

MaCHO Short for *massive compact halo object*. MaCHOs include brown dwarfs, white dwarfs, and black holes and are candidates for dark matter. Compare *WIMP*. (Ch. 19)

magma Molten rock, often containing dissolved gases and solid minerals. (Ch. 8)

magnetic field A field that can exert a force on a moving electric charge. Compare *electric field*. (Ch. 5)

magnetic force The force exerted on electrically charged particles such as protons and electrons, arising from their motion. Compare *electric force*. See also *electromagnetic force*. (Ch. 5)

magnetosphere The region surrounding a planet that is filled with relatively intense magnetic fields and plasmas. (Ch. 8)

magnitude A system used by astronomers to describe the brightness or luminosity of stars. The brighter the star, the lower its magnitude. (Ch. 13)

major axis The long axis of an ellipse. (Ch. 3)

main asteroid belt See *asteroid belt*. (Ch. 12)

main sequence The strip on the H-R diagram where most stars are found. Main-sequence stars are fusing hydrogen to helium in their cores. (Ch. 13)

main-sequence lifetime The amount of time a star spends on the main sequence, fusing hydrogen into helium in its core. (Ch. 16)

main-sequence turnoff The location on the main sequence of an H-R diagram made from a population of stars of the same age (such as a star cluster) where stars are just evolving off the main sequence. That location is determined by the age of the population of stars. (Ch. 17)

mantle The solid portion of a rocky planet that lies between the *crust* and the *core*. (Ch. 8)

mare (pl. maria) A dark region on the Moon composed of basaltic lava flows. (Ch. 8)

mass 1. Inertial mass: the property of matter that determines its resistance to changes in motion. Compare *weight*. 2. Gravitational mass: the property of matter defined by its attractive force on other objects. According to general relativity, the two are equivalent. (Ch. 3, 4)

mass-luminosity relationship An empirical relationship between the luminosity (L) and mass (M) of main-sequence stars; for example, $L \propto M^{3.5}$. (Ch. 13, 16)

mass transfer The transfer of mass from one member of a binary star system to its companion. Mass transfer occurs when one of the stars evolves to the point that it overfills its Roche lobe, so that its outer layers are pulled toward its binary companion. (Ch. 16)

massive element Also called *heavy element*. Any element more massive than helium. (Ch. 13)

matter 1. Objects made of particles that have mass, such as protons, neutrons, and electrons. 2. Anything that occupies space and has mass. (Ch. 5)

Maunder Minimum The period from 1645 to 1715 during which very few sunspots were observed. (Ch. 14)

medium The substance that a wave, such as light, travels through—for example, air or glass. Compare *vacuum*. (Ch. 5)

medium-mass star A star with a main-sequence mass between 3 and 8 solar masses (M_{Sun}). Compare *high-mass star* and *low-mass star*. (Ch. 16)

megaparsec (Mpc) A unit of distance equal to 1 million parsecs, or 3.26 million light-years. (Ch. 19)

meridian The imaginary arc in the sky running from the horizon at due north through the zenith to the horizon at due south. The meridian divides the observer's sky into eastern and western hemispheres. Compare *equator*. (Ch. 2)

mesosphere The layer of Earth's atmosphere immediately above the stratosphere, extending from an altitude of 50 kilometers (km) to about 90 km. Compare *troposphere*, *stratosphere*, and *thermosphere*. (Ch. 9)

meteor A meteoroid that enters and burns up in a planetary atmosphere, often leaving an incandescent trail. Compare *meteorite* and *meteoroid*. (Ch. 8)

meteor shower A larger-than-normal display of meteors, occurring when Earth passes through the orbit of a disintegrating comet, sweeping up its debris. Compare *sporadic meteor*. (Ch. 12)

meteorite A piece of rock or other fragment of material (a meteoroid) that survives to reach a planet's surface. Compare *meteor* and *meteoroid*. (Ch. 7, 8)

meteoroid A small cometary or asteroidal fragment, ranging in size from 100 microns (μm) to 100 meters. When entering a planetary atmosphere, the meteoroid creates a *meteor*. Compare *meteor* and *meteorite*; also *planetesimal* and *zodiacal dust*. (Ch. 8)

micrometer (μm) See *micron*. (Ch. 5)

micron (μm) One-millionth (10^{-6}) of a meter; a unit of length used for the wavelength of infrared light. (Ch. 5)

microwave radiation Electromagnetic radiation with frequencies, photon energies, and wavelengths between those of infrared radiation and radio waves. (Ch. 5)

Milky Way Galaxy The galaxy in which the Sun and Solar System reside. (Ch. 1)

minor planet See *asteroid*. (Ch. 16)

minute of arc See *arcminute*. (Ch. 13)

molecular cloud An interstellar cloud composed primarily of molecular hydrogen. (Ch. 15)

molecular-cloud core A dense clump within a molecular cloud that forms as the cloud collapses and fragments. Protostars form from molecular-cloud cores. (Ch. 15)

molecule Generally, the smallest particle of a substance that retains its chemical properties and is composed of two or more atoms. (Ch. 5)

momentum The product of the mass and velocity of a particle. Possible units include kilograms times meters per second (kg m/s). (Ch. 7)

moon A less massive satellite orbiting a more massive object. Moons are found around planets, dwarf planets, asteroids, and Kuiper Belt objects. The term is usually capitalized when referring to Earth's Moon. (Ch. 7)

Mpc See *megaparsec*. (Ch. 19)

multiverse A collection of parallel universes that together make up all that exists. (Ch. 22)

mutation In biology, an imperfect reproduction of self-replicating material. (Ch. 24)

N

N See *newton*. (Ch. 3)

nadir The point on the celestial sphere located directly below an observer, opposite the *zenith*. (Ch. 2)

nanometer (nm) One-billionth (10^{-9}) of a meter; a unit of length used for the wavelength of visible light. (Ch. 5)

natural selection The process by which forms of structure, ranging from molecules to whole organisms, that are best adapted to their environment become more common than less well-adapted forms. (Ch. 24)

NCP See *north celestial pole*. (Ch. 2)

neap tide An especially weak tide that occurs around the time of the first or third quarter Moon, when the gravitational forces of the Moon and the Sun on Earth are at right angles to each other. Compare *spring tide*. (Ch. 4)

near-Earth asteroid An asteroid whose orbit brings it close to the orbit of Earth. See also *near-Earth object*. (Ch. 12)

near-Earth object (NEO) An asteroid, comet, or large meteoroid whose orbit intersects Earth's orbit. (Ch. 12)

nebula (pl. nebulae) A cloud of interstellar gas and dust, either illuminated by stars (for bright nebulae) or seen in silhouette against a brighter background (for dark nebulae). (Ch. 7)

nebular hypothesis The first plausible theory of the formation of the Solar System, proposed by Immanuel Kant in 1755, which stated that the Solar System formed from the collapse of an interstellar cloud of rotating gas. (Ch. 7)

NEO See *near-Earth object*. (Ch. 12)

neutrino A very low-mass, electrically neutral particle emitted during beta decay. Neutrinos interact with matter only very feebly and so can penetrate great quantities of matter. (Ch. 6, 14)

neutron capture The process in which an atomic nucleus forms a heavier nucleus after colliding with a neutron (Ch. 17)

neutrino cooling The process in which thermal energy is carried out of the center of a star by neutrinos rather than by electromagnetic radiation or convection. (Ch. 17)

neutron A subatomic particle having no net electric charge and a rest mass and rest energy nearly equal to that of the proton. Compare *electron* and *proton*. (Ch. 5)

neutron star The neutron-degenerate stellar core left behind by a Type II supernova. (Ch. 17)

new Moon The phase of the Moon in which the Moon is between Earth and the Sun; from Earth, we see only the side of the Moon not being illuminated by the Sun. Compare *full Moon*. See also *first quarter Moon* and *third quarter Moon*. (Ch. 2)

newton (N) The force required to accelerate a 1-kilogram (kg) mass at a rate of 1 meter per second per second (m/s²). Unit: kilograms multiplied by meters per second squared (kg m/s²). (Ch. 3)

Newton's first law of motion The law, formulated by Isaac Newton, stating that an object will remain at rest or will continue moving along a straight line at a constant speed until an unbalanced force acts on it. (Ch. 3)

Newton's laws The three physical laws of motion that Isaac Newton formulated. (Ch. 3)

Newton's second law of motion The law that Isaac Newton formulated, stating that if an unbalanced force acts on a body, the body will accelerate in proportion to the unbalanced force and in inverse proportion to the object's mass: $a = F/m$. The acceleration will be in the direction of the unbalanced force. (Ch. 3)

Newton's third law of motion The law that Isaac Newton formulated, stating that for every force an equal force in the opposite direction exists. (Ch. 3)

nm See *nanometer*. (Ch. 5)

normal matter See *luminous matter*. (Ch. 19)

north celestial pole (NCP) The northward projection of Earth's rotation axis onto the celestial sphere. Compare *south celestial pole*. (Ch. 2)

North Pole The location in the Northern Hemisphere where Earth's rotation axis intersects Earth's surface. Compare *South Pole*. (Ch. 2)

nova (pl. novae) A stellar explosion that results from runaway nuclear fusion in a layer of material on the surface of a white dwarf in a binary system. (Ch. 16)

nuclear burning See nuclear fusion. (Ch. 14)

nuclear fusion The combination of two less massive atomic nuclei into a single more massive atomic nucleus. (Ch. 14)

nucleosynthesis The formation of more massive atomic nuclei from less massive nuclei, either in the Big Bang (Big Bang nucleosynthesis) or in the interiors of stars (stellar nucleosynthesis). (Ch. 17)

nucleus (pl. nuclei) 1. The dense, central part of an atom. 2. The central core of a galaxy, comet, or other diffuse object. (Ch. 5)

O

objective lens The primary optical element in a telescope or camera that produces an image of an object. (Ch. 6)

oblateness The flattening of an otherwise spherical planet or star caused by its rapid rotation. (Ch. 10)

obliquity The inclination of a celestial body's equator to its orbital plane. (Ch. 2)

observational uncertainty The fact that real measurements are never perfect; all observations are uncertain by some amount. (Ch. 13)

Occam's razor The principle that the simplest hypothesis is the most likely, named after William of Occam (circa 1285–1349), the medieval English cleric to whom the idea is attributed. (Ch. 1)

Oort Cloud A spherical distribution of comet nuclei stretching from beyond the Kuiper Belt to more than 50,000 astronomical units (AU) from the Sun. (Ch. 12)

opacity A measure of how effectively a material blocks the radiation going through it. (Ch. 14)

open cluster A loosely bound group of a few dozen to a few thousand stars that formed together in the disk of a spiral galaxy. Compare *globular cluster*. (Ch. 17)

open universe An infinite universe with a negatively curved spatial structure (much like the surface of a saddle) such that the sum of the angles of a triangle is always less than 180 degrees. Compare *closed universe* and *flat universe*. (Ch. 22)

orbit The path taken by one object moving around another object under the influence of their mutual gravitational or electric attraction. (Ch. 4)

orbital resonance A situation in which the orbital periods of two objects are related by a ratio of small integers. (Ch. 11)

orbiter A spacecraft placed in orbit around a planet or moon. Compare *flyby*. (Ch. 6)

organic Containing the element carbon. (Ch. 7)

P

P wave See *primary wave*. (Ch. 8)

pair production The creation of a particle-antiparticle pair from a source of electromagnetic energy. (Ch. 22)

paleoclimatology The study of changes in Earth's climate throughout its history. (Ch. 9)

parallax Also called *parallactic angle*. The displacement in the apparent position of a nearby star caused by the changing location of Earth in its orbit. (Ch. 13)

parent element A radioactive element that decays to form more-stable *daughter products*. (Ch. 8)

parsec (pc) Short for *parallax second*. The distance to a star with a parallax of 1 arcsecond (arcsec) using a base of 1 astronomical unit (AU). One parsec is approximately 3.26 light-years. (Ch. 13)

partial lunar eclipse An eclipse that occurs when the Moon is partially in Earth's shadow. (Ch. 2)

partial solar eclipse The type of eclipse that occurs when Earth passes through the penumbra of the Moon's shadow, so that the Moon blocks only a portion of the Sun's disk. Compare *annular solar eclipse* and *total solar eclipse*. (Ch. 2)

pc See *parsec*. (Ch. 13)

peculiar velocity The motion of a galaxy relative to the overall expansion of the universe. (Ch. 23)

penumbra (pl. penumbrae) 1. The outer part of a shadow, where the source of light is only partially blocked. Compare *umbra* (definition 1). 2. The region surrounding the umbra of a sunspot. The penumbra is cooler and darker than the surrounding surface of the Sun but is not as cool or dark as the umbra. Compare *umbra* (definition 2). (Ch. 2, 14)

penumbral lunar eclipse A lunar eclipse in which the Moon passes through the penumbra of Earth's shadow. Compare *total lunar eclipse*. (Ch. 2)

perihelion (pl. perihelia) The point in a solar orbit that is closest to the Sun. Compare *aphelion*. (Ch. 12)

period The time that a regularly repetitive process takes to complete one cycle. (Ch. 17)

period-luminosity relationship The relationship between the period of variability of a pulsating variable star, such as a Cepheid or RR Lyrae variable, and the luminosity of the star. Longer-period pulsating variable stars are more luminous than shorter-period ones. (Ch. 17)

Perseids A prominent August meteor shower associated with the dust debris left by comet Swift-Tuttle. (Ch. 12)

phase One of the various appearances of the sunlit surface of the Moon or a planet caused by the change in viewing location of Earth relative to both the Sun and the object. Examples include crescent phase and gibbous phase. (Ch. 2)

photino A hypothetical subatomic particle related to the photon. One of the candidates for cold dark matter. (Ch. 23)

photochemical Resulting from light acting on chemical systems. (Ch. 11)

photodisintegration An atomic nucleus absorbs an energetic gamma ray and emits some particles (Ch. 17)

photodissociation The breaking apart of molecules into smaller fragments or individual atoms by the action of photons. Compare *recombination* (definition 1). (Ch. 11)

photoelectric effect The emission of electrons from a substance illuminated by electromagnetic radiation greater than a certain critical frequency. (Ch. 18)

photon Also called *quantum of light*. A discrete unit or particle of electromagnetic radiation. The energy of a photon is equal to Planck's constant (h) multiplied by the frequency (f) of its electromagnetic radiation: $E_{photon} = h \times f$. The photon is the carrier of the electromagnetic force. (Ch. 5)

photosphere The apparent surface of the Sun as seen in visible light. Compare *chromosphere* and *corona*. (Ch. 14)

photosynthesize The process by which plants and algae absorb sunlight, water, and carbon dioxide, and release oxygen. (Ch. 9, 24)

physical law A broad statement that predicts a particular aspect of how the physical universe behaves and that is supported by many empirical tests. See also *theory*. (Ch. 1)

pixel The smallest picture element in a digital image array. (Ch. 6)

Planck era The early time, just after the Big Bang, for which the universe as a whole must be described with quantum mechanics. (Ch. 22)

Planck spectrum Also called *blackbody spectrum*. The spectrum of electromagnetic energy emitted by a blackbody per unit area per second, which is determined only by the temperature of the object. (Ch. 5)

Planck's constant (h) The constant of proportionality between the energy and the frequency of a photon. This constant defines how much energy a single photon of a given frequency or wavelength has. Value: 6.63×10^{-34} joule-second. (Ch. 5)

planet 1. A large body that orbits the Sun or other star that shines only by light reflected from the Sun or star. 2. In the Solar System, a body that orbits the Sun, has enough mass for self-gravity to overcome rigid body forces so that it assumes a spherical shape, and has cleared smaller bodies from the neighborhood around its orbit. Compare *dwarf planet*. (Ch. 7)

planet migration The theory that a planet can move to a location away from where it formed, through gravitational interactions with other bodies or loss of orbital energy from interaction with gas in the protoplanetary disk. (Ch. 7)

planetary nebula The expanding shell of material ejected by a dying asymptotic giant branch star. A planetary nebula glows from fluorescence caused by intense ultraviolet light coming from the hot, stellar remnant at its center. (Ch. 16)

planetary system A system of planets and other smaller objects in orbit around a star. (Ch. 7)

planetesimal A primitive body of rock and ice, 100 meters or more in diameter, that combines with others to form a planet. Compare *meteoroid* and *zodiacal dust*. (Ch. 7)

plasma A gas composed largely of charged particles but that also may include some neutral atoms. (Ch. 10, 14)

plate tectonics The geological theory concerning the motions of lithospheric plates, which in turn serves as the theoretical basis for *continental drift*. (Ch. 8)

positron A positively charged subatomic particle; the antiparticle of the *electron*. (Ch. 14)

power Energy per unit time. Possible units include watts (W) and joules per second (J/s). (Ch. 15)

precession of the equinoxes The slow change in orientation between the ecliptic plane and the celestial equator caused by the wobbling of Earth's axis. (Ch. 2)

pressure Force per unit area. Possible units include newtons per square meter (N/m²) and bars.

primary atmosphere An atmosphere, composed mostly of hydrogen and helium, that forms at the same time as its host planet. Compare *secondary atmosphere*. (Ch. 7)

primary mirror The principal optical mirror in a reflecting telescope. The primary mirror determines the telescope's light-gathering power and resolution. Compare *secondary mirror*. (Ch. 6)

primary wave Also called *P wave*. A longitudinal seismic wave, in which the oscillations involve compression and decompression parallel to the direction of travel. Compare *secondary wave*. (Ch. 8)

principle A general idea or sense about the universe that guides us in constructing new scientific theories. Principles can be testable theories. (Ch. 1)

prograde motion 1. Rotational or orbital motion of a moon that is in the same direction as the planet it orbits. 2. The counterclockwise orbital motion of Solar System objects as seen from above Earth's orbital plane. Compare *retrograde motion*. (Ch. 3)

prominence An archlike projection above the solar photosphere often associated with a sunspot. (Ch. 14)

proportional See *proportionality*. (Ch. 1, Appendix 1)

proportionality A relationship between two things whose ratio is a constant. (Ch. 1, Appendix 1)

proton (p or p^+) A subatomic particle having a positive electric charge of 1.6×10^{-19} coulomb (C), a rest mass of 1.67×10^{-27} kilogram (kg), and a rest energy of 1.5×10^{-10} joule (J). Compare *electron* and *neutron*. (Ch. 5)

proton-proton chain One way in which hydrogen fusion can take place. That path is the most important for hydrogen fusion in low-mass stars such as the Sun. Compare *carbon-nitrogen-oxygen cycle*. (Ch. 14)

protoplanetary disk The remains of the accretion disk around a young star from which a planetary system may form. Sometimes called *circumstellar disk*. (Ch. 7)

protostar A young stellar object that derives its luminosity from converting gravitational energy to thermal energy rather than from nuclear reactions in its core. (Ch. 7, 15)

pulsar A rapidly rotating neutron star that beams radiation into space in two searchlight-like beams. To a distant observer, the star appears to flash on and off. (Ch. 17)

pulsating variable star A variable star that undergoes periodic radial pulsations. (Ch. 17)

Q

QCD See *quantum chromodynamics*. (Ch. 22)

QED See *quantum electrodynamics*. (Ch. 22)

quantized Existing as discrete, irreducible units. (Ch. 5)

quantum chromodynamics (QCD) The quantum theory describing the strong nuclear force and its mediation by gluons. Compare *quantum electrodynamics*. (Ch. 22)

quantum efficiency The likelihood that a particular photon falling on a detector will actually produce a response in the detector. (Ch. 6)

quantum electrodynamics (QED) The quantum theory describing the electromagnetic force and its mediation by photons. Compare *quantum chromodynamics*. (Ch. 22)

quantum mechanics The branch of physics that deals with the quantized and probabilistic behavior of atoms and subatomic particles. (Ch. 5)

quantum of light See *photon*. (Ch. 5)

quark The building block of protons and neutrons. (Ch. 22)

quasar Short for *quasi-stellar radio source*. The most luminous of the active galactic nuclei, seen only at great distances from the Milky Way. (Ch. 19)

R

radial velocity The component of velocity directed toward or away from the observer. (Ch. 5)

radian The angle at the center of a circle subtended by an arc equal to the length of the circle's radius; 2π radians equals 360°, and 1 radian equals approximately 57.3°. (Appendix 1)

radiant The point in the sky from which the meteors in a meteor shower appear to come. (Ch. 12)

radiation Waves or particles of energy traveling through space or a medium. (Ch. 14)

radiation belt A toroidal ring of high-energy particles surrounding a planet. (Ch. 9)

radiative transfer The transport of energy from one location to another by electromagnetic radiation. (Ch. 14)

radiative zone A region within a star where energy is transported outward by radiation. Compare *convective zone*. (Ch. 14)

radio galaxy A type of elliptical galaxy that has an active galactic nucleus at its center and very strong emission (10^{35} to 10^{38} watts [W]) in the radio part of the electromagnetic spectrum. Compare *Seyfert galaxy*. (Ch. 19)

radio telescope An instrument for detecting and measuring radio frequency emissions from celestial sources. (Ch. 6)

radio wave Electromagnetic radiation in the extreme long-wavelength region of the spectrum, beyond the region of microwaves. (Ch. 5)

radioisotope A radioactive element. (Ch. 8)

radiometric dating Use of the radioactive decay of elements to measure the ages of materials such as minerals. (Ch. 8)

ratio The relationship in quantity or size between two or more things. (Ch. 1, Appendix 1)

ray 1. A beam of electromagnetic radiation. 2. A bright streak emanating from a young impact crater. (Ch. 6; 8)

recombination 1. The combining of ions and electrons to form neutral atoms. Compare *photodissociation*. 2. An event early in the evolution of the universe in which hydrogen and helium nuclei combined with electrons to form neutral atoms. The removal of electrons caused the universe to become transparent to electromagnetic radiation. (Ch. 11, 21)

red giant A low-mass star that has evolved beyond the main sequence and is now fusing hydrogen in a shell surrounding a degenerate helium core. (Ch. 16)

red giant branch A region on the H-R diagram defined by low-mass stars evolving from the main sequence toward the horizontal branch. (Ch. 16)

reddening The effect by which stars and other objects, when viewed through interstellar dust, appear redder than they actually are. Reddening occurs because blue light is more strongly absorbed and scattered than red light. (Ch. 15)

redshift The Doppler shift toward longer (redder) wavelengths of light from an approaching object. Compare *blueshift*. (Ch. 5)

reflecting telescope A telescope that uses mirrors to collect and focus incoming electromagnetic radiation to form an image in their focal planes. The size of a reflecting telescope is defined by the diameter of the primary mirror. Compare *refracting telescope*. (Ch. 6)

reflection The redirection of a beam of light that strikes, but does not cross, the surface between two media having different refractive indices. If the surface is flat and smooth, the angle of incidence equals the angle of reflection. Compare *refraction*. (Ch. 6)

refracting telescope A telescope that uses objective lenses to collect and focus incoming electromagnetic radiation to form an image. Compare *reflecting telescope*. (Ch. 6)

refraction The redirection or bending of a beam of light when it crosses the boundary between two media having different refractive indices. Compare *reflection*. (Ch. 6)

refractory material Material that remains solid at high temperatures. Compare *volatile material*. (Ch. 7)

regular moon A moon that formed together with the planet it orbits. Compare *irregular moon*. (Ch. 11)

reionization A period after the Dark Ages during which objects formed that radiated enough energy to ionize neutral hydrogen, at redshift $6 < z < 20$. (Ch. 23)

relative humidity The amount of water vapor held by a volume of air at a given temperature compared (stated as a percentage) to the total amount of water that could be held by the same volume of air at the same temperature. (Ch. 9)

relative motion The difference in motion between two individual frames of reference. (Ch. 2)

relativistic Describing systems that travel at nearly the speed of light or are located near very strong gravitational fields. (Ch. 18)

relativistic beaming The effect created when material moving at nearly the speed of light beams the radiation it emits in the direction of its motion. (Ch. 19)

relativistic speed A speed high enough that special relativity, rather than Newtonian physics, is needed to describe the motion. Speeds greater than about 10% the speed of light are relativistic. (Ch. 18)

remote sensing The use of images, spectra, radar, or other techniques to measure the properties of an object from a distance. (Ch. 6)

resolution The ability of a telescope to separate two point sources of light. Resolution is determined by the telescope's aperture and the wavelength of light it receives. (Ch. 6)

rest wavelength The wavelength of light that is seen coming from an object at rest with respect to the observer. (Ch. 5)

retrograde motion 1. Rotation or orbital motion of a moon that is in the opposite direction to the rotation of the planet it orbits. 2. The clockwise orbital motion of Solar System objects as seen from above Earth's orbital plane. Compare *prograde motion*. 3. Apparent retrograde motion is a motion of the planets with respect to the "fixed stars," in which the planets appear to move westward for a time before resuming their normal eastward motion. (Ch. 3)

revolve Motion of one object in orbit around another. (Ch. 2)

right ascension A measure, analogous to *longitude*, that tells you the angular distance of a celestial body eastward along the celestial equator from the vernal equinox. Compare *declination*. (Ch. 2)

ring An aggregation of small particles orbiting a planet or star. The rings of the four giant planets of the Solar System are composed variously of silicates, organic materials, and ices. (Ch. 11)

ring arc A discontinuous, higher-density region within an otherwise continuous, narrow ring. (Ch. 11)

ringlet A narrowly confined concentration of ring particles. (Ch. 11)

Roche limit The distance at which a planet's tidal forces exceed the self-gravity of a smaller object—such as a moon, asteroid, or comet—causing the object to break apart. (Ch. 4)

Roche lobes The hourglass-shaped or figure eight–shaped volume of space surrounding two stars, which constrains material that is gravitationally bound by one or the other. (Ch. 16)

rotation curve A plot showing how the orbital velocity of stars and gas in a galaxy changes with radial distance from the galaxy's center. (Ch. 19)

rover A remotely controlled instrumented vehicle designed to move and explore the surface of a terrestrial planet or moon. Compare *lander*. (Ch. 6)

RR Lyrae variable A variable giant star whose regularly timed pulsations are good predictors of its luminosity. RR Lyrae variables are used in measuring the distance to globular clusters. (Ch. 17)

S

S-type asteroid An asteroid made of material that was once part of the outer layer of a larger, differentiated body that has since broken into pieces. Compare *C-type asteroid* and *M-type asteroid*. (Ch. 12)

S wave See *secondary wave*. (Ch. 8)

S0 galaxy A galaxy with a bulge and a disk-like spiral, but smooth in appearance like ellipticals. Compare *barred spiral galaxy*, *elliptical galaxy*, *irregular galaxy*, and *spiral galaxy*. (Ch. 19)

satellite An object in orbit around a more massive body, for example, a moon of any planet or an artificial satellite. (Ch. 4)

scale factor (R_U) A dimensionless number proportional to the distance between two points in space. The scale factor increases as the universe expands. (Ch. 21)

scattering The random change in the direction of travel of photons, caused by their interactions with molecules or dust particles. (Ch. 9, 16)

Schwarzschild radius The distance from the center of a nonrotating, spherical black hole at which the escape velocity equals the speed of light. (Ch. 18)

scientific method The formal procedure—including hypothesis, prediction, and experiment or observation—used to test (that is, to attempt to falsify) the validity of scientific hypotheses and theories. (Ch. 1)

scientific notation The standard expression of numbers with one digit (which can be zero) to the left of the decimal point and multiplied by 10 to the exponent required to give the number its correct value. Example: $2.99 \times 10^8 = 299{,}000{,}000$. (1, Appendix 1)

SCP See *south celestial pole*. (Ch. 2)

second law of thermodynamics The law stating that the entropy or disorder of an isolated system always increases as the system evolves. (Ch. 24)

second of arc See *arcsecond*. (Ch. 6, 13)

secondary atmosphere An atmosphere that forms—as a result of volcanism, comet impacts, or another process—sometime after its host planet has formed. Compare *primary atmosphere*. (Ch. 7)

secondary crater A crater formed from ejecta thrown from an *impact crater*. (Ch. 8)

secondary mirror A small mirror placed on the optical axis of a reflecting telescope that returns the beam back through a small hole in the *primary mirror*, thereby shortening the mechanical length of the telescope. (Ch. 6)

secondary wave Also called *S wave*. A transverse seismic wave, which involves the sideways motion of material. Compare *primary wave*. (Ch. 8)

sediment Solid material such as rock or sand that has been affected by erosion and weather, and transported by water, wind, or ice. (Ch. 8)

seismic wave A vibration due to an earthquake, a large explosion, or an impact on the surface that travels through a planet's interior. (Ch. 8)

seismometer An instrument that measures the amplitude and frequency of seismic waves. (Ch. 8)

self-gravity The gravitational attraction among all parts of the same object. (Ch. 4)

semimajor axis Half of the longer axis of an ellipse. (Ch. 3)

SETI The Search for Extraterrestrial Intelligence project, which uses advanced technology combined with radio telescopes to search for evidence of intelligent life elsewhere in the universe. (Ch. 24)

Seyfert galaxy A type of spiral galaxy with an active galactic nucleus at its center; first discovered in 1943 by Carl Seyfert. Compare *radio galaxy*. (Ch. 19)

shepherd moon A moon that orbits close to rings and gravitationally confines the orbits of the ring particles. (Ch. 11)

shield volcano A volcano formed by very fluid lava flowing from a single source and spreading out from that source. Compare *composite volcano*. (Ch. 8)

short gamma-ray burst Bursts of gamma rays lasting a few milliseconds to a few minutes, thought to originate in neutron star mergers. (Ch. 17)

short-period comet A comet with an orbital period of less than 200 years. Compare *long-period comet*. (Ch. 12)

sidereal day Earth's period of rotation with respect to the stars—about 23 hours 56 minutes—which is the time Earth takes to make one rotation and face the same star on the meridian. It differs from the *solar day* because of Earth's motion around the Sun. (Ch. 2)

sidereal period An object's orbital or rotational period measured with respect to the stars. Compare *synodic period*. (Ch. 2)

silicate One of the family of minerals composed of silicon and oxygen in combination with other elements. (Ch. 7)

singularity The point at which a mathematical expression or equation becomes meaningless, such as a fraction whose denominator approaches zero See also *black hole*. (Ch. 18)

solar abundance The relative amount of an element detected in the atmosphere of the Sun, expressed as the ratio of the number of atoms of that element to the number of hydrogen atoms. (Ch. 5)

solar day The day in common use—24 hours, which is Earth's period of rotation that brings the Sun back to the same local meridian where the rotation started. Compare *sidereal day*. (Ch. 2)

solar eclipse An eclipse that occurs when the Moon partially or entirely blocks the Sun. Compare *lunar eclipse*. (Ch. 2)

solar flare An explosion on the Sun's surface associated with a complex sunspot group and a strong magnetic field. (Ch. 14)

solar maximum (pl. **maxima**) The time, occurring about every 11 years, when the Sun is at its peak activity, meaning that sunspot activity and related phenomena (such as prominences, flares, and coronal mass ejections) are at their peak. (Ch. 14)

solar neutrino problem The historical observation that only about a third as many neutrinos as predicted by theory seemed to be coming from the Sun. (Ch. 14)

Solar System The gravitationally bound system made up of the Sun, planets, dwarf planets, moons, asteroids, comets, and Kuiper Belt objects, along with their associated gas and dust. (Ch. 1)

solar tide A tide on Earth caused by the differential gravitational pull of the Sun. Compare *lunar tide*. (Ch. 4)

solar wind The stream of charged particles emitted by the Sun and that flows at high speeds through interplanetary space. (Ch. 9)

solstice Literally, "Sun standing still." 1. One of the two most northerly and southerly points on the ecliptic. 2. Either of the two times of year (the *summer solstice* and *winter solstice*) when the Sun is at one of these two positions. Compare *equinox*. (Ch. 2)

south celestial pole (SCP) The southward projection of Earth's rotation axis onto the celestial sphere. Compare *north celestial pole*. (Ch. 2)

South Pole The location in the Southern Hemisphere where Earth's rotation axis intersects Earth's surface. Compare *North Pole*. (Ch. 2)

spacetime A concept that combines space and time into a four-dimensional continuum with three spatial dimensions plus one dimension of time. (Ch. 18)

special relativity See *special theory of relativity*. (Ch. 18)

special theory of relativity Sometimes referred to as simply *special relativity*. Einstein's theory explaining how the fact that the speed of light is a constant affects nonaccelerating frames of reference. Compare *general theory of relativity*. (Ch. 18)

spectral type A classification system for stars that is based on the presence and relative strength of absorption lines in their spectra. Spectral type is related to the surface temperature of a star. (Ch. 13)

spectrograph Also called *spectrometer*. A device that spreads out the light from an object into its component wavelengths. (Ch. 6)

spectrometer See *spectrograph*. (Ch. 6)

spectroscopic binary A binary star system whose existence and properties are revealed to astronomers only by the Doppler shift of its spectral lines. Most spectroscopic binaries are close pairs. Compare *eclipsing binary* and *visual binary*. (Ch. 13)

spectroscopic parallax Use of the spectroscopically determined luminosity and the observed brightness of a star to determine the star's distance. (Ch. 13)

spectroscopy The study of an object's electromagnetic radiation in terms of its component wavelengths. (Ch. 6)

spectrum (pl. **spectra**) Waves sorted by wavelength. See also *electromagnetic spectrum*. (Ch. 5)

speed The rate of change of an object's position with time, without regard to the direction of movement. Possible units include meters per second (m/s) and kilometers per hour (km/h). Compare *velocity*. (Ch. 3)

spherically symmetric Describing an object whose properties depend only on distance from the object's center, so that the object has the same form viewed from any direction. (Ch. 4)

spin-orbit resonance A relationship between the orbital and rotation periods of an object such that the ratio of their periods can be expressed with simple integers. (Ch. 11)

spiral density wave A stable, spiral-shaped change in the local gravity of a galactic disk that can be produced by periodic gravitational kicks from neighboring galaxies or from nonspherical bulges and bars in spiral galaxies. (Ch. 20)

spiral galaxy A galaxy of Hubble type "S" class, with a discernible disk in which large spiral patterns exist. Compare *barred spiral galaxy*, *elliptical galaxy*, *irregular galaxy*, and *S0 galaxy*. (Ch. 19)

spoke One of several narrow radial features seen occasionally in Saturn's B Ring. Spokes appear dark in backscattered light and bright in forward, scattering light, indicating that they are composed of tiny particles. Their origin is not well understood. (Ch. 11)

spreading center A zone from which two tectonic plates diverge. (Ch. 8)

spring tide An especially strong tide that occurs around the time of a new or full Moon, when lunar tides and solar tides reinforce each other. Compare *neap tide*. (Ch. 4)

stable equilibrium An equilibrium state in which the system returns to its former condition after a small disturbance. Compare *unstable equilibrium*. (Ch. 5)

standard candle An object whose luminosity either is known or can be predicted in a distance-independent way, so its brightness can be used to determine its distance via the inverse square law of radiation. (Ch. 19)

standard model The theory of particle physics that combines electroweak theory with quantum chromodynamics to describe the structure of known forms of matter. (Ch. 22)

star A luminous ball of gas held together by gravity. A normal star is powered by nuclear reactions in its interior. (Ch. 1)

star cluster A group of stars that all formed at the same time and in the same general location. (Ch. 15)

static equilibrium A state in which the forces within a system are all in balance so that the system does not change. Compare *dynamic equilibrium*. (Ch. 5)

Stefan-Boltzmann constant (σ) The constant of proportionality that relates the flux emitted by an object to the fourth power of its absolute temperature. Value: 5.67×10^{-8} W/(m² K⁴) (W = watts, m = meters, K = kelvins). (Ch. 5)

Stefan-Boltzmann law The law, formulated by Josef Stefan and Ludwig Boltzmann, stating that the amount of electromagnetic energy emitted from the surface of a body (flux), summed over the energies of all photons of all wavelengths emitted, is proportional to the fourth power of the temperature of the body: $\mathcal{F} = \sigma T^4$. (Ch. 5)

stellar evolution The stages of development that a star goes through over its life cycle. Astronomers observe many stars at the same time to build a picture of how stars change. (Ch. 16)

stellar-mass loss The loss of mass from the outermost parts of a star's atmosphere during the star's evolution. (Ch. 16)

stellar occultation An event in which a planet or other Solar System body moves between the observer and a star, eclipsing the light emitted by that star. (Ch. 10)

stellar population A group of stars with similar ages, chemical compositions, and dynamic properties. (Ch. 17)

stereoscopic vision The way an animal's brain combines the different information from its two eyes to perceive the distances to objects around it. (Ch. 13)

stony-iron meteorite A meteorite consisting of a mixture of silicate minerals and iron-nickel alloys. Compare *iron meteorite* and *stony meteorite*. (Ch. 12)

stony meteorite A meteorite composed primarily of silicate minerals, similar to those found on Earth. Compare *iron meteorite* and *stony-iron meteorite*. (Ch. 12)

stratosphere The atmospheric layer immediately above the troposphere. On Earth, it extends upward to an altitude of 50 kilometers (km). Compare *troposphere*, *mesosphere*, and *thermosphere*. (Ch. 9)

string theory See *superstring theory*. (Ch. 22)

stromatolite A structure created by living or fossilized cyanobacteria. (Ch. 24)

strong nuclear force The attractive short-range force between protons and neutrons that holds atomic nuclei together. One of the four fundamental forces of nature, along with the *electromagnetic force*, the *weak nuclear force*, and *gravity* (definition 1). The force is mediated by the exchange of gluons. (Ch. 14, 22)

subduction zone A region where two tectonic plates converge, with one plate sliding under the other and being drawn downward into the interior. (Ch. 8)

subgiant A giant star that is smaller and lower in luminosity than normal giant stars of the same spectral type. Subgiants evolve to become giants. (Ch. 16)

subgiant branch A region of the H-R diagram defined by stars that have left the main sequence but have not yet reached the red giant branch. (Ch. 16)

sublimation The process by which a solid becomes a gas without first becoming a liquid. (Ch. 12)

summer solstice 1. One of two points where the Sun is at its greatest distance from the celestial equator. 2. The day on which the Sun appears at that location, marking the first day of summer (about June 20 in the Northern Hemisphere and December 21 in the Southern Hemisphere). Compare *winter solstice*. See also *autumnal equinox* and *vernal equinox*. (Ch. 2)

Sun The star at the center of the Solar System. (Ch. 1)

sungrazer A comet whose perihelion is within a few solar diameters of the surface of the Sun. (Ch. 12)

sunspot A cooler, transitory region on the solar surface produced when loops of magnetic flux break through the surface of the Sun. (Ch. 14)

sunspot cycle The approximate 11-year cycle during which sunspot activity increases and then decreases. This is one-half of a full 22-year cycle, in which the magnetic polarity of the Sun first reverses and then returns to its original configuration. (Ch. 14)

supercluster A large conglomeration of galaxy clusters and galaxy groups, typically more than 100–300 megaparsecs (Mpc) in size and containing tens of thousands to hundreds of thousands of galaxies. Compare *galaxy cluster* and *galaxy group*. (Ch. 1, 23)

super-Earth An extrasolar planet with about 2–10 times the mass of Earth. (Ch. 7, 24)

superior planet A Solar System planet that orbits farther from the Sun than Earth does. Compare *inferior planet*. (Ch. 3)

superluminal motion The appearance (though not the reality) that a jet is moving faster than the speed of light. (Ch. 21)

supermassive black hole A black hole of 1,000 solar masses (M_{Sun}) or more that resides in the center of a galaxy and whose gravity powers active galactic nuclei. (Ch. 19)

supernova (pl. **supernovae**) A stellar explosion resulting in the release of tremendous amounts of energy, including the high-speed ejection of matter into the interstellar medium. See also *Type Ia supernova* and *Type II supernova*. (Ch. 16)

supernova remnant The material ejected from the outer layers of a star after a supernova explosion. (Ch. 16)

supersonic Moving within a medium at a speed faster than the speed of sound in that medium. Compare *subsonic*. (Ch. 15)

superstring theory The theory that conceives of particles as strings in 10 dimensions of space and time; the current contender for a theory of everything. (Ch. 22)

surface brightness The amount of electromagnetic radiation emitted or reflected per unit area. (Ch. 6)

surface wave A seismic wave that travels on the surface of a planet or moon. (Ch. 8)

symmetry In theoretical physics, the properties of physical laws that remain constant when certain things change, such as the symmetry between matter and antimatter even though their charges may be different. (Ch. 22)

synchronous rotation The case that occurs when a body's rotation period equals its orbital period around another body. A special type of spin-orbit resonance. (Ch. 2)

synchrotron radiation Radiation from electrons moving at close to the speed of light as they spiral in a strong magnetic field, so named because that kind of radiation was first identified on Earth in particle accelerators called synchrotrons. (Ch. 10)

synodic period An object's orbital or rotational period measured with respect to the Sun. Compare *sidereal period*. (Ch. 2)

T

T Tauri star A young stellar object that has dispersed enough of the material surrounding it to be seen in visible light. (Ch. 15)

tail A stream of gas and dust swept away from the coma of a comet by the solar wind and by radiation pressure from the Sun. See also *ion tail* and *dust tail*. (Ch. 12)

tectonism Deformation of the lithosphere of a planet. (Ch. 8)

telescope The basic tool of astronomers. Working over the entire range from gamma rays to radio waves, astronomical telescopes collect and concentrate electromagnetic radiation from celestial objects. (Ch. 6)

temperature A measure of the average kinetic energy of the atoms or molecules in a gas, solid, or liquid. (Ch. 5)

terrestrial planet An Earth-like planet, made of rock and metal and having a solid surface. In the Solar System, the terrestrial planets are Mercury, Venus, Earth, and Mars. Compare *giant planet*. (Ch. 7)

theoretical model A description of the properties of a particular object or system in terms of known physical laws or theories; often, a computer calculation of predicted properties based on such a description. (Ch. 1)

theory A well-developed idea or group of ideas that are tied solidly to known physical laws and make testable predictions about the world. A very well-tested theory may be called a *physical law*, or simply a fact. Compare *hypothesis*. (Ch. 1)

theory of everything (**TOE**) A theory that unifies all four fundamental forces of nature: strong nuclear, weak nuclear, electromagnetic, and gravitational forces. (Ch. 22)

thermal conduction See *conduction*. (Ch. 14)

thermal energy The energy that resides in the random motion of atoms, molecules, and particles, by which we measure their temperature. (Ch. 5)

thermal equilibrium The state in which the rate of thermal-energy emission by an object is equal to the rate of thermal-energy absorption. (Ch. 5)

thermal motion The random motion of atoms, molecules, and particles that gives rise to thermal radiation. (Ch. 5)

thermal radiation Electromagnetic radiation resulting from the random motion of the charged particles in every substance. (Ch. 5)

thermosphere The layer of Earth's atmosphere at altitudes greater than 90 kilometers (km), above the mesosphere. Near its top, at an altitude of 600 km, the temperature can reach 1000 K. Compare *troposphere*, *stratosphere*, and *mesosphere*. (Ch. 9)

third quarter Moon The phase of the Moon in which only the eastern half of the Moon, as viewed from Earth, is illuminated by the Sun. It occurs about one week after the full Moon. Compare *first quarter Moon*. See also *full Moon* and *new Moon*. (Ch. 2)

tidal bulge Distortion of a body resulting from tidal stresses. (Ch. 4)

tidal force A force caused by the change in the strength of gravity across an object. (Ch. 4)

tidal locking Synchronous rotation of an object caused by internal friction as the object rotates through its tidal bulge. (Ch. 4)

tide On Earth, the rise and fall of the oceans as Earth rotates through a tidal bulge caused by the Moon and the Sun. See also *lunar tide*, *neap tide*, *solar tide*, and *spring tide*. (Ch. 4)

time dilation The relativistic "stretching" of time. Compare *general relativistic time dilation*. (Ch. 18)

TOE See *theory of everything*. (Ch. 22)

Topography The study of the natural physical features of the surface of a planet or moon (Ch. 8)

tornado A violent rotating column of air, typically 75 meters across with 200-kilometer-per-hour (km/h) winds. Some tornadoes can be more than 3 km across, and winds up to 500 km/h have been observed. (Ch. 9)

torus (pl. **tori**) A three-dimensional, doughnut-shaped ring. (Ch. 10, 19)

total lunar eclipse A lunar eclipse in which the Moon passes through the umbra of Earth's shadow. Compare *penumbral lunar eclipse*. (Ch. 2)

total solar eclipse The type of eclipse that occurs when Earth passes through the umbra of the Moon's shadow, so that the Moon blocks the disk of the Sun. Compare *annular solar eclipse* and *partial solar eclipse*. (Ch. 2)

transit method A method of detecting extrasolar planets by measuring the decrease in light from a star as its orbiting planet passes in front of the star as viewed from Earth. (Ch. 7)

trans-Neptunian object See *Kuiper Belt object*. (Ch. 12)

transverse wave A wave that oscillates perpendicular to the direction of the wave's propagation. Compare *longitudinal wave*. (Ch. 8)

triple-alpha process The nuclear fusion reaction that combines three helium nuclei (alpha particles) into a single nucleus of carbon. (Ch. 16)

Trojans A group of asteroids orbiting in the L_4 and L_5 Lagrangian points of Jupiter's orbit. (Ch. 12)

tropical year The time between two crossings of the vernal equinox. Because of the precession of the equinoxes, a tropical year is slightly shorter than the time Earth takes to orbit once around the Sun. Compare *year*. (Ch. 2)

Tropics The region on Earth between latitudes 23.5° south and 23.5° north, where the Sun appears directly overhead twice during the year. (Ch. 2)

tropopause The top of a planet's troposphere. (Ch. 9)

troposphere The convection-dominated layer of a planet's atmosphere. On Earth, the atmospheric region closest to the ground within which most weather phenomena take place. Compare *stratosphere*, *mesosphere*, and *thermosphere*. (Ch. 9)

turbulence The random motion of blobs of gas within a larger cloud of gas. (Ch. 15)

Type Ia supernova A supernova explosion with a calibrated peak luminosity that occurs as a result of runaway carbon fusion in a white dwarf star that accretes mass from a companion and approaches the Chandrasekhar mass limit of 1.4 M_{Sun}. (Ch. 16)

Type II supernova A supernova explosion in which the degenerate core of an evolved massive star suddenly collapses and rebounds. (Ch. 17)

U

ultrafaint dwarf galaxy A dim dwarf galaxy with only 1,000–100,000 times the Sun's luminosity. Ultrafaint dwarf galaxies differ from globular clusters in that they are composed of large amounts of dark matter. (Ch. 23)

ultraviolet (**UV**) **radiation** Electromagnetic radiation having frequencies and photon energies greater than those of visible light but less than those of X-rays and having wavelengths shorter than those of visible light but longer than those of X-rays. (Ch. 5)

umbra (pl. **umbrae**) 1. The darkest part of a shadow, where the source of light is blocked. Compare *penumbra* (definition 1). 2. The darkest, innermost part of a sunspot. Compare *penumbra* (definition 2). (Ch. 2, 14)

unbound orbit An orbit in which an object is no longer gravitationally bound to the body it was orbiting. An unbound orbit's velocity is greater than the escape velocity. Compare *bound orbit*. (Ch. 4)

uncertainty principle See *Heisenberg uncertainty principle*. (Ch. 22)

unified model of AGN A model in which many types of activity in the nuclei of galaxies are all explained by accretion of matter around a supermassive black hole. (Ch. 19)

uniform circular motion Motion in a circular path at a constant speed. (Ch. 4)

unit A fundamental quantity of measurement. The meter is an example of a metric unit; the foot is an example of an English unit. (Ch. 1, Appendix 1)

universal gravitational constant (G) The constant of proportionality in the universal law of gravitation. Value: 6.67×10^{-11} meters cubed per kilogram second squared [$m^3/kg\ s^2 = N\ m^2/kg^2$]. (Ch. 4)

universal law of gravitation The law, formulated by Isaac Newton, stating that the gravitational force between any two objects is proportional to the product of their masses and inversely proportional to the square of the distance between them: $F_{grav} = G \times \dfrac{m_1 \times m_2}{r^2}$ (Ch. 4)

universe 1. All of space and everything contained therein. 2. Our own universe in a collection of parallel universes that together make up all that exists. (Ch. 1)

unstable equilibrium An equilibrium state in which a small disturbance will cause a system to move away from equilibrium. Compare *stable equilibrium*. (Ch. 5)

UV Ultraviolet. See *ultraviolet radiation*. (Ch. 5)

V

vacuum A region of space devoid of matter. In quantum mechanics and general relativity, however, even a perfect vacuum has physical properties. (Ch. 5, 22)

variable star A star with varying luminosity. Many periodic variables are found in the instability strip on the H-R diagram. (Ch. 17)

velocity The rate and direction of change of an object's position with time. Possible units include meters per second (m/s) and kilometers per hour (km/h). Compare *speed*. (Ch. 3)

vernal equinox 1. One of two points where the Sun crosses the celestial equator. 2. The day on which the Sun appears at that location, marking the first day of spring (about March 20 in the Northern Hemisphere and September 22 in the Southern Hemisphere). Compare *autumnal equinox*. See also *summer solstice* and *winter solstice*. (Ch. 2)

virtual particle A particle that, according to quantum mechanics, comes into existence only momentarily. According to theory, fundamental forces are mediated by the exchange of virtual particles. (Ch. 18)

visual binary A binary system in which the two stars can be seen individually from Earth. Compare *eclipsing binary* and *spectroscopic binary*. (Ch. 13)

void A region in space containing little or no matter. Examples include regions in cosmological space that are largely empty of galaxies. (Ch. 23)

volatile material Generally called *ice* in its solid form. Material that remains gaseous at moderate temperature. Compare *refractory material*. (Ch. 7)

volcanism A form of geological activity on a planet or moon in which molten rock (magma) erupts onto the surface. (Ch. 8)

W

W See *watt*. (Ch. 5)

waning Describing the changing phases of the Moon as it becomes less fully illuminated between full Moon and new Moon as seen from Earth. Compare *waxing*. (Ch. 2)

waning crescent Moon The phases of the Moon between third quarter and new Moon (Ch. 2)

waning gibbous Moon The phases of the moon between full Moon and third quarter. (Ch. 2)

water cycle The flow of water on, above, and through Earth's surface. (Ch. 9)

watt (W) A measure of *power*. Unit: joules per second (J/s). (Ch. 5)

wave A disturbance moving along a surface or passing through a space or a medium. (Ch. 5)

wavefront The imaginary surface of an electromagnetic wave, either plane or spherical, oriented perpendicular to the direction of travel. (Ch. 6)

wavelength The distance on a wave between two adjacent points having identical characteristics. The distance a wave travels in one period. Possible units include meters (m). (Ch. 5)

waxing Describing the changing phases of the Moon as it becomes more fully illuminated between new Moon and full Moon as seen from Earth. Compare *waning*. (Ch. 2)

waxing crescent Moon The phases of the Moon between new and first quarter. (Ch. 2)

waxing gibbous Moon The phases of the Moon between first quarter and full. (Ch. 2)

weak nuclear force The force underlying some forms of radioactivity and certain interactions between subatomic particles. It is responsible for radioactive beta decay and for the initial proton-proton interactions that lead to nuclear fusion in the Sun and other stars. One of the four fundamental forces of nature, along with the *electromagnetic force*, the *strong nuclear force*, and *gravity* (definition 1). The force is mediated by the exchange of W and Z particles. (Ch. 22)

weather The state of an atmosphere at any given time and place. Compare *climate*. (Ch. 9)

weight The gravitational force acting on an object, that is, the force equal to the mass of an object multiplied by the local acceleration due to gravity. In general relativity, the force equal to the mass of an object multiplied by the acceleration of the frame of reference in which the object is observed. Compare *mass*. (Ch. 4)

white dwarf The stellar remnant left at the end of the evolution of a low-mass star. A typical white dwarf has a mass of 0.6 solar mass (M_{Sun}) and a size about equal to that of Earth; it is made of non-fusing, electron-degenerate carbon. (Ch. 16)

Wien's law A law, named for Wilhelm Wien, stating that location of the peak wavelength in the electromagnetic spectrum of an object is inversely proportional to the temperature of the object. (Ch. 5)

WIMP Short for *weakly interacting massive particle*. A hypothetical massive particle that interacts through the weak nuclear force and gravity but not with electromagnetic radiation. WIMPs are candidates for dark matter. Compare *MaCHO*. (Ch. 19)

winter solstice 1. One of two points where the Sun is at its greatest distance from the celestial equator. 2. The day on which the Sun appears at that location, marking the first day of winter (about December 21 in the Northern Hemisphere and June 20 in the Southern Hemisphere). Compare *summer solstice*. See also *autumnal equinox* and *vernal equinox*. (Ch. 2)

X

X-ray Electromagnetic radiation having frequencies and photon energies greater than those of ultraviolet (UV) light but less than those of gamma rays and having wavelengths shorter than those of UV light but longer than those of gamma rays. (Ch. 5)

X-ray binary A binary system in which mass from an evolving star spills over onto a collapsed companion, such as a neutron star or black hole. The material falling in is heated to such high temperatures that it glows brightly in X-rays. (Ch. 17)

Y

year The time Earth takes to make one revolution around the Sun. A solar year is measured from equinox to equinox. A sidereal year, Earth's true orbital period, is measured relative to the stars. Compare *tropical year*. (Ch. 2)

Z

zenith The point on the celestial sphere located directly overhead of an observer. Compare *nadir*. (Ch. 2)

zero-age main sequence The strip on the H-R diagram plotting where stars of all masses in a cluster begin their lives. (Ch. 15)

zodiac The 12 constellations lying along the plane of the ecliptic. (Ch. 2)

zodiacal dust Particles of cometary and asteroidal debris smaller than 100 microns (μm) that orbit the inner Solar System close to the plane of the ecliptic. Compare *meteoroid* and *planetesimal*. (Ch. 12)

zodiacal light A band of light in the night sky caused by sunlight reflected by zodiacal dust. (Ch. 12)

zonal wind The east–west component of a wind. (Ch. 9)

Selected Answers

These pages contain selected answers for all 24 chapters in the complete volume of *21st Century Astronomy*, Sixth Edition. The Solar System Edition does not include Chapters 15–23. The Stars and Galaxies Edition does not include Chapters 8–12.

Chapter 1

CHECK YOUR UNDERSTANDING
1.1 (d, a, e, c, f, b)
1.2 (b)
1.3 (c)

TEST YOUR UNDERSTANDING
4. (a)
7. (c)
10. (d)
11. (c)
14. (b)

THINKING ABOUT THE CONCEPTS
18. 8.3 minutes (see Figure 1.3).
23. A *theory* is generally understood to mean an idea, regardless of any proof, evidence, or way to test it. A *scientific theory* is an explanation for an occurrence in nature that must be based on observations and data and make testable predictions.
29. Some scientific fields rely heavily on math, but not all. Following the scientific method is the important part, not use of mathematics.

APPLYING THE CONCEPTS
31. (a) 7×10^9 (b) 3.46×10^{-3} (c) 1.238×10^3
36. 5,333 km/hr. 6.7 times faster than a jet plane.

Chapter 2

CHECK YOUR UNDERSTANDING
2.1A (b)
2.1B (d)
2.2 (d)
2.3 A full Moon is rising as the Sun sets. At noon, a third-quarter Moon will be setting.
2.4 If the synodic period of the Moon phases were precisely 30 days, and Earth took precisely 12 months to orbit the Sun, the current "wandering" holidays would always fall on the same days from year to year.
2.5 (b)

TEST YOUR UNDERSTANDING
1. (c)
4. (b)
6. (e)
9. (d)
14. (b)

THINKING ABOUT THE CONCEPTS
16. Magellan could not use the North Star (Polaris) for navigation because he was in the Southern Hemisphere; thus, Polaris was never above the horizon.
18. If Gemini is high in the night sky in winter, it is high in the daytime sky in summer, and the Sun's light blots out all light from the stars.
23. (a) The Earth takes 24 hours to complete one rotation (about its axis) with respect to the Sun—23 h 56 m with respect to the stars. (b) The Earth takes 26,000 years to complete one "wobble."

APPLYING THE CONCEPTS
31. 12,795 km.
34. 25° N latitude. Hawaii, southern Florida.

Chapter 3

CHECK YOUR UNDERSTANDING
3.1A Ancient civilizations knew planets were different from stars because planets wandered through the sky—hence the name planet, from the Greek *planetes,* meaning "wanderer."
3.1B (c)
3.2 (c, b, d, a)
3.3 (c, d)
3.4 (b)

TEST YOUR UNDERSTANDING
1. (d)
5. (a)
9. (a)
12. (c)
14. (b)

THINKING ABOUT THE CONCEPTS
16. During normal motion, Mars moves toward the east. Retrograde motion for Mars is motion toward the west. In this image, Mars was in retrograde motion for about 2.5 months, or about 75 days. Mars observers would have seen Earth in retrograde motion as well.
18. Saturn takes 29.5 years to orbit the Sun, so after 1 Earth orbit it has not moved far in its orbit. Earth catches up to and passes Saturn is just over one Earth year.
27. We wear seat belts to protect us from acceleration. We travel at the same speed as the vehicle, but because we have inertia of our own, we will keep moving forward (Newton's first law) until something stops us (that is, the seat belt) if the vehicle suddenly decelerates.

APPLYING THE CONCEPTS
33. 354 years.
42. 11 days.

Chapter 4

CHECK YOUR UNDERSTANDING
4.1A (d)
4.1B (c), (a)
4.2 (c, a, b)
4.3 (d, b, a, c)
4.4 (a)

TEST YOUR UNDERSTANDING
3. (c)
5. (c)
8. (a)
10. (b)
13. (b)

THINKING ABOUT THE CONCEPTS
17. Circular velocity is the speed needed to orbit a body in a circular path. Escape velocity is the speed needed to escape the gravitational force of a body. Escape velocity will always be larger because more energy is needed to escape the gravitational force of an object.

20. The comet in an elliptical orbit will eventually return to the immediate vicinity of the Sun, depending on the size of its orbit. The comet in an unbound orbit will escape the Sun entirely and head out into interstellar space, never to return.
28. Solid bodies can be tidally distorted, and tidal distortion occurs on Earth. That distortion of the planet's landmasses is dissipated in frictional heating of rock.

APPLYING THE CONCEPTS

32. 5.4 km/s.
37. 9.81 m/s^2

Chapter 5

CHECK YOUR UNDERSTANDING

5.1A (e, c, b, d, a)
5.1B (c)
5.2 The spectral features correspond to the composition of elements present in the stars' atmospheres. Each element—even its ions and isotopes—has a unique set of lines.
5.3 (d)
5.4 (d, b, a, c, e)
5.5 (c)

TEST YOUR UNDERSTANDING

2. (b)
6. (d)
8. (a)
11. (c)
15. (b)

THINKING ABOUT THE CONCEPTS

16. Light can travel at a speed slower than 3×10^5 km/s when it is traveling in a medium. Light, like sound, travels at a certain speed in each kind of medium.
19. A beam of red light can have as much energy as a blue beam if the red beam has many more photons.
23. Redshifted objects are moving away from us; blueshifted objects are moving toward us.

APPLYING THE CONCEPTS

33. 472.4 Hz.
36. 0.005 percent of its brightness as seen from Earth.

Chapter 6

CHECK YOUR UNDERSTANDING

6.1A (a, b, c)
6.1B (a)
6.2 (c)
6.3 (d)
6.4 (c)
6.5 (a, b, d)

TEST YOUR UNDERSTANDING

4. (b)
7. (a)
10. (b)
13. (d)
15. (b)

THINKING ABOUT THE CONCEPTS

16. Simple lenses have chromatic aberration; they bend different colors by different amounts. The resulting image is not nearly as crisp as in telescopes with multiple lenses.
19. Different types of astronomical objects are at different temperatures and thus emit light at different peak wavelengths. Many astronomical sources that look normal at visible wavelengths may be much brighter or more complex in other regions of the spectrum.
25. The stars are so far away that a telescope in space is only infinitesimally closer to the stars than a telescope on Earth. Telescopes in space avoid atmospheric effects and can observe at wavelengths or energies that Earth's atmosphere blocks.

APPLYING THE CONCEPTS

33. 16 times more light. Stars 1/4 as bright, or stars of 1 to 2 magnitudes dimmer.
40. (a) 0.35 milliarcseconds
    (b) 1.1 milliarcseconds

Chapter 7

CHECK YOUR UNDERSTANDING

7.1 (a, b, d)
7.2 (a)
7.3 (b)
7.4 (a)
7.5 (c)

TEST YOUR UNDERSTANDING

1. (a, b, h, c, e, d, f, g)
3. (b)
8. (b)
14. (c)
15. (d)

THINKING ABOUT THE CONCEPTS

16. The hydrogen and most helium atoms that make up the Sun and the rest of the Solar System all came from the Big Bang. Everything else was formed in previous generations of stars.
20. Scientists will have to propose a formation mechanism consistent with all other evidence for planet formation. Scientists will consider all the relevant physics, propose the most likely scenarios, and then propose tests of those scenarios and search to validate or invalidate them.
29. Many of the initial discoveries of exoplanets were made using the radial velocity method, which at the time could do no better than resolve massive planets in orbits close to their stars with periods of about a week.

APPLYING THE CONCEPTS

31. 30 m/s, about 70 mph. 0.1 percent
34. 0.003

Chapter 8

CHECK YOUR UNDERSTANDING

8.1 (b)
8.2 (d)
8.3A (d)
8.3B (c)
8.4 (c)
8.5 (b)
8.6 (a, c, d, e)

TEST YOUR UNDERSTANDING

3. (e)
6. (c)
7. (c)
12. (a)
15. (a)

THINKING ABOUT THE CONCEPTS

18. Rock ages can be measured with radiometric dating. Rocks get older as one goes deeper into the canyon.
20. The Moon is smaller than Earth, so the Moon contained less thermal energy in its core to begin with, and a smaller core cools faster than a larger one. The core of Earth also contains more radioactive material than the core of the Moon.
23. The impactor theory of the Moon is supported by the compositional similarities of Earth and the Moon, the lack of volatiles on the Moon, and the inclination of Earth's spin axis. However, evidence of the impacting object's own composition has yet to be observed.

APPLYING THE CONCEPTS

34. Figure 8.18 shows a relatively smooth surface with lots of small craters. Figure 8.23 also shows a very smooth surface with fewer small craters, suggesting that this region is younger.

41. (a) 6 half-lives old.
    (b) 34,200 years old.

Chapter 9

CHECK YOUR UNDERSTANDING
9.1 (a, b, c, d)
9.2 (c)
9.3A (c, a, e, b, d)
9.3B (a)
9.4 (a, d, c, b)
9.5 (a, b, c)

TEST YOUR UNDERSTANDING
1. (d, b, c, f, a, e)
4. (c)
7. (b)
10. (a)
11. (c)

THINKING ABOUT THE CONCEPTS
17. After the primary atmospheres were stripped from the terrestrial planets, the planets outgassed volatile material from their interiors. The gas expelled from planetary interiors combined with those materials delivered by comet impacts to produce secondary atmospheres.
22. The troposphere is the lowest layer of the atmosphere; that is where ozone is a pollutant. In the higher layers (mainly the stratosphere), ozone absorbs UV photons, protecting the planet from that harmful radiation.
27. The slow rotation of Venus means that strong wind currents do not occur across the planet's longitude, only in latitude, so hot gas at the equator is constantly mixed with cooler gas at the poles to make a uniform temperature. Venus has a thick atmosphere which traps heat efficiently.

APPLYING THE CONCEPTS
31. (a) The axes are linear.
    (b) The variations of CO_2 and temperature are correlated.
    (c) The highest CO_2 level was about 300 ppm, but it is 410 ppm today.
36. 1.8×10^5 N.
    (b) ≈ 20,700 kg.
    (c) ≈ 350 kangaroos on a person's head.

Chapter 10

CHECK YOUR UNDERSTANDING
10.1 (a)
10.2 (b)
10.3 (b)
10.4 (c)
10.5 (a, d, b, c)

TEST YOUR UNDERSTANDING
2. (b)
6. (b)
8. (a)
13. (d)
15. (d)

THINKING ABOUT THE CONCEPTS
20. Uranus is discovered → Hypothesis: We can predict its orbit → Prediction: On the basis of its distance from the Sun, we predict Uranus's orbit → Test: Uranus's orbit does not match predictions → FAIL → New hypothesis: Something might be pulling on Uranus → New prediction: Another planet might exist → New test: Find the planet → Success: We found Neptune.
25. They measure the positions of clouds as they move across the surface of the planet, and see how they vary in latitude. They also compare these speeds with the rotation speed of the interior of the planet, according to its magnetic fields.
30. Migration of the outer giant planets could leave inner ones alone completely as they eject objects orbiting nearby, or they could completely disrupt their orbits and send them flying out into space.

APPLYING THE CONCEPTS
33. The interiors of Jupiter and Saturn reach very high pressures and temperatures over 10,000 K, violent and hot enough to strip the electrons off the hydrogen atoms.
44. 0.09 atm/km, 0.04 atm/km, 1.33 atm/km, 4 atm/km.

Chapter 11

CHECK YOUR UNDERSTANDING
11.1 (a, b, c)
11.2A (a)
11.2B (a, d, b, c)
11.3 (a)
11.4 (b)

TEST YOUR UNDERSTANDING
1. (a)
4. (d)
8. (a)
11. (a)
13. (d)

THINKING ABOUT THE CONCEPTS
17. Cryovolcanism is volcanic activity with eruptions of materials that are liquid at very low temperatures, such as water, ammonia, methane. Seen on Enceladus, Titan, Ganymede and others.
20. Titan resembles Earth with the nitrogen composition of its atmosphere and surface features similar to rivers and lakes. On Titan, methane appears to play the same role that water plays in the hydrologic cycle on Earth.
27. The discovery of kinks and knots in Saturn's rings meant that either our understanding of rings was completely wrong or that we had an incomplete theory. Because scientific understanding is provisional, our understanding of rings *was* correct *until* we learned more about them and required a theory that incorporated more physics. Thus, our understanding grew with more observations to become a more complete explanation.

APPLYING THE CONCEPTS
31. (a) 2.5 kilometers per second (km/s).
    (b) 2.5 times faster than the typical speed of gas escaping from Io's vents. That means most of the vented material will fall back on the moon rather than escape into space.
40. 14.3 hours

Chapter 12

CHECK YOUR UNDERSTANDING
12.1 According to the IAU definitions, those objects are dwarf planets and not moons because they orbit the Sun, not a planet.
12.2 (c)
12.3A (c)
12.3B (c)
12.4 (a, b, c, e)
12.5 The origins of a meteorite come from its precollision orbit as measured by multiple observations, as well as its structure, and its chemical makeup.

TEST YOUR UNDERSTANDING
2. (b)
5. (c)
8. (a)
11. (b)
15. (d)

THINKING ABOUT THE CONCEPTS
16. Pluto does not have a nearly circular orbit and does not orbit in the plane of the other planets, which suggests that its formation was different from that of the eight "classical" planets. Pluto also is not massive enough to have gravitationally swept out the space around it.

19. *Meteoroids* are particles of space debris smaller than about 100 meters in diameter. *Meteors* are ionized trails left by the passage of meteoroids and interplanetary dust through the upper atmosphere. *Meteorites* are chunks of meteoroids that survive their atmospheric passage to land on the surface.
28. Ion tails are pushed directly away from the comet by the charged particles of the solar wind, whereas the dust tails are pushed away both by solar wind and solar radiation. The two tails need not point in the same direction because the shape of the dust tail depends on the particle makeup of the dust and how it is accelerated away from the comet by the Sun's light.

APPLYING THE CONCEPTS
32. 340 km/hr
41. (a) 29 orbits.
    (b) 9×10^{12} kg lost.
    (c) 4 percent of its total mass.

Chapter 13

CHECK YOUR UNDERSTANDING
13.1 (b)
13.2 (d)
13.3 (a, b, d)
13.4 (b, d)

TEST YOUR UNDERSTANDING
1. (b)
5. (b)
10. (d)
12. (a)
15. (a)

THINKING ABOUT THE CONCEPTS
19. Every blackbody emits light at all wavelengths. A star with a 2500 K blackbody emits mostly in the red and infrared but still gives off a measurable amount of blue light.
20. Betelgeuse is cooler than Rigel. Nothing can be said about their luminosity or size without knowing their distance.
24. Binary stars can transit each other just like planets, so any mission looking for regular variability in the light from the stars is going to discover binary stars (and variable stars).

APPLYING THE CONCEPTS
33. The range in luminosities is about a hundred-billion (10^{11}); extremely hard to display in a linear plot. Logarithmic.
41. Betelgeuse; Achenar.

Chapter 14

CHECK YOUR UNDERSTANDING
14.1A (c)
14.1B (b)
14.2 (c)
14.3 (b)
14.4 (d)

TEST YOUR UNDERSTANDING
1. (a)
9. (d)
10. (c)
13. (c)
14. (d)

THINKING ABOUT THE CONCEPTS
17. To fuse two nuclei, very high temperatures (millions of degrees), pressures, and densities are required so the two nuclei can overcome their mutual repulsion and fuse. The oceans do not satisfy those conditions.
24. The Sun is made of gas and plasma only, so it has no actual surface at all. The photosphere is as deep into the Sun as we can see because it becomes opaque at the base of that layer.
30. The solar wind blows against the interstellar medium and clears out an area like the inside of a bubble. This area, the heliosphere, protects us from galactic high-energy cosmic rays as the Sun and Solar System move through our galaxy.

APPLYING THE CONCEPTS
31. (a) The height axis increases by adding 1,000 km to each tick mark; thus, it is linear.
    (b) The density axis increases by a multiplicative factor of 100 for each tick mark; thus, it is logarithmic.
40. 4.3 days.

Chapter 15

CHECK YOUR UNDERSTANDING
15.1 (a, d)
15.2 (c)
15.3 (a)
15.4 (b)

TEST YOUR UNDERSTANDING
4. (c)
5. (d)
8. (c)
10. (c)
15. (a)

THINKING ABOUT THE CONCEPTS
17. Star formation occurs in the dense, cold, and dusty interstellar medium, where dust blocks visible light. Infrared radiation is less affected by interstellar dust. Also protostars emit significantly in the infrared.
26. Smaller regions (clumps) can be denser and contract into individual stars. As the cloud collapses, the central regions of the protostars in those clumps become more massive, more compressed, and hotter. Eventually, the center of each protostar gets hot and dense enough for fusion to start, and a single star is born.
28. Some brown dwarfs in the ~10–80 $M_{Jupiter}$ mass range can temporarily have low levels of nuclear fusion in their cores; others cannot. Anything larger than about 10–20 $M_{Jupiter}$ is usually considered a planet.

APPLYING THE CONCEPTS
34. 6300 K, 4030 K. Dust has a significant effect on dimming and reddening light from stars.
45. ~1/112,000 L_{Sun}, so 112,000 brown dwarfs would produce the luminosity of the Sun.

Chapter 16

CHECK YOUR UNDERSTANDING
16.1 (a)
16.2 (b)
16.3 (a)
16.4 (a)
16.5 (d)

TEST YOUR UNDERSTANDING
2. (a, e, d, c, f, b, g)
5. (a)
8. (d)
9. (d)
12. (c)

THINKING ABOUT THE CONCEPTS
17. The energy generated would increase, which would be used to lift the atmosphere of the star, making it grow larger and cool down. That would drop the core temperature, returning the star to its original size and luminosity.
20. The brightness we measure won't change, so if all Type Ia supernovae are shown to be more luminous than previously thought, they must be farther away.
26. A white dwarf is just a cooling blackbody of constant size. As the white dwarf emits radiation and cools, its temperature will decrease, and hence its luminosity drops as well.

APPLYING THE CONCEPTS
37. 37 km/s.
44. $1.8 \times 10^{13} L_{Sun}$, $1.8 \times 10^5 L_{Sun}$, $1.8 \times 10^{-3} L_{Sun}$, $1.8 \times 10^{-11} L_{Sun}$

Chapter 17

CHECK YOUR UNDERSTANDING
17.1 (b, d)
17.2 (b)
17.3 (c)
17.4 (b)

TEST YOUR UNDERSTANDING
2. (e, a, f, b, c, d)
3. (d)
7. (e)
12. (a)
15. (b)

THINKING ABOUT THE CONCEPTS
19. In each successive fusion cycle, fewer and fewer nuclei exist to be fused. Also, fusion of heavier nuclei requires much higher energy, so the collisions happen more often.
21. New generations of stars and their planets incorporate the heavy elements fused in supernova and scattered into the interstellar medium. Also the shock wave from the expanding supernova bubble can compress interstellar clouds, triggering their collapse into new stellar systems.
30. Although AGB stars create some C and O, most of the heavy elements in the universe were made in supernovae and distributed into space in their ejecta. Later generations of stars like the Sun (and its planets) formed from this enriched material.

APPLYING THE CONCEPTS
40. (a) 3.8×10^{23} kg; (b) 5.25 M_{Moon}/minute
43. 160 times the radius of a pulsar (10 km)

Chapter 18

CHECK YOUR UNDERSTANDING
18.1 (d)
18.2 (b)
18.3 (d)
18.4 (c)

TEST YOUR UNDERSTANDING
2. (c)
10. (a)
11. (c)
12. (d)
15. (c, d, f)

THINKING ABOUT THE CONCEPTS
18. A photon is a massless particle, and all massless particles travel at exactly the speed of light in a vacuum.
21. Moving clocks run slowly, so the person's pulse rate will appear extremely slow to you.
23. The astronaut is moving with the ship and is in the same reference frame; thus, he will see no length contraction. An outside viewer who is *not* moving at that high speed would observe length contraction.

APPLYING THE CONCEPTS
40. 10^{-25} meters
43. 3×10^8 meters per second

Chapter 19

CHECK YOUR UNDERSTANDING
19.1A (d)
19.1B (c)
19.2 (c)
19.3 (a)
19.4 (c)

TEST YOUR UNDERSTANDING
2. (c)
7. (d)
10. (d)
11. (c)
15. (b)

THINKING ABOUT THE CONCEPTS
17. Cepheids that Hubble found in the Andromeda Galaxy (M31) showed that M31 was too far away to be part of our own Milky Way.
20. A Type Ia supernova is a standard candle only if we know exactly how luminous each one is. For that, we have to know the type of its progenitor.
30. In the crowded center of galaxies, several tidal encounters among stars could disrupt planetary orbits. Even if the center of a galaxy is not an active AGN, an overabundance of radiation harmful to life may still be present.

APPLYING THE CONCEPTS
32. 10^{12} galaxies
33. (a) 0.143
    (b) 4.28×10^7 m/s
    (c) 612 Mpc

Chapter 20

CHECK YOUR UNDERSTANDING
20.1 Answers will vary; may include comparisons, content, observations, motions, star formation.
20.2 The disk (and bulge) contains old and young stars; the halo contains only old stars.
20.3 (c)
20.4 Andromeda and the Milky Way are moving toward each other because of their mutual gravitational attraction.

TEST YOUR UNDERSTANDING
2. (a)
5. (a)
7. (c)
10. (b)
15. (d)

THINKING ABOUT THE CONCEPTS
18. Open clusters are young and therefore will generally be found near spiral arms and regions of recent star formation. Globular clusters are found primarily in the galactic halo.
23. Synchrotron radiation causes the disk of the galaxy to glow and it comes from cosmic rays, mostly electrons, spiraling through the galaxy's magnetic field.
30. Astronomers exclude regions of rapid star formation (both because of the UV light and frequent supernovae), the galactic center (high radiation), close stellar encounters, and regions that do not have enough heavy elements to form rocky planets.

APPLYING THE CONCEPTS
33. 1.8×10^9 yr
42. 2.1×10^3 km/s

Chapter 21

CHECK YOUR UNDERSTANDING
21.1 (b, a)
21.2 (d)
21.3A (c)
21.3B (b)
21.4 (a)

TEST YOUR UNDERSTANDING
1. (b)
2. (d)
8. all
11. (d)
12. (c)

THINKING ABOUT THE CONCEPTS
17. Space itself is expanding. Imagine space filled with a grid: the space between gridlines gets larger. See Figure 21.5.
21. Objects held together by their own self-gravity or the gravitational attraction of its contents do not expand. The expansion happens only on the largest scales, in which gravitational attraction is weaker than the expansion rate.

24. Regardless of your position in the universe, all distant galaxies recede from you as the universe expands. If you observe a distant galaxy moving rapidly away, an observer in that distant galaxy would say the Milky Way is rapidly moving away from them. The Big Bang took place everywhere; all observers, regardless of location, think they are at the center.

APPLYING THE CONCEPTS
34. ~12 billion years
39. (a) 4,100 Mpc; (c) 0.147

Chapter 22

CHECK YOUR UNDERSTANDING
22.1 (c)
22.2 (c)
22.3 Inflation solves the horizon and flatness problems.
22.4 (d)
22.5 (a)

TEST YOUR UNDERSTANDING
2. (c)
3. (a)
5. (a, f, j, b, e, g, d, i, h, c)
7. (c)
14. (b)

THINKING ABOUT THE CONCEPTS
17. When we look at distant galaxies and observe Type Ia supernovae, we see that their recessional velocity increases with distance faster than the linear relationship of Hubble's law, suggesting a nonzero cosmological constant. Einstein brought in the cosmological constant to explain why the universe didn't collapse in on itself.
20. Hubble time does not take into account the changing dynamics that result from having different cosmological densities and a cosmological constant (dark energy).

23. The early universe was *hot* and *dense*, which are situations that might be replicated in particle accelerators. At high temperatures and densities, all sorts of exotic forms of matter can exist, which may have substantially influenced how the early universe evolved.

APPLYING THE CONCEPTS
39. 6.1×10^{-10}
45. 1.2×10^{26}

Chapter 23

CHECK YOUR UNDERSTANDING
23.1 (c, b, d, a)
23.2 (b)
23.3 (c)
23.4 (b)

TEST YOUR UNDERSTANDING
3. (b)
5. (c)
8. (d)
11. (d)
14. (d)

THINKING ABOUT THE CONCEPTS
17. If voids were filled with considerable dark matter, we would expect to see gravitational lensing of galaxies behind those voids. We also would have expected galaxies to have formed inside the dark matter voids. So it is much more likely that they are fairly empty of any mass.
28. Hot dark matter could not be gravitationally bound enough to form the halos needed to contain galaxies and to explain galactic rotation curves.
30. The nuclear forces are too short ranged to work over cosmic distances, and electromagnetic forces work only for charged particles; most matter in the universe is neutral. That leaves only gravity.

APPLYING THE CONCEPTS
31. 1.05×10^7 m/s
36. (a) $10^{13}\, M_{Sun}$; (b) 10^{14} to $10^{15}\, M_{Sun}$; (c) 10^{16} to $10^{18}\, M_{Sun}$

Chapter 24

CHECK YOUR UNDERSTANDING
24.1 (d)
24.2 (a)
24.3A (d)
24.3B (a)
24.4 (b)

TEST YOUR UNDERSTANDING
4. (c)
7. (d)
12. (b)
14. (a)
15. (d)

THINKING ABOUT THE CONCEPTS
21. A burst of biodiversity around 500 million years ago. Possibly caused by a decrease in predators, a warming of Earth, or oxygen reaching high enough levels and with an adequate ozone layer for modern life to evolve.
26. All the hydrogen and most of the helium are from the Big Bang. All the other atoms in our bodies are from dying stars, especially supernovae.
30. Main-sequence stars slowly become more luminous. In a few billion years, the increase in the Sun's luminosity will cause the oceans to evaporate and generally scorch the Earth.

APPLYING THE CONCEPTS
33. 1,024 organisms
38. About 17 generations

Credits

These pages contain credits for all 24 chapters in the complete volume of *21st Century Astronomy*, Sixth Edition. The Solar System Edition does not include Chapters 15–23. The Stars and Galaxies Edition does not include Chapters 8–12.

Photo

Frontmatter

(in chronological order): NASA/JPL-Caltech/MSSS; © Damian Peach; NASA; robertharding/Alamy Stock Photo; Caltech/NASA/JPL; Paul Fearn/Alamy Stock Photo; NASA/JPL-Caltech/Univ. of Arizona; Stockli, Nelson, Hasler, Goddard Space Flight Center/NASA; NASA/JPL/Space Science Institute; ESA/Rosetta/NavCam, CC BY-SA IGO 3.0; Visuals Unlimited/Getty Images; NASA/SDO/AIA; NASA and ESA; NASA/CXC/SAO; NASA/ESA; Brad Goldpaint/Aurora Open/Alamy; NASA/JPL/Space Science Institute; Seth Shostak/SETI Institute; Caltech/NASA/JPL; Courtesy of Stacy Palen.

Author bio photos

Zac Williams; Courtesy of Laura Kay; Courtesy of George Blumenthal.

Chapter 1

Pages 2–3 NASA **5 both** NASA **6 second to bottom right** Ryan Somma, CC BY-SA **6 bottom right** NASA **11** Courtesy of the Archives, California Institute of Technology **12** Matheisl/Getty Images **15** ESA/NASA.

Chapter 2

Pages 22–23 Science Photo Library/Science Source **24 left** Roger Ressmeyer/Getty Images **24 center** J Gerard Sidaner/Getty Images **24 right** Henry Westheim Photography/Getty Images **28 right** Pekka Parviainen/Science Source **28 left** David Nunuk/Science Source **38** Arnulf Husmo/Getty Images **44** akg-images/Rabatti - Domingie/The Image Works **46** Laura Kay **47 top** Heinz-Peter Bader/Reuters/Newscom **47 bottom** Laura Kay and Lisa Rand **49 left** knickohr/Getty Images **49 right** © Wang Letian.

Chapter 3

Pages 58–59 © Damian Peach **60 top** Image Select/Art Resource, NY **60 bottom** © Tunc Tezel **61 center** Forum/UIG/AgeFotostock **61 bottom** Bettmann/Getty Images **64** Kean Collection/Getty Images **65 top** De Mundi Aetherei Recentioribus Phaenomenis, 1588. The Tycho Brahe Museum, Sweden. www.tychobrahe.com **65 center** Science & Society Picture Library/Getty Images **67 top** Universal Images/Superstock **67 top center** Iberfoto/Superstock **67 bottom center** Iberfoto/Superstock **67 bottom** Heinz-Dieter Falkenstein/age fotostock/Superstock **70 top** Leemage/Getty Images **70 bottom** The Granger Collection, NYC **71 top** Jamie Cooper/Getty Images **71 bottom** DEA/G. Nimatallah/Getty Images.

Chapter 4

Pages 82–83 NASA **91 top right** NASA **91 bottom right** NASA **91 bottom left** JSC/NASA **91 center left** JSC/NASA **91 top left** Heinz-Dieter Falkenstein/age fotostock/Superstock **97 top** Eric Carr/Alamy Stock Photo **97 bottom** Education Images/UIG via Getty Images **101** NASA.

Chapter 5

Pages 108–109 robertharding/Alamy Stock Photo **115 top left** Paul Fearn/Alamy Stock Photo **115 top right** Heinz-Dieter Falkenstein/age fotostock/Superstock **115 bottom left** Iberfoto/Superstock **115 bottom right** Underwood Photo Archives/Superstock **133 all** NASA **134** ESO/Y. Beletsky.

Chapter 6

Pages 140–141 Rick Peterson/W.M. Keck Observatory **142** ASU Physics Instructional Resource Team **143** Yerkes Observatory photographs: Richard Dreiser **145 top** ASU Physics Instructional Resource Team **145 bottom** Jim Sugar/Getty Images **147 top** Enrico Sacchetti/Science Source **147 center (both)** The Space Telescope Science Institute **147 bottom** ASU Physics Instructional Resource Team **149 bottom (both)** NASA, ESA, HEIC, and The Hubble Heritage Team (STScI/AURA) **150 top** ESO/L. Calada/ACe Consortium **150 bottom** NASA/NOAA **151 top** Hulton-Deutsch Collection/CORBIS/Corbis via Getty Images **151 bottom** Jean-Charles Cuillandre (CFHT) **153 top** Rick Peterson/W.M. Keck Observatory **153b** NASA **153c** NASA **153d** NASA **153e** NASA **153f** Joint Astronomy Center in Hilo, Hawaii **153g** Photo by Dave Finley, courtesy National Radio Astronomy Observatory and Associated Universities, Inc. **153h** NRAO/AUI, James J. Condon, John J. Broderick, and George A. Seielstad **153i** NASA/Courtesy of the Arecibo Observatory, a facility of the NSF **154 top** Roger Ressmeyer/Getty Images **154 center** NASA/Courtesy of the Arecibo Observatory, a facility of the NSF **154 bottom** Photo by Dave Finley, courtesy National Radio Astronomy Observatory and Associated Universities, Inc. **155 top** National Geographic Creative/Alamy Stock Photo **155 center** NASA Photo / Carla Thomas **bottom right** NASA/Tom Tschida **155 bottom** NASA Photo/Carla Thomas **bottom right** NASA/Tom Tschida **157 top** NASA/JPL-Caltech/MSSS **157 bottom** NASA/CXC/NGST **158 top** NASA/JPL-Caltech/MSSS **158 bottom** NASA/JPL-Caltech/MSSS **159 top** NASA/JPL/ESA, D. Ducros **159 bottom** NASA **160 top** CERN **160 bottom** Jim Haugen/NSF **161 center** LIGO Laboratory/CalTech/U.S. National Science Foundation **161 left** Volker Steger/Science Source **161 right** EADS Astrium/ESA **162 all** Patrik Jonsson, Greg Novak & Joel Primack, UC Santa Cruz, 2008.

Chapter 7

Pages 170–171 Caltech/NASA/JPL **173 top left** Science Source **173 top right** S. Andrews (Harvard-Smithsonian CfA); B. Saxton (NRAO/AUI/NSF); ALMA (ESO/NAOJ/NRAO) **173 bottom** Massimo Brega/Science Source **top** NASA/JPL-Caltech/T. Pyle **175 bottom** John Schults/Reuters/Newscom **178** NASA **182** NASA/Johns Hopkins University Applied Physics Laboratory/Carnegie Institution of Washington **187** NASA **189** NASA Ames/SETI Institute/JPL-Caltech.

Chapter 8

Pages 196–197 NASA/JPL-Caltech/Univ. of Arizona **198** NASA **199 all** RGB Ventures/SuperStock/Alamy Stock Photo **200 top** D.J. Roddy and K.A. Zeller, USGS, Flagstaff, AZ **200 bottom** NASA/Goddard/MIT/Brown **201** NASA/JPL/Caltech **202** NASA Goddard **205 both** NASA **209** Spencer Grant/Science Source **210** Francois Gohier/Science Source **214 top** NASA **214 bottom left** NASA/JSC **214 bottom right** NSSDC/NASA **215** NASA/Magellan Image/JPL **216 top** Tomas Janececk **216 bottom** Jukurae/Shutterstock **217 top** NASA/JSC **217 center** NASA/JSC/Arizona State University **217 bottom** NASA/Johns Hopkins University Applied Physics Laboratory/Arizona State University/Carnegie Institution of Washington. Image

C-2 CREDITS

reproduced courtesy of Science/AAAS. **218 top** Corbis via Getty Images **218 center and bottom** NASA/JPL/ESA, D. Ducros **219** NASA **220 top** NASA/JPL/University of Arizona **220 bottom** NASA **221 both** NASA **223** Joe Tucciarone/Science Source.

Chapter 9

Pages 230–231 Stockli, Nelson **240** NOAA **244 top** NASA **244 bottom** All Canada Photos/Alamy Stock Photo **247 top** JAXA/ISAS/DARTS/Damia Bouic **247 bottom** JAXA/ISAS/DARTS/Damia Bouic **248 top** NASA/JPL-Caltech/MSSS **248 bottom** NASA/JPL/Caltech **249 top** NASA/STScI **249 bottom** NASA/JPL-Caltech/UA **256 left** JAXA/ISAS/DARTS/Damia Bouic **256 center and right** NASA **257** NASA's Goddard Space Flight Center.

Chapter 10

Pages 264–265 NASA/SwRI/MSSS/Gerald Eichstadt/Sean Doran **266 all** NASA **267 all** NASA **268 top left** JPL/NASA **268 bottom right** JPL/NASA **269 top** NASA/JPL/Space Science Institute **269 bottom** NASA/JPL **270** NASA, ESA, and the Hubble Heritage Team (STScI/AURA) **271 top** NASA/SwRI/MSSS/Gerald Eichstadt/Sean Doran **271 bottom** NASA/JPL-Caltech/SwRI/ASI/INAF/JIRAM **272 top left** NASA **272 top right** NASA, ESA **272 bottom left** NASA/JPL/Caltech. **272 center right** NASA, ESA, and M.H. Wong and J. Tollefson (UC Berkeley) **272 bottom right** NASA **276** NASA/JPL/Caltech. **277** NASA, ESA, L. Sromovsky and P. Fry (University of Wisconsin), H. Hammel (Space Science Institute), and K. Rages (SETI Institute) **281** freenaturepictures.com/Wikimedia Commons **282 bottom left** NASA Goddard Photo and Video **282 both insets** NASA Goddard Photo and Video **282 bottom right** NASA Goddard Photo and Video **284** NASA's Goddard Space Flight Center/S. Wiessinger **286** NASA/JPL-Caltech/SwRI/MSSS/Jason Major **291 top** NASA/JPL **291 bottom (both)** NASA/JPL.

Chapter 11

Pages 292–293 NASA/JPL/Space Science Institute **294** NASA/JPL/Caltech **295 right** ESA/DLR/FU Berlin (G. Neukum) **295 left** NASA/JPL/University of Arizona **297 top** NASA **297 bottom** NASA/JPL/Caltech **297 bottom insets** NASA/JPL/Caltech **298 all** NASA/JPL-CalTech **299 bottom** NASA/JPL Caltech/UCLA/ MPS/DLR/IDA/ **299 top** NASA/JPL/Space Science Institute **300 top** NASA/JPL/Space Science Institute **300 bottom** NASA/JPL/Space Science Institute **300 center** NASA/JPL/University of Arizona/University of Idaho **301 top** NASA **301 bottom** NASA/JPL/USGS **302** NASA, ESA and A. Feild/STScI **303 top** NASA/JPL/Space Science Institute **303 bottom** NASA/JPL/Ted Stryk **304** NASA/JPL/Space Science Institute **306** NASA/JPL/Space Science Institute **307** NASA/JPL/Space Science Institute **308 top** NASA/JPL/Space Science Institute **308 bottom** NASA/JPL/Space Science Institute **309 bottom** NASA **309 center** NASA/JPL/Space Science Institute **309 top** Stephanie Swartz/Dreamstime **310 top** NASA **310 center** NASA/JPL/Space Science Institute **310 bottom** NASA/JPL/Space Science Institute **311 left** NASA/JPL-Caltech **311 center** NASA/JPL/Space Science Institute **311 right** NASA/JPL/Space Science Institute **312 bottom** NASA/JPL/Caltech. **312 top** NASA/JPL/Space Science Institute **313** NASA **319 all** NASA/JPL/Caltech.

Chapter 12

Pages 320–321 ESA/Rosetta/NavCam, CC BY-SA IGO 3.0 **322 center** NASA/JHUAPL/SwRI **322 bottom** NASA/JHUAPL/SwRI **322 top** NASA/JPL **323 center** JPL/NASA **323 top** NASA/JPL-Caltech/UCLA/MPS/DLR/IDA **323 bottom** NASA/JPL-Caltech/UCLA/MPS/DLR/IDA/ **324 right** JPL/NASA/Johns Hopkins University Applied Physics Laboratory/Southwest Research Institute **324 center** JPL/NASA/Johns Hopkins University Applied Physics Laboratory/Southwest Research Institute **324 left** NASA/JPL-Caltech **325** NASA **330 bottom** NASA/JPL-Caltech/UCLA/MPS/DLR/IDA **330 top** NASA/JPL/USGS **331 top** NASA/JPL-Caltech/UCLA/MPS/DLR/IDA **331 bottom** NASA/JPL-Caltech/UCLA/MPS/DLR/IDA **336 bottom** Courtesy of Terry Acomb **336 top** Science Source **337 top** NASA/JPL/Caltech **337 bottom left** NASA/JPL/Caltech/UMD **337 bottom right** NASA/JPL/Caltech/UMD. **338** NASA/JPL-Caltech/UMD **340 top** Alexandar/Shutterstock **340 bottom** Barrie Rokeach/Getty Images **341 top left** Courtesy of Ron Greeley **341 top right** Courtesy of Ron Greeley **341 center left** Courtesy of Ron Greeley **341 center right** Courtesy of Ron Greeley **341 bottom** NASA/JPL Caltech/LANL/CNES/IRAP/ LPGNantes/CNRS/IAS/MSSS **342** ESO/Y. Beletsky **343 bottom** Lunin Gleb/ITAR-TASS Photo/Avalon/Photoshot **343 top** NASA **343 center** Science Photo Library/Science Source **351 all** Larry Marschall and Christy Zuidema, Gettysburg College.

Chapter 13

Pages 352–353 Visuals Unlimited/Getty Images **367** ESA/Hubble **371** Norman Lockyer Observatory, Sidmouth **372 left** GL Archive/Alamy Stock Photo **372 top right** Science History Images/Alamy Stock Photo **372 center right** Paul Fearn/Alamy Stock Photo **372 top center** GL Archive/Alamy Stock Photo **372 center right** Interfoto/Alamy Stock Photo **375** Norman Lockyer Observatory, Sidmouth.

Chapter 14

Pages 382–383 NASA/SDO/AIA **390** Hinode JAXA/NASA/PPARC **394 both** Brocken Inaglory, 2006; http://creativecommons.org/licenses/by-sa/3.0/deed.en **395 top** Nigel Sharp, NOAO/NSO/Kitt Peak FTS/AURA/NSF **395 right** Westend61 GmbH/Alamy Stock Photo **395 center bottom** John Chumack/Science Source **395 center left** Hinode JAXA/NASA/NASA **396** NASA **398 left** NASA/SDO/Solar Dynamics Observatory **398 right** NASA/SDO/Solar Dynamics Observatory **399** SOHO/ESA/NASA **400** NASA/SDO **401 top left** SOHO/ESA and NASA **401 top center** NASA SDO **401 top right** SOHO/ESA and NASA **401 bottom** SOHO/ESA and NASA **403** JPL/NASA.

Chapter 15

Pages 410–411 ESO **412 top** Axel Mellinger **412 both** NASA/JPL-Caltech/IRAS/2MASS/COBE **413** U.S. Coastguard **415 bottom** IRAS/COBE (NASA) **415 top** NASA/JPL-Caltech/UCLA **416 bottom** Dr. Douglas Finkbeiner **416 top** Masashi Kimura/MAXI/SSC **417 top center** Courtesy of Emmanuel Mallart www.astrophoto-mallart.com & www.axisinstruments.com, **417 top right** ESO **417 top center** Jean-Charles Cuillandre (CFHT), Hawaiian Starlight, CFHT. **417 bottom** NRAO/AUI **418 top** ESO **418 bottom** ESO **420** ESA/Planck Collaboration **423 left** ESA/Herschel/PACS/SPIRE/Hill, Motte, HOBYS Key Programme Consortium **423 center** MPG/ESO **423 right** VLT/ISAAC/McCaughrean & Andersen/AIP/ESO **426** NASA/JPL-Caltech/UCB **429 bottom** ESO/M. McCaughrean **429 top** NASA/JPL-Caltech/UIUC **431 bottom** NASA/JPL-Caltech/UIUC **431 top** Science Photo Library/Alamy Stock Photo.

Chapter 16

Pages 438–439 NASA and ESA **453 all** NASA, ESA and The Hubble Heritage Team **460 all** NASA.

Chapter 17

Pages 468–469 NASA/CXC/SAO **474** NASA/CXC/GSFC/M.Corcoran et al.; Optical: NASA/STScI **478** NASA/CXC/SAO **479 center** Alexandra Angelich (NRAO/AUI/NSF); NASA Hubble; NASA Chandra **479 bottom** NASA/CXC/SAO/PSU **479 top** NASA/CXC/SAO/PSU **482 bottom center** Emilio Segré Visual Archives/American Institute of Physics/Science Source **482 top center** PA Images/Alamy Stock Photo **482 bottom left** Romolo Paradisi/ESO **482 top left** Science and Society/Superstock **482 bottom right** Science Photo Library/Superstock **482 top right** Science Photo Library/Superstock **485 bottom** Hillary Mathis/REU Program/NOAO/AURA/NSF/Science Source **485 top** NASA **486 right** Barbara Mochejska Hedberg, J. Kaluzny (Copernicus Astronomical Center) **486 left** European Space Agency & NASA.

Chapter 18

Pages 496–497 NASA/ESA **510 right** Heinz-Dieter Falkenstein/age fotostock/Superstock **510 left** Underwood Photo Archives/Superstock **511 bottom**

CREDITS

ESA/Hubble & NASA **511 top** NASA/ESA **515** NASA/CXC/M.Weiss **517 top** NASA/CXC/Cal; Chandra X-ray Observatory Center **517 bottom** NASA/CXC/Cal; Chandra X-ray Observatory Center.

Chapter 19

Pages 524–525 NASA, ESA, and The Hubble Heritage Team (STScI/AURA) **526** NASA, ESA, and the Hubble Heritage Team (STScI/AURA) **527 top left** ESA/Hubble & NASA **527 bottom left** ESA/Hubble & NASA **527 bottom right** ESA/Hubble & NASA; Acknowledgement: Judy Schmidt **527 top right** ESO/INAF-VST/A. Grado/L. Limatola/INAF-Capodimonte Observatory **527 bottom center** NASA **527 top center** NASA/Hubble Heritage Team **528 top left** NOAO/AURA/NSF **528 top center left** NASA/JPL-Caltech **528 top center right** NASA/JPL-Caltech **528 top right** NASA/JPL-Caltech **528 center (all)** NASA/JPL-Caltech **528 bottom (all)** NASA/JPL-Caltech **529 top** NOAO/AURA/NSF **529 bottom** NOAO/AURA/NSF **530** NASA and The Hubble Heritage Team (STScI/AURA) **531 top both** European Southern Observatory **531 bottom right** NASA, ESA, Hubble Heritage Team (STScI/AURA) **531 bottom left** NOAO/AURA/NSF/Science Source **533 top** BJ Fulton/Bryne Observatory at Sedgewick Reserve and the Palomar Transient Factory **533 bottom** NASA/Swift/Peter Brown, Univ. of Utah **535 both** NASA **538 top** NASA/CXC/Penn State/G. Garmire; Optical: NASA/ESA/STScI/M. West **538 bottom** NASA/CXC/Penn State/G. Garmire; Optical: NASA/ESA/STScI/M. West **540 top** HST **540 bottom** NASA/CXC/CfA/R.Kraft et al; Radio: NSF/VLA/Univ.Hertfordshire/M.Hardcastle; Optical: ESO/WFI/M.Rejkuba et al. **544** X-ray (NASA/CXC/KIPAC/N. Werner, E. Million et al); Radio (NRAO/AUI/NSF/F. Owen **546** NASA/CXC/CfA/R.Kraft et al; Radio: NSF/VLA/Univ.Hertfordshire/M.Hardcastle; Optical: ESO/WFI/M.Rejkuba et al. **548** NASA, ESA, and J. Lotz and the HFF Team (STScI) **553** NASA, ESA, and the Hubble Heritage Team (STScI/AURA).

Chapter 20

Pages 554–555 Brad Goldpaint/Aurora Open/Alamy **556 bottom** C. Howk and B. Savage (University of Wisconsin); N. Sharp (NOAO)/WIYN, Inc. **556 top** NG Images/Alamy Stock Photo **557 top** NASA/JPL-Caltech/ESO/R. Hurt **557 center** NOAO/AURA/NSF **557 bottom left** NASA/JPL-Caltech **557 bottom right** Robert Gendler/Corbis/Getty Images **558 bottom** Richard Rand, University of New Mexico **558 top** Todd Boroson/ NOAO/AURA/NSF **561 top left** Emilio Segré Visual Archives/American Institute of Physics/Science Source **561 bottom left** John Irwin Collection/AIP/Science Source **561 center right** NASA Goddard **561 top right** NASA/JPL-Caltech/ESA/Institute of Astrophysics of Andalusia, University of Basque **561 bottom right** NASA/JPL-Caltech/ESA/Institute of Astrophysics of Andalusia, University of Basque **564 center** ESA/Gaia/DPAC **564 bottom** ESA/Gaia/DPAC **565** NASA **566 top** ASPERA/G.Toma/A.Saftoiu **566 bottom** Courtesy of Pierre Auger Observatory **567** DPAC/ESA **568 both** NASA **569** UCLA Galactic Center Group - W.M. Keck Observatory Laser Team **570** NASA's Goddard Space Flight Center Image of Fermi all-sky map, with bubbles outlined courtesy of NASA/GSFC/DOE/Fermi LAT/D.Finkbeiner et al. **571 bottom** Amanda Smith/University of Cambridge **571 center** ESA/Gaia/DPAC **571 top** Royal Observatory, Edinburgh/Science Source **572 top** NASA **572 bottom (all)** NASA, ESA, Z. Levay and R. van der Marel (STScI), T. Hallas, and A. Mellinger **574** ESA/Gaia/DPAC.

Chapter 21

Pages 580–581 ESA and the Planck Collaboration **586 bottom** The LIFE Images Collection./Getty Images **586 top** Underwood Photo Archives/Superstock **594** Lucent Technologies, Bell Labs **595 top** Emil Oprisa/Alamy Stock Photo **595 bottom** Sasho Bogoev/Alamy Stock Photo **596 bottom** ESA and the Planck Collaboration **596 top** NASA COBE Science Team **596 top center** NASA COBE Science Team **596 bottom center** NASA COBE Science Team **603 all** Courtesy of Stacy Palen.

Chapter 22

Pages 604–605 NASA/JPL/Space Science Institute **610 bottom** The LIFE Images Collection./Getty Images **610 top** Underwood Photo Archives/Superstock **633** CERN PhotoLab.

Chapter 23

Pages 634–635 NASA, ESA, H. Teplitz and M. Rafelski (IPAC/Caltech), A. Koekemoer (STScI), R. Windhorst (Arizona State University), and Z. Levay (STScI) **636 top** NASA/ESA/ESO/NAOJ/G. Paglioli **636 bottom** R. Brent Tully (U. Hawaii) et al., SDvision, DP, CEA/Saclay **637 both** Max Tegmark/SDSS Collaboration **638 both** ASA/CXC/MIT/E.-H Peng et al; Optical: NASA/STScI **638** NASA, ESA, Jennifer Lotz and the HFF Team (STScI) **639 both** NASA, A. Fruchter and the ERO team (STScI, ST-ECF). NASA, ESA **643** NASA/WMAP Science Team **644** NASA/JPL-Caltech/ESA/Institute of Astrophysics of Andalusia, University of Basque **645** Nicolle R. Fuller, National Science Foundation **646 both** NASA/JPL-Caltech/GSFC **647 right** NASA , ESA, and B. Salmon (STScI) **647 left** NASA, ESA, and P. Oesch (Yale University) **649 both** Guedes, Javiera, et al. The Astrophysical Journal, Volume 742, Issue 2, article id. 76 (2011). **650 bottom** NASA/CXC/SAO/P **650 top** NASA/Hubble Space Telescope Cosmic Assembly Near-infrared Deep Extragalactic Legacy Survey **651 top, both** NASA, ESA, and N. Pirzkal (STScI/ESA); NASA, ESA, J. Lotz (STScI), M. Davis (University of California, Berkeley), and A. Koekemoer (STScI) **651 bottom left** NASA/CXC/CfA/M.Markevitch et aL. **651 bottom right** NASA/STScI; Magellan/U.Arizona/D. Clowe et al. **652 top, all** Courtesy Joel Primack and George Blumenthal **652 bottom, all** Courtesy Joel Primack and George Blumenthal **653 top, both** Courtesy Joel Primack and George Blumenthal **653 bottom** NASA, ESA, and A. Feild (STScI).

Chapter 24

Pages 660–661 Seth Shostak/SETI Institute **662 top** Dr. Michael Perfit, University of Florida, Robert Embley/NOAA **662 bottom** Woods Hole Oceanographic Institution, Deep Submergence Operations Group, Dan Fornari **663 center** Chris Boydell/Getty Images **663 top** Gaertner/Alamy **663 bottom** National Park Service Photo by Jim Peaco **664 bottom** Dr. Robert Calentine/Getty Images **664 top** Peter Essick/Aurora Photos **666** Courtesy of National Science Foundation **667** Lucas Oleniuk/Getty Images **670** NASA/JPL-Caltech/MSSS **671 top** NASA/JPL-Caltech/MSSS **671 center** NASA/JPL-Caltech/UA/USGS **671 bottom** PHL@UPR Arecibo, NASA **674** NASA/Wikimedia Commons/Joshua Krissansen-Totton **675 bottom** F. Drake (UCSC) et al., Arecibo Observatory (Cornell, NAIC) **675 top** NASA **677** F. Drake (UCSC) et al., Arecibo Observatory (Cornell, NAIC) **679** NASA/JPL-Caltech/Space Science Institute.

Line Art

Figure 7.17: Graph: "Radial Velocity/ Year" from Exoplantets.org. Reprinted by permission of the Department of Terrestrial Magnetism, Carnegie Institution of Washington.

Figure 10.18: Figure from "First Earth-Sized Planets Found," Harvard-Smithsonian Center for Astrophysics/ David A. Aguilar (CfA), December 20, 2011. Reprinted with permission.

Figure 18.21: "Black Holes with Known Masses." Courtesy Caltech/MIT/LIGO Laboratory.

Figure 23.10: Figure from "The First Galaxies," by V. Bromm & N. Yoshida. Reproduced with permission of Annual Review of Astronomy & Astrophysics, Volume 49 (373-407) © by Annual Reviews, http://www.annualreviews.org

Figure 24.12: "The Origin of the Solar System Elements," Graphic created by Jennifer Johnson. Astronomical Image Credits: ESA/NASA/AASNova. Licensed under a Creative Commons Attribution-ShareAlike 4.0 International License.

Figure 24.17: "Oceans of the Solar System," from PHL @UPR Arecibo. Reprinted by permission of PHL@UPR Arecibo.

Figure 24.18: Figure 1 from "Summary of the Limits of the New Habitable Zone." PHL.UPR.edu, March 23, 2013. Reprinted by permission of PHL@UPR Arecibo.

Index

This index contains page references for all 24 chapters in the complete volume of *21st Century Astronomy*, Sixth Edition. The Solar System Edition does not include Chapters 15–23 (pp. 410–659). The Stars and Galaxies Edition does not include Chapters 8–12 (pp. 196–351).

Page numbers in *italics* refer to illustrations and tables.

Abell 1689, *638*
Abell 2218, *639*
aberration of starlight, 498
absolute magnitude, 358
absolute zero, 128
absorption, 121
absorption lines
 absorption lines, overview, *120*, 121–122, 128
 Doppler shift and, 125
 and stellar composition, 362, *374*
 and stellar temperature and classification, 360–362, *374*
absorption spectra, 120–121, *361*, 540
acceleration
 defined, *72*, 73, 74
 equivalence principle, 506–508
 gravitational acceleration (g), 84, 85
 Newton's second law of motion and, *72*, 73, 74, 84
accretion disks
 angular momentum and, 175–177, 179, 649
 bipolar outflow, 177, 429–430
 black holes, 496–497, 515, 542, 645
 energy and temperature in, 178–179, 181–183, 542
 formation, *176*, 176–177, 422
 galaxies, 643, 649
 inner disk, 178–179
 large object formation in, 177–178, 182–183
 neutron star in binary system, 484, 542
 outer disk, 178–179
 planets, 295
 supermassive black holes, 542–544
 white dwarf in binary system, 457, 542
achondrites, *329*, 340, 341
active comets, 331, 334–336
active galactic nuclei (AGNs)
 accretion disks around supermassive black holes, 542–544
 blazars, 544, *545*
 emission lines in spectra, 540
 interaction of galaxies, 556
 jets, 540, 542, 544, *545*
 merging galaxies, 556
 normal galaxies and AGNs, 544, 556
 quasars as, 540
 radio galaxies, 540, *545*
 relativistic beaming, 544
 Seyfert galaxies, 540, *545*
 size, density, and power, 543
 size comparable to Solar System, 541–542
 synchrotron radiation, 540, 544
 torus of gas and dust, 544
 unified model of AGN, 544, *545*
Adams, John Couch, 267
adaptive optics, 148–149, 187
AGB. *see* asymptotic giant branch
AGNs. *see* active galactic nuclei
albedo, defined, 132, 325
Alexander, Claudia, 346
ALH84001 meteorite, 341
Allen, Paul, 677
Allen Telescope Array (ATA), *660–661*, 677
ALMA. *see* Atacama Large Millimeter/submillimeter Array
Alpha Canis Majoris (Sirius), 34, 44, 358, 454
Alpha Centauri, 357, *367*, 376, 587
alpha particles, 447
Alpher, Ralph, 592
Altwegg, Kathrin, 346
amino acids, 313, 662, 669
ammonia
 in atmospheres of giant planets, 273, 279
 on Callisto, 303
 from comets, 662
 decomposition by ultraviolet radiation, 233–234, 236
 photodissociation of, 300
 on Titan, 300
Amor asteroids, 328
amplitude, 112
ancient astronomy
 archaeological evidence, 24
 Beijing Ancient Observatory in China, *24*
 calendars and, 44
 estimating size of Earth, 31
 importance to agriculture, 13, 24, 43
 Mayan El Caracol at Chichén Itzá, *24*
 prediction of eclipses, 45
 Stonehenge, *24*, 45
 using the stars for navigation, 30–31
Andromeda Galaxy
 approaching Milky Way Galaxy, 572, 585
 blueshifted spectrum, 572, 585
 Cepheid variable stars, *526*, 527
 collision with Milky Way Galaxy in future, 572, 653, 678
 dark matter, 561
 distance, 5, *526*, 572
 in Local Group, 570–572
 spiral arms, 557
angular momentum
 accretion disk formation and, 175–177, 179, 649
 conservation of angular momentum, 175–177, 186, 649
 defined, 175
 factors affecting, 175–176
 orbital angular momentum, 177, 181, 188
 spin angular momentum, 177
annular solar eclipse, 46, *46*, 48
Antarctic Circle, *35*, 39
ANTARES experiment, 160
anthropic principle, 626–627
anticyclonic motion, 245, 272, 276
antimatter, 388, 619, 621, 623
antiparticles, 514, 619–621, 623
antiprotons, 619–621
Ant Nebula, *453*
aperture
 defined, 142
 diffraction and, 147–148
 focal ratio and, 143
 magnification and, 144, 186
aphelion, defined, 326
Aphrodite Terra, 215
Apollo asteroids, 325, 328
Apollo missions, 224
 Apollo 8, *2–3*, 159
 Apollo 10, *214*
 Apollo 13, 159
 Apollo 15, *91*, *217*
 Apollo 17, 210
 gravitational constant G measurements, 102
 lunar rock samples, 159, 203–204, *205*, 210, *211*
 magnetic field measurements, 209
 overview, 159
 seismometers left on the Moon, 207
apparent daily motion, 25, *25*, 26
apparent magnitude, 358
appeal to belief, 21
Archaea, 664, *665*
arcminute (arcmin), 142, 148, 356
arcsecond (arcsec), 148, 356
Arctic Circle, *35*, 39
Arecibo Observatory (Puerto Rico), *153*, 154, 675
Ariel, *294*, 303
Ariel 1 (British satellite), 155
Aristarchus of Samos, 60
Aristotle, 24
Arrhenius, Svante, 252
Artemis, 40
asteroid belt
 defined, 322
 Kirkwood gaps, 327–328
 location, *172*, 182, 322
 main asteroid belt, 182, 322, 326, 327, *328*
 orbital resonances, 327–328

I-1

asteroids, 326–327
  Amor asteroids, 328
  Apollo asteroids, 325, 328
  Aten asteroids, 328
  Ceres classified as, 323
  composition and classification, 328–329
  C-type (carbon) asteroids, 328
  Dactyl, moon of Ida, 330
  discovery, 326–327
  distribution of, 327–328
  Eros, 330
  formation from planetesimals, 328
  Gaspra, 330
  Ida, 94, 330
  Itokawa, 330, 344
  M-type (metal) asteroids, 330, 341
  near-Earth asteroids, 328
  as planetesimals from early Solar System, 327, 341
  spacecraft missions, 330–331
  S-type (stony) asteroids, 328–329, 343
  Trojan asteroids, 328
  24 Themis, 329
  2005 YU55, 325
  Vesta, 330–331
astrobiology, 669–675
  defined, 15, 660
  Drake equation, 675–677
  Earth Similarity Index (ESI), 674
  interconnections between branches of science, 668
  liquid water on Enceladus, 282, 299, 308, 313, 671
  liquid water on Europa, 298–299, 313, 671
  Planetary Habitability Index, 674
  planetary orbits and development of life, 75
  search for intelligent life, 675–678
  sending messages, 675
  tidal forces and development of life, 101
  see also life
astrometric method for extrasolar planets, 186
astronomer's periodic table of the elements, 123
astronomical parallax. see parallax
astronomical seeing, 148
astronomical tools and instruments, 142–169
  ancient instruments, 24
  charge-coupled devices (CCDs), 151–152, 160, 186, 189
  interferometric arrays, 154–155
  optical detectors and instruments used with telescopes, 150–153
  orbiting observatories, 155–157
  photographic plates, 151
  photography in astronomy, 151–152
  planetary spacecraft, 157–159
  see also telescopes
astronomical unit (AU), 63, 64, 368, 532, 543
astronomy, defined, 2
asymptotic giant branch (AGB)
  defined, 450
  fate of planets, 450
  high-mass stars, 472, 480, 489
  low-mass stars, 450–452, 454–456
  medium-mass stars, 472
  stellar-mass loss, 450–451
  see also H-R diagram

ATA (Allen Telescope Array), 660–661, 677
Atacama Cosmology Telescope (Chile), 163
Atacama Desert, 155, 155, 313, 663
Atacama Large Millimeter/submillimeter Array (ALMA)
  cosmic microwave background radiation (CMB), 163, 640
  Herbig-Haro object HH 46/47, 429
  radio dishes, 155, 155
  Supernova 1987A, 479
  W3 IRS5 star-forming complex, 431
Aten asteroids, 328
atmospheres
  atmosphere retention, 235
  atmosphere structure on giant planets, 273–275
  exospheres, 249
  formation, 180–181, 232–234
  Mars, 257
  planetary mass effects, 234–236
  primary atmospheres, 180–181, 182, 282–283
  secondary atmosphere evolution, 234–238
  secondary atmosphere formation, 180–182, 232, 233–234, 662
  see also Earth's atmosphere; Sun's atmosphere
atmospheres of terrestrial planets, 230–263
  ammonia decomposition by sunlight, 233–234
  atmosphere retention, 235
  comets and, 201, 232, 233
  comparison of atmospheric greenhouse effect, 237–238, 254, 256
  formation and loss of primary atmospheres, 180–182, 282–283
  physical properties and composition, 234
  planetary mass effects, 234–236
  see also Earth's atmosphere; terrestrial planets
atmospheric pressure
  Earth's atmosphere, 240–241
  Mars, 236, 248, 256, 674
  Venus, 236, 246–247
atmospheric probes, 158, 273, 300–301, 671
atmospheric windows, 153
atomic energy levels, 117–121
  absorption spectra and, 120–121
  decay, 118, 123–124
  emission spectra and, 118–119
  excited states, 118, 119, 123–124
  ground state, 118
  hydrogen, 122
  spectral fingerprints of atoms and, 121–123
atomic mass number, 388
atomic nucleus, 117, 385
atomic structure, 117
AU. see astronomical unit
AU Microscopii, 183
auroras, 244, 282, 283, 396
autumnal equinox, 38
axes, of graphs, 14
axion, 641

Baade, Walter, 481
Baker, Daniel, 404
balloon observations of millimetric extragalactic radiation and geophysics (BOOMERANG), 163
bar (pressure), 240

barred spiral galaxies
  bulges, 528
  characteristics, 528
  spiral arms, 558
  stellar motions, 539
  structure, 524–525, 528, 529
  see also Milky Way Galaxy; spiral galaxies
Barringer Crater, 200
Barrish, Barry, 516
Barstow, Jo, 462
Barstow, Martin, 462
Bay of Fundy, tides in, 97
begging the question, 21
Beijing Ancient Observatory in China, 24
Bell, Jocelyn, 482
Bell Laboratories, 151, 153, 163, 594, 598
Bessel, F. W., 356–357
Beta Pictoris b, 187
biased sample, 21
Big Bang
  age of universe, 102, 587–590
  Big Bang nucleosynthesis, 547, 596–597, 623, 640–641, 645, 654
  defined, 15, 142, 589
  recombination of the universe, 595, 595, 640, 643
  separation of forces in the cooling universe, 623–624
  see also cosmic microwave background radiation; expanding universe
Big Crunch, 612
Big Freeze or the Big Chill, 612
Big Rip, 612, 654
binary stars, 365–368, 456–460
  center of mass, 365, 367, 456, 515
  defined, 365
  Doppler shift, 365–367
  eclipsing binary stars, 367–368, 374
  evolution of close binary system, 457
  extrasolar planets around, 431
  low-mass stars, evolution, 456–460
  masses of stars, measuring, 367–368, 374, 515
  mass transfer, 456–457
  novae, 457–458
  orbits, 365–367
  Roche lobes, 456–457, 484
  spectroscopic binary stars, 367–368
  Type Ia supernovae, 458, 459–460, 513, 533
  visual binary systems, 367, 368
  X-ray binary systems, 484, 514, 515, 542, 645–646
binding energy, 474–475
bipolar outflow, 177, 429–430
blackbodies
  blackbody laws, 129–130
  cosmic microwave background radiation spectrum, 592, 595
  defined, 129
  effective temperature, 394, 514
  Planck spectra or blackbody spectra, 129, 163, 359, 360, 514
  scale factor and, 592
  Stefan-Boltzmann law, 129, 132
  stellar radiation and, 359
  Wien's law, 129–130, 359, 592

blackbody spectra (Planck spectra), 129, 163, 359, 360, 514
black holes, 512–516
   accretion disks, *496–497*, 515, 542, 645
   Cygnus X-1, *496–497*, 514–515
   defined, 513
   discovery of, 11
   effective temperature, 514
   event horizon, 513–514, 543
   formation of, 512–513
   Hawking radiation, 514
   mass of, 480–481
   mergers of black holes, 646
   properties, 513–514
   Schwarzschild radius, 513, 543
   seeing black holes, 514–515
   singularities, 512
   winds from the disk around a black hole, 515
   X-ray binary systems, 484, 514, 515, 542, 645–646
   see also Chandrasekhar limit; supermassive black holes
blazars, 544, *545*
blue-green algae (cyanobacteria), 239, 663–664
blueshifted light, 125
blue supergiant stars, 474
Bohr, Niels, 118
Bohr model, 118
Boltzmann, Ludwig, 129
Boltzmann constant (*k*), 235, 620
BOOMERANG (balloon observations of millimetric extragalactic radiation and geophysics), 163
bound orbits, 93
Bradley, James, *115*, 498
Brahe, Tycho, 61, 64–66, *67*, 331
brightness
   apparent magnitude, 358
   defined, 130, 359
   luminosity and distance effects, 130–133, *132*, 358–359
   magnitude system, 358
   standard candles, 102, *526*, 532–534, 585
   wave amplitude and, 112
Brown, Anthony, 574
brown dwarfs, 184, 187, 426, 443
bubble chambers, 633
bulges
   barred spiral galaxies, 528
   elliptical galaxies, 528
   Milky Way Galaxy, 557, 560, *564*, 568, *649*, 650
   S0 galaxies, 533
   spiral galaxies, 528, 539
   supermassive black holes in, 556, 650
   tidal bulges, 96–97, 99–100
Bullet Cluster, 651
Butterfly Nebula, *453*

Cabeus Crater, 222
Cadonatt, Laura, 518
calendars, 43–45
Callisto, *294*, 303–304, 313
Cambrian explosion, 664, 673
Cancer, *34*, 39
Cannon, Annie Jump, 361, 372
Capricornus, *34*, 39

carbon
   in biological compounds, 667, 669
   carbon-12 from helium fusion, 447, 449, 475, 667
   carbon cores in stars, 449–451
   as catalyst in CNO cycle, 470
   in organic compounds, 180, 662
carbonaceous chondrites, *329*, 340, 345
carbon dioxide
   from burning fossil fuels, 252
   in Earth's atmosphere, 233, 240, *251*, 252
   and Earth's temperature, *251*, 252
   greenhouse gas, 237, 251
   on Mars, 221, 233, 236, 237, 248, 670
   sources and sinks on Earth, 240
   on Venus, 232, 233, 236, 237
   from volcanoes, 181, 234, 662
carbon fusion, 461, 472, 474, 476
carbon-nitrogen-oxygen (CNO) cycle, 470–472, 645
Carrington Event, 404
Carter, Jimmy, 675
Casertano, Stefano, 376
Cassini, Jean-Dominique, 308
Cassini Division, *305*, 306, 308–309
Cassini mission
   avoiding contamination of Saturn's moons, 314
   clouds on Saturn, *264–265*
   Earth as seen from Saturn, *679*
   Enceladus and Saturn's E Ring, *307*
   Huygens atmospheric probe of Titan, 158, 300–301, 672
   lightning on Saturn, *272*, 275
   Mimas image, *303*
   Pandora and Prometheus and Saturn's F Ring, *309*
   radar imaging of Titan, 300
   Saturn images, *264–265*, *292–293*
   Saturn's Encke Gap, *310*
   spokes in Saturn's B Ring, *310*
   structure of Saturn's rings, *308*
   water detected on Enceladus, 282, 299, 671
CATE (Citizen Continental-America Telescopic Eclipse), 52
Cat's Eye Nebula, *149*, 453
CCDs. *see* charge-coupled devices
celestial equator, 25, *27*, 28–30, 35–36, 38–39
celestial sphere
   celestial equator, 25, *27*, 28–30, 35–36, 38–39
   ecliptic, 25, 34, 35
   meridian, 25
   motion of stars and celestial poles, 28–30
   nadir, 25
   north celestial pole (NCP), 25–31, 39
   right ascension, 25
   south celestial pole (SCP), 25–28, 30
   using the stars for navigation, 30–31
   viewed from different latitudes, 27–31
   view from the poles, 26–27
   zenith, 25, 26
Centaurus A (radio galaxy), 540
Center for High Angular Resolution Astronomy (CHARA) array, 155
center of mass
   binary stars, 365, 367, *456*, 515
   defined, 94, 365
   molecular clouds, *420*

   Pluto and Charon, 94, 296
   Sun and Jupiter, 184, 185
centripetal force, 90
Cepheid variables
   in Andromeda Galaxy, *526*, 527
   characteristics, 472–473
   distance measurement, 376, 472, *526*, 527
   period-luminosity relationship, 472, *473*, 527, 560
   standard candles, *526*, 533
Ceres, *69*, *172*, 182, 322, 323, 346
CFCs (chlorofluorocarbons), 239
Chandrasekhar limit
   black hole formation and, 512–513, 515
   Type Ia supernovae and, 459, 513
   white dwarfs and, *458*, 459, 513, 533
Chandra X-ray Observatory, *153*, 155, *156*, 568–569, *650*, 651
Chandrayaan-1 spacecraft, 210, 222
chaotic encounters, 180
CHARA (Center for High Angular Resolution Astronomy) array, 155
charge-coupled devices (CCDs), 151–152, 160, 186, 189
Charon, 94, 100, 270, *294*, 322–323
Chelyabinsk meteoroid, 343–344
chemical elements. *see* elements
chemosynthesis, 313
chlorofluorocarbons (CFCs), 239
chondrites, *329*, 340
chondrules, 340
chromatic aberration, 144
chromosphere, 394–396, 400, 401
circular velocity (v_{circ}), 89, 90, 94, 95
circumpolar stars, *27*, 28, 30
Citizen Continental-America Telescopic Eclipse (CATE), 52
Clementine mission, 221
climate, defined, 249
climate change
   anthropogenic (human-caused) change, 253
   computer models, 251, 253
   on Earth, 251–253
   effects of, 253
   global climate models, 251, 252
   greenhouse gases and, 249–254
   ice ages, 51, 250–252
   mechanisms controlling climate, 250–251
   Milankovitch cycles, 250, 252
   role of life in, 251, 254, 256
closed universe, 613
CMB. *see* cosmic microwave background radiation
CMEs. *see* coronal mass ejections
CNO cycle, 470–472, 645
COBE telescope. *see* Cosmic Background Explorer telescope
cold dark matter, 640–643, *652*
color
   color index of stars, 360, *374*
   false-color images, 152, *200*, *215*, *217*
   and surface temperature of stars, 360
   temperature and color of light emission, 129, 360
coma, 331, 334–335
Coma Cluster, 553, 636

comet nuclei
  disintegration, 338
  frozen state in outer Solar System, 331, 334
  impact energy, 344
  in Kuiper Belt, 322, 332
  number in Solar System, 331
  in Oort Cloud, 332–333
  as planetesimals from early Solar System, 183, 327, 331
  structure and composition, 334
comets, 331–338
  active comets, 331, 334–336
  atmosphere formation and, 201, 232, 233, 662
  coma, 331, 334–335
  Comet 67P/Churyumov-Gerasimenko, 320–321, 338, 346
  Comet Hale-Bopp, 333, 336
  Comet Halley, 333, 334, 336–337
  Comet Hartley 2, 338, 346
  Comet Hyakutake, 344
  Comet Ikeya-Seki, 336
  Comet McNaught, 336
  Comet Shoemaker-Levy 9, 342–343
  Comet Swift-Tuttle, 340
  Comet Tempel 1, 337
  Comet Tempel-Tuttle, 340
  Comet Wild 2, 337
  defined, 181, 331
  dust tails, 334, 335, 336
  early records of, 331
  head, 335
  ion tails, 334, 335, 336
  jets of gas and dust from, 336–338
  in Kuiper Belt, 233, 234
  long-period comets, 333, 334, 336, 344
  meteor showers and, 339–340
  in Oort Cloud, 233, 332–333
  orbits of, 333–334
  as possible source of Earth's water, 338, 345, 346
  short-period comets, 333–335, 337, 340
  source of volatile compounds on planets, 181, 201, 233, 345
  spacecraft visits, 336–338
  sungrazers, 336
  tails, 335
  see also comet nuclei
comparative planetology, 198–199, 251, 266
composite volcanoes, 216
compound lens, 145
computers
  adaptive optics, 148–149
  analysis of astronomical data, 160, 162, 312
  climate models, 251, 253
  computer modeling, 162
  extrasolar planets, 284, 285
  giant planet models, 270, 279
  large-scale structure simulations, 162, 556, 572, 645, 649, 652–653
  moon models, 298
  pole reversal simulation, 209
  protoplanetary disk simulation, 181
  supercomputers, 162, 312, 639, 649, 652
conduction, 389

conjunction, 62–63, 76
Conselice, Christopher, 16
conservation of angular momentum, 175–177, 186, 649
constellations, 24, 34
continental drift, 211. see also plate tectonics
continuous spectrum, 128, 153
convection
  atmospheric convection on Earth, 237, 241–242, 244–245
  atmospheric convection on giant planets, 273, 275, 278
  convective zone of the Sun, 388, 390, 393, 394
  in high-mass stars, 471, 476, 489
  mantle convection, 211–213, 215–216
  in protostars, 423, 426, 429
  in the Sun, 389, 390
Convection Rotation and Planetary Transits (COROT) space telescope, 156, 186
Copernicus, Nicolaus
  biography, 61
  Copernican Revolution, 61–62
  De revolutionibus orbium coelestium, 61, 62, 71
  heliocentric theory, 61–62, 64, 65, 67, 68, 70
  scale of the Solar System, 62–63
core accretion–gas capture, 182–183
cores
  carbon cores in stars, 449–451
  defined, 207
  Earth's inner core, 15, 207
  Earth's outer core, 206, 207, 209
  giant planets, 182, 269, 277–280
  helium cores in stars, 443–444, 446–448, 450
  iron cores of high-mass stars, 472, 476–477
  molecular-cloud cores, 421–422
  radioactive heating in, 208
  Sun, 385, 387, 389, 438, 442
  terrestrial planets, 180, 207
  thermal radiation from, 207–208
Coriolis effect, 32–33, 245–247, 275, 276
corona, 394–397, 401, 416
coronae (Venus), 215
coronal holes, 396
coronal loops, 396, 400
coronal mass ejections (CMEs), 401, 403, 404
COROT (Convection Rotation and Planetary Transits) space telescope, 156, 186
Cosmic Background Explorer (COBE) telescope, 163, 415, 595, 596, 639
cosmic inflation. see inflation
cosmic microwave background radiation (CMB)
  defined, 163, 594
  discovery of, 594, 598, 615
  flatness of the universe, 614–615
  horizon problem, 615–616
  images, 580–581, 596
  measuring temperature of, 594–595
  Planck blackbody spectrum, 592, 595
  recombination of the universe and, 595, 595, 640, 643
  variations in, 595–596, 640–641
  see also Big Bang
cosmic rays, 250, 403, 504, 517, 565–566

cosmological constant, 609, 611, 612, 626–627, 642
cosmological principle, 582–587
  defined, 10, 61, 580, 582
  homogeneous universe, 582–583, 585, 591, 637
  Hubble's law and, 582–585
  isotropic universe, 582–583, 591, 637
  large-scale structure of the universe and, 637
  as testable hypothesis, 10, 582
cosmological redshift, 593, 647
cosmology, 604–633
  defined, 582, 604, 618
  fate of the universe, 606, 611–612, 653–654
  multiverses, 625–626
  particle physics and, 618–624
  superstring theory, 624–625
  see also Big Bang; expanding universe; universe
Crab Nebula, 468–469, 484–485, 540
craters and impacts. see impacts and craters
Cretaceous-Paleogene (K-Pg) boundary, 223
critical density, 606–608, 611, 626–627
crust, 207
Crux, the Southern Cross (constellation), 30
cryovolcanism, 299, 300–302, 671
C-type (carbon) asteroids, 328
Curiosity rover, 76, 158, 220, 341, 671
Curtis, Heber D., 526
cyanobacteria (blue-green algae), 239, 663–664
cyclonic motion, 245, 246, 272, 276
Cygnus X-1, 496–497, 514–515

Dactyl, moon of Ida, 330
Dark Ages, 643, 653, 654
dark energy, 628
  accelerating universe, 611–612, 625
  defined, 4, 611
  discovery, 102
  evolution over time, 611
  fate of the universe and, 611–612
  universe expansion and, 376, 590, 611–612, 625, 654, 655
Dark Energy Survey, 611
dark matter, 628
  cold dark matter, 640–643, 652
  composition of, 538–539
  defined, 4, 536–538
  detection and measurement, 537–538
  distribution of regular and dark matter in early universe, 641, 654
  elementary particles, 539, 641
  and formation of galaxies, 640–643, 654
  in galaxies, 536–538, 561, 654
  in galaxy groups and clusters, 637–639
  hot dark matter, 641–643
  Lambda-CDM (Lambda Cold Dark Matter) theory, 640–642, 649, 652
  Milky Way Galaxy, 566–570
  minihalos and formation of first stars, 645, 646, 648
dark matter halo
  galaxies, 538, 539, 647, 648, 651
  minihalos and formation of first stars, 645, 646, 648
  simulations, 652, 653
daughter products, 203, 204

Dawn spacecraft, 330, 346
decay (energy states), 118, 123–124
declination, defined, 25
Deep Impact spacecraft, 337–338
degeneracy pressure, 443
Deimos, 295
Delta Cephei, 472
deoxyribonucleic acid. *see* DNA
deuterium
   from Big Bang nucleosynthesis, 596–597, *597*, 623
   deuterium fusion in brown dwarfs, 426
   deuterium fusion in protostars, 422–423
   in heavy water, 222, 346
   from proton-proton chain, *387*, 388
   see also hydrogen
Diana, 40
Dicke, Robert, 592
differential rotation, 398
differentiation, 207
diffraction
   defined, 147
   diffraction grating, 152
   diffraction limit, 148
   prisms and spectra, *119*, 120–121, 128, 152
Diffuse X-ray emission from the Local Galaxy (DXL) instrument, 490
Dione, *294*, 299, 302
direct imaging of extrasolar planets, 186–187
dispersion, 145
distance ladder, 376, 532–533
distance measurement
   Cepheid variables, 376, 472, *526*, 527
   distance ladder, 376, 532–533
   H-R diagram and, 371, 373, *374*
   light-year, defined, 5, 111, 356
   to nearest stars, 354–357, 359, 364, 370, *374*, 376
   parsec (pc), 356
   redshift, 596, 637
   spatial scanning technique, 376
   spectroscopic parallax, 371, 373, *374*
   stereoscopic vision and, 354, *355*, 359
   Type Ia supernovae, 376, 459, 533, 534, 609
   in universe, 4–7
   see also parallax
DNA (deoxyribonucleic acid)
   composition, 313, 669
   damage by radiation, 517, 674
   genetic code and, 663, 666–667, 669
   mutation, 666
   phylogenetic tree, 664, *665*
   prebiological evolution, 662, 663
   in prokaryotes and eukaryotes, 664, *665*
   sequencing and analysis, 460, 664, 665
Doppler effect, 124–125
Doppler shift
   binary stars, 365–367
   blueshifted light, 125
   formula, 126
   helioseismology, 393
   radial velocity (v_r) and, 125, 126, 184–185, 187, 189
   redshifted light, 125, 593
   rest wavelength (λ_{rest}) and, 125, 126

Tully-Fisher relation and distance estimates, 533
21-cm radiation of neutral hydrogen, 417–418
   see also Hubble's law
30 Doradus, 417
Drake, Frank, 675, 677
Drake equation, 675–677
Draper, John W., 151
dust
   dust storms on Mars, 220, *247*, 248, 249
   dust tails of comets, *334*, 335, *336*
   giant dust ring around Saturn, 309
   in protoplanetary disks, 173, 177–178, 180, 182, *183*
   in spiral galaxies, 530
   stardust, in human body, 15, 24
   zodiacal dust, 341–342
   see also interstellar dust
dust devils, 246, 249
dust tails of comets, *334*, 335, *336*
dwarf galaxies, 529
   dark matter in, 538, 571
   in Local Group, 570–571, 647
   mass, 531
   Segue 1, 647
   Small Magellanic Cloud, 571
   ultrafaint dwarf galaxies, 648
dwarf novae, 458
dwarf planets, 322–325
   Ceres, *69*, *172*, 183, 322, 323, 346
   comparison to Earth and the Moon, *325*
   Haumea, 294, 322, 325–326
   in Kuiper Belt, 183
   Makemake, 322, 325–326
   periods and semimajor axes of orbits, *69*
   see also Eris; Pluto
DXL (Diffuse X-ray emission from the Local Galaxy) instrument, 490
dynamo theory and magnetic fields, 209

Eagle Nebula, *423*
Earth
   age of, 6, 7, 203, 460
   composition of interior, 205–206
   density, 205, 270
   destruction of impact craters, 200, 203
   distance from Sun, 5
   estimating size of, 31
   fate of life on Earth, 678–679
   impact craters, 200, 203, 223
   K-Pg boundary and mass extinction, 223
   layers of, 240–243
   location of, 4
   magnetic field, 209–210, 243–244
   magnetosphere, 209, 243–244
   measuring mass of, 205
   model of interior, 206–207
   origin of life on Earth, 662–664
   physical properties, *198*
   as seen from Saturn, *679*
   sidereal period, *69*
   small number of impact craters, 200
   solar activity effects, 401–402
   view from space, *2–3*, *230–231*, *679*
   when life began on Earth, 460, 663–664

Earth, orbit of
   average speed, 498
   circumference, 498
   eccentricity (*e*), 66, 75
   Milankovitch cycles, 250, 252
   seasons and, 32
   semimajor axis of orbit, *69*
   tilt of axis and, 34–36, 51, 250
Earth, rotation of, 24–33
   apparent daily motion of astronomical objects, *25*, 25–30
   Coriolis effect and, 32–33, 245–246
   effect of tidal forces, 99–101
   motion of stars and celestial poles, 27–31
   tilt of axis and, 35, 51, 250
   view from the poles, *25*, 26–27
   see also celestial sphere
earthquakes, 205–206, 212, 213
Earth's atmosphere
   atmospheric convection, 237, 241–242, 244–245
   atmospheric greenhouse effect, 237
   atmospheric pressure, 240–241
   comets and, 201, 662
   composition, 234, *234*
   electromagnetic radiation blocked by, 153, 155–157
   ionosphere, 243
   layers of, 240–243
   life effect on composition, 234, 237–240, 254, 256
   light scattering, 49
   limitations on telescope resolution, 147–149
   magnetosphere and, 209, 243–244
   mass of, 240
   mesosphere, 242
   meteorites and, 199–200, 339–340
   molecular nitrogen in, 234, 236
   ozone, 239–240, 242, 416, 517, 664–665
   possible effects of supernovae, 517
   primary atmosphere, 182
   role of life in, 234, 237
   secondary atmosphere formation, 182, 233–234, 662
   stratosphere, 242
   thermosphere, 242, 273
   troposphere, 241–242
   wind and weather, 244–246
   see also specific gases
eccentricity (*e*), 65, *65*, 75, 325
eclipses, 45–51
   annular solar eclipse, 46, *46*, 48
   defined, 45
   eclipse seasons, 49–51
   gravitational lensing during, 509, 511
   line of nodes and, 50–51
   lunar eclipses, 48–50
   partial lunar eclipse, 49
   partial solar eclipse, 46, 47, 64
   paths of solar eclipses, 48
   penumbra, 46, 49
   penumbral lunar eclipse, 49
   solar eclipses, 46–48
   total lunar eclipse, 48
   total solar eclipse, 46, 48, 49
   umbra, 46, 48, 49

eclipse seasons, 49–51
eclipsing binary stars, 367–368, *374*
ecliptic, 25, 34, 35
ecliptic plane, defined, 36
Eddington, Arthur Stanley, 509, 511
E-ELT (European Extremely Large Telescope), *147*, 149–150
effective temperature, 394, 514
Einstein, Albert
   cosmological constant, 609, 611
   on falsifiable theories, 8
   formulation of spacetime, 609
   mass-energy equation, 386, 503, 543, 620
   Nobel Prize in Physics, 502
   on photoelectric effect, 114, 502
   photograph, *11*
   quantum mechanics and, 12
   scientific revolutions and, 11–12, 506
   speed of light as constant, *115*, 500–502
   thought experiments, 501–502
   see also general theory of relativity; special theory of relativity
Einstein ring, 511
ejecta, *198*, 199
El Caracol, *24*
electric fields, 111
electric force, 111
electromagnetic force, 618, 619, 621, 623
electromagnetic radiation
   atmospheric windows, 153
   frequency (f), 113–114, 116
   frequency–wavelength relationship, 113, 116
   properties of electromagnetic waves, 111–112
   visible light, 113
   visible spectrum, 113, *113*
   wavelength, 113–114, 116
   see also light; radiation
electromagnetic spectrum, 111–114
   defined, 113
   gamma rays, 113–114
   infrared (IR) radiation, 114, 155
   microwave radiation, 114–116
   photon energy and, 114, 116
   radio waves, 114, 154
   ultraviolet (UV) radiation, 114
   visible light, 113
   visible spectrum, 113, *113*
   X-rays, 113
   see also light
electron-degenerate matter, 443–444, 447, 450, 451, 453
electron neutrino, 391
electrons, 116
electroweak theory, 619
elementary particles, 641. *see also specific particles*
elements
   astronomer's periodic table, *123*
   creation in massive dying stars, 489, 547, 645–646
   defined, 118
   heavy elements, 15, 270, 362, 547, 645–646, 667
   in living organisms, 667, 669
   massive elements, 362
   origins of, 15

   parent element, 203, 204
   periodic table, *123*
   relative abundance in Solar System, *480*
   relative abundance in universe, 596–597
   solar abundance, 123, *363*
   in Sun's atmosphere, 363
   see also nucleosynthesis; *specific elements*
ellipses
   defined, 65
   eccentricity (*e*), 65, *65*, 75, 325
   focus, 65, *65*
   Kepler's first law, 65–66, 71
   semimajor axis, 66, 68, 89, 95, 185
   see also Kepler's laws
elliptical galaxies
   bulges, 528
   characteristics, *528*
   color, 530, 547
   distance estimates and surface brightness, 533
   extrasolar planets in, 547
   giant and dwarf elliptical galaxies, 531
   hot gas in, 530
   mass, 538
   older stars in, 530, 547
   radio galaxies, 540, *545*
   stellar motions, 529–530
   structure, 528, *529*
Ellis, Richard, 628
El Niño, 253
emission lines
   from atmosphere of stars, 360, *361*, 362
   defined, 119
   Hα emission line, 395, 400, 416, 537, *558*
   from interstellar gas, 416
   from molecular clouds, 418
   overview, 119, 121–122, 125, 128
   from planetary nebula, 453
emission spectra, 118–119
empirical laws, 71, 84, 89, 93
empirical science, 64
Enceladus
   cryovolcanism, 299, 671
   escape velocity, 308
   evidence of liquid water oceans, 282, 299, 308, 313, 671
   icy geysers, 282, *299*, 308, 313
   mass and radius, 308
   methane, 313
   orbital resonance with Dione, 299
   possibility of life, 313, 671
   and Saturn's E Ring, 307
   tidal flexing, 299
   torus of plasma associated with, 282
Encke Gap, *305*, *306*, 310
energy
   binding energy, 474–475
   defined, 111
   Einstein's mass-energy equation, 386, 503, 543, 620
   gravitational potential energy, 178–179
   law of conservation of energy, 179
   of light, 111, 114, 116
   thermal energy, defined, 111, 207, 422
   see also kinetic energy

energy transport in the Sun, 389
Englert, François, 619
epicycles, 61, 62, *67*
EPOXI spacecraft, 338
equator, 25, 36
equilibrium, 126–127
   dynamic equilibrium, 126–127
   hydrostatic equilibrium, 206, 384, 420, 425
   stable equilibrium, 127
   static equilibrium, 126–127
   temperature of planet interiors, 208
   temperature of planets, 127, *127*, 132, 133, 236, 256
   thermal equilibrium, 127, *127*
   unstable equilibrium, 127
equinoxes, 38
equinoxes, precession of, 39
equivalence principle, 506–508
Eratosthenes, 31
Eris
   characteristics, 323, 325
   classification as a dwarf planet, 322
   discovery, 323
   Dysnomia, moon, 294, 323
   location, *172*
   orbit, *69*, 323, 325
   as planetesimal, 183
Eros (asteroid), 330
erosion, 219–220
escape velocity (v_{esc})
   black holes, 513
   defined, 93
   and gas molecules in atmosphere, 232, 234, 235, 673
   neutron stars, 481
   surface of astronomical bodies, 94, 308, 451, 606, 673
   terrestrial planets, *198*
   universe, 606–608
ESI (Earth Similarity Index), 674
ESO. *see* European Southern Observatory
Eta Carinae, *15*, 474
eternal inflation, 625–626
Eukarya, 664, *665*
eukaryotes, 664, *665*
Europa
   evidence of liquid water oceans, 298–299, 313, 671
   possibility of life, 313, 671
   tidal flexing, 299, 671
   variable magnetic field, 298, 313
Europa Clipper, 314
European Southern Observatory (ESO), 134, 149, 155
European Space Agency, 76, 157, 163, 187, *214*, 215
event, defined, 500
event horizon, 513–514, 543
Event Horizon Telescope, 154
excited states, 118, 119, 123–124
exoplanets. *see* extrasolar planets
exospheres, 249
expanding universe, 580–603
   accelerating universe, 608–614, 625
   age of universe, 587–590
   critical density, 606–608, 611, 626–627
   dark energy and, 376, 590, 611–612, 625, 654, 655
   flatness problem, 614–617

galaxies and, 589–590
gravity and, 606–608
horizon problem, 615–616
inflation, 616–617, 623, 625–626
inflationary model, 616–617, 623, 625–626
matter density parameter (Ω_m), 607–608, 611–614, 616
rate estimation by spatial scanning technique, 376
shape in spacetime, 613–614
see also Big Bang; cosmological principle; Hubble's law; universe
exponential growth, 666
exponential processes, 14
extrasolar planets, 183–188, 284–285
around binary stars, 431
astrometric method, 186
atmospheres, spectroscopy, 187
vs. brown dwarfs, 184, 187
defined, 184
direct imaging, 186–187
discovery of, 180, 184–186, 197–198
Earth Similarity Index (ESI), 674
in elliptical galaxies, 547
estimation of orbit size, 185
estimation of planet mass, 184–186, 189
estimation of planet radius (size), 185, 186
gas dwarfs, 284
habitable zone, 189, 285, 375, 547, 672–675
hot Jupiters, 197–198, 284
Kappa Andromedae b, 284
Kepler 10c, 284
KOI-314c, 284
microlensing, 186, 188
mini-Neptunes, 188, 284–285
moons and rings, detection methods, 312
properties and orbits, 10, 189
puffy Jupiters, 284
radial velocity (v_r) method, 184–185, 187, 189
search for, 184–187, 189
super-Earths, 188, 284–285
tidal forces, 101
transit method, 186–189, 312
see also planetary system formation
Extremely Large Telescope (ELT), 147, 149–150
extremophiles, 313, 663, 671, 678
eye, human
anatomy, 142
angular resolution of, 142, 148
focal length, 147
integration time, 150–151
planets visible to naked eye, 44, 60
quantum efficiency, 151
schematic view of, 142
sensitivity to visible light wavelengths, 113, 116
stars visible to naked eye, 34
stereoscopic vision, 354, 355
eyepiece, 143

fact, defined, 8
Faherty, Jackie, 574
false-color images, 152, 200, 215, 217
falsified hypothesis, 7–8, 9, 34–35, 37, 202
falsified theories, 8, 67

Far Ultraviolet Spectroscopic Explorer, 462
faults, 212, 213
Fermi, Enrico, 677
Fermi Gamma-ray Space Telescope, 153, 156, 569, 570
Fermi paradox, 677
filaments, 637, 639, 652
filters, optical, 360
first quarter Moon, 41, 95
flatness problem, 614–617
flat universe, 613–615, 625
flux ($\mathcal{F}$), 129, 130
flux tubes, 283, 297, 396, 400
flyby missions, 157–158
focal length, 143–145, 147
focal plane, 143, 145, 148–149
focal ratio, 143
focus of ellipse (plural, foci), 65, 66
Fomalhaut, 187, 462
forces
    centripetal force, 90
    defined, 72
    electric force, 111
    electromagnetic force, 618, 619, 621, 623
    four fundamental forces of nature, 618
    grand unified theories (GUTs), 621–624
    magnetic force, 111
    Newton's laws of motion, 72–75, 84
    quantum chromodynamics (QCD), 619
    separation in the cooling universe, 623–624
    strong nuclear force, 385–387, 447, 478, 518–519, 626–627
    theory of everything (TOE), 619, 622, 624–626
    weak nuclear force, 618, 619, 621, 623
    see also gravity; tidal forces
formaldehyde as prebiotic molecule, 573
fossil fuels, 252
fossils, 223, 251, 663, 664, 667
frame of reference
    cosmic reference frame and CMB, 595–596
    equivalence principle, 506–508
    inertial reference frame, 72, 499, 500, 503, 507
    and relative motion, 31–33, 38, 61, 72, 498
    and relativistic time dilation, 502–504
free fall, 90
frequency (f), 113–114, 116
Friedmann, Alexander, 582, 607
full Moon, 40, 41, 42, 95, 98

Gaia space observatory, 156, 186, 371, 574
galactic fountains, 563, 564
galaxies, 16, 524–553
    accretion disks, 643, 648–649
    bottom-up models of galaxy formation, 639, 646, 647, 648, 650
    classification and types, 527–528, 650
    CMB variation and galaxy formation, 640
    collision of, computer modeling, 162
    color, 530
    dark matter and formation of, 640–643, 654
    dark matter halo, 538, 539, 647, 648, 651
    dark matter in, 536–538, 654
    dark matter in galaxy groups and clusters, 637–639

    defined, 526
    discovery of, 524, 526–527
    distance estimation methods, 532–536, 637
    distance ladder, 532–533
    dust and gas in, 530
    dwarf galaxies, 531, 538
    Einstein ring, 511
    evolution, 556, 649–654
    expanding universe and, 589–590
    first galaxies, 645–648, 654
    gas between galaxies, 638, 639
    giant galaxies, 531, 571, 636
    gravitational lensing, 511, 638
    habitable zones in, 547, 573, 674
    hierarchical clustering and growth, 639, 646, 649–652
    interaction of, 556, 585
    irregular galaxies, 528, 530–531
    luminosity, 530
    mass, 531, 536–537, 638
    mass distribution, 537
    mass of a cluster of galaxies, 638
    merging galaxies, 162, 547, 556, 571, 573, 650–652
    merging galaxies, biological effects, 573
    parallels between galaxy and star formation, 418, 420–422, 648–649
    peculiar velocity, 637
    redshift surveys, 637
    rotation curves, 537–538, 567, 569, 638
    stellar motions and galaxy shape, 529–530
    supermassive black holes in most distant galaxies, 650
    supermassive black holes in normal galaxies, 544, 556
    tidal forces and, 101, 572
    types of galaxy structures, 636
    ultrafaint dwarf galaxies, 648
    see also active galactic nuclei; barred spiral galaxies; dwarf galaxies; elliptical galaxies; large-scale structure of the universe; S0 galaxies; spiral galaxies; specific galaxies
galaxy clusters, 636, 638, 651
galaxy groups, 5, 570, 636–639
Galeazzi, Massimiliano, 490
Gale Crater, 221, 671
Galilei, Galileo
    Dialogo sopra i due massimi sistemi del mondo, 71
    experiments with motion and gravity, 70, 74, 84–86, 90, 97
    heliocentric theory and, 61, 71
    law of inertia, 70–72
    Saturn rings observation, 304
    telescopic observations, 70, 142, 304
Galileo spacecraft, 286
    Callisto's cratered surface, 303
    Europa's magnetic field, 298
    Gaspra (asteroid), 330
    Ida (asteroid) and its moon, Dactyl, 330
    Jupiter probe, 158, 273
    Jupiter's rings, 310
    volcanoes on Io, 297, 298, 319
Galle, Johann Gottfried, 267
gamma-ray bursts (GRBs), 458, 516–517, 573, 645

gamma rays (gamma radiation)
  absorption by Earth's atmosphere, 153–154
  from center of Milky Way, 569–570
  from CNO cycle, 470–471
  in electromagnetic spectrum, 113–114
  from novae, 458
  from proton-proton chain, *387*, 388
  solar flares, 400–401
  solar radiation, 390, 401–402
  thermal gamma-ray photons, *477*, 479
  from triple-alpha process, *447*
gamma-ray telescopes, 153–154, *156*
Gamow, George, 592
Ganymede, *294*, 295, 296, 302
Garrick-Bethell, Ian, 224
gas dwarfs, 284
gas giants, defined, 269. *see also* giant planets; Jupiter; Saturn
Gaspra (asteroid), 330
Gemini telescopes, *146*, 149
general relativistic time dilation, 511–512
general theory of relativity, 506–512
  black holes and, 512–516
  cosmological constant, 609, 611
  equivalence principle, 506–508
  failure to describe early universe, 622
  general relativistic time dilation, 511–512
  geodesics, 506–508
  gravitational lensing, 509, 511, 538, 638, 651
  gravitational redshift, 511–512, 514, 593, 596, 641
  gravitational waves, 515–516
  gravity as distortion of spacetime, 506–512, 589
  and Newton's universal law of gravitation, 508–509
  observable consequences of, 509, 511–512
  precession of elliptical orbits, 509
  scientific method and, 8
  spacetime distortion by mass, 506, 508, 509, 589, 613, 637–639
  see also cosmological principle; gravity
genetic code, 663, 666–667, 669
geocentric model, 60–62, 70, 71
geodesics, 506–508
giant galaxies, 531, 571, 636, 638
Giant Magellan Telescope (GMT), *146*, 149
giant molecular clouds, *410*–*411*, 418
giant planets, 264–291
  atmosphere structure, 273–275
  atmospheric convection, 273, 275, 278
  auroras, *282*, 283
  characteristics of, 264, 266–267, 269
  clouds, 269–273
  composition of, 269–270
  core accretion–gas capture, 182–183
  cores, 182, 269, 270, 277–280
  days and seasons, 270–271
  defined, 266–267
  densities, 270
  formation of, 182–183, 279
  gas giants, 269
  helium, 269, 270, 279
  hot Jupiters, 197–198, 284
  hydrogen, 269, 270, 278–279
  ice giants, 269
  images of, *266*–*267*

  magnetic fields, 279–283
  magnetospheres, 279–282
  physical properties of, *266*
  planet migration, 181, 285
  puffy Jupiters, 284
  radiation belts, 282
  super-Jupiters, 284
  synchrotron radiation, 281–282
  tidal forces and rings of, 101
  tropospheres, 273, *274*
  winds and weather, 275–277
  see also Jupiter; Neptune; Saturn; Uranus
Giotto spacecraft, 336
global circulation, 245, 247
Global Oscillation Network Group (GONG), 393
globular clusters
  ages, 485–486, *486*, 562, 571, 612
  defined, 485
  Milky Way Galaxy, 559–560, 562, 563, 565, 571, 650
  stellar evolution in, 485–489
gluons, 619, 621, 622
GMT (Giant Magellan Telescope), *146*, 149
Gold, Thomas, *482*
GONG (Global Oscillation Network Group), 393
Gopalswamy, Nat, 404
grand unified theories (GUTs), 621–624
Gran Telescopio CANARIAS (GTC), *146*
graphs, reading, 14
gravitational lensing, 509, 511, 538, 638, 651
gravitational potential energy, 178–179
gravitational redshift, 511–512, 514, 593, 596, 641
gravitational waves, 162, 163, 484, 515–516, 518
gravity, 82–107
  defined, 84, 618
  equivalence principle, 506–508
  expansion of the universe and, 606–608
  and formation of large-scale structure, 639–643, 655
  free fall and, 90
  grand unified theories and, 622
  gravitational acceleration (g), 84, 85
  gravitational redshift, 511–512, 514, 593, 596, 641
  inverse square law, 86, *87*
  on a neutron star, 481
  orbits and, 89–90
  as a quantum phenomenon, 622
  self-gravity, 88, 95, 100, 172–173, 176, 420–421
  universal gravitational constant (G), 86, 102, 296, 607
  universal law of gravitation, 86–89, *91*, 95, 508–509
  variation on and near a sphere, 88
  weight and, 84, 88, 90
  see also general theory of relativity
GRBs. *see* gamma-ray bursts
Great Attractor, 637, 655
Great Dark Spot, 272, 275
greatest elongation, 63, 64
Great Red Spot, 271–272, 275, 286
Green, Jim, 257, 314
greenhouse effect
  atmospheric greenhouse effect, 236–237, 246, 252, 254, 256
  comparison of terrestrial planets, 237–238, 254, 256
  defined, 236–237

greenhouse gases
  carbon dioxide, 236, 251
  climate change and, 249–254
  defined, 236
  increase in Earth's atmosphere, 252–253
  methane, 236, 251
  water vapor, 237, 240
Gregorian calendar, 44, 45
ground state, 118
GTC (Gran Telescopio CANARIAS), *146*
Gum, Colin, 134
Gum 41 nebula, 134
Guth, Alan, 616
GUTs (grand unified theories), 621–624
GWI70104, 518

habitable zones
  defined, 189, 254, 256, 375
  extrasolar planets in, 189, 285, 375, 547, 672–675
  galactic habitable zones, 547, 573, 674
  in Milky Way Galaxy, 573, 674
  Solar System moons, 313, 674
  Solar System planets, 254, 256, 285, 455–456, 672–675, 678
Hadley circulation, 244, 247
Hale, George Ellery, 399
half-life, 203–204
Halley, Edmund, 334
halo
  dark matter halo, 538, 539, 647, 648, 651
  dark matter halo simulations, *652*, *653*
  inner halo, *560*, *564*, 564–565
  MaCHOs (massive compact halo objects), 538–539, 638, 641
  Milky Way Galaxy, 560, 562–565
  minihalos and formation of first stars, 645, 646, 648
  outer halo, *560*, 564–565
halogens, 239
harmonic law, 69. *see under* Kepler's laws
"Harmony of the Worlds" (harmonic law). *see under* Kepler's laws
Haumea, *294*, 322, 325–326
Hawaiian Islands, 216
Hawking, Stephen, 514
Hawking radiation, 514
Hayabusa, 330
Hayashi, Chushiro, 427
Hayashi track, *427*, 427–430
HD 226868, 514–515
heavy elements, 15, 270, 362, 547, 645–646, 667
heliocentric theory, 61–62, 64, 65, *67*, 68, 71
helioseismology, 391, 393
Helios II spacecraft, 502
heliosphere, 403
helium
  from Big Bang nucleosynthesis, 596–597, *597*
  discovery in solar spectrum, 395
  emission spectrum, *122*
  on giant planets, 269, 270, 278, 279
  helium cores in stars, 443–444, 446–448, 450
  helium shell fusion, 457, 472
  in interstellar medium, 412
  in primary atmospheres, 232
helium flash, defined, 449

helium fusion
  in binary stars, 457
  in cores of red giant stars, 447–451
  helium flash, 448–449
  helium shell fusion, 457, 472
  in high-mass stars, 471, 475, 480
  horizontal branch, 449–451, 454–455, 457, 471
  medium-mass stars, 472
  triple-alpha process, 447–448, 451, 475, 480
  white dwarf formation from post-AGB star, 454
Helix Nebula, *453*
hematite, 220
Herbig-Haro objects (HH objects), 429
heredity, 666–667
Herman, Robert, 592
Herschel, Caroline, *326*, 526
Herschel, William, 267, 303, 326, 526, 670
Herschel crater on Mimas, 303
Herschel Space Observatory, 157, 332, 338, 431
hertz (Hz), 112
Hertzsprung, Ejnar, 369, 372
Hertzsprung-Russell diagram. *see* H-R diagram
HET (Hobby-Eberly Telescope), *146*
Hewish, Anthony, *482*
HH (Herbig-Haro) objects, 429
hierarchical clustering, 639, 646, 649–652
Higgs, Peter, 619
Higgs boson, 619
Higgs field, 619
high-mass stars
  blue supergiant stars, 474
  convection in, 471, 476, 489
  defined, 440, 470
  first stars, 643–649, 654
  iron cores, 472
  red supergiant stars, 471, 474
  spectral types, 440
  yellow supergiant stars, 472
high-mass stars, evolution, 468–495
  asymptotic giant branch (AGB), 472, 480, 489
  blue supergiant stars, 474
  carbon fusion, 472
  CNO cycle, 470–471
  core collapse, 476–479
  final days of, 475–476
  helium fusion, 471, 475, 480
  helium shell fusion, 472
  hydrogen fusion, 470–471
  hydrogen shell fusion, 471
  layered structure, 472, 474
  leaving the main sequence, 471–472
  low-density winds and mass loss, 473–474
  luminous blue variable (LBV) stars, 474, 480
  red supergiant stars, 471, 474
  triple-alpha process, 475, 480
  yellow supergiant stars, 472
H II regions ("H two"), 416–417, 452, 563
Hipparchus, 358
Hipparcos satellite, 357, 371
Hobby-Eberly Telescope (HET), *146*
Hogg, David, 574
Homestake Mine experiment, 391, *392*
homogeneous universe, 582–583, 585, 591, 637
horizon, defined, 25

horizon problem, 615–616
horizontal branch, 449–451, 454–455, 457, 471, 473, 560
Horsehead Nebula, *417*
hot dark matter, 641–643. *see also* neutrinos
hot Jupiters, 197–198, 284
hot spots, 213, 215–217
H-R diagram (Hertzsprung-Russell diagram)
  close binary systems, 457
  distance estimation, 371, 373, *374*
  evolutionary track of protostars, 427–428
  giants, *370*, 373
  Hayashi track, *427*, 427–428
  horizontal branch, 449–451, 454–455, 457, 471, 473, 560
  instability strip, 472–474, **560**
  luminosity class, 373–375
  main sequence, 369–371, 373
  overview, 369–371
  red giant branch, 444–446, 448, 450, 471
  spectroscopic parallax, 371, 373, *374*
  star clusters, 485–488
  stars not on the main sequence, 373–375
  subgiant branch, 445
  supergiants, *370*, 373–375
  white dwarfs, *370*, 373, 453–455
  see also asymptotic giant branch
HST. *see* Hubble Space Telescope
Hubble, Edwin P.
  Andromeda Galaxy's blueshift, 572
  Cepheid variables, 526
  classification of galaxies, 527–528, *649*
  discovery of Hubble's law, 534–535, 593
  expanding universe, 534, 582, *586*, 632–633
  Local Group identified, 570
Hubble constant (H_0), 534–535, 585, 587, 588, 607, 612
Hubble flow, defined, 585
Hubble's law
  age of universe estimation, 587–590
  cosmological principle and, 582–585
  defined, 534, 582–585
  discovery of, 534–535
  distance measurement and redshift, 585, 637
  Hubble constant (H_0), 534–535, 585, 587, 588, 607, 612
  Hubble time, 587, 589, *589*, 612
  look-back time, 587, 593
  mapping the universe, 636–637
  see also Doppler shift; expanding universe
Hubble Space Telescope (HST), 286
  auroral rings on Jupiter and Saturn, *282*
  Cassini Division in rings of Saturn, *306*
  Cepheid variables as standard candles, 533
  Coma Cluster of galaxies, 553
  Comet Shoemaker-Levy 9 collision with Jupiter, 342, 343
  diameter, 156
  disks around newly formed stars, *173*
  dust storm on Mars, *249*
  evidence of liquid water oceans on Europa, 298, 671
  extrasolar planet imaging, 187
  filters and false colors, 152
  gravitational lensing, 511, *639*

  image, *153*
  Jupiter, *282*
  low-magnitude star detection, 358
  M104 or Sombrero Galaxy, *530*
  merging galaxies, *651*
  Neptune, 272
  observing high-redshift objects, 647
  optical limitations, 358
  orbit above atmosphere, 148, 155
  parallax measurement, 376
  planetary nebulae, *453*
  quasars, *540*, 556
  resolution, 148
  Ring Nebula, *438–439*
  rings of Uranus, 304, 310, 312
  Saturn, *269*, *270*, *282*
  servicing missions, 157
  spatial scanning technique, 376
  Ultra Deep Field 2014, *634–635*, 653
Hubble time, 587, 589, *589*, 612
Huygens, Christiaan, 304, 308
hydrogen
  atomic energy levels, 122
  comparison to oxygen, 232, 235, 238
  emission spectrum, *122*
  on giant planets, 269, 270, 278–279
  H⁻ ion and stellar temperature, 427–428
  Hα emission line, 395, 400, 416, 537, *558*
  in interstellar medium, 412
  isotopes, relative abundance of, 596–597
  metallic hydrogen, *277*, 279
  in primary atmospheres, 232
  21-cm radiation of neutral hydrogen, 417–418, *419*, 556–557, 559, 567
hydrogen fusion
  in binary stars, 457, 458
  CNO cycle, 470–472, 645
  high-mass stars, 470–471
  hydrogen shell fusion, 444–446, 471
  in main-sequence stars, 385–386, 440, 442, 446
  in novae, 458
  overview, 385–386
  proton-proton chain, 387–388, 390, 442, 470, 471, 619
  in protostars, 427, 430
  in the Sun, 430, 438, 442, *443*
  white dwarf formation from post-AGB star, 454
hydrogen shell fusion, 444, 445, 446, 471
hydrostatic equilibrium, 206, 384, 420, 425
hydrothermal vents, 313, *662*, 663, 671
Hyperion, 295, 303, 304
hypernovae, or Type Ib or Ic supernovae, 517
hypothesis
  defined, 7
  falsified, 7–8, 9, 34–35, *37*, *202*

Iapetus, *294*, 303, *309*
IAU (International Astronomical Union), 183
ice, defined, 180
ice ages, 51, 250–252
IceCube neutrino observatory, 160
ice giants, defined, 269. *see also* giant planets; Neptune; Uranus
Iceland, volcanism, 216

Ida (asteroid), 94, 330
idea, defined, 8
Igaluk, 40
impacts and craters
  central peak, *198*, 199
  Chelyabinsk meteoroid, 343–344
  Comet Shoemaker-Levy 9 collision with Jupiter, 342–343
  death of the dinosaurs, 223, 345
  ejecta, *198*, 199
  impact energy of a comet nuclei, 344
  K-Pg boundary (Cretaceous-Paleogene boundary), 223
  late heavy bombardment, 201, 217, 663
  meteoroids, 199, 200
  micrometeoroids, 200, 219
  planets reshaped by giant impacts, 201
  radiometric dating, 203–204
  secondary craters, *198*, 199, 200
  on surfaces of terrestrial planets, 182, 183, 196–202
  Tunguska event, 343
  use as geological clock, 203
index of refraction (*n*), 143, 144
inertia, defined, 72
inertial frame of reference, 72, 499, 500, 503, 507
inferior conjunction, *62*, 63
inferior planets, 62–64
inflation, 616–617, 623, 625–626
Infrared Astronomical Satellite (IRAS), 415
infrared (IR) radiation
  absorption by Earth's atmosphere, 149, 153, 155, 242
  absorption by greenhouse gases, 237, 246
  atmosphere of Venus and, 215, 218, 246
  penetration of interstellar dust, *412*, 413
  properties, 114
  as thermal radiation, 237, 241, 414–415
infrared telescopes, 153, 155, 187
instability strip, 472–474, 560
integration time, 150–151
intercloud gas, 415–418
interferometer, 154
interferometric arrays, 154–155
International Astronomical Union (IAU), 183
International Space Station, *82–83*, 100, 111, 155
Interstellar Boundary Explorer spacecraft, 397
interstellar clouds
  accretion disk formation, 175–176
  defined, 415
  gravity effects, 88, 175–176
  nebular hypothesis and, 172–173
  properties of, *415*
  see also molecular clouds; star formation
interstellar dust
  effects on light, 412–415, 420, *430*
  formation of, 412
  interstellar extinction, 413, 414
  in molecular clouds, 418
  penetration by infrared radiation, *412*, 413
  penetration by radio waves, 413, *417*, 420
  reddening, 414
  Stardust robotic probe, 432
interstellar extinction, 413, 414

interstellar medium, 412–420
  composition, 412
  defined, 396, 412
  density and properties, 412, 415–418
  H II regions ("H two"), 416–417, 452, 563
  intercloud gas, 415–418
  properties of components, *415*
  regions of cool, dense gas, 418, 420
  solar wind boundary with, 397
  temperature, *415*, 416
  see also interstellar clouds; interstellar dust
inverse square law, 86
Io
  flux tube, 283, 297
  orbital resonance of Ganymede, Europa, and Io, 296
  Pillan Patera, 297
  plasma associated with, 282, 297
  Prometheus Patera, *297*, *319*
  semimajor axis and period of orbit, 296
  tidal flexing, 297
  volcanic activity, 282, 297–298, 307, *319*
ionize, 243, 416
ionosphere, 243
ion tails of comets, *334*, *335*, *336*
IRAS (Infrared Astronomical Satellite), 415
iron
  binding energy, 474–475
  in cores of high-mass stars, *472*, 476–477
  in cores of planetesimals, 328–329
  in cores of terrestrial planets, 15, 180, 207, 489
  density, 205
  end product of nuclear fusion, 474–476
  in Europa's core, 298
  in Ganymede's core, 302
  iron-60 in rocks, 517
  magnetic properties, 209–210
  in meteorites, 199, *200*, 205, 339, 341
  in rocks and minerals, 219, 220, 238, 248, 517
  from supernova explosions, 459, 489
  in Vesta's core, 330
iron meteorites, *329*, 341, *341*
IR radiation. *see* infrared radiation
irregular galaxies, 528, 530–531
irregular moons, 295, 301, 303
Ishtar Terra, 215
isotopes, 118, 203–204, 388
isotropic universe, 582–583, 591, 637
Itokawa (asteroid), 330, 344

Jakosky, Bruce, 257
James Webb Space Telescope, 156–157, 640
jansky (Jy), 154
Jansky, Karl, 154
jets
  from active galactic nuclei (AGNs), 540, 542, 544, 545
  bipolar outflow, 177, 429, *429*
  of gas and dust from comets, 336–338
  from newly formed stars, *173*, 177
  from radio galaxies and quasars, 540
  relativistic beaming, 544
  from supermassive black holes, *540*, 542

  of water on Europa, 671
  of water vapor and ice on Enceladus, 282, *299*, 308, 313, 671
Jovian planets. *see* giant planets
Juno mission, 273, 286
Jupiter
  ammonia in atmosphere, 273
  angular momentum in orbit, 177
  atmosphere structure, 273, *274*, 278–279
  auroras, *282*, 283
  axial tilt, *266*, 270
  chemical composition, 270
  clouds, 269, 271–273
  Comet Shoemaker-Levy 9 collision, 342–343
  core, 182, 270
  core accretion–gas capture, 182
  diameter, *184*, 266
  formation, 182, 279
  Galileo atmospheric probe, 158, 273
  Galileo spacecraft, 273
  gravitational effect on Mars, 250
  gravitational effect on Sun, 184, 185
  Great Red Spot, 271–272, 275
  heavy elements, 270
  helium, 270, 278, 279
  hydrogen, 269, 270, 278–279
  images, *58–59*, 266–267
  interior, 278–279
  internal thermal energy, 278
  Juno mission, 273, 286
  lightning, 275
  magnetic field, 279
  magnetosphere size and shape, 280, 281
  mass, 184, 266, 269, 279, 296
  mass of atmosphere, 182
  methane in atmosphere, 273
  orbital angular momentum, 177
  orbital period, 184
  physical properties of, *266*
  planet migration, 285
  primary atmosphere formation, 181
  radiation belts, 282
  rings, composition of, 307
  rings, discovery of, 304
  rings, formation and evolution, 307
  rings, structure, *305*, 310
  semimajor axis of orbit, 69
  sidereal period, 69
  tidal forces, 100, 101
  troposphere, 273
  winds and weather, 275
Jupiter's moons
  Adrastea, 307
  Amalthea, 307
  Callisto, *294*, 303–304, 313
  Galilean moons, *58–59*, 70, 282, 295, 296, 303
  Ganymede, *294*, 295, 296, 302
  major moons, images, *294*
  Metis, 307
  orbital motions and speed of light measurement, 110, *115*
  orbital resonance of Ganymede, Europa, and Io, 296

orbits of, 295
Thebe, 307
see also Europa; Io
JWST. *see* James Webb Space Telescope

Kaguya mission, 222
Kant, Immanuel, 172, 526
Kappa Andromedae b, 284
Keck Observatory (Hawaii), *140–141*, 144, 147, 149, 153, *647*
kelvin (K), 128
Kelvin scale, 128
Kepler, Johannes, 61, 65–69, 71, 86, 95
Kepler 10c, 284
Kepler-11 system, 186, *186*
Kepler-186 system, *189*
Kepler's laws
    as empirical laws, 71, 84, 89, 93
    influence of Sun on planetary orbits, 86
    Kepler's first law, 65–66, 71
    Kepler's second law (law of equal areas), 66, 68, 93
    Kepler's third law (harmonic law), 68, *69*, 70, 94, 205
    overview, 65
    see also planetary motion
Kepler space telescope
    Earth-sized planets, 189
    eclipsing binary stars, 367
    heavy elements and planet formation, 547
    launch year, *156*
    mini-Neptunes or super-Earths, 188
    planets in elliptical galaxies, 547
    planets in habitable zone, 189, 375
    transit method, 186, 189, 312, 375
kinetic energy (E_K)
    defined, 111, 179
    equation, 128
    gas temperatures and, *127*, 179, 232, 235
Kirkwood, Daniel, 327
Kirkwood gaps, 327–328
knowledge
    hierarchy of, 8
    as power, 24
    scientific revolution and, 11–12
    scientific theory and, 8
KOI-314c, 284
K-Pg boundary (Cretaceous-Paleogene boundary), 223
Kuiper, Gerard, 331
Kuiper Belt
    comets in, 233, 322
    defined, 322, 332
    dwarf planets in, 182, 322
    Kuiper Belt objects (KBOs), 322, 332
    location, *172*, 233, 332
    possible source of Earth's water, 338, 346
    Quaoar, 332
    trans-Neptunian objects (TNOs), 322

Laboratory for Atmospheric and Space Physics (LASP), 257
LADEE (Lunar Atmosphere and Dust Environment Explorer), 249
Lagoon Nebula, 134
Lambda-CDM (Lambda Cold Dark Matter) theory, 640–642, *649*, 652
Lambda Centauri Nebula (Running Chicken Nebula), 134
landers, 158, 219
Laniakea Supercluster, 4, 5, 636, 637, 655
La Niña, 253
Laplace, Pierre-Simon, 172
Large Binocular Telescope (LBT), *146*
Large Hadron Collider, 160, 539, 619, 641
Large Magellanic Cloud (LMC)
    distance, 417, 478
    in Local Group, 571
    Milky Way Galaxy halo and, *565*
    orbital motion, 567
    Supernova 1987A, 478, *479*
    30 Doradus, 417
large-scale structure of the universe, 635–659
    bottom-up models of galaxy formation, 639, *646*, 647, 648, 650
    computer simulations, *162*, 556, 572, 645, 649, 652–653
    cosmological principle and, 637
    dark matter and formation of structure, 640–643
    dark matter in galaxy groups and clusters, 637–639
    distribution of regular and dark matter in early universe, 641, 654
    filaments, 637, 639, 652
    first galaxies, 645, 646–648, 654
    galaxy clusters, 636, 638, 651
    galaxy groups, 5, 570, 636–639
    gas between galaxies, 638, *639*
    gravitational instabilities and, 639, 641, 648
    gravity's role in formation of, 639–643, 655
    Great Attractor, 637, 655
    hierarchical clustering, 639, 646, 649–652
    Lambda-CDM (Lambda Cold Dark Matter) theory, 640–642, 649, 652
    large-scale structure, defined, 637
    mass of a cluster of galaxies, 638
    parallels between galaxy and star formation, 418, 420–422, 648–649
    predictions about the deep future, 653–654
    superclusters, 4–5, 636, 637, 655
    types of galaxy structures, 636
    voids, 636, 637, 639, 652
    walls, 637, 639, 652
Large Synoptic Survey Telescope (LSST), *146*, 149
Laser Interferometer Gravitational-Wave Observatory (LIGO), 162, 163, 484, 515–516, 518
La Silla Observatory (Chile), 134
LASP (Laboratory for Atmospheric and Space Physics), 257
latitude, defined, 25
law, defined, 8
law of conservation of angular momentum, 175–177, 186, 648
law of conservation of energy, 179
law of equal areas. *see under* Kepler's laws
LBT (Large Binocular Telescope), *146*
LBV (luminous blue variable) stars, 474, 478

LCROSS (Lunar Crater Observation and Sensing Satellite), 222
leap years, 45
Leavitt, Henrietta, 472
Leibniz, Gottfried, 71
Leiden Observatory, 16
Lemaître, Georges, 589
Leonid meteor shower, 340, *340*
Le Verrier, Urbain-Jean-Joseph, 267
life, 660–685
    anthropic principle, 626–627
    and atmospheric oxygen, 234, 239
    Cambrian explosion, 664, 673
    chemical processes, 667–669
    chemosynthesis, 313
    definitions, 662
    effect of supernovae and gamma ray bursts, 516–517
    effect on Earth's atmosphere, 234, 237–240, 254, 256
    elements in living organisms, 667, 669
    evolution, 666–667
    exponential growth, 666
    extremophiles, 313
    fate of life on Earth, 678–679
    fossils, 223, 251, 663, 664, *667*
    habitability of galaxies, 547
    habitable zones, 189, 254, 256, 285, 375, 455–456, 672–675
    hydrothermal vents and, 313, *662*, 663, 671
    increasing complexity, 664–665
    mutation, 666–667
    natural selection, 667
    origin of life on Earth, 662–664
    phylogenetic tree, 664, *665*
    radiation and, 547
    role in climate change, 251, 254, 256
    search for intelligent life, 675–678
    search for life in Solar System, 670–672
    search for life on Mars, 670–672
    "seeding" hypothesis, 663
    solar wind and, 403
    star formation and, 431
    stellar lifetimes and biological evolution, 460
    stellar nucleosynthesis and, 596–597, 626–627, 667, 669
    tidal forces and development of life, 101
    tilt of Earth's axis and, 51
    Urey-Miller experiment, 662
    when life began on Earth, 460, 663–664
    see also astrobiology
light, 108–143
    as electromagnetic wave, 111–112
    energy of, 111, 114, 116
    frequency (f), 113–114, 116
    as a particle, 114, 116
    prisms and spectra, *119*, 120–121, 128, 152
    visible spectrum, 113, *113*
    wavelength, 113–114, 116
    wave properties, 110
    see also electromagnetic radiation; electromagnetic spectrum; luminosity; speed of light

light-year, defined, 5, 111, 356
LIGO. *see* Laser Interferometer Gravitational-Wave Observatory
LIGO Scientific Collaboration (LSC), 518
limb darkening, 394
line of nodes, 50–51
Lippershey, Hans, 142
lithosphere, 210, 211. *see also* plate tectonics
lithospheric plates, 211–213, 215, 216. *see also* plate tectonics
Liu, Ying D., 404
LMC. *see* Large Magellanic Cloud
"Local Bubble" of gas from neighboring supernovae, 416, 490
Local Group
  Andromeda Galaxy in, 570–572
  defined, 4
  diameter, 5, 570–571
  dwarf galaxies in, 570–571, 647
  galaxies in, 570–571
  in Laniakea Supercluster, 4, 5, 636
  Large Magellanic Cloud in, 571
  mergers and collisions of galaxies, 571
  Milky Way Galaxy in, 4, 5, 570–572
  Sagittarius Dwarf Galaxy in, 571
  Small Magellanic Cloud in, 571
  tidal interactions between galaxies, 572
  Triangulum Galaxy in, 571, 572
logarithms, 14
logical fallacies, 21
longitude, defined, 25
longitudinal wave, 205
look-back time, 587, 593
Lorentz, Hendrik, 505
Lowell, Percival, 220, 670
low-mass stars
  defined, 440
  diameter, comparison to Sun, *184*
  RR Lyrae variables, 473, 559–560
  spectral types, 440
low-mass stars, evolution, 438–467
  asymptotic giant branch (AGB), 450–452, 454–456
  binary stars, 456–460
  changes in structure, 442–446
  effect of mass on main-sequence lifetime, 430–443, 646
  electron-degenerate matter in cores, 443–444, 447, 450, 451, 453
  evolutionary stages of a solar-type mainsequence star, *454*
  fate of planets around evolved stars, 443, 455–456, 460, 462, 678–679
  helium flash, 448–449
  helium fusion in core, 447–450, 451
  horizontal branch, 449–451, 454–455, 457, 473, 560
  hydrogen fusion, 440, 442, 446
  hydrogen shell fusion, 444–446
  leaving the main sequence, 443–446
  nebula formation, *438–439*
  planetary nebula, 451–453
  red giant branch, 444–447, *448*, 451
  stars in close binary system, 457
  subgiant branch, 445
  subgiant stars, 444–445
  triple-alpha process, 447–448, 451
LRO. see Lunar Reconnaissance Orbiter
LSC (LIGO Scientific Collaboration), 518
LSST (Large Synoptic Survey Telescope), *146*, 149
Luhmann, Janet G., 404
luminosity
  absolute magnitude, 358
  brightness and distance and, 130–133, 358–359
  comparison to Sun in solar units (L_{Sun}), 359
  defined, 128, 130
  distribution of, 359
  formula, 359, *374*, 429
  H-R diagram and, 369–371, *370*, 373–375
  mass-luminosity relationship for main-sequence stars, 371, 373, *374*, 441, 442
  period-luminosity relationship, 472–473, 527, 560
  protostars, 427, 429
  Stefan-Boltzmann law, 129, 132, 363–364, 394
  of the Sun, 384, 385, 456
  surface temperature and radius and, 364, 427, 429, 444, 473
  temperature and, 128
  see also light
luminosity class, 373–375
luminosity-temperature-radius relationship, 364, 427, 429, 444, 473
luminous blue variable (LBV) stars, 474, 478
luminous matter (normal matter), 536–539
Luna 3 (Soviet flyby mission), 157
Lunar Atmosphere and Dust Environment Explorer (LADEE), 249
lunar calendars, 43, 45
Lunar Crater Observation and Sensing Satellite (LCROSS), 222
lunar eclipses, 48–50
Lunar Laser Ranging Experiment, 102
Lunar Prospector mission, 222
Lunar Reconnaissance Orbiter (LRO), 200, 217, 222
lunar tides, 95–97
lunisolar calendars, 44, 45

M13 (star cluster), 675
M55 (globular cluster), 486, 488
M87 (radio galaxy), 543, 544
M104 or Sombrero Galaxy (spiral galaxy), *530*
M109 (barred spiral galaxy), *557*
MaCHOs. *see* massive compact halo objects
Magellan I and II telescopes, *146*
Magellan mission to Venus, 215, 247
magma, 215, *216*, 217
magnetic fields, 224
  defined, 111
  dynamo theory and, 209
  Earth, 209–210, 243–244
  Europa, 298, 313
  flux tubes, 283, 297, 396, 400
  giant planets, 279–283
  Jupiter, 279
  Mars, 201, 210, 236
  Mercury, 210
  Milky Way Galaxy, 565–566
  Moon, 209–210
  Neptune, 279
  recorded in lava, 209, 211
  reversal of Sun's magnetic field, 400
  Saturn, 279
  solar activity and, 396–397
  solar wind and, 210, 234, 236, 281, 396, 403
  terrestrial planets, 209–210, 236
  Uranus, 279
  Venus, 210
magnetic force, 111
magnetospheres
  auroras, 244, *282*, 283
  defined, 209
  Earth, 209, 243–244
  giant planets, 279–282
  interaction with solar wind, 243, 244, 280–282, 396, 403
  Jupiter, 280, 281
  Mercury, 210
  Moon, 209–210
  Neptune, 280–281
  radiation belts, 244, 282
  radio waves from, 276, 281
  Saturn, 280, 281
  size, shape, and orientation, *209*, 280–282
  Uranus, 280, 281
magnitude, 358
main asteroid belt, 182, 322, 326, 327, *328*
main sequence
  defined, 371
  evolution of protostars and, 425–431
  H-R diagram and, 369–371, 373
  leaving the main sequence, 443–446, 471–472
  main-sequence turnoff, 487–488
  mass and location on main sequence, 371, 373, 374
main-sequence stars
  changes in structure, 442–443
  evolution before main sequence, 427–431
  hydrogen fusion, 440, 446
  main-sequence lifetime, defined, 441
  main-sequence lifetime, effect of mass on, 430–443, 645–646
  main-sequence lifetime, estimating, 441, 442
  mass-luminosity relationship for, 371, 373, *374*, 441, 442
  properties of, by spectral type, *371*
main-sequence turnoff, 487–488
Major, Jason, 286
major axis, 66
Makemake, 322, 325–326
mantle, 207
mantle convection, 211–213, 215–216
maria (singular, mare), 217
Mariana Trench, 213
Mariner 10 mission, 217
Mars
  atmosphere, 248–249, 257
  atmospheric composition, 233, 236
  atmospheric greenhouse effect, 236, 256
  atmospheric pressure, 236, 248, 256, 674
  atmospheric water vapor, 248
  canali ("channels"), 220, 670

carbon dioxide in atmosphere, 233, 236, 237, 248
carbon dioxide in polar caps, 221, 670
density, 270
different terrain of northern and southern hemispheres, 183
distance from Sun, 75
dust devils, 249
dust storms, 220, *247*, 248, 249
early giant impacts, 201
eccentricity (*e*) of orbit, 75
erosion, 219
evidence of liquid water in the past, 200, 220–221, 238, 249, 256, 671
Gale Crater, 221, 671
hematite, 220
impact craters, 200–201, 203
inequality of seasons, 75
lack of ozone, 248
magnetic field, 201, 210, 236
Martian meteorites on Earth, 158, 341
model of interior, *207*
Newton Crater, *196–197*
Nili Patera dune field, *219*
Olympus Mons, 217–218, 248
Phobos and Deimos, moons, 295
physical properties, *198*
retrograde motion, *60*
search for life on, 670–672
search for water, 220, 670–672
seasons, 248–249
secondary atmosphere formation, 182, 233–234
semimajor axis of orbit, *69*
shield volcanoes, *214*, 217
sidereal period, *69*
solar conjunction, 76
subsurface water ice, 221, 238, 256, 670–672
tectonic features, 214
temperatures, 248
tilt of axis, 250
Valles Marineris, 214
Viking landers, 249
volcanism, 217–218
water in polar ice caps, 221
winds, 219–220, 249
Mars Atmosphere and Volatile EvolutioN (*MAVEN*) mission, 236, 257, 671
Mars Express orbiter, 76, *214*, 220, 221
Mars Odyssey orbiter, 76, 220
Mars Reconnaissance Orbiter (*MRO*), 76, *196–197*, *219*, 220–221, 249, *295*
Mars spacecraft and exploration
  communications blackout during Mars solar conjunction, 76
  Curiosity rover, 76, 158, 220, *341*, 671
  Mars Atmosphere and Volatile EvolutioN (*MAVEN*) mission, 236, 257, 671
  Mars Express orbiter, 76, *214*, 220, 221
  Mars Odyssey orbiter, 76, 220
  Mars Reconnaissance Orbiter (*MRO*), 76, *196–197*, *219*, 220–221, 249, *295*
  Opportunity rover, 76, 220–221
  Phoenix lander, 221, 670–672
  Spirit rover, 220, *248*

Viking landers, 249, 670
Viking Orbiter, 214
mass
  distortion of spacetime, 506–512, 589, 613, 637–639
  Einstein's mass-energy equation, 386, 503, 543, 620
  inertia and, 73, 507
  measurement using Newton's laws, 93–95, 269
  vs. weight, 84–85
mass-energy equation, 386, 503, 543, 620
massive compact halo objects (MaCHOs), 538–539, 638, 641
massive elements, 362
mass-luminosity relationship, 371, 373, *374*, 441, 442
mass transfer, 456–457
Mather, John, 163, 595
matter, 116
matter density parameter (Ω_m), 607–608, 611–614, 616
Mauna Kea Observatories (MKO), *140–141*, 147, 149, 155
Maunder Minimum, 400
MAVEN mission. *see Mars Atmosphere and Volatile EvolutioN* mission
Maxwell, James Clerk, 111–113, *115*, 308
Maxwell's equations, 111–112, 500
medium, defined, 112
medium-mass stars
  CNO cycle, 472
  defined, 440, 470, 472
  evolution, 472
  helium fusion, 472
  spectral types, 440
Mercury
  core, 210
  erosion, 219
  evidence of water ice in polar regions, 222
  exosphere, 249
  giant impact, 201
  impact craters, *182*, 183, 200, 201, 203
  lack of atmosphere, 219, 233, 256, 670
  magnetic field, 210
  magnetosphere, 210
  Mariner 10 mission, 217
  Messenger mission, 203, 214, 217, 222, 249
  model of interior, *207*
  physical properties, *198*
  precession of orbit, 509
  semimajor axis of orbit, *69*
  sidereal period, *69*
  tectonism, *69*
  volcanism, 217
mercury (element), emission spectrum, *122*
meridian, 25
messages from Earth to space, 675
Messenger mission, 203, 214, 217, 222, 249
Messier, Charles, 526
Meteor Crater in Arizona, 199–200
meteorites, 338–342
  achondrites, *329*, 340, 341
  ALH84001, 341
  from asteroids, 328, 338–341
  carbonaceous chondrites, *329*, 340, 345

Chelyabinsk meteoroid, 343–344
chondrites, *329*, 340
chondrules in, 340
from comets, 339
defined, 173, 199
Earth's atmosphere and, 199–200, 339–340
information about early Solar System, 173, *174*, 205, 341
iridium in, 223
iron meteorites, *329*, 341, *341*
Martian meteorites on Earth, 158, 341
radiometric dating, 203, 455
stony-iron meteorites, *329*, 341, *341*
stony meteorites, 340
meteoroids, 199, 200, 339–341, 343–344
meteors, 199, 338–340
meteor showers, 339–340
methane
  from biological activity, 251, 252
  from comets, 233, 313, 662
  and Earth's temperature, 251, 252
  on Enceladus, 313
  as greenhouse gas, 236, 251
  photodissociation by ultraviolet radiation, 233, 300
  on Titan, 300–301, 671
microlensing, 186, 188
micrometeoroids, 200, 219
micron (μm), 114
microwave radiation
  atmospheric window, 153
  in electromagnetic spectrum, 114–116
  see also cosmic microwave background radiation
microwave telescopes, 162–163
Mid-Atlantic Ridge, 211, 216
Milankovic, Milutin, 250
Milankovitch cycles, 250, 252
Milky Way Galaxy, 574
  age and chemical composition of stars, 560–566
  all-sky images, *412*, 413, *415*, *420*
  approaching Andromeda Galaxy, 572, 585
  barred spiral classification, 528, 557
  bulge, 557, 560, *564*, 568, *649*, 650
  chemical evolution of stars, 560–566
  collision with Andromeda Galaxy in future, 572, 653, 678
  computer simulation of formation, *649*, *649*
  cosmic rays, 565–566
  dark matter, 561, 566–570
  dark matter halo, 539
  diameter of, 5
  disk of, *4*, 564
  in Earth's night sky, *554–555*, 556, 559
  formation of, 571, *649*, 650
  future of, 571–572
  galactic fountains, 563, *564*
  globular clusters, 559–560, 562, 563, 565, 571, 650
  habitable zones in, 573, 674
  halo, 560, 562–565
  H II regions, 563
  inner halo, *560*, *564*, 564–565
  interstellar dust blocking view of, *412*, 413
  in Local Group, 4, 5, 570–572
  location of Sun in, 4

Milky Way Galaxy (*continued*)
   magnetic fields, 565–566
   mass, 566–568
   outer halo, *560*, 564–565
   Perseus arm, 557
   rotation curve, 567, 569
   Sagittarius A* (Sgr A*), 568–569
   Scutum-Centaurus arm, 557
   size estimates, 526, 559–560
   spiral arms, 557–559
   spiral structure, 556–557
   star formation, 557–559, 562, 564
   Sun's location in, 560
   supermassive black hole in, 566, 568–570
   see also barred spiral galaxies
Miller, Stanley, 662
millibar (pressure), 240
Mimas, 183, *294*, 299, 303, 306
mini-Neptunes, 188, 284–285
Miranda, *294*, 303
MKO. *see* Mauna Kea Observatories
MMT (Multiple Mirror Telescope), *146*
molecular-cloud cores, 421–422
molecular clouds
   accretion disk formation, 176
   collapse of, 420–422
   dust in, 418
   giant molecular clouds, *410–411*, 418
   molecular-cloud cores, 421–422
   properties and composition, 418, 420
   self-gravity in, 420–421
   star formation, *410–411*, 412, 418, 420–422
   water detected in W3 IRS5 star-forming complex, 431
molecules, 118
Molyneux, Samuel, 498
month, 42
Moon
   age, 203–204
   apparent magnitude of full Moon, 358
   Cabeus Crater, 222
   core, 207, 210
   diameter, 4, *5*, 356
   distance from Earth, 4, *5*
   erosion, 219
   evidence of water ice at poles, 222
   exosphere, 249
   far side, 40, 43, 157, 165
   fault valleys, 214
   giant impact theory of origin, 183, 201, 202, 296
   impact craters, *198*, 200, 203
   lack of atmosphere, 219, 234, 256, 670
   lunar calendars and, 43, 45
   lunar eclipses, 48–49
   lunar tides and, 95–97
   magnetic field, 209–210, 224
   magnetosphere, 209–210
   maria (singular, mare), 217
   model of interior, *207*
   phases of, 40–43
   physical properties, *198*
   Rima Ariadaeus, *214*
   search for water, 220
   seismic data, 207
   sidereal period, 40, 42
   synchronous rotation, 43, 99, 100, 101
   synodic period, 42
   tidal locking, 99–100, 295
   tides and, 95–97
   volcanism, 216–217
   water from solar wind, 222
moons, 294–304
   defined, 182–183
   distribution, 294–295
   formation, 182–183
   formerly active moons, 303
   geological activity and water, 297–304
   geologically dead moons, 303–304
   irregular moons, 295, 301, 303
   major moons, images, *294*
   number in Solar System, 294
   orbits of, 295–296
   regular moons, 295–296
   shepherd moons, 306, 307, 309, 310, *311*
   use to compute mass of planets, 296
   see also specific moons and planets
Moon spacecraft and probes
   Chandrayaan-1 spacecraft, 210, 222
   Clementine mission, 221
   Kaguya mission, 222
   Luna 3 (Soviet flyby mission), 157
   Lunar Atmosphere and Dust Environment Explorer (LADEE), 249
   Lunar Crater Observation and Sensing Satellite (LCROSS), 222
   Lunar Prospector mission, 222
   Lunar Reconnaissance Orbiter (LRO), 200, 217, 222
   see also Apollo missions
Mould, Jeremy, 102
MRO. *see* Mars Reconnaissance Orbiter
M-type (metal) asteroids, 330, 341
Multiple Mirror Telescope (MMT), *146*
multiverses, 625–626
muon neutrino, 391
muons, 504
mutation, 666–667

nadir, 25
nanometer (nm), 113
National Aeronautics and Space Administration (NASA). *see specific spacecraft and missions*
National Solar Observatory (NSO), 52
natural selection, 667
navigation, stars used for, 30–31
NCP (north celestial pole), 25–31, 39
neap tides, 98
near-Earth asteroids, 328
near-Earth objects (NEOs), 328, 343, 344
NEAR Shoemaker spacecraft, 330
nebular hypothesis, 172–173, *174*
neon, emission spectrum, *122*
NEOs. *see* near-Earth objects
Neptune
   ammonia in atmosphere, 279
   atmosphere structure, *274*
   axial tilt, *266*, 270–271
   clouds, 270, 272, 273, 276
   core accretion–gas capture, 182
   density, 270
   diameter, 267
   discovery, 267, *268*
   distance from Sun, 5, 68
   Great Dark Spot, 272, 275
   helium, 279
   hydrogen, 279
   images, *266–267, 269, 272*
   interior, 279
   lightning, 275
   magnetic field, 279
   magnetosphere size and shape, 280–281
   mass of atmosphere, 182
   methane in atmosphere, 270, 272, 273, 275, 279
   physical properties of, *266*
   planet migration, 181, 285
   rings, 304, 307, 310–312
   semimajor axis of orbit, 68, *69*
   sidereal period, 68, *69*
   Triton, moon, 270, *294*, 295, 301–302
   troposphere, 273, 275
   winds and weather, 275, 276, 278
neutrinos
   carbon fusion, 476
   CNO cycle and, 470–471
   from collapsing core, *477*, 479
   defined, 160
   efforts to detect, 160, 390–391, 480
   electron neutrino, 391
   Homestake Mine experiment, 391, *392*
   muon neutrino, 391
   neutrino cooling, 476
   from proton-proton chain, 387, 388, 390
   release during core collapse, *477*, 479
   silicon fusion, 476
   solar neutrino problem, 391, *392*
   speed of light velocity, 388
   from Sun's core, 390–391
   tau neutrino, 391
   from Type II supernovae, 479
neutrons, 117
neutron stars, 480–484
   accretion disk in binary system, 484
   black hole formation, 512–513
   discovery of signals from, *482*
   escape velocity, 481
   general relativistic time dilation, 511–512
   gravitational redshift, 511–512
   gravity on, 481
   supernovae and, 480–484
   see also pulsars
New Horizons spacecraft, 286, 323, 332
new Moon, 40–42, 50, 95, 98
newton (N), 74
Newton, Isaac
   calculus invention, 71
   cannonball "thought experiments," 89–90, 92, 94
   development of classical physics, 11
   heliocentric theory and, 61
   law of gravity development, 85
   reflecting telescope, 145
Newton Crater, *196–197*
Newton's first law of motion, 72, *74*, 90
Newton's law of gravity, 85–86

Newton's laws
  general theory of relativity and, 508–509
  measuring mass using Newton's laws, 93–95, 269
  Newton's first law of motion, 72, *74*, 90
  Newton's law of gravity, 85–86
  Newton's second law of motion, 72–74, 84, 87
  Newton's third law of motion, 73–75, 85
  Newton's universal law of gravitation, 86–89, *91*, 95, 508–509
  orbit shape, predicted by Newton's laws, 89, 92–93
  special theory of relativity and, 499–503, 505, 509
  testable predictions, *268*
Newton's second law of motion, 72–74, 84, 87
Newton's third law of motion, 73–75, 85
Newton's universal law of gravitation, 86–89, *91*, 95, 508–509
NGC 891 (spiral galaxy), *556*
NGC 1132 (elliptical galaxy), *538*
NGC 1300 (barred spiral galaxy), *524–525*
NGC 1365 (spiral galaxy), *604–605*
NGC 2362 (star cluster), *488*
NGC 3198 (spiral galaxy), *537*
NGC 6240 (galaxy with two supermassive black holes), *650*
NGC 6530 (open cluster), *486*
nickel
  in cores of planetesimals, 329
  in cores of terrestrial planets, 15, 180, 207, 489
  in meteorites, 199, *200*, 341
  from supernova explosions, 459, 489
Nili Patera dune field on Mars, *219*
nitrogen in Earth's atmosphere, 236, 238
normal matter (luminous matter), 536–539
north celestial pole (NCP), 25–31, 39
Northern Hemisphere
  constellations, 34
  Coriolis effect in, 32
  seasons in, *12*, 35–36
  star maps in each season, *12*
  tilt of Earth's axis and, 35
  using the stars for navigation, 30–31
  view of the sky from, *12*, 28, 30
North magnetic pole, 209
North Pole, 24, *25*, 26–28, 36, 38–39
novae, 457–458
NSO (National Solar Observatory), 52
nuclear fusion
  carbon fusion, 461, 472, 474, 476
  defined, 385
  deuterium fusion in protostars, 422–423
  properties, 388, 641–643
  proton-proton chain, 387–388, 390, 442, 470, 471, 619
  in the Sun, 385–387, 390
  see also helium fusion; hydrogen fusion
nucleic acids, 662. *see also* DNA; RNA
nucleosynthesis
  Big Bang nucleosynthesis, 547, 596–597, 623, 640–641, 645, 654
  cosmic ray production, 565
  life and stellar nucleosynthesis, 596–597, 626–627, 667, 669
  stellar nucleosynthesis and the interstellar medium, *562*
  from supernovae, 479–480, 485, 489, 645–646, 669
  see also elements
nucleus, 116

Oberon, *294*, 303
objective lens, 143–144
observable universe, defined, 4, 589, 624
Occam's razor, 10, *482*
ocean floor rifts, 211, 212
Olympus Mons, 217–218
Oort, Jan, 331
Oort Cloud, 233, 332–333
opacity, 389–390, 393
open clusters, 485, *486*, 562
Opportunity rover, 76, 220–221
opposition, *62*, 63, 64
optical axis, 143, *144*
optical filters, 360
optical telescopes, 142–150
  adaptive optics, 148–149
  astronomical seeing, 148
  focal length, 143–145, 147
  focal plane, 143, 145, 148–149
  objective lens, 143–144
  observatory locations, *146*, 149–150
  optical and atmospheric limitations, 147–149
  optical axis, 143, *144*
  refracting telescopes, 142–145
  resolution, 147–149, 152
  use by Galileo, 70, 142
  world's largest optical telescopes, *146*
  see also specific telescopes
orbital resonances, 296, 299, 304–306, 327–328
orbiters, 157–158
  Clementine mission, 221
  Lunar Reconnaissance Orbiter (LRO), *200*, 217, 222
  Magellan mission to Venus, 215, 247
  Mars Express orbiter, 76, *214*, 220, 221
  Mars Odyssey orbiter, 76, 220
  Mars Reconnaissance Orbiter (MRO), 76, *196–197*, *219*, 220–221, 249, *295*
  Messenger mission to Mercury, 203, 214, 217, 222, 249
  Venus Express (European Space Agency), 215, 218, 238, *247*, 248
  Viking Orbiter, 214
orbiting observatories, 155–157
orbits
  binary stars, 365–367
  bound orbits, 93
  circular velocity (v_{circ}), 89, 90, 94, 95
  defined, 90
  free fall, 90
  gravity and, 89–90
  Kepler's first law, 65–66, 71
  Kepler's second law (law of equal areas), 66, 68, 93
  Kepler's third law (harmonic law), 68, *69*, 70, 94, 205
  of moons, 295–296
  Newton's cannonball "thought experiments," 89–90, 92, 94
  precession of elliptical orbits, 509
  shape, predicted by Newton's laws, 89, 92–93
  unbound orbits, 93
  uniform circular motion and, 90
  see also planetary motion
organic compounds, defined, 180, 662
Orion (constellation), *352–353*, 364, 431
Orion Nebula, *352–353*, 417
oxygen
  amount in Earth's atmosphere, *234*, 238
  comparison to hydrogen, 232, 235, 238
  life and atmospheric oxygen, 234, 239
  reaction with rocks and minerals, 219, 238, 239
ozone, 239–240, 242, 416, 517, 664–665

Pacini, Franco, *482*
pair production, 620–621
paleoclimatology, 251
parallax
  defined, 356, 376
  distance to nearest stars, 354–357, 359, 364, 370, *374*, 376
  spatial scanning technique, 376
  spectroscopic parallax, 371, 373, *374*
parent element, 203, 204
Parkes radio telescope (Australia), *154*
parsec (pc), 356, 526
partial lunar eclipse, 49
partial solar eclipse, 46, 47, 64
particle accelerators
  Large Hadron Collider, 160, 539, 619, 641
  mass and energy of particles, 503, 566, 622, 624
  overview, 160
  pair production, 620
  quark-gluon plasma, 621
  relativistic effects, 502, 504, 505
particle physics
  alpha particles, 447
  antimatter, 388, 619, 621, 623
  antiparticles, 514, 619–621, 623
  axion, 641
  bubble chambers, 633
  and cosmology, 618–624
  defined, 618
  electromagnetic force, 618, 619, 621, 623
  electroweak theory, 619
  grand unified theories (GUTs), 621–624
  Higgs boson, 619
  pair production, 620–621
  photino, 641–642
  pions, 505
  positrons, 387, 388, 470, *483*, 619–620
  quantum chromodynamics (QCD), 619
  quantum electrodynamics (QED), 518–519
  quarks, 619, 621
  standard model of particle physics, 618, 619, 621
  strong nuclear force, 385–387, 447, 478, 518–519, 626–627
  theory of everything (TOE), 619, 622, 624–626
  virtual particles, 514, 618, *619*
  weak nuclear force, 618, 619, 621, 623
  see also neutrinos
Payne-Gaposchkin, Cecilia, 372
peculiar velocity, 637
penumbra, 46, 49, 398
penumbral lunar eclipse, 49

Penzias, Arno, 163, 594, *594*, 595, 598
perihelion, defined, 326
periodic table, *123*
period-luminosity relationship, 472–473, 527, 560
Perlmutter, Saul, 609
Perseid meteor shower, 340
phases of the Moon, 40–43
    first quarter Moon, 41, 95
    full Moon, 40, 41, *42*, 95, 98
    new Moon, 40–42, 50, 95, 98
    third quarter Moon, 42, 95
    waning crescent Moon, *41*, 42
    waning gibbous Moon, *41*, 42
    waxing crescent Moon, 40–41
    waxing gibbous Moon, 41
Phobos, 295
Phoenix lander, 221, 670–672
photino, 641–642
photodissociation, 300, 313
photoelectric effect, 114
photography in astronomy, 151–152
photons
    defined, 114
    energy of, 114, 117, 620
    virtual photons, 618, *619*
photosphere, 394–396, 398, 401, 423, 427
photosynthesis, 239, 251, 517, 664
phylogenetic tree, 664, *665*
phytoplankton, 251, 517, 664
Piazzi, Giuseppe, 323, 326
Pillan Patera on Io, *297*
Pinwheel Galaxy, *533*, 534
Pioneer spacecraft, 248, 282, 675
pions, 505
pixels, 151–152, 160
Planck, Max, 129
Planck era, 622, *623*
Planck's constant (*h*), 114, 116, 620
Planck space observatory
    all-sky image of carbon monoxide distribution, 420
    cosmic microwave background radiation, 163, *580–581*, 596, 596–597, 639
    Hubble constant measurement, 535
    launch date, 156
Planck spectra (blackbody spectra), 129, 163, 359, 360, 514
planetary moons. *see* moons
planetary motion, 58–81
    conjunction, 62–63
    geocentric model, 60–62, 70, 71
    greatest elongation, 63, 64
    heliocentric theory, 61–62, 64, 65, *67*, 68, 71
    inferior conjunction, *62*, 63
    opposition, *62*, 63, 64
    prograde motion, 60
    retrograde motion, 60–63
    sidereal period, 63, 64, 68
    superior conjunction, *62*, 63
    synodic period, 63, 64, 68
    as viewed from above North Pole, *25*
    see also Kepler's laws; orbits
planetary nebula, 451–453

planetary spacecraft, 157–159, 220
    atmospheric probes, 158, 273, 300–301, 671
    flyby missions, 157–158
    landers, 158, 219
    rovers, 76, 158, 220–221
    sample returns, 159
    see also orbiters; *specific missions*
planetary system formation, 170–195
    accretion disks, *176*, 176–179, 181–183
    chaotic encounters, 180
    disk formation in nebulas, 172, 173, *174*
    evidence from planetary scientists, 173
    formation around stars, 172–175
    formation of large objects, 177–178
    nebular hypothesis, 172–173, *174*
    planetary system, defined, 172
    planet migration, 179–180, 188, 284, 285, 455, 678
    protoplanetary disks, 175, 177–178, 180–183, 188
    protostars, 175–179, 182
    self-gravity and, 172–173, 176
    see also extrasolar planets; planet formation
planetesimals
    comet nuclei, 183, 327, 331
    composition, 180
    defined, 178
    fate in early Solar System, 328
    formation, 178, 182
    growth to planet-sized bodies, 182, 322
    heavy bombardment in early Solar System, 201
    remaining planetesimals in Solar System, 182–183, 320
    see also asteroids; comets; dwarf planets
planet formation
    in accretion disk, 177–178
    atmosphere formation, 181
    compositions of planets, 179–180
    core accretion–gas capture, 182–183
    disk instability and, 182
    evolution of planetary interiors, 207–208
    giant planets, 182–183
    terrestrial planets, 181–182
    see also planetary system formation
planet migration, 179–180, 188, 284, 285, 455, 678
planets
    compositions of, 179–180
    defined, 172
    fate of planets around evolved stars, 443, 455–456, 460, 462
    visible to naked eye, 44, 60
    see also dwarf planets; extrasolar planets; giant planets; planet formation; terrestrial planets; *individual planets*
plasma, defined, 282, 400–401
plate tectonics, 210–215
    and climate, 250–251
    faults, *212*, 213
    hot spots, 213, 215–217
    lithospheric plates, 211–213, 215, 216
    magnetic record and, 211
    mantle convection, 211–213, 215–216
    ocean floor rifts, 211, 212

    on other planets, 214–215
    radiometric dating evidence, 211
    spreading centers, 211, *212*, 213, 216
    subduction zones, 212–213
    tectonism, defined, 210
    tectonism on other planets, 214–215
    theory of, 211
Pleiades or "Seven Sisters," 430
Pluto
    carbon monoxide ice, 323
    characteristics, 322–323
    Charon, moon, 94, 100, 270, *294*, 322–323
    classification as a dwarf planet, 183, 322–324
    discovery, 322
    eccentricity (*e*) of orbit, 66
    moons, 294, 296, 322
    New Horizons images, *322*
    orbit, *69*, 322
    orbital resonance of moons, 296
    as planetesimal, 183
    sidereal period, *69*, 322
    tidal locking with Charon, 100, 296, 322
Polaris, the "North Star," 30, 35, 39
Porter, F. Scott, 490
positrons
    antiparticle of electrons, 388, 619, 620
    around neutron stars, *483*
    from CNO cycle, 470
    defined, 387
    from proton-proton chain, 387, 388, 619
post hoc ergo propter hoc, 21
precession of the equinoxes, 39
primary atmospheres, 180–182, 232–233
primary waves (P waves), 205, 206
principles, scientific, 8
prisms and spectra, *119*, 120–121, 128, 152
Process of Science
    active galactic nuclei (AGNs), *545*
    agreement between fields, *115*
    astrobiology, *668*
    causes of seasonal change, *37*
    convergence of evidence, *174*
    cosmological constant, *610*
    dark matter, evidence for, *644*
    data as ultimate authority, *586*
    falsified hypothesis, *202*
    falsified theories, *67*
    finding the common thread, *545*
    following up on the unexpected, *311*
    general theory of relativity, *510*
    gravitational waves, *161*
    heliocentric theory of Copernicus, *67*
    interconnections between branches of science, *419*, *668*
    learning from failure, *392*
    multiple streams of evidence, *644*
    nebular hypothesis, *174*
    new science encompassing old, *510*
    Occam's razor, *482*
    orbits of particles in Saturn's rings, *311*
    revisiting old ideas in the light of new data, *610*
    scientific method, overview, 7–8, *9*
    solar neutrino problem, *392*

speed of light, 115
symbiosis of science and technology, 161
testable predictions of scientific laws, 268
theories must fit facts, 37
theories of origin of Moon, 202
thinking about complex issues, 255
21-cm radiation of neutral hydrogen, 419
universality of Newton's law of gravitation, 91
see also science; scientific method
prograde motion, 60
prokaryotes, 664, 665
Prometheus Patera on Io, 297, 319
prominences, 400–401
proportionality, 13
proton-proton chain, 387–388, 390, 442, 470, 471, 619
protons, 117
protoplanetary disks
    atmosphere formation from, 180–181, 232
    defined, 175
    dust in, 173, 177–178, 180, 182, 183
    extrasolar planet formation, 188, 284
    formation of large objects, 177–178
    giant planet formation, 264, 279
    Solar System formation, 175, 182–183
    temperature variations, 180
    terrestrial planet formation, 196
protostars
    beginning of hydrogen fusion, 425
    bipolar outflow, 177, 429–430
    classification, 430
    convection in, 423, 426, 429
    defined, 175
    deuterium fusion, 422–423
    evolution, 423–426, 427–431
    formation, 175–177, 176, 422–423
    Hayashi track, 427, 427–430
    Herbig-Haro objects (HH objects), 429
    H⁻ ion and stellar temperature, 427–428
    influence of mass, 427, 430
    luminosity, surface temperature, and radius of, 427, 429
    main sequence and, 425, 427–431
    in minihalos in early universe, 645
    photosphere, 423, 427
    protostellar wind, 430
    proto-Sun, 175–176, 180, 182
    radiated energy, 178–179, 423
    thermal energy from gravitational collapse, 422–423, 425
    T Tauri stars, 430
    see also star formation; stars
Proxima Centauri, 6, 126, 332, 357, 505
Ptolemy (Claudius Ptolemaeus), 34, 39, 60–61, 65
puffy Jupiters, 284
pulsars, 468–469, 482, 483, 484, 485. *see also* neutron stars
pulsating variable stars, 472

QED (quantum electrodynamics), 518–519
quadrature, 62, 63, 64
quantum chromodynamics (QCD), 619
quantum efficiency, 151
quantum electrodynamics (QED), 518–519
quantum mechanics
    defined, 12, 114
    in early universe, 615–616
    Einstein's problems with, 12
    quantum chromodynamics (QCD), 619
    quantum electrodynamics (QED), 518–519
    uncertainty principle, 615
Quaoar (Kuiper Belt object), 332
quarks, 619, 621
quasars (quasi-stellar radio sources)
    as active galactic nuclei, 540
    discovery of, 539–540
    in early universe, redshift of, 645, 646, 650
    effect of viewing angle, 545
    emission lines in spectra, 540
    size comparable to Solar System, 541–542
    3C 273, 540

radial velocity (v_r), 125, 126, 184–185, 187, 189
radial velocity method for extrasolar planets, 184–185, 187, 189
radiant, 340
radiation
    blocked by Earth's atmosphere, 153, 155–157
    habitability and, 547, 674
    Hawking radiation, 514
    solar radiation, 389–390, 393, 395, 401–403
    synchrotron radiation, 281–282, 485
    thermal radiation, 127, 128, 130, 133, 414–415
    21-cm radiation of neutral hydrogen, 417–418, 556–557, 559, 567
    see also cosmic microwave background radiation; electromagnetic radiation
radiation belts, 244, 282
radiative zone of the Sun, 388, 389–390
radio galaxies, 540, 545
radioisotopes, 203–204
radiometric dating, 203–204, 205, 209, 211
radio telescopes, 144–145
radio waves
    atmospheric window, 153
    penetration of interstellar dust, 413, 417, 420
    from planetary magnetospheres, 276, 281
    from pulsars, 481
    from radio galaxies, 540
    reflection by ionosphere, 242
    wavelengths, 114, 154
rainbows, 108–109, 112
ratios, 13
reading a graph, 14
Reber, Grote, 154
recombination of the universe, 595, 595, 640, 643
reddening, 414
red dwarf stars, 443, 488, 489
red giant branch, 444–447, 448, 451
red giant stars
    changes in structure, 444–446
    defined, 445
    escape velocity (v_{esc}), 451
    helium flash, 448–449
    helium fusion in core, 447–450
    hydrogen shell fusion, 444, 445, 446
    stellar winds and interstellar dust formation, 412
    structure, 445
    Sun, in the future, 678
redshift
    age of universe and, 640
    cosmological redshift, 593, 647
    distance measurement, 585, 637
    Doppler shift, 125, 593
    due to changing scale factor, 591–592
    general theory of relativity, 511–512, 514, 593
    gravitational redshift, 511–512, 514, 593, 596, 641
    observing high-redshift objects, 647
    redshift surveys of galaxies, 637
red supergiant stars, 471, 474
reflecting telescopes
    adaptive optics, 148–149
    concave mirrors, 145
    defined, 142
    focal length, 145, 147
    focal plane, 145, 148–149
    Isaac Newton's, 145
    secondary mirrors, 145
reflection, 145, 145
refracting telescopes
    compound lenses, 145
    defined, 142
    focal length, 143–144
    focal plane, 143
    objective lens, 143–144
    shortcomings of, 143–144
refraction, 142, 143
refractory materials, 180
regular moons, 295–296
reionization, 643
relative motion
    aberration of starlight and, 498
    defined, 31
    and frame of reference, 31–33, 38, 61, 72, 498
    relative speeds close to speed of light, 499–500
relativistic beaming, 544
relativistic effects, defined, 502
relativistic speeds, 500, 502, 542
relativity, 496–523
    Galilean relativity, 499, 500
    implications of, 502–506
    relative speeds close to speed of light, 499–500
    relativistic speeds, 500, 502, 542
    see also general theory of relativity; special theory of relativity
Reltze, David, 518
resolution, 142, 147–149, 152, 154–155
rest wavelength (λ_{rest}), 125, 126
retrograde motion, 60–63, 295
retrograde rotation of Venus, 201
revolve, defined, 33
Rhea, 294
ribonucleic acid. *see* RNA
Riess, Adam, 376, 609
right ascension, 25
Rima Ariadaeus, 214
ring arcs, 312
Ring Nebula, 438–439, 453

rings, planetary, 304–314
    composition of ring material, 307
    defined, 304
    discovery of, 304–305
    formation and evolution, 306–307
    orbits of ring particles, 304–306
    ring arcs, 312
    Roche limit, 306–307
    shepherd moons, 306, 307, 309, 310, *311*
    size and complexity around giant planets, *305*, 308–314
    tidal forces and, 101
    see also specific planets and rings
RNA (ribonucleic acid), 313, 662, 663, 674
Roche limit, 100–101, 306–307, 455
Roche lobes, 456–457, 484
rocks
    iron in, 219, 220, 238, 248, 517
    lunar, 159, 203–204, *205*, 210, *211*
    oxygen reaction with, 219, 238, 239
    radiometric dating of, 203–204, *205*, 209, 211
Rømer, Ole, 110, *115*
Rosetta spacecraft, 338, 346
rotation curves, 537–538, 567, 569, 638
rovers
    Curiosity rover, 76, 158, 220, *341*, 671
    Opportunity rover, 76, 220–221
    Spirit rover, 220, *248*
RR Lyrae variables, 473, 559–560
Rubin, Vera, 537, 561
Running Chicken Nebula (Lambda Centauri Nebula), 134
Russell, Henry Norris, 369, 372

S0 galaxies
    bulges, 533
    color, 530, 547
    in Coma Cluster, 636
    defined, 528
    distance, 533, 650
    heavy elements in, 547
    mass, 531
Sagittarius A* (Sgr A*), 568–569
Sagittarius Dwarf Galaxy, 571
Saha, Meghnad, 372
SALT (South African Large Telescope), *146*
San Andreas Fault, 213
satellite, 92
Saturn
    ammonia in atmosphere, 273
    atmosphere structure, *274*, 278–279
    auroras, *282*
    axial tilt, *266*, 270–271
    clouds and storms, *264–265*, 269, 272, 273
    core accretion–gas capture, 182
    density, 270
    diameter, 4–5, 266–267
    formation, 182, 279
    heavy elements, 270
    helium, 269, 278, 279
    hydrogen, 269, 278–279
    images, *264–265*, 270, *282*, *292–293*
    interior, 278–279
    internal thermal energy, 278
    lightning, 272, 275–276
    magnetic field, 279
    magnetosphere, 280, 281
    mass, 279
    methane in atmosphere, 273
    planet migration, 285
    winds and weather, *264–265*, 272, 275, 276
Saturn's moons, 314
    Dione, *294*, 299, 302
    Herschel crater on Mimas, 303
    Hyperion, 295, 303, 304
    Iapetus, *294*, 303, 309
    major moons, images, *294*
    Mimas, 183, *294*, 299, 303, 306
    Oberon, *294*, 303
    orbital resonances, 296, 299, 304–306
    Pandora, 309
    Prometheus, 309
    Rhea, *294*
    shepherd moons, 309, 310, *311*
    Tethys, *294*, 302
    see also Enceladus; Titan
Saturn's rings
    Cassini Division, *305*, 306, 308–309
    composition of, 307, 308
    discovery of, 304
    Encke Gap, *305*, *306*, 310
    formation and evolution, 308
    giant dust ring around Saturn, 309
    magnetosphere interaction with, 281
    orbits of particles, 304–306, *311*
    shepherd moons, 309, 310, *311*
    size, structure, and complexity, *292–293*, *305*, *306*, 308–312
    spokes in Saturn's B Ring, *Voyager 1*, 310
scale factor (R_U)
    and age of universe, 612
    cosmological constant and, *611*, 612
    defined, 590
    expanding universe and, 590–591, 592, *611*
    Planck blackbody spectrum, 592
    redshift and, 591–592
Schiaparelli, Giovanni, 220, 670
Schmidt, Brian, 102, 609
Schmidt, Maarten, 539
Schwarzschild, Karl, 11, 513
Schwarzschild radius, 513, 543
science
    arts compared with, 11
    as creative activity, 11
    mathematics as language of, 13
    patterns and, 12–13
    as way of viewing the universe, 7–12
    see also Process of Science
scientific method, 7–11
    defined, 7
    facts, 8
    falsified hypothesis, 7–8, *9*, 34–35, *37*, 202
    hypothesis, defined, 7
    ideas, 8
    laws, 8
    repeatable observations, 8
    reproducible observations, 7
    testable predictions, 8, *9*, 11, 12, *268*, 625
    theoretical models, 8, 10–11
    see also Process of Science; theory
scientific notation, 13
scientific principles, defined, 8
scientific revolutions, 11–12, 61–62
Scott, David, 97
SCP (south celestial pole), 25–28, 30
SDO. *see* Solar Dynamics Observatory
SDSS. *see* Sloan Digital Sky Survey
Search for Extraterrestrial Intelligence (SETI), 677
seasons
    amount of sunlight and, 36, *37*
    changes in the landscape, *12*, 22–23
    changing patterns in the sky, 12–13
    distance from sun and, 34–35, *37*, 75
    Earth's poles, 36, 38–39
    length of days and, 36
    near equator, 36
    seasonal changes in sky, 12–13
    tilt of Earth's axis and, 34–36, *37*, 51
secondary atmosphere
    evolution, 234–238
    formation, 180–182, *232*, 233–234, 662
secondary craters, *198*, 199, 200
secondary waves (S waves), 205–206
Sedna, in Oort Cloud, 332
Segue 1 (ultrafaint dwarf galaxy), 647
seismic waves, 205–206, 212
seismometers, 206, 207
selection effects, 187
semimajor axis, 66, 68, 89, 95, 185
SETI (Search for Extraterrestrial Intelligence), 677
Seyfert, Carl, 540
Seyfert galaxies, 540, *545*
Shapley, Harlow, 526, 559–560, 579
shepherd moons, 306, 307, 309, 310, *311*
shield volcanoes, *214*, 216–218
Shoemaker, David, 518
short gamma-ray bursts, defined, 484
sidereal day, 44
sidereal period, 42, 63, 64, 68
significant digits, 13
silicates, defined, 179
singularity, 512
Sirius (Alpha Canis Majoris), 34, 44, 358, 454
SKA (Square Kilometre Array), 154–155
Slipher, Vesto, 534
slippery slope, 21
Sloan Digital Sky Survey (SDSS), 628, 637, 653
Sloan Great Wall, 637
slope, 14
Small Magellanic Cloud (SMC), 565, 571
Smoot, George, 163, 595
"Snowball Earth" hypothesis, 664
sodium emission spectrum, *122*
SOFIA (Stratospheric Observatory for Infrared Astronomy), 155, *431*
SOHO (Solar and Heliospheric Observatory), *401*, 402

solar abundance, 123
Solar and Heliospheric Observatory (*SOHO*), *401*, 402
solar day, 44
Solar Dynamics Observatory (SDO), *156*, *382–383*, 398, *401*, 402
solar eclipses, 46–48
solar flares, 396, 400–401, 403
solar maxima, *399*, 400, *402*
solar neutrino problem, 391, *392*
solar prominences, 400–401
solar radiation, 389–390, 393, 395, 401–403
Solar System
　age, 203, 455
　compositions of planets, 179–180
　distances in, 5
　general description, 4, 172
　geocentric model, 60–62, 70, 71
　giant planets, formation of, 182–183
　heliocentric theory, 61–62, 64, 65, *67*, 68, 71
　location, 4
　scale of, 62–63, 172
　search for life in Solar System, 670–672
　terrestrial planets, formation of, 181–182
　volcanism in, 216–218
Solar System formation, 170, 172–183
　accretion disk, *176*, 176–179, 181–183
　cataclysmic collisions, 183
　disk formation in nebula, 172, 173, *174*
　evidence from planetary scientists, 173
　formation of large objects, 177–178
　nebular hypothesis, 172–173, *174*
　protoplanetary disk, 175, 177–178, 180–183
　protostar, 175–179, 182
Solar TErrestrial Relations Observatory (STEREO), 404
solar tides, 97–98
solar wind
　boundary with interstellar medium, 397, 403
　charge exchange between ions, 490
　and comets, 335, 337, 396
　from coronal holes, 396
　defined, 234
　and Earth's atmosphere, 242
　and Earth's magnetosphere, *243*, 244, 396, 403
　interactions with exospheres, 249
　and life, 403
　and loss of atmosphere, 234
　magnetic fields and, 210, 234, 236, 281, 396, 403
　magnetospheres of giant planets, 280–282
　and Moon's water, 222
　particle speeds, 396
　and planetary magnetic fields, 210, 234, 236, 281
　structure and properties, 397
solstices, *35*, 36, 38–39
Sombrero Galaxy (spiral galaxy), *530*
South African Large Telescope (SALT), *146*
south celestial pole (SCP), 25–28, 30
Southern Hemisphere
　constellations, 34
　Coriolis effect in, 33
　seasons in, *35*, 36
　tilt of Earth's axis and, *35*, 36
　using the stars for navigation, 30–31
　view of the sky from, 28
South magnetic pole, 209
South Pole, 25–28, 36, 38–39
South Pole Telescope (Antarctica), 163
space observatories, 155–157
spacetime
　defined, 502
　distortion by mass, 496, 506, 508, 509, 589, 613, 637–639
　expanding universe shape in, 613–614
　gravity as distortion of, 506–512, 589
　special theory of relativity, 502, 506
Space Very Long Baseline Interferometer (SVLBI), 154
spatial scanning technique, 376
special theory of relativity, 496, 500–506
　event, defined, 500
　implications of, 502–506
　inertial reference frames, 499, 500, 503, 507
　length contraction, 504
　Lorentz factor (γ), 504, *504*, 505
　mass-energy equation, 386, 503, 543, 620
　and Newton's laws of motion, 499–503, 505, 509
　scientific method and, 8
　spacetime, 502, 506
　space travel and, 505–506, 678
　speed of light as ultimate speed limit, 503, 506, 593
　speeds close to speed of light, 496, 499–500
　time and relativity, 500–506
　time dilation, 502–506, 678
　twin paradox, 504
spectral types of stars, 360–362, *371*
spectrographs, 152–153
spectrometers, 152
spectroscopic binary stars, 367–368
spectroscopic parallax, 371, 373, *374*
spectroscopy
　absorption spectra, 120–121, *361*, 540
　atmospheric windows, 153
　continuous spectrum, 128, 153
　defined, 152
　emission spectra, 118–119
　spectral fingerprints of atoms, 121–123
　spectral types of stars, 360–362
　spectrographs, 152–153
　spectrometers, 152
　21-cm radiation of neutral hydrogen, 417–418, *419*
　see also absorption lines; Doppler shift; emission lines
spectrum, 113
speed of light (*c*)
　change in a medium, 143
　constant value for all observers, *115*, 499–500, 502, 506
　in Einstein's mass-energy equation, 386
　measurement of, 110–111, *115*
　as ultimate speed limit, 503, 506, 593
　value, 5, 111, 499–500
　wavelength and frequency relationship, 112, 116
spin, of subatomic particles, 417
spiral density waves, 558–559

spiral galaxies
　bulges, 528, 539
　characteristics, 528, *528*
　color, 530
　dust in, 530
　Seyfert galaxies, 540, *545*
　spiral density waves, 558–559
　star formation in, 530, 547, 557–559
　stellar motions, 539
　Tully-Fisher relation and distance estimates, 533
　see also barred spiral galaxies
Spirit rover, 220, *248*
Spitzer Space Telescope
　description, *156*
　giant dust ring around Saturn, 309
　Hubble constant measurement, 535
　image, *153*
　jets from newly formed stars, *429*
　Milky Way Galaxy structure, 557, 568
　observing high-redshift objects, 647
　planetary nebulae, *453*
　solar orbit, 155–157
　water detected in W3 IRS5 star-forming complex, *431*
spreading centers, 211, *212*, 213, 216
spring balances, 424
spring tides, 98
Sputnik 1, 92
Square Kilometre Array (SKA), 154–155
stable equilibrium, 127
standard candles, 102, *526*, 532–534, 585
standard model of particle physics, 618, 619, 621
star clusters
　age, color, and chemical compositions, 488–489, 562, 563
　cluster ages and H-R diagrams, 485–488
　defined, 430
　M55 (globular cluster), 486, 488
　main-sequence turnoff, 487–488
　NGC 6530 (open cluster), 486
　open clusters, 485, *486*, 562
　Pleiades or "Seven Sisters," 430
　stellar evolution in, 485–489
　stellar population, 488
　30 Doradus, 417
　see also globular clusters
stardust, in human body, 15, 24
Stardust robotic probe, 337, 432
star formation
　Dark Ages, 643, *653*, 654
　first stars, 643–649, 654
　and galaxy formation, parallels between, 418, 420–422, 648–649
　giant molecular clouds, *410–411*
　influence of mass, 430
　and life, 431
　Milky Way Galaxy, 557–559, 562, 564
　molecular clouds, *410–411*
　rate of, changes over time, 650
　red glow of nebulas in star-forming regions, 134
　in spiral galaxies, 530, 547, 557–559
　time scale, 410
　see also protostars

stars
- apparent daily motion, 27–31
- classification by surface temperature, 360–362
- color index of, 360, *374*
- defined, 183
- first stars, 643–649, 654
- H⁻ ion and stellar temperature, 427–428
- measuring mass in binary systems, 367–368, *374*, 515
- number visible to naked eye, 34
- spectra and chemical composition, 362, *374*
- spectral types of stars, 360–362
- Stefan-Boltzmann law and size of stars, 363–364, *374*, 427
- very low-mass stars, 440
- *see also* star formation; stellar evolution

static equilibrium, 126–127
Stefan, Josef, 129
Stefan-Boltzmann constant (σ), 129, 398
Stefan-Boltzmann law
- defined, 129, 364
- energy and temperature, 279, 363–364, 427
- luminosity and temperature, 129, 132, 363–364, 394
- size of stars, 363–364, *374*, 427

stellar evolution
- binary stars, 456–460
- defined, 440
- effect of mass on main-sequence lifetime, 430–443, 645–646
- electron-degenerate matter in cores, 443–444, 447, 450, 451, 453
- evolutionary track of protostars, 427–428
- fate of planets around evolved stars, 443, 455–456, 460, 462
- Hayashi track, *427*, 427–430
- influence of mass on protostars, 427, 430
- leaving the main sequence, 443–446, 471–472
- before main sequence, 425, 427–431
- observation and modeling, 440
- protostars, 423–426, 427–431
- in star clusters, 485–489
- stellar lifetimes and biological evolution, 460
- *see also* high-mass stars, evolution; low-mass stars, evolution

stellar-mass loss, 450–451, 453
stellar occultation, 267
stellar population, 488
STEREO (Solar TErrestrial Relations Observatory), 404
stereoscopic vision, 354, *355*, 359
Stonehenge, *24*, 45
stony-iron meteorites, *329*, 341, *341*
stony meteorites, 340
stratosphere, 242
Stratospheric Observatory for Infrared Astronomy (SOFIA), 155, *431*
stromatolites, 663
strong nuclear force, 385–387, 447, 478, 518–519, 626–627
structure of the universe. *see* large-scale structure of the universe
S-type (stony) asteroids, 328–329, 343
Subaru Telescope, *146*

subduction zones, 212–213
subgiant branch, 445
subgiant stars, 444–445
sublimation, 335
summer solstice, *35*, 36, 38, 39
Sun, 382–410
- astronomer's periodic table of the elements, *123*
- changes in structure and chemical composition, 442, *443*
- circumpolar motion at high latitudes, 38–39
- convective zone, *388*, 390, 393, *394*
- core, 385, 387, 389, 438, 442
- differential rotation, 397–398
- distance from Earth, 5
- energy transport from interior, 388–390
- evolutionary stages of a solar-type, main-sequence star, *454*, 678
- future evolution, 678
- gravitational effect of Jupiter, 184, 185
- helioseismology, 391, 393
- hydrogen fusion, 430, 438, 442, *443*
- hydrostatic equilibrium, 384
- Hα emission line, 395, 400
- lifetime, 386
- location of, 4, 560
- luminosity, 384, 385, 456
- main-sequence lifetime, 438, 678
- mass of, 95
- nuclear fusion, 385–387, 390
- opacity, 389–390, 393
- pressure, density, and temperature in interior, 384
- as protostar, 175–179, 182
- radiation, 389–390, 393, 395, 401–403
- radiative zone, *388*, 389–390
- reversal of Sun's magnetic field, 400
- rotation, 397–398
- solar activity effects on Earth, 401–402
- spectrum and absorption lines, 395
- spin angular momentum, 177
- temperature, 427
- ultraviolet image of, *382–383*

sungrazers, 336
Sun's atmosphere, 394–402
- chemical composition of atmosphere, 363
- chromosphere, 394–396, 400, 401
- corona, 394–397, 401, 416
- coronal holes, 396
- coronal loops, 396, 400
- coronal mass ejections (CMEs), 401, 403, 404
- limb darkening, 394
- magnetic fields and solar activity, 396–397
- photosphere, 394–396, 398, 401, 423, 427
- solar flares, 396, 400–401, 403
- solar prominences, 400–401
- sunspot cycle, 398–399, 401
- sunspots, 394, 396, 397–410

sunspot cycle, 398–399, 401
sunspots, 394, 396–410
superclusters, 4–5, 636, 637, 655
super-Earths, 188, 284–285
supergiant stars
- blue supergiant stars, 474
- H-R diagram, *370*, 373
- red supergiant stars, 471, 474
- yellow supergiant stars, 472

superior conjunction, *62*, 63
superior planets, 62–64
super-Jupiters, 284
supermassive black holes
- accretion disks, 542–544
- in bulges, 556, 650
- defined, 542
- evolution, 650
- jets, *540*, 542
- mass, 556
- merging galaxies, 547, 556, 650
- in Milky Way Galaxy, 566, 568–570
- in most distant galaxies, 650
- in normal galaxies, 544, 556
- size, density, and power, 543
- unified model of AGN, 544, *545*

Supernova 1987A, 478, *479*
supernovae, 459–460
- defined, 459
- energetic and chemical legacy of, 479–480
- first stars, 645
- heating of intercloud gas, 415–416
- hypernovae, or Type Ib or Ic supernovae, 517
- Kepler's supernova, 459, *460*
- "Local Bubble" of gas from neighboring supernovae, 416, 490
- neutrinos released, 479
- neutron stars and, 480–484
- nucleosynthesis, 479–480, 485, 489, 645–646, 669
- possible effects on Earth's atmosphere, 517
- SN 1006, *460*
- SN 1987A, *479*
- SN 2011fe, *533*
- Supernova 1987A, *479*
- Tycho's supernova, 459, *460*
- Type Ia supernovae and Chandrasekhar limit, 459, 513
- Type Ia supernovae and distance measurement, 376, 459, 533, 534, 609
- Type Ia supernovae and light curves, 478
- Type Ia supernovae and white dwarfs, 102, 376, *458*, 459–460, 484, 513
- Type Ia supernovae in binary systems, *458*, 460, 484, 533
- Type II supernovae, 477, 479
- unchanging value of gravitational constant (*G*) and, 102

supernova remnants, 416, 460, *478*
superstring theory, 624–625
surface waves, 205
SVLBI (Space Very Long Baseline Interferometer), 154
S waves (secondary waves), 205–206
Swift Gamma-Ray Burst Mission, *156*, 546
synchronous rotation, 43, 99–101
synchrotron radiation, 281–282, 485, 540, 544, 566
synodic period, 42, 63, 64, 68

tau neutrino, 391
Taurus, *34*, 39
tectonism, 210, 214–215. *see also* plate tectonics

telescopes
  adaptive optics, 148–149
  defined, 142
  focal length, 143–145, 147
  focal plane, 143, 145, 148–149
  gamma-ray telescopes, 153–154, *156*
  infrared telescopes, 153, 155, 187
  microwave telescopes, 162–163
  objective lens, 143–144
  observatory locations, *146*
  optical, 142–150
  optical and atmospheric limitations, 147–149
  optical axis, 143, *144*
  optical detectors and instruments used with, 150–153
  orbiting observatories, 155–157
  radio telescopes, 144–145
  refracting telescopes, 142–145
  resolution, 147–149, 152, 154–155
  use by Galileo, 70, 142
  world's largest optical telescopes, *146*
  X-ray telescopes, 11, 153, 155, *156*
  see also specific telescopes
Tempel, Elmo, 655
temperature, 127–128
  absolute zero, 128
  blackbody laws and, 129–130
  classification of stars by surface temperature, 360–362
  color and surface temperature of stars, 360
  and color of light emission, 129
  continuous spectra and, 128, 153
  defined, 127
  Fahrenheit scale, 128
  H-R diagram and surface temperature, 369–371, *370*, 373–375
  Kelvin scale, 128
  luminosity and, 128
  luminosity and radius and, 364, 427, 429, 444, 473
  Planck spectra or blackbody spectra and, 129, 163, 359, 360, 514
  of planets, at equilibrium, 127, *127*, 132, 133, 236, 256
  Stefan-Boltzmann law, 128, 129, 132, 279, 363–364, 394, 427
  thermal motions of atoms and, 127
  Wien's law and, 129–130, 359, 363, 414–415, 452, 592
terrestrial planets, 196–229
  atmospheric greenhouse effect, comparison of, 237–238, 254, 256
  comparative planetology, 198–199, 254, 256
  core, 180, 207
  defined, 181–182
  densities, 270
  evolution of planetary interiors, 207–208
  formation of, 181–182
  geological evidence for water, 218–222
  magnetic fields, 209–210, 236
  models of interiors, 207
  physical properties, *198*
  radioactive heating in core, 208
  secondary atmospheres, 181–182

  tectonism, 210–215
  thermal radiation from core, 207–208
  volcanism, 215–218
  see also atmospheres of terrestrial planets; Earth; Mars; Moon; Venus
Tethys, *294*, 302
theoretical models, 8, 10–11
theory
  defined, 8
  falsified theories, 8, *67*
  testable predictions of, 8, *37*
theory of everything (TOE), 619, 622, 624–626
thermal energy, defined, 111, 207, 422
thermal equilibrium, 127, *127*
thermal motions, 127
thermal radiation, 127, 128, 129, 133, 414–415
thermosphere, 242, 273
third quarter Moon, 42, 95
30 Doradus, 417
Thirty Meter Telescope (TMT), *146*, 149
Thorne, Kip, 516
thought experiments, 89–90, 92, 94, 501–502
tidal bulges, 96–97, 99–100
tidal forces
  calculating, 98
  defined, 96, 98
  and development of life, 101
  effects on solid bodies, 99–101
  Enceladus, tidal flexing, 299
  Europa, tidal flexing, 299, 671
  extrasolar planets, 101
  and galaxies, 101, 572
  on many scales, 100–101
  Moon and, 95–98
  rings of giant planets, 101
  Roche limit, 100–101, 306–307, 455
  Sun and, 98
  tidal bulges and, 96–97, 99–100
  tidal locking, 99–101, 295
tidal locking, 99–101, 295, 296
tides, 95–98
  defined, 95
  high tides, 96–99, 101
  low tides, 96–99
  lunar tides, 95–97
  Moon and, 95–97
  neap tides, 98
  solar tides, 97–98
  spring tides, 98
Tikoo, Sonia, 224
time dilation
  general relativistic time dilation, 511–512
  special theory of relativity, 502–506, 678
  twin paradox, 504
Titan
  ammonia, 300
  atmosphere, 244, 295, 300–301, 313, 671
  cryovolcanism, 300
  evidence of liquid water, 301, 671
  Huygens atmospheric probe, 158, 300–301, 671
  images, *270*, *294*, *300*
  methane, 300–301, 671
  oceans, 300–301

  organic molecules on, 300–301, 313, 671
  reconnaissance spacecraft, 158
  size, 295
  surface, 300–301
Titania, *294*, 303
TMT (Thirty Meter Telescope), *146*, 149
TNOs (trans-Neptunian objects), 322
TOE. *see* theory of everything
Tombaugh, Clyde W., 322
tools of astronomers. *see* astronomical tools and instruments
total lunar eclipse, 48
total solar eclipse, 46, 48, 49
trade winds, 245, 276
transit method for extrasolar planets, 186–189, 312, 375
trans-Neptunian objects (TNOs), 322
transverse waves, 205
Triangulum Galaxy, 571, 572
triple-alpha process, 447–448, 451, 475, 480
Triton, 270, *294*, 295, 301–302
Trojan asteroids, 328
tropical year, 44
Tropic of Cancer, *31*, *35*, 39
Tropic of Capricorn, *35*, 39
tropics, 39
tropopause, 242
troposphere, 241–242, 273, *274*
T Tauri stars, 430
Tully, R. Brent, 655
Tully-Fisher relation, 533
Tunguska event, 343
24 Themis (asteroid), 329
twin paradox, 504
2005 YU55 (asteroid), 325
Type Ia supernovae
  in binary systems, *458*, 460, 484, 533
  Chandrasekhar limit, 459, 513
  distance measurement, 376, 459, 533, 534, 609
  from double-degenerate systems, 533
  in the galaxy NGC 1365, *604–605*
  indication of accelerating universe, 609, 612
  and light curves, 478
  luminosity, 533
  observed brightness as a function of redshift, 609
  and white dwarfs, 102, 376, *458*, 459–460, 484, 513
Type Ib or Ic supernovae, or hypernovae, 517
Type II supernovae, *477*, 479

Uddin, Syed, 102
ultrafaint dwarf galaxies, 648
ultraviolet (UV) radiation
  absorption by atmospheric ozone, 416, 517, 664
  ammonia decomposition, 233–234, 236, 300
  in electromagnetic spectrum, 114
  interstellar medium and, 412, 413, 416–417
  methane photodissociation, 233, 273, 300
  ultraviolet image of the Sun, *382–383*, 396, 401
umbra, 46, 48, 49, 398
Umbriel, *294*, 303, 304
unbound orbits, 93
uncertainty principle, 615
unified model of AGN, 544, *545*

uniform circular motion, 90
universal gravitational constant (G)
   in bubble universes, 626–627
   defined, 86
   supernovae and unchanging value of G, 102
   value of, 94, 185, 296, 568, 607, 638
universal law of gravitation, 86–89, *91*, 95, 508–509
universe
   age of, 7, 15, 102, 587–590, 612, *613*, 624
   amount of ordinary matter in, 654
   closed universe, 613
   cosmic timeline, *7*
   critical density, 606–608, 611, 626–627
   Dark Ages, 643, *653*, 654
   dark energy and fate of the universe, 611–612
   dark matter in, 4, 654
   density of matter in, 596–597
   distribution of regular and dark matter in early universe, 641, 654
   Earth's place in, 4
   "escape velocity," 606–608
   evolution of the universe, 623–624
   flatness problem, 614–617
   flat universe, 613–615, 625
   horizon problem, 615–616
   inflationary model, 616–617, 623, 625–626
   mass distribution, 606
   measuring distances in, 4–7
   observable universe, defined, 4, 589, 624
   open universe, 613
   origins of, 2, 15
   predictions about the deep future, 653–654
   reionization, 643
   shape in spacetime, 613–614
   size of, 5, *6*, 172
   see also expanding universe; large-scale structure of the universe
unstable equilibrium, 127
uranium, 204
Uranus
   ammonia in atmosphere, 279
   axial tilt, 183, *266*, 270
   clouds, 270, 272, 273, 276–277
   core accretion–gas capture, 182
   diameter, 267
   discovery, 267, *268*
   helium, 279
   hydrogen, 279
   images, *266–267*, *272*, *277*
   impacts, 183, 270
   interior, 279
   lightning, 275
   magnetic field, 279
   magnetosphere size and shape, 280, 281
   mass of atmosphere, 182
   methane in atmosphere, 270, 272, 273, 275, 279
   physical properties of, *266*
   planet migration, 285
   rings, 304–305, 307, 310, *312*
   semimajor axis of orbit, 69
   sidereal period, 64, *69*
   synodic period, 64
   tidal forces, 100, 101
   troposphere, 273, 275
   winds and weather, *275*, 276–277
Uranus's moons
   Ariel, *294*, 303
   major moons, images, *294*
   Miranda, *294*, 303
   Oberon, *294*, 303
   Titania, *294*, 303
   Umbriel, *294*, 303, 304
Urey, Harold, 662
Urey-Miller experiment, 662
UV radiation. *see* ultraviolet radiation

vacuum, 110–112, 611
Vallenari, Antonella, 574
Valles Marineris, 214
variable stars
   Cepheid variables, 376, 472, *473*, 474
   defined, 472
   luminous blue variable (LBV) stars, 474, 480
   period-luminosity relationship, 472–473, 527, 560
   pulsating variable stars, 472
   RR Lyrae variables, 473, 559–560
   yellow supergiant stars, 472
Vega, *39*, 462
Vega spacecraft, 336
velocity, defined, 73
Venera 14 mission, *247*
Venus
   Aphrodite Terra, 215
   apparent magnitude, 358
   atmosphere, *215*, 246–248
   atmospheric composition, 234, 236, 246
   atmospheric greenhouse effect, 237, 246, 256
   atmospheric pressure, 236, 246–247
   atmospheric probes, 158
   atmospheric sulfur compounds, 218, 237, 246, 247
   atmospheric temperature, 246–248
   atmospheric water vapor, 221, 238, 256
   carbon dioxide in atmosphere, 232, 234, 236, 237
   coronae, 215
   destruction of impact craters, 200, 203
   erosion, 219
   evidence of water in the past, 221, 238
   giant impact and retrograde rotation, 201
   global circulation, 247
   global Hadley circulation, 247
   hot spots and coronae, 215, 218
   impact craters, 200
   Ishtar Terra, 215
   lack of magnetic field, 210
   lightning, 248
   Magellan mission, 215, 247
   model of interior, *207*
   phases and apparent size, 70
   physical properties, *198*
   Pioneer spacecraft, 248
   primary atmosphere, 182
   retrograde rotation, 201, 247
   secondary atmosphere formation, 182, 233–234
   semimajor axis of orbit, *69*
   shield volcanoes, 218
   sidereal period, 69
   small number of impact craters, 200
   surface features, *215*, 247
   tectonic features, 214–215
   Venera 14 mission, 247
   Venus Express (European Space Agency), 215, 218, 238, *247*, 248
   volcanism, 218
   windstorms, 219
vernal equinox, 38
Very Large Array (VLA), *153*, 154
Very Large Telescope (VLT), *146*, 155
Very Large Telescope Interferometer (VLTI), 155
Very Long Baseline Array (VLBA), 154
very low-mass stars, 440
Vesta (asteroid), 330–331
Viking landers, 249, 670
Viking Orbiter, 214
Virgo Supercluster, 5, 636, 655
virtual particles, 514, 618, *619*
visible spectrum, 113, *113*
visual binary systems, 367, 368
VLA (Very Large Array), *153*, 154
VLBA (Very Long Baseline Array), 154
VLT (Very Large Telescope), *146*, 155
VLTI (Very Large Telescope Interferometer), 155
voids, 636, 637, 639, 652
volatile materials (volatiles), 180, 273
volcanism, 215–218
   composite volcanoes, 216
   cryovolcanism, 299–302, 671
   defined, 215
   and geologically active planets, 213, 215–218
   hot spots, 213, 215–217
   Iceland, 216
   Io, 282, 297–298, 307, *319*
   magma, 215, *216*, 217
   magnetic fields recorded in lava, 209, 211
   Mars, *214*, 217
   Mercury, 217
   Moon, 216–217
   seismic waves, 206
   shield volcanoes, *214*, 216–218
   in the Solar System, 216–218
   source of secondary atmosphere, 180–181
   at tectonic plate boundaries, 213, 215–216
   Venus, 218
Voyager spacecraft, 286
   clouds and jet stream on Jupiter, *Voyager 2*, 272
   cryovolcanism on Triton, *Voyager 2*, 301
   evidence of liquid water oceans on Europa, 298
   filled impact scars on Ganymede, *302*
   Great Red Spot on Jupiter, *Voyager 1*, *271*
   interstellar space, *Voyager 1*, 398
   Jupiter's magnetosphere, *Voyager 1*, 282
   lightning on Jupiter, *Voyager 1*, 275
   lightning on Uranus and Neptune, *Voyager 2*, 275, 276
   rings of Jupiter, 304, 310
   rings of Neptune, *Voyager 2*, 304, 312
   rings of Uranus, *Voyager 2*, 304–305, 310
   routes of travel, 157
   Saturn's rings and Cassini Division, *Voyager 1*, 308–310, *311*

spokes in Saturn's B Ring, *Voyager 1*, 310
Uranus flyby, *Voyager 2*, 272, 275, 276–277
volcanoes on Io, 297

walls, 637, 639, 652
waning crescent Moon, *41*, 42
waning gibbous Moon, *41*, 42
water
    amount on Earth, 238
    comets as possible source of Earth's water, 338, 345, 346
    detected in W3 IRS5 star-forming complex, 431
    erosion and, 219–220, 238
    evidence of liquid water on Mars in the past, 200, 220–221, 238, 249, 256, 671
    evidence of water ice at poles on Moon, 222
    geological evidence on terrestrial planets, 218–222
    heavy water, 222, 346
    isotope signature in, 338, 346
    Kuiper Belt as possible source of Earth's water, 338, 346
    liquid water on Enceladus, 282, 299, 308, 313, 671
    liquid water on Europa, 298–299, 313, 671
    in polar ice caps on Mars, 221
    requirement for life, 219, 345, 670, 672
    in Solar System, search for, 220–223
    subsurface water ice on Mars, 221, 238, 256, 670–672
    Venus, evidence of liquid water in the past, 221, 238
water cycle, 245–246, 256
water vapor
    in atmosphere of Mars, 248
    in atmosphere of Venus, 221, 238, 256
    in Earth's atmosphere, 240
    as greenhouse gas, 237, 240
    infrared radiation blocking, 155
    from volcanoes, 236, 238, 662
wavelength, 113–114, 116

waves
    amplitude, 112
    characterizing, 112
    frequency (f), 113–114, 116
    frequency–wavelength relationship, 112, 116
    interaction with matter, 413
    light as electromagnetic waves, 111–112
    velocity, 112
    wavelength, 113–114, 116
    wave properties of light, 110
waxing crescent Moon, 40–41
waxing gibbous Moon, 41
weakly interacting massive particles (WIMPs), 539, 641
weak nuclear force, 618, 619, 621, 623
weather, 244–246
weathering, 219
Weber bar, *161*
Wegener, Alfred, 211
weight, 84, 88, 90
Weiss, Rainer, 516
Weldon, Marcus, 598
Westphal, Andrew, 432
white dwarfs
    in binary systems, 456–460
    carbon fusion, 461
    Chandrasekhar limit, *458*, 459, 513, 533
    defined, 102, 453
    formation, *452*, 453–455, 472
    H-R diagram, *370*, 373, 453–455
    novae, 457–458
    pollution with other elements, 455, 462
    Type Ia supernovae, 102, 376, *458*, 459–460, 484, 513
Wide-field Infrared Survey Explorer (WISE), 328, 415, 426
Wien's law, 129–130, 359, 363, 414–415, 452, 592
Wilkinson, Aaron, 16
Wilkinson Microwave Anisotropy Probe (WMAP), 163, 596, *596*, 609, 639

Wilson, Robert, 163, 594, *594*, 595, 598
WIMPs (weakly interacting massive particles), 539, 641
winds
    anticyclonic motion, 245, 272, 276
    Coriolis effect, 32–33, 245–247, 275, 276
    cyclonic motion, 245, *246*, 272, 276
    from the disk around a stellar black hole, 515
    on giant planets, 275–277
    global circulation, 245, 247
    low-density winds of high-mass stars, 473–474
    on Mars, 219–220, 249
    protostellar wind, 430
    trade winds, 245, 276
    on Venus, 219
    and weather on Earth, 244–246
    zonal winds, 245, 275
    see also solar wind
winter solstice, *35*, 38, 39
WISE. *see* Wide-field Infrared Survey Explorer
WMAP. *see* Wilkinson Microwave Anisotropy Probe

X-ray binary systems, 484, 514, 515, 542, 645–646
X-rays in electromagnetic spectrum, 113
X-ray telescopes, 11, 153, 155, *156*

year, defined, 33
Yellowstone National Park, 216, 313
yellow supergiant stars, 472
Yerkes 1-meter refractor, 143–144

zenith, 25, 26
zodiac, 34
zodiacal dust, 341–342
zodiacal light, 342
Zolensky, Mike, 432
zonal winds, 245, 275
Zwicky, Fritz, 481, 561